Communications
in Computer and Information Science 142

Vinu V Das   Janahanlal Stephen
Yogesh Chaba (Eds.)

Computer Networks and Information Technologies

Second International Conference on Advances in Communication, Network, and Computing, CNC 2011
Bangalore, India, March 10-11, 2011
Proceedings

Springer

Volume Editors

Vinu V Das
Engineers Network
Trivandrum, Kerala, India
E-mail: vinuvdas@gmail.com

Janahanlal Stephen
ILAHIA College of Engineering
Kothamangalam, INDIA
E-mail: drlalps@gmail.com

Yogesh Chaba
Guru Jambheshwar University
of Science and Technology
Hisar, Haryana, India
E-mail: yogeshchaba@yahoo.com

ISSN 1865-0929 e-ISSN 1865-0937
ISBN 978-3-642-19541-9 e-ISBN 978-3-642-19542-6
DOI 10.1007/978-3-642-19542-6
Springer Heidelberg Dordrecht London New York

Library of Congress Control Number: 2011922133

CR Subject Classification (1998): C.2, H.3.5, F.1, I.2.3, I.2, I.4-5

Typesetting: Camera-ready by author, data conversion by Scientific Publishing Services, Chennai, India

Printed on acid-free paper

Springer is part of Springer Science+Business Media (www.springer.com)

Preface

The Second International Conference on Advances in Communication, Network, and Computing – CNC 2011 – is organized every year to bring together in a common forum innovative academics and industrial experts in the field of computer science, information technology, computational engineering, communication and network. The conference will be held every year to make it an ideal platform for people to share views and experiences in information, telecommunication, computing techniques and related areas. CNC 2011 was organized by The Association of Computer Electronics and Electrical Engineers (ACEEE) during March 10–11, 2011 in Bangalore, Karnataka, India.

This volume contains 130 papers selected through a rigorous reviewing process. The contributions reflect the richness of research on topics within the scope of the conference and represent several important developments, specifically focused on theoretical foundations and methods for information technology and computer networks.

Organizing a conference like this one is not possible without the assistance and continuous support of many people and institutions. Particularly, I would like to express my gratitude to the organizers of sessions on dedicated topics that took place during the conference.

I also thank Stefan Goeller, Janahanlala Stephen, R. Vijay Kumar, and Yoghesh Chaba for the constant support and guidance. I would like to express my gratitude to the Springer LNCS-CCIS editorial team, especially Ms. Leonie Kunz, for producing such a wonderful quality proceedings book.

January 2011 Vinu V Das

CNC 2011 – Organization

Technical Chair

Janahan Lal	Mangalam College of Engineering, India

Technical Co-chair

R. Vijayakumar	NSS College of Engineering, India
Ford Lumban Gaol	University of Indonesia

Organizing Chair

P.M. Thankachan	Mar Ivanious, India

Organizing Co-chair

Srinivasa K G	M. S. Ramaiah Institute of Technology, India
HATIM A ABOALSAMH	King Saud University, Riyad

General Chair

Vinu V Das	Tabor Engineers Network P. Ltd., India

Publicity Chair

Mohammad Hammoudeh	University of Wolverhampton, UK

Publication Chair

Mohamed Jamaludeen	SRM University, India

Advisory Committee

Sudarshan TSB	BITS Pilani, India
Sumeet Dua	Louisiana Tech University, USA
Ansari Nirwan	

Program Committee

Shelly Sachdeva	Jaypee Institute o Information and Technology University, India
Pradheep Kumar K.	SEEE, India
Rupa Ashutosh Fadnavis	Yeshwantrao Chavan College of Engineering, India
Shu-Ching Chen	Florida International University, USA
Stefan Wagner	Fakultät für Informatik Technische Universität München, Germany
Juha Puustjärvi	Helsinki University of Technology, Finland
Selwyn Piramuthu	University of Florida, USA
Werner Retschitzegger	University of Linz, Austria
Habibollah Haro	Universiti Teknologi Malaysia
Derek Molloy	Dublin City University, Ireland
Anirban Mukhopadhyay	University of Kalyani, India
Malabika Basu	Dublin Institute of Technology, Ireland
Tahseen Al-Doori	American University in Dubai
V.K. Bhat	SMVD University, India
Ranjit Abraham	Armia Systems, India
Naomie Salim	Universiti Teknologi Malaysia
Abdullah Ibrahim	Universiti Malaysia Pahang
Charles McCorkell	Dublin City University, Ireland
Neeraj Nehra	SMVD University, India
Muhammad Nubli	Universiti Malaysia Pahang
Zhenyu Y Angz	Florida International University, USA
Keivan Navi	Shahid Beheshti University, Iran

Table of Contents

Full Papers

Poster Papers

Short Papers

Erratum

Solar Cell – A Novel Process Control Transducer

Snehlata Mishra, Vandana Neha Tigga, and S.N. Singh

Department of Electronics and Communication Engineering
NIT, Jamshedpur (India) – 831014
{snehlatamsr12,vntjsr,snsnitjsr}@gmail.com

Abstract. This paper deals with a novel instrumentation scheme for measurement of process control parameters using an optoelectronic solar cell device. The I-V reverse bias characteristic of the device has been used to measure output differential voltage when it is exposed to sun radiation. The uniform constant intensity of light is exposed on the cell and the moving object is placed in between them. Due to masking of light intensity caused by a moving object gives the variation in output as differential voltage which is calibrated in terms of displacement or its derivative parameter. It can also be used for measuring process parameters like pressure, level and temperature etc in an industrial process control environment. The experimental results validated with computed value of the displacement of moving object as a primary parameter shows a good conformity.

Keywords: Solar Cell, Displacement transducer, Sensitivity, Reliability.

1 Introduction

An instrumentation system consists of sensor as input, signal conditioning module as interfacing device and indicator or display unit as output device. The signal conditioning unit, which include amplifiers and filters circuit module etc, takes the input signal(s) obtained through transducer and conditioned it to a standard format in terms of voltage or current i.e. 0-5V/0-10V or i.e. 0-20mA/4-20mA respectively [1,3,4]. The uniform light intensity falling on the cell delivers varying output current across the load. Its relatively large power output, small physical size, and its quick response to change in light intensity are some of the unique features of this device. It can be used to sense and measure physical movements of objects that may be small in magnitude or occur too fast for visual observations [2]. It can be used in locations that are inaccessible to other sensors. Thus the constrained faced in conventional transducers like LVDT etc are overcome with this transducer. Typical dimension of a solar cell used as a transducer is shown in Fig.1 as 28mm x 8mm x 2mm. The advantages of its small size senses very small changes in light intensity caused by either due to small mechanical movement of object that cast a shadow on the cell or movement of light causing variation in light intensity.

V.V. Das, J. Stephen, and Y. Chaba (Eds.): CNC 2011, CCIS 142, pp. 1–8, 2011.

2 Solar Cell Device: Fabrication and I-V Electrical Characteristic

A typical silicon solar cell is a single crystal of the silicon base semiconductor material, having diffused onto its surface a thin film of similar material having P and N type impurity to form a P-N junction. The main fabrication steps include (a) growth of a single crystal, (b) doping of P type impurity material (i.e. a few parts per billion) using boron followed by N type impurity (phosphorus) to form P-N junction in a diffusion furnace under controlled temperature and for a predefined period, and (c) applying metal contacts to the P and N regions to form a cell. The other semiconductor materials that are being used to form a cell include selenium, gallium arsenide, indium arsenide, and indium antimonite. The photons (sun-radiation) falling on the surface of cell produces electron hole pair and give rise to power generating property. The electrons from the P region and holes from the N region diffuse across the junction and constitute current in the cell. In order to use this property of a solar cell, some load is connected to it in an external circuit. A simple circuit containing a solar cell and its load is diagrammed in Fig.2(a)

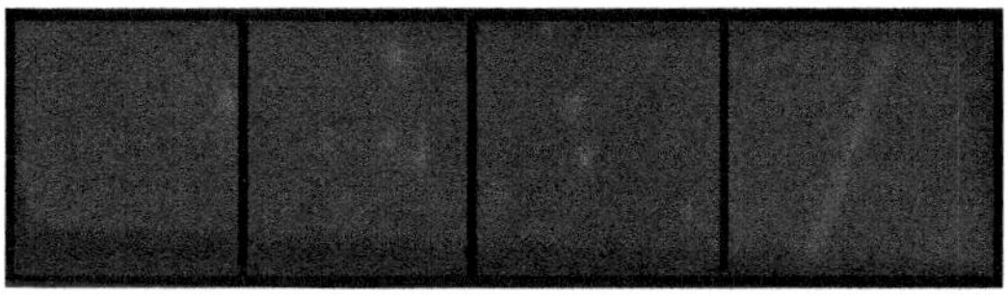

Fig. 1. A solar cell – Transducer Device (2.8cm long)

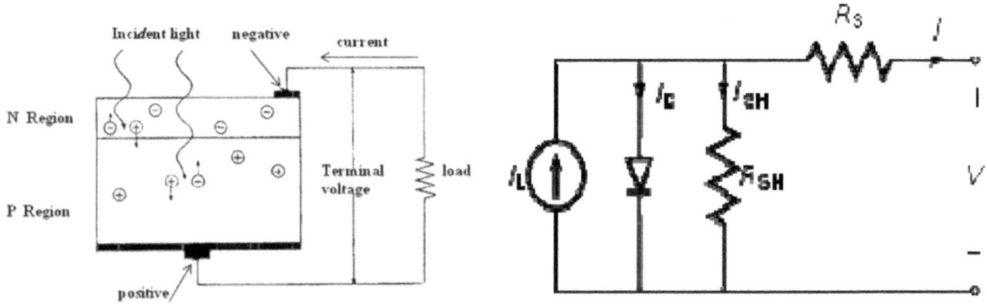

Fig. 2. (a) Solar cell with load (b) Equivalent electrical circuit model of a solar cell

The equivalent electrical circuit model of solar cell is shown in Fig. 2(b). The voltage and current relation of the cell is governed by the equation (1). And its I-V characteristic is shown in Fig.3.

$$I = I_L - I_D(e^{K_0(V+R_S I)}-1) - V/R_{SH}. \quad (1)$$

Where,

$K_0 = q/AKT$ (q = electronic charge, A = completion factor, K = Boltzmann constant, T = absolute factor), I_D = reverse saturation current, R_{SH} = shunt resistance, R_S = series resistance, I_L = photo voltaic current I = current through load.

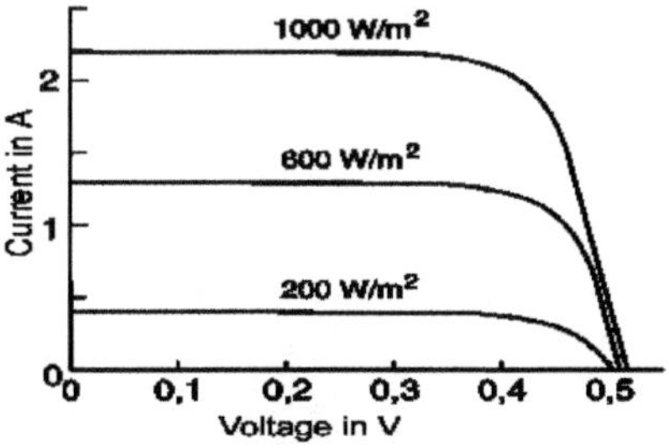

Fig. 3. I-V Characteristic of Solar Cell

3 Solar Cell as a Process Control Transducer

In process control applications, solar cell can be used for measurement of a few parameter(s) directly (such as displacement, temperature, radiation,velocity etc) and few other parameters indirectly such as pressure level etc in terms of displacement.

3.1 Temperature Measurement

The variation of open circuit voltage (V_{OC}) with temperature (T) is represented in silicon cells by:

$$dV_{OC} / dT = - 0.00288 \text{ volt /degree centigrade} \tag{2}$$

Since the output power of the device varies linearly with Voc, the power decreases with temperature at a rate greater than 0.5 percent per degree centigrade (Fig.4).

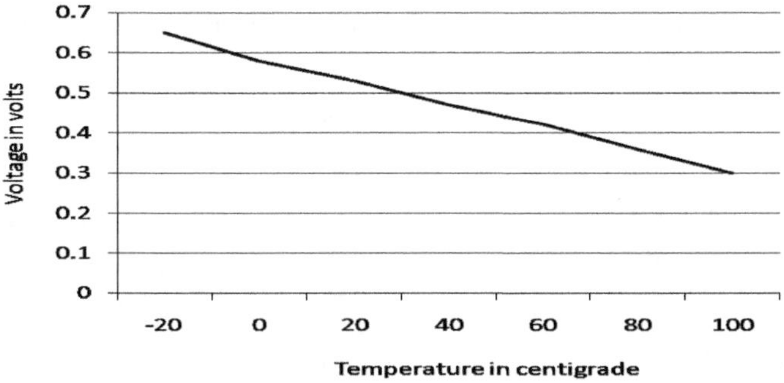

Fig. 4. Voltage across one cell as a function of temperature

3.2 Displacement Measurement

An experimental study of the behavior of a typical cell and its voltage variation across the cell operating under varying exposed level of light intensity (0-1000W/m2), corresponding to displacement (mm) can be carried out as shown in Fig.5.

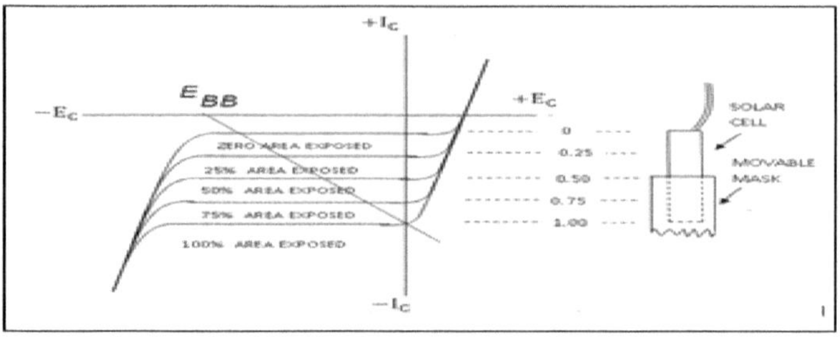

Fig. 5. Solar cell as a transducer

3.3 Speed/Velocity of Moving Object

The solar cell transducer can be used to measure the speed or velocity of the movable object. As velocity is defined as the rate of displacement so it can be calculated by taking the differentiation of the displacement with respect to time . A velocity transducer is formed by having two solar cells mounted side by side, and covered with an opaque mask having two small holes at measurable distance apart so that each hole coincides with one cell (Fig.6(a)). When the shadow of the moving part moves across the holes, each circuit will switch as the edge of the shadow passes. If the output from the two circuits is fed into an oscilloscope, the time from the edge of the shadow to move from one hole to another can be displayed. From this the velocity can be calculated. The circuit used is shown in Fig 6(b) To improve the output signal, a differentiator RC circuit has been added. This can be called a peaking circuit, for it will result in a slightly peaked output on the oscilloscope when there is a step type change in the voltage drop across the transistor. The output peaks corresponding to the shadow passing the holes will permit a more accurate determination of the average velocity

Let V be the velocity of the moving mask attached to object, S be the displacement (mm) and T is the time (sec), then,

$$V = dS/dT \quad \text{mm/sec} \tag{3}$$

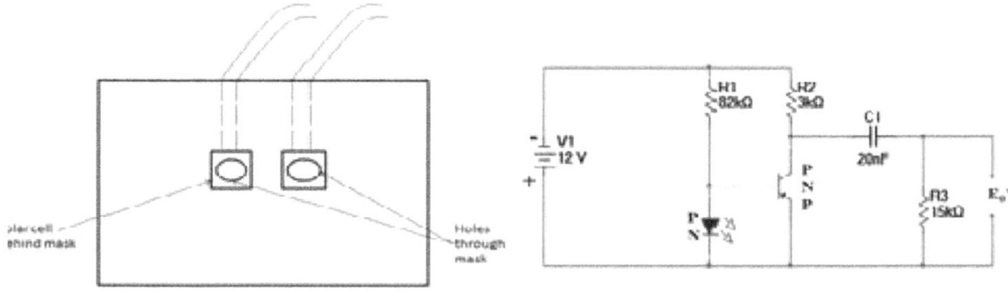

(a) (b)

Fig. 6. (a) Schematic of velocity transducer (b) Circuit associated with solar cell velocity transducer

3.4 Measurement of Pressure and Level

An attachment is added to primary transducer such as bourdon tube or diaphragm to convert pressure (Kg/m2) into displacement (mm) and so on as shown in Fig.7.

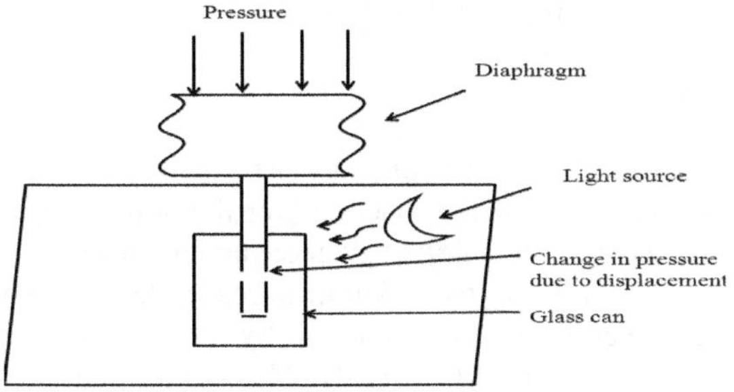

Fig. 7. Pressure measured using a solar cell

4 Sensitivity and Reliability

4.1 Sensitivity

The sensitivity of solar cell transducer are supposed to be consistent during the period of measurement but due to some inherent problems (such as the nonlinear behaviour of output current produced due to light intensity falling on the cell), it need to be calibrated for a specific installation of the cell. Reduced sensitivity also results in regions of higher light intensity. Its response to variations in light intensity also affects the sensitivity parameter. For example, when it is used to detect the motion of an object, a source of constant light intensity is required. An AC lamp source could produce an oscillating output from the cell, caused by small variations in its intensity occurring at the frequency of the ac source. Such variations in the output might be mistaken for measurements of small movements. It is important to preclude variations of the source that are comparable in amplitude to those caused by the system being measured. For this reason, a dc supply for the light source will be desirable in nearly all applications of the solar cell as a sensor/ transducer.

4.2 Reliability

A critical requirement of a reliable sensor discriminates against unwanted input signals, or “noise.” The sensor should be able to differentiate between different signals. It should refuse to mix two responses or signals. Let us take an example of a transducer which is sensitive to both insolation and temperature. When it is used to sense light intensity then it must be shielded from temperature and vice versa. It is not always possible to eliminate the response of a transducer to more than one variable but its construction and application must permit the one of interest to predominate. Extraneous input signals must be subdued to tolerable limits. Since there is no shortage of principles and physical effects that could be used as the basis for a transducer, calibration of the installation is necessary before using it to make measurements.

5 Experimental Investigation: Displacement Measurement and Its Validation with Computed Value

5.1 Data Acquisition

Experimental investigation has been carried out to measure commonly used primary parameter (i.e. displacement) in a process control application. Methodology of an instrumentation scheme was developed to measure displacement by connecting the cell and exposed from minimum to maximum area under uniform constant light intensity. This was achieved by masking the cell by moving object placed between cell and source generating light intensity (Fig.8). The voltage and current thus produced by the cell was decided by the load line on I-V characteristic of cell. The upper and lower point of the load line was measured at dark (fully masked) and fully illuminated (no mask). Depending on the requirements, the static condition of the solar cell was positioned in between these two extreme points. This had yielded the largest possible dynamic output, by allowing a maximum excursion of the shadow across the cell.

Fig. 8. Experimental set-up for displacement Measurement

The instrumentation scheme and experimental set-up for the linear displacement of a moving object under dynamic and static condition has been shown in Fig.8.The uniform constant light intensity had been produced in the experiment by illuminating a filament lamp powered by a AC/ DC regulated power supply.

5.2 Calibration

The gain of the Instrumentation amplifier cascaded with the solar cell device is set at maximum value and the cell is exposed to minimum area at a uniform intensity of light. The open circuit voltage (V_{OC}) thus obtained is noted down and marked as maximum position of displacement. The solar cell was exposed to its maximum area of light at a particular intensity. The no load output voltage of the cell corresponds to minimum displacement / position of movement of object this process was initialized before the start of recording the observations.

5.3 Results and Discussion

The measured value of displacement of moving object as recorded has been shown in Table (1). It can be concluded from data as recorded during investigation that output voltage produced by the solar cell gives desirable linear characteristic. The sensitivity was found as greatest when the largest change of exposed area was used to measure the dynamic conditions.

Table 1. Measured value of displacement of a typical object

S. No.	*Displacement (mm)*	*Output (Volt)*	***Light intensity* (W/m^2)**
1	2.8	0.1	Constant uniform Light intensity (1000W/m^2) Filament Source of Light
2	5.6	0.6	
3	8.4	0.6	
4	11.2	0.8	
5	14.0	1.0	
6	16.8	1.2	
7	22.4	1.8	
8	28.0	2.0	

6 Conclusions

An energy conversion optoelectronics device has been used in the instrumentation scheme to measure process parameter such as displacement of an object. Its main advantages are its small physical size, permitting measurements in confined spaces, and its response to changes in illumination, when they occur as a varying area of shadow on the cell. As it has some advantage it has some disadvantages too like its nonlinear current vs. light intensity characteristic prevents a linear response to measured movements, requiring prior calibration of a system at the start of taking measurement every time. The necessity for shielding against unwanted "noise"-extraneous light sources and variations of temperature-creates other problems that we face during the working of the transducer. In spite of these entire problems it is immune to noises due to large amplitude of output produced by this sensor. Further it is used as non contact type and does not require any specific attachment to mount it on the moving object to measure mechanical movements and hence finds its wide application in process control industry.

References

1. Puckett, R.E., Thompson, L.W.: The solar cell - A Novel Transducer. University of Kentucky, Lexington (1965)
2. Wolf, M.: Limitations and possibilities for improvement of photovoltaic solar energy converters. Pro. IRE 48, 1246–1263 (1960)
3. Puri, M., Jha, J., Kumar, S.: Elements of electronic instrumentation, pp. 143–152. Narosa Publishing House, Bombay (1996)
4. Kalsi, H.S.: Electronic instrumentation, p. 423. Tata Mc-Graw Hill Publication Ltd. (1995)
5. Mishra, S., Tigga, V.N., Singh, S.N.: The Solar Cell- A Novel Static Transducer. In: IEEE Students Technology Symposium, IIT Kharagpur (2010)

A Hybrid Aggregation Technique for Continuous-Monitoring in Wireless Sensor Networks

R. Rajkamal and P. Vanaja Ranjan

Department of Electrical and Electronics Engineering,
College of Engineering Guindy, Anna University, Chennai –25
rajkamalr@gmail.com

Abstract. Network lifetime is the one of the most important issues in WSN and is fundamental to any design and development effort. In continuous monitoring application, the Wireless sensor Network (WSN) is generated data continuously and transmit to base station with a predefined frequency. The dominant energy consumption in the WSN occurs in radio transceiver that is responsible for transmission and reception of data. In this work, a hybrid (combined data and information) aggregation scheme is proposed to achieve energy efficiency by exploring the impact of heterogeneity of network for continuous Power Quality (PQ) monitoring application.

Keywords: Continuous Monitoring, Wireless Sensor Network, Radio Transceiver, Network lifetime, Power Quality monitoring.

1 Introduction

Wireless Sensor Network (WSN) is a system that relies on the collective effort of several micro nodes that promise fine-grain monitoring in wide variety of environments. The ability to communicate not only allows information and control to be communicated across the network of nodes, but also nodes to cooperate in performing more complex tasks, like statistical sampling, data aggregation [1], system health and status monitoring. WSNs can be broadly categorized in to event driven monitoring, continuous monitoring and hybrid monitoring. A Sensor node (generally known as mote) is made up of four basic components namely sensing unit, processing unit, transceiver and power unit. Sensing units are usually composed of two sub units: one or more number of sensors and analog-to-digital converters (ADCs). The processing unit is generally associated with a small storage unit, manages the procedures that make the sensor node collaborate with the other nodes to carry out the assigned sensing tasks. The transceiver unit connects the node to the network. One of most important components of a sensor node is the power unit. The sensor node is being a micro electronic device; the power unit is treated as most important unit [2]. Therefore, sensor node lifetime shows a strong dependence on battery lifetime.

Energy consumption of a sensor node can be divided into three domains namely sensing, data processing and communication. The dominant energy consumption in the WSN occurs in the radio transceivers (i.e. Communication domain of a sensor node) [3], [4], [5], [6]. The energy consumption of the other components in the sensor node is

V.V. Das, J. Stephen, and Y. Chaba (Eds.): CNC 2011, CCIS 142, pp. 9–14, 2011.

very small. For this reason, many researches address this issue in WSN and advanced to reduce the wireless transmission.

In many applications of WSNs, data aggregation is a critical operation needed for extracting useful information from the operating environment and is regarded as an effective technique to reduce energy consumption [8]. Data aggregation attempts to collect the most critical data from sensors and make it available to the sink in an energy efficient manner with minimum latency [7]. The design of efficient data aggregation algorithm is an inherently challenging task. There has been intense research in the recent past on data aggregation in WSNs. Most of the existing work like Directed diffusion [8], SPIN [9], LEACH [10], HEED [11], CLUDDA [12], EADAT [13], PEGASIS [14], EDAP [15] and SAG [16] has mainly focused on the communication domain, especially in the aspects of routing, Medium Access Control (MAC) and collaborative mechanism for data aggregation in WSNs. However the performance of the data aggregation protocol is strongly coupled with the infrastructure of the network. There has not been significant research on exploring the impact of heterogeneity on the performance of the data aggregation protocols. This motivates us to explore the impact of heterogeneity of the WSN for hybrid (information and data) aggregation scheme to achieve energy consumption of the radio of a WSN.

We wish to clarify the term information and data; especially in the context of WSNs. Data are plain facts. When data are processed, organized and structured or presented in a given context so as to make them useful, they are called Information. For instance, consider a query, SELECT Temperature, FROM Temp. Sensors, WHERE region Z, HAVING Temp >30°c, DURATION for 5 mins, EVERY 1 hour which can produce the enormous raw data from sensors employed in region Z. This may not be interpreted any meaningful information to the user who sends the query. The enormous data from sensors is to be processed and presented in the form of useful information to user for further steps to be taken in the monitoring applications. The rest of the paper is organized as follows. Section 2 describes Network Model, Proposed Scheme and Application while Section 3 presents results and discussions followed by concluding remarks in Section 4.

2 Network Model, Proposed Scheme and Application

2.1 WSN Model

In a heterogeneous network, two or more different types of nodes with different battery energy and functionality are used. The motivation being that the more complex hardware and extra battery energy can be embedded in few cluster head (CH) nodes, thereby reducing hardware cost of the rest of the network. This impact of the heterogeneous networks enabled us to design the network for PQ monitoring with a Base Station (Sink - type 0), clusters having Cluster Heads (CH) (type-1) and number of Sensor Nodes (type-2) with the following capabilities.

Cluster Head Node (type-1)

- Receive/ Transmit raw data or queries from/to type-2 nodes in the cluster.
- Transmit/ Receive raw data / information from / to type1 nodes of one cluster to other clusters in the network.

- Ability to classify (Neural Network) the events from received raw data.
- Decision making capability to send /not to send the information / data to Sink (type-0).

Sensor Node (type-2)

- Transmit raw data to type-1 nodes in the cluster.
- Receive queries from type –1 node in the cluster.
- Ability to compare the sensed data to stored data.

The main task on the motes in WSN applications is to carry out wireless communication through the radio transceiver. Generally the power consumption measurement for radio is considered the following different radio operation: receive (RX)/listen, transmit (TX) under different transmission power thresholds, sleep and idle state. When a radio is in sleep or idle mode, its RF power rating is irrelevant. The RF power affects power consumption only when the radio is TX / RX mode. In this work, all type –2 nodes deal only with its locally sensed data (D_1) and compares with stored data (D_0) that is in memory of the node permanently. Data aggregates to type –1 node, only when the value is changed with respect to the stored data. The MICAz - MPR2400CA is having measurement (serial) flash memory of 512K; part of this memory can be used to store the data of pure sine wave.

Since the application under consideration is continuous monitoring, data is generated continuously with a predefined frequency. The battery life can be extended significantly with modifying the transmission rate but type –2 nodes have to perform long transmissions to base station, most of the power consumption taken place at this point.

2.2 Proposed Scheme

By restricting the unnecessary data transmission in the continuous monitoring environment the considerable energy saving in the radio transceiver section of type-1 and type-2 nodes is achieved through the following algorithm.

Algorithm in Type-2 node

1.Start timer
2.if (timer expired)
3. Start to sense the environment (for PQ disturbance)
4. Compare present sensed data (D_1) with the stored data (D_0)
5.if (change in value of comparison (D_0 and D_1))
6. Sensor nodes transmit D_0 to CH
7.else
8. No transmission
9.Go to Step 1

Algorithm in Type –1 node

1.if (Node receives data)
2. Classify the data for Disturbance event (Neural Network Classifier)
3. Disturbance event information (not data) will be sent to sink
4.else
5. Do nothing

2.3 Application -- Power System Model Consideration

Power Quality (PQ), or more specifically, a Power Quality disturbance, generally defines as any change in power (voltage, current or frequency) that interferes with normal operation of electrical equipments. The application under consideration is Power Quality monitoring that monitors the power system for an occurrence of Power Quality disturbances. The Power System model is considered from our previous work [18].

3 Results and Discussion

Computer simulated waveforms are generated for some of the IEEE defined PQ disturbances such as notch, transient and outage from the power system model using MATLAB is shown in Fig. 1.

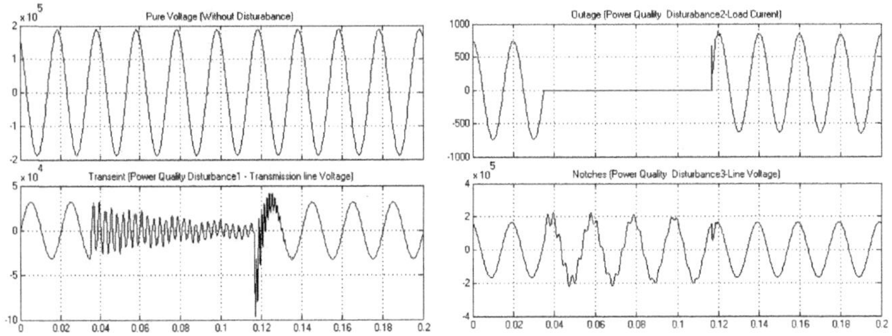

Fig. 1. Without PQ Disturbance and With PQ Disturbance (Transient, Outage and Notches)

We have considered continuous monitoring WSN employed for PQ disturbance monitoring, type-2 nodes sense the environment continuously with the frequency f_0 and type-1 node classify the PQ disturbance event then sending information (not data of the event) to type-0 node. Table. I shows the energy consumption/bit of a node in different modes of transceiver operation.

Table 1. Energy Consumption / bit for Different Modes of Radio

Mode of Operation @ Power Threshold	Energy = Current * Voltage * Time
Receive	19.7 mA * 3 V * 4 * 10^{-6} sec =0.23 μJ / Bit
Transmit @ -10dBm	0.13 μJ / Bit
Transmit @ -5dBm	0.16 μJ / Bit
Transmit @ -0dBm	0.21 μJ / Bit

The proposed algorithm is employed in type-1 and type-2 nodes to achieve considerable amount of savings in energy with the restricted TX/RX operation of radio transceiver that extends the lifetime of battery. Type –2 nodes are sensing for PQ disturbance with the predefined frequency but it transmit the data only when it detects the disturbance in the sensed data. This ability of type-2 nodes operates radio for the transmission of only disturbance data in PQ monitoring. It is not only ensured the better continuous monitoring but also save considerable amount of energy. Type- 2 nodes employed with the ability to classify the data according to the PQ disturbance category and send only the information about the disturbance to sink not the entire data of the PQ event. The communication from the type-1 nodes to base station is only few packets carries information not the entire raw data of the disturbance event. This nature of type –1 node can further enormously improve the energy saving on transmission that extends the lifetime of battery.

4 Conclusion

Detailed knowledge of each functional module's current demands of a WSN mote is essential when striving to achieve energy consumption. An energy efficient hybrid aggregation scheme targeted to radio transceiver of the WSN by exploring the impact of heterogeneity of the network is presented. The design of network nodes with different capabilities presented in this work can assist the proposed hybrid aggregation scheme to achieve the considerable savings in energy. In this work, the continuous monitoring application of WSN – PQ disturbance monitoring is considered and the proposed algorithm is employed to achieve considerable savings of energy that extends network lifetime.

References

1. Intanagonwitat, C., Estrin, D., Govindan, R., Heidemann, J.: Impact of Network Density on Data Aggregation in wireless sensor Networks. In: ICDCS 2002 (2002)
2. Fakildiz, I., Su, W., Sankarasubramaniam, Y., Cayirci, E.: A Survey on Sensor Networks. IEEE Communications Magazine 40(8), 102–114 (2002)
3. Min, R., et al.: Energy-centric enabling technologies for wireless sensor networks. In: Proceedings of IEEE Wireless Communications, pp. 28–39 (2002)
4. Sohrabi, K., et al.: Protocols for self-organization of a wireless sensor network. In: Proceedings of IEEE Personal Communications, vol. 7, pp. 16–27 (2000)
5. Estrin, D., et al.: Next century challenges: scalable coordination in sensor networks. In: Proceedings of MOBICOMM, pp. 263–270 (1999)
6. Raghunathan, V., et al.: Energy-Aware Wireless Sensor Networks. Proceedings of IEEE Signal Processing 19, 40–50 (2002)
7. Rajagopalan, R., Varhney, P.k.: Data Aggregation Techniques in Sensor Networks: A Survey. IEEE Communications Survey and Tutorials 8(4), 48–62 (2006)
8. Intanagonwiwat, C., Govindan, R., Estrin, D., Heidemann, J., Silva, F.: Directed Diffusion for Wireless Sensor Networking. ACM/IEEE Transactions on Networking 11(1), 2–16 (2003)
9. Heinzelman, W.R.: Application-Specific Protocol Architectures for Wireless Networks, Ph.D. thesis, Massachusetts Institute of Technology (2000)
10. Younis, O., Fahmy, S.: HEED: a Hybrid, Energy-Efficient, Distributed Clustering Approach for Ad Hoc Sensor networks. IEEE Trans. Mobile Computing 3(4), 366–379 (2004)

11. Chatterjea, S., Havinga, P.: A Dynamic Data Aggregation Scheme For Wireless Sensor Networks. In: Proc. Program for Research on Integrated Systems and Circuits, Veldhoven, The Netherlands (2003)
12. Lindsey, S., Raghavendra, C., Sivalingam, K.M.: Data Gathering Algorithms in Sensor Networks Using Energy metrics. IEEE Trans. Parallel and Distributed Systems 13(9), 924–935 (2002)
13. Ding, M., Cheng, X., Xue, G.: Aggregation Tree Construction in Sensor Networks. In: IEEE 58th Vehicle. Tech. Conf., vol. 4(4), pp. 2168–2172 (2003)
14. Tan, H.O., Korpeoglu, I.: Power Efficient Data Gathering and Aggregation in Wireless Sensor Networks. SIGMOD Record 32(4), 66–71 (2003)
15. Lindsey, S., Raghavendra, C.: PEGASIS: Power Efficient Gathering in Sensor Information Systems. In: IEEE Aerospace Conference, vol. 3, pp. 3-1125–3-1130 (2002)
16. Azim, M.A., Moad, S., Bouabdallah, N.: SAG: Smart Aggregation Technique for Continuous-Monitoring in Wireless Sensor Networks. In: IEEE Conference on Communications, pp. 1–6 (2010)
17. IEEE. Wireless medium access control (MAC) and physical layer (PHY) specifications for low rate wireless personal area networks (LR-WPANs). The Institute of Electrical and Electronics Engineers, New York, NY, USA (2003)
18. Rajkamal, R., Vanaja Ranjan, P.: Soft Computing Feature Extraction Algorithm with Minimum Entropy. International Journal of Distributed Energy Resources 6(3), 253–261 (2010)

FLD-SIFT: Class Based Scale Invariant Feature Transform for Accurate Classification of Faces

B.H. Shekar[1], M. Sharmila Kumari[2], Leonid M. Mestetskiy[3], and Natalia Dyshkant[3]

[1] Department of Computer Science, Mangalore University, Karnataka, India
[2] Department of Computer Science and Engineering,
P A College of Engineering, Mangalore, Karnataka, India
[3] Department of Computational Mathematics and Cybernetics,
Moscow State University, Moscow, Russia
bhshekar@gmail.com, sharmilabp@gmail.com, mestlm@mail.ru,
natalia.dyshkant@gmail.com

Abstract. In this paper, a new model called FLD-SIFT is devised for compact representation and accurate recognition of faces. Unlike scale invariant feature transform model that uses smoothed weighted histogram and massive dimension of feature vectors, in the proposed model, an image patch centered around the keypoint has been considered and linear discriminant analysis (FLD) is employed for compact representation of image patches. Contrasting to PCA-SIFT model that employs principal component analysis (PCA) on a normalized gradient patch, we employ FLD on an image patch exists around the keypoints. The proposed model has better computing performance in terms of recognition time than the basic SIFT model. To establish the superiority of the proposed model, we have experimentally compared the performance of our new algorithm with $(2D)^2$-PCA, $(2D)^2$-FLD and basic SIFT model on the AT&T face database.

Keywords: Linear discriminant analysis, Local descriptor, Face classification.

1 Introduction

In this digital world, automated face recognition (AFR) systems are widely used and plays a critical role in biometric systems. The AFR system has numerous applications including visual surveillance and security, person authentication etc. The design of AFR system is very challenging because of its interclass recognition problem and distinctiveness of face is quite low when compared to other biometrics [1]. In addition, changes caused by expressions, illumination, pose, occlusion etc impose further challenges on accurate face recognition.

Several models have been proposed ranging from appearance based approaches to sophisticated systems based on thermal information, high resolution images or 3D models. A comprehensive survey of face recognition algorithms is given by [18]. Principal Component Analysis (PCA) and Fisher's Linear Discriminant Analysis (FLD) are the two most widely used and successful appearance based models for face

V.V. Das, J. Stephen, and Y. Chaba (Eds.): CNC 2011, CCIS 142, pp. 15–21, 2011.

recognition. The currently existing appearance based recognition models date back to the earlier works of Sirovich and Kirby [9] and Kirby and Sirovich [11]. Turk and Pentland [13] popularized the use of PCA for face recognition. Although PCA ensures least reconstruction error, it may not be optimal from a discrimination stand point. Belhumeur et al., [17] have proposed the FLD based model that extract features which posses the best discrimination capability. Additionally, the effect of the non-uniform illumination conditions on the face recognition performance is studied and the superiority of the FLD method over the PCA method has been shown in [17].

On the other hand, local descriptors [4, 6, 15] are commonly employed in a number of real-world applications such as object recognition [4, 6] and image retrieval [15] as they can be computed efficiently, are resistant to partial occlusion, and are relatively insensitive to changes in viewpoint. Mikolajczyk and Schmid [14] presented a comparative study of several local descriptors including differential invariants [6], moment invariants [10], Scale Invariant Feature Transform [4], and cross-correlation of different types of interest points [2, 15]. A more distinctive representation for local image descriptors called PCA-SIFT [19] is proposed by Ke and Sukthankar and they argued that the representation of gradient patch in a compact eignspace is better for image retrieval applications. However, as discussed above, it is a well known fact that linear discriminant analysis is a better alternative when class-wise information is available and hence is suitable for classification problems. In this context, we proposed better representation scheme for image patches that are subsequently used in face recognition based applications. Like SIFT, in the proposed model, descriptors encoded the salient aspects of the image patch in the feature point's neighborhood; however, instead of using SIFT's smoothed weighted histograms, we apply linear descriminant analysis to the normalized image patch. Our experiments demonstrate that the FLD based local descriptors are more distinctive and more compact than the standard SIFT representation.

The remainder of this paper is organized as follows. The proposed model (FLD-SIFT) is given in section 2. Section 3 presents experimental results along with comparative analysis. Conclusion is given in section 4.

2 FLD-Based SIFT Descriptors

Scale Invariant Feature Transform (SIFT) is an approach for detecting and extracting local feature descriptors that are reasonably invariant to changes in illumination, image noise, rotation, scaling, and small changes in viewpoint. The SIFT transforms image data into scale-invariant coordinates relative to local features that consists of four major phases: (1) scale-space peak selection; (2) keypoint localization; (3) orientation assignment; (4) keypoint descriptor.

The proposed model for keypoint descriptors (called FLD-SIFT) consists of two major stages namely *extraction of keypoints* from training samples and the *construction of fisherspace* for compact representation of feature space. In the first stage, the information associated with keypoints is extracted which is nothing but the information obtained due to standard SIFT keypoints: the sub-pixel location, scale, and dominant orientations of the keypoint. In the second stage, we build the fisherspace using the 39x39 image patches at the given scale, centered over the keypoint, and rotated to align its dominant orientation to a canonical direction and subsequently use this fisherspace to derive a compact feature vector for all the images used for training purpose. During recognition, similar procedure is employed to obtain feature vector and the Euclidean distance between two feature vectors is used to determine whether the two vectors correspond to the same keypoint in different images or not.

2.1 Computation of Patch Fisherspace

FLD explicitly attempts to model the difference between the classes of data and enables us to linearly-project high-dimensional samples onto a low-dimensional feature space. FLD is also closely related to principal component analysis that looks for linear combinations of variables which best explain the data. PCA on the other hand does not take into account any difference in class. The input vector is created by obtaining the image patch of size 39x39 centered at the keypoint. Thus, the input vector has 39x39=1521 elements. This vector is subjected to normalization process to have unit magnitude in order to minimize the impact of variations in illumination. It shall be observed here that the patches considered here are centered on a local maximum in scale-space; rotated so that (one of its) dominant gradient orientations is aligned to be vertical; and contains information for the scale appropriate to this keypoint. In order to construct our fisherspace, the first three stages of the SIFT algorithm on a set of training face images is considered and each was processed as described above to create a 1521-element vector, and FLD was applied to the covariance matrix of these vectors. The matrix consisting of the top d eigenvectors was stored on disk and used as the projection matrix for FLD-SIFT. The face images used in building the fisherspace were discarded and not used in any of the recognition experiments. The construction of fisherspace is briefly described below.

Given a set of samples of each class, the proposed FLD-SIFT extracts most informative features which could establish a high degree of similarity between samples of the same class and a high degree of dissimilarity between samples of two different classes.

Formally, let there be C number of classes each with T_i, i=1..C, number of training images and each T_i contains varying number of patches, say p_i^j, i=1..C, j=1..T_i. Therefore we have totally $N = \sum_{i=1..C \text{ and } j=1..T_i} p_i^j$ number of training image patches. Let p_i^j be an image patch of size m x n representing jth image patch in ith class. Let $\overline{m_i}$ be the average image of all p_i training image patches of the ith class, arranged in a vector form (Note here that the input vector is created by converting 39x39 image patch centered at the keypoint into a vector form). Let m be the average image patch of all N training image patches. The between-class scatter matrix G_b and the within-class scatter matrix G_w are computed as

$$G_b = \frac{1}{N}\sum_{i=1}^{C} T_i(\overline{m_i} - m)^T(\overline{m_i} - m) \tag{1}$$

$$G_w = \frac{1}{N}\sum_{i=1}^{C}\sum_{j=1}^{T_i}(p_i^j - \overline{m_i})^T(p_i^j - \overline{m_i}) \tag{2}$$

Once G_b and G_w are computed, it is recommended to find the optimal projection axis X so that the total scatter of the projected samples of the training image patches is maximized. For this purpose, the Fisher's criterion given by

$$J(X) = \frac{X^T G_b X}{X^T G_w X} \tag{3}$$

is used. It is a well-known fact that the eigenvector corresponding to the maximum eigenvalue of $G_w^{-1}G_b$ is the optimal projection axis which maximizes *J(X)*. Generally, as it is not enough to have only one optimal projection axis, we usually go for *d* number of projection axes, say $X_1, X_2, \ldots, X_d$, which are the eigenvectors corresponding to the first *d* largest eigenvalues of $G_w^{-1}G_b$. Once these $X_1, X_2, \ldots, X_d$ are computed, each patch p_i^j is then projected onto these X's to obtain the feature vector f_i^j. So, during training, for each training image patch p_i^j, a corresponding feature vector f_i^j is constructed and stored in the knowledge base for matching at the time of recognition. We have used Euclidean distance measure for classification purpose. In order to classify a given face image, we project image patches centered around the keypoint onto fisherspace and the test feature vectors are compared against the knowledge base to decide the class, which is based on the maximum number of matches.

3 Experimental Results

This section presents the results of the experiments conducted on the standard face database to corroborate the success of the proposed model. We have conducted experimentation on AT&T face dataset. We have specifically chosen this dataset as this dataset is used by many researchers as a benchmark dataset to verify the validity of their proposed face recognition models. All experiments are performed on a P-IV 2.99GHz Windows machine with 1GB of RAM.

The AT&T face dataset contains images from 40 individuals, each providing 10 different images of size 112x92. In our experiment, we have considered alternate five samples per class for training and the remaining samples for testing. Similarly, we have conducted experiments considering 160, 120 and 80 faces as training faces of the AT&T database choosing 4, 3 and 2 faces respectively from each person and the recognition performance has been obtained considering the remaining faces as test faces. The recognition performance of proposed model with varying dimension of feature vectors is given in Table 1.

Table 1. Recognition accuracy of the proposed model for AT&T face dataset

No. of Training faces	No. of Test faces	% of recognition accuracy for varying dimensions of feature vectors			
		10	15	20	25
200	200	97.5	98	99	100
160	240	95.41	96.66	97.5	97.5
120	280	93.92	94.64	94.64	95.00
80	320	80.00	85.00	86.56	87.18

Table 2. Recognition accuracy of $(2D)^2$-PCA for AT&T face dataset

No. of Training faces	No. of Test faces	% of Recognition accuracy for varying dimensions of feature vectors			
		6x6	8x8	10x10	12x12
200	200	95.5	95.5	95.5	95.5
160	240	95.83	96.66	95.83	95.83
120	280	91.00	92.50	92.85	92.14
80	320	79.37	79.37	79.68	79.68

Table 3. Recognition accuracy of $(2D)^2$-FLD for AT&T face dataset

No. of Training faces	No. of Test faces	% of Recognition accuracy for varying dimensions of feature vectors			
		3x3	4x4	5x5	6x6
200	200	97.25	98.25	98.75	98.75
160	240	95.25	97.75	98.00	99.25
120	280	94.25	94.50	96.00	95.75
80	320	82.50	84.00	86.50	87.50

For the purpose of establishing the superiority of the proposed model, we have considered bi-directional appearance based face recognition models: $(2D)^2$-PCA [3] and $(2D)^2$-FLD [16] for comparative study. We performed four tests with varying number of training faces. Here too, we have considered alternate five samples per class for training and the remaining samples for testing. Similarly, we have conducted experiments considering 160, 120 and 80 faces as training faces of the AT&T database choosing 4, 3 and 2 faces respectively from each class and the recognition performance has been obtained considering the remaining faces as test faces. The recognition performance with varying dimension of feature vectors is given in Table 2 and Table 3 respectively for $(2D)^2$-PCA [3] and $(2D)^2$-FLD [16] techniques. We have also compared our proposed methodology with other well known subspace analysis models to reveal the success of the proposed model considering the standard AT&T face database. In this experimental set-up, we have considered alternate five samples per class for training and the remaining samples for testing. The recognition performance of 2DPCA [7], 2D-FLD [5, 8], alternative 2DPCA [3], 2DLDA [12] with varying dimension of feature vectors is given in Fig. 1(a) and the generalized versions of

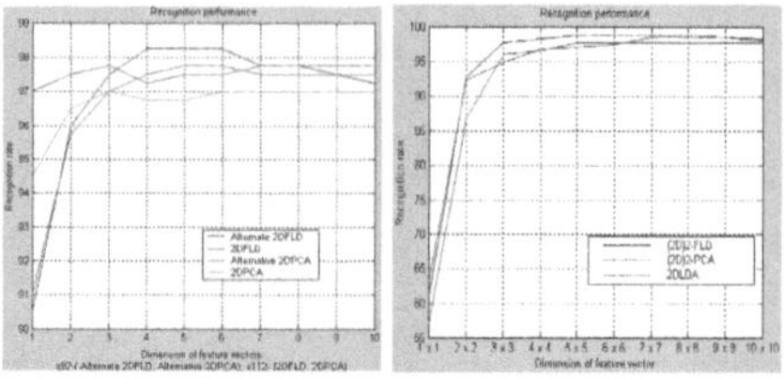

(a) (b)

Fig. 1. Recognition performance of different subspace models with varying dimension of feature vectors on AT&T face database

bi-directional models called $(2D)^2$PCA [3], $(2D)^2$ FLD [16] is given in Fig. 1(b). Table-4 summarizes the comparisons of methods [3, 5, 7, 8, 12, 16] with respect to the best recognition rate along with the corresponding dimension of the feature vectors. It shall be observed from Table-4 that the proposed FLD-SIFT achieve the best recognition rate with reduced dimension of feature vector among all the approaches.

Table 4. Recognition accuracy of subspace based methods

Method	Dimension of feature vector	Best recognition rate (%)
2DPCA [3]	3x112	97.00
2DFLD [5,8]	3x112	97.75
Alternate-2DPCA[3]	5x92	97.75
Alternate-2DFLD (16)	4x92	98.25
2DLDA[12]	8x8	98.75
$(2D)^2$PCA[3]	5x5	97.75
$(2D)^2$FLD [16]	5x5	98.75
FLD-SIFT (Proposed Model)	25	100.00

4 Conclusion

An alternate representation for local image descriptors for the SIFT algorithm is presented which exploits class-wise information for compact representation and accurate classification of faces. When compared to the standard SIFT representation and the subspace based models, the proposed representation is both more distinct and more compact leading to significant improvements in recognition accuracy (and speed) and hence is good for real time applications. We are currently extending our representation based on the kernel approaches, designed for classification type algorithms and exploring ways to apply the ideas behind FLD-SIFT to other keypoint algorithms.

Acknowledgements

The authors would like to acknowledge the support provided by DST-RFBR, Govt. of India to carry out this research work vide No. INT/RFBR/P-48 dated 19.06.2009.

References

1. Jain, A.K., Ross, A., Prabhakar, S.: An introduction to biometric recognition. IEEE Transactions on Circuits and Systems for Video Technology 14(1), 4–20 (2004)
2. Harris, C., Stephens, M.: A combined corner and edge detector. In: Alvey Vision Conference, pp. 147–151 (1988)
3. Zhang, D., Zhou, Z.-H. $(2D)^2$PCA: 2-directional 2-dimensional PCA for efficient face representation and recognition. Journal of Neurocomputing 69(1-3), 224–231 (2005)
4. Lowe, D.G.: Distinctive image features from scale-invariant keypoints. International Journal of Computer Vision 60(2), 91–110 (2004)
5. Xiong, H., Swamy, M.N.S., Ahmad, M.O.: Two-dimensional FLD for face recognition. Pattern Recognition 38(7), 1121–1124 (2005)
6. Koenderink, J., van Doorn, A.: Representation of local geometry in the visual system. Biological Cybernetics 55, 367–375 (1987)
7. Yang, J., Zhang, D., Frangi, A.F., Yang, J.: Two-Dimensional PCA: A new approach to appearance based face representation and recognition. IEEE Transactions on Pattern Analysis and Machine Intelligence 26(1), 131–137 (2004)
8. Yang, J., Zhang, D., Yang, X., Yang, J.: Two-dimensional discriminant transform for face recognition. Pattern Recognition 38(7), 1125–1129 (2005)
9. Sirovich, L., Kirby, M.: Low dimensional procedure for the characterization of human faces. Journal of the Optical Society of America A: Optics, Image science, and Vision 4(3), 519–524 (1987)
10. Van Gool, L., Moons, T., Ungureanu, D.: Affine/photometric invariants for planar intensity patterns. In: Buxton, B.F., Cipolla, R. (eds.) ECCV 1996. LNCS, vol. 1065. Springer, Heidelberg (1996)
11. Kirby, M., Sirovich, L.: Application of the Karhunen-Loeve procedure for the characterization of human faces. IEEE Transactions on Pattern Analysis and Machine Intelligence 12(1), 103–107 (1990)
12. Li, M., Yuan, B.: 2D-LDA: A statistical linear discriminant analysis for image matrix. Pattern Recognition Letters 26(5), 527–532 (2005)
13. Turk, M., Pentland, A.: Eigenfaces for recognition. Journal of Cognitive Neuroscience 3(1), 71–86 (1991)
14. Mikolajczyk, K., Schmid, C.: A performance evaluation of local descriptors. In: Proceedings of Computer Vision and Pattern Recognition (June 2003)
15. Mikolajczyk, K., Schmid, C.: Indexing based on scale invariant interest points. In: Proceedings of International Conference on Computer Vision, pp. 525–531 (July 2001)
16. Nagabhushan, P., Guru, D.S., Shekar, B.H. $(2D)^2$ FLD: An efficient approach for appearance based object recognition. Journal of Neurocomputing 69(7-9), 934–940 (2006)
17. Belhumeur, P.N., Hespanha, J.P., Kreigman, D.J.: Eigenfaces vs. Fisherfaces: Recognition using class specific linear projection. IEEE Transactions on Pattern Analysis and Machine Intelligence 19(7), 711–720 (1997)
18. Zhao, W., Chellappa, R., Phillips, P.J., Rosenfeld, A.: Face Recognition: A Literature Survey. ACM Computing Survey 35(4), 399–458 (2003)
19. Ke, Y., Sukthankar, R.: PCA-SIFT: A more distinctive representation for local image descriptors. In: Computer Vision and Pattern Recognition (2004)

Understanding the Performance of Multi-core Platforms

V.V. Srinivas and N. Ramasubramaniam

Department of Computer Science and Engineering,
National Institute of Technology - Tiruchirappalli
{206110022,nrs}@nitt.edu

Abstract. Recent advancements in multi-core systems have created interest among many research groups to develop efficient algorithms for multi-core platforms. The performance evaluation of such multi-core architectures becomes essential before the design of such high end, cutting edge systems. One important question is how to evaluate the performance of such multi-core systems? In this paper, the authors have made a preliminary study of simulating the various architectural components of a typical multi-core environment using Super ESCalar (SESC) simulator and evaluated their performance using lmbench.

Keywords: multi-core, performance, simulator, SESC, lmbench.

1 Introduction

1.1 Motivation

Multi-core technology plays a significant role in faster execution of user and kernel processes. These form the essential component of any real-time computing system. Effective utilization of multiple processors is the key to increase the computing power in case of multi-core platforms [5], [1]. It can be seen that, the currently implemented algorithms and policies will not scale [6], as the number of cores and on-chip memory units increase. In this paper, we take into consideration some of the previous analytical insights on performance in multi-core environment; refer [2], [3], [4].

The evolution in multi-core technology has lead to exploiting the processing power of multiple cores on a single chip. An important conclusion of our paper is that, proper evaluation of performance, of each architectural component inside a multi-core chip would help greatly, in harnessing the power of multi-core platforms. The evaluation of performance depends on numerous parameters. As compared to the single core systems without pipelining and onboard cache, testing of multi-core systems with a number of onboard components is more difficult. The advancement in transistor technology has resulted in, embedding a large number of onboard components. Hence, the performance of such components has to be evaluated using micro-benchmarking suite.

V.V. Das, J. Stephen, and Y. Chaba (Eds.): CNC 2011, CCIS 142, pp. 22–26, 2011.

1.2 Contribution

In this paper, the authors have evaluated the performance of different architectural components, using lmbench on MIPS instruction set based SESC simulator. The results are reported for a dual core system model.

The rest of this paper is organized as follows: Section 2 discusses the various tools and experimental setup used, followed by Section 3 dealing with the benchmark suite. In Section 4, the results obtained from the experiments are summarized. Section 5 concludes with the prospects for future research.

2 Tools Used

The set of tools namely: SESC (Super ESCalar Simulator), lmbench and condor described here have been used for analyzing the performance. We describe below the details regarding the tools.

Super ESCalar Simulator is a multiprocessor environment developed by i-acoma research group at UIUC in collaboration with other research groups. It is an event-driven, full system simulator [8]. MIPS processor is emulated using SESC simulator. The working of SESC consists of executing the instructions on a MIPS module, which returns the result to the SESC simulator. Further details are available in [9]. The compilation of SESC tool requires gcc version 3.4. The details related to the installation of gcc-3.4 along with gcc-4.X, and the installation of SESCutils is available in [7]. In our experiments we use condor [11], a tool for submitting the configuration scripts and the input to the SESC simulator. LMbench [12] is a set of binaries that helps in evaluating the performance of the system parameters. We have taken lmbench version-3 for running our experiments on MIPS architecture based simulator.

Table 1. Parameters used in SESC experiments

Parameter	Value
Hardware parameters	
Number of memory strides	16, 32, 64, 128, 256, 512, 1024
Number of processors	2
System parameters	
Fetching policy	Out-of-order
Presence of instruction cache	True
Chip parameters	
Clock frequency	2.1 GHz
Chip frequency	45 nm

(a)

Parameter	Value
Processor configuration	
Instruction queue size	8
Integer unit configuration	
Latency for load / store	1
Multiplication latency	2
Division latency	12
Floating unit configuration	
ALU latency	1
Multiplication latency	4
Division latency	12

(b)

3 Benchmarks

The lat_mem_rd measures the memory read latency for different memory sizes. A wide range of memory sizes from 32 MB, 64 MB, 128 MB, 256 MB, 512 MB and 1024 MB have been used for conducting experiments. The results show that as the memory strides increase, the latency for accessing the memory stride also increases; see Fig. 1.a. The lat_fs benchmark helps in measurement of the speed with which files in the current file system can be created and destroyed. The output obtained from the benchmark consists of file size and the number of files deleted per second. Based on the result, the creation and deletion latencies of a single file are computed see Fig. 2.a.

The lat_ctx benchmark measures the context switching time. It takes processes of size 0, 16, 32, 64, 128, 256, 512, 1024 kB as input. Process size of 0 kB denotes, the process does not perform any operation, it simply switches context from one process to another see Fig. 1.b. In Fig. 2.d, the run time latency for operations involving 32 and 64 bit integer and floating point addition, subtraction, multiplication and division are shown.The bw_mem_rd and bw_mem_wr benchmarks calculate the memory read and write rates in terms of Mbps. These benchmarks take file size as input. The former estimates the time for reading zeros for the specified input file size while the later estimates the time for writing zeros. Fig. 2.b and Fig 2.c shows the read rate and write rate for a wide range of file sizes. Using bw_pipe the latency for transferring 50 MB of data between two processes through 64 kB segmented pipe was calculated as 6.1289 microseconds. MHz benchmark finds the clock speed of the processor along with the delay to obtain the clock rate. The processor clock rate obtained was 2095 MHz with delay of 0.4773 nanoseconds.

4 Summary of Results and Avenues for Future Research

Study of lmbench on SESC has helped in analyzing the performance of various architectural components in a multi-core system. Evaluating the performance of simulators using standard benchmarks would provide us a novel and much deeper insight into the performance that one could achieve while designing high end systems.

This research was just a preliminary step towards understanding of multi-core platforms using simulators. Future avenues would include, incorporating scheduling mechanisms and running standard benchmarks to find the performance measures. Another area would be to find whether over-provisioning processor cores in a single machine would always increase performance.

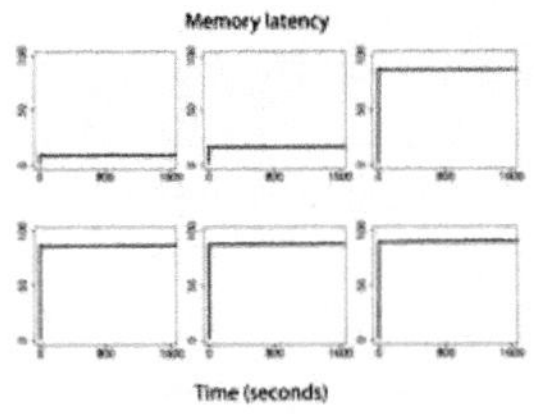

(a) Memory latency with stride values 32, 64, 128, 256, 512, 1024

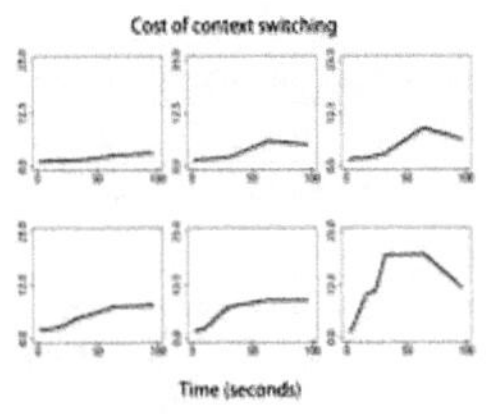

(b) Cost of context switching when the file size is 0, 4, 8, 16, 32, 64 kB

Fig. 1.

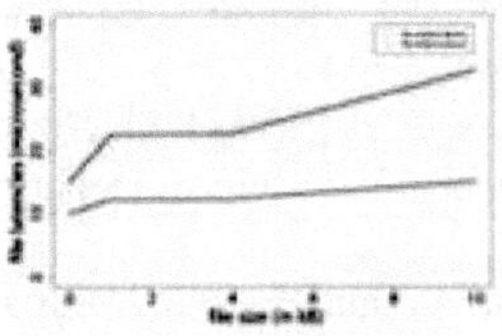

(a) File access latencies for different file size

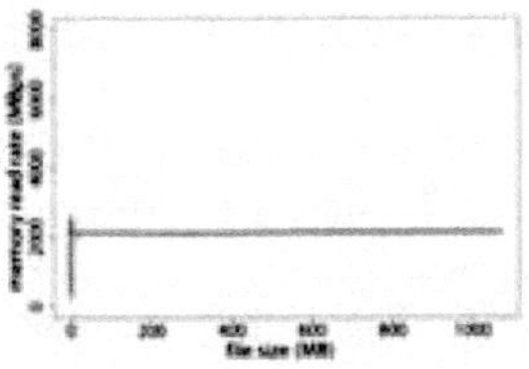

(b) Memory read rate

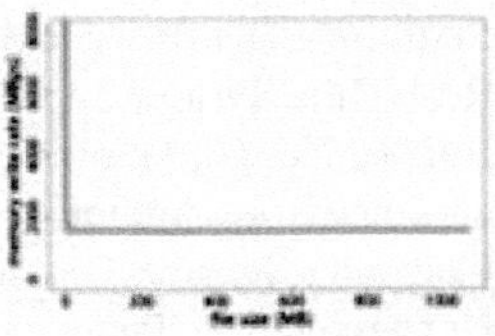

(c) Memory write rate

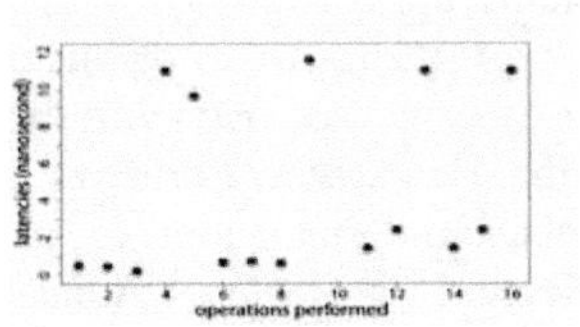

(d) Integer and floating point operational latencies

Fig. 2.

In Fig. 2.d. the values 1 to 16 plotted as dots along the xy plane indicates the various operations performed. 1-int sub, 2-int add, 3-int mul, 4-int div, 5-int modulo, 6-64 bit int sub, 7-64 bit int add, 8-64 bit int mul, 9-64 bit int div, 10-float add, 11-float mul, 13-double add, 15-double mul, 16-double div.

5 Conclusion

In this work, the authors have made a successful attempt in understanding the performance of various architectural components on a multi-core platform using SESC simulator and lmbench benchmark suite. The performance evaluated using SESC showed, results that qualitatively resemble the present day systems. The performance evaluation done on a simulator would certainly help improve the simulator and also to analyze multi-core systems in greater detail.

Acknowledgement

The authors would like to acknowledge National Institute of Technology, Tiruchirappalli for extending its facilities for this research.

References

1. Calandrino, J.M., Anderson, J.H.: On the Design and Implementation of a Cache Aware Multicore Real-Time Scheduler. In: Proceedings of 21st Euromicro Conference on Real-Time Systems, pp. 194–204 (2009)

2. Calandrino, J.M., Anderson, J.H.: Cache-Aware Real-Time Scheduling on Multi-core Platforms: Heuristics and a Case Study. In: Proceedings of 20th Euromicro Conference on Real-Time Systems, pp. 299–308 (2008)
3. Bui, J., Xu, C., Gurumurthy, S.: Understanding Performance Issues on both Single Core and multicore Architecture. In: Technical Report, University of Virginia, Department of Computer Science, Charlottesville (2007)
4. Hennessy, J.L., Patterson, D.A.: Computer Architecture: A Quantitative Approach, 4th edn. Elsevier Inc., Amsterdam (2007)
5. SESC, http://manio.org/sesc-tutorial-1-install-whole-sesc-step-by-step577.html#how-to-install-gcc3.4
6. Ortego, P.M., Sack, P.: Super ESCalar simulator. In: Proceedings of 17th Euromicro Conference on Real-Time Systems, pp. 1–4 (2004)
7. SESC, http://iacoma.cs.uiuc.edu/~paulsack/sescdoc/
8. LMbench, http://bitmover.com/lmbench/lmbench.html
9. Condor, http://www.cs.wisc.edu/condor/tutorials/
10. McVoy, L., Staelin, C.: lmbench: Portable tools for performance analysis. In: Proceedings of the 1996 Annual Conference on USENIX Annual Technical Conference, pp. 23–39 (1996)

Improvisation of E-Learning System Using Web Services Composition

J. Jayapradha, B. Muruganantham, and G. Archana

SRM University

Abstract. Many web based systems are adopting asynchronized way, the teacher publishes learning content statically on the internet and the student obtain static learning materials at different times where, real classroom learning follows a synchronized learning process ,the students in and out of the classroom can listen to the live instructions while the tutor gives the lectures. So, the synchronized learning process is much more attractive than the asynchronized learning process. We build a prototype application called Real Open Smart Classroom build on our software infrastructure using Web Services Composition Technology. Real Open Smart Classroom tries to connect two Smart Classroom together to give a novel experience for the students and the tutor in the class. Simple sharing of the video and voice are inadequate. It is necessary for the Real Open Smart Classroom to provide a new mechanism for the tutor and the students to learn and interact in their Web Based Learning Environment. A Web Based Communication Scheme is established, deployed in an open network for the classroom to make use of the abundant dedicated services using Web Services Composition.

Index Terms: Asynchronized, Synchronized, Web Services Composition, Web Based Learning.

1 Introduction

Distance education system promise wider access to education and support for lifelong learning. These systems are either asynchronized or synchronized. Asynchronized systems are relatively simple where as synchronized are very effective. Real Open Smart Classroom therefore plays an indispensible significance in distance learning. Even though the sharing of videos and documents are lively done, there should be an improvised mechanism where the students in the remote area can clarify the doubts at that moment with the tutor.

To carry out this process a multichat application using Langrid Blackboard [1] is introduced. The Real open smart classroom is equipped with several cameras, to monitor the smart spaces. There are many learning systems embedded with these features such as Active Class [2], Class Talk [3], Interactive Workspace [4], which successfully seek a better searching and learning mode based on traditional classroom learning. These are called classroom based e-Learning system.

V.V. Das, J. Stephen, and Y. Chaba (Eds.): CNC 2011, CCIS 142, pp. 27–32, 2011.

2 New Challenges and Requirements

The classroom based e-learning system helps both the tutor and the student in the learning process and have achieved successful results. However, with the development of Real Open Smart Classroom new services are required for the full-fledged performance of the task. The main aspect of this issue is illustrated below in detail.

2.1 Web Services Composition to Increase the Performance

The proposed approach is Web Services Composition, which is the process of creating new services by composing already existing services. Web Services Composition technology is introduced to build an effective e-learning system. In Web Services Composition by deploying more services, the performance of the task will be increased. When the services invoked by the tutor or students cannot perform in a full-fledged manner, it will get the help of the other services. By adopting Web Services Composition, it facilitates the creation of innovative services that suits the end-users needs with a reduced time-market and better component reuse.

3 Proposed Technique

Addressing the challenge mentioned above, we propose a prototype system called Real Open Smart Classroom. Real Open Smart Classroom is a classroom based e-learning system. We built two classrooms, one is with the tutor and other one is in the remote area. Classroom where the tutor and students are present has four wall size projector screens, smart board and in remote classroom LCD screens are present and with the equipped cameras. Several natural human-computer interfaces are integrated in classroom E.g.: Touch Sensitive SmartBoard. Connected by open internet, two classrooms and multiple software modules in them provide the classroom functionality.

All software modules run on our software infrastructure called Real Open Smart Platform, to communicate and coordinate with each other. Thus the Real Open Smart Platform is not only to build Real Open Smart Classroom but also has important significance in the research of Web Services, contributing to the research of web Services Composition as well.

3.1 Real Open Smart Classroom

Open Smart Classroom [1] was a generic classroom with many modules connected to perform a specified task. In MAS (Multi Agent System) model, only one service was accomplished by specified task. Modules in Real Open Smart Classroom are loosely coupled in order to maintain a robust system, and embedded technologies are involved to remove the computing devices from people sight.

In Open Smart Classroom, Service-Oriented Communication channel between inside and outside of classroom was created [1] to bridge services, help the system to manage and utilize resources. Web Services Composition is implemented in Open Smart Classroom to invoke required services for greater performance, results in Real Open Smart Classroom.

3.2 Real Open Smart Platform

After presenting the motivation of Real Open Smart Classroom, we are going for Real Open Smart Platform. Real Open Smart Platform is developed based on Open Smart Platform such as MAS architecture and Ds-Container-Agent Hierarchy. Web Services Composition is the key improvement to provide more scalability.

3.2.1 Architecure

We inherit the Multi Agent Architecture of Open Smart Platform because it is suitable for Real Open Smart Classroom, where the modules are dynamic and autonomous.

The three system modules:

1. Web-Service-Wrapper-Agent(WSWA),
2. Smart-Platform-Agent Webservice(SPAW),
3. Open-Smart-Platform Gateway (OSPG).

WSWA: WSWA is an agent, which invokes an outside Web service based on the message received from other agents and returns the reply from the Web service to the agents. Agents in Smart Platform can invoke outside Web services by sending messages to this agent.

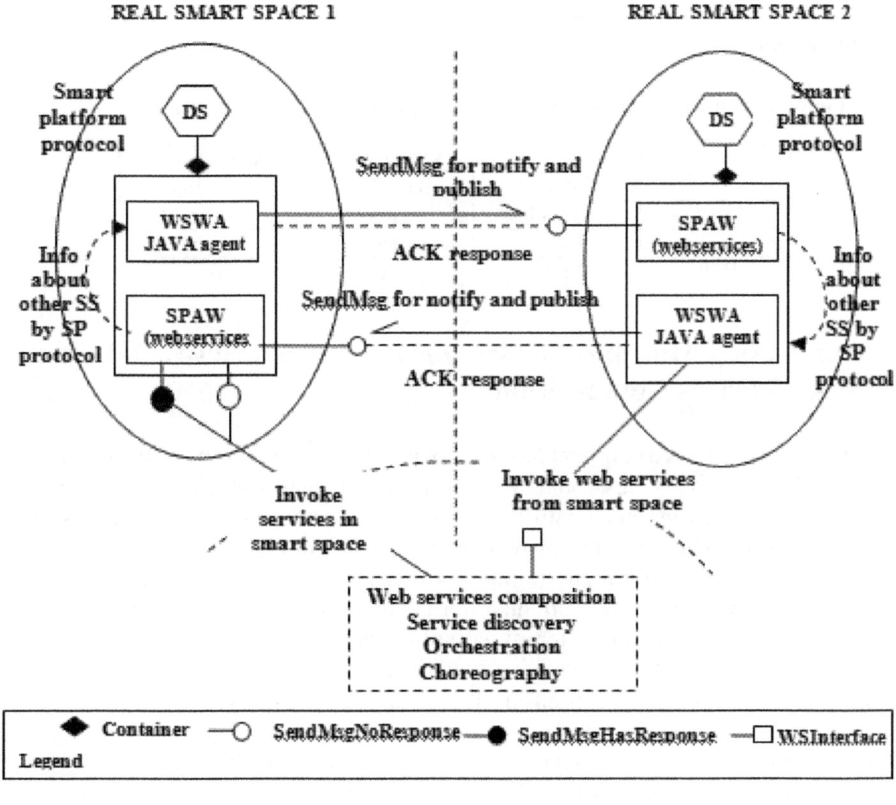

Fig. 1. Real Open Smart Platform runtime architecture for two Smart

SPAW: SPAW is a Web service deployed on Axis[7]. It receives messages from outside systems, transforms the messages to the protocol used in Smart Platform, creates a SPAW-agent to dispatch the message, and returns the reply from the agent to outside systems. Through this mechanism, outside systems can interact with agents in Smart Platform as Web services. SPAW provides two Web service interfaces: sendMsgNoResponse and sendMsgHasResponse. The former is used if the services requester is NOT concerned about the result from the services. Otherwise, the latter is used to get the result from the services.

OSPG: OSPG is a Web application serving as the proxy between mobile devices and Smart Space. It consists of a servlet deployed on Tomcat [6] and several Web services deployed on Axis [7]. Specifically, the servlet builds Web pages for mobile devices to use their browsers to interact with the services inside Smart Space; the Web services is to receive necessary messages from SPAW[1].

Taking two Smart Spaces as the simplest Smart Community for example, the two Real Open Smart Platform runtime architecture for Smartspaces is shown in Fig. 1

4 New Features

In Real open Smart Platform, there are three communication schemes that correspond to a different QoS requirement, which are suitable for multiple smart space communication.

4.1 Three Communication Schemes in Real Open Smart Platform

	Amount of data	Sensitive to data drop	Sensitive to real time
Message	Small	Yes	Yes
Stream	Large	No	No
Bulk	large	Yes	No

5 Connecting Multiple Classrooms in an Open Network Using Web Services Composition

Real Open Smart Classroom emphasizes openness a new experience for the tutors and students. Enabling two classrooms to have class together, especially for those in different countries has greater significance in intercultural learning. It is necessary for Real Open Smart Classroom to provide new mechanism for the tutors and the student to learn in their intercultural environment. Current focus is mainly on the composition of services. Additionally, the administrator is able to modify the workflow of the task in open network to meet the different requirements.

In Open Smart Platform [1] if particular language translation, like Chinese language translation service is required, then it has to be deployed in an open network and if any other language translation service is needed like Japanese language then, Chinese language translation service has to be removed and Japanese language translation service has to be deployed. To overcome the issues, we are going for web services composition where the services can be deployed permanently, and according to the needs the services can be dynamically selected. So, the Real Open Smart Platform has a service management module to address the issues such as dynamical service composition.

6 Advantages of Web Services Composition for Real Open Smart Classroom

Using web services composition there are several advantages for Real Open Smart Classroom. It enables the administrator to manage the system easily. For example, when a new classroom joins in the system, it is very easy to choose its services to invoke, according to the language or any other services to cope up with the new classroom.

The Web services Composition requires the interfaces of agents in the classroom to built in a standard way and thus can be reused as the services resources by other classroom. Reusability plays a major role here. Another is the decision engine of smart cameraman [5], which uses the observed content of the Real Open Smart Classroom as the input to give the appropriate view of the class, as the result.

7 System Analysis

Currently, Real Open Smart Platform which is the software infrastructure for building Real Open Smart Classroom which provides many services that can be used by the tutors and students. New students before joining the classroom should register their account and that will be verified by the administrator and finally registry is updated. New classroom with the same infrastructure and interfaces works effectively in Real Open Smart Platform.

8 Related Work

8.1 Existing Learning System

Several projects work on improving the experience of traditional classroom–based learning using Smart Space Technologies. Active Class [2] encourages in class participation using personal wireless devices, where the students give feedback on the class using their wireless devices, but it lacks support for mobile devices to control and interact with the classroom and thus has few natural human-computer interaction interfaces to enhance the experience of class.

EClass [8] at Georgia tech is another project with automated capture of live experience for late access and it works only for single classroom. iRoom at Stanford[4] explores new possibilities for people to work and learn together in technology rich spaces. However iRoom also workout only for single classroom.

Smart classroom at Tsinghua University [5] is similar to iRoom but supports remote student's interaction and communication but, it lacks a mobile device communication without any preinstalled modules and limited to utilizing useful services outside the classroom. Open Smart Classroom provided a standard communication interface for mobile devices to interact with the other modules and it provided an Open service-oriented channel for easy deployment of different requirements and for improving system capability and it supports the multiple classrooms but it lacks in service management module to address the issue such as dynamical service composition.

9 Conclusion and Future Work

In informal user evaluation with several tutors from different departments at our university, our prototype system received positive comments. Especially for its rich and intuitive interactive channels for working with remote students. Real Open Smart Platform as an upgrade of open smart platform is still MAS integrated with several extended modules. It enables better support in

- Service Management module to address the issues of dynamical service composition.

Currently, the Real Open Smart Platform does not have service management module to address the issue such as dynamical service arbitration.

References

1. Suo, Y., Miyata, N., Morikawa, H., Ishida, T.: Open Smart Classroom: Extensible and Scalability Learning System in Smart Space using Web Service Technology. IEEE Transactions on Knowledge and Data Engineering 21(6) (June 2009)
2. Ratto, M., Shapiro, R.B., Truong, T.M., Griswold, W.: The ActiveClass Project: Experiments in Encouraging Classroom Participation. In: Proc. Computer Supported Collaborative Learning Conf. (CSCL 2003), pp. 477–486 (2003)
3. Dufresne, R.J., Gerace, W.J., Leonard, W.J., Mestre, J.P., Wenk, L.: Classtalk: A Classroom Communication System for Active Learning. J. Computing in Higher Education 7(2), 3–47 (1996)
4. Johanson, B., Fox, A., Winograd, T.: The Interactive Work-spaces Project: Experiences with Ubiquitous Computing Rooms. IEEE Pervasive Computing 1(2), 159–168 (2002)
5. Shi, Y., et al.: The Smart Classroom: Merging Technologies for Seamless Tele-Education. IEEE Pervasive Computing 2(2), 47–55 (2003)
6. The Apache Tomcat (2008), `http://www.apache.org/`
7. Web Services Axis (2008), `http://ws.apache.org/axis/`
8. Abowd, G.D.: Classroom 2000: An Experiment with the Instru-mentation of a Living Educational Environment. IBM Systems J. 38(4), 508–530 (1999)

Artificial Neural Network Based Contingency Ranking

Mithra Venkatesan[1] and Bhuvaneshwari Jolad[2]

[1] Department of Electronics & Telecommunication Padmashree Dr.D.Y.Patil Institute of Engineering & Technology, Pune-411018
[2] Department of Electronics, Padmashree Dr. D.Y.Patil Institute of Engineering & Technology, Pune-411018
mithrav@rediffmail.com, b_jolad@yahoo.com

Abstract. Increased power demand without appropriate expansion of transmission lines has resulted in exploitation of the existing power transmission system. In view of this, the problem of voltage instability leading to voltage collapse is attracting more and more attention. Continuous monitoring of the system status through on-line contingency analysis and contingency ranking is therefore a necessary requirement. Due to its ability to learn off-line and produce accurate results on-line, Artificial Neural Network (ANN) is widely applied for on-line ranking of critical contingencies. Therefore this paper proposes an Artificial Neural Network based approach for fast voltage contingency ranking. The off-line load flow studies are adopted to find the Minimum Singular Value (MSV), which reflects the degree of severity of the contingencies in terms of voltage stability margin of the system, and the results from load flow study are used to train the multilayered ANN for estimating the MSV. The effectiveness of the proposed method is demonstrated by applying it to line outage contingency ranking under different loading conditions for a practical 22-bus Indian system. Once trained the neural network gives fast and accurate ranking for unknown patterns and classifies the contingency considered into groups in accordance to their severity based on the predicted value of MSV. The developed model is found suitable for on-line applications at load dispatch centers.

Keywords: Artificial Neural Networks, Voltage Collapse, Singular value decomposition, Contingency Ranking.

1 Introduction

The interconnections of bulk power systems, brought about by economic and environmental pressures and tendency towards maximizing economic benefits, have led to increasingly more complex systems that must operate close to their stability limits. This has necessitated the study of stressed or heavily loaded situation of the power system network to accomplish efficient energy management without losing its reliability. The system instability, which occurs when the system is heavily loaded, turns into voltage collapse. Hence on-line monitoring of power system stability has become a vital factor for electric utilities. An effective method for real time monitoring of system status and voltage collapse prediction and analysis becomes necessary. Contingency

V.V. Das, J. Stephen, and Y. Chaba (Eds.): CNC 2011, CCIS 142, pp. 33–38, 2011.

ranking is one of the important components of on-line system security assessment. The purpose of contingency ranking is to arrange contingencies according to their severity and pick out those contingencies, which are potentially more harmful to the system, so as to reduce the number of contingencies that require detailed analysis. Suitable preventive control actions can be implemented considering contingencies that are likely to affect power system performance. The main requirement that the processing of all contingencies cases be done in a very short time imposes a heavy computational burden. Hence it is advisable to consider only few severe contingencies for analysis. To further ease the computational burden, the problem of assessing the voltage stability in electrical power system is addressed through Artificial Neural Network approach. Hence, the growing demand causes system to operate close to their stability limit sometimes leads to voltage instability and collapse. And this problem can be prevented or mitigated by ANN based voltage collapse contingency ranking.

Though different algorithms have been applied for the above problem, very few methods yield good speed and accuracy. Hence, to perform ANN based voltage contingency ranking with good efficiency has generated interest and this has motivated the attempts in this paper. This developed voltage collapse based contingency ranking has been applied to a 22-bus practical SREB real system.

2 Voltage Collapse

Voltage collapse is a non-linear phenomenon, occurring over a period of few seconds to hours, characterized by fall in voltage and shortage of reactive power supply. Voltage collapse may be total (blackout) or partial. To use the theory of Singular Value Decomposition on power systems, a linearised relation between the active and reactive powers at nodes versus voltage magnitudes and node angles has to be found which is established by the power flow Jacobian matrix. This relationship has been proposed by Lof.P.A et al (1992).

$$\begin{bmatrix} \Delta P \\ \Delta Q \end{bmatrix} = \begin{bmatrix} H & N \\ J & L \end{bmatrix} \begin{bmatrix} \Delta\Theta \\ \Delta V/V \end{bmatrix}$$

Hence the jacobian matrix essentially establishes relation between the unknowns (voltage magnitude and angles) and the knowns (active and reactive power at nodes). So when the inverse of the jacobian doesn't exist because of matrix singularity, the relation between knowns and unknowns fails to exist. Hence, following inability to establish the relationship, the system becomes unstable and this leads to voltage collapse. Therefore, as the Minimum Singular Value (MSV) is a measure of power flow jacobian's closeness to singularity, it is used as a measure to find nearness to voltage collapse. Hence the analysis of voltage collapse is done in this paper by Minimum Singular Value Decomposition technique where the distance of MSV from zero to an operating point is a measure of proximity to voltage collapse.

Due to the complex network topology of the power system, various equipments and unpredictable loads are subjected to different kinds of abnormalities such as faults, failures, overloading, under loading etc. Together these are termed as 'CONTINGENCIES' on the power system. Contingency analysis is essential for evaluating the system performance and determining measures to improve performance. An exhaustive contingency analysis involves identification and simulation of all possible outages. However, such an analysis would prove ineffective because of the cost and time involved. Therefore, contingencies having high severity and probability of occurrence are identified and termed as critical contingencies and are ranked resulting in the development of contingency ranking algorithm.

In this work, voltage collapse based contingency ranking is carried out. The analysis of voltage collapse is done by minimum singular value decomposition technique.

3 Artificial Neural Networks

The proposed ANN model uses a three-layer network, which is trained, by using error back propagation algorithm. Power system states can be characterized either by bus injections, bus voltages, network configuration or the line power flows.

The following are the neurons in the input layer for desired Input information:

1. Voltage: Voltage stability is the ability of a power system to maintain acceptable voltages at all buses in the system under normal operating conditions and after being subjected to a disturbance. Furthermore, voltage instability results in a progressive and uncontrollable decline in voltage. Therefore voltage profiles are direct indicators of voltage stability. Hence voltage magnitudes at all buses are part of the input.
2. Active and Reactive power of the load: With the increase in load, generation must correspondingly increase to meet the increased demand. However, it is not possible to increase the reactive power generation in generators beyond an extent, as it will damage the field windings. Hence there is always a shortage in reactive power supply, which results in decrease in voltage. Cumulative decrease of voltage leads the system to voltage collapse. Hence active and reactive power of the load is used as additional information to be included in the input.
3. Line Configuration: The presence or absence of a line is likely to take the system either towards or away from collapse point. Hence line configuration forms an integral part of the input data. 1 or 0 represents all lines present in the system, where 0 indicated the occurrence of an outage and line being removed, while 1 indicates the presence of the line. Hence, the input vector of the neural network comprise of

P = [V1, V2.........Vn, Pd1, Pd2...Pdn, Qd1, Qd2............Qdn, 1 or 0.... p times]

Where n is the number of buses in the power system

V is the voltage of each bus

Pd, Qd are the active and reactive power of the load

P is the number of lines connecting any two buses, in the power system

There are no rules for the selection of the number of middle layers and number of neurons in each layer. The number of neurons in each layer is upto the designer. While in general, the more neurons in middle layer, the better the network can fit the targets, too many neurons in middle layer can result in over fitting. Thus choice of number of neurons in the middle layer is based on experimentation and simulation. The output vector [o] of the proposed ANN contains one element, the minimum singular value. In the study MSV is used as a voltage stability index to rank the contingencies. The training set is the representative of the different operating states of a power system. Off-line power calculation results are used to construct the training set.

The proposed neural network was trained by using the training patterns and error back propagation algorithm. In order to evaluate the performance of the proposed ANN, a set of test patterns, which are not part of the training patterns, were generated. The evaluation of the ANN performance includes the sum-squared error of the neural network and the generalization ability of the neural network for new operating conditions.

4 Case Study and Results

22-bus Southern Region Electricity Board Real System is used as working examples in this study. The present study is intended for on-line environment.

A. Algorithm of developed ANN model

The whole strategy of the developed online ANN model is summarized below:

Step 1: Check the type of contingency and accordingly change the input system data for load flow studies.

Step 2: Normalize the result obtained from the load flow studies to form input data for the ANN.

Step 3: Build up the input vector for the current operating condition, with normalized input data and line configuration corresponding to the line outage caused.

Step 4: Present the input vector to the trained MLP and get the value of Minimum Singular Value (MSV)

Step 5: Denormalise the obtained MSV and check the range of the obtained MSV

Step 6: According to the obtained range, classify the contingency as most severe, moderately severe and less severe.

The 35 contingencies are considered. For varying system loads from 10% of base load to base load condition, MSVs are calculated. The obtained MSV and contingency ranking based on the MSV are found.

B. ANN model for 22-bus system

Training patterns are generated using the 35 different contingencies. For each contingency 10 different operating conditions are taken, making a total of 350 different conditions. The total inputs to the ANN are 93, comprising 22 bus voltages, active and reactive power of the load (22+22) and 27 lines present in the system. In each hidden layer 30 and 15 neurons are found to be optimum. Hence the ANN structure for contingency ranking is 93-30-15-1. The training after normalization is carried out till error convergence is obtained. The results of the obtained ANN are tabulated. The performance of the trained neural network was tested on 25 unseen patterns. The performance of neural network for these unseen patterns is shown in fig 5.1 below.In the below figures., straight lines indicate ANN output, while dotted lines indicate actual output The difference between the straight lines and dotted lines is very less, indicating minimum error obtained among the tested unseen patterns. To facilitate easier understanding of results, the contingencies are grouped in accordance to contingency ranking.

Group 1: CRITICALLY SEVERE, Rank 1-5
Group 2: MODERATELY SEVERE, Rank 6-16
Group 3: LESS SEVERE, Rank 17-35

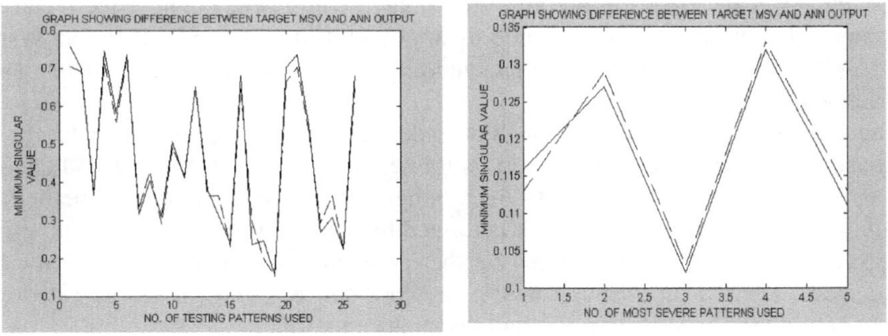

Fig. 5.1 & 5.2. Performance of neural network for unseen patterns and for severe contingencies

C. Observations on ANN results
The ANN results were tabulated. Based on these data that show the error between the target MSV and ANN output graphs were plotted. Subsequently, graphs (Fig. 6.2,6.3 a, b, c, d) illustrate the difference between target MSV and ANN output for the test data as well as when the contingencies are classified into groups. It is observed that the error is minimum. This proves the generalization ability of the trained neural network. The ANN once trained and evaluated provides fast contingency ranking. In fact, the minimization of error by the network proves the generalization ability of the network.

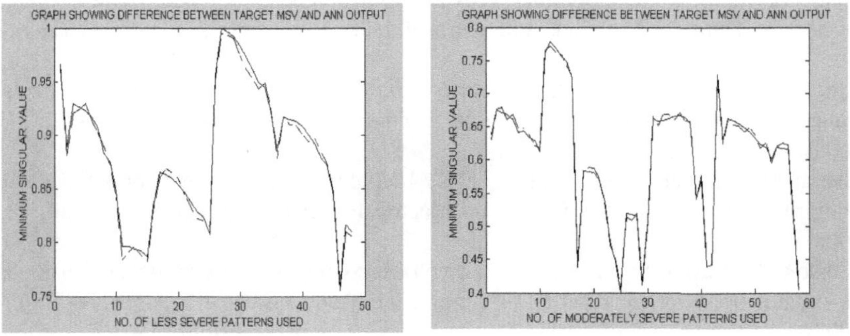

Fig. 5.3 & 5.4. Performance of neural network for moderately severe and less severe contingency

5 Conclusion

This paper proposes an ANN based method for voltage collapse based contingency ranking. By using the results of off-line load flow calculations, Minimum Singular

Value (MSV) is found, which indicates the proximity to voltage collapse. The multilayered artificial neural network is trained to estimate the MSV by using error back propagation learning rule. Hence, the proposed method can provide the information about voltage stability assessment corresponding to both pre-fault and post-fault status for any load condition and any line outage contingency. The effectiveness of the proposed approach is demonstrated by applying it to 22-bus SREB Indian system. The results show that the proposed technique is well suited to be used in on-line environment.

The proposed ANN based approach provides satisfactory calculation accuracy and good analysis speed. Once the network is trained and evaluated, this method can provide a fast contingency ranking. By using singular value decomposition method, the multilayer perceptron neural network is trained to classify the contingency into groups in accordance to their severity based on the obtained value of MSV. Hence the proposed method based on artificial neural network can be used for online ranking of critical contingencies to be used in load dispatch centers.

References

1. Schmidt, M.P.: Application of artificial neural networks to dynamic analysis of the voltage stability problem. IEE Proc. on Generation, Transmission, Distribution 144(4), 371–376 (1997)
2. Srivastava, L., Singh, S.N., Sharma, J.: Knowledge-bases neural network for voltage contingency selection and ranking. IEE Proc. on Generation, Transmission, Distribution 146(6), 649–656 (1999)
3. Pandit, M., Srivastava, L., Sharma, J.: Contingency ranking for voltage collapse using parallel self-organising hierarchical neural network. International Journal on Electric Power and Energy Systems 23, 369–379 (2001)
4. Srivastava, L., Singh, S.N., Sharma, J.: A hybrid neural network model for fast voltage contingency screening and ranking. International Journal on Electric Power and Energy Systems 22, 35–42 (2000)
5. Wan, H.B., Ekwue, A.O.: Arificial neural network based contingency ranking method for voltage collapse. International Journal on Electric Power and Energy Systems 22, 349–354 (2000)
6. Devaraj, D., Yegnanarayana, B., Ramar, K.: Radial basis function networks for fast contingency ranking. International Journal on Electric Power and Energy Systems 24, 387–395 (2002)
7. Manjaree, P., Laxmi, S., Jaydev, S.: Fast voltage contingency selection using fuzzy parallel self-organizing hierarchical neural network. IEEE Trans. on Power Systems 18(2), 657–664 (2003)
8. Moghavvemi, M., Faruque, M.O.: Power system security and voltage collapse: a line outage based indicator for prediction. International Journal on Electric Power and Energy Systems 21, 455–461 (1999)

Detection and Tracking of Moving Object in Compressed Videos

E. Jayabalan[1] and A. Krishnan[2]

[1] Assistant Professor of Computer Science, Government Arts College (Autonomous), Salem-7, Tamilnadu, India
[2] Dean (Academic), KSR College of Engineering, Tiruchengode, Tamilnadu, India
ejksrcas@rediffmail.com, ejksrcas@gmail.com, ejksrcas@yahoo.co.in

Abstract. Object tracking in video sequences is a very important topic and has various applications in video compression, surveillance, robot technology, etc. In many applications, the focus is on tracking moving object. Object tracking is the problem of generating an inference about the motion of an object from the sequence of frames in a video. In this paper, we propose a method for tracking moving objects in video sequences.

Keywords: Object, Tracking, Video, frames, Compression.

1 Introduction

A recently there has been a great demand for the effective and efficient organization of video data [1]. In the view of real world application like content based video retrieval, the major need is to store more number of videos in the database. The advance technology makes better compression ratio that can effectively utilize digital video. The necessity of better video compression is achieved by using standard formats like MPEG through the use of motion estimation and compensation. Today achieving the motion analysis and motion tracking in compressed video format is still remains an open problem.

Detection and tracking involves real time extraction of moving objects from video and continuous tracking over time to form persistent object trajectories. To achieve this objective, the first step is to represent and understand the detected objects present in the scene is known as *object extraction*. The next step is to analyze how these detected objects are moving with respect to each other. These problems are referred in the Computer Vision community as *Object Tracking*. In general, object tracking has five different states that denote the state of objects at any given time instant. These states are new object appearance, an object disappearance, moving object, a stationary object and an occluded object. The conditions for these transitions are marked along the edges. Tracking is one of the most challenging issues of Computer Vision, Robotics, Pattern recognition, artificial intelligence and multimedia. Tracking is used in wide range of applications like Surveillance system, Content based retrieval system, Monitoring systems and Navigation systems, Human motion modeling etc [2][3][4][5][6].

V.V. Das, J. Stephen, and Y. Chaba (Eds.): CNC 2011, CCIS 142, pp. 39–43, 2011.

2 Proposed System

2.1 Tracking of Moving Objects in Compressed Videos

The implementation of multiple objects tracking is achieved in two phases. The first phase focuses on Object detection. To achieve this step, preprocessing is to extract the motion vectors from an MPEG2 stream and by using extracted motion vectors Objects are detected. Second phase is to track the objects by using extract features from the detected objects.

3 Experimental Results

3.1 Object Detection

The proposed system takes the motion vectors at the time of decoding the compressed video stream. In general, considered test videos are having consecutive P-frames separated by two or three B-frames are still similar and would not vary too much. Besides, it must be noted that B-frames are just "interpolating" frames that hinge on the hard motion information provided in P-frames and therefore using them for the concatenation of displacements would be redundant. However, the motion vectors of P-frames or B-frames in MPEG2 compression standard may not actually represent actual motions in a frame. These detected motion vectors have to be stored in an order that will allow the motion from frame to frame to be calculated. Then the algorithm starts with reordering the bit stream order to the display order followed by storage of motion vectors. Then transform the motion vector information into discrete form as given below.

```
for each macro block m_t (x,y) do
If  m_t (x,y)>0 or m_t (x,y)<0 then store_t (x,y)=1
    else store_t (x,y)=0
```

In the above transformation $store_t\,(x,y)$ consists discrete information of each frame in a video.

3.2 Tracking Based on Multiple Feature Approach

One of the easiest ways of finding tracking of objects in compressed domain is by calculating the use of motion estimation and compensation. For calculating the motion estimation, corresponding macro blocks are compared between the frames. This search method was mainly used to achieve the regions wherever the object motion was presented. There by extracting the features like centroids, dispersion rate for each cluster, tracking is maintained effectively.

The tracking results depend on object trajectories like object position etc. This object position is defined by the centroid of the object (C_x, C_y) as refer to the equation (1)

$$C_x = \left(\sum_{(i,j)\in O} P_{i,j} \times i\right) \Big/ \left(\sum_{(i,j)\in O} P_{i,j}\right)$$

$$C_y = \left(\sum_{(i,j)\in O} P_{i,j} \times j\right) \Big/ \left(\sum_{(i,j)\in O} P_{i,j}\right) \qquad (1)$$

Where O is the set of coordinates of an object area and $P_{i,j}$ is the intensity value of the frame at the position (i, j).

Generally the object trajectory drawn from the centroids of the objects in the adjacent frames are close to straight lines. So that, Object acceleration rate is constant for the adjacent frames. By using the initial velocity and acceleration rate, it is easier to predict the locations in the current frame. By comparing the predicated positions and real positions, it is possible to achieve trajectory based tracking. If tracking is performed only with centroids, it's difficult to find the situations like new tracks, track collisions. So that more features like object dispersion rate, gray scale information and textures etc are used for efficient tracking. The limitation of using these features is that the sizes of the objects should not change dramatically between adjacent frames. The dispersion of the object *disp* as refer to equation (2).

$$disp = \left(\sum_{(i,j)\in O} \sqrt{(i-C_x)^2 + (j-C_y)^2} \times P_{i,j} \right) \Big/ \left(\sum_{(i,j)\in O} P_{i,j} \right) \qquad (2)$$

When the dispersion rate is computed for each object region in all frames, it is easy to track the objects by comparing them.

In case of compressed domain, implementation focuses on extraction of the motion vectors from an MPEG2 stream. This information is used to calculate the motion of each and every object from one frame to another. The implementation is done in four phases. In the first phase, implemented algorithm has two threads called scan thread and animated thread. In the second phase, MPEG decoder is used to extract and decode the motion vectors. In third phase, the objects are segmented based on clustered information obtained from object detection algorithm. The fourth phase describes about feature based tracking for multiple objects. If the detected objects are more than 4 objects, then those objects are tracked by using graph theoretical approach. The results obtained for video object segmentation and it's tracking up to frame 32 as shown in Fig. 3 to Fig.4.

Inference: In test video shown Fig. 1 to Fig. 4 the detected objects were moving from left to right.

Fig. 1. Tracking of a Single Object (ball)

Fig. 2. Multiple object tracking upto 15 frames

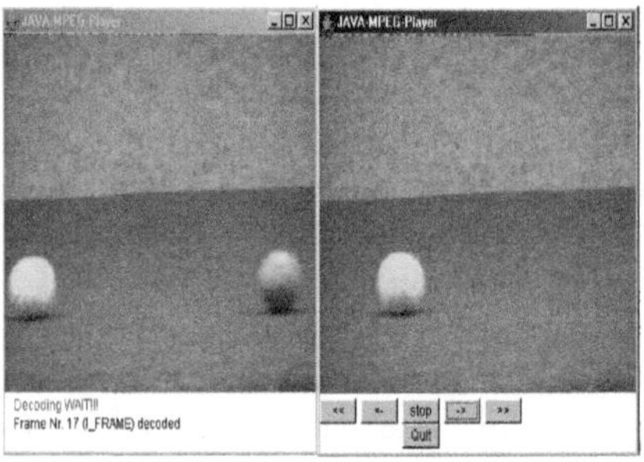

Fig. 3. Detection of *Collision* at the 16th frame

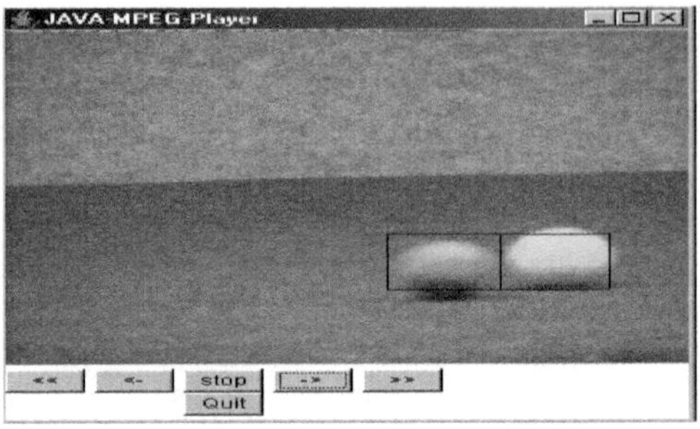

Fig. 4. Detection of *stationary* object

Graphical Analysis and Interpretation

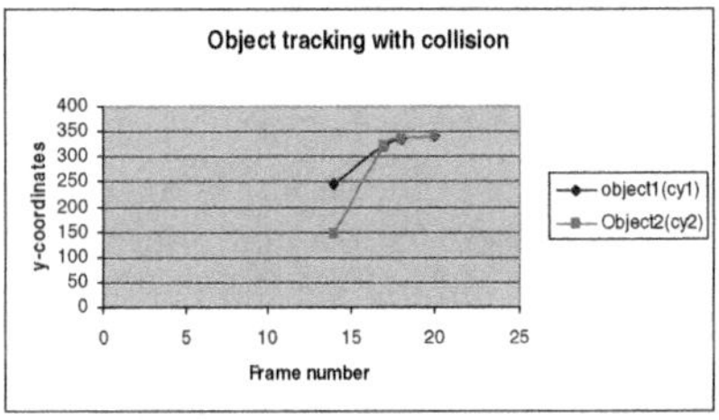

Fig. 5. Multiple Object Tracking in normal conditions

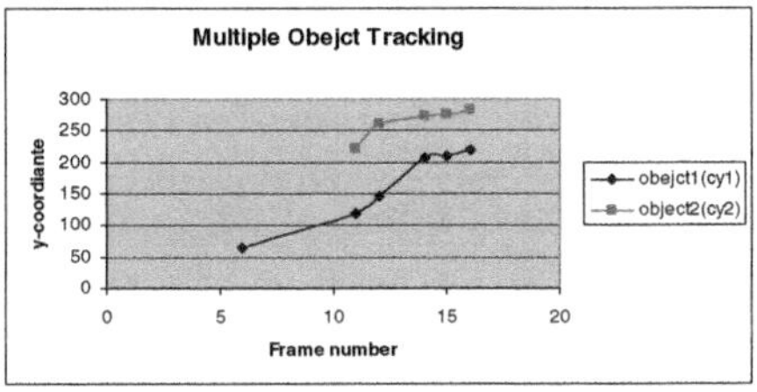

Fig. 6. Multiple Object tracking with Collision

Interpretation

Fig. 5. shows tracking in case of normal conditions (where the objects are moving from left to right), y-coordinate of both objects gradually increases with constant x-coordinate, where as in Fig. 6. shows the state of collision where the y-coordinate are matching from 17th frame onwards with constant x-coordinate.

4 Conclusion

In this paper, we analyze the constraint underlying general object motion including non-rigid cases. In case of compressed domain, implementation focuses of extraction of the motion vectors for an MPEG stream. This work estimates those motion and its positions of moving objects of all frames in a video.

References

1. Kim, N.W., Kim, T.Y., Choi, J.S.: Motion Analysis using the Normalization of Motion Vectors on MPEG Compressed Domain. In: International Technical Conference on Circuits/Systems, Computers and Communications, Phuket, Thailand, vol. 3, pp. 408–1411 (2002)

2. Zang, Q., Klette, R.: Object Classification and Tracking in Video Surveillance. Technical report 128, Communication and Information Technology Resarch, The University of Aukland (2003)
3. Zhou, Q., Aggrwal, J.K.: Tracking and Classifing Moving Objects from Video. In: 2nd IEEE Workshop on Performance Evaluation of Tracking and Surveillance, Kauai, Hawaii, USA ((2001)
4. Cohen, I., Medioni, G.: Detecting and Tracking Moving Objects for Video Surveillance. In: IEEE Computer Society Conference on Computer Vision and Pattern Recognition, Fort Collins, Colorado, vol. 2, pp. 2319–2325 (1999)
5. Achler, O., Trivedi, M.M.: Camera Based Vehicle Detection, Tracking, and Wheel Baseline Estimation Approach. In: 7th International IEEE Conference on Intelligent Transportation Systems, Washington, D.C, pp. 743–748 (2004)
6. Rosenberg, Y., Werman, M.: Real-Time Tracking from a Moving Video Camera: A Software Approach on a PC. In: 4th IEEE Workshop on Applications of Computer Vision, Princeton, NJ, USA, pp. 238–239 (1998)
7. Porikli, F.: Real-Time Video Object Segmentation for MPEG Encoded Video Sequences. In: 8th SPIE Conference on Real-Time Imaging, San Jose, vol. 5297, pp. 195–203 (2004)
8. Symes, P.: Video Compression Demystified. McGraw Hill, International Edition (2001)
9. Yokoyama, T., Iwasaki, T., Watanabe, T.: Motion Vector Based Moving Object Detection and Tracking in the MPEG Compressed Domain. In: 7th IEEE Workshop on Content-Based Multimedia Indexing, Chania, pp. 201–206 (2009)
10. Yokoyoama, T., Ota, S., Watanabe, T.: Noisy MPEG Motion Vector Reduction for Motion Analysis. In: 6th IEEE International Conference on Advanced Video Signal Based Surveillance, Genova, pp. 274–279 (2009)

An Enhanced Round Robin Policy for Real Time Traffic

Prosanta Gope[1], Ajit Singh[2], Ashwani Sharma[2], and Nikhil Pahwa[2]

[1] Member IACSIT, M.Tech. in Computer Science & Engineering, NIT Durgapur, West Bengal, Zip-713709, India
prosanta.dgp@gmail.com
[2] TIT&S, Bhiwani
ajit713@gmail.com, er_ashwani@in.com, nikiii.pahwa@gmail.com

Abstract. The amount of traffic carried over the Internet has dynamically increase with the tremendous popularity of World Wide Web because of that it is mandatory to have a router or high speed switch that can deals with the network traffic in an efficient way. This one is also applicable for a computer system of multiprogramming environment where multiple process are required to have the support of CPU to perform the required operations but the CPU will handle then in a proper manner by using an efficient CPU scheduling algorithm. Here in this paper we have introduced an enhanced cost effective model of Round Robin Technique which is applied on a real time traffic based on policy based routing. Apart from this we have also analysis the enhanced model of Round Robin with some valuable case studies.

Keywords: Round Robin, Quantum, Preemptive, BGP Router, Context Switches, CPU cycles.

1 Introduction

The world went through a long period (late 80's, early 90's) in which most popular operating system (DOS, Mac) has no sophisticated CPU scheduling algorithms. They were single threaded and ran one process at a time until a user direct them to run another process. More recent system (Window NT) has back to having more sophisticated CPU scheduling algorithm. When more than one process is runnable, the operating system must decide which one to run first and which one will be the next. The part of the operating system which makes this decision is schedule; the algorithm it uses is called the scheduling algorithm. Different CPU scheduling algorithms have different properties and may favor one class of processes over another. The algorithm which can maximize the CPU utilization and throughput and can minimize the Turnaround Time, Waiting Time and Response Time is known to be an efficient one.

There are some scheduling algorithms those are preemptive in nature and others are not preemptive. The strategy of allowing process that are logically runnable to be temporally suspended is called preemptive scheduler and ease to contrast to the

V.V. Das, J. Stephen, and Y. Chaba (Eds.): CNC 2011, CCIS 142, pp. 44–50, 2011.

run to completion method of early batch systems. Run to completion is also called non-preemptive scheduling. The Table 1 is representing the functionalities of different CPU scheduling algorithms along with their nature. Here in this paper we will discuss some methodologies through which the performance of Round Robin techniques can be made efficient one. Thus a new enhanced version of Round Robin scheduling algorithm has been designed which we have applied on a real time traffic environment with some valuable case studies. This algorithm is suitable for an environment where router is connected with different autonomous systems deal with the packet of variable size mostly the larger one with different characteristics.

Table 1. Comparative studies of several Scheduling Algorithms

Algorithm	Nature	Features	Drawbacks
FCFS	Non-preemptive	The process that requests first is allocated the CPU	Convoy-effect
SJF	Preemptive/Non-preemptive	Shortest job get the CPU first	Difficult to know the length of next CPU request.
Priority Scheduling	Preemptive	CPU is allocated on basis of priority	Infinite blocking (Starvation)
Round Robin	Preemptive	Designed on the basis of Time quantum, applicable for time sharing	Sometimes it can behaves same as FCFS policy
Proposed one	Preemptive	Variable time quantum and efficient	Bit complexity

2 Round Robin Scheduling

It is one of the oldest, simplest, fairest and most widely used scheduling algorithms, designed especially for time-sharing systems. A small unit of time called time slice or quantum[3] is defined. All runnable processes are kept in a circular queue. The CPU scheduler goes around this queue, allocating the CPU to each process for a time interval of one quantum. New processes are added to the tail of the queue. The CPU scheduler picks the first process from the queue, sets a timer to interrupt after one quantum, and dispatches the process. If the process is still running at the end of the quantum, the CPU is preempted and the process is added to the tail of the queue. If the process finishes before the end of the quantum, the process itself releases the CPU voluntarily. In either case, the CPU scheduler assigns the CPU to the next process in the ready queue. Every time a process is granted the CPU, a counter context switch occurs, which adds overhead to the process execution time, Round-Robin Analysis. If there are **n** processes in the ready queue and the time slice is **q**, then each process ideally would get **1/n** of the CPU time in chunks of **q** time units, and each process would wait no longer than **nq** time units until its next quantum. A more realistic formula would be **n(q+o)**, where **o** is the context switch overhead. So, for practical purposes, it is desirable that the context switch be negligible compared to the time slice. The performance of the Round-Robin algorithm depends heavily on the size of the quantum. If the quantum is very large, the Round-Robin algorithm is similar to the **First- Come, First-Served** algorithm. If the quantum is very small, the Round-Robin approach is called *processor sharing*.

3 Proposed Algorithm

Step 1: The individual number of cycles (INC) of each process as (CPU Burst Time)/(Original time Quantum) is evaluated.
Step 2: Calculate total number of Context Switches CPU Cycles according to the simple Round Robin Scheduling algorithm.
Step 3: Compare the average number of Context Switches or Average number of CPU Cycles with the INC of each process.
Step 4: The average numbers of Context Switches or Average Total numbers of cycles are verified with the INC of each process and self adjusted time quantum is made according to CPU burst of a particular process as per the requirement.
Step 5: Now variable number of time Quantum has been introduced.

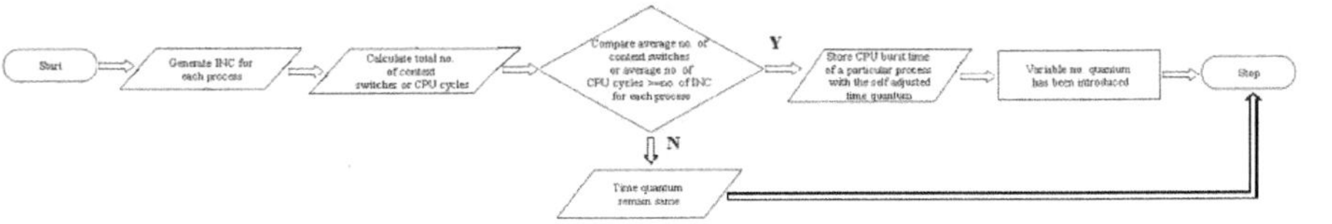

4 Case Studies

4.1 For Simple Round Robin Algorithm

Process ID	CPU Burst Time (ms)	INC(Individual no. of CPU Cycles)
P1	25	7
P2	5	2
P3	15	4
P4	8	2
P5	10	3

Gantt Chart

P1	P2	P3	P4	P5	P1	P2	P3	P4	P5	P1	P3	P5	P1	P3	P1	P1	P1

No. of context switches=15; No. of CPU cycles=18

Waiting time
WT1=38; WT2=20; WT3=39; WT4=25; WT5=37
Total Waiting Time =159ms; Avg. Waiting Time=31.8ms;
Avg. Turnaround Time =45ms

4.2 For Proposed Algorithm

Process ID	CPU Burst Time	Number of Cycle Req.	Self Adjusted Time Quantum	New no. of Cycles
P1	25	7	4	7
P2	5	2	5	1
P3	15	4	4	4
P4	8	2	8	1
P5	10	3	10	1

Gantt Chart

P1	P2	P3	P4	P5	P1	P3	P1	P3	P1	P3	P1	P1	P1

Number of Context Switches =11; Number of CPU Cycles =14

Waiting Time
WT1=38; WT2=04; WT3=39; WT4=13; WT5=21
Total Waiting Time=115ms; Avg. Waiting Time=23ms; Avg. Turnaround Time=35.6ms

Now from the above result it has been proved the performance of the new enhanced model of Round Robin Scheduling is efficient with respect to Context Switch rate, number of CPU Cycles, Average Waiting Time and Average Turnaround Time.

5 Enhanced Model for Policy Based Routing

In today's high performance internetworks organization need the freedom to implement packet forwarding and routing to their own defined policies in a way that goes beyond traditional routing protocol concerns, where administrative issues dictate the traffic be routed through specific path .By using policy based routing customer can implement policies that selectively cause packets to take different paths. Policy routing also provides a mechanism to mark packets so that certain kind of traffic differentiated. Now by applying the enhanced model of Round Robin Technique on a certain environment shown below where the Packet from different departments of several sizes arrived at Router 1 those destination are the BGP Router .Under this circumstances we are considering that the Enhanced Round Robin Technique is running on the Router 1,where the time quantum will be varied on the basis of packet sizes arrived from each and every department.

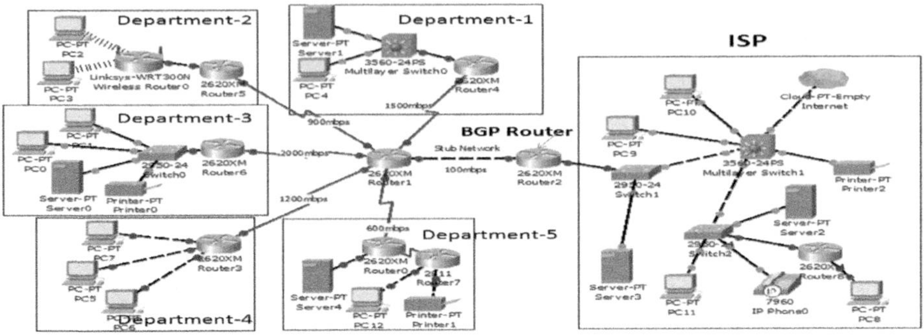

the Router 1
Router>enable
Router#config t
Enter configuration commands, one per line.End with CNTL/Z.
//Now configuring every router and their interfaces//
Router(config)#int ser1/1
Router(config-if)#ip address 182.168.21.1 255.255.255.0
Router(config-if)#description line -1500mbps have 15 time quantum
Router(config-if)#clockrate 1500000
Router(config-if)#show controller s1/1 | in DCE/DTE
Router(config-if)#no shut
Router(config-if)#exit
Router(config)#int ser1/2
Router(config-if)#ip address 182.168.31.1 255.255.255.0
Router(config-if)#description line -900mbps have 9 time quantum
Router(config-if)#clockrate 900000
Router(config-if)#show controller s1/2 | in DCE/DTE
Router(config-if)#no shut
Router(config-if)#end
//Similarly Configuring the every interface of the router 1 and giving the quantum value to each interface//

Now applying simple RR approach on router R1 we have

Dept. ID	Dept. Data transfer required(mb)	Time required(sec.)	No. of cycles
1	1500	15	4
2	900	09	3
3	2000	20	5
4	1200	12	3
5	600	6	2

Let the time quantum = 4sec; No. of context switches=15; No. of CPU cycles=17; Total waiting time=186; Average waiting time=37.2
Applying the proposed algorithm on R1 we have

Dept. ID	Time req.	cycles acc. to quantum	Self-adjusted time	cycle acc. to proposed algorithm
1	15	4	04	04
2	09	3	09	01
3	20	5	04	05
4	12	3	12	01
5	06	2	06	01

No. of context switches=10; No. of the cycles according to proposed algorithm=12
Total waiting time=131; Average waiting time=26.2

Hence once again it has been proved that the newer enhanced RR policy is the better than the simple one.

6 Conclusion and Future Work

Here in this paper we have introduced an enhanced version of Round Robin scheduling technique by analyzing some specific parameters like context switching and CPU cycles. Apart from this we have also seen how an efficient algorithm can change the effect on packet switched network, where a router which is receiving packet from different autonomous systems can take some valuable decisions before it forwards to other network on the basis of specific parameters. The proposed directions for the future work could be to analyze following factors:

1. Make the enhanced model more effective by introducing some other parameters.
2. Introduce the enhanced one on more complex network traffic like ISP, etc.

Acknowledgement. The authors would like to acknowledge and thank their parents, god and all the human being with great heart for their support and encouragement as well.

References

[1] Chaskar, H.M., Madhow, U.: Fair Scheduling with tunable latency:a round robin approach. IEEE/ACM Trans. Netw. A247, 529–551 (1955)
[2] Katevenis, M., Sidoropojlos, S., Courcoubetis, C.: Weighted round robin cell multiplexing in a general-purpose ATM switch chip. IEEE J. Sel. Areas Commun. (1991)
[3] Shreedhar, M., Varghese, G.: Efficent fair queueing using deficit Round Robin Technique. IEEE/ACM Trans. Netw. (1996)
[4] Saha, D., Mukherjee, S., Tripathi, S.K.: Carry-over round robin: a simple cell scheduling mechanism for ATM networks. IEEE/ACM Trans. Netw. (1998)
[5] Saha, D., Mukherjee, S., Tripathi, S.K.: Carry-over round robin: a simple cell scheduling mechanism for ATM networks. IEEE/ACM Trans. Netw. (1998)

[6] Zhang, L.: Virtual Clock: a new traffic control algoritm for packet switching networks. ACM SIGCOMM Comput. Commun. Rev. (1990)
[7] Garg, R., Chen, X.: RRR: recurcive round robin scheduler. Int. J. Comput. Telecommun. (1999)
[8] Bae, J.J., Suda, T., Simha, R.: Analysis of a finite buffer queue with heterogenous markow modulated arrival process: A study of a the effects of traffic burstiness on individual packet loss. In: Proceeding of the IEEE INFOCOM
[9] Saito, H.: Optimal queuing discipline for real time traffic at ATM switching nodes. IEEE Transaction on Communications
[10] Tanenbaum, A.S.: Operating System Desin and Implementation. Princeton Hall, Englewood Cliffs (1987)
[11] Gueria, R., Ahmadi, H., Naghahineh, M.: Equivalent Capacity and its application to bandwidth allocation is high-speed networks. IEEE Journal on selected Areas in Communication
[12] Boedihardjo, A.P., Liang, Y.: Hierarchical smoothed round robin scheduling in high speed networks, `http://www.ietdl.org`
[13] Silerschatz, A., Galvin, P.B., Gagne, G.: Operating System Concepts

A Resultant Output

(a) According to Simple Algorithm

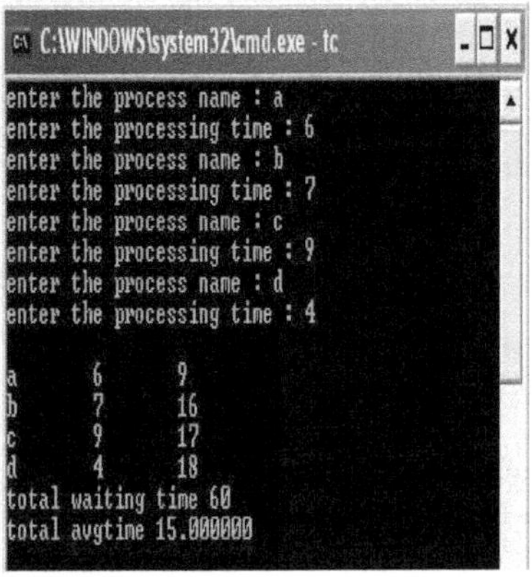

(b) According to Proposed Algorithm

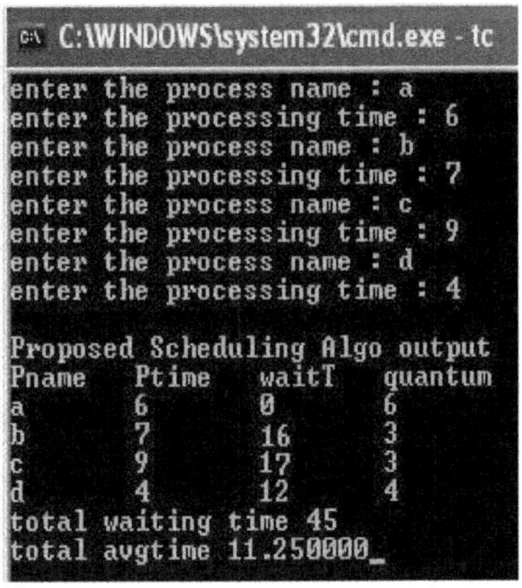

A Novel Reconfigurable Video Burst Processor Offering High QoS

B. Bag, A.K. Jana, and M.K. Pandit

Department of Electronics & Comm Engg., Haldia Institute of Technology (HIT)
C/o ICARE (Indian Centre for Advancement of Research and Education)
Haldia -721657 WB India
mkp10011@yahoo.com

Abstract. Real-time video processing is till today a formidable task thanks to the strict requirement on latency control and packet loss minimization. Burst processing has come to the rescue by offering buffer less operation and separation of control and data information. We report here for the first time a novel statistical embedded burst scheduling method suitable for processing class-differentiated video channels. The method is based on optimized Markov chains where the initial scheduled Markov transition probabilities are subsequently adaptively reconfigured by the scheduler to maintain the best system QoS.

Keywords: Embedded systems, Markov process, Video communication, QoS, Reconfigurable computing.

1 Introduction

In today's world the Internet is seen to be the most promising approach to build the next generation communication infrastructure. In the next generation Internet one must address the QoS issue for mission critical and real-time applications that require low transmission delay and low packet losses.

Yoo et al [1] have studied QoS for embedded systems. But the QoS considered is specific for an individual process. S. J. Yoo [2] suggested some new approaches towards scheduling but they are not *QoS-aware*. Safavian [3] and Tomoyoshi et al [4] have studied QoS [5-6] in embedded systems but none of them considered Markov chains. The concept of *dynamically-optimized* system-QoS based on Markov statistics, is for the first time we have worked out here to our best of knowledge.

2 Methodology

In our system we assume there are three process of different priority level based on their traffic rate and required service rate such that waiting time is negligible or very less. High priority is class1, medium priority is class 2 and low priority is class 3. To serve the incoming packet depending on the traffic rate we applied the concept of software *threads*.

V.V. Das, J. Stephen, and Y. Chaba (Eds.): CNC 2011, CCIS 142, pp. 51–55, 2011.

3 Video Burst

The burst length should neither be very small nor be very large. If the burst length is small, throughput will be less, while if the burst length is very large, delay will grow up. This is due to higher wait times suffered when packets are formed into larger bursts - this causes additional overall delay. The burst length should therefore be optimized for the best QoS.

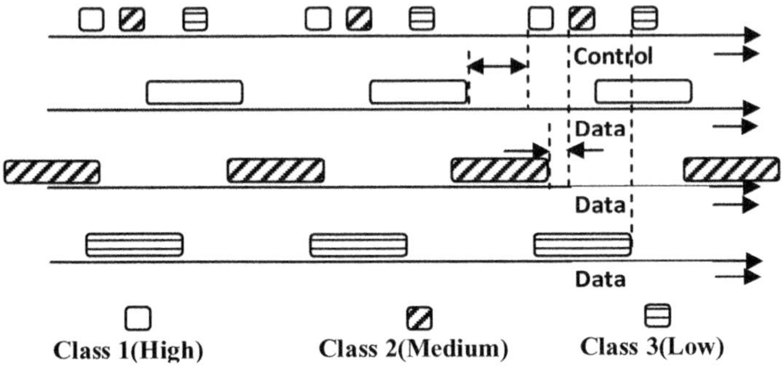

Fig. 1. Video Channel Bursts with assigned priorities

3.1 Markov Process

The video bursts are assumed to arrive with a Poisson probability distribution. The high-priority classes have a longer offset time (separating control and data burststhan the low-priority classes as shown in Fig. 1. For this system, we calculate the total burst loss probability and the burst loss probability for each priority class.

We use *three* types of video traffic to evaluate the performance of the embedded video processor. The first one is NTSC *hard real-time* video signal which has extremely tight timing requirements and so treated as class I. Second one is *real-time* video which has slightly less strict timing requirements and considered as class II traffic. Third one is streaming video signal and assigned class III priority.

3.2 Optimization Parameters

We analyze the lower bounds on the burst loss probability. The results obtained from analysis and simulations are described. The number of average burst length (L) and the number of threads (k) for different classes are considered as the key design parameters in the system. The class I channel generates bursts according to an exponentially distributed arrival rate with average λ_i and with an exponentially distributed service rate with average μ_i where $\mu_i = 1/L_i$.

The normalized traffic intensity of class i is $\rho_i = (\lambda_i / \mu_i) /k = r_i / k$ where traffic intensity $r_i = \lambda_i / \mu_i$. The processes in the embedded processor are modeled as Markovian. The lower bound on blocking probability is determined by assuming that there is *minimal buffer*. It is given according to Markov model as

$$\mathrm{Pb_l} = \frac{\mathrm{r^D}}{\mathrm{k^{D-k}\,.\,k!}} P_0 \tag{1}$$

$$\text{where } \mathrm{P_0} = \left[\sum_{n=0}^{k-1} \frac{\mathrm{r^n}}{\mathrm{k!}} + \sum_{n=k}^{D} \frac{\mathrm{r^n}}{\mathrm{k^{n-k}.k!}} \right]^{-1} \tag{2}$$

where the total number of bursts that can be processed by the embedded processor is $D = k + k.N$, N = physical memory block size allocated to each thread.

The performance of this model is assessed by evaluating the blocking probability. The processing nodes are assumed to follow Poisson traffic model. The results of burst loss probability vs. normalized traffic Intensity for all classes using equation (2) have been shown in Fig 2.

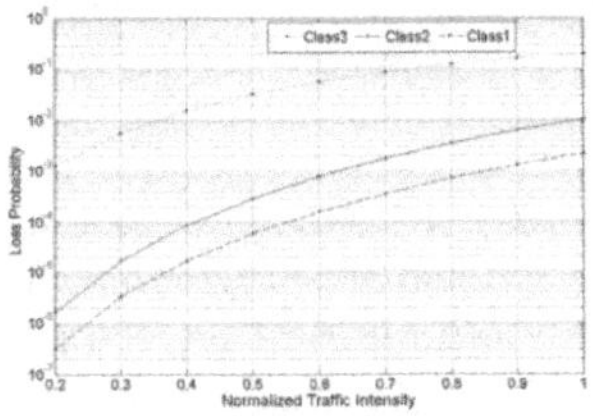

Fig. 2. Loss probabilities for different classes of traffic

We observe that class 3 traffic suffers the highest loss and as the traffic escalates loss probability also rises. From Figure 2 it is clear that up to 0.3 unit traffic load, class 3 burst loss probability grows rapidly and at 0.3 unit, it is approximately 5×10^{-3}.

The system QoS is calculated using *Markov chain* analysis where the different processes are treated for their state transition probabilities. In our model we consider three processes assigned to a single processor. Considering a finite state machine (FSM), the three states $<P_1>$, $< P_2>$, $< P_3>$ define the execution of three corresponding processes by the CPU.

4 Simulation Model

For the sake of simulation, we consider two matrices 1) A, a state transition matrix for Markov model assumed (here three state model) and 2) B, an error probability matrix (at normalized traffic intensity of 0.6).

$$A = \begin{bmatrix} 0.90 & 0.08 & 0.02 \\ 0.10 & 0.80 & 0.10 \\ 0.20 & 0.20 & 0.60 \end{bmatrix} \quad B = \begin{bmatrix} 0.9998 & 0.9990 & 0.9500 \\ 0.0002 & 0.0010 & 0.0500 \end{bmatrix} \tag{3}$$

The simulation is done by MATLAB version 7.0. We perform simulation for calculating error vector which give error positions in 20000 sequences (iterations).

4.1 Simulation Results

At normalized traffic intensity of 0.6, the probabilities of processes 1, 2, and 3 are found to be 0.59, 0.24 and 0.17 respectively and the system (loss) error probability 0.03. It is seen from Fig 3 that the system loss probability grows up with traffic. Consequently it keeps changing with varying *traffic pattern*.

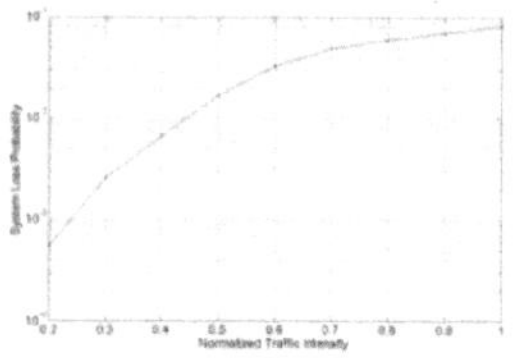

Fig. 3. The system loss probability as a function of traffic

4.2 Reconfigurability

As the traffic pattern changes, the pre-allocated state transition probabilities of matrix A are *unable* to maintain the QoS at its *maximum*. We solve this by *re-configuring* matrix A with the help of reconfiguration parameters Δ_1, Δ_2 and Δ_3 as shown in Equation 4.

$$A_{reconfig} = \begin{bmatrix} 0.90\Delta 2\Delta_1 & 0.08+\Delta_1 & 0.02+\Delta_1 \\ 0.10+\Delta_2 & 0.80-2\Delta_2 & 0.10+\Delta_2 \\ 0.20+\Delta_3 & 0.20+\Delta_3 & 0.60-2\Delta_3 \end{bmatrix} \quad (4)$$

This brings the probability of error to a minimum 0.007 and hence QoS back to *permissible maximum* as shown in Fig 5.

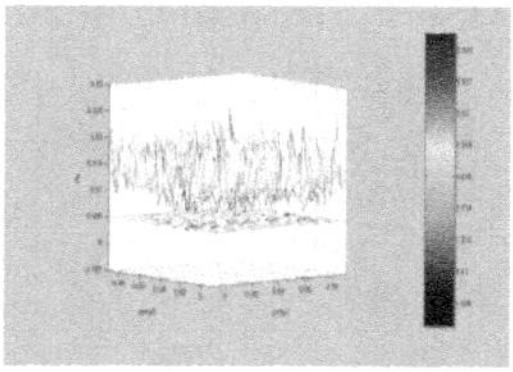

Fig. 5. Pe vs Δ 1, Δ3 ($\Delta 2 = 0$)

5 Conclusion

We have proposed and designed a novel embedded scheduler using Markov statistics.

References

1. Yoo, M., Qiao, C., Dixit, S.: QoS performance of optical burst switching in IP-over-WDM networks. IEEE/OSA J. Lightwave Technology 18(10), 2062–2071 (2000)
2. Ben Yoo, S.J.: Optical packet and burst switching for the future photonic internet. IEEE/OSA J. Lightwave Technology 24(12), 4468–4492 (2006)

3. Rasoul Safavian, S.: How to dimension wireless networks for packet data services with guaranteed QoS. Bechtel Telecommunications Technical Journal 3(1) (March 17-19, 2008)
4. Tomoyoshi, S., Kosuke, T.: Table-based QoS control for embedded real-time systems. In: Proc of the ACM SIGPLAN 1999 Workshop on Languages, Compilers, and Tools for Embedded Systems, pp. 65–72 (1999)
5. Jain, R., Hughes, C.J., Adve, S.V.: Soft Real-time Scheduling on Simulataneous multithreaded processors. In: Proc. of 23rd Real-time Systems Symposium (RTSS-23), pp. 134–135. IEEE Press, Los Alamitos (2002)
6. Snaverly, A., Tullsen, D.M., Voelker, G.: Symbiotic job scheduling with priorities for a simultaneous multithreaded processor. In: Proc. of 9th International Conf. on Architectural Support for Programming Languages and Operating Systems, pp. 234–244. ACM Press, New York (2000)

Public Key Cryptography Based Approach for Securing SCADA Communications

Anupam Saxena, Om Pal, and Zia Saquib

Centre for Development of Advanced Computing, Mumbai, India
{anupam,ompal,saquib}@cdacmumbai.in

Abstract. SCADA (supervisory Control and Data Acquisition) security has become a challenge in today's scenario because of its connectivity with the world and remote access to the system. Major challenge in the SCADA systems is securing the data over the communication channel.

PKI (public key infrastructure) is an established framework to make the communication secure. SCADA systems, normally have limited bandwidth and low computation power of RTUs (Remote Terminal Units); through the customization of general PKI we can reduce its openness, frequent transfer of certificates and possibility of DOS (Denial of Service) attacks at MTUs (Master Terminal Units) and RTUs.

This paper intends for securing data over communication channel in SCADA systems and presents the novel solutions pivoted on key distribution and key management schemes.

Keywords: Broadcasting, Key Distribution and Management, Multicasting, Broadcast, Public Key Infrastructure, SCADA security.

1 Introduction

Supervisory Control and Data Acquisition (SCADA) systems facilitates management, supervisory control, monitoring of process control and automation systems through collecting and analyzing data in real time. Primarily these systems were not build to operate within the enterprise environment, this lead to inability within SCADA components to deal with the exposure to viruses, worms, malware etc.

Internet connectivity of SCADA systems increased the risk of cyber attacks, and hence security of these systems became a challenging issue. Technology become vulnerable to attacks and this may cause damage on critical infrastructures like electric power grid, oil gas plant and water management system. Protection of such Internet connected SCADA systems from intruders is a new challenge for researchers and therefore, it makes necessary to apply information security principles and processes to these systems.

SCADA system consists of a human-machine interface (HMI), a supervisory system (controller or MTU), remote terminal units (RTUs), programmable logic controllers (PLCs) and a communication infrastructure connecting the supervisory system to the RTUs.

V.V. Das, J. Stephen, and Y. Chaba (Eds.): CNC 2011, CCIS 142, pp. 56–62, 2011.

With the development of SCADA industry, vendors started adopting open standards, which reduced the total number of SCADA protocols to a smaller number of protocols that were popular and were being promoted by industry. These protocols include MODBUS, Ethernet/IP, PROFIBUS, ControlNet, InfiNET, Fieldbus, Distributed Network Protocol (DNP), Inter-Control Center Communications Protocol (ICCP), Tele-control Application Service Element (TASE) etc. The most widely used communication protocols in SCADA system are DNP3 (Distributed Network Protocol version 3.0), IEC 62351 and Modbus. Initially due to isolation of SCADA system from outside world, cyber security was not an issue. With the interconnection of system to the outside world, the necessity of securing the system is increased.

Cryptographic techniques are widely used for providing many security features of higher security, reliability, and availability of control systems etc. to the SCADA systems. There is a need of establishment of secure keys before application of cryptographic techniques. In this paper first we discuss key distribution and key management issues and then we present our PKI based approach for securing the SCADA system; this scheme is designed to fulfill the essential requirement of availability along with integrity in the SCADA systems.

In next section, we discuss key challenges and related work, In Section 3 we present our proposed key distribution and key management technique with Conclusions in Section 4 and References at the end.

2 Key Challenges and Related Work

Connectivity of SCADA system to the Internet, leads to emergence of many security threats, like unauthorized access of devices, capturing and decoding network packets and malicious packet injection in the network.

To protect SCADA system from these threats, certain security requirements are needed to be considered:

1. Authentication: It is to ensure that the origin of an object is what it claims to be.
2. Integrity: The manipulation of messages between nodes and insertion of new nodes can be hazardous. A malicious attacker may cause physical damage if it has the ability to alter or create messages.
3. Confidentiality: It is to ensure that no one can read the message except the intended receiver.
4. Availability of resources: Resources should be available for legitimate users. It insures that the information is there when needed by those having authorization to access or view it.

To protect any SCADA system, these challenges need to be considered along with installation and configuration limitations of the system. Ludovic Piètre-Cambacédès[3] has pointed out the some constraints of SCADA system:

1. Limited computational capacity: The most of the RTUs have low computational capabilities.
2. Limited Space Capacity: Memory available in the most of the RTUs is low.
3. Real-time processing: Lack of timely transmission and processing of data in SCADA can cause latency problems.

4. Key freshness: In the absence of key freshness entities would keep reusing an 'old' key for longer time, which might have been compromised, so there is a need of key freshness for eliminating this possibility.
5. Small number of messages: Due to low bandwidth, number of messages exchanged between nodes need to be minimum and also length of messages need to be also small.
6. Broadcasting: There should be facility of common message announcement to all devices.

These constraints are needed to keep in mind before building any security mechanism for the system. Efforts have been made in this area, but still there exists a wide scope for improvements.

A cryptographic key management and Key Establishment approach for SCADA (SKE) was proposed by Sandia National Laboratories [1] in 2002, which divides the communication into two categories: first is 'controller to subordinate (C-S) communication' and second is 'subordinate to subordinate (S-S) communication'. The C-S is a master-slave kind of communication and is ideal for symmetric key technique. The C-C is a peer-to-peer communication and it can use asymmetric key approach.

Information Security Institute, Queensland University of Technology, Australia [2] also proposed a Key Management Architecture for SCADA systems (SKMA). In this scheme a new entity 'Key Distribution Center (KDC)' was introduced, which is maintains long term keys for every node.

In this paper, we concentrate on accomplishment of fundamental security goals of communications. We are using the PKI approach in an optimized manner, which reduces the openness of Public Key in such a way that it makes the communication secure, and provides broadcasting and multicasting (for any group of RTUs) in the limited environment of SCADA system.

3 Proposed Methodology

Our scheme assumes that messaging takes place among two entities namely MTU and RTU. Scheme assumes that MTU has high computing capability to take most of the computational load in order to provide security in SCADA systems.

Scheme uses asymmetric key approach while putting restriction on accessibility of keys. Scheme uses 'Key and CounterKey pair'. Though these keys are analogues to Private-Public keys ('Key' corresponds to Private Key, and 'CounterKey' corresponds to Public key), but are not same in the true sense, because in the scheme, the 'CounterKeys' are not publicly accessible to all.

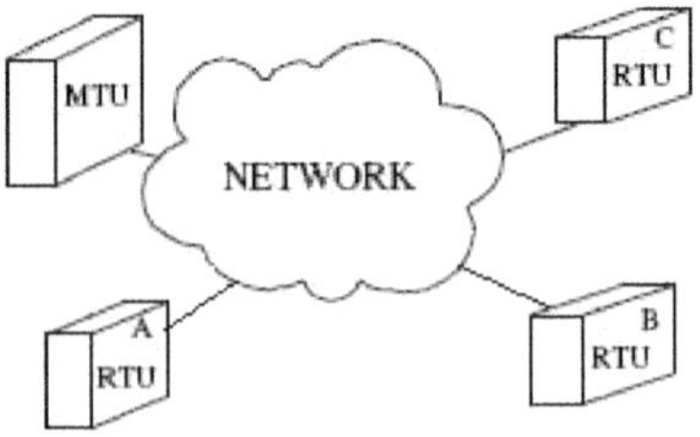

Fig. 1. SCADA System

Scheme assumes that, for a sub section, there is an MTU (with high computational power) and n number of RTUs (with low computational power). MTU and RTUs are attached to each other.

Initially long term keys are stored manually at each node, 'n' unique keys stored at

MTU (corresponding to each RTU), and one at each RTU which belongs to that corresponding RTU.

RTU R_i has key L_i where i = 1,2,...n. The MTU passes 'Key' (K) to corresponding RTUs by using the pre shared-keys (L_i, where i = 1,2,...'n'). Also it passes MTU's CounterKey (CK_{MTU}) to each RTU to each RTU.

For maintaining key freshness, there is a provision of re-distribution of "Key-CounterKey pair" after certain period of time using long term keys; and also these long term keys would be replaced manually after a long (fixed) time period. This fixed time can be adjusted depending on the requirement of system.

Each RTU maintains a database of counter keys of other RTU to which communicates and if the required CounterKey is not available in its database then it requests to MTU for obtaining required CounterKey before initiation of communication. All keys and CounterKeys are confidential, any node can request to MTU to obtain the CounterKey of other node only when it wants to communicate to that node. After the initial communication, both parties (sender and receiver) store the CounterKeys of each other in their database for future use.

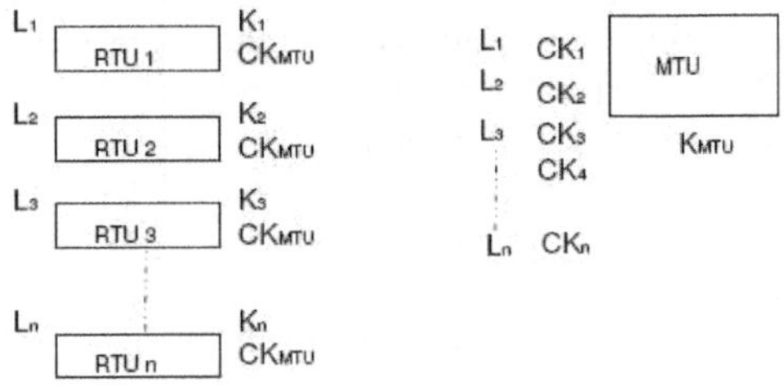

Fig. 2. Initial Key Setup

In general PKI architectures, a CA authenticates any node in the network by issuing certificate to that node but in this case, where MTU takes the responsibilities of a CA, extra computational overhead of certificate might be an issue for some RTUs because of their low processing power. In such cases we have two options, first one is to reduce the certificate as much as possible by removing unnecessary extra fields from it [3], and second is to replace certificate value with a single unique value like "MAC Address" of the corresponding entity.

In the proposed scheme, node encrypts hash of its own MAC address with its own key and forwards this value to other node, to prove its authenticity.

For load balancing, and to avoid an MTU to become a single point of failure, scheme also recommends the deployment of distributed MTUs.

Proposed scheme categorizes the communication into three categories as follows

A RTU to RTU communication
B MTU to RTU communication
C RTU to MTU communication

A RTU to RTU Communication

If an RTU wants to communicate with another RTU, it will not store CounterKeys of all RTUs but it will only store CounterKey of other RTU at run time if it is needed, which it takes from MTU.

If RTU A wants to communicate with RTU B then it checks the CounterKey of RTU B in its database if it is there then RTU A encrypts the message with CounterKey of RTU B and sends to RTU B. If CounterKey of RTU B is not available in database of RTU A then it calculates Hash of its own MAC address, signs it by its own Key, and then encrypts the resulting block (along with address of RTU B) by MTU's CounterKey, and sends to MTU. When MTU gets the request, it decrypts it by its own Key, and then checks the signature of RTU A with the help of Hash Function and by using CounterKey of RTU A. After checking the validity of the RTU B, MTU prepares a response. If both RTUs are genuine, then MTU sends this response (which contains CounterKey of corresponding RTU B) to RTU A by signing the Hash of MAC address of MTU by its own key and encrypting the resulting block by the CounterKey of RTU A.

After getting the response of MTU, RTU A decrypts the Block by its own Key and after decrypting hash value by CounterKey of MTU, compares hash of MAC address of MTU with received hash value from MTU, if values match, then RTU A stores the CounterKey of RTU B in its database.

Table 1. Counterkey Storage in Databases of Communicating Rtus

States	RTU A	RTU B	Encryption
Initial State	CK_{MTU}, CK_A	CK_{MTU} CK_B	
State after Message 1	"	"	Message encrypted with CK_{MTU}
State after Message 2	CK_{MTU}, CK_A, CK_B	"	Message encrypted with CK_A
State after Message 3	"	"	Message encrypted with CK_{MTU}
State after Message 4	"	"	Message encrypted with CK_{MTU}
State after Message 5	"	CK_{MTU}, CK_B, CK_A	Message encrypted with CK_B

Now RTU A sends message to RTU B by encrypting the message with CounterKey of RTU B. After receiving it, the RTU B starts communication if it has the CounterKey of RTU A in its database, otherwise it does not respond.

If the RTU A does not get reply from RTU B then it waits for a fixed time (this time duration may vary from system to system and will depend on requirements), after that it sends a communication initiation request to RTU B, by calculating hash of its own MAC address, signing it by its Key, and then encrypting the resulting block (along with address of RTU A) by MTU's CounterKey, and sends to RTU B. After receiving the request from RTU A, RTU B passes this request to MTU. When MTU gets the request, it decrypts it by its own Key, and then checks the signature of RTU A with the help of Hash function and by using CounterKey of RTU A. After checking the validity of the sender (who initiated the request), the MTU prepares a response. If initial sender A is genuine, then MTU sends a response (which contains CounterKey of RTU A) to RTU B by signing the hash of MAC address of MTU by its own key and with encrypting the resulting block by the CounterKey of RTU B.

After getting the response of MTU, RTU B decrypts the Block by its own Key and after decrypting hash value by CounterKey of MTU, compares the hash, if values match then RTU B stores the CounterKey of RTU A in its database. Now the communication can start.

Table 2. Counterkey Storage in Databases of Communicating Mtu And Rtu

States	MTU	RTU B	Encryption
Initial state	CK_{MTU}, CK_B	CK_{MTU}, CK_B	
Message 1	"	"	Message encrypted with CK_B

B MTU to RTU Communication

When an MTU wants to communicate to an RTU then it fetches the CounterKey of that RTU from its database, then MTU encrypts the message with CounterKey of that RTU and sends to it.

As both the MTU and RTU B contain CounterKeys of each other in their databases, so they can always communicate with each other.

C RTU to MTU Communication

When an RTU wants to communicate to MTU then it sends a message to MTU, by encrypting it with CounterKey of MTU (which is known to every RTU). After receiving it, the MTU starts communication as it also has the CounterKey of that RTU in its database.

Table 3. Counterkey Storage in Databases of Communicating Rtu and Mtu

States	**MTU**	**RTU B**	**Encryption**
Initial state	CK_{MTU}, CK_B	CK_{MTU}, CK_B	
Message 1	"	"	Message encrypted with CK_{MTU}

Dynamic Arranged Database for Optimal Key Storage

Due to low memory, MTUs and RTUs can store a limited number of counter keys in their databases. This limited storage of keys can cause an extra overhead at run time, if required key is not available in database of MTU/RTU.

To overcome this problem, scheme uses 'dynamic arranged database for optimal key storage'. Each RTU/MTU stores CounterKey of MTU in first row of database. Always new Counter Key will be stored in second row of database and all keys will shift downward by one row. Key at the bottom row is removed if database is already full. If any CounterKey is used by the node from its database then this used key will be shifted to second row and all CounterKeys (which were above in database from used counter key) will shifted downward by one row. The place of CounterKeys those are at lower position from used Counterkeys, will be unchanged.

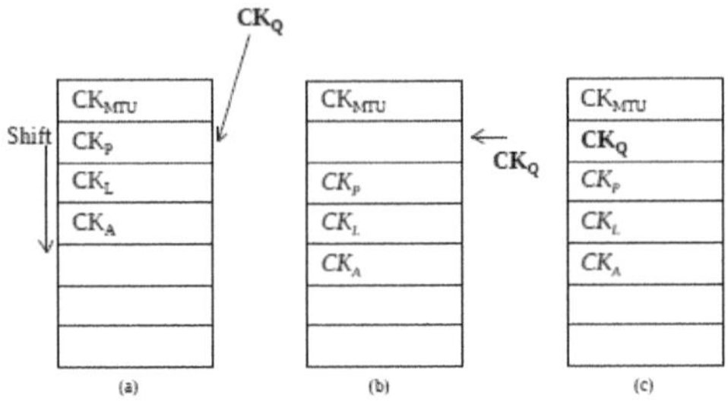

Fig. 3. Insertion of new CounterKey, when database is partially filled

4 Conclusion

In SCADA system, due to limited bandwidth and rare communications among some RTUs (Remote Terminal Units), there is a need of efficient use of general PKI, to reduce the openness of Public Key, frequent transfer of certificates and reduction in DOS (Denial of Service) attacks at MTUs (Master Terminal Units) and RTUs.

We have discussed a variety of existing issues, challenges, and schemes on cryptographic key distribution and key management for SCADA systems. We have devised and proposed this new scheme to emphasize on key distribution and key management in constrained environment using the strength of PKI.

Our attempt is to address the issues of securing the data over the communication channel in the constrained environment and to present a novel solution for key distribution and key management schemes. Proposed scheme supports broadcasting of messages and also provides multicasting as an additional feature for SCADA system.

References

1. Beaver, C. L., Gallup, D.R., NeuMann, W. D., Torgerson, M.D. : Key Management for SCADA (SKE): printed at Sandia Lab (March 2002)
2. Dawson, R., Boyd, C., Dawson, E., Manuel, J., Nieto, G.: SKMA – A Key Management Architecture for SCADA Systems. In: Fourth Australasian Information Security Workshop AISW-NetSec (2006)
3. Piètre-Cambacédès, L., Sitbon, P.: Cryptographic Key Management for SCADA Systems, Issues and Perspectives. In: Proceedings of International Conference on Information Security and Assurance (2008)
4. Hentea, M.: Improving Security for SCADA Control Systems: Interdisciplinary. Journal of Information, Knowledge, and Management 3 (2008)
5. Lee, S., Choi, D., Park, C., Kim, S.: An Efficient Key Management Scheme for Secure SCADA Communication. In: Proceedings Of World Academy Of Science, Engineering And Technology, vol. 35 (November 2008)
6. Wang, Y., Chu, B.-T.: sSCADA: Securing SCADA Infrastructure Communications (August 2004)
7. Paukatong, T.: SCADA Security: A New Concerning Issue of an Inhouse EGAT-SCADA: Electricity Generating Authority of Thailand, 53 Charan Sanit Wong Rd., Bang Kruai, Nonthaburi 11130, Thailand

gBeL: An Efficient Framework for e-Learning Using Grid Technology

M.B. Geetha Manjusha[1], Suresh Jaganathan[2], and A. Srinivasan[3]

[1] M.E. Final Year
Department of Computer Science and Engineering
Sri Sivasubramania Nadar College of Engineering
Chennai, Tamilnadu, India
geethamtech0911@gmail.com
[2] Assistant Professor,
Department of Computer Science and Engineering
Sri Sivasubramania Nadar College of Engineering
Chennai, Tamilnadu, India
whosuresh@gmail.com
[3] Professor,
Department of Computer Science and Engineering
Misrimal Navajee Munoth Jain Engineering College
Chennai, Tamilnadu, India
asrini30@gmail.com

Abstract. E-learning is the process of stretching learning or carrying instructional resource sharing gateways, to locations away from a classroom to another classroom, by using multimedia communications. The process of e-learning will cause some needs, such as i) high storage capacity systems for storing e-learning materials, ii) high network throughput for faster and efficient access of materials and iii) efficient streaming of e-learning materials which are in multimedia format. The first and second needs can be tackled by grid solutions and last one by embedding efficient video compression algorithm in grid. In this paper we proposed a framework called gBeL [Grid Based E-Learning] framework, which can fulfill the above said needs by adopting the grid technology and efficient streaming by reducing the video size by using video compression algorithm and embedding it in gBeL framework.

Keywords: e-learning, multimedia streaming, video compression, grid technology.

1 Introduction

E-learning [4] is a term that can be used to refer to computer-dependent learning. It makes use of computer dependent guidance and online meetings, teaching supplies, email, discussion forums, computer-aided evaluation, and other related methods. Most of the experts are concerned in the application of technology in education recognize the significance of e-learning where the trainer also has some good interaction with

V.V. Das, J. Stephen, and Y. Chaba (Eds.): CNC 2011, CCIS 142, pp. 63–68, 2011.

the students. As the e-learning comprises lots of multimedia files which consume a large amount of storage space and larger bandwidth while streaming, for reducing the size of multimedia file and to avoid wastage of bandwidth, the multimedia file can be compressed using an efficient video compression algorithm [2,3]. As the huge volume of raw multimedia data entails rigorous bandwidth requirements on the network, compression is widely employed to attain transmission effectiveness. As the video takes larger bandwidth (56 kbps-15 Mbps) than audio (8-128 kbps), video can be used for amendment in order to meet QoS requirements. Hence, the focus on characteristics of video compression becomes useful for amendment. The intention of video compressing is for optimizing the video excellence over bit-rate range instead of a specified bit-rate and also the specified bit stream can be moderately decodable at any bit rate range to renovate the optimized quality. Rest of the paper is organized as, in Section 2 sample e-Learning use case diagram is explained, in Section 3 gBeL framework is explored, in Section 4 shows possible system scenario and results and in last Section 5concludes the paper.

2 Use-Case Diagram for E-Learning

Figure 1 explains the procedure how e-learning can be used in engineering colleges and the use cases involved in it. It shows the various actors and the relationship between various use cases available in the e-learning.

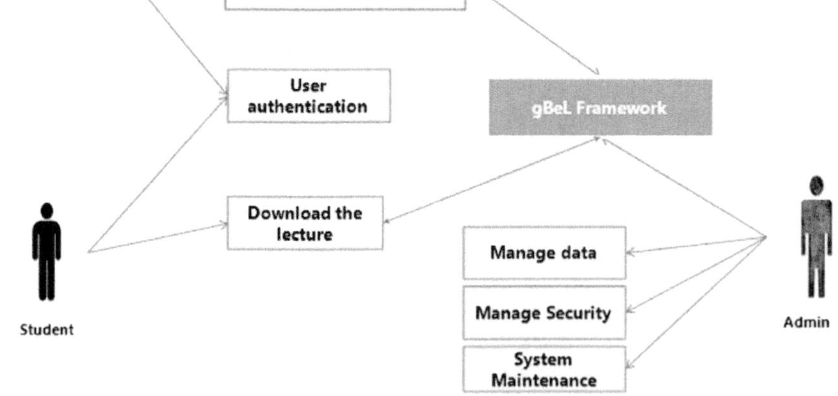

Fig. 1. Use Case Diagram for e-Learning

3 Grid Based E-Learning [gBeL]

The proposed framework fulfills the needs of e-learning and in the coming sections the needs are explained.

3.1 gBeL Component Model

Figure 2 presented below illustrate the general schema adopted in this framework and the software component architecture. The component for the physical interconnection architecture is a class room, this is the physical environment the lecture is broadcasted and from which the students might join. In addition to the physical facilities for housing everyone attending the lecture, it has all the necessary equipment to integrate it into a "virtual lecture room". Equipment includes a computer along with a projector to which a video-camera is connected, and these are managed by a specialized operator.

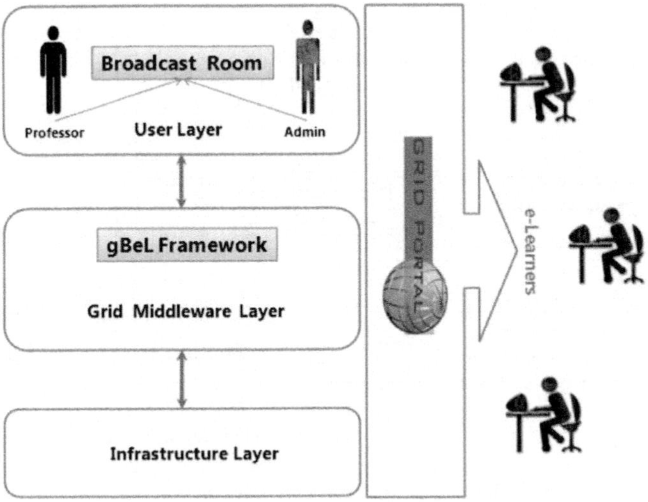

Fig. 2. gBeL Component Model

Individual site, this denotes the places in which the students may join the conference using personal computing resources. This might be widely scattered and will be having all the facilities for participating in the lecture. Broadcasting room is the place from which the professor broadcasts the lecture to all the students sitting in classrooms or individual sites. The technical resources are the same as those found in classrooms, with the exception that a video camera is imperative. An operator assists the professor with the management of such equipment. Grid portal is the portal by which users perform activities such as submitting lectures, subscribing to lectures, and uploading and downloading the content. Our gBeL framework sits in the center and describes the technique that can be used to give high quality of service for the transmission of lectures from professors to students.

3.2 gBeL Framework

Figure 3 presented below illustrate the gBeL framework. First, a video clip is taken and 3D DCT compression algorithm [1] is applied to the video clip. The reason for going to 3D DCT method is less computational time and high compression ratio. In this method, the video clip is decomposed using in spatial and temporal manner and combined to form a representation called "accordion representation", which is a compressed version. Then to decompress the video clip at the receiver end inverse operation is done. This framework is embedded into grid middleware using globus XIO framework. The Globus XIO framework [5] manages the I/O operation requested by an application via the user API. This framework does not work to deliver the data in an I/O operation nor does it manipulate the data. The framework's job is to manage requests and map them to the drivers interface and it is up to the driver to manipulate it and transport the data.

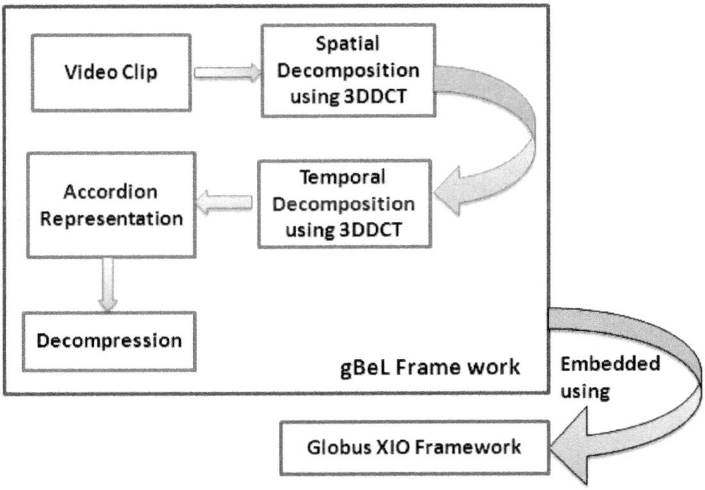

Fig. 3. gBeL Framework

The Table 1 illustrates the performance of three variants of DCT based compression algorithms and the from Figure 4, it can be noticed that the Peak Signal to Noise Ratio (PSNR) for 3D-DCT has a linear increase then compared to other DCT variants.

Table 1. Comparison of DCT variants

	DCT	2D-DCT	3D-DCT
Bit Rate	PSNR[dB]	PSNR [dB]	PSNR [dB]
500KB	25.1546	29.483314	32.893331
1MB	26.8324	30.110880	33.584134
2MB	27.9364	32.29383	34.84368
5MB	28.0315	33.483314	35.65412
10MB	28.9141	34.310880	38.484134

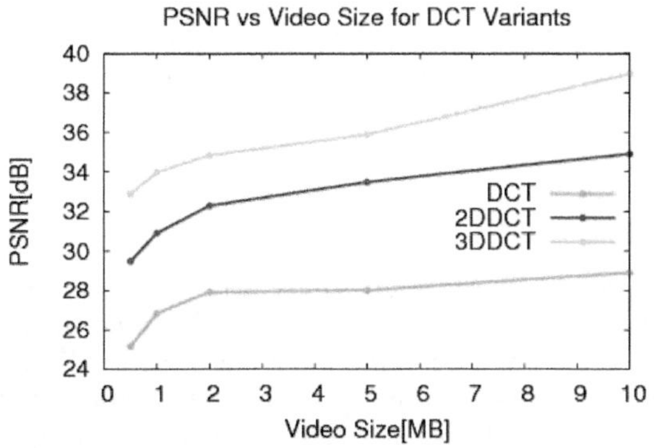

Fig. 4. PSNR comparisons for DCT variants

4 Implementation Scenario

The Figure 5 shows the implementation scenario of the proposed framework in which the network uses 100Mbits/s LAN. The actors are Professors and Students. Professors upload the lessons of different subjects in the grid server. A student has to login (using grid portal) into the student login page in order to download the lessons of different subjects, for example Component Based Technology (CBT), Computer Network Management (CNM) and Grid Computing (GC). Student has two options for the downloading of lessons, first direct downloading i.e. directly from the professor's system and the second is using third party downloading i.e. using grid server, and both the option adopts the *globus_url_copy* command. When the student opts for

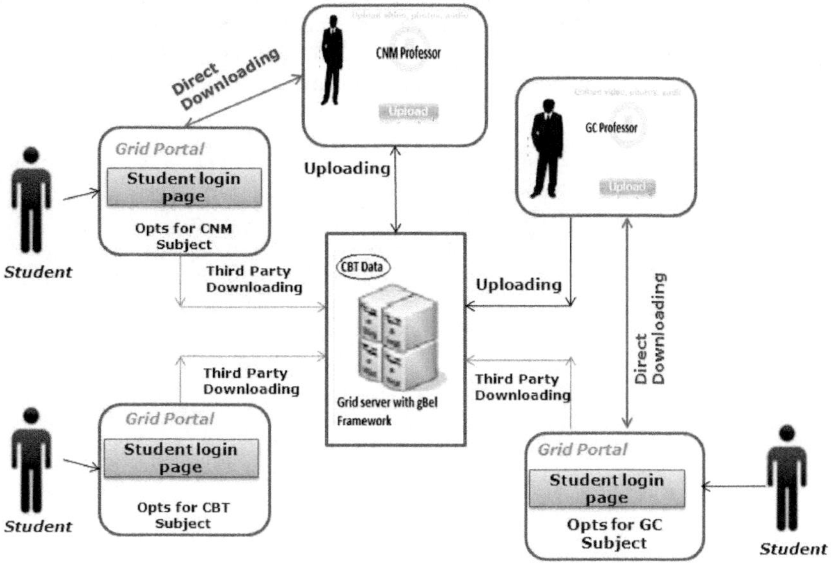

Fig. 5. Implementation Scenario

second option, the video is compressed using our framework and uploaded. From the results got, our gBeL framework solves the lower bandwidth problems because of using more efficient 3D-DCT compression algorithm and embedding into grid technology and this leads in increase of throughput up to 95% and fairness to 92.3%.

5 Conclusion

In this paper, we proposed a new framework for efficient e-learning system using grid technology, which satisfies the important needs of e-learning such as storage and saving bandwidth while streaming e-learning materials which will be in multimedia format. The proposed framework uses the grid for high storage, high throughput and QoS and provides solution for low-bandwidth problems. By using this framework we can efficiently utilize the available bandwidth, which gives reliable quality of video results and reduced processing time. Improvements for this framework can be done in two areas: i) when lots of small numbers of files are streamed and ii) when the number of parallel streams increased in *globus_url_copy*.

References

1. Ouni, T., Ayedi, W., Abid, M.: New low complexity DCT based video compression method. In: TeleCommunication ICT 2009, pp. 202–207 (2009), doi:10.1109/ICTEL2009.5158644
2. Das, A., Hazra, A., Banerjee, S.: An Efficient Architecture for 3-D Discrete Wavelet Transform. IEEE Transactions on Circuits and Systems (2010), doi:10.1109/TCSVCT.2009.2031551
3. Lodge, N.K.: Video compression techniques. IEEE Colloquium on Applications of Video Compression in Broadcasting 2/1, 214 (2002)
4. Kawamura, T., Sugahara, K., Kinoshita, S., Nakatani, R., Motomura, S.: Backup and recovery scheme for distributed E-learning system, Integration of Knowledge Intensive Multi-Agent Systems, KIMAS2007,doi:10.1109/KIMAS.2007.369788,73-78 (2007)
5. http://www.globus.org/toolkit/docs/4.0/common/xio/developer-index.html

An Algorithm Design to Evaluate the Security Level of an Information System

Sunil Thalia[1], Asma Tuteja[2], and Maitreyee Dutta[3]

[1] ME Student, NITTTR, Chandigarh
[2] Assistant Professor, MITS University, Rajasthan
[3] Associate Professor, NITTTR, Chandigarh
sunilthalia@rediffmail.com, asmatuteja@yahoo.co.in,
d_maitreyee@yahoo.co.in

Abstract. Measuring the security of an Information System has become a critical issue in the era of Information Technology. As any other process, security can not be improved, if it can not be measured. The need of security metrics is important for assessing the current security status. Since all systems and organizations are different, there is no single set of metrics that is generally applicable. This paper presents an algorithm to develop the necessary security metrics for assessing the information system in a structured way and a quantitative evaluation model with qualitative decision based on Analytic Hierarchy Process (AHP) to measure the security level of the Information System. At last, a test case is given to illustrate the algorithm and effectiveness of this model.

Keywords: Information system, Security metrics, Analytic hierarchy process.

1 Introduction

Security, one of the most discussed subjects in computer science today, is considered as an issue of vital importance for all Information Technology (IT) based systems. The purpose of a security assessment is to give an indication of how well a system meets specified criteria. Although perfect security is the ultimate goal for a system, an assessment can not guarantee any level of security, but it can provide a basis for confidence in the assessed system. However, since security is an abstract, subjective and intangible property, properly assessing the security of an information system is hard and, currently, there are no methods for efficient, reliable, and valid security assessment. Security metrics is a central concept, in order to perform the measurements of security assessment.

The main contribution of this paper is to introduce an algorithm to develop the necessary security metrics for assessing the information system in a structured way and a quantitative evaluation model with qualitative decision based on Analytic Hierarchy Process (AHP) [5] to measure the security level of the Information System. The rest of this study is organized in following way: Section 2 analyze the background and related work, section 3 presents proposed algorithm, Section 4 gives a test case and finally, Section 5 gives conclusions and discusses future work.

V.V. Das, J. Stephen, and Y. Chaba (Eds.): CNC 2011, CCIS 142, pp. 69–75, 2011.

2 Background

In 1883 Lord Kelvin wrote if you can not measure it, you can not improve it. Classical measurement theory [3] tells us that a "valid" measure is a mapping from an empirical domain to a numerical range such that the empirical relations are preserved by the numerical relations. The most recognized security objectives are Confidentiality, Integrity and Availability (CIA) [2], often referred to as the "CIA model". Wang and Wulf [6] describe a general-level framework for measuring system security based on a decomposition approach. A better definition, in that sense, was given at the WISSRR Workshop on Security management [9]. It defines the Information Security (IS) metric as follows: An IS metric/measure is a value, selected from a partially ordered set by some assessment process that represents an IS-related quality of some object of concern. It provides, or is used to create, a description, prediction, or comparison, with some degree of confidence. Besides, they are increasingly treated in national and international standards (NIST800-55[4], ISO27004 [12]). Yet, what is often unclear is a concise definition of an IS metric. Vaughn, Henning and Siraj [8] focused the attention on the area of metrics or measures for systems that had security or assurance as a requirement, but were not successful in coming to agreement on a set of measures to be used or even finding consensus in any particular approach. In 2005 the issues of system-wide IT security assessment such as specification of requirements, design of system security assessment methods, categorization and evaluation of current and future methods were addressed by Hallberg, Hunstad and Peterson [7]. R. Savola et al. [1] present a systematic and holistic method for security metrics development for software intensive systems.

3 Proposed Algorithm

Our proposed algorithm is based on the decomposition methodology. In the first part of algorithm, a security standard such as ISO/IEC2700-series [10], [11], [12], NIST800-series are taken as a target goal for the information system, business process or an organization. Now, this target goal is decomposed into several more specific sub goals, recursively and a goal tree is generated in a top down manner.

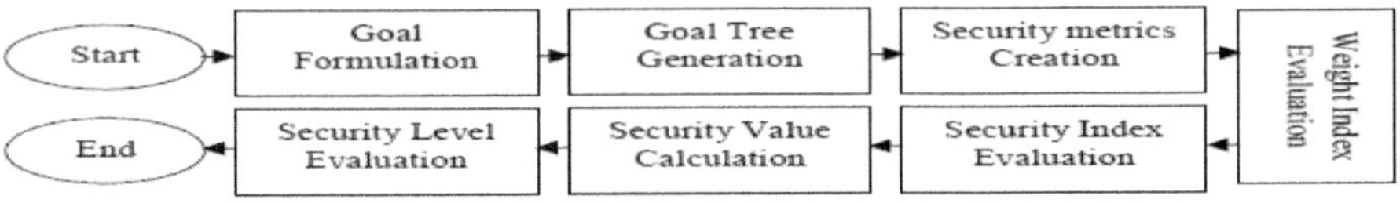

Fig. 1. Steps of Proposed Algorithm

In next part, all the nodes of this tree are inserted into metrics templates to design a set of security metrics for the system under consideration. At last, an evaluation model is given in which, first we decide the relative importance of all the sub goals using AHP, after that security value of all the metrics are calculated quantitatively and finally, security level of whole system is evaluated in a sequential manner (see Fig.1.).

1. *Target Goal Formulation:* Make a high level information security goal such as NIST-800, ISO/IEC 27000-series etc. for the information system under consideration.
2. *Goal Tree Generation:* Decompose the target Goal into several sub-goals to generate a goal tree in following manner:

i. If target goal is specific enough so that a set of simple measurements can show that it is fulfilled.
Then stop the decompositions; *Else* go to step (ii).
ii. Decompose the target goal in to sub goals that are more specific than the original goal {maintain a record of the parent /child relation between goals.}
iii. Do step (i) for each of the sub goals from step (ii).

3. *Security Metrics Creation:* Design security metrics for the information system in following manner:

i. Traverse the goal tree from left to right in top-down manner and create Internal Node List (INL) for all the internal nodes present in the tree & Terminal Node List (TNL) for all the terminal nodes present in the tree. {Except Root Node}
ii. Design security metrics templates i.e. Security Metrics Template for Terminal Nodes (SMTTN) and Security Metrics Template for Internal Node (SMTIN). {SMTTN contains following fields: MetricsId, Name, Goal, Scale, Range, Formula, Implementation Evidence, Target Value, Measured value, Validity, Collection Frequency, Reporting Frequency, data Sources, Information Owner, Information Collector, Information Receiver, Reporting Time, Parent Node, Latest Revision, and Update History. while SMTIN contains all the fields present in the SMTTN except Implementation Evidence. SMTIN also contains one more field i.e. child nodes}
iii. Insert all nodes of INL & TNL into SMTIN & SMTTN respectively and fill all the fields of the templates.
iv. Generate measurement questionnaires ("yes"/ "no" type) for each node of TNL.

4. *Weight Index (W_i) Evaluation:* Apply Analytic Hierarchy Process (AHP) [20] to evaluate weight index for each node (except root node) of goal tree.
i. Take root node (parent node) as a final goal and make a pair wise comparison among its child nodes and decide the relative importance by ranking them as shown in Table-1 in following manner:

Table 1. Relative Importance (Ranking)

value a_{ij}	Comparison Description
1	Child "i" and "j" are of equal importance
3	Child "i" is weakly more important that "j"
5	Child "i" is strongly more important that "j"
7	Child "i" is very strongly more important that"j"
9	Child "i" is absolutely more important that "j"

a) Compare child "i" with child "j". {Where "i" is assumed to be at least as important as "j"}
b) Select a value (a_{ij}) from Table-1
c) Let a_{ij} = k, then set a_{ji} = 1/k and a_{ii} = 1.
d) If n = number of children nodes. *Then* no. of comparisons among them is given by: {n × (n-1)/2}

ii. Create an (n × n) comparison matrix with respect to the parent node as follows:

a) Let a_{ij} = element of ith row and j^{th} column, of the matrix.

b) Then $a_{ji} = 1/ a_{ij}$ and $a_{ii} = 1$.

c) Create (n × n) comparison matrix as shown below:

$$\begin{bmatrix} a_{11} & a_{12} & \dots & a_{1n} \\ a_{21} & a_{22} & \dots & a_{2n} \\ . & . & \dots & . \\ a_{n1} & a_{n2} & \dots & a_{nn} \end{bmatrix}$$

iii. Calculate weight index (W_i) for each child node as follows:

a) Sum each column of comparison matrix & get T_1, T_2,T_n as:

$$T_1 = a_{11} + a_{21} + \dots + a_{n1} \quad (1)$$
$$T_2 = a_{12} + a_{22} + \dots + a_{n2} \quad (2)$$
$$T_n = a_{1n} + a_{2n} + \dots + a_{nn} \quad (3)$$

b) Divide each element of matrix with the sum of its column and obtain another (n × n) matrix with elements "b_{ij}".

c) Measure (W_i) of each child by averaging across the row,as:{$\Sigma (W_i)_n = 1.$}

$$(W_i)_1 = (b_{11} + b_{12} + \dots + b_{1n})/n \quad (4)$$
$$(W_i)_2 = (b_{21} + b_{22} + \dots + b_{2n})/n \quad (5)$$
$$(W_i)_n = (b_{n1} + b_{n2} + \dots + b_{nn})/n \quad (6)$$

iv. Calculate Principal Eigen Value (λ_{max}) for comparison matrix as follows:

$$\lambda_{max} = \sum_{k=1}^{n} T_k \times (W_i)_k \quad (7)$$

v. Calculate Consistency Index (C_i) for the matrix as:

$$C_i = \{(\lambda_{max} - n) / (n - 1)\} \quad (8)$$

vi. Calculate Consistency Ratio (C_r) for the matrix as follows:

a) Select appropriate Random Consistency Index (R_i) from Table-2 for above matrix. {if n<3; then skip this step & go to step (viii)}

b) Measure Consistency Ratio (C_r) by following formula:

$$C_r = (C_i / R_i) \quad (9)$$

vii. If ($C_r < 0.1$); *Then* go to step (viii); *Else* revise the subjective judgment.
viii. Assign weight index (W_i) to each child node, as measured in step (iii).
ix. Do steps (i) to (viii) for each internal node exist in INL.

Table 2. Random Consistency Index

n	3	4	5	6	7	8	9
R_i	0.58	0.9	1.12	1.24	1.32	1.41	1.45

5. *Security Index (S_i) Evaluation:* Evaluate security index for all terminal nodes in following manner:

i. Evaluate the answers (either in "yes" or "no") of all measurement questionnaires (related to the information system) for each node exists in TNL.

ii. Calculate security index for each terminal node as follows:
Security Index (S_i) = No. of questions answered by "yes" / Total no. of questions.

6. *Security Value (S_v) Calculation:* Calculate security value for all nodes in tree.

i. Calculate security value for each terminal node exists in TNL as follows:

$$(S_v)_{terminal} = (W_i)_{terminal} \times (S_i) \quad \text{(1)}$$

$(S_v)_{terminal}$ = Security Value of Terminal Node, $(W_i)_{terminal}$ = Weight Index of Terminal Node, (S_i) = Security Index of Terminal Node
ii. Calculate security value for each internal node (with "n" child nodes) exists in INL as follows:

$$(S_v)_{internal} = \sum_{j=1}^{n} (W_i)_{internal} \times (S_v)_j \quad \text{(2)}$$

$(S_v)_{internal}$ = Security Value of Internal Node, (W_i)internal = Weight Index of internal node, (S_v)j = Security Values of child nodes
iii. Calculate security value for root node (with "m" child nodes) as follows:

$$(S_v)_{root} = \sum_{k=1}^{m} (S_v)_k \quad \text{(3)}$$

$(S_v)_{root}$ = Security Value of Root Node, $(S_v)_k$ = Security Values of child nodes

7. *Security Level Evaluation:*

i. Evaluate degree & security level of the information system from the Table-3.

Table 3. Security Levels

Security Level	Degree	Range of Security Values
Level 1	Excellent	0.90~1.00
Level 2	High	0.80~0.89
Level 3	Medium	0.70~0.79
Level 4	Low	0.60~0.69
Level 5	Worst	0.00~0.59

4 Test Case

A test case is given in this section to explain the above algorithm. First, select a security standard as a target goal for the information system and generate a goal tree as shown in Fig.2. Next, prepare INL (i.e. A, and C) and TNL (i.e. B, Z, X, and Y) and inserted into SMTIN & SMTTN respectively. After that, generate security related questionnaires for TNL. Now, apply AHP, to obtain weight indices for all the nodes

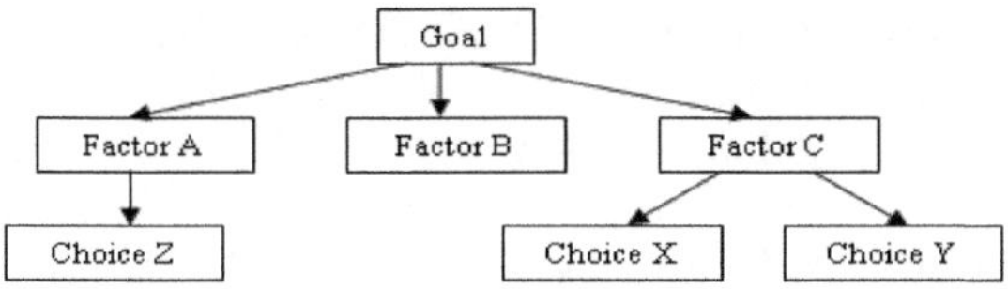

Fig. 2. Goal Tree

as follows: Let we assume that at level-1, Factor A is weakly more important than Factor B; Factor C is strongly more important than Factor A; and Factor C is very strongly more important than Factor B. Therefore, the comparison matrix for the Level-1 is generated as follows:

	A	B	C
A	1	3	1/5
B	1/3	1	1/7
C	5	7	1

The weight index of each node at the level-1 is calculated as following manner:

$$(Wi)_A= (3/19 + 3/11 + 7/47)/3= 0.19$$
$$(Wi)_B= (1/19 + 1/11 + 5/47)/3 = 0.08$$
$$(Wi)_C= (15/19 + 7/11 + 35/47)/3= 0.73$$

Thus, Principal Eigen Value λ_{max} is calculated as:

$$\lambda_{max} = ((19/3)*0.19) + 11*0.08 + ((47/35)*0.73)) = 3.04$$

In next step, the value of Consistency Index (C_i) & Consistency Ratio (C_r) is calculated as: {as value of random index (R_i) is 0.58 for n=3; see Table-2}

$$C_i = ((3.04 -3)/ (3-1)) =0.02$$
$$C_r = 0.02/0.58 = 0.03$$

As $C_r < 0.1$, we can say that this much inconsistency is acceptable. Now, at level-2, because of there is only one child of Factor A i.e. Choice Z. So Weight Index $(W_i)_Z$ of Choice Z is 1. But, Factor C has two choices i.e. X & Y and we assume that choice X is weakly more important than choice Y. Therefore, the comparison matrix for the Level-2 is generated as follows:

	X	Y
X	1	3
Y	1/3	1

Because, it's a (2×2) matrix (i.e. n<3), so there is no need to measure λ_{max} , C_i and C_r.

In this manner, after calculating the security index of each node in the goal tree, security value of every node is measured by applying the formula (mentioned in the algorithm) as follows:

Security Value $(S_v)_B$ of B = 0.08*8/10=0.064
Security Value $(S_v)_Z$ of Z = 1*7/10=0.7
Security Value $(S_v)_X$ of X = 0.75*7/10=0.525
Security Value $(S_v)_Y$ of Y = 0.25*9/10=0.225
Security Value $(S_v)_A$ of A = 0.19*0.7=0.133
Security Value $(S_v)_C$ of C = 0.73*0.525 + 0.73*0.225 = 0.5465
Security Value $(S_v)_{Root}$ of Root = $(S_v)_A + (S_v)_B + (S_v)_C$
Security Value $(S_v)_{Root}$ = 0.133 + 0.064 + 0.5475 = 0.7445

At last, security level of the information system is evaluated on behalf of Table-3. Thus, the security level of above information systems' is Level 3 i.e. Medium, as Security value of Root Node lies between 0.70~0.79.

5 Conclusion and Future Work

In this paper we have introduced an algorithm to evaluate the security level of an information system. In the beginning of this algorithm a high level information security goal are transformed in to low level security metrics using decomposition of goals that are then inserted into metrics templates to develop a perfect set of metrics aligned with the goals of the information system. In the middle, Analytic Hierarchy Process (AHP) is applied to identify the relative importance among sibling components and to evaluate their weight index. Finally, a quantitative approach is used to evaluate the security level of the information system.

In our future work, we intend to standardize a method to prove the validity of above algorithm based on measurement theory and to develop a set of automated tools for data gathering, qualitative and quantitative analysis.

Acknowledgments. The work presented in this paper has been carried out in partial fulfillment of the requirement for the award of Master Degree to Panjab University, Chandigarh, INDIA.

References

1. Savola, R.: A Security Metrics Development Method for Software Intensive Systems. In: ISA 2009. CCIS, vol. 36, pp. 11–16. Springer, Heidelberg (2009)
2. Parker, D.B.: Computer Security Management. Reston Publishing Company, Reston (1981)
3. Roberts, F.: Measurement Theory, with Applications to Decision-Making, Utility, and the Social Sciences. Addison-Wesley, Reading (1979)
4. Swanson M., Nadya B., Sabato J., Hash J., Graffo L.: Security Metrics Guide for Information Technology Systems, National Institute of Standards and Technology Special Publication #800-26NIST 800-55 (2003)
5. Saaty, T.: The Analytic Hierarchy Process. McGraw-Hill, New York (1980)
6. Wang, C., Wulf, W.A.: Towards a Framework for Security Measurement. In: 20th National Information Systems Security Conference, Baltimore, MD, USA, pp. 522–533 (October 1997)
7. Hallberg, J., Hunstad, A., Peterson, M.: A Framework for System Security Assessment. In: Proceedings of Sixth Annual IEEE SMC Information Assurance Workshop, IAW 2005, 224–231 (2005)
8. Vaughn Jr., R.B., Henning, R., Siraj, A.: Information Assurance Measures and Metrics - State of Practice and Proposed Taxonomy. In: 36th Annual Hawaii International Conference on System Sciences Proceedings, p. 10 (2003)
9. WISSRR Workshop Proceedings, Security System Scoring and Ranking, ACSA (May 2001)
10. ISO27002: The ISO 27001 and ISO 27002 Directory, `http://www.27002.net`
11. Introduction to ISO 27002 / ISO27002, `http://www.27000.org/iso-27002.htm`
12. Introduction to ISO 27004 / ISO27004, `http://www.27000.org/iso-27004.htm`

Multi-agent Based Dynamic Data Integrity Protection in Cloud Computing

S. Venkatesan and Abhishek Vaish

Indian Institute of Information Technology Allahabad
India
venkalt_s@yahoo.co.in, abhishek@iiita.ac.in

Abstract. Cloud computing is the emerging worldwide technology to mitigate the burden of the users by providing different types of services. There are numerous of critical questions in data protection to answer before adopting this new cloud technology. Data integrity protection is one of the prime challenge in the cloud computing because the data is located in the third party server. This paper proposes an efficient multi-agent based static and dynamic data integrity protection by periodically verifying the hash value of the files stored in the enormous date storage. In order to provide optimized result we propose to take an adequate chunk of data in a randomized fashion, instead of the whole cumulative hash value to the data owner for integrity check. The matching process of the hash value by the agent can identify the malicious change on the data stored in the remote server and it will recover the data efficiently through its backup storage.

Keywords: Cloud Computing, Data Integrity, Multi-Agent systems.

1 Introduction

Cloud computing is the technique of distributed computing to handle large sets of computational problems with different services with speed and reliability. In spite of the fact of various advantages, cloud still appears gray in many aspects. One of the grayest area of the environment is the integrity control of the data. The only way to trust the cloud computing service providers is through (i) maintaining high standards, (ii) evidence for secure certifications, (iii) Subject to inspection by auditors and (iv) placing it under higher scrutiny than typical in-house security teams. Even then the service providers fixing these kinds of things, peoples always feel for the security of their data's [12]. The reason is the service provider himself can act as an adversary against the data's. This is a critical issue to the data owners to detect or identify the attack. To protect the data from the adversaries (customer or service provider), we proposed the multi-agent based data integrity model for the cloud environment.

The remaining section of this paper is organized as follows: Section 2 gives the brief description over the related works and their impact in the intended environments. Section 3 describes the proposed architecture for protecting the integrity of the data stored in cloud environment. Section 4 describes the validation of the proposed model. Section 5 concludes the paper with the directions of future enhancements.

V.V. Das, J. Stephen, and Y. Chaba (Eds.): CNC 2011, CCIS 142, pp. 76–82, 2011.

2 Related Works

To protect the integrity for the data reside in the remote locations, various works were carried out by the researchers. Like, Juels and Burton [6] developed the model of POR (Proofs of Retrievability) for large files. In their first approach, they involved the key hash function of the file. In this approach, prior to archiving a file in the remote location, the verifier or owner of the data computes and stores a hash value along with the secret and random key. To check the integrity, the verifier will release the key and ask the prover (remote location) to generate the hash value for the file and send the hash value to it. It will provide strong proof to the data integrity. But the drawbacks of this model is storing the multiple keys and multiple hash values, it requires more computational cost. Also the prover has to read the entire file for every proof of the file. In their second approach, they will store only one cryptographic key for verifying the integrity. Also the prover has to access only a small portion of the large file to prove its integrity. However this is applicable only for the static data but not for the dynamic data.

Lillibridge et al [2] developed a P2P backup scheme in which blocks of a data file are isolated across $m+k$ peers using an $(m+k,m)$-erasure code. Peers can request random blocks from their backup peers and verify the integrity using separate keyed hashes attached on each block. Their scheme can detect data loss from unchanged peers. In case all the peers are changed, then it is not able to identify the modification and also it is not applicable for dynamic data's [11]. Filho et al [4] proposed a model to verify data integrity using RSA-based hash to demonstrate uncheatable data possession in peer-to-peer file sharing networks. However, their proposal requires exponentiation over the entire data file, which is clearly impractical for the server whenever the file is large [11]. Shah et al [5] proposed a model to allow a TPA to keep online storage honest by first encrypting the data then sending a number of pre-computed symmetric-keyed hashes over the encrypted data to the auditor. However, their scheme only works for encrypted files and auditors must maintain long-term state. Schwarz et al [3] model ensure file integrity across multiple distributed servers, using erasure-coding and block-level file integrity checks. However, their scheme only considers static data files. Apart from these, Bowers et al[8][10] and Ateniese et al [7] also proposed the model to protect the data integrity only when the data is static not for dynamic. Next the Ateniese et al [9] and Cong et al [11] proposed the model for dynamic data integrity protection.

Ateniese et al [9] described a PDP scheme that uses only symmetric key cryptography. This method has lower-overhead than their previous scheme [7] and allows for block updates, deletions and appends to the stored file (dynamic data). However, their scheme focuses on single server scenario and does not address small data corruptions, leaving both the distributed scenario and data error recovery issue unexplored [11]. Next cong et al [11] proposed the model for dynamic data integrity protection model, where data is stored simultaneously in n servers. If n_c servers are misbehaving or the data of that servers modified. It is compared with the others servers and detecting the probability of data modification. The drawbacks of this model are the storage and the easy compromise. It needs more storage space to store the n copies of the data in different servers. It is also easy to compromise all the n servers residing in the single cloud. To overcome all this issues and protect the dynamic data (including static data), we proposed our multi-agent based dynamic data integrity protection model with less overhead in the cloud environment.

3 Proposed Model

The proposed data integrity model is based on the multi-agent system. The reason for embedding the agent concept is known, that is the agent is having capability of autonomous, persistence, social ability and etc [1]. The proposed architecture has multiple agents to monitor and maintain the data integrity also the architecture includes three entities (respectively customer, service provider and the data owner) as shown in Fig.1.

- Customer – is the client who is going to make use of the data available with the service provider. The customers are allowed to read and modify the data[1].
- Service Provider – is the server who is going to deliver the data to the customer world. Service provider is the owner of the cloud.
- Data Owner – is the originator of the data like Wikipedia (they are having the content; any one can read or modify the content).

With this architecture of cloud computing, customer who is accessing the respective data's from the storage server is always pass through the screening agent. The modification (update or new insertion) of the data passing through the screening agent is considered as the legal modification and the modification (update or new insertion) done through bypassing the screening agent is referred to as the malicious modification (this modification is may be by the customer or by the compromised service providers). This paper focused to protect the data integrity against this malicious modification of the data. The description of this architecture is as follows. The customer can send its query through the screening agent (SA) to the data storage and they will fulfill their needs. The SA is always to monitor the query of the customer and also it will help to provide the necessary service to the customer. The query from the customer may be to read, update or insert the data. In case of the read query, SA will simply process the query and satisfy the customer needs. In case of the update or insert query, the SA will forward the query to the modifier agent and process customer needs.

The modifier agent will modify the backup storage (whether backup may be with the data owner or it will be stored in some other clouds) data as per the customer query. Hence both the backup storage and the data storage in the service provider will get the same update.

Next the penetration agent residing inside the data owner is responsible to check the robustness of the screening agent. The reason for checking the robustness is, the SA is residing in the remote location and processing for data owner. There is no guarantee the SA agent will always running properly (agent may be altered by the adversary to misbehave). For this test, the penetration agent will periodically send the queries like the customer and it will get the respective response. The dynamic integrity agent is temporary in nature and generates hash for different set of block as per

[1] Authentication for modifying the data is not concentrated in this paper. The main motive of this paper is to maintain the data integrity for the data stored in the remote location.

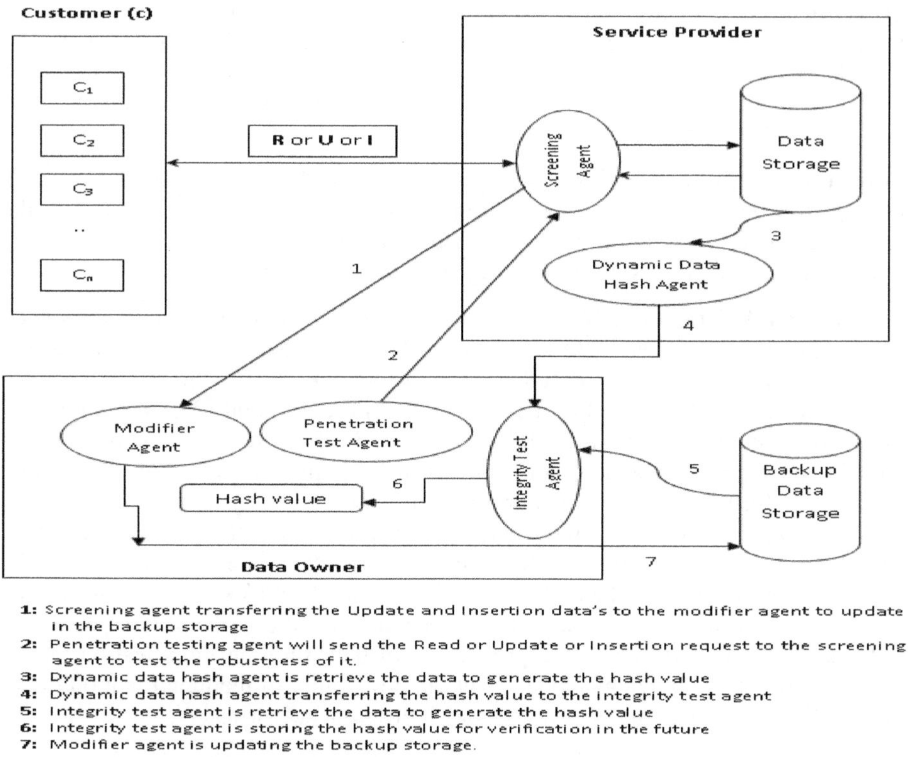

Fig. 1. Multi-agent based Data Integrity protection Architecture[2]

the requirement of the integrity agent. For optimized result, the hash value would not be generated for the entire data but a part of it.

The part of data will divide further in small part and hash is generated for the randomized small part of the data's. The other agent of the architecture is the Integrity agent. It is having two functionalities: one is preprocessing and the second is checking the integrity of the data's. The pre-processing refers to generate the hash value and store it in the hash value storage for different set of blocks whenever the data gets modified by the customer only in case of legal modification. The integrity check is to compare the hash value from the dynamic data hash agent and the hash value available with the hash value storage. If the value of the two is not matched then it is identified that the data is modified in the remote location and it to be replaced.

4 Concept Validation

The proposed multi-agent based dynamic data integrity protection model has lot of advantages than the existing models.

---

[2] The numbers given in the architecture is not the sequence. Every module can perform its task in any order based on the requirements (modification of data's).

Multiple location storage: We are keeping the backup data's away from the regular data storage. The place where the backup might be residing is the agent owner or some other clouds. In case of agent owner no false rate. In case of other cloud, false rate will be there but very low because compromising the two clouds (remote location) is a complicated task.

Robustness: As discussed in previous section the model has capabilities to its own audit agent through the penetration testing module. To achieve the robustness, the agent owner query will be sent to the SA and based upon the response the necessary action will be taken either to replace the agent with new one or keep the old.

Data Integrity Verification: In our model, only the particular set of data's has been taken for integrity verification as same as the previous model instead of the whole data. In set of 100% data in the remote storage we are taking only the 20% of data for integrity check. The 20% (*K*%) of data is also from the different blocks (randomized data's). For example consider that 100% (5*20%) of data is stored in the database in *m* row. Existing models are taking only any of the 20% data for integrity verification. But we have further divided that portion in matrix *m***n* format and choosing the blocks for data integrity verification. After that we are taking the 20% data for hash verification in randomized manner from different blocks. If we compare our model with the existing model [6], the probability of detecting the attack is 100%, in case the modification of data is in that particular block otherwise the probability of detecting the attack is 0%. In our model, the probability of detecting the modification on the total data is 1/5 as the previous models. But the probability of the identification of the flaw modification of the data is as follows with different scenarios.

Existing Model: In case of the 20% ({0,0}{0,1}{0,2}{0,3}{0,4}) of the data is taken for integrity verification, then consider that the modification is in the portion {0,0}, then the probability of flaw modification identification is 100%. In case of the modification in any other portion of ({1,0}{1,1}{1,2}{1,3}{1,4}), then the probability of flaw modification identification is 0%.

Proposed Model: In case of the 20% ({0,4}{1,4}{2,4}{3,1}{4,1}) of our randomized model. Here also we will get the same % of probability result as the previous one, but with the randomized check, there is a possibility to improve the probability percentage of the detection rate in each partition as in Table.1. Also for certain time period, the set of randomized data is verified, after that the other set of randomized data will be verified, it will continue for every periodical time. Hence, the modification on any portion will be detected in periodical times. The periodical time should be decided by the agent owner based on the frequent detection of the flaw modification. The portion of data selected in our model is decided by the integrity test agent.

Dynamic Data support: Customers are allowed to insert or update the data. It never affects the integrity identification. Whenever, the customer changing the data of the remote storage locations, the same thing will be reflected in the backup storage with the help of the screening agent and the modifier agent. After the modification of the data, the integrity test agent will generate the new hash value for that affected portion and then it will compare with the hash value from the dynamic hash agent. In case of 20% content, the single change should need the new hash value for all the 20%. In

Table 1. Probability Percentage of Detecting the Modification

Data partition	Single block Data modification Identification probability	Sub-partitioned block Data modification Identification probability
20%	100%	20% (1/5)
20%	0%	20% (1/5)
20%	0%	20% (1/5)
20%	0%	20% (1/5)
20%	0%	20% (1/5)
Total achievement for 100% data	20% (1/5)	20% (1/5)

case of our model, the change is in only small partition; hence we need to generate the hash value for that partition only.

5 Conclusion

Data integrity is the major trust issue in storing the data in remote locations. The data owner will ask when my data is on my server there is no need to worry about that when it is in some other location, how do I know it is protected? [13] The answer for this question from our proposed model is, every owner can able to check their data integrity through the multiple agents as like as the data residing in their machine. The proposed model provides integrity for the dynamic data with the hash value generated for the randomized set of blocks. Also there is no need to worry about the robustness of the screening agent because the periodic penetration test is generated by the penetration testing agent. With all these, proposed model is able to protect the integrity of the dynamic data in the remote location with efficient manner (randomized integrity check) than the other existing models. The future work of this paper is to authenticate the customer to modify the data and encrypting the hash value transferred from the dynamic hash agent to the integrity agent with minimum number of secret keys.

References

1. Wilson, L.F., Burroughs, D.J., Kumar, A., Sucharitaves, J.: A Framework for Linking Distributed Simulations Using Software Agents. Proceedings of the IEEE 89(2), 186–200 (2001) (invited paper)
2. Lillibridge, M., Elnikety, S., Birrell, A., Burrows, M., Isard, M.: A Cooperative Internet Backup Scheme. In: Proceedings of the 2003 USENIX Annual Technical Conference (General Track), pp. 29–41 (2003)
3. Schwarz, T.S.J., Miller, E.L.: Store, Forget, and Check: Using Algebraic Signatures to Check Remotely Administered Storage. In: Proceedings of ICDCS 2006, p.12 (2006)
4. Filho, D. L. G., Barreto, P. S. L. M.: Demonstrating Data Possession and Uncheatable Data Transfer. Cryptology ePrint Archive, Report 2006/150 (2006), `http://eprint.iacr.org/`

5. Shah, M.A., Baker, M., Mogul, J.C., Swaminathan, R.: Auditing to Keep Online Storage Services Honest. In: Proceedings of 11th USENIX Workshop on Hot Topics in Operating Systems (HOTOS 2007), pp. 1–6 (2007)
6. Juels, J., Kaliski, B.S.: PORs: Proofs of Retrievability for Large Files. In: Proceedings of CCS 2007, pp. 584–597 (2007)
7. Ateniese, G., Burns, R., Curtmola, R., Herring, J., Kissner, L., Peterson, Z., Song, D.: Provable Data Possession at Untrusted Stores. In: Proceedings Of CCS 2007, pp. 598–609 (2007)
8. Bowers, K. D., Juels, A., Oprea, A.: Proofs of Retrievability: Theory and Implementation. Cryptology ePrint Archive, Report 2008/175 (2008), http://eprint.iacr.org/
9. Ateniese, G., Pietro, R.D., Mancini, L.V., Tsudik, G.: Scalable and Efficient Provable Data Possession. In: Proceedings of SecureComm. 2008, pp. 1–10 (2008)
10. Bowers, K. D., Juels, A., Oprea, A.: HAIL: A High-Availability and Integrity Layer for Cloud Storage. Cryptology ePrint Archive, Report 2008/489 (2008), http://eprint.iacr.org/
11. Wang, C., Wang, Q., Ren, K., Lou, W.: Ensuring Data Storage Security in Cloud Computing. In: Proceedings of 17th International Workshop on Quality of Service, IWQoS, pp. 1–9 (2009)
12. Urquhart, J.: The Biggest Cloud- 2. Computing Issue of 2009 is Trust. C-Net News (2009), http://news.cnet.com/830119413_3-10133487-240.html
13. Whitney, L.: Air Force taps IBM for secure cloud, http://news.cnet.com/8301-1009_3-1447483-83.html

Design of Efficient Reversible Parallel Binary Adder/Subtractor

H.G. Rangaraju[1], U. Venugopal[2], K.N. Muralidhara[3], and K.B. Raja[2]

[1] Department of Electronics and Communication Engineering,
Government Engineering College, Chamarajanagar, Karnataka, India
rang_raju@yahoo.com
[2] Department of Electronics and Communication Engineering,
University Visvesvaraya College of Engineering, Bangalore, Karnataka, India
venu.ubaradka@gmail.com
[3] Department of Electronics and Communication Engineering,
P E S College of Engineering, Mandya, Karnataka, India

Abstract. In recent years, Reversible Logic is becoming more and more prominent technology having its applications in Low Power CMOS, Quantum Computing etc. Reversibility plays an important role when energy efficient computations are considered. In this paper, Reversible 8-bit Parallel Binary Adder/Subtractor with Design I, Design II and Design III are proposed. In all the three design approaches, the full Adder and Subtractors are realized in a single unit as compared to only full Subtractor in the existing design. The performance analysis is verified using number gates, Garbage inputs/outputs and Quantum Cost. It is observed that Reversible 8-bit Parallel Binary Adder/Subtractor with Design III is efficient compared to Design I and II.

Index Terms: Reversible Logic, Garbage Input/output, Quantum Cost, Reversible Parallel Binary Adder/Subtractor.

1 Introduction

Energy loss is an important consideration in digital design. Higher level of integration and use of new fabrication processes have dramatically reduced the heat loss over the last decades. The power dissipation in a circuit can further be reduced by the use of reversible logic. Landauer's [1] principle states that logic computations that are not reversible necessarily generate heat energy of KTln2 joules for every bit of information that is lost, where '*K*' is Boltzmann's constant and *T* the absolute temperature. Bennett [2] showed that zero energy dissipation is possible only if a computation is carried out in Reversible logic, as the amount of energy dissipated in a system is directly related to the number of bits erased during computation.

Contribution: In this paper, novel three design types viz., Design I, Design II and Design III Reversible 8-bit Parallel Binary Adder/Subtractor are proposed. The Reversible gates such as Fredkin (F), Feynman (FG), Peres (PG) and TR are used. *Organization:* The paper is organized into the following sections. The Background work

V.V. Das, J. Stephen, and Y. Chaba (Eds.): CNC 2011, CCIS 142, pp. 83–87, 2011.

is described in section 2. Section 3 is the proposed design, Result analysis of the proposed design is presented in section 4 and conclusions are contained in section 5.

2 Literature Survey

Majid Mohammadi et al., [3] presented modular synthesis method to realize a Reversible Binary Coded Decimal Adder/Subtractor optimized in terms of Garbage Input/outputs and Quantum Cost. William Hung et al., [4] proposed an approach to synthesize Quantum circuits by reachability analysis resulting minimum Quantum Cost and speed. The proposed approach is applicable to small Reversible functions like half and full adders. Thapliyal et al., [5] proposed a Reversible Half Subtractor using Fredkin and Feynman Gates. Using this gate Reversible Full Subtractor was designed and later extended for Reversible Parallel Binary Subtractor. Thapliyal and Srinivas [6] proposed Reversible TSG gate to design Adder. The Reversible Full Adder is implemented using single TSG gate. Saiful Islam and Rafiqul Islam [7] presented a survey on design of Reversible Adders which are efficient in terms of Gate counts, Garbage outputs and Quantum Cost. Hafiz Md. Hasan Babu et al., [8] introduced a Reversible Full Adder that requires three Reversible gates and produce least number of Garbage outputs. Bruce et al., [9] discussed an efficient Adder using Fredkin gate which has lower hardware complexity.

3 Proposed Model

3.1 Full Adder/Subtractor – Design I

The Design I Reversible Full Adder/Subtractor with five FG, two F and TR gate is shown in the Figure 1. The three inputs are A, B and Cin and the outputs are Sum/Difference (S/D) and Carry/Borrow (C/B). The Control (Ctrl) input differentiates the Addition and Subtraction functionalities. For Ctrl value *zero*, the circuit performs addition and Subtraction for Ctrl value *one*. The numbers of Garbage inputs are 3 represented by logical *zero*. The Garbage outputs are 5 represented by g_1 to g_5. The Quantum Cost for five FG gates are five, for two F gates is ten, one TR gate costs six and total design Quantum Cost is 21.

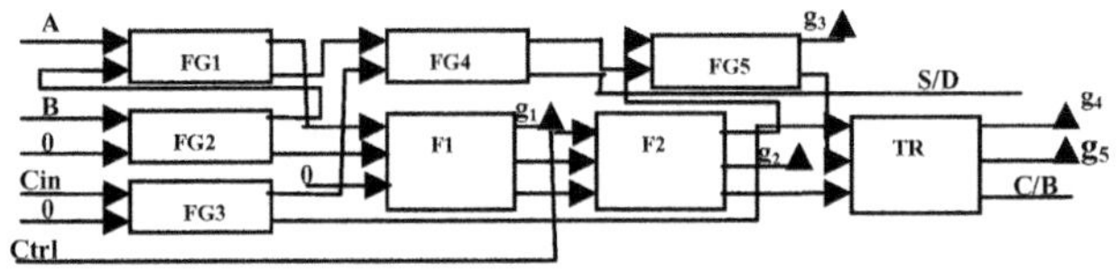

Fig. 1. Reversible Full Adder/Subtractor – Design I

3.2 Full Adder/Subtractor – Design II

Two TR gates and two FG gates are used to realize Deign II Reversible Full Adder/Subtractor unit as shown in Figure 2. The Quantum Cost for the design is 14. A Quantum Cost advantage of 7 is obtained when compared to Design I.

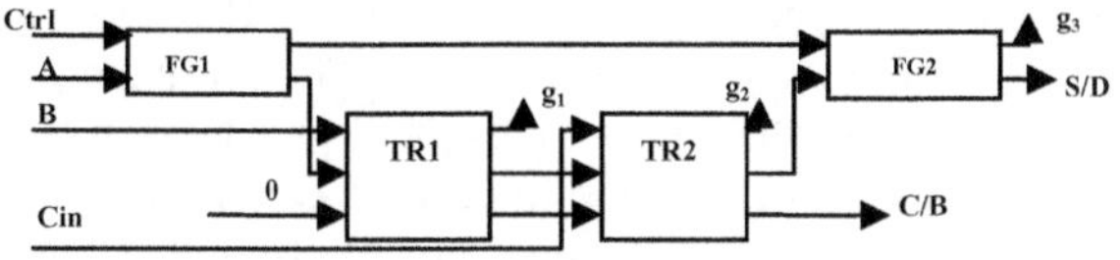

Fig. 2. Reversible Full Adder/Subtractor – Design II

3.3 Full Adder/Subtractor – Design III

The Reversible Full Adder/Subtractor Design III consists of two FG, two PG gates, and their interconnections are shown in the Figure 3. The numbers of Garbage inputs are 1, Garbage outputs are 3. The Quantum Cost for the design is 10. A Quantum Cost advantage of 11 is obtained when compared to Adder/Subtractor Design I and of 4 when compared to Adder/Subtractor Design II.

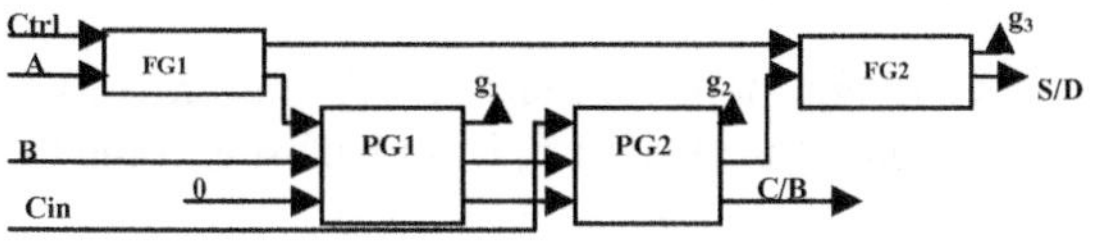

Fig. 3. Reversible Full Adder/Subtractor – Design III

3.4 Reversible Eight-Bit Parallel Binary Adder/Subtractor

The Full Adder/Subtractor Design I, II and III are used to construct Reversible 8-bit Parallel Binary Adder/Subtractor is shown in the Figure 4. The ctrl input is used to differentiate 8-bit addition and subtraction functions. The two 8-bit binary numbers are A0 to A7 and B0 to B7. The outputs Sum/Difference and Carry/Barrow are shown as S_D0 to S_D7 and C_B8 respectively. The implementation requires 7 Full Adder/Subtractors and one Half Adder/Subtractor.

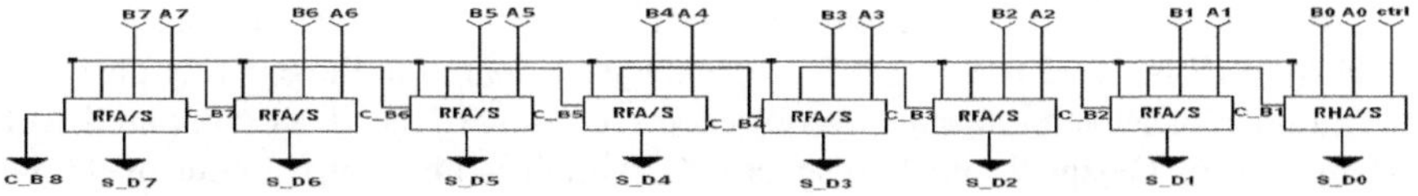

Fig. 4. Reversible eight-bit Parallel Binary Full Adder/Subtractor

4 Results

4.1 Reversible Full Adder/Subtractor

The comparison of Reversible full Adder/Subtractor Design I, II and III in terms of the number gates, number of Garbage inputs/outputs and Quantum Cost of the logics is as shown in the Table 1.

The number of Reversible gates required for Design III is only 4 as compared to 8 and 4 in the cases of Design I and II respectively. The Garbage outputs are 5 in the case of Design I, whereas 3 in the case of Design II and Design III. The Garbage inputs are 3 in the case of Design I and one in case of Design II and Design III. Quantum Cost of Design III, Design II and Design I are 21, 14 and 10 respectively.

Table 1. Comparison of Reversible Full Adder/Subtractor

	Reversible Gates	Garbage outputs	Garbage inputs	Quantum Cost
Add/Sub–Design I	08	05	03	21
Add/Sub–Design II	04	03	01	14
Add/Sub–Design III	04	03	01	10

4.2 Reversible Eight-Bit Parallel Binary Adder/Subtractor

The number of gates, Garbage inputs/outputs and Quantum Cost for Reversible 8-bit Parallel Binary Adder/Subtractor Design I, II and III are compared as shown in the Table 2.

Table 2. Comparison of Reversible eight-bit Parallel Binary Adder/Subtractor

	Reversible Gates	Garbage outputs	Garbage inputs	Quantum Cost
Add/Sub– Design I	60	38	23	159
Add/Sub–Design II	31	23	08	106
Add/Sub- Design III	31	23	08	76

The existing Binary full Subtractor based on Reversible gate [10] requires Quantum Cost of 12, Garbage input of *1* and Garbage outputs of *2*. The proposed Reversible 8-bit Parallel Binary Adder/Subtractor Design III is better compared to the existing design in terms of Quantum Cost, Garbage inputs and Garbage outputs. Hence Design III is better in terms of performance compared to the existing designs.

The number of Reversible gates required for Design III is 31 as compared to 60 and 31 in case of Design I and II respectively, which is an improvement of 93.54% compared to Design I. The Garbage outputs are 38 in case of Design I, whereas 23 in case of Design II and III, yields an improvement of 65.21% in Design III compared to Design I. The Garbage inputs are 23 in case of Design I and 8 in Design II and III, resulting 187.50% improvement in Design III compared to Design I. Quantum Cost of Design III, II and I are 76, 106 and 159 respectively, hence an improvement of Design III over Design II and I are 39.47% and 109.21% respectively.

5 Conclusions

In this paper, we proposed an efficient Reversible Parallel Binary Adder/Subtractor unit. Three designs are used to implement full Adder/Subtractor. The Design III implementation of 8-bit Parallel Binary Adder/Subtractor has better performance as compared to Design I and Design II in terms of number of gates used, Garbage inputs/outputs and Quantum Cost, hence can be used for low power applications. The full Adder/Subtractor is implemented in a single unit as compared to only full Subtractor in the existing design. In future, the design can be extended to any number of bits and also for low power reversible ALUs, Multipliers and Dividers.

References

1. Landauer, R.: Irreversibility and Heat Generation in the Computational Process. IBM Journal of Research and Development 5(3), 183–191 (1961)
2. Bennett, C.H.: Logical Reversibility of Computation. IBM Journal of Research and Development 17(6), 525–532 (1973)
3. Mohammadi, M., Eshghi, M., Haghprast, M., Bahrololoom, A.: Design and Optimization of Reversible BCD Adder/Subtractor Circuit for Quantum and Nanotechnology Based Systems. Journal of Word Applied Science, 787–792 (2008)
4. Hung, W.N.N., Song, X., Yang, G., Yang, J., Perkowski, M.: Optimal Synthesis of Multiple Output Boolean Functions using a Set of Quantum Gates by Symbolic Reachability Analysis. IEEE Transactions on Computer-Aided Design of Integrated Circuits and Systems 25(9), 1652–1663 (2006)
5. Thapliyal, H., Srinivas, M.B., Arabnia, H.R.: Reversible Logic Synthesis of Half, Full and Parallel Subtractors. In: Proceedings of the International Conference on Embedded Systems and Applications, pp. 165–181 (2005)
6. Thapliyal, H., Srinivas, M.B.: A New Reversible TSG gate and its Application for Designing Efficient Adder Circuits. In: International Symposium on Representations and Methodology of Future Computing Technologies (2005)
7. Islam, S., Islam, R.: Minimization of Reversible Adder Circuits. Asian Journal of Information Technology 4(12), 1146–1151 (2005)
8. Babu, H.M.H., Islam, M.R., Chowdhury, S.M.A., Chowdhury, A.R.: Synthesis of Full Adder Circuit using Reversible Logic. In: International Conference on VLSI Design, pp. 757–760 (2004)
9. Bruce, J.W., Thornton, M.A., Shivakumaraiah, L., Kokate, P.S., Li, X.: Efficient Adder Circuits Based on a Conservative Reversible Logic Gate. In: Proceedings of the IEEE Computer Society Annual Symposium on VLSI, pp. 83–88 (2002)
10. Thapliyal, H., Ranganathan, N.: Design of Efficient Reversible Binary Subtractors Based on a New Reversible Gate. In: Proceedings of the IEEE Computer Society Annual Symposium on VLSI, pp. 229–234 (2009)

A Computationally Simplified Numerical Algorithm for Evaluating a Determinant

S.N. Sivanandam[1] and V. Ramani Bai[2]

[1] Professor and Head, Department of Computer Science and Engineering, PSG College of Technology, Coimbatore, India 641 004
sns@mail.psgtech.ac.in, snsivanandam@yahoo.co.in
[2] Associate Professor, Department of Computer Science and Engineering, SAINTGITS College of Engineering, Kottayam, India 686 532
ramanibaiv@saintgits.org, ramanisivan@gmail.com

Abstract. A computationally simplified new procedure is presented in this paper to evaluate the determinant of a matrix A [nxn], where A may be ill-conditioned. The proposed method reduces the nth order determinant using the elementary row operations into a sequence of column vectors and then the determinant is evaluated by multiplying the elements of all column vectors. We improve the condition of ill-conditioned determinant first and then evaluate the well-conditioned one. This procedure is direct and simple in application compared to Gauss reduction method. Both the procedures are applied to illustrative examples and the comparison is also reported.

Keywords: Determinant, Ill-Conditioned Determinant, Well-Conditioned Determinant, Column Vectors; Computational complexity.

1 Introduction

Triangularization [1], is also known as Gauss reduction method [3], [6], is one of the computational methods to evaluate a determinant. This method converts a given determinant into a lower or upper triangular form and then the value of the determinant is obtained by multiplying the diagonal elements of the lower or upper triangular form. In this method, if any of the diagonal elements is zero then further reduction is not possible, because divisions by a diagonal element are required at every stage of triangular formation. Also if any diagonal element of the determinant is small, then the division leads to the loss of significant digits. To prevent the loss of significant digits, the pivoting techniques [4], [7] are used.

The present work follows the direction of row reduction but involves a simple procedure to convert the given determinant into a sequence of column vectors using only elementary column operations. Also the proposed method improves the given ill-conditioned determinant first and then evaluates the well-conditioned determinant. Therefore, the pivoting technique is not required and thus makes this method simpler in application.

V.V. Das, J. Stephen, and Y. Chaba (Eds.): CNC 2011, CCIS 142, pp. 88–95, 2011.

2 A New Numerical Method for Evaluating a Determinant

Consider a determinant of a matrix A of order n as

$$\Delta A = \begin{vmatrix} a_{11} & a_{12} & a_{13} & \dots & a_{1n} \\ a_{21} & a_{22} & a_{23} & \dots & a_{2n} \\ a_{31} & a_{32} & a_{33} & \dots & a_{3n} \\ \dots & \dots & \dots & \dots & \dots \\ a_{n1} & a_{n2} & a_{n3} & \dots & a_{nn} \end{vmatrix} \tag{1}$$

The proposed method reduces the determinant (1) into a sequence of n column vectors $C_0, C_1, C_2 \dots C_{n-1}$ and then the value of determinant (1) is obtained by multiplying all the column vectors as $\Delta A = C_0 \times C_1 \times C_2 \times \dots \times C_{n-1}$ i.e,

$$\Delta A = \overset{(0)}{A}[1:n,1:1] \times \overset{(1)}{A}[1:(n-1),2:2] \times \overset{(2)}{A}[1:(n-2),3:3] \times \dots \times \overset{(n-1)}{A}[1:1,n:n] \tag{2}$$

The procedure starts from the 0th level of the given determinant (1), proceeds to 1st level, from 1st level to 2nd level and so on up to $(n-1)^{th}$ level with reduction of one row from each level to extract the column vectors $C_0, C_1, C_2 \dots C_{n-1}$ respectively as

$$C_0 = \begin{vmatrix} \overset{(0)}{a}_{11} \\ \overset{(0)}{a}_{21} \\ \overset{(0)}{a}_{31} \\ \overset{(0)}{a}_{41} \\ \vdots \\ \overset{(0)}{a}_{n1} \end{vmatrix}, \quad C_1 = \begin{vmatrix} \overset{(1)}{a}_{12} \\ \overset{(1)}{a}_{22} \\ \overset{(1)}{a}_{32} \\ \vdots \\ \overset{(1)}{a}_{(n-1)2} \end{vmatrix}, \quad C_2 = \begin{vmatrix} \overset{(2)}{a}_{13} \\ \overset{(2)}{a}_{23} \\ \vdots \\ \overset{(2)}{a}_{(n-2)3} \end{vmatrix}, \quad \dots, \quad C_{n-1} = \begin{vmatrix} \overset{(n-1)}{a}_{1n} \end{vmatrix} \tag{3}$$

The 0th level of the determinant (1) is written as

$$\overset{(0)}{a}_{ij} = a_{ij} \qquad ; i, j = 1,2,\dots,n \tag{4}$$

We can extract the column vector C_0 from determinant (4) as

$$C_0\,[1:n, 1:1] = \overset{(0)}{a}_{ij} \qquad ; i = 1,2, \dots, n-1 \,;\, j = 1 \tag{5}$$

Thus, a new 1st level of the determinant (4) is formed as

$$\overset{(1)}{a}_{ij} = \frac{\overset{(0)}{a}_{(i+1)j}}{\overset{(0)}{a}_{(i+1)1}} - \frac{\overset{(0)}{a}_{ij}}{\overset{(0)}{a}_{i1}} \qquad ; i = 1,2,\dots, n-1 \,;\, j=2,3,\dots,n \tag{6}$$

From the 1st level determinant (6), the column vector C_1 is extracted as

$$C_1\ [1{:}n\text{-}1, 2{:}2] = a_{ij}^{(1)} \qquad ; i = 1,2, \dots, n\text{-}1 \ ; j = 2 \tag{7}$$

Finally, at the (n-1)[th] level a new determinant is formed and its corresponding column vector C_{n-1} is extracted as

$$a_{ij}^{(n-1)} = \frac{a_{(i+1)j}^{(n-2)}}{a_{(i+1)(n-1)}^{(n-2)}} - \frac{a_{ij}^{(n-2)}}{a_{i(n-1)}^{(n-2)}} \qquad ; i = 1,2 \ ; j = n\text{-}1,n \tag{8}$$

$$C_{n-1}\ [1{:}1, n{:}n] = a_{ij}^{(n-1)} \qquad ; i = 1 \ ; j = n \tag{9}$$

In general, the transformation of the determinant from (1) into the determinant (8) is formulated as

$$a_{ij}^{(k)} = \frac{a_{(i+1)j}^{(k-1)}}{a_{(i+1)k}^{(k-1)}} - \frac{a_{ij}^{(k-1)}}{a_{ik}^{(k-1)}} \qquad ;i=1,2,\dots,(n\text{-}k)\ ;j=k+1,\dots,n\ ;k=1,2,\dots,n\text{-}1 \tag{10}$$

Then the determinant of A is evaluated from (10) as

$$(a_{11}^{(0)} \text{ x } a_{12}^{(0)} \text{ x} \dots \text{x } a_{1n}^{(0)}) \text{ x } (a_{21}^{(1)} \text{ x } a_{22}^{(1)} \text{ x} \dots \text{x } a_{1(n-1)}^{(1)}) \text{ x} \dots \text{x } (a_{1n}^{(n-1)}) \text{ i.e.,}$$

$$\Delta A = \prod a_{ij} \qquad ; i=1,2,\dots,n\ ;\ j=1,2,\dots,(n+1\text{-}i) \tag{11}$$

If any element in the given determinant is zero, then it shows the ill-conditioning of the given determinant (1). Without the loss of generality and for simplicity, we introduce a scheme for converting an ill-conditioned determinant into a well-conditioned determinant by filling up the elements of the matrix A (1) as

$$\Delta A = \begin{vmatrix} a_{11}+a_{12} & a_{12}+a_{13} & a_{13}+a_{14} & \cdots & a_{1n}+a_{11} \\ a_{21}+a_{22} & a_{22}+a_{23} & a_{23}+a_{24} & \cdots & a_{2n}+a_{21} \\ a_{31}+a_{32} & a_{32}+a_{33} & a_{33}+a_{34} & \cdots & a_{3n}+a_{31} \\ \cdots & \cdots & \cdots & \cdots & \cdots \\ a_{n1}+a_{n2} & a_{n2}+a_{n3} & a_{n3}+a_{n4} & \cdots & a_{nn}+a_{n1} \end{vmatrix} \tag{12}$$

The elements of two consecutive columns of the determinant (1) are added to form the new columns of the determinant (12). The nth column of the determinant (12) is formed by adding the nth column of the determinant (1) and the first column of the determinant (12). Now the determinant (1) becomes the determinant (12). The above procedure is repeated until all the elements of the determinant (12) become non zero. Then the proposed procedure is applied for the newly generated determinant (12) to find its value.

3 Numerical Algorithm for Proposed Method

In this method, the given determinant of the matrix A of order n is reduced into a sequence of column vectors $C_0, C_1, C_2 \ldots C_{n-1}$ and the determinant A is evaluated by multiplying the elements of all column vectors.

Algorithm 3.1 ProposedMethod_EvaluateDeterminant

1. *Read* the matrix $A = [a_{ij}]$; $i, j = 1, 2, \ldots, n$
2. *If* $a_{ij}^{(0)} = 0$ *then* ; $i, j = 1,2,\ldots,n$

   Call **Algorithm3.2 IllConditionedDeterminant**
3. *Set* k = 1 and Product =1
4. *Compute* Product = Product x $a_{ik}^{(k-1)}$; $i = 1,2,\ldots,(n+1-k)$
5. *Compute* a_{ij}'s using the expression

$$a_{ij}^{(k)} = \frac{a_{(i+1)j}^{(k-1)}}{a_{(i+1)k}^{(k-1)}} - \frac{a_{ij}^{(k-1)}}{a_{ik}^{(k-1)}}$$

;$i = 1,2,\ldots,(n-k)$;$j = k+1,\ldots,n$;$k = 1,2,\ldots,n-1$

6. Increase k by 1
7. *If* k < n *then* go to Step 4

   //Extracting column vectors

   Otherwise Print ΔA = Product and *Stop.*

***End* ProposedMethod_EvaluateDeterminant**

If the given determinant (1) is found ill conditioned in step 2 of the algorithm 3.1 ProposedMethod_EvaluateDeterminant when any a_{ij} is zero, algorithm 3.2 IllConditionedDeterminant is called. This algorithm fills up all the elements of the determinant (1) by repeated addition.

Algorithm 3.2 IllConditionedDeterminant

1. *Compute* a_{ij}'s using the expression

$$a_{ij} = a_{ij} + a_{i(j+1)} \qquad ;i=1,2,\ldots,n\ ;j=1,2,\ldots,n-1$$

2. *Set* $a_{in} = a_{in} + a_{i1}$;i=1,2,…,n
3. *If* $a_{ij} = 0$ *then* go to Step 1 ;i,j=1,2,3,…,n

End **IllConditionedDeterminant**

3.1 Complexity of the Algorithm

Step 5 in algorithm 3.1 ProposedMethod_EvaluateDeterminant requires approximately $(n-1)+(n-1)^2$ divisions and 2n+(n-1) subtractions only. If the given determinant (1) is ill conditioned, then to convert it into a well-conditioned determinant, the algorithm 3.2 IllConditionedDeterminant requires only $n+n^2$ additions. For evaluating the determinant, the step 4 in algorithm 3.1 ProposedMethod_EvaluateDeterminant requires approximately 2n+(n-1) multiplications only. Hence the proposed method is comparatively effective than other high complexity direct methods.

4 Numerical Illustrations

The proposed algorithm 3.1 is simple and straightforward to be implemented using 'C' language. The program is executed in Intel Core2 Duo, 1.6GHz, 32 bit CPU for the following three examples i.e. non singular, tridiagonal and ill-conditioned determinants. The values of the determinants are obtained with the accuracy of 10^{-12} and the program is executed six times and average execution time is measured. Also the execution of the proposed method is compared with the Gauss reduction method and the comparison is reported.

4.1 Illustration 1

Consider the given determinant

$$\Delta A1 = \begin{vmatrix} 1 & 2 & -12 & 8 \\ 5 & 4 & 7 & -2 \\ 6 & -12 & 8 & 3 \\ 3 & -7 & -9 & -5 \end{vmatrix} \qquad (13)$$

The above determinant (13) is non-singular but not diagonally dominant. The determinant (13) is evaluated using the algorithm 3.1 and the results are tabulated in Table 1. It is observed from the execution times of proposed method and Gauss reduction method that the proposed method evaluates the determinant and yields the result in the average time of 0.119 milliseconds and faster than the Gauss reduction method which finds the value of the determinant (13) in the average time of 0.150 milliseconds.

Table 1. Evaluation of the determinant A1 (13)

Solution	$\Delta A1 = -12441.00000000000$ (10^{-12} Accuracy) Six Execution Times (milliseconds)						Average Exe.Time (milliseconds)
Proposed Method	0.126	0.119	0.120	0.116	0.123	0.111	**0.119**
Gauss Reduction Method	0.162	0.134	0.132	0.196	0.129	0.149	**0.150**

4.2 Illustration 2

Consider the given tridiagonal determinant

$$\Delta A2 = \begin{vmatrix} 2 & -1 & 0 & 0 & 0 \\ -1 & 2 & -1 & 0 & 0 \\ 0 & -1 & 2 & -1 & 0 \\ 0 & 0 & -1 & 2 & -1 \\ 0 & 0 & 0 & -1 & 2 \end{vmatrix} \tag{14}$$

The above determinant (16) is a tridiagonal determinant. Since the tridiagonal determinant (16) contains many zeros, the algorithm 3.1 calls algorithm 3.2 to fill all coefficients to make it a well-conditioned determinant.

The algorithm 3.1 evaluates the transformed determinant without the loss accuracy of the value and the results are tabulated in Table 2. The proposed method evaluates the determinant (16) in an average time of 0.129 milliseconds which is well comparable with the execution times of Gauss reduction method.

Table 2. Evaluation of the determinant A2 (14)

Solution	$\Delta A2 = 6.000000000000$ (10^{-12} Accuracy) Six Execution Times (milliseconds)						Average Exe.Time (milliseconds)
Proposed Method	0.121	0.128	0.153	0.128	0.120	0.124	0.129
Gauss Reduction Method	0.136	0.136	0.129	0.129	0.132	0.137	0.133

4.3 Illustration 3

Consider the given ill-conditioned determinant

$$\Delta A3 = \begin{vmatrix} 1 & -2 & 0 & 0 & 4 \\ 0 & 3 & 2 & -1 & 0 \\ 2 & 1 & -4 & 3 & 0 \\ 0 & 0 & 1 & 1 & 1 \\ 1 & -1 & 1 & -5 & 1 \end{vmatrix} \tag{15}$$

The above determinant (15) is a sparse determinant and has many zeros. The algorithm 3.1 calls algorithm 3.2 to fill all coefficients to make it a well-conditioned determinant. Then algorithm 3.1 evaluates the converted determinant and the results are tabulated in Table 3. The proposed method evaluates the determinant (15) in an average time of 0.134 milliseconds which is well comparable with the execution times of Gauss reduction method.

Table 3. Evaluation of the determinant A3 (15)

Solution	$\Delta A3 = 130.000000000000$ (10^{-12} Accuracy) Six Execution Times (milliseconds)						Average Exe.Time (milliseconds)
Proposed Method	0.163	0.120	0.129	0.124	0.136	0.133	0.134
Gauss Reduction Method	0.174	0.134	0.135	0.132	0.212	0.189	0.163

5 Result Analysis and Discussion

The observations made from the illustrations 4.1, 4.2 and 4.3 are discussed here. The execution times measured in illustration 4.1 show that the proposed method is straightforward in solving the non-singular determinant (13) which has 20.67% progress than Gauss Reduction method using partial pivoting. From the illustrations 4.2 and 4.3, it is observed that the execution time of the proposed method is well comparable with Gauss Reduction method. Since the determinants (14) and (15) contain many zeros, the proposed method converts them into well-conditioned determinants and evaluates them with the accuracy of 10^{-12}. And the procedure does not follow any further conversions to obtain the actual results.

For the above examples 4.1, 4.2 and 4.3, the procedures are repeated six times to consider the average execution time and the number of repetition can be extended as per the application.

6 Conclusion

The proposed method is reliable in evaluating any determinant which may be singular or ill-conditioned with a chosen accuracy of 10^{-12}. The proposed method follows a simple procedure for column reduction to convert the given determinant into a sequence of column vectors. Since this procedure does not use partial pivoting like Gauss reduction method, the complexity involved in this algorithm is reduced compared to other methods. This proposed procedure is most suitable for ill-conditioned determinants in which the method converts the ill-conditioned into a suitable well-conditioned determinant and evaluates it. The procedure is suitably modified to evaluate the determinant having complex values. The illustrative examples show that the method is direct and simple in application to be implemented using any programming language.

References

1. Rice, J.R.: Matrix Computations and Mathematical Software. McGraw-Hill, New York (1981)
2. Press, W.H.: Numerical Recipes: The Art of Scientific Computing, Cambridge (1992)
3. Hoffman, J.D.: Numerical Methods for Engineers and Scientists. McGraw-Hill, New York (1992)
4. Mucha, M., Sankowski, P.: Maximum matching via Gaussian elimination. In: 45th Annual IEEE Symposium on Foundations of Computer Science (accepted 2004)
5. Sankowski, P.: Dynamic Transitive Closure via Dynamic Matrix Inverse. In: Proceedings of the 45th Annual IEEE Symposium on Foundations of Computer Science, FOCS 2004 (2004)
6. Bunch, J., Hopcroft, J.: Triangular factorization and inversion by fast matrix multiplication. Math. Comp. 28, 231–236 (1974)
7. Gloub, G.H.: Matrix Computations. John Hopkins University Press, London (1989)
8. Broyden, C.G.: A class of Methods for Solving Nonlinear simultaneous equations. Mathematics of Computation 19(22), 577–593 (1965)
9. Aho, A.V., Hopcroft, J.E., Ullman, J.D.: The Design and Analysis of Computer Algorithms. Addison-Wesley Longman Publishing Co., Inc., Amsterdam (1974)
10. Percival, W.S.: Improved Matrix and Determinant methods for Solving Networks. Monograph No. 96. Radio Section, pp. 258–266 (1964)
11. Clarkson, K.L.: A short method for evaluating determinants and solving system of linear equations with real and complex coefficients. AIEE Transactions 60, 1235–1248 (1941)
12. Bronnimann, H., Yvinec, M.: Efficient Exact Evaluation of Signs of Determinants. Algorithimica 27, 21–56 (2000) doi: 10.10007/s004530010003,

Deriving Association between Learning Behavior and Programming Skills

S. Charles[1], L. Arockiam[2], and V. Arul Kumar[3]

[1] Assistant Professor, Department of Computer Science, St Joseph's College, Tiruchirappalli
[2] Associate Professor, Department of Computer Science, St Joseph's College, Tiruchirappalli
[3] Research Scholar, Department of Computer Science, St Joseph's College, Tiruchirappalli
charli_e@hotmail.com, larockiam@yahoo.com, varulkumar@ymail.com

Abstract. Data mining is the process of discovering patterns from large amount of data stored in student database. It predicts the facts from the dataset and interpreted as useful information to the society. The learning behavior of the student and programming skill are dissimilar in a classroom and programming lab. Data mining techniques are used to find association between the Learning behavior and programming skills using students' dataset. Clustering is used to determine the similarity in the students' dataset based on the nature of the learning behavior and programming skills. Each cluster reveals the identity based on its learning behavior of the student. Likewise the programming skill is categorized based on descriptive modeling. Multilayer Perceptron (MLP) technique classifies the learning behavior of students and their programming skills based on learning by example. The task is to determine the association between learning behavior and programming skills through descriptive and predictive modeling using mapping or function. It reveals that there is a positive correlation between student learning behavior and programming skills. This analysis could help the staff members to provide right training to the students for their improvement of programming skill.

Keywords: Multilayer Perceptron, K-Means Clustering, Criterion Reference Model, Learning Behavior.

1 Introduction

The process of translating raw information from the data set to valuable information can be used to inform design verdict. Information is useful if it can be significantly interpreted in the appropriate context. The chance of finding meaningful patterns, leads to the extension of the mining process with domain and problem specific. Cluster analysis is an unsupervised learning method that is used to investigate the inter-relationships among a collection of student objects, by organizing them into homogeneous clusters. K-means clustering technique is used to group the student objects based on the Learning behavior into classes of similar students' objects [1]. Similarly the programming skills are classified as good or fair.

The MLP is used for subcategory classification in the students' dataset. It is one of the approaches for discerning a pattern of subcategories. It estimates the useful

V.V. Das, J. Stephen, and Y. Chaba (Eds.): CNC 2011, CCIS 142, pp. 96–103, 2011.

information through mining and determines the interesting patterns of the students' learning style. Similarly it classifies the programming skills of students. This paper focuses on deriving association between learning behavior of the students and their programming skills based on supervised and unsupervised learning process.

2 Motivation

2.1 Learning Behavior

The learning style or behavior is defined as "the way each person absorbs and retains information and skills". Each learner possesses an individual learning style, which is preferential mode of learning. Learning style in college education varies among individual students and group of students. The patterns of repetitive and consistent learning behavior in the classrooms are observed. Such patterned behaviors are characteristics of the styles of learning. The learning style characteristics are 1) Serious, Analytical learners 2) Active, Practical learners 3) Observation-Centered Learner 4) Passive Accepting Learner 5) Concrete, Detail, Fact Oriented Learner and 6) Non-Adaptive Struggling learner [2].

The problem of discrimination and learning behavior classification is mainly concerned with differentiating between g (g>=2) mutually exclusive students and with classifying learning behavior on the basis Multivariate observations [3, 4]. The mathematical problem of discrimination is to divide the observation.space R^P into g mutually exclusive and exhaustive regions $R_1 \ldots R_g$ so that if an observation space x falls in R_i then it is classified as belonging to the population. The sub category problem is a mutually exclusive population according to their attributes and classifying an unseen instance into its sub categories using multiple features. In this study, the subcategories are experimental unit variables and observed features are observational unit variables.

Learning styles are characterized based on abstractness and correctness in learning style or motivation and responsibility. The proposed model determines the learning style and is able to characterize, how the mind functions are while learning.

2.2 Criterion-Referenced Model

A criterion-referenced model is an assessment model where the purpose is to assess the extent to which a student has achieved the goals of a course. The summative assessment is carried out aligned with specified criteria. Results are expressed in terms of how well a given student's performance matches set criteria. The standards are set before teaching takes place. The result is assigned on the basis of the standard that the student has achieved on the criteria [5, 18].

2.3 K-Means Approach

The K-means algorithm is one of the methods, where each cluster is represented by mean value of the objects in the cluster. It takes the input parameter **k**, and partitions a set of **n** students' data objects into **k** clusters so that the resulting intracluster similarity is high but the intercluster similarity is low. Cluster similarity is measured with regard to the mean value of the student data object in a cluster, which can be viewed as the cluster centroid [6, 7].

2.4 Multi-layer Perceptron

The Multilayer Perceptron is one of the techniques for classification task. It is a network of simple neurons called perceptrons. The perceptron computes a single output from multiple real-valued inputs by forming a linear combination according to its input weights and then possibly putting the output through some nonlinear activation function [7].

3 DALP Approach

Our approach can be divided into four phases which is depicted in Fig 1. In Phase I, the research questions are designed by the expert based on learning behavior and pretest the questionnaire. After pretesting a questionnaire, the learning behavior analysis test has been conducted in Post Graduate Course. The written and practical marks are obtained from the semester exam, which is viewed as programming skills (good (Marks>=60) or fair (Marks<60)) [5, 8, 10, 18] based on Criterion-Referenced Model. The students' dataset contains the test results of Learning Behavior and performance of the students in the written and practical test in programming subject.

In Phase II and III, K-Means clustering and Multi Layer Perceptron techniques are employed to find the association between the Learning Behavior and programming skills. K-Means clustering reveals the identity of the students' Learning behavior and designates the cluster as 1) Serious, Analytical learners 2) Active, Practical learners 3) Observation-Centered Learner 4) Passive Accepting Learner 5) Concrete, Detail, Fact Oriented Learner and 6) Non-Adaptive Struggling learner. The same method is employed for program skill based clustering [11, 12, 13]. The MLP technique is used to

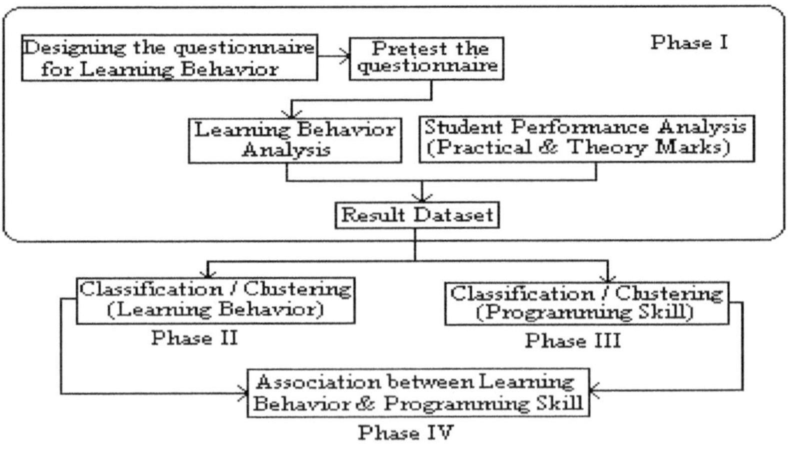

Fig. 1. DALP approach

classify the learning behavior of the students. The six sigmoid nodes are used as experimental variables, the weights are assigned to each node, and six output layers are classified based on the category of learning behavior. Similarly the skill-based classification can be carried out based on the programming skills [14]. In Phase IV, the association between learning and behavior of the students is analyzed using interesting measures.

4 Results and Discussion

The dataset containing 408 students' objects is used to find the association between learning behavior and their programming skills of the student. Fig 2 shows that various clusters and corresponding learning behavior are categorized. It reveals that Active and practical learner objects are silhouetted in cluster 0, Observation centered learner objects are silhouetted in cluster 1, Concrete detail, fact oriented learner objects are silhouetted in cluster 2, passive accepting learner objects are silhouetted in cluster 3 and Serious Analytical learner objects and Non passive struggling learner objects are silhouetted in cluster 4. The observed values are shown in table 1.

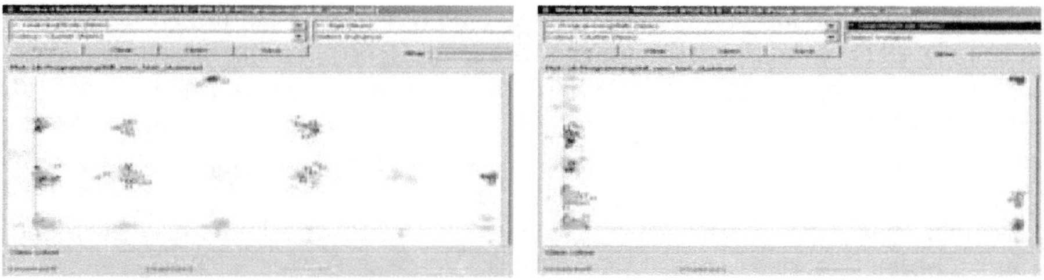

(a) Clustering based on learning behavior (b) Clustering based on programming skills

Fig. 2. Clustering based on learning behavior and programming skills

In Cluster 4, Serious Analytical learner objects are merged with Non-Passive struggling learner. Similarly the MLP classifies the students' objects with a support of the class variable as learning behavior and experimental variables as demographic information of objects and theory and practical examination mark. The confusion matrix reveals that various classes to identify the six learning styles. A misclassification occurs in the classification based on the nature of students' learning styles.

Table 1. Cluster assisgments and Learning behavior

C0	C1	C2	C3	C4	LB
63	17	36	45	53	A
18	8	18	27	0	C
0	9	9	18	0	P
17	18	0	17	9	O
0	0	0	9	0	N
8	0	0	9	0	S

Table 2. Confusion Matrix for Learning Behavior

Cross validation		Predicted						
		a	b	c	d	e	f	Learning Behavior
Actual	a	209	0	0	0	5	0	Active and Practical
	b	11	51	0	9	0	0	Concrete, fact-Oriented
	c	4	6	25	1	0	0	Passive, Accepting
	d	1	8	0	52	0	0	Observation centered
	e	5	0	0	0	4	0	Non - Passive
	f	0	0	0	0	0	17	Serious, Analytical

Table 3. Learning Style Preference of Student

Learning Style Characteristics	Response	Agreement %
Serious, Analytical Learner	17	4
Active and Practical Learner	214	52
Observation Centered Learner	61	15
Concrete Detail, Fact-Oriented Learner	71	18
Passive, Accepting Learner	36	9
Non-Passive, Struggling Learner	9	2

Table 4. Confusion Matrix for Classification Programming Skill

Validation Data		
Actual Category	Predicted Category	
	G	F
G	375	0
F	11	22
Mean absolute error	0.0515	
Root mean squared error	0.1764	

Table 3 clearly indicates that the general ways of learning, which majority of Post Graduate Students preferred. Approximately 4% of the students have viewed themselves as Serious, Analytical Learners. 52% of the students are Active learners, 15% of the students are Observation centered learners, 18% of the students are Concrete Detailed learners, 9% of the students are Passive learners and 2 % of the students are Non-Passive Learners, who agreed with the preferred learning styles.

Table 4 shows that the classification of programming skill using the experiment variables. It reveals that 375 student objects are observed as good programmers, 22 students' objects are observed as fair programmers and 11 students' objects are misclassified and viewed as Good programmers but the objects are existed in Fair programming students' objects. Similarly the learning styles are categorized using clustering.

The Student objects are classified based on the leaning behavior and programming skill. 52% of Active, Practical learners, 18% of Detail fact oriented learners 16% observation centered learners and 9% passive accepting learners are good in programming. Fig 3 reveals that the Active and practical learners having good programming skill other than other learner. It is evident that the Practical Learners are good in programming.

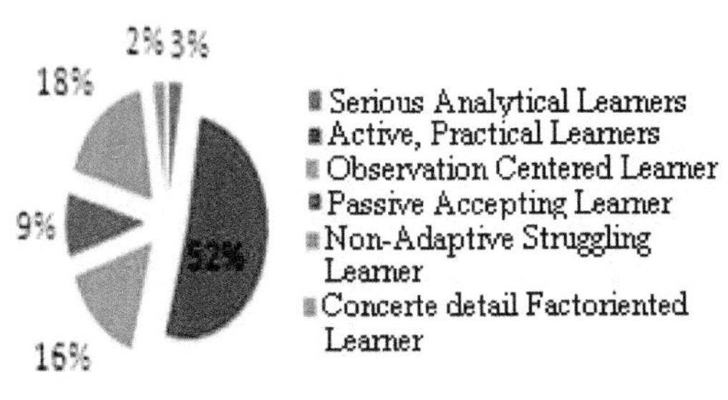

Fig. 3. Classification of Learning behavior and Good Programmers

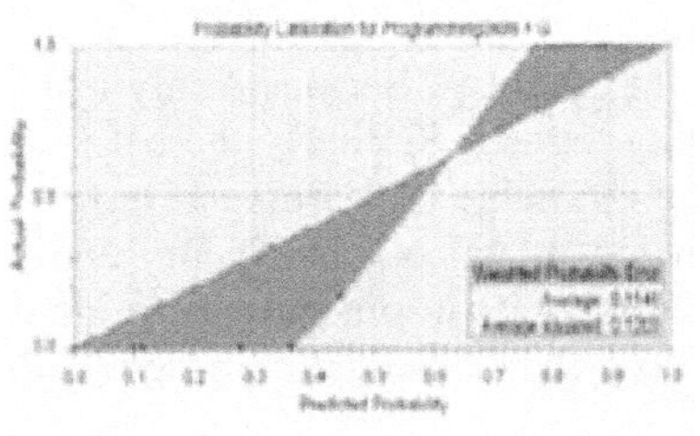

Fig. 4. Probality Calibration Chart for Good Programmers

Fig 4 shows that the probability of students having good programming skill. The Probability Calibration Chart shows how the predicted probability for a target category is distributed and provides a means for gauging the accuracy of predicted probabilities. The average predicted probability for cases between 0.6 and 0.7 about 0.65; the actual probability based on the rate of occurrence for those cases were about 0.87.

If the predicted probabilities match the actual probabilities, the points fall on the diagonal line. The red shaded area shows the error, which is the difference between the predicted and actual probabilities. The average predicted probability for those rows is 0.1481, which is about the midpoint of the range. The weighted probability error value is low, which expose that the model is perfect.

Fig.5 shows Receiver Operating Characteristic (ROC) chart and probability threshold is high [16]. It means that most of the students' programming skill is good. The sensitivity is the proportion of the students having a good or fair programming skill who are identified by the experiment.

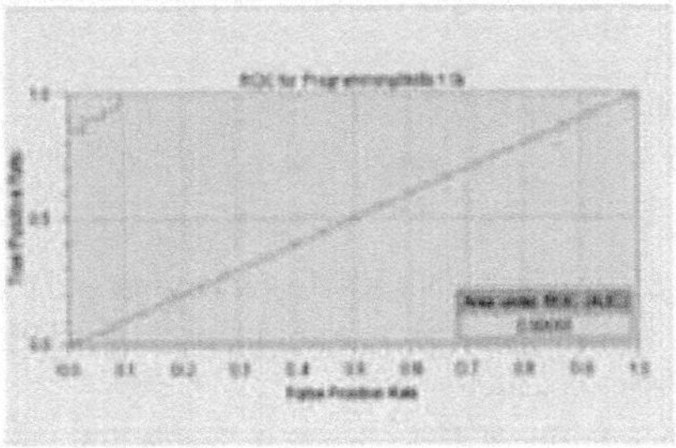

Fig. 5. ROC chart for Good Programmers

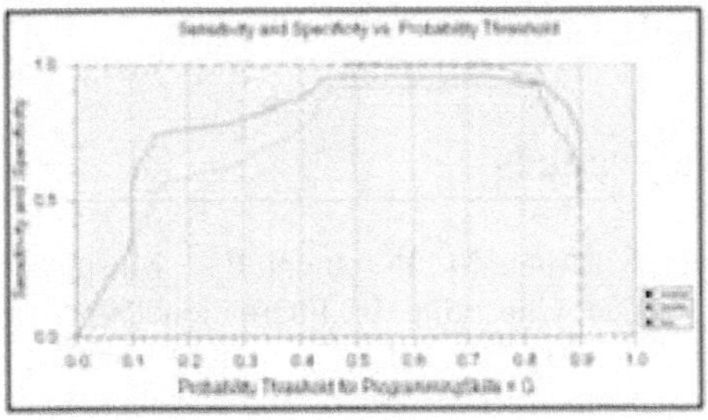

Fig. 6. Sensitivity and Specificity curve for Programming Skill

The specificity is the proportion of the students who do not have good or fair programming skill. Sensitivity and Specificity would be both 0.9. Fig.6 shows the sensitivity curve, which shows that the students having an active, practical learning style, concrete detail fact oriented learner, observation centered learning style and passive accepting learning style are good in programming. Only few students are fair in programming in all learning styles. The prediction can also be viewed as a mapping or function Y = **f (X)** where **X** is the input and the output Y is a categorical value [16]. The mapping or function model shows the association between X and Y. The lift and cosine measure are used to find the correlation. The correlated value of X and Y are 1.079, which is positively correlated, meaning that the X and Y are associated with each other. The cosine value of X and Y are 0.512, meaning that there is a positive correlation between X and Y. It makes known that Learning behavior have a strong relationship with programming skill.

5 Conclusion

In this Paper, Clustering technique and Multi-Layer perceptron are used to find the association between Learning behavior and programming skills. Learning by observation method is an effective means for classifying the objects such as Serious, Analytical learners, Active, Practical learners, observation-Centered Learner, Passive Accepting Learner, Concrete, Detail, Fact Oriented Learner and Non-Adaptive Struggling learner and models each cluster. It reveals the identity of the learning behavior.

Likewise, the programming skills based clusters are observed and modeled as good and fair in programming. The classifier reveals that, each individual student's learning behavior related with the programming skill. The students having an Active and Practical Learner are good in programming, some students having a Observation Centered Learning style and Concrete Detail, Fact-Oriented Learning style are also good in programming, Passive accepting style learner, Serious analytical learner and Non-Adaptive struggling learner are fair in programming. Passive accepting style learner, serious analytical learner and Non-Adaptive struggling learner are need to be improved in their programming skill. The correlation analysis unveils that the learning style and programming skills have strong association. The findings provide an insight into developing the programming skills of students to improve it through practical oriented teaching.

References

[1] Khalilian, M., Boroujeni, F.Z., Mustapha, N., Sulaiman, M.N.: K-Means Divide and Conquer Clustering. In: International Conference on Computer and Automation Engineering, pp. 306–309. IEEE Computer Society, Los Alamitos (2009)

[2] Kolb, D.A.: Experimental learning, Experience as the source of the learning and development. Prentice Hall Inc., Englewood Cliffs (1984)

[3] Choi, S.C., Rodin, E.Y.: Statistical Methods of Discrimination and Classification. In: Advances in Theory and applications (1986)

[4] Duda, R.O., Hart, P.E., Stork, D.G.: Pattern Classification, 2nd edn. John. Wiley & sons Inc., Chichester (2000)

[5] Khamis, I., Idris, S.: Issues and Solutions in Assessing Object-oriented programming Skills in the Core Education of Computer Science and Information Technology. In: 12th WSEAS International Conference on Computers, Heraklion, Greece, July 23-25 (2008)

[6] Wu, X., Kumar, V., Ross Quinlan, J., Ghosh, J., Yang, Q., Motoda, H., McLachlan, G.J., Ng, A., Liu, B., Yu, P.S., Zhou, Z.-H., Steinbach, M., Hand, D.J., Steinberg, D.: Top 10 algorithms in data mining. Knowl. Inf. Syst. 14, 1–37 (2008)

[7] Ayers, E., Nugent, R., Dean, N.: Skill Set Profile Clustering Based on Student Capability Vectors Computed from Online Tutoring Data. In: Baker, R.S.J.d., Barnes, T., Beck, J.E. (eds.) Proceedings of 1st International Conference on Educational Data Mining 2008, Montreal, Quebec, Canada, June 20-21, pp. 210–217.

[8] Pavlik Jr., P.I., Cen, H., Wu, L., Koedinger, K.R.: Using Item-type Performance Covariance to Improve the Skill Model of an Existing Tutor. In: 1st Proceedings of International Conference on Educational Data Mining, Canada, June 20-21, pp. 77–86 (2008)

[9] Green, T.M., Jeong, D.M., Fisher, B.: Using Personality Factors to Predict Interface Learning Performance. In: 43rd Hawaii International Conference on System Sciences, Honolulu, HI, January 5-8, pp. 1–10. IEEE Computer Society, Los Alamitos (2010)

[10] Marshall, L., Austin, M.: The relationship between software skills and subject specific knowledge, theory and practice. Learning and Teaching Projects (2002/2003)

[11] McCracken, M., Almstrum, V., Diaz, D., Guzdial, M., Hagen, D., Kollokant, Y., Lazer, C., Thomas, L.A., Utting, I.: A Luti national Study of Assessment of Programming Skills of First Year CS students. SIGC, SE Bulletin 33, 125–140 (2001)

[12] Weka Tool (3.6.2),

[13] `http://www.cs.waikato.ac.nz/ml/weka/index_downloading.html`

[14] Chiu, C.: Cluster Analysis for Cognitive Diagnosis:Theory and Applications. Ph.D.Dissertation, Educational Psychology, University of Illinois at Urbana Champaign (2008)
[15] Ayers, E., Nugent, R., Dean, N.: A Comparison of Student Skill Knowledge Estimates. In: Proceedings of 2nd International Conference on Educational Data Mining 2009, Cordoba, Spain, July 1-3, pp. 1–10 (2009)
[16] Nghe, N.T., Janecek, P., Haddawy, P.: A Comparative Analysis of Techniques for Predicting Academic Performance. Paper presented at 37th ASEE/IEEE Frontiers in Education Conference, Milwaukee, WI, October 10-13 (2007)
[17] Ramaswami, M., Bhaskaran, R.: A Study on Feature Selection Techniques in Educational Data Mining. Journal of Computing 1(1) (December 2009)
[18] Han, J., Kamber, M.: Data Mining Concepts and Techniques, 2/e. Morgan Kaufmann Publishers, San Francisco (2006)
[19] Arockiam, L., Charles, S., Arul Kumar, V.: Deriving Association between Personality Traits and Programming Skills. In: International Conferences on Advances in Communication, Network and Computing (October 2010)

Analysis of a New Base Station Receiver for Increasing Diversity Order in a CDMA Cellular System

Harish Kumar[1], Member IEEE, V.K. Sharma[1], Pushpneel Verma[1], and G. Venkat Babu[2]

[1] Deptt.of ECE, Bhagwant Institute of Technology, MuzaffarNagar,U.P (India)
[2] Hindustan University, Chennai (India)
hc78@rediffmail.com, viren_krec@yahoo.com, pushpneelverma@gmail.com

Abstract. In this paper the analysis a new base station receiver is proposed and analyzed for a code-division multiple-access (CDMA) cellular system. The proposed receiver can achieve remarkable diversity gain by increasing diversity order with reasonable cost and complexity. From the numerical results, it is confirmed that the proposed receiver structure can be a practical solution for enhancing reverse-link capacity and improving performance in CDMA cellular system operations. The result in the letter can find its applications to legacy IS-95/cdma 2000 1xbase stations with simple modifications.

Keywords: Code-division Multiple access (CDMA), Diversity, Multipath fading, Receiver complexity.

1 Introduction

Several diversity techniques have been studied and found practical use in many communication systems in order to combat multipath fading and improve performance. Among the techniques, spatial diversity has been commonly used at cellular base station (BS) receivers due to its simplicity in implementation [1].

Although it is well known that higher order diversity can improve the receiver performance, the second-order spatial diversity with two receiving antennas at a BS has been most popular. Because the higher order spatial diversity at a BS requires additional cost associated with strict zoning requirements. The antenna elements for the spatial diversity need to be separated at least ten times wavelength in order to obtain signals that fade independently [1]. The possibility of using polarization diversity has been studied with a motivation that the polarization diversity using a dual-polarized antenna can reduce cost and space for installation compared with the traditional spatial diversity. The previous investigations on polarization diversity have revealed that the polarization diversity is able to achieve comparable performance to the spatial diversity [2]-[5].

V.V. Das, J. Stephen, and Y. Chaba (Eds.): CNC 2011, CCIS 142, pp. 104–110, 2011.

Exploring the polarization as an additional source of diversity has been considered in [6] and [7] Particularly [7] investigated the benefit of increased diversity order using spatially separated polarized antennas at the BS receiver of a code-division multiple access (CDMA) cellular system. The paper proved that the four branch diversity receiver combining spatial and polarization diversity could considerable improve receiver performance and enhance reverse-link capacity. However implementation of the four-branch diversity receiver using spatially separated cross-polarized antennas requires high cost and considerable modification, compared with the conventional dual-diversity receiver. The four-branch diversity receiver also needs additional hardware units and space for increased diversity branches.

2 Proposed Receiver Structure

In sectorized cellular CDMA systems, a typical BS supports three sectors, and the structure for the four-branch diversity receiver using spatially separated dual-polarized antennas in [7] is shown in Fig. 1(a). In each sector, the polarized antennas form four diversity branches and the signals from each branch are down-converted and digitized in the radio frequency (RF) and the intermediate frequency (IF) circuits. Then, they are transferred to the baseband RAKE receiver for matched filtering and combining.

The four-branch diversity receiver requires additional hard-ware units for the increased diversity branches, compared with the conventional dual-diversity receiver, because the signals from antenna elements have the same pseudo noise (PN) code offsets and suffer independent fading.

In cellular CDMA systems, through the transmitted signal from a mobile station (MS) is spread by a user-specific PN code, the received signals from the MS have different PN offsets due to the multipath components. The RAKE receiver in the baseband modern resolves the multipath signals and combines them [8]. The RAKE receiver can resolve the combined signals based on the PN code offsets. With this motivation, we propose a new BS receiver that can increase diversity order with low cost and complexity. The structure of the proposed receiver is shown in Fig. [1].

The RF signal from one antenna element of a cross-polarized antenna is internally delayed by the amount of a predetermined value and added by the RF signal from the other antenna element. In the proposed receiver, M/2 branches are for normal branches and the remaining M/2 branches are for the delay branches among the total M-diversity branches [M=4 for Fig. 1 (b)]. Then the combined signal in the RF level is input to the RF and IF units for down-conversion and digitization, and transferred to the baseband RAKE receiver. All of the signals from M diversity branches can be resolved in the RAKE receiver, because the sufficient delay can make the fading statistics appear independent even for the same user and assure the resolvability as well.

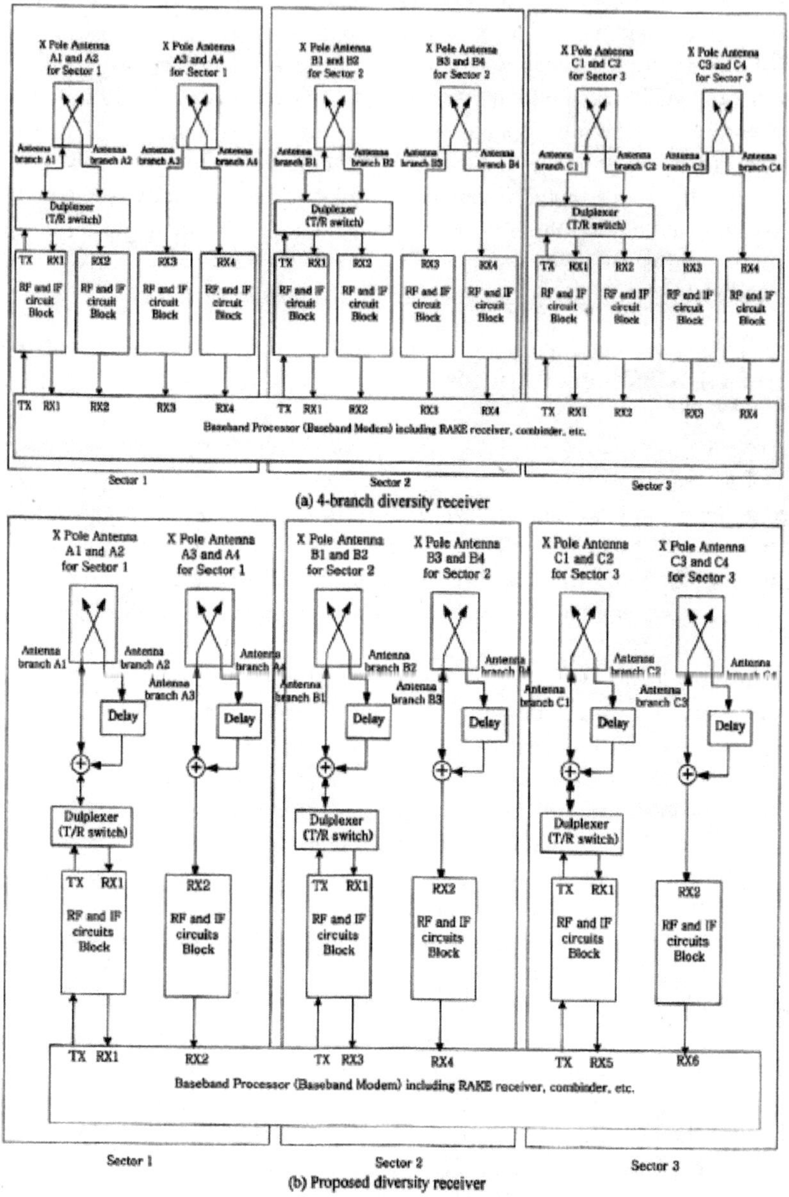

Fig. 1. BS receiver structures. (a) Four branch diversity receiver (b) Proposed diversity receiver.

3 Performance Analysis

In this section, the outage probability and BEP are derived for each type of receiver by modifying the results of [9] and [10]. Although the synchronous assumption is not valid in commercial CDMA systems a synchronous reverse link is assumed in the performance analysis because it allows a complete comparative analysis of the diversity techniques.

A. Conventional M-Branch Diversity Receiver

For the M-branch diversity receiver, matched-filter (MF) output for a user 0 can be given in a vector form by

$$Z_0(n) = 8_0(n)A_0 + \sum_{i=1}^{N} s_i(n)A_i + n(n) \tag{1}$$

Where each element of Z_0 (n) represents the MF output corresponding to each diversity branch. $s_i(n)$ $(i=0.1...N)$ is a transmitted data sequence of user i. $A_1(i = 0.1....N)$ is an M-dimensional vector which represents filtered outputs of the attenuated signals for the user i, N is the number of interfering users, and $n(n)$ is an M-dimensional Gaussian zero-mean vector with covariance matrix $\sigma^2 I$.

In an MRC system, the signal of each branch is weighted by a corresponding complex-valued channel gain before combining. Thus, the weighting vector for the MF output given in (1) becomes A_o Then. the output signal-to-interference-plus-noise ratio of the MRC system (SINR_0) is given by

$$\gamma = \frac{A_0^H A_0 A_0^H A_0}{A_0^H C_{nn} A_0} = \frac{x_0}{\sum_{i=1}^{N} z_i + \sigma^2} \tag{2}$$

where the superscript ^{H} denotes complex conjugate transpose, and

$C_{nn} = \sum_{i=1}^{N} A_i A_i^H + \sigma^2 I$, $x_0 = A_0^H A_0$ and $z_i = | A_0^H A_i |^2 / x_0$.

For the MRC, the BEP can be obtained by GA for multiple access interference (MAI). In (1) the second term represents the MAI and can be approximated as a Gaussian random variable from the result of [11] then the MAI plus additive white Gaussian noise (AWGN) is also a Gaussian random variable and the BEP can be given by [3].

$$P_e(\gamma) = \left[\frac{1}{2}(1-\mu)\right]^M \sum_{k=0}^{M-1} \binom{M-1+k}{k} \left[\frac{1}{2}(1+\mu)\right]^k \tag{3}$$

where $\mu = \sqrt{\gamma/(1+\gamma)}$ and $\overline{\gamma}$ is an average SINR per branch (SINR_br).

B. Proposed M. Branch Diversity Receiver

The proposed receiver, the RAKE receiver resolves the signals of each diversity branch from the coupled signal. The MF output for a user 0 can be divided into two parts, which are the resolved MF outputs for the signals from normal and delay branches, respectively, and given by

$$Z_0(n) = \begin{bmatrix} Z_0^{(1)}(n) \\ Z_0^{(2)}(n) \end{bmatrix} \tag{4}$$

and thus, intentional delay should be larger than T_c but smaller than the shifted value for user distinction. The proposed scheme may insert sufficient delay to make the fading statistics appear independent even for the same user and assure the resolvability as well.

In the matched filtering for user 0's signals from the normal branches, user 0' signals from the delay branches act as an other MAI source given in the fourth term is statistically equivalent to the cross-correlated interference between different users. The cross-correlation appears stationary, and the signals from the diversity branches are independent. It can be effectively approximated to

$$Z_0^{(1)}(n) \cong 8_0(n) A_0^{(1)} + \sum_{i=1}^{2N+1} 8_i(n) A_i^{(1)} + n^{(1)}(n) \tag{5}$$

$$Z_0^{(2)}(n) \cong 9_0(n) A_0^{(2)} + \sum_{i=1}^{2N+1} 8_i(n) A_i^{(2)} + n^{(2)}(n) \tag{6}$$

From (5) and (6), the MF output for a user 0 can be given in an M-dimensional vector.

C. On the Validity of Gaussian Approximation

The method to improve the inaccuracy of GA in this situation remains our further study. However, considering the fact that the purpose of this letter is to investigate the feasibility of the proposed receiver, the derived BEP with the GA can be useful in comparative analysis by it-self. For the conventional diversity receiver, the BEP with the GA can be regarded as only an optimistic loss bound because of the inaccuracy of the GA. But for the proposed receiver, the inaccuracy of the GA is obviously improved, since the proposed receiver causes the effect of increasing the number of interfering user $(N \rightarrow 2N+1)$.

4 Numerical Results

Fig.2 shows the BEP versus average SINR_br in a Rayleigh fading channel environment in the general case, including the situation where the MAI is not dominant compared with AWGN. For a BEP of 10^{-3}, the four-branch receiver has about 1-dB gain over the proposed four-branch receiver, while the proposed four-branch receiver has about 5-dB gain over the conventional two-branch receiver. Though the four-branch receiver shows better performance than any other receivers considered, it requires high cost and considerable complexity, compared with the conventional two-branch receiver.

Fig. 3 shows the outage probability versus SNR $(\Omega_0 = P_0 / \sigma^2)$ as another performance measure. Five interfering users, 5-DB outage threshold SINR and 128 length user specific codes are assumed for all the types of receivers for performance comparison. For an outage probability of 10^{-2}, the gap of the required SNR between the proposed four-branch receiver and the conventional two-branch receiver is about 15 dB. That means the received power level of the conventional two-branch receiver should be 15 dB larger than that of the proposed four-branch receiver at the same level of AWGN to achieve the same outage probability of 10^{-2} Thus, transmit power of the MS in the proposed four-branch receiver can be 15 dB less than that of the conventional two-branch receiver.

From the BEP and outage performance shown in Figs, 2 and 3 the proposed four-branch diversity receiver can achieve remarkable gains over the conventional two-branch diversity receiver with low cost and complexity, and the gains can directly be linked to a capacity enhancement.

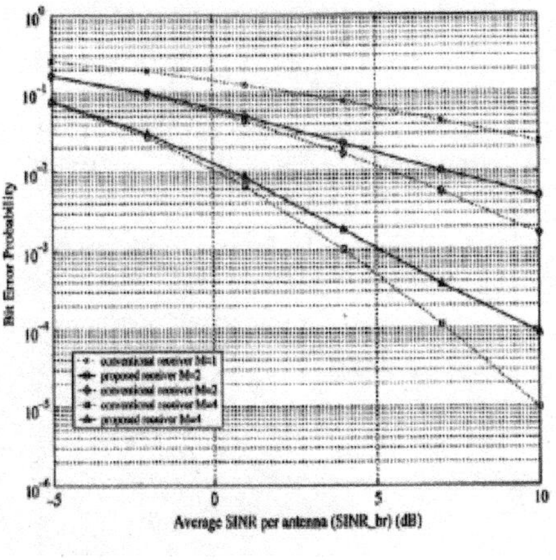

Fig. 2. BEP versus average SINR_br

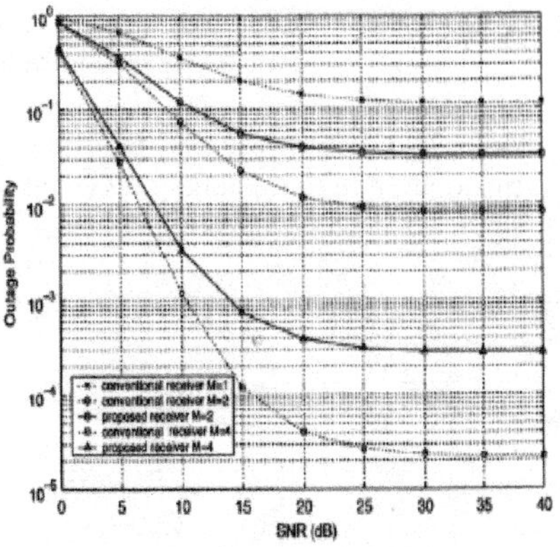

Fig. 3. Outage probability versus SNR

5 Conclusion

In this paper the new BS receiver has been proposed and analyzed in order to increase diversity order in the CDMA cellular system. The BEP and outage probability are evaluated and compared for the conventional and the proposed diversity receivers. From the numerical results, it has been demonstrated that the proposed receiver can achiever remarkable diversity gain with reasonable cost and complexity. The proposed receiver is expected to provide a practical solution for enhancing reverse link capacity and improving performance in the CDMA cellular system operations. The proposed receiver can be applied to the legacy IS-95/ cdma 2000 1x BSs with simple modifications.

References

[1] Jakes, W.C.: Microwave Mobile Communications. Wiley, New York (1974)

[2] Lee, W.C.Y., Yeh, Y.S.: Polarization diversity system for mobile radio. IEEE Trans. Commun. COM 26, 912–923 (1972)

[3] Turkmani, A.M.D., Arowojolu, A.A., Jefford, P.A., Kellen, C.J.: An experimental evaluation of the performance of two-branch space and plarization diversity scheme at 1800 MHz. IEEE Trans. Veh. Tehnol. 44, 318–326 (1995)

[4] LIndmark, B., Nilsoon, M.: On the available diversity gain from different dual-polarized antennas. IEEE J. Select Areas Commun. 19, 287–294 (2001)

[5] Siquiera, G.I., et al.: Combined use of space and polarization diver-sity on mobile cellular network. In: Proc. IEEE Globecom, Rio de Janeiro, Brazil, pp. 863–867 (December 1999)

[6] Turin, G.L.: Introduction to spread spectrum ant multipath techniques and their application to urban digital radio. Proc. IEEE 68, 328–353 (1980)

[7] Aydin, L., Esteves, E., Padovani, R.: Reverse link capacity and coverage improvement for CDMA cellular systems using polarization and spatial diversit. In: Proc. IEEE Int. Conf. Communications, New York, NY, pp. 1887–1892 (April 2002)

[8] Cut, J., Sheikh, A.U.H.: Outage probability of cellular radio systems using maximum ratio combining in the presence of multiple interferers. IEEE Trans. Commun. 47, 1783–1787 (1999)

[9] Aalo, V.A., Chayawan, C.: Outage probability of cellular radio systems using maximal ratio combining in Rayleigh fading channel with multiple interferes. IEEE Electron. Lett. 36, 1314–1315 (2000)

An Improved SRS Document Based Software Complexity Estimation and Its Robustness Analysis

Sharma Ashish[1] and D.S. Kushwaha[2]

[1] Department of Computer Science & Engineering, GLA Inst. of Tech. & Mgt., Mathura
[2] Department of Computer Science & Engineering, MNNIT, Allahabad, India
sharma.asb@gmail.com, dsk@mnnit.ac.in

Abstract. Complexity measures are used to predict critical information about reliability and maintainability of software systems from analysis of the source code. Complexity measures also provide continuous feedback during a software project to help control the development process. During testing and maintenance, it provides detailed information about software modules to help pinpoint areas of potential instability. The various complexity measures established so far are based on code and it is too late to perform this activity because a major part of investment has already taken place. Although it is established that a high quality SRS is pre requisite for high quality software, but software complexity estimation based on SRS document has not been properly researched and we find little proposals. Hence the work presented in this paper attempts to empirically demonstrate that the complexity of the code to be produced can be closely estimated based on IEEE software requirement specification document (IEEE 830-1998). Results obtained shows that the complexity values obtained from improved requirement based complexity (I-RBC) are comparable with other established measures and hence the complexity of the software to be produced could be computed from its SRS document. Its robustness is established by strictly evaluating and comparing proposed measure against Weyuker properties.

Keywords: Requirement Based Complexity, Input Output Complexity, Product Complexity, Personal Complexity Attributes, Interface Complexity, User Location Complexity, Requirement Complexity.

1 Introduction and Related Work

Complexity analysis has an extremely high payoff for the investment. Moving from counting lines of code to calculating Cyclometric complexity has immediate, measurable benefits in terms of risk management; reliability prediction, cost containment, project scheduling, and improving overall software quality have been proposed in the past. Most of the software complexity measures are based on code, but when we use the code for computing software complexity it is too late. We can make use of the requirement engineering document and properly analyze each and every component of this document and further derive the various attributes that contributes towards estimation of complexity measure based on SRS [7]. As an effect, the code and design decisions can be made much in advance.

V.V. Das, J. Stephen, and Y. Chaba (Eds.): CNC 2011, CCIS 142, pp. 111–117, 2011.

Maurice Halstead proposed the measure [2] based on the principle of Count of Operators and Operand, The problem with this method is that, they are difficult to compute. It is not suited when we want fast and easy computation Cyclometric Complexity [3] developed by Thomas J Mc Cabb in 1976. The difficulty lies within a conditional statement is never acknowledged. Also there is no penalty for embedded loops versus a series of single loops; both have the same complexity. Klemola and Rilling [5] proposed KLCID based complexity measure. This method can become very time consuming when comparing a line of code with each line of the program. Wang and Shao [6] have proposed to measure the cognitive complexity using cognitive weights and basic control structures (BCS). CICM (Cognitive Information Complexity measure) [4] uses weighted information count of the software and cognitive weights of Basic Control Structure (SBCS) of the software. Limitation with these measures are that, they all uses code and hence they are code dependent measures, which itself is a problem as stated earlier.

2 Introduction to Improved Requirement Based Complexity Measure (I-RBC)

The proposed I-RBC measure is based on the framework derived as shown in figure 1 from SRS Document [1]. The computation method for the proposed measure is explained below:

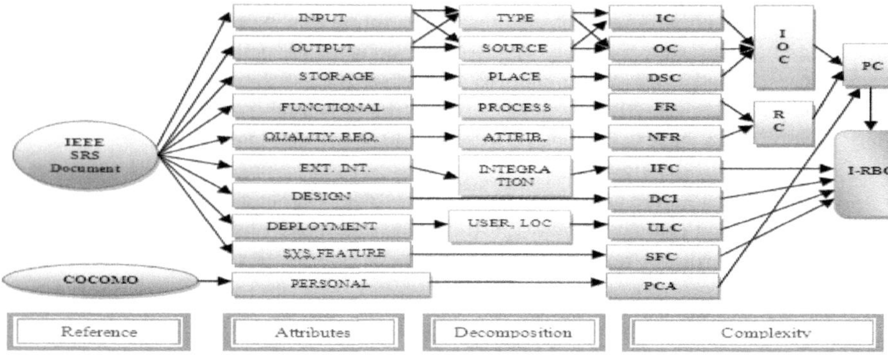

Fig. 1. Frame-work for the computation of Improved RBC

a. Input Output Complexity (IOC)

IOC is a combination of Input Complexity (IC). Output Complexity (OC) and Data Storage Complexity (DSC). For IC it is absolutely necessary to consider the source and type of input.

Input source refers the location of input entering to the system and these sources/ locations are mentioned in table 1:

Table 1. Types of input-output information

Parameters	Type	Count
Input/ Output Type	Text, character, integer, float etc.	1
	Image, picture, graphics, animation	2
	Audio, video etc.	3

Table 2. Sources of input-output information

Parameters	Source	Count
Input/ Output Source	External I/O through devices	1
	Files, database, process, S/w, H/w.	2
	Outside system	3

Further the type of input entering to the system, which will have a heavy impact on input output complexity. It can be described as shown in table 2

Upon considering the attributes we can compute the Input Complexity IC as:

$$\text{Input Complexity} = \sum_{i=1}^{3}\sum_{j=1}^{3} Nij\ X\ (Source)i\ X\ (Type)j$$

where N is the number of inputs. Similarly output complexity can be calculated based on output produced by the system:

$$\text{Output Complexity} = \sum_{i=1}^{3}\sum_{j=1}^{3} Oij\ X\ (Source)i\ X\ (Type)j$$

where O is the number of Outputs. Further the data storage we can have from table 3.

Table 3. Types of Data Storage & relative count

Parameters	Type	Count
Data Storage	Local data storage	1
	Remote data storage	2

$$\text{Storage Complexity} = \sum_{i=1}^{2}(Number\ of\ Storage)i\ X\ Associated\ Count$$

Now, Input Output Complexity can be expressed as:

$$IOC = Input\ Complexity + Output\ Complexity + Storage\ Complexity$$

b. Functional Requirement (FR)

A functional requirement defines a function of a software system or its component. It may be appropriate to partition the functional requirement into sub-functions or sub-processes.

$$FR = No.\ of\ Functions * \sum_{i=1}^{n} SPFi$$

where SPF is Sub Process or Sub-functions received after decomposition of functionality.

c. Non Functional Requirement (NFR)

This refers to the Quality related requirements for the software apart from functionality. These qualities related requirements can be mathematically expressed as:

$$NFR = \sum_{i=1}^{6}\sum_{j=1}^{m}(Attribute)i\ X\ (Sub\ attribute)j$$
$$1 \qquad otherwise\ for\ inbuilt\ quality\ attribute$$

Where m is number of sub attribute for corresponding attribute.

d. Requirement Complexity (RC)
It refers to the sum of all requirements i.e. functional requirement and non-functional requirement which can be stated as: $RC = FR + NFR$

e. Product Complexity (PC)
It refers to the overall complexity based on the functionality of the system. It can be defined as product of Requirement Complexity and Input Output Complexity. $PC = IOC\ X\ RC$

f. Personal Complexity Attributes (PCA)
It refers to technical expertise based on personal capability [13] Mathematically PCA can be described as: $PCA = \sum_{i=1}^{5} Multiplying\ factors$

g. Design Constraints Imposed (DCI)
It refers to number of constraints that are to be considered for the development of software product by external body. This metrics is mathematically defined as: $DCI = \sum_{i=0}^{n} C_i$

Where C_i is Number of Constraints and value of C_i will vary from 0 to n.

$$C_i = \begin{cases} 0 & If\ Blind\ Development \\ Value & If\ Constraints\ exists \end{cases}$$

h. Interface Complexity (IFC)
It is used to define number of external integration/external interfaces for the proposed module/ program/ system. $IFC = \sum_{i=0}^{n} EIi$

Where EI_i is number of external interfaces and value of EI_i will vary from 0 to n

$$EIi = \begin{cases} 0 & If\ no\ external\ interface \\ n & Number\ of\ external\ interface\ exists \end{cases}$$

i. Software Deployment Location Complexity (SDLC)
This measure refers to the number of user class as shown in table 4 for accessing the system and locations on which the system is to be deployed.

$$SDLC = \sum_{i=1}^{4} (User\ Class)_i\ x\ Location$$

j. System Feature Complexity (SFC)
This refers to the specific features to be added to enhance look and feel feature of the system

$$SFC = (Feature1 * Feature2 * \ldots\ldots\ldots\ldots * Feature\ n)$$

Table 4. User classes and their description

S	User Class	Description
1	Casual End Users	Occasionally access the software
2	Naïve or Parametric Users	Deals with a sizeable portion of the database
3	Sophisticated Users	Implemented for complex requirement
4	Standalone Users	Uses readymade program package

k. Improved Requirement Based Complexity (I-RBC)
Finally the improved requirement based complexity can be mathematically expressed as:

$$\textbf{I-RBC} = ((\textbf{PC X PCA}) + \textbf{DCI} + \textbf{IFC} + \textbf{SFC}) \textbf{ X SDLC}$$

3 Results

This section analyses the result of applying I-RBC on 20 SRS's and generating their source code too as shown in table 5. The Plot shows strict comparison of IRBC. It can be observed that measured complexity values follow the same trend with all the categories as shown in figure 2, 3 and 4.

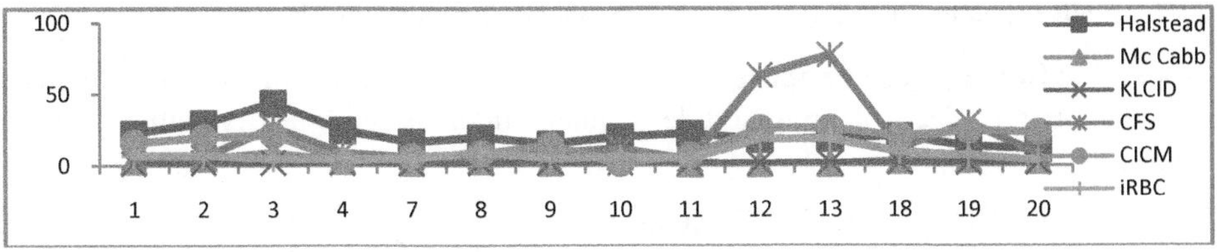

Fig. 2. I-RBC versus Other established code and cognitive information based measures

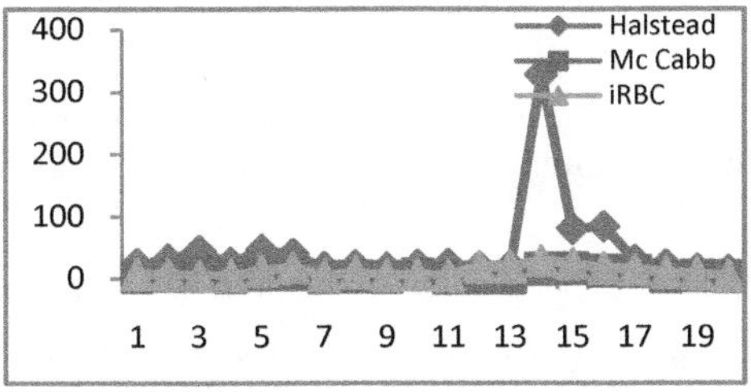

Fig. 3. I-RBC versus code based measures

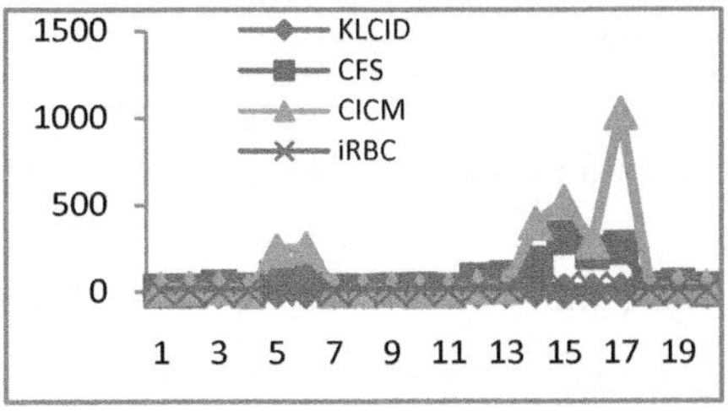

Fig. 4. I-RBC versus Cognitive measures

4 Robustness Analysis with Complexity Properties

In this section we attempt to verify the proposed measure with control flow complexity measure.

Property 1: **Functional Dependency**: If we add or delete functional requirement then the complexity value of the module will get change. As module 7 & 8 are having different functionality with different values shown in table 5.

Property 2: **Language Independency**: Since IRBC is not a code based measure, so, adding or deleting any functional statement does not affect it.

Property 3: **Simplicity of Calculation**: The proposed measures considers input- output, storage, functionality, quality related requirements, constraints and computes complexity value simply.

Property 4: **Non-negativity**: Our measure is always true for this property.

Property 5: **Ranking of the basic control constructs**: Control constructs, controls the execution of the program. Functionally equivalent is decomposition of requirement as leveled it will be that much better the calculation will also be. see module 9 and 10 as shown in table 5.

Property 6: **Additive property**: Let us consider a module for sorting using selection and insertion sort. So the total complexity value for the module will be individual sum of the modules as shown in module 5 & 6 i.e. 29.2 = 11.7 + 17.5

Property 7: **Sensitivity of Nesting**: It shows equivalence as decomposition of functionality of requirement and further classify them, then if module having 2 level decomposition (module 8) will be showing different complexity value then single level decomposition as module 7 i.e. $9.36 \neq 4.68$. Hence property holds.

Property 8: **Substitution Property**: Consider module 3, 4, 7 & 8. If we substitute any segment on the place of any module segment then the complexity value will get change.

Property 9: **CASE Complexity**: Since proposed measure is based on functionality and requirement and using n way case construct or if then construct is used in core programming. Hence this property is not applicable to proposed measure.

Property 10: **Structuring, Transformation and Modularization** Measures holds it completely

Hence proposed measure holds all the properties of control flow complexity measure except for property 9, which is based on specific content of code.

Table 5. Complexity values of programs obtained from code, cognitive and proposed measure

Sl.	Program	Halstead	Mc Cabe	KLCID	CFS	CICM	RBC
1	Program to calculate Factorial of a given number	22.5	3	1.4	6	16.01	7.02
2	Program to obtain Fibonacci series of number	29.7	5	2	6	19.79	7.02
3	Comparing two strings based on string length	43.7	9	1.93	27	21.57	4.68
4	Program for String Concatenation	24.6	3	1.46	9	3.28	4.68
5	Program for Insertion Sort	45	7	3.3	84	237.4	11.7
6	Program for Selection Sort	38.2	7	3.14	84	254.3	17.5
7	Program to find the length of a given string	16.5	2	1.16	6	6.2	4.68
8	Program for appending of string	19.5	3	1.45	6	8.34	9.36
9	Program to find the greater between two numbers	15.1	2	2	9	14.2	4.68
10	Add., Sub., Mul. and Division of Two No.	20.0	6	1.5	12	1.25	5.85
11	Swap two numbers without using third variable	22.1	1	1.6	4	7.04	4.68
12	To read and display of information of employee	18.0	1	1	63	25.92	18.7
13	To read and display marks of students using C++	18.7	1	1.5	77	26.23	18.7
14	Implementation of Round Robin Scheduling	327.	1	3.47	16	393.2	28.0
15	Implementation of Binary Search Tree Traversal	79.8	1	2.4	31	521.6	23.4
16	Implementation of Shortest Job First Scheduling	83.3	1	3.22	22	284.4	16.3
17	Program for implementation of Dijkstra's	27.6	1	3.45	25	1025.	14.0
18	Program to find sum and average	20.4	3	1.81	12	20.42	9.36
19	Program to check for prime number	13.5	4	1.83	30	24.48	7.02
20	Program for Temperature Conversion	11.4	3	1.87	8	23.31	3.51

5 Conclusions

The proposed work computes the improved requirement based complexity based on IEEE-830: 1998 standard. It can be said to be a comprehensive method because it not only validates all nine Weyuker's properties but also verifies properties of control flow complexity measures. Further it encompasses all major parameters and attributes that are required to find out complexity of yet to be produced software. Result validates that IRBC is also comparable with prominent code based and cognitive complexity measures. The proposed measure is carried out at a very early stage of software development and this will aid the developer and practitioner in evaluating the software complexity in order to reduce the rework.

References

[1] IEEE Computer Society: IEEE Recommended Practice for Software Requirement Specifications, New York (1994)
[2] Halstead, M.H.: Elements of Software Science. Elsevier North, New York (1977)
[3] Mc Cabe, T.H.: A Complexity measure. IEEE Transactions on Software Engineering SE-2(6), 308–320 (1976)
[4] Kushwaha, D.S., Misra, A.K.: A Modified Cognitive Information Complexity Measure of Software. ACM SIGSOFT Software Engineering Notes 31(1) (January 2006)
[5] Klemola, T., Rilling, J.: A Cognitive Complexity Metric based on Category Learning. In: IEEE International Conference on Cognitive Informatics (ICCI 2004) (2004)
[6] Wang, Y.: Measurement of the Cognitive Functional Complexity of Software. In: IEEE International Conference on Cognitive Informatics (2003)
[7] Ashish, S., Kushwaha, D.S.: A Complexity measure based on requirement engineering document, JCSE UK (May 2010)
[8] Boehm, B.: Cost Models for Future Software Life Cycle Processes. In: Annals of Software Engineering Special Volume on Software Process and Product Measurement, Netherlands (1985)
[9] Lakshmanan, K.B., Jayaprakash, S., Sinha, P.K.: Properties of control flow complexity measures. IEEE Transactions on Software Engineering 17(12), 1289–1295 (1991)
[10] Weyuker, E.J.: Evaluating Software complexity measures. IEEE transactions on software engineering 14(14), 1357–1365 (1988)

A Novel Model for Software Effort Estimation Using Exponential Regression as Firing Interval in Fuzzy Logic

J.N.V.R. Swarup Kumar[1], T. Govinda Rao[2], M. Vishnu Chaitanya[3], and A. Tejaswi[4]

[1] IEEE MEMBER, Asst.Prof in CSE Dept., Gudlavalleru Engineering College, Gudlavalleru, India
swarupjnvr@yahoo.co.in
[2] Asst.Prof in BS&H Dept., Gudlavalleru Engineering College, Gudlavalleru, India
govind.tangudu@gmail.com
[3] Asst.Prof in School of IT and Engg., VIT University, Vellore, India
vishnu.chaitanya@vit.ac.in
[4] Asst.Prof in IT Dept, Gudlavalleru Engineering College, Gudlavalleru, India
tejaswi_arja@yahoo.co.in

Abstract. Software effort estimation is the process of estimating the cost and time required to develop a software system. It plays a prominent role in software project decisions like resource allocation and bidding which are major parts of planning where as the substratal goals of planning are to scout for the future, to diagnose the attributes that are essentially done for the consummation of the project successfully. So, the effective Software cost estimation is one of the most challenging and important activities in Software development. This paper articulates the new model using fuzzy logic to estimate effort required in software development. We use MATLAB for tuning the parameters of famous various cost estimation models. The performance of model is evaluated on published software projects data. Comparison of results from our model with existing ubiquitous models is done.

Keywords: Fuzzy Logic, Effort Estimation, KLOC, COCOMO, Fuzziness, Membership Function.

1 Introduction

Software effort estimation is the process of predicting the amount of time (Effort) required to build a software system. In order to perform cost-benefit analysis, cost estimation is to be performed by client or developer. Cost Estimation is achieved in terms of person-months (PM), which can be translated into actual dollar cost. Estimation carries inherent risk and this risk leads to obscurity. The concept of software cost estimation has been growing rapidly due to practicality and demand for it. Today the people are expecting high quality software with a low cost; So many popular cost estimation models like COCOMO81, COCOMOII, SLIM, FP, Delphi, Halsted Equation, Bailey-Basili, Doty, and Anish Mittal Model have come into existence. These models are created as a result of regression analysis methods applied to historical data. A recent review of surveys on software cost estimation found that software

V.V. Das, J. Stephen, and Y. Chaba (Eds.): CNC 2011, CCIS 142, pp. 118–127, 2011.

projects have cost overruns. Today most of the software companies follow COCOMOII for estimating the cost of products; but we found some variations in this model [2], [5-10]. There are several reasons like "unrealistic over-optimum", "complexity", "and overlooked tasks" [11], [12].

From 1990's. Researchers particularly have turned their attention to a set of approaches that are soft computing based. These include artificial neural networks, fuzzy logic models and genetic algorithms. Fuzzy logic with its features of a powerful linguistic representation can signify imprecision in inputs and outputs, while providing a more expert knowledge based approach to model building. The fuzzy logic model uses the fuzzy logic concepts introduced by Lofti A. Zadeh [3], [4], [13].

1.1 Membership Functions

Fuzzy numbers are one of the ways to describe data vagueness, obscurity and imprecision. A fuzzy number is an extension of a regular number in the sense that it does not refer to one single value but rather to a connected set of possible values, where each possible value has its own weight between '0' and '1'. This weight is called the membership function. The membership function is increasing towards the mean and decreasing away from it. The Fuzzy number can be of three types 1) Triangular fuzzy Number 2) Trapezoidal fuzzy number 3) Bell shaped fuzzy number.

Fuzziness in a fuzzy set is characterized by its membership functions. A membership function (MF) is a curve that defines how each point in the input space is mapped to a membership value (or degree of membership) between 0 and 1. The graphical representations may include different shapes. There are certain restrictions regarding the shapes used, which is an important criterion that has to be considered. There are different methods to form membership functions.

A. S Function (Figure 1)

Defined by its lower limit a, it's upper limit b, and the value m or point of inflection so that a <m <b. A typical value is: m = (a + b) / 2. Growth is slower when the distance a – b increases.

$$S(x) = \begin{cases} 0 & if\ x \leq a \\ 2\left\{\frac{x-a}{b-a}\right\}^2 & if\ x \in (a,m) \\ 1-2\{\frac{x-b}{b-a}\}^2 & if\ x \in (m,b) \\ 1 & if\ x \geq b \end{cases} \quad \ldots \quad (1)$$

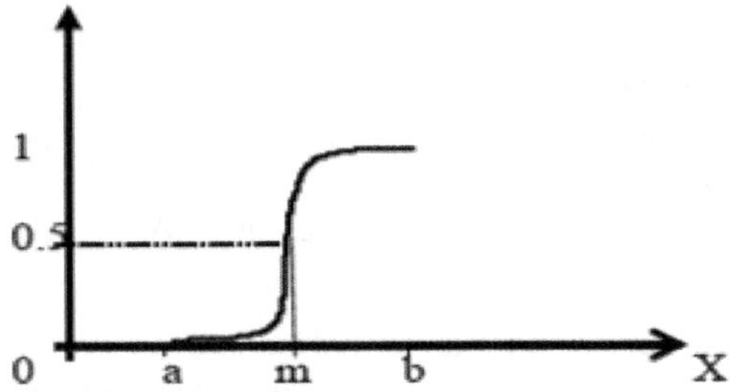

Fig. 1. S Fuzzy Set

2 Cost Estimation Models Literature Review

Within last few decades, to improve the accuracy of cost estimation many software cost estimation models [2], [6-10] were introduced. It seems to be impractical because of the inherent obscurity in software development projects and the impact of software development cost use. Still, it is likely that the estimation can be improved because software development cost estimates are systematically overoptimistic and very inconsistent. The primary objective of the software engineers has been to develop required models using which software cost can be accurately estimated. Estimation models use KDLOC (Thousands of Delivered Lines of Code) as the primary input. This input is not sufficient for accurately estimating the cost of products. Several other parameters have to be considered.

2.1 COCOMO81 Model

Boehm described COCOMO as a collection of three variants: basic model, intermediate model, detailed model [11].

1. Basic model
The basic COCOMO model computes effort as function of program size, and it is same as single variable method.

$$\text{Effort} = a * \text{size}^b \tag{2}$$

2. Intermediate model
An intermediate COCOMO model effort is calculated using a function of program size and set of cost drivers or effort multipliers.

$$\text{Effort} = (a * \text{size}^b) * \text{EAF} \tag{3}$$

3. Detailed model
In detailed COCOMO the effort is calculated as function of program size and a set of cost drivers given according to each phase of software life cycle. The phases used in detailed COCOMO are requirements planning and product design, detailed design, code and unit test, and integration testing.

$$\text{Effort} = (a * \text{size}^b) * \text{EAF} * \text{sum}(W_i). \tag{4}$$

2.2 COCOMO II Model

Boehm and his colleagues have refined and updated COCOMO called as COCOMO II. This consists of application composition model, early design model, post architecture model [11].

1. The Early Design Model
It uses to evaluate alternative software system architectures where unadjusted function point is used for sizing.

$$\text{Effort} = a * \text{KLOC} * \text{EAF} \tag{5}$$

2. The Post Architectural Model
It is used during the actual development and maintenance of a product. The post architecture model includes a set of 17 cost drivers [11] and a set of 5 factors determining the projects scaling component.

$$\text{Effort}=(a*\text{size}^{b})*\text{EAF} \tag{6}$$

There are various effort estimation equations based on KLOC Such As:

2.3 Doty: [1]

$$\text{Effort} = 5.288(\text{KLOC})^{1.047} \tag{7}$$

2.4 Mittal Model: [1]

$$\text{Fuzzification:}\quad u(E) = \begin{cases} 0 & \text{if } E \le a \propto^{b} \\ \frac{(\frac{E}{a})^{1/b} - \propto}{m - \propto} & if\ a \propto^{b} \le E \le am^{b} \\ \frac{\beta - (\frac{E}{a})^{\frac{1}{b}}}{\beta - m} & if\ am^{b} \le E \le a\beta^{b} \\ 0 & if\ E \ge a\beta^{b} \end{cases} \tag{8}$$

$$\text{Defuzzification:}\quad E=\frac{\left(w1(a\propto^{b})+w2(am^{b})+w3(a\beta^{b})\right)}{(w1+w2+w3)} \tag{9}$$

3 Proposed Model Effort Estimation Using Fuzzy

3.1 Fuzzification

We Use S function Fuzzy number T(N) which is defined as Follows:

$$T(n) = \begin{cases} 0 & if\ n \le \propto \\ 2\,[\frac{n-\propto}{\beta-\propto}]^{2} & if\ \propto < n < m \\ 1 - 2[\frac{n-\beta}{\beta-\propto}]^{2} & if\ m < n < \beta \\ 1 & if\ n \ge \beta \end{cases} \tag{10}$$

Where n is size as input. E is effort as output $\propto, m\ and\ \beta$ are the parameters of membership function T(n), m is the modal value, $\propto$ and β are the left and right boundaries respectively.

From S function,

$$\text{Modal value } m = (\propto + \beta)/2 \tag{11}$$

As by definition of fuzziness,

$$F = (\beta - \propto)/\,2m \tag{12}$$

So by solving (11) and (12), we get

$$\propto = (1 - F) * m \tag{13}$$

$$\beta = (1 + F) * m \tag{14}$$

Similarly, the SFN μ(E) is defined as,

$$\mu\ (E)=\begin{cases} 0 & if\ \ E \le ae^{b\propto} \\ 2\ [\frac{\left(\frac{1}{b}\right)log\left(\frac{E}{a}\right)-\propto}{\beta-\propto}]^2 & if\ ae^{b\propto} < E < ae^{bm} \\ 1-2[\frac{\left(\frac{1}{b}\right)log\left(\frac{E}{a}\right)-\beta}{\beta-\propto}]^2 & if\ ae^{bm} < E < ae^{b\beta} \\ 1 & if\ E \ge ae^{b\beta} \end{cases} \quad \ldots \qquad (15)$$

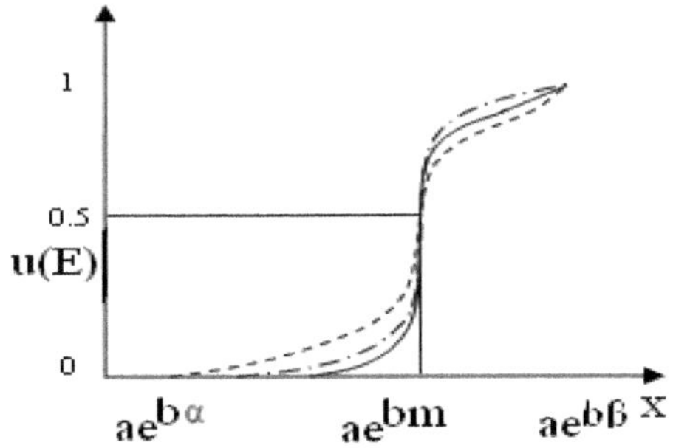

Fig. 2. Representation of u(E)

$$\mu(E) = SFN(ae^{b\propto}, ae^{bm}, ae^{b\beta}), F = 0.1$$
$$\mu(E) = SFN(ae^{b\propto}, ae^{bm}, ae^{b\beta}), F = 0.2$$
$$\mu(E) = SFN(ae^{b\propto}, ae^{bm}, ae^{b\beta}), F = 0.3$$

Table1, gives values of ∝ and β for F=0.1, 0.2 and 0.3 for various values using equations (13) and (14), where m, size estimate in KLOC.

Table 1. Values of ∝ and **β** for various values of F

F	∝	β
0.1	0.9m	1.1m
0.2	0.8m	1.2m
0.3	0.7m	1.3m

3.2 Defuzzification

The output, fuzzy estimate of E, can be computed as a weighted average of the optimistic ($ae^{b\propto}$), most likely (ae^{bm}) and pessimistic estimate ($ae^{b\beta}$). Fuzzy effort estimate (E_m) is given as

$$E_m=(w1(ae^{b\propto})+w2(ae^{bm})+w3(ae^{b\beta}))/(w1+w2+w3) \qquad \ldots(16)$$

Where w1, w2 and w3 are weights of the optimistic, most likely and pessimistic estimate respectively. Maximum weight should be given to the most expected estimate.

Effort $E_m = (w1(ae^\wedge(b \propto)) + w2(ae^\wedge bm) + w3(ae^\wedge b\beta)) / (w1+w2+w3)$ (17)

Where a=70.737, b=0.004 obtained from exponential regression method by using MATLAB, and m represents the size in KLOC

$$\propto = (1 - F) * m \quad (18)$$

$$\beta = (1 + F) * m \quad (19)$$

Here F, w1, w2 and w3 are arbitrary constants. The effort is estimated in man months (MM).

4 Research Methodology

The performance of proposed software effort estimation model is evaluated by comparing against various software cost estimation models. The methodology used in empirical evaluation is described as follows:

- For each model, using MRE we evaluate the impact of estimation accuracy using (MRE, MARE) evaluation criteria.
- Criterion for measurement of software effort estimation model performance.

$$\text{MARE}(\%) = \sum_{i=1}^{n}(|\frac{estimate_i - actual_i}{actual_i}|) * 100 \quad \ldots (20)$$

Where $estimate_i$ is the estimated effort (E) from the model, $actual_i$ ($\hat{E}$) is the actual effort and n is the number of projects.

5 Experimental Result Analysis

The Data is taken from [2] and given in Table 2

Table 2. Data for Experimental Study

SL.NO	Project No	KLOC	Actual
1	9	39	72
2	2	40.5	82.5
3	6	50	84
4	11	128.6	230.7
5	12	161.4	157
6	13	164.8	246.9
7	7	200	130.3
8	4	214.4	86.9
9	1	253.6	287
10	10	254.2	258.7
11	8	289	116
12	5	449.9	336.3
13	3	450	1107.31

Let F = 0.3, then from Table 1, $\alpha = 0.7m$, $\beta = 1.3m$. We have taken W1 = 6, W2 = 2 and W3=1 for our Model. Table 3 below gives the experimental results.

Table 3. Effort Estimated Of Various Models

Proj.No	KLOC	Actual Effort	COCOMO Basic Model	COCOMO Inter(Nom)	Detailed (Nom)	Early Design Model (High)	post Arch Model (H - H)	Doty Model	Mittal Model	Swarup Model
1	39	72	292.1	227.2	245.4	107.7	211.9	245	68.34707	80.59925
2	40.5	82.5	305.7	237.7	256.8	111.8	221.3	254.9	70.08411	81.00658
3	50	84	393.6	306.1	330.6	138.1	282	317.8	80.62635	83.63769
4	128.6	230.7	1222.9	951.1	1027.2	355.1	835.7	854.4	151.1149	109.23106
5	161.4	157	1606.2	1249.2	1349.2	445.6	1085.3	1083.8	175.7591	122.2722
6	164.8	246.9	1646.8	1280.9	1383.3	455	1111.6	1107.8	178.2126	123.71601
7	200	130.3	2077.5	1615.8	1745.1	552.2	1388.8	1356.7	202.6968	139.77894
8	214.4	86.9	2258.3	1756.4	1896.9	592	1504.4	1459.1	212.2885	146.97896
9	253.6	287	2762.4	2148.5	2320.4	700.2	1824.8	1739.5	237.3675	168.66038
10	254.2	258.7	2770.2	2154.6	2327	701.9	1829.8	1743.9	237.7408	169.01766
11	289	116	3231.3	2513.2	2714.3	798	2120.7	1994.6	258.9161	191.18887
12	449.9	336.3	5495.9	4274.6	4616.6	1242.2	3528	3170.3	347.5234	342.81432
13	450	1107.31	5497.4	4275.8	4617.8	1242.5	3528.9	3171.1	347.5748	342.94131

Figure 3 below show the comparison of experimental estimated effort from Swarup model, Mittal model and Actual effort.

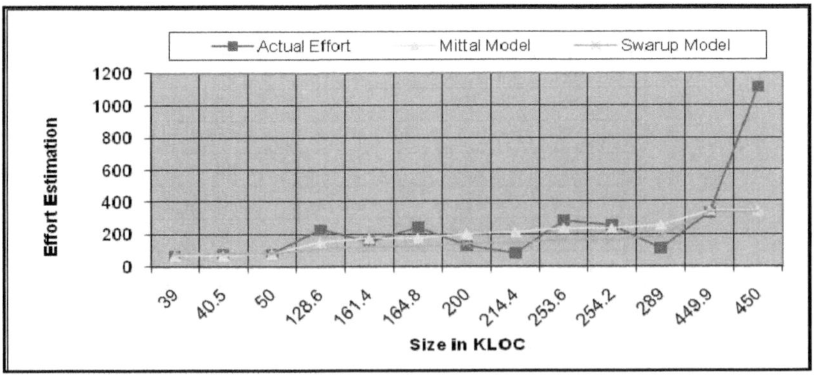

Fig. 3. Effort Model vs. Mittal Model and Actual Effort

Figure 4 below show the comparison of effort from various models versus estimated effort.

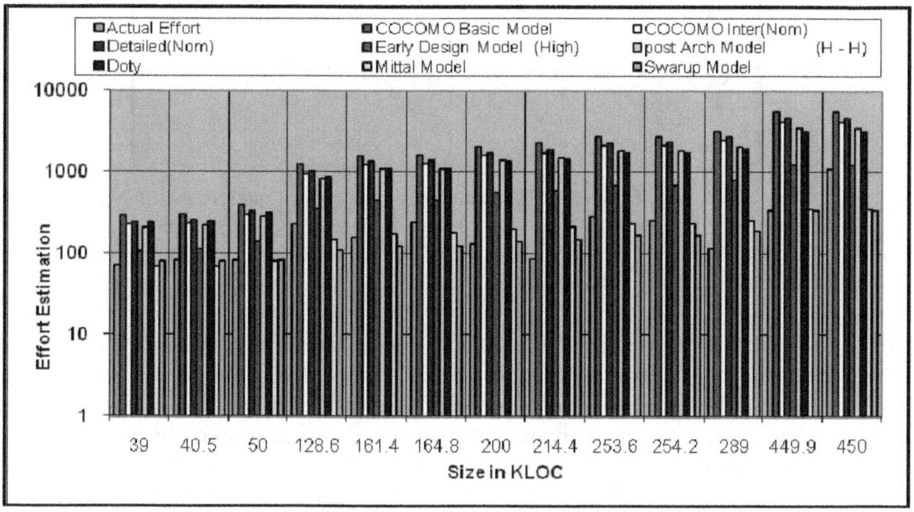

Fig. 4. Comparison of Effort Estimated of various models

Comparison of various models on the basis of various performance criterions for software cost estimation is given in below Table 4. Figure 5 below shows the Mean Absolute Relative Error (%) comparison of various models.

Table 4. Various Models MARE % Values

Model Type	MARE (%)
COCOMO Basic Model	13307.7
COCOMO Inter(Nom)	10061.5
Detailed(Nom)	10970.6
Early Design Model (High)	2561.3
post Arch Model (H - H)	8438
Doty	8186.1
Mittal Model	518.7
Swarup Model	426.923

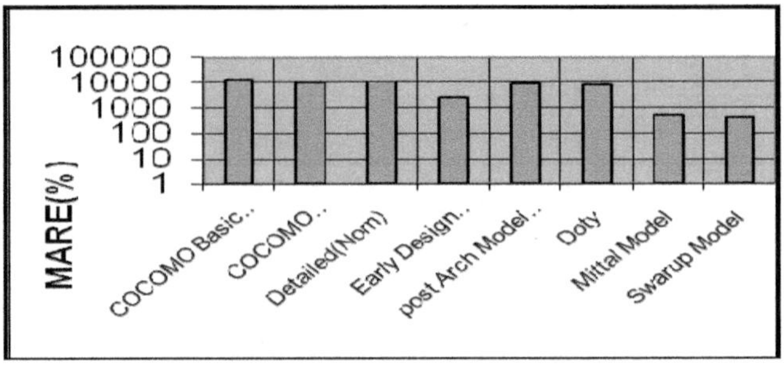

Fig. 5. MARE (%) Comparison of various models

6 Conclusion and Future Scope

The important issue for project managers is to estimate the software development effort required accurately and reliably in the early stages of software development lifecycle so that the resources allocation can be done perfectly. In this paper we postulated fuzzy software cost estimation model that handles ambiguousness, obscurity and then compared with other popular software cost estimation models. From the empirical evaluation we concluded that: proposed model showed better software effort estimates in view of the MARE evaluation criteria as compared to the traditional estimation models. The above results demonstrate that applying fuzzy logic method to the software effort estimation is an expedient approach to address the problem of obscurity and vagueness existed in software effort drivers. Hence, the utilization of fuzzy logic for other applications in the software engineering field can also be explored in the future.

References

1. Mittal, A., Parkash, K., Mittal, H.: Software Cost Estimation using fuzzy logic. ACM SIGSOFT Software Engineering Notes 35(1) (November 2010)
2. Zmud, R.W., Kemerer, C.F.: An Empirical Validation of Software Cost Estimation Models. Communication of the ACM 30(5) (May 1987)
3. Zadeh, L.A.: Fuzzy sets. Info and Control 8, 338–353 (1965)
4. Galindo, J.: Handbook of Research in Fuzzy Information Processing in Databases. In: Information Science Reference (2008)
5. Johnson, K.: Software Cost Estimation: Metrics and Models, pp. 1–17. Dept of Computer Science, University of Calgary, Alberta, Canada
6. Hari, C.V.M.K., et al.: Identifying the Importance of Software Reuse in COCOMO81, COCOMOII. International Journal on Computer Science and Engineering 1(3), 142–147 (2009); ISSN: 0975-3397
7. Baiely, Basili, j.w.: A Metamedel for Software Development Resource Expenditure. In: Proc. Intl. Conference Software Egg., pp. 107–115 (1981)
8. Boehm, B.: Software Engineering Economics. Prentice Hall, Englewood Cliffs (1981)
9. Boehm, B.: Cost Models for Future Life Cycle Process: COCOMO2. Annals of Software Engineering (1995)

10. jalote, P.: An Integrated Approach for Software Engineering, 3rd edn., ISBN: 978-81-7319-702-4
11. Jorgensen, M., Grimstad, S.: Over- Optimism in Software Development Projects: The Winner's Curse. In: CONIELECOMP 2005. Simula Research Laboratory, Norway (2005)
12. Menzies, T., Port, D., Chen, Z., Hihn, J., Stukes, S.: Validation Methods for calibrating software effort models. In: ICSE 2005: Proceedings of the 27th International Conference on Software Engineering, pp. 587–595. ACM Press, New York (2005)
13. Lotfi Zadeh, A.: Fuzzy Logic, Neural Networks and Soft Computing. Communication of ACM 37(3), 77–84 (1994)

Fault Informant Distributed Deadlock Detection Using Colored Probes

V. Geetha[1] and N. Sreenath[2]

[1] Dept. of Information Technology
Pondicherry Engineering College
Puducherry – 605014
[2] Dept. of Computer Science & Engg.
Pondicherry Engineering College
Puducherry – 605014
vgeetha@pec.edu, nsreenath@pec.edu

Abstract. Several probe based algorithms have been proposed for deadlock detection and resolution in distributed environment. These algorithms work only in fault free environment. This paper proposes an enhanced probe algorithm by using colored probes to work under faulty environment also. The objective of this algorithm is that the initiator is always informed about the status of the probe, deadlock or otherwise. This increases the reliability.

1 Introduction

In a distributed system, resources and transactions are remotely placed. Any transaction can request for any resource. A transaction may need one or more resources to complete execution. In a truly distributed system, the transactions and resource managers have to communicate through messages. The most popular algorithm for distributed deadlock detection and resolution is probe based algorithm by Chandy and Mishra[6]. In this algorithm, the transaction that suspects deadlock sends probe messages along the wait for edges in wait for graph. If the probe returns back to the initiator, it indicates the presence of deadlock. If the probe does not return back to initiator, then initiator will deduce that it is a live lock condition. Deduction of site failure as live lock is a serious problem. The initiator will assume that it is a finite wait for resources and will wait for time out. After time out, it will send probe messages and this repeats forever. So these algorithms will not work in faulty distributed system. Priority based algorithms [3, 5] have been proposed to detect phantom deadlocks. All these algorithms expect the underlying system model to be fault free. In Ozu and Valduriez [2], it is stated that the failures in distributed systems could be categorized as: Transaction failure- bug in code, Site failure- processor failure, Link failure - communication link failure.

In Li and Mcmillin[1], a totally distributed probe based fault tolerant DDDR algorithm is proposed using fault diagnosis model for site failure. In this, the processors are categorized into faulty and non faulty. All non faulty processors will certify the other processors as faulty or non faulty. A fault vector is attached as part of the probe in which each bit represents a processor in the system. 0 represents non faulty and 1

V.V. Das, J. Stephen, and Y. Chaba (Eds.): CNC 2011, CCIS 142, pp. 128–132, 2011.

represents faulty state of the processors. This paper has the following drawbacks: All the processors are diagnosed periodically by the other non faulty processors. If the period is very small, the non faulty processors need to spend more time in diagnosing other processors than executing their transactions. This will reduce throughput of the system. On the other hand, if the period is more then reliability reduces. Hence the success of this algorithm lies in choosing ideal diagnosis interval. Moreover, fault diagnosis is a function of the underlying network model. However fault information needs to be known for better deadlock detection and resolution. Message complexity is more in propagating updated processors' status and clean messages. It can identify only 1 processor failure per deadlock cycle. Hansdah et al. [7] discusses about link failure, where grant messages are lost or delayed. Brzezinski et al. [8] offers solution for asynchronous messaging system, where the messages are not delivered in FIFO basis. However these algorithms also assume no site failures.

2 Motivation

Though there are several algorithms for providing fault tolerance in distributed systems, generally it is not expected by deadlock detection algorithms to be fault tolerant. However any probe based DDDR algorithm should facilitate for the initiator to infer whether it is due to livelock or site failure (where deadlock may be present), that the probe has not come back to it. If it is due to livelock, then the transactions can wait for finite time and then start sending resources' requests again. If the probe does not come back due to site failure, then the system has to be reconfigured to exclude faulty sites for future requests. Since there is no ideal fault tolerance algorithm for distributed systems and site failures are bound to happen, the initiator needs to get the probe back always. Hence this paper proposes a fault-informant algorithm to send colored probes to initiator about the status of sites. It can detect at most two site failures per deadlock cycle. The proposed algorithm uses the following colors in probe messages to indicate the probe status.

RED: Indicates presence of deadlock and there is no site failure.
ORANGE: Indicates site failure. In this deadlock/ livelock status is unknown due to site failure.
WHITE: Indicates livelock and there is no site failure.

In this section, definitions and descriptions related to the proposed scheme are given:

Definitions

Definition 1: Wait for Graph (WFG (N, E)) is a directed graph where nodes N represent transactions currently participating in the system and E is a finite set of edges representing the dependency of transactions on resources. Ti-→Tj ϵ E where Ti is waiting on Tj for the resource held by Tj. So Tj is successor of Ti and Ti is predecessor of Tj.A **Deadlock** is identified by a directed cycle in the WFG.

Definition 2: Forward probe (*Forward_Probe (Initiator, Sender, Receiver, Forward_Probe_Color)*) is a traversal of dependency edges in WFG from initiator and propagates until it reaches back initiator or terminates when there is no dependency edge for a transaction in the path i.e. T_I →T1→T2…Tn, where {Tn = T_I or Tn has no dependency edge | T_I ,T1,T2….Tn ϵ N}.

Definition 3: Backward probe (*Backward_Probe (Initiator, Sender, Receiver, Backward_Probe_Color))* is a traversal from initiator and propagates backwards along the dependency edges in WFG.

Definition 4: A **Fault Vector** (*FaultVector*) V = S1S2…Sn, where S1, S2 …Sn denotes the N sites participating in the system domain. Si = 1, if site i is faulty; Si = 0, if site i is non faulty.

Definition 5: Victim is the lowest priority transaction which will be aborted to break the deadlock cycle. Here initiator is the victim.

Descriptions

Returnprobe (*Return_Probe (Initiator,Sender, Forward/Backward_Probe_Color, FaultVector, FaultSiteID,))* is the probe forwarded by the site Si holding transaction Ti to the initiator about its successor faulty site Sj holding transaction Tj, where Ti→Tj ϵ E. This probe updates the status of site Sj in fault vector and sends it to the initiator. Initially the color of the probe is RED and it stays unchanged until the *awake* message is received from the faulty site Sj. This is done during forward probe. In backward probe, if predecessor Tj is faulty, then this return probe is forwarded by the successor. The return probe will have *FaultVector* and *FaultSiteID* only under faulty environment.

Acknowledgement message (*Ack_msg (Receiver, Sender)):-* Every site on receiving either forward or backward probe message from its sender should send an acknowledgement message to its neighbors. If this message is not received by the sender before timeout period, it assumes that receiver is faulty. Then sender sends return probe to initiator updating fault vector about this faulty site.

Clean message broadcast all the sites to clean the probes sent by victim in faulty site.

Awake message (*awake_mesg(SiteID)*) is sent by all sites on startup or after fault recovery. It is needed to update its status in fault vector and include it for future transaction requests.

3 Proposed Scheme

In the proposed algorithm, initiator uses forward and backward probe messages to detect the reason for not getting the resource granted. Initially the color of the probe message is RED. The definitions of all the messages used in the proposed algorithm are given in section 3. The probe messages travel both along the wait for edges and on the opposite direction along the wait for edges by traversing along the nodes in WFG. A node that receives a probe message should send an acknowledgement to its sender. This is used to inform the awake status of receiver to the sender. If sender does not receive acknowledgement message from the receiver before timeout, it infers that the receiver site is faulty. Then sender will change the color of sent forward/backward probe message into ORANGE after updating the corresponding bit of the faulty site in fault vector as faulty. The sender will send a return probe which is addressed always to the initiator with the color of forward/ backward probe along with updated faulty vector and fault site ID. The initiator will broadcast the faulty state of this site to all the sites. This faulty status is modified only when the faulty site broadcasts awake message, after recovering from site failure .If the faulty site ID in both forward and backward probe messages is same, it can be inferred that it is single site failure. If they are different, then it indicates failure of two sites. Then initiator will invoke the reconfiguration phase. If there is no site failure, both the forward and backward probe messages will reach the initiator and by the RED color of the probe, the initiator will infer the presence of deadlock and will start deadlock resolution phase. If there is no site failure, but some node in the wait for graph does not have any wait for edge, then receiver will not be able to send message any further. This indicates live lock condition. But it will send acknowledgement message to sender indicating that it is active. Then receiver will change the color of the forward/backward probe received into WHITE, and send return probe with updated probe color to initiator. When the initiator realizes it is live lock, it will wait for some more time, before trying again. The algorithms are not given for want of space.

4 Correctness of the Algorithm

The algorithm is proved correct under the following assumptions:

1. Transactions use single request model for requesting the resource.
2. No transaction in deadlock aborts unless it is victimized in resolution phase.
3. There are atmost 2 site failures in a cycle.
4. Initiator alone is allowed to send probe messages to eliminate phantom deadlocks.

Theorem 1: The algorithm detects deadlock only if there is a deadlock.

Proof: This algorithm detects a deadlock only when initiator receives both forward and backward probes and their colors are RED. In that case, there was no site failure when the probe was traversing. If there is any site failure, then it must have happened only after the probe message had passed this site. If the site had failed after forwarding the probe message, then this failure will be known only on next deadlock cycle detection through return probes. So the probe message forwarded by this site must have reached the initiator before the return probe. Therefore it is impossible that the transaction in a faulty site is aborted before the deadlock is detected.

Theorem 2: In a deadlock cycle of size greater than two, the failure of a process i is identified by its successor j and predecessor k, if $k \rightarrow i \rightarrow j$ is part of the cycle.

Proof: In a deadlock cycle with $k \rightarrow i \rightarrow j$ as part of the cycle, when transaction i fails, its predecessor k will not receive acknowledgement for its backward probe message and successor j will not receive acknowledgement for its forward probe message. The initiator will receive return probe messages from both k and j on faulty status of i. The color of the probe will be changed to ORANGE. Since fault site ID of both return probes will be same as the site ID of i, as given in the procedure executed by all transactions, this site failure is identified in deadlock cycle.

Theorem 3: In a deadlock cycle of size two, site failures can be identified.

Proof: In [10], it is given that $2t+1$ processors are needed to detect t failures using fault diagnosis model. However in our case, it can be inferred that t failures can be detected with minimum of one processor.

In our algorithm at most 2 site failures can be detected. Hence there are only two values for t, where $t=1$ or $t=2$. Since our algorithm is based on messaging, one processor itself is enough to deduce the faults in both the cases when $t= 1$ and $t=2$.

Theorem 4: The algorithm identifies whether the wait state is due to live lock or deadlock in fault free environment. Proof: In fault free environment, both forward as well as backward probe messages will reach initiator. If probe does not reach initiator in fault free environment, it is because of live lock. Some node in the wait for graph does not have any wait for edges. Then it cannot send forward or backward probes to its neighbors. So it will inform its status to initiator by sending return probe with WHITE color as mentioned in the algorithm for transactions.

5 Conclusion

A new fault tolerant DDDR algorithm is proposed with the following features. Initiator always knows the status of probe whether deadlock or livelock or site failure. In [1] every non-faulty site tests other sites periodically for site failures. In the proposed algorithm the site failure is decided by acknowledgement messages. This improves the throughput of non-faulty sites. Checking whether faulty sites are rectified is known by *awake* message. This situation is not handled in [1]. Only one site failure is handled in [1]. This paper however handles at most 2 site failure which improves fault tolerance. The probe colors are used to indicate the status of the system. The worst case message complexity is 4n where n is the number of transactions in the deadlock cycle. This occurs when there is no site failure and deadlock occurs. The 4 messages are the forward and backward probe messages to next nodes and acknowledgement messages to senders. Further fault identification is better than fault diagnosis model which needs 2t+1 processors to identify t failures. In messaging mechanism, 1 processor is enough to identify t failures, where t = 1 or 2.

References

1. Li, P.-y., McMillin, B.: Fault-tolerant Distributed Deadlock Detection/Resolution. IEEE Transactions on Parallel and Distributed Systems, 224–230 (1993)
2. Ozsu, M.T., Valduriez, P.: Principles of Distributed Database Systems. Pearson Education, London (1999)
3. Chowdhary, A.N., et al.: A modified priority based probe algorithm for distributed deadlock detection and resolution. IEEE Trans. Software Eng., SE-15, 10–17 (1989)
4. Mitchell, D.P., Merrit, M.J.: A distributed algorithm for deadlock detection and resolution. In: Proc. 3rd ACM Symp. Principles of Distributed Computing, Canada, pp. 282–284 (August 1984)
5. Sinha, M.K., Natarajan, N.: A priority based distributed deadlock detection algorithm. IEEE Trans. Software Eng. SE-11, 67–80 (1985)
6. Mani Chandy, K., Mishra, J.: Distributed Deadlock Detection. ACM Transactions on Computer Systems 1(2), 144–156 (1983)
7. Hansdah, R.C., et al.: A Fault Tolerant distributed Deadlock Detection Algorithm. In: Das, S.K., Bhattacharya, S. (eds.) IWDC 2002. LNCS, vol. 2571, pp. 78–87. Springer, Heidelberg (2002)
8. Brzezinski, J., et al.: Deadlock models and a General Algorithm for Distributed Deadlock Detection. Journal of Parallel and Distributed Computing 31, 112–125 (1995)
9. Preparata, et al.: On the connection assignment problem of diagnostic systems. IEEE Transactions on Electronic Computers EC-16 (1967)

Experiments with a Behavior-Based Robot

Barnali Das[1], Dip Narayan Ray[2], and Somajyoti Majumder[2]

[1] Assistant Professor, B.I.E.T., Birbhum, West-Bengal, India
[2] Scientist, C.M.E.R.I., Durgapur, West-Bengal, India
{barnalidas1985,dipnarayan.ray}@gmail.com

Abstract. Behavior based robot up lifted today's science into a glorious position of progression. The essence of behavior based robotics is to develop robotic systems which can exhibit behaviors normally found in the nature. This paper is focused on a methodology to model the control architecture that optimizes the behavior rules using Subsumption architecture, the logical representation, the experimental result and the simulation result of the system. The results were very promising and the knowledge gathered will be used to develop an upgraded version of this system. This work concludes the significance of Subsumption Architecture and Hybrid Architecture for navigation of behavior-based autonomous robots.

Keywords: Behavior-based system, Subsumption architecture, Control model, robot intelligence.

1 Introduction

In 1953, W.Grey Walter created first Autonomous Robot named as Grey Walter's tortoise. It could effectively exhibit certain behaviors such as backing away from strong light and heading toward weak light [1]. After Walter in the year 1977 Hilare invented another autonomous robot whose weight was 400kg [2]. Moravec constructed a small cylindrical robot named CMU Rover in 1983 [3].Though the system was moderately reliable, but it was very slow. In the mid-1980's Rodney Brooks touched off a firestorm of interest in autonomous robots with the introduction of the Subsumption architecture. In his paper, Brooks (1986) advocated the use of a layered control system, embodied by the subsumption architecture but layered along a different dimension than what traditional research was pursuing [4].In the year 1992, A.Ram, R.C.Arkin, K.Moorman, R.J Clark discussed a case based method for dynamic selection and modification of behavioral assemblages for navigational systems [5]. In 1998, Mataric introduce a behavior-based system focuses on functional modeling [6]. In 2001 Abreu and Correia, [7] in2004, Yang, S.X., Hao Li, Meng, M.Q.-H., Liu [8] and in 2005 Hansberger, Okon, Aghazarian, Robinson constructed multi behavior autonomous robot using the concept of fuzzy controller [9]. In 2006 D. N. Ray discussed about the concept of Behavior Based Robotics and differences with the classical/conventional robotics [10]. In the year 2008 Dr. Reuven Granot defined the structure of autonomous robot architecture where different types of control model are described [11]. B. Das, A. Mondal, D. N. Ray, S. Majumder in 2009 introduced an

V.V. Das, J. Stephen, and Y. Chaba (Eds.): CNC 2011, CCIS 142, pp. 133–141, 2011.

autonomous robot based on intelligent behavior [12]. In this paper the system aims to address these limitations of preceeding works and implementation of subsumption architecture for behavior-based robot. The weight of the system is very less to move freely and the hybrid architecture won't slow down the system.

2 Architecture of the System

Maja J. Mataric defines the purpose of architecture as "a principled way of organizing a control system". In the field of robotics, 'control architectures' are methodologies that supply structure and impose constraints on the way that robots are controlled [11]. The basic architecture control systems are follows:

1. Deliberative control – looks ahead, think and plan, then act.
2. Reactive control – no look ahead, react (time-scale).
3. Behavior-based control – distribute thinking over acting.
4. Hybrid control – combine reactive with deliberative, think slowly, react quickly.

2.1 Hybrid Control

"Hybrid control is a combination of reactive and deliberative control. Hybrid robotic architecture states that a union of deliberative and reactive approaches can potentially yield the best of both worlds" [13]. A modern hybrid system typically consists of three components: a reactive layer, a planner and a layer that puts the two together [14]. Subsumption architecture is a methodology for developing artificial intelligence that is heavily associated with behavior based robotics [4] [15] [16]. Subsumption has been widely influential in autonomous robotics and elsewhere in real-time AI. It is a way of decomposing complicated intelligent behavior into many "simple" behavior modules, which are in turn organized into layers. Each layer implements a particular goal of the agent, and higher layers are increasingly abstract. After wandering at the external environment the system follows three layers as per the priority basis in the Fig. 1 (a).

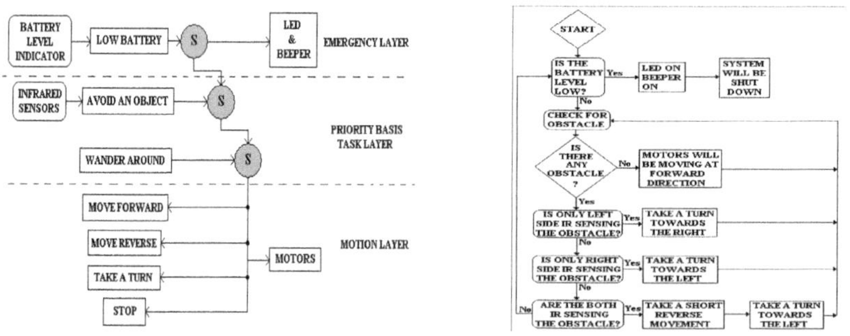

Fig. 1(a). Hybrid Control Architecture of The System based on Subsumption Architecture [left] and **(b)** Logical Representation of the System

Emergency layer deals with urgent purpose. For example, the system is avoiding an obstacle; at that time if on board power supply goes down then system will be shut down. That means low battery charge (hunger) is the most important among all the tasks. So this layer is called Emergency Layer. After this layer there is task layer. This layer presents how the behaviors have been prioritized by the programmer to the exploration of the system. The last layer of the architecture is motion layer. The speed and direction of the motor for various behaviors are decided from this layer.

The system is controlled by the user defined logic or programming. It uses two DC geared motors to drive two wheels for its movement [17]. Two Infrared sensors are used as its sensing elements and a small wireless (RF) video camera, mounted on the front, is used for monitoring/viewing the area in front of the system. One LED and one Beeper are used to indicate the low onboard battery power. If the obstacles are on the right side, then the system takes a left turn and vice versa for avoiding the obstacles on the left side [18]. But when the obstacle is completely in front of the system, initially it moves in reverse direction and then takes a short left turn.

3 Probable Mathematical Model

The system creates a hypothetical boundary using sensors. Any object or interference within this boundary generates a stimulus response. This will however be presented by the following relations:

$$R(\theta) : S(\mathrm{K}) \rightarrow v_t \tag{1}$$

Where, $R(\theta) =$ Responsivity of the system, $\theta \in 0,1,2,\ldots\ldots\ldots\mathrm{N}$ (Where, $\mathrm{N} =$ any integer), $S(\mathrm{K}) =$ Sensitivity of the sensor, $\mathrm{K} \in 0,1,2,\ldots\ldots\ldots\mathrm{N}$, $V_{avg} = v_t$ =average linear velocity of the system

$V_t = (V_l + V_r)/2$ $\quad t \in 1,2,3,\ldots\ldots\ldots\mathrm{N}$

Now, we know that linear velocity $= V = r\omega$

Where, $r =$ Radius of the wheel, $\omega =$ Angular velocity of the wheel of the system

Now, for the two wheels, ω_l =Angular velocity of the left wheel, ω_r =Angular velocity of the right wheel

So, the average linear velocity of the system is, $V_{avg} = r(\omega_l + \omega_r)/2$ (2)

Now, instantaneous velocity, $V_t^+ = r(\omega_l^{t^+} + \omega_r^{t^+})/2$ (3)

Where, ('+') sign indicates the forward direction/clockwise direction of the wheels.

$$V_t^- = r(\omega_l^{t^-} + \omega_r^{t^-})/2 \tag{4}$$

Where, ('-') sign indicates the reverse direction/counter clockwise direction of the wheels. $2\Psi =$ Heading or heading angle of the system, $2\Psi' = \gamma_H$ =Rate of change of heading angle of the system

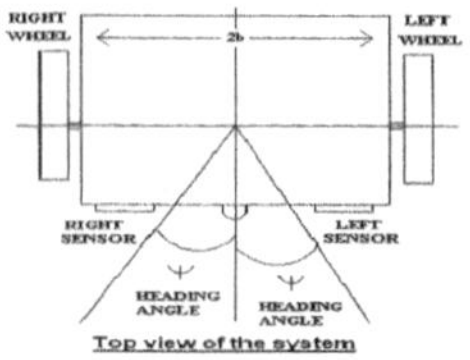

Fig. 2. Schematic Model of the System

Now, $\Psi = \tan^{-1}(||V_l| - |V_r||)t/2b$, Where, V_r=Velocity of right wheel, V_l=Velocity of left wheel, $2b$=Distance of the two wheels of the system. Now, we are assuming that the change of heading angle $(\Delta\Psi)$ is very small in a very small time (Δt) respect to the whole system. So we can write,

$$\tan\Psi = (|V_l - V_r|)t/2b = r(|\omega_l - \omega_r|)t/2b \tag{5}$$

Or, $\tan(\Delta\Psi) = [r(|\omega_l - \omega_r|)t/2b]\Delta t$ (6)

For very small value of (Δt), $\tan(\Delta\Psi) \rightarrow (\Delta\Psi)$ (7)

Or, $\Delta\Psi = [r(|\omega_l - \omega_r|)t/2b]\Delta t$ (8)

Therefore, the rate of change of the heading angle,

$$\gamma_H = \partial\Psi/\partial t = \Psi' = r(|\omega_l - \omega_r|)t/2b \tag{9}$$

Case 1

When there is no such obstacle in front of the system, then the linear instantaneous velocity of the system will be, $V_t^+ = r(\omega_l^{t^+} + \omega_r^{t^+})/2$ (10)

As there is no obstacle so the angular velocity of the two wheels will be same.

That is, $\omega_l^{t^+} = \omega_r^{t^+}$ (11)

Therefore, $\gamma_H = r(|\omega_l^{t^+} - \omega_r^{t^+}|)/2b = 0$ (12)

Therefore, responsivity of the system

$$R(\theta): S(\mathrm{K}) \rightarrow V_t^+ = r(\omega_l^{t^+} + \omega_r^{t^+})/2 \tag{13}$$

Case 2

When there is an obstacle in front of right sensor. The angular velocity of the left wheel will be zero. That is, $\omega_l^{t^+} = 0$ (14)

Therefore, the instantaneous velocity of the system is,

$$r(0 + \omega_r^{t^+})/2 = r(\omega_r^{t^+})/2 \tag{15}$$

As there is an obstacle in front of the right sensor, the rate of change of heading angle of the system will be towards left.

Therefore, $\gamma_H = r(|0-\omega_r^{t^+}|)/2b = r(|\omega_r^{t^+}|)/2b$ (16)

We are assuming that the turning towards right is the positive turning. ('-') sign is indicating the turning to the opposite side (towards left sensor). But here we can ignore the negative sign because of only the magnitude is concerned.

Therefore responsivity of the system, $R(\theta): S(\mathrm{K}) \rightarrow V_t^+ = r(\omega_r^{t^+})/2$ (17)

Case 3

When there is an obstacle in front of left sensor. The angular velocity of the right wheel will be zero. That is, $\omega_r^{t^+} = 0$ (18)

Therefore, the instantaneous velocity of the system is,

$$r(\omega_r^{t^+} + 0)/2 = r(\omega_l^{t^+})/2 \quad (19)$$

As there is an obstacle in front of the right sensor, the rate of change of heading angle of the system will be towards left.

Therefore, $\gamma_H = r(|\omega_l^{t^+} - 0|)/2b = r(|\omega_l^{t^+}|)/2b$ (20)

We are assuming that the turning towards right is the positive turning.
Therefore responsivity of the system,

$$R(\theta): S(\mathrm{K}) \rightarrow V_t^+ = r(\omega_l^{t^+})/2 \quad (21)$$

Case 4

When there is an obstacle within the hypothetical boundary of both sensors, the linear velocity of the system will be in a discrete form.

$$\sum_{t=1}^{\mathrm{N}/4} V_t^- \Delta t = r(\omega_l^{t^-} + \omega_r^{t^-})/2 \quad (22)$$

$t \in 1,2,3,\ldots\ldots\ldots\ldots\ldots\mathrm{N}/4$ (^{t} Belongs to the total time that is N)

Here, we are assuming that the system will move in the reverse direction for ¼ of the total time. Rest of the time will be for turning towards left.

$$\sum_{t=\mathrm{N}/4}^{\mathrm{N}} V_t^+ \Delta t = r(0 + \omega_r^{t^+})/2 = r(|\omega_r^{t^+}|)/2 \quad (23)$$

As, $\omega_l^{t^+} = 0$, Therefore,

$$\sum_{t=1}^{\mathrm{N}} V_t \Delta t = \sum_{t=1}^{\mathrm{N}/4} V_t^- \Delta t + \sum_{t=\mathrm{N}/4}^{\mathrm{N}} V_t^+ \Delta t = r[(\omega_l^{t^-} + \omega_r^{t^-}) + (\omega_r^{t^+})]/2 \quad (24)$$

Now, the rate of change of heading angle of the system will also be discrete,

For, $t = 1,2,3,\ldots\ldots\ldots\ldots\ldots\ldots\ldots N/4$

$$[\gamma_H]_{1/4} = r(|\omega_l^{t^-} + \omega_r^{t^-}|)/2b \tag{25}$$

For, $t = N/4\ldots\ldots\ldots\ldots\ldots\ldots\ldots N$

$$[\gamma_H]_{3/4} = r(|0-\omega_r^{t^+}|)/2b = r(|\omega_r^{t^+}|)/2b \tag{26}$$

Therefore, responsivity of the system,

$$R(\theta): S(\mathrm{K}) \to \sum_{t=1}^{N} V_t \Delta t = r[(\omega_l^{t^-} + \omega_r^{t^-}) + (\omega_r^{t^+})]/2 \tag{27}$$

4 Experiments

"When simple behaviors work together, they can create what appear to be complex behaviors to a naive observer," noted by Bryan Adams. Different sub parts, used in the system, were tested separately and finally the whole integrated system was tested in the rough terrains as well as in the areas with full of obstacles to check the reliability and efficiency of the system. The exploration by the system was observed by changing the pattern of the obstacles, forming once a round shape, then a U-shape, and next time a zigzag shape. At the beginning of the experiment all the behaviors of the system were assigned to zero.

Fig. 3. (a) Round Shape (b) "U" Shape (c) Zigzag Shape

The first pattern was Round shaped tested area, where the system was avoiding the obstacle with proper deviation and following the perimeter. Second pattern was 'U' shaped area, where the system successfully avoiding the obstacle by following the convenient path. The last experiment was done creating a 'Zigzag' shaped area, which was most challenging. The test result on this pattern proved that how much the system is reliable, efficient and sensitive in autonomous navigation.

5 Results and Discussion

5.1 Simulation Result

Y-axis is representing the Y-component of distance and X- axis is representing the X-component of distance. The positive side of the Y axis is representing the left turn and the negative side for right turn. The positive side of the X axis is representing the forward direction and positive side for reverse direction.

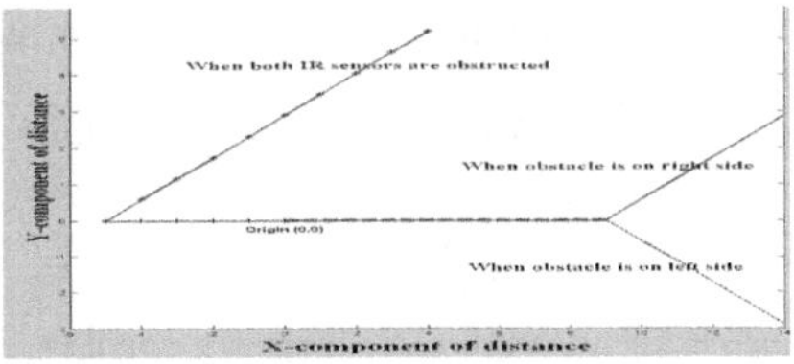

Fig. 4. Movement of the system when sensor (s) is (are) obstructed

5.2 Experimental Result

This pattern shows the system is being tested in the "Round" shaped area enclosed completely. It is moving into that trap of obstacles and avoiding them intelligently.

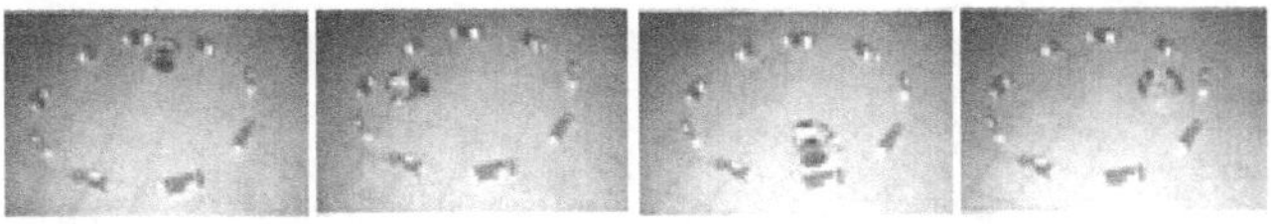

Fig. 5. Rotation inside the circle: Following a circular pattern of obstacle the system first sensing the obstacle one by one then avoiding the obstacle taking a turn

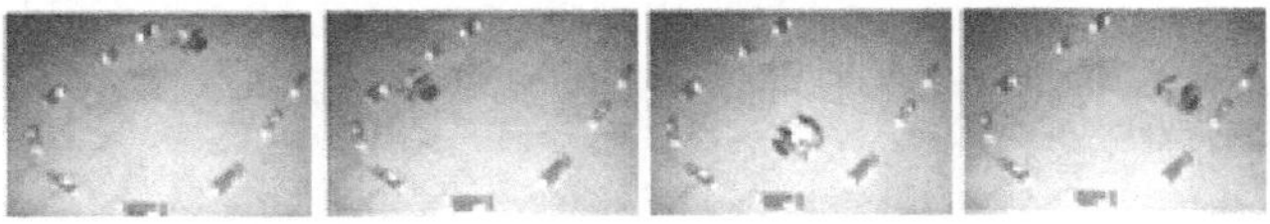

Fig. 6. Entering into the 'U' shaped pattern, avoiding obstacles by taking a turn then coming out from 'U' Shaped pattern

These sequences of pictures depict system's actions during moving in the "U" shaped area. The sequence proceeds from left to right avoiding the obstacles to achieve its goal.

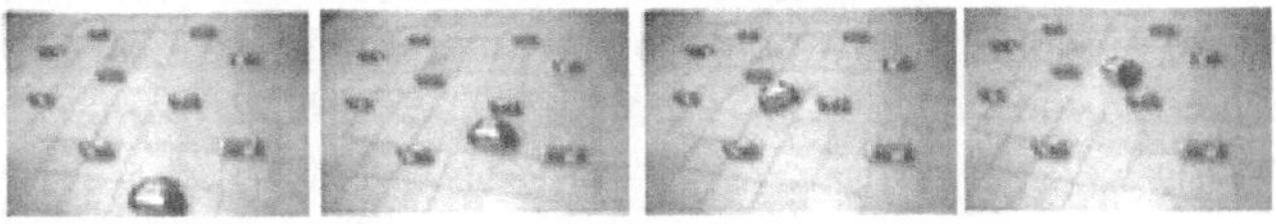

Fig. 7. Entering into the Zigzag pattern, sensing the obstacles shaped, avoiding that obstacle coming out from there

The system is being tested in the "Zigzag" shaped area in the Fig. Though the obstacle is placed in the haphazard sequences, the system can avoid the obstacle easily.

5.3 Discussion

The following table shows the results of the experiments (only five frames). The table consists of distance along x-direction, distance along y-direction, resulting distance and

Table 1. Analyzed data captured during the performance of the system

Frames Number	Distance (meter) along x-axis	Distance (meter) along y-axis	Resultant Distance (meter)	Absolute Velocity (mm/frames)
1.	2.24	1.66	2.79	0
2.	2.21	1.66	2.76	0
3.	2.17	1.65	2.73	35.1
4.	2.14	1.65	2.7	35.69
5.	2.1	1.64	2.66	36.1

the absolute velocity of the system. When the system wanders freely, its position and velocity varies with respect to the frame. The graphs of the resulting distance and the absolute velocity of the system with respect to the origin of the frame are shown below.

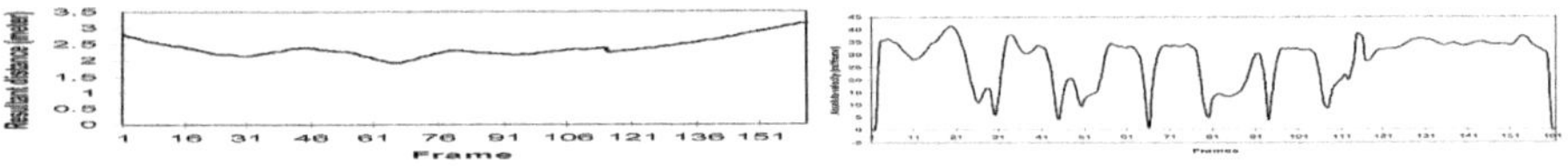

Fig. 8. (a) Graph of Resulting Distance vs. Frame and (b) Graph of Absolute Velocity vs. Frame

6 Conclusions

This work attempts to describe the hybrid architecture along with the Subsumption architecture. A control system for behavior based robot should be decomposed into three functional elements: a sensing system, a planning system, and an execution system. From the experimental field we have seen that when the system is wandering at the external environment the information is gathered by the sensors. The system uses this information to act against any environmental change and simultaneously monitors the area in front of it. The system is highly sensitive to response the obstacle within few microseconds. It is compatible in indoor as well as outdoor. Here we used a reprogrammable microcontroller which provides a set of behavioral primitives for further improvement of the system. It is simple, robust, reliable, low cost and capable for intelligent motion.

Acknowledgement

Authors are grateful to Director, CMERI and other CMERI personals who have directly or indirectly helped to complete the project successfully.

References

1. Walter, W.G.: Grey Walter's Tortoise (1953), `http://www.ias.uwe.ac.uk/Robots/gwonline/gwonline.html` (accessed January 26, 2010)
2. Arkin, R.C.: Behavior-Based Robotics. MIT Press, Cambridge (1998)

3. Moravec, H.P.: The Stanford Cart and the CMU Rover. Proceedings of the IEEE 71, 872–884 (1983)
4. Brooks, R.: A Robust Layered Control System for a Mobile Robot. IEEE Journal of Robotics and Automation RA-2, 14–23 (1986)
5. Ram, A., Arkin, R.C., Moorman, K., Clark, R.J.: A Case Based Reactive Navigation: A Case Based Method for On-Line Selection and Adoptation of Reactive Control Parameters in Autonomous Robotic Systems. IEEE Transactions on Systems, Man and Cybernetics Part B 27(3), 376–394 (1992)
6. Mataric, M.J.: Behavior-Based Robotics as a tool for Synthesis of Artificial Behaviors and Analysis of Natural Behavior. In: Trends in Cognitive Science, pp. 82–87 (March 1998)
7. Abreu, A., Correia, L.: A fuzzy behavior-based architecture for decision control in autonomous vehicles. In: IEEE International Symposium on Intelligent Control, Issue 2001, pp. 370–375 (2001)
8. Yang, S.X., Li, H., Meng, M.Q.-H., Liu, P.X.: An Embedded fuzzy Controller for a Behavior-Based Mobile Robot with Guaranteed Performance. IEEE Transactions on Fuzzy systems 12(4) (2004)
9. Stroupe, A., Hunsberger, T., Okon, A., Aghazarian, H., Robinson, M.: Behavior-Based Multi-Robot Collaboration for Autonomous Construction Tasks. In: IEEE International Conference on Intelligent Robot & Systems, Edmonton, Canada, vol. 2(4) (2005)
10. Ray, D.N.: Behavior Based Robotics: Steps towards Intelligence. The Bulletin of Engineering and Science 1(2) (2006)
11. Granot, R., Ina, G.: AuRA – Architecture of Autonomous System. Mathematics and Computer science Department at Haifa University, Israel (2008)
12. Mondal, A., Das, B., Ray, D.N., Majumder, S.: ARBIB: Development of an Autonomous Robot based on Intelligent Behavior. Published in National Conference on Machine and Mechanism 2009, pp. 195–200 (2009)
13. Mataric, M.J.: Integration of Representation into Goal-Driven Behavior-Based Robots. IEEE Transactions on Robotics and Automation 8(3) (June 1992)
14. Low, K.H., Leow, W.K., Marcelo Jr., H.A.: A Hybrid Mobile Robot Architecture with Integrated Planning and Control. In: 1st AAMAS 2002, Bologna, Italy, July 15-19, pp. 219–226 (2002)
15. Brooks, R.: Planning is just a way of Avoiding Figuring out What to Do Next. Working Paper 303, MIT Al Laboratory (September 1987a)
16. Brooks, R.: New Approaches to Robotics. Science 253, 1227–1232 (1991b)
17. Badal, S., Ravela, S., Draper, B., Hanson, A.: A Practical Obstacle detection and Avoidance System. In: Proceedings of the Second IEEE Workshop on Applications of Computer Vision, Sarasota, FL, pp. 97–104 (December 1994)
18. Braitenberg, V.: Vehicles: Experiments in Synthetic Psychology. MIT Press, Cambridge (1984)

Investigation of Channel Capacity of Railway Power Lines for Track and Level Cross Monitoring Application

Anish Francis and Geevarghese Titus

Amal Jyothi College of Engineering, Kanjirappally, Kottayam, 686518, Kerala, India
anishfrancis@amaljyothi.ac.in

Abstract. Security and monitoring of remote railway tracks and level crosses is an important issue these days in view of terrorist attacks and accidents. This paper analyzes the channel capacity of the railway power lines as broad band carrier for developing monitoring applications for railways. The BPLC technology has underwent significant advancements through developments in coding and modulation. The paper applies multi conductor transmission line theory to model the power lines and estimate channel capacity.

Keywords: Power lines, ABCD parameters, channel capacity, frequency response, video transmission.

1 Introduction

Currently systems available for monitoring the railway tracks are limited, which are mainly based on radio wave technologies like GPS as in [1]. The accidents due to unguarded level crosses, and failure of the mechanical components of the track etc. indicates the need of an advanced monitoring system for railways.

A conceptual system based on video cameras placed at electric poles in the railway track, is an example of an advanced system that can tackle the above mentioned issues. Since railway network is vast, the usage of existing railway power cable as channel for data communication for the development of the system is a good option. Hence, a study of channel capacity of the railway powerlines, and how it responds to the different frequency ranges of audio and video, based on the existing modulation schemes has got a high priority. It has been found that, OFDM is best suited for broadband video transmission through indoor powerlines [8], [9]. The objective of the paper is to model the power line using a multi conductor transmission theory approach and estimate the channel capacity. The modeling of the channel and specification of transfer function is required in order to apply the Shannon's theory for estimation of channel capacity.

2 Electric Power Distribution System of Indian Railways

This work focuses on Indian railways. Indian railway network is the largest railway network with more than 60% route is electrified [4].The Indian railway had adopted a

V.V. Das, J. Stephen, and Y. Chaba (Eds.): CNC 2011, CCIS 142, pp. 142–146, 2011.

25kv,50Hz,A.C. supply system for traction purposes according to[4],[5].The power supplies are derived from 220 KV/132 KV 3-phase transmission system from various grids.

Power from the 3- phase grid is transported to Grid substations, from where it is transferred to traction substations. The traction substations supply power for 35-40 km track length [4], by over head single phase lines operating at 50 Hz. The A.C power from the lines is collected by pantograph and converted to D.C by loco-transformer and D.C series motors in the train. [4].This medium voltage, 25 kv,50/60 Hz supply system is also employed in U.S,U.K ,Russia and several other countries[6]. The average distances between substations in that case is 20-40 miles [5]. The network of the above described overhead single phase power line is subjected to study for monitoring purposes.

3 Analytical Model of Railway Powerlines

Railway overhead lines uses copper, copper alloys and aluminium based catenary cables. In some parts of the world, insulated wire is also used depending on the voltage. In India, we have single phase double lines for over head contact, mostly bare catenary cables are used. In the present work, we tested both copper and aluminium, for varying distances of separation. Also the simulations tested for different frequencies and different lengths of wires.

To model the overhead lines, we followed the multiconductor transmission line theoretical approach [1].This is a bottom down approach and the lines are viewed as a two port network. The two port network is specified by the ABCD matrix which is also called transmission matrix. Every uniform transmission line can be modeled as a two port network from the basic transmission line theory. Now the electrical characteristics between the lines are characterized by the transmission matrix. The ABCD coefficients are derived from the primary line equations for characteristic impedance and propagation constant

$$Z_C = \sqrt{(R + j\omega L)/(G + j\omega C)} \tag{1}$$

$$\gamma = \sqrt{(R + j\omega L)(G + j\omega C)} \tag{2}$$

In the above equations R, L, C & G can be found using standard formulae. These parameters depend on radius of cable, distance between cables, wave frequency, permeability and conductivity of conducting material. The parameters can be computed using the simulation program by inputting data relevant to the railway powerlines. After finding the above parameters, we can model the transfer function of overhead lines using ABCD parameters. Transmission matrix is given by

$$T \begin{vmatrix} A & B \\ C & D \end{vmatrix} \tag{3}$$

$$\left.\begin{array}{l} A=D=\cosh(\gamma l) \\ B=Z_0\sinh(\gamma l) \\ C=(\frac{1}{Z_0})\sinh(\gamma l) \end{array}\right\} \quad (4)$$

where γ, l and Z_0 are the propagation constant, the length and characteristic impedance of the powerline cable. The channel between the source and load is represented by the ABCD matrix. The transfer function of the channel is represented by the function

$$H(f)=\frac{Z_l}{AZ_l+B+CZ_sZ_l+DZ_s} \quad (5)$$

By cascading the transmission matrices, the transfer function of systems with multiple transmitters can be computed using, chain rule in [2]. In that case, the overall transmission matrix can be computed by taking the product of individual transmission matrices. For example as in our present work, if the video transmitters are connected at electric poles, that support the overhead catenary, the individual section can be modeled as a transmission matrix. If there are n sections between the source and receiving centers, each having a transmission matrix T (n1),T(n2)...etc. Then the over whole transmission matrix can be found by

$$T = T(n_1)T(n_2)T(n_3)\dots T(n_n) \quad (6)$$

Channel gain can be computed by plotting frequency versus H(f) function found from equation (5). From that gain and phase variation, with respect to frequency, length, diameter and distance between the cables, can be found. The channel capacity can be estimated using the Shannon's channel capacity equations in [3].

$$C=\int_{f1}^{f2} log_2(1+\frac{S(f)}{N(f)})\, df \quad [bits/s] \quad (7)$$

The signal power can be computed from H(f) using the equations in [3] which require the estimation of noise power and PSD for railway overhead lines. We assumed a -40 dB noise power and -50dB PSD for the present work, based on [3].The load and source impedance were assumed to be 100Ω.The characteristic impedance also varies from 100 Ω to 1M Ω.

4 Results and Findings

The simulations were done in MatLab. The data used for analyzing is included in Table 1 which was compiled from [6],[10]. In this we assume a distance of 0.125 - 0.75 meter of separation between the cables. The simulations were done to compute channel gain vs. frequency, channel capacity vs. diameter of the cable and distance transmitter to receiver. In the present work we are not considering the capacitive effect between the lines.

For a one meter length, 50 mm diameter, the gain response is plotted, for frequency up to 50 Mhz as in Fig.1a, from which we can see that, attenuation happens at high frequencies, approximately greater than 10MHz. As diameter of the cable is varied keeping the frequency constant, it is observed that gain increases for higher diameters. The input diameter was in the range of practically used power cable standards. The gain response in Fig.1b is plotted for a length of one meter, and frequency 1 MHz.

Table 1. Data specification for railway powerlines

Parameters	Copper	Aluminium
Diameter(d) [mm]	44.2,40.6,24.8	53.0,40.2,28.3,19.4,14.3-35.5 (sector shaped)
Conductivity (σ) [siemens/m]	1.58E+7 to 5.58E+7	4.58E+7 to 5.56E+7
Permeability(μ) [H/m]	1.256×10-6	1.256×10-6

The channel capacity was plotted for frequencies up to 100Mz for a 50mm diameter, one meter length, of cable which is shown in Fig.1c. It shows that capacity increases at low frequency, reaches a peak value and then decreases thereafter. A maximum value of 10 Mbps was observed.

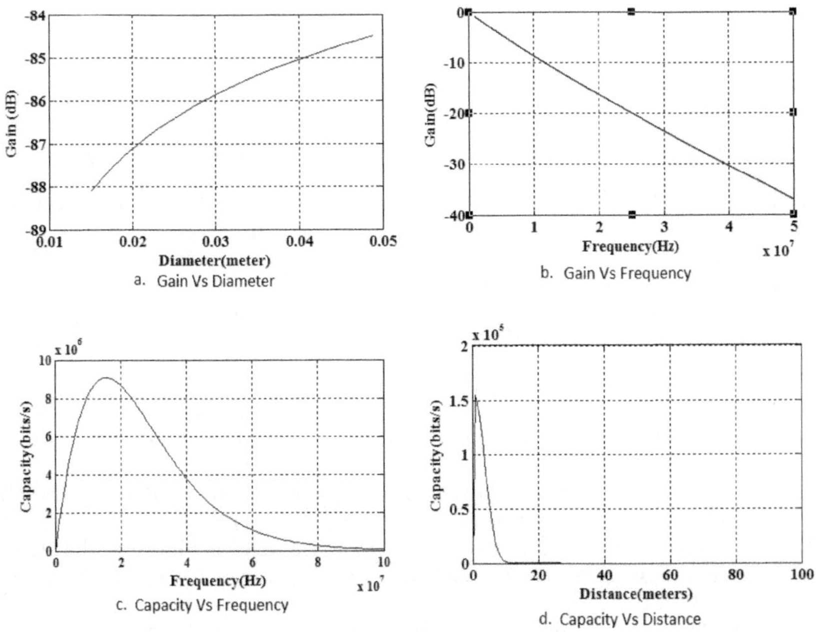

Fig. 1(a-d). Various simulations in MatLab

This capacity is enough for video transmission across the cable. But as the distance increases, the channel capacity reduces considerable, which suggests the requirement of repeaters, across the channel for robust video transmission. Channel gain drops after a 10 meter distance between the source and load as per the simulation for a 1 MHz, 40 mm diameter cable as in Fig.1d.

But for the case of audio frequency, the channel gain drops only after a significant distance. We found that channel attenuates audio frequency completely after a distance of around 45 meters. The phase is non-linear at distance near the transmitter and tends to be linear near the receiver. This non-linearity suggests the need of FIR filters for audio regeneration.

5 Conclusion

In this paper we analyzed the channel model of railway overhead lines. The study suggested the possibility of video and audio transmission through the cables, with repeaters included in the communication link. The present work is made based on ABCD parameters, which should be remodeled using s parameters for considering the capacitive effects and multipath effects. Also the transfer function has to be computed for branching taking into account of reflections in the power line. The study opens the challenges of suitable coding techniques and coupler designs to implement the system.

References

1. Kim, Z.W., Kohn, T.E.: Pseudo realtime activity detection for rail road grade crossing safety
2. Paul, C.R.: Analysis of Multiconductor Transmission Lines. Wiley Publications, Chichester
3. Galli, S., Banwell, T.: A Novel Approach to the Modeling of the Indoor Powerline Channel—Part II: Transfer Function and its Properties. IEEE Trans. Power Del. 20(3), 1869–1878 (2005)
4. Motilal Nehru National Institute of Technology, http://www.mnnit.ac.in/departments/eed/iee_sem/sem_proced/vol1/ElectricTractionSupplySystemonIndianRailways.pdf
5. Indian Railway, http://www.indianrailways.gov.in/indianrailways/codesmanual/ACTraction-II-P-I/ACTractionIIPartICh1_data.htm
6. Bhargava, B.: Railway Electrification Systems and Configurations. IEEE paper no: 0-7803-5569 (1999)
7. Goldring, A.G.: Development of overhead equipment for British Railways 50 Hz A.C. Electrification since 1960. Proc. IEEE 118(8) (August 1971)
8. Zhang, J., Meng, J.: Robust Narrrowband Interference Rejection for Powerline Communication System using IS-OFDM, 2036179, IEEE paper no: 0885-8977 (2009), doi: 10.1109/ TPWRD
9. Zhang, J., Meng, J.: Noise Resistant OFDM for Power-Line Communication Systems. IEEE Transactions on Power Delivery 25(2) (April 2010)
10. Wester Railway, http://www.wester-rail.co.uk

A Distributed Weighted Cluster Based Routing Protocol for MANETs

Naveen Chauhan, L.K. Awasthi, Narottam Chand, and Ankit Chugh

Department of Computer Sc. & Engineering
National Institute of Technology, Hamirpur (H.P.)-177005
{naveen,lalit,nar}@nitham.ac.in, ankitnitham@yahoo.co.in

Abstract. MANETs or Mobile ad-hoc networks are a form of wireless networks which do not require a base station for providing network connectivity. Mobile ad-hoc networks have many characteristics which distinguish them from other wireless networks which makes routing in mobile ad-hoc networks a challenging task. Cluster based routing scheme is one of the routing schemes for MANETs in which various clusters of mobile nodes are formed with each cluster having its own clusterhead which are responsible for routing between clusters. In this project we have designed an implementation of Distributed Weighted Clustering Algorithm. This approach is based on combined weight metric that takes into account several system parameters like the node degree, transmission range, energy and mobility of the nodes. After implementation we have evaluated the performance of our cluster based routing scheme in various network situations.

Keywords: MANETs, clustering, routing, wireless communication, distributed clustering.

1 Introduction

MANETs or Mobile ad-hoc networks are a form of wireless networks which do not require a base station for providing network connectivity. In this networking concept, mobile devices form a temporary community without any planned installation, or human intervention. Each node acts as a host and a router at the same time. This means that each node participating in a MANET commits itself to forward data packets from a neighboring node to another until a final destination is reached. Mobile ad-hoc networks have many characteristics which distinguish them from other wireless networks. These factors are no fixed network infrastructure, dynamic network configuration, node mobility and frequent node failure, low battery power etc which make routing in MANETs quite a challenging task. Various routing protocols have been proposed for MANETs with varying performance in different network conditions. Cluster based routing scheme is one of the routing schemes for MANETs in which various clusters of mobile nodes are formed with each cluster having its own clusterhead which are responsible for routing between clusters. Clustering of nodes saves energy and communication bandwidth in ad-hoc networks.

In this paper we discuss an implementation of Distributed Weighted Clustering Algorithm. This approach is based on combined weight metric that takes into account

V.V. Das, J. Stephen, and Y. Chaba (Eds.): CNC 2011, CCIS 142, pp. 147–151, 2011.

several system parameters like the ideal node degree, transmission range, energy and mobility of the nodes. Depending on specific applications some of these parameters can be used in the metric to determine the cluster-head. However more cluster-heads lead to extra number of hops for a packet when it has to be routed from source to destination. On the other hand we can choose to have minimum number of cluster-heads to maximize the resource utilization. Various parameters and their respective weights can be determined to arrive at an efficient weighted cluster based routing scheme.

The remainder of this paper is organised as follows. Section II gives description of proposed algorithm. Section III gives performance analysis of proposed algorithm. Finally, Section IV concludes the paper.

2 Proposed Algorithm

In this section we present our improved distributed weighted clustering algorithm. This algorithm can be divided into three phases. These phases are described as:

2.1 Formation of Clusters

Initially each node is assigned a random ID value. It broadcasts its ID value to its neighbours and builds its neighbourhood table. Each node calculates its own weight based on the following factors:

- Node connectivity: The number of nodes that can communicate directly with the given node i.e. that are in its transmission range.
- Battery Power: The power currently left in each node. The energy is consumed by sending and receiving of messages.
- Mobility: Running average of speed of each node. If mobility is less, the node is more suitable to become clusterhead.
- Distance: Sum of distance of the node from all its neighbours.

After finding its own weight, each node broadcasts its weight to its neighbours. If it has maximum weight among its neighbours, it sets the clhead variable to 1 otherwise, the clhead variable is set to 0. The node with maximum weight broadcasts clhead message to other nodes. On receiving a clhead message a node checks all the nodes from which it receives clhead message. The node with maximum weight becomes the clusterhead of that node. The node sends a unicast message to the clusterheads.

2.2 Running Mobility

Then we assign a random value between 0 and 1 to each node and a threshold value is taken. If the random value assigned to the node is greater than threshold value, the node becomes mobile, otherwise it remains stationary. Mode is made mobile by changing its x and y position in a random manner in a random direction. E.g. if a node is to move in east direction, its position will change as

$$x=x+1;\ y=y+0;$$

The direction and the duration the node moves in a particular direction is random. On reaching new destination, it joins the nearest clusterhead.

2.3 Cluster Maintenance

Cluster maintenance is required when a node moves out of the range of its clusterhead, if a new node is added or the clusterhead fails. First case has been discussed earlier in the mobility handling section. In case a new node is added, it calculates its weight as discussed earlier. However, even if its weight is more than the clusterhead, it does not immediately become the clusterhead and instead makes the current clusterhead its clusterhead. This is to reduce unnecessary overhead in selection of the clusterhead each time a new node is added. In case the clusterhead fails, the nodes attached to that clusterhead recalculate their weights and select new clusterhead with the maximum weight.

The algorithm can be given as follows:

Weight Calculation Algorithm

1. Find connectivity c for each node which is the number of neighbours of each node.
2. Find the energy remaining, e for each node.
3. Compute the mobility M for each node which is the running average of the speed until the current time t.

$$M_v = \frac{1}{T}\sum_{t=1}^{T}\sqrt{(X_t - X_{t-1})^2 + (Y_t - Y_{t-1})^2}$$

4. Compute the sum of distances with all its neighbours, d for each node.
5. Calculate the combined weight W as

$$W = w_1 * c + w_2 * e - w_3 * M + w_4 * d$$

3 Performance Analysis

In this section we present the performance of the proposed algorithm obtained by simulation using Omnet++. The measured performance of the proposed algorithm was compared with that of DWCA. The mobile network model assumed in this paper consists of random number of mobile nodes with each node having fixed energy and random mobility.

3.1 No of Clusterheads Formed vs. No of Nodes

Figure 3 depicts the average number of clusters formed with respect to the total number of nodes in the Ad-hoc If the node density is increased, our algorithm produced constantly less clusters in comparison with the original algorithm. As a result, our algorithm gave better performance in terms of the number of clusters when the node density and node mobility in the network are high.

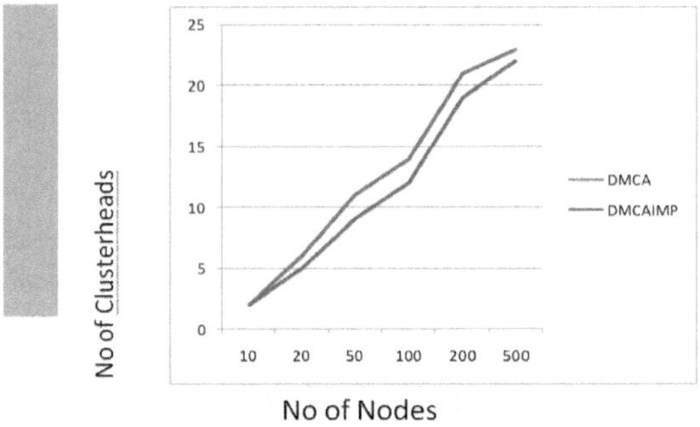

Fig. 3. No of clusterheads vs no of nodes

3.2 No. of Control Messages vs. No. of Nodes

Figure 4 shows the overhead of packets generated per node during the initial clustering set up phase. The overhead increased as the number of nodes increases. Each node independently chose one of its neighbors with the highest score to be its cluster head and thus the cluster head selection was performed in a distributed manner with most recently gathered information of current state of the neighbors.

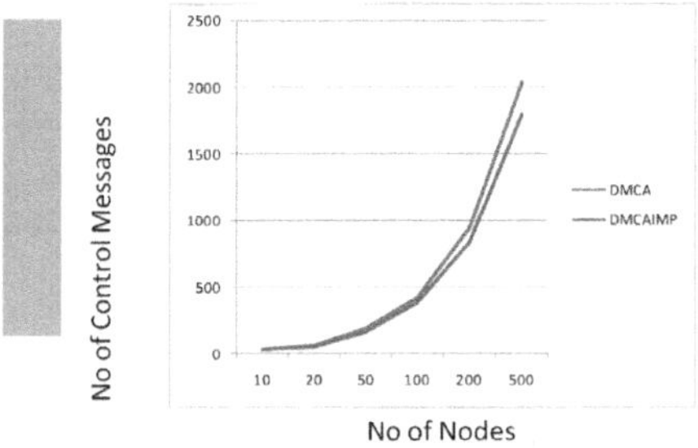

Fig. 4. No of control messages vs no of nodes

4 Conclusions

In this paper we have studied various clusterbased routing schemes and implemented a novel distributed weighted clustering algorithm making some modifications and improvements on existing algorithms[1, 2].

As demonstrated, our algorithm reduces the clusterhead formation and control messages overhead thus improving overall performance and reducing energy utilization. Since energy utilization is the most important criteria in cluster based routing schemes, our protocol provides better results than existing distributed clustering algorithms.

References

1. Adabi, S., Jabbehdari, S., Rahmani, A., Adabi, S.: A Novel Distributed Clustering Algorithm for Mobile Ad-hoc Networks. Journal of Computer Science, 161–166 (2008)
2. Choi, W., Woo, M.: A Distributed Weighted Clustering Algorithm for Mobile Ad Hoc Networks. In: Proceedings of the Advanced International Conference on Telecommunications and International Conference on Internet and Web Applications and Services (2006)
3. Basagni, S.: Distributed clustering for ad hoc networks. In: Proceedings of I-SPAN 1999, pp. 310–315 (1999)
4. Chatterjee, M., Das, S., Turgut, D.: WCA: a weighted clustering algorithm for mobile ad hoc networks. Journal of Cluster Computing (Special Issue on Mobile Ad hoc Networks) 5, 193–204 (2002)
5. Brust, M.R., Andronache, A., Rothkugel, S.: WACA: A Hierarchical Weighted Clustering Algorithm optimized for Mobile Hybrid Networks. In: Proceedings of the Third International Conference on Wireless and Mobile Communications, ICWMC 2007 (2007)
6. Elhdhili, M.E., Azzouz, L.B., Kamoun, F.: Lowest Weight: Reactive Clustering Algorithm for Adhoc Networks. IEEE Journal on Mobile Communications (2006)
7. Ramachandran, L., Kapoor, M., Sarkar, A., Aggarwal, A.: Clustering algorithms for wireless ad hoc networks. In: Proceedings of Workshop on Discrete Algorithms and Methods for Mobile Computing and Communications, Boston, pp. 54–63 (August 2000)
8. Wang, Y., Chen, H.R., Yang, X.Y., Zhang, D.Y.: WACHM: Weight Based Adaptive Clustering for Large Scale Heterogeneous MANET. IEEE Journal (2008)
9. Jiang, M., Li, J., Tay, Y.C.: Cluster based routing protocol (cbrp). draft-ietf-manet-cbrp-spec-01.txt, IETF, Internet draft version 01 (July 1999)
10. Tao, Z., Wu, H., Huang, C., Cui, L.: Hybrid Cluster Routing: An Efficient Routing Protocol for Mobile Ad Hoc Networks. In: IEEE International Conference Communications, vol. 8, pp. 3554–3559 (2006)
11. Lin, C.R., Gerla, M.: Adaptive Clustering for Mobile Wireless Networks. IEEE J. Select. Areas Commun. 15, 1265–1275 (1997)
12. Baker, F.: An outsider's view of MANET draft-baker manet review. Network Working Group, March 17 (2002)
13. Perkins, C., Das, S.: Ad hoc On-Demand Distance Vector (AODV) Routing. Network Working Group (July 2003)
14. Dhurandherl, S.K., Singh, G.V.: Power Aware Clustering Technique in Wireless Ad Hoc Networks. IEEE Journal on Mobile Communications (2006)

ReeRisk –A Decisional Risk Engineering Framework for Legacy System Rejuvenation through Reengineering

Anand Rajavat[1] and Vrinda Tokekar[2]

[1] Computer Science & Engineering, Shri Vaishnav Institute of Technology and Science, Indore, M.P., India
[2] Information Technology, Institute of Engineering and Technology(DAVV), Indore, M.P., India

Abstract. Maintaining legacy system to meet continual changing user and business needs is difficult. However organization must consider modernizing these legacy systems to remain viable. Variety of solutions is available for legacy system modernization. Reaching a decision about how to evolve a legacy system cannot be made spontaneously; rather it requires understanding the strengths and weaknesses of each modernization technique. Over the past few years reengineering has emerged as a popular modernization technique. Decision about when to evolve a legacy system through reengineering cannot be made spontaneously; rather it should require risk assessment of legacy application from system, managerial and technical point of view. We present a decision driven risk assessment framework ReeRisk that examines system, managerial and technical domain of legacy system in accordance with requirements of target system. The result of ReeRisk framework is used to take decision about when evolution of a legacy system through reengineering is successful.

Keywords: Domain, Perspective, Risk cluster, ReeRisk.

1 Introduction

Legacy systems [1] were developed before the widespread use of modern software engineering [2] [3] methods and have been maintained to accommodate changing business and user requirements. Most of the legacy systems we use have complex design structure have inefficient coding and incomplete documentation. Modernizing legacy system to meet continual changing user and business needs is difficult. A legacy system may evolve in a number of ways, depending on factors such as its technical condition, its managerial value and the characteristics of the organization involved in maintaining and operating the system. Continued maintenance, reengineering [4] and replacement are the general evolution strategies of which one or a combination may be an appropriate way of evolving a legacy system. [5]

Software risk management [6] [7] generally focuses on goals relating to identification and controlling of cost, schedule and quality risk within the software development lifecycle. Nevertheless certain goals such as risk identification and risk assessment of legacy system to support system evolution activities have gained

V.V. Das, J. Stephen, and Y. Chaba (Eds.): CNC 2011, CCIS 142, pp. 152–158, 2011.

importance recently. Though there are several contributions in the area of legacy system assessment, still a lot need to be done for risk engineering in legacy system evolution process. We summaries the following research questions.

- How risk engineering can systematically integrate at early reengineering stage to reach a decision about how best to evolve a legacy system through reengineering.
- What are the main areas require to be attained during early stage of system evolution to select reengineering strategy.
- How risk components and risk factors that obstruct the reengineering decision identified and assess from system, managerial and technical domain of legacy system.

To answer these questions, the aim of this research is to propose a decision driven risk engineering framework to select reengineering [8] [9] as a system evolution strategy.

The research contributes a decisional risk engineering framework ReeRisk to identify and assess risk in reengineering process of software systems. The main focus is to identify and assess risk from system, managerial and technical domain of legacy system in accordance with target system requirements.

2 ReeRisk (Reengineering Risk) Framework Architecture

The enterprise ReeRisk framework consists of three domains that are building block for a successful risk engineering process for the selection of reengineering as a system modernization technique. Each domain covers different perspectives of software development process [10] that are essential for developing risk engineering framework. For each perspective one or more risk clusters are identified which includes risk component and the risk measurement model. Critical risk factors for each risk component are identified to support measurement model, which is used to measure the effect of particular risk component.

A simplified conceptual view of the elements in the ReeRisk framework is presented in Fig. 1. ReeRisk framework comprises with Domain, Perspective, risk cluster and risk factors.

A domain is a field of study that defines a set of common Perspectives, Risk Cluster, and relative risk factors to identify and measure cumulative risk impact in reengineering process of legacy system. Domain represents a subset of ReeRisk framework.

Perspective is a viewpoint according to which different risk clusters are identified and measured using different risk measurement model.

Risk cluster covers risk component and the risk measurement model, which is used to measure the effect of particular risk component on system evolution decision.

Risk component contain different types of negative outcomes from system, managerial and technical domain of legacy system.

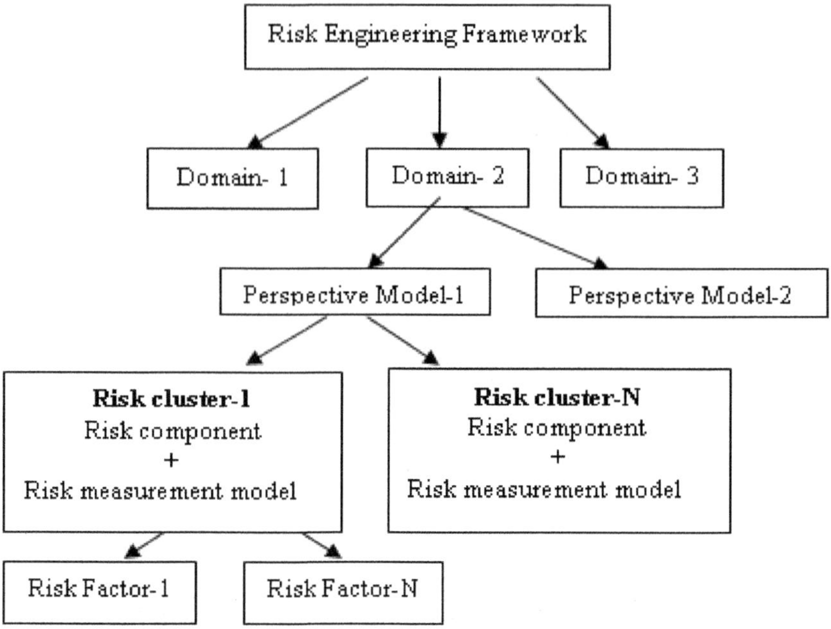

Fig. 1. Conceptual view of ReeRisk framework

Risk measurement model measures different types of risk components from system, managerial and technical domain of legacy system in accordance with desired state of target system and reengineering strategy.

Risk factor encompasses sources of risk component from system, managerial and technical domain of legacy system.

3 ReeRisk Process

The ReeRisk process overview is presented in Fig.2 which illustrates the role of the framework elements and their interrelationship with legacy and target system.

The ReeRisk process begins with analyzing legacy system to identify current state of legacy system and desired state of the target system and ends with the management decision concerning selection of reengineering strategy for legacy system evolution.

Management decision for the selection of reengineering as a system modernization strategy needs identifying requirements of target system as well as migration strategy use to transform legacy system to target system.

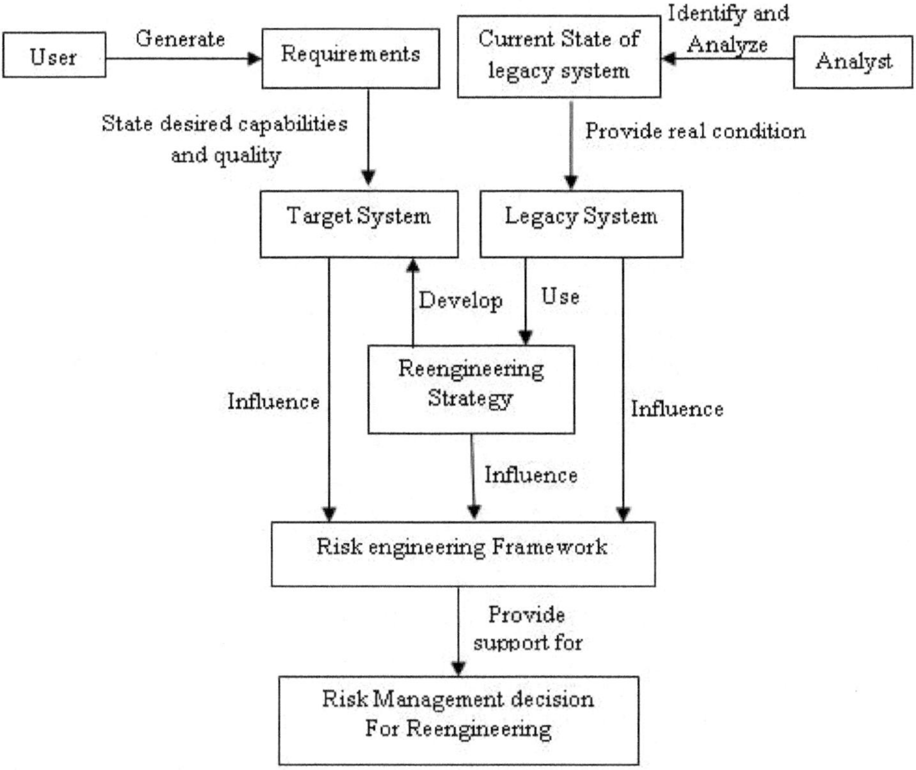

Fig. 2. ReeRisk process

4 Components of ReeRisk Framework

ReeRisk framework is a logical modal that shows hypothesized links between different domains, perspectives, risk components and factors of legacy system in concern with requiring state of target system. An analytical model expands on a specific portion of logical model and is used to measure cumulative impact of different risk components.

The framework serves to draw out important risk issues early in the development cycle. Framework provide insight for developing and implanting risk engineering by concentrating on system, managerial and technical aspects of legacy and Target system to realize successful reengineering effort.

ReeRisk framework focus on the following issues to take decision about when evolution of a legacy system through reengineering is likely to succeed and when they are likely to fail.

- Enhancement of application
- Benefits of new technology
- Better user experience

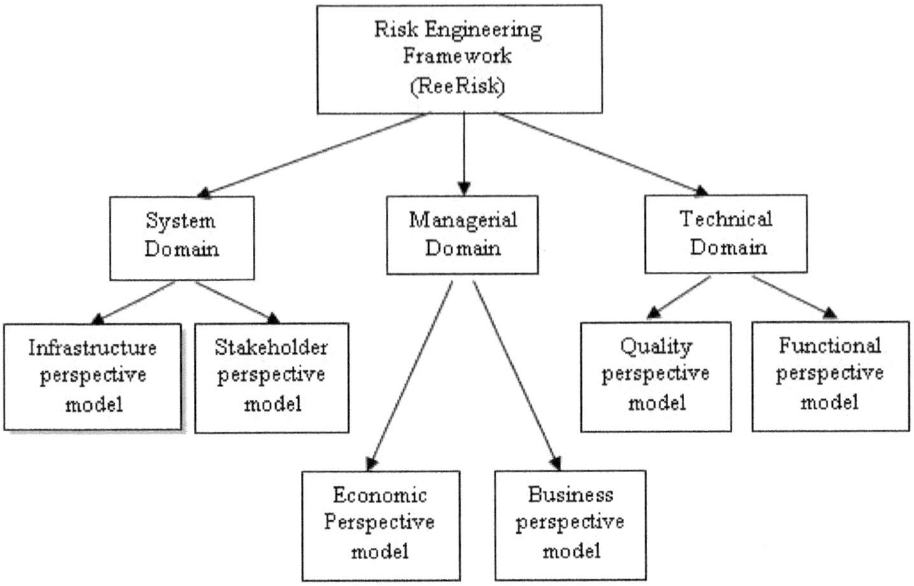

Fig. 3. Components of ReeRisk Framework

- Solution for the existing limitation
- Cost effective system functionality

Fig.3 shows relationship between domain and perspectives of Reerisk framework. The main domains of ReeRisk Framework are:

4.1 System Domain

In ReeRisk framework risk engineering begins with system domain. The term "System domain" denotes a structural unit that is responsible for maintaining a system that provides products and services to its customers. In this context, the System is responsible for planning and structuring the system Infrastructure efforts, organizing the stakeholder's tasks and ensures that the products and services fulfill the organization's goals and objectives. The element of the System domain consists of Infrastructure perspective model and Stakeholder perspective model. The system domain addresses the requirements of the customer, organization's strategic goals and objectives and operational environment of the enterprise. The system domain concentrates on the current legacy system and their operational environment as well as how the proposed system will be affected by (or affect) elements of system domain. System domain is characterized in terms of a fundamental set of risk component and factors that are indicative of the present state of legacy and desired state of Target System. In system domain two types of perspective model i.e. Infrastructure perspective Model and Stakeholder Perspective Model is developed by analyzing existing states of legacy system and desired state of target system.

4.2 Managerial Domain

Managerial domain of ReeRisk framework covers managerial issues related to system evolution process .Impact of market factors and effect of competitive products, on quality [11] & cost of target system are measured within the context of managerial domain. Domain identifies and measure organizational economic value to support system evolution activities. The element of the managerial domain consists of business value perspective model and economic value perspective model. Managerial domain is characterized in terms of a fundamental set of risk components and factors that are indicative for economics of present state of legacy and desired state of Target System.

4.3 Technical Domain

Technical domain of ReeRisk framework has a significant impact on software functionality and software quality. Technical domain helps in analyzing and testing the legacy system to better understand the function's capabilities and quality features and assess the impact of the proposed changes .The technical domain covers the technologies of legacy system as compared to technologies being considered for the proposed (target) system. The element of the technical domain consists of functional perspective model and Quality perspective model. Technical domain is characterized in terms of a fundamental set of risk components and factors that are indicative for function's capabilities and quality attributes of legacy and Target System.

The ReeRisk framework presented in this paper essentially involves measuring the system, managerial and technical domain of legacy system. The ReeRisk framework provides guidance on deriving and interpreting risk measurement process for each domain. The cumulative result can be used as the basis to take decision about when evolution of a legacy system through reengineering is successful.

5 Conclusion

In this paper, we have presented an overview of decision driven risk engineering framework to take decision concerning when evolution of legacy system through reengineering is successful. The ReeRisk framework has been designed to perform comprehensive assessment of legacy system in accordance with requirements of target system. ReeRisk framework assesses system, managerial and technical domain of legacy system to take decision about when evolution of a legacy system through reengineering is likely to succeed and when they are likely to fail. ReeRisk framework identify and measure critical risk factors from each domain. The result of ReeRisk framework is a level of understanding for the selection of reengineering as a system evolution strategy. Future work can be done to identify different risk clusters and relative risk factors for system, managerial and technical domain of ReeRisk framework.

References

1. Brodie, M.L., Stonebraker, M.: Migrating Legacy Systems: Gateways, Interfaces, & the Incremental Approach. Morgan Kaufmann Publishers, Inc., San Francisco (1995)
2. Sommerville, I.: Software Engineering, 7th edn. Pearson Education, London (2006)
3. The management of software engineering, Part I: Principles of software engineering. IBM Systems Journal 19(4), 414–420, ISSN: 0018-8670, doi: 10.1147/sj.194.0414
4. Sneed, H.M.: Risks Involved in Reengineering Projects. In: WCRE: Proceedings of the 6th IEEE Conference on Reverse Engineering, p. 204 (1999)
5. Ransom, J., Somerville, I., Warren, I.: A Method for Assessing legacy systems for evolution. In: Proceedings of the Second Euromicro Conference on Software Maintenance and Reengineering (1998) ISBN: 0-8186- 8421-6, doi:10.1109/CSMR
6. Boehm, B.W.: Software Risk Management: Principles and Practices. IEEE Software 8(1), 32–41 (1991)
7. Kulik, P.: What is Software Risk Management, 1st edn. white paper KLCI, Inc. (1996)
8. Briden, P.: Software Re-engineering process. Tessella Support Services PLC Technical report, Issue V2.R1.M1 (2000)
9. Feiler, P.H.: Reengineering: An Engineering Problem. Technical Report Software Engineering Institute Carnegie Mellon university Pittsburgh Pennsylvania 15213, CMU/SEI-93-SR-5 (1993)
10. Russ, M.L., McGregor, J.D., Korson-McGregor, Clemson, S.C.: A software development process for small projects. IEEE Software, 96–101, ISSN : 0740-7459, doi:10.1109/52.877874
11. Boehm, B., Chulani, S., Verner, J., Wong, B.: Fifth Workshop on Software Quality. In: 29th International Conference ICSE 2007 Companion on Software Engineering - Companion, pp. 131–132 (2007), ISBN: 0-7695-2892-9, doi:10.1109/ICSECOMPANION.2007.38

Quantitative and Qualitative Comparison Study of Reference Based Time Synchronization Approach in Wireless Sensor Network

Surendra Rahamatkar[1], Ajay Agarwal[2], and Praveen Sen[1]

[1] Department of Computer Science, Nagpur Institute of Technology, Nagpur, India
[2] Department of MCA, Krishna Institute of Engg. & Technology, Ghaziabad, India

Abstract. Time synchronization is an important research issue in wireless sensor networks (WSNs). Many applications based on these WSNs assume local clocks at each sensor node that need to be synchronized to a common notion of time. Some inherent properties of sensor networks such as limited energy, storage, computation, and bandwidth resources, combined with potentially high density of nodes make conventional synchronization methods unsuitable for these networks. Therefore, an increasing research focus on designing synchronization algorithms is required. In this paper we explore various time synchronization protocols and present theoretical analysis of protocols based on quantitative and qualitative criteria with the proposed Tree Structured Time Synchronization Protocol. The comparative study shows that it is an excellent conciliation among synchronization accuracy, computational complexity, and convergence time.

Keywords: Ad Hoc Tree Structure; Clock synchronization; Wireless sensor networks, Hierarchical sensor network.

1 Introduction

WSNs can be applied to a wide range of applications in domains as diverse as medical, industrial, military, environmental, scientific, and home networks [3]. Since the sensors in a WSN operate independently, their local clocks may not be synchronized with one another. This can cause difficulties when trying to integrate and interpret information sensed at different nodes. Some inherent properties of sensor networks such as limited energy, storage, computation, and bandwidth resources, combined with potentially high density of nodes make conventional synchronization methods unsuitable for these networks. Therefore, an efficient time synchronization approach is required. Hence, this is the motivation for developing software based approaches to achieve in network time synchronization.

This paper is organized in several Sections. Section 2 contains the review of existing synchronization protocols. We present brief description of proposed Tree Structured Time Synchronization scheme in Section 3. The comparative study of the proposed scheme with existing work is given in Section 4. Finally, Section 5 contains the conclusion of the paper.

V.V. Das, J. Stephen, and Y. Chaba (Eds.): CNC 2011, CCIS 142, pp. 159–163, 2011.

2 Existing Approaches to Time Synchronization

Time synchronization algorithms providing a mechanism to synchronize the local clocks of the nodes in the network have been extensively studied in the past. The most widely adapted protocol used in the internet domain is the NTP devised by Mills [4].

In RBS [5], a reference message is broadcasted. The receivers record their local time when receiving the reference broadcast and exchange the recorded times with each other. In the case of multi hop synchronization, the RBS protocol would lose its accuracy. Santashil PalChaudhuri et al [9] extended the RBS protocol to handle multi hop clock synchronization in which all nodes need not be within single-hop range of a clock synchronization sender. The TPSN [6] algorithm first creates a spanning tree of the network and then performs pair wise synchronization along the edges. Each node gets synchronized by exchanging two synchronization messages with its reference node one level higher in the hierarchy. TinySeRSync [10] protocol works with the ad hoc deployments of sensor networks. This protocol proposed two asynchronous phases: Phase I –secure single-hop pair wise synchronization, and Phase II–secure andresilient global synchronization to achieve global time synchronization in a sensor network. Van Greunen et al [12] *Lightweight Tree-based Synchronization* (LTS) protocol is a slight variation of the network-wide synchronization protocol of Ganeriwal et al. [11]. A disadvantage is that the accuracy of synchronization decreases linearly in the depth of the synchronization tree.

3 Proposed Time Synchronization Scheme

The proposed Tree Structured Referencing Time Synchronization (TSRT) scheme proposed by [1], [2], is considered for comparative study, which aims to minimize the complexity of the synchronization. The synchronization accuracy is assumed to be given as a constraint, and the target is to devise a synchronization algorithm with minimal complexity to achieve given precision. The scheme works on two phases. First phase used to construct an ad hoc tree structure and second phase used to synchronize the local clocks of sensor nodes.

The goal of the TSRT is to achieve a network wide synchronization of the local clocks of the participating nodes. We assume that each node has a local clock exhibiting the typical timing errors of crystals and can communicate over an unreliable but error corrected wireless link to its neighbors. The TSRT synchronizes the time of a sender to possibly multiple receivers utilizing a single radio message time-stamped at both the sender and the receiver sides. MAC layer time-stamping can eliminate many of the errors, as observed in [7]. However, accurate time-synchronization at discrete points in time is a partial solution only. Compensation for the clock drift of the nodes is inevitable to achieve high precision in-between synchronization points and to keep the communication overhead low. Linear regression is used in TSRT to compensate for clock drift as suggested in [5].

In proposed approach, before synchronization, each sensor node efficiently floods the network to form a logical hierarchical structure of network from a designated source point.

The synchronization phase is initiated by the root node's syn_begin message. On receiving this message, nodes of level 1 initiate a two-way message exchange with the root node. Before initiating the message exchange, each node waits for some random time, in order to minimize collisions on the wireless channel. Once they get back a reply from the root node, they adjust their clocks to the root node. Level 2 nodes, overhearing some level 1 node's communication with the root, initiate a two-way message exchange with a level 1 node, again after waiting for some random time to ensure that level 1 nodes have completed their synchronization. This procedure eventually gets all nodes synchronized with reference to the root node or reference node of subsequent level.

4 Evaluation and Comparison

In this section we compare the proposed scheme TSRT with existing synchronization protocols. Before evaluating the various protocols, fist we define the quantitative and quality criteria in detail.

4.1 Quantitative Evaluation

Here, the protocols differ broadly in terms of their computational requirements, energy consumption; precision of synchronization results, and communication requirements. In [8] authors suggested, there is no protocol clearly performs the others in all possible applications of wireless networks. Relatively, it is quite likely that the choice of a protocol will be based on the characteristics and requirements of each application. The quantitative comparisons of the various protocols are presented in Table 1.

Table 1. Quantitative performance comparison of synchronization protocols

Protocols	Precision	Piggy backing	Complexity	Convergence time	Network size	Sleep mode
TSRT [1] [2]	16.9 μs	Yes	Low	Unknown	150-300 Nodes	Yes
RBS [5]	1.85 ± 1.28 μs	N/A	High	N/A	2-20 Nodes	Yes
TS/MS[15]	945 μs	No	Low	High (Multi-hop)	N/A	Yes
TDP [13]	100 μs	No	High	High (Multi-hop)	200 Nodes	Yes
AD [14]	Unknown	No	High	High (Multi-hop)	200–400 Nodes	Yes

In the case of Computational complexity, RBS [5], asynchronous diffusion [14], and TDP [13] have higher computational and storage complexity as compared to TSRT protocol and the Tiny Sync and Mini Sync [15]. In Convergence time, RBS does not emphasize on low convergence time because it is based on reactive routing. The protocols Tiny Sync and Mini Sync [15] can tolerate high convergence times for multi-hop networks. As per as the Network size is concern, some authors presented empirical evaluations of synchronization protocols on actual sensor networks. Although this information is not commonly available for most of the protocols that we have considered for studies, the TSRT time synchronization protocol is significant in this regard. Compatibility with sleep mode is expressed in the term of ability of a node to be in low-power (sleep) mode can be vital to meeting the node's energy requirements. RBS highlights this feature by way of post-facto synchronization, TSRT and other protocols Tiny [15] also support this feature as well.

4.2 Qualitative Evaluation

In this section we evaluate the protocols based on overall quality criteria presented by [8]. In contrast to Section 5.1 above, here, features of each protocol extended and discussed based on various qualitative criteria. Table 2 compares the various protocols in terms of the following qualitative criteria.

Most authors have not measured scalability in their experiments; Ganeriwal et al. [7] have used a network of 150-300 nodes to test scalability. In addition, Su and Akyildiz used a net with 200 nodes to test the scalability of TDP [13]. RBS [5] typically synchronizes 3-20 nodes in a neighborhood; however, this protocol works well even in much larger networks using gateways between neighborhoods.

Some fault-tolerant protocols [6, 13] address message loss to some extent, but others have not addressed this issue. Consequently, it is unclear how sensitive their protocols are to message loss. This is somewhat troublesome because handling message loss can result in significant overheads and performance degradation during synchronization.

Table 2. Qualitative Performance Comparisons of Synchronization Protocols

Protocols	Accuracy	Energy Efficiency	Overall Complexity	Scalability	Fault Tolerance
TSRT[1] [2]	High	Average	Low	Good	Yes
RBS [5]	High	High	High	Good	No
TS/MS [15]	High	High	Low	N/A	Yes
TDP [13]	High	Average	High	Good	Yes
AD [14]	Unknown	Low	High	N/A	Yes

5 Conclusion

In this paper we have proposed a Tree Structure referencing Time Synchronization approach. The proposed approach is able to produce tight, deterministic synchronization with only few message exchanges. While the proposed approach is suitable for any type of network, it is especially useful in wireless sensor networks which are typically constrained on the available computational power and bandwidth and have some of the most exotic needs for high precision synchronization. The proposed approach was designed to switch between Timing-sync Protocol for Sensor Networks and the Reference Broadcast Synchronization algorithm. These two algorithms allow all the sensors in a network to synchronize themselves within a few microseconds of each other, while at the same time using the least amount of resources possible. Performance of proposed work is verified theoretically with the existing results and compared with existing protocols, which shows that proposed approach is an excellent conciliation among synchronization accuracy, computational complexity and convergence time.

References

1. Rahamatkar, S., Agarwal, A.: An Approach towards Lightweight, Reference Based, Tree Structured Time Synchronization Scheme in WSN. In: COSIT 2011. CCIS, vol. 131, pp. 189–198. Springer, Heidelberg (2011)
2. Rahamatkar, S., Agarwal, A.: Tree Structured Time Synchronization Protocol in Wireless Sensor Network. J. Ubi. Comp. & Comm. 4, 712–717 (2009)
3. Rhee, K., Lee, J., Wu, Y.C.: Clock Synchronization in Wireless Sensor Networks: An Overview. Sensors 9, 56–85 (2009)
4. Mills, D.L.: Internet Time Synchronization: The Network Time Protocol. IEEE Trans. Comm. 39(10), 1482–1493 (1991)
5. Elson, J.E., Girod, L., Estrin, D.: Fine-Grained Network Time Synchronization using Reference Broadcasts. In: OSDI, pp. 147–163 (2002)
6. Ganeriwal, S., Kumar, R., Srivastava, M.B.: Timing-Sync Protocol for Sensor Networks. In: First ACM Conference SenSys, pp. 138–149 (2003)
7. Woo, C.D.: A Transmission Control Scheme for Media Access in Sensor Networks. In: Mobicom, pp. 221–235 (2001)
8. Rahamatkar, S., Agarwal, A.: Analysis and Comparative Study of Clock Synchronization Schemes in Wireless Sensor Networks. Int. J. C. S. E. 2(3), 523–528 (2010)
9. PalChaudhuri, S., Saha, A.K., Johnson, D.B.: Adaptive clock synchronization in sensor networks. In: IPSN, USA, pp. 340–348 (2004)
10. Sun, K., Ning, P., Wang, C.: TinySeRSync: Secure and Resilient time synchronization in wireless sensor networks. In: ACM CCS, USA, pp. 264–277 (2006)
11. Ganeriwal, S., Kumar, R., Adlakha, S., Srivastav, M.: Network-wide Time Synchronization in Sensor Networks. Technical Report, E.E. Dept., UCLA (2003)
12. Greunen, J., Rabaey, J.: Lightweight Time Synchronization for Sensor Networks. In: 2nd ACM Int. Workshop on WSN & Applications, California, pp. 11–19 (2003)
13. Akyildiz, I.F., Su, I.W.: Time-Diffusion Synchronization Protocols for Sensor Networks. IEEE/ACM Transactions on Networking (2005)
14. Li, Q., Daniela, R.: Global clock synchronization in sensor networks. In: IEEE InfoCom (2004)
15. Sichitiu, M.L., Veerarittiphan, C.: Simple, accurate time synchronization for wireless sensor networks. In: IEEE WCNC (2003)

Efficient Video Streaming in Wireless P2P Network Using HIS-MDC Technique

Jeevan Eranti[1], Suresh Jaganathan[2], and A. Srinivasan[3]

[1] M.E. Final Year
Department of Computer Science and Engineering
Sri Sivasubramania Nadar College of Engineering
Chennai, Tamilnadu, India
jeevan4007@gmail.com
[2] Assistant Professor,
Department of Computer Science and Engineering
Sri Sivasubramania Nadar College of Engineering
Chennai, Tamilnadu, India
whosuresh@gmail.com
[3] Professor,
Department of Computer Science and Engineering
Misrimal Navajee Munoth Jain Engineering College
Chennai, Tamilnadu, India
asrini30@gmail.com

Abstract. In this paper, we propose a technique which enhances continuous flow of streaming despite packet loss, error resilience and quality of service using multiple descriptions coding (MDC). The main aim of MDC is to encode a video into multiple frames called descriptions, supporting multiple stages of decoding and when compared to layered coding techniques, each description can be decoded separately to estimate the model. In this proposed work we improve the error resilience and user heterogeneity using new multiple description coding method called Hybrid Interpolation Scalable MDC (HIS-MDC) technique and using both spatial and temporal interpolation schemes corrupted portions of a frame are concealed. Experimental results show that, proposed approach attains reasonable quality of video performance compared with previous multiple description methods.

Keywords: interpolation, peer-to-peer, wireless network, multiple description coding, video streaming.

1 Introduction

Streaming is a technology used for providing multimedia content distribution between clients on wireless network. With this technology, the client can playback the media content without waiting for the entire media file to arrive. Contrast with conventional data communication, delivery of multimedia data has more severe impact on network bandwidth, delay and loss. P2P is a term that normally refers to a file distribution

V.V. Das, J. Stephen, and Y. Chaba (Eds.): CNC 2011, CCIS 142, pp. 164–169, 2011.

network and supports a decentralized communication model. It does not include any central server and all computers in the network serve as both server and client. Peer-to-peer (P2P) systems are more popular due to their capability to hand over large amounts of multimedia data at a reduced deployment cost.

Multiple description coding (MDC) [1] is developed to deal with data transmission over error-prone networks. MDC reduces the adverse effect of packet losses by transmitting different descriptions along different paths and is useful to improve the continuity of service in P2P live video streaming systems. Interpolation is defined as approximating a missing block from a previous frame. Spatial interpolation is defined as a standard of missing pixels, approximated from the nearby pixels of a similar frame. Temporal interpolation is based on the equivalent areas of the reference frames. In temporal error concealment, association between current decoded frame and previous decoded frames is exploited. Hybrid Interpolation is applied when macro block mode information is reliable and concealed neighbors decide which concealment method to use whether spatial or temporal neighbors. Rest of the paper is organized as, Section 2 reveals related works on MDC, Section 3 explains proposed system, Section 4 shows possible system scenario and results and last Section 5 concludes the paper.

2 Related Works on MDC

One challenging problem for video communications over wireless networks is to provide error resilience for reliable communications. There are different techniques have been proposed to improve the error robustness of video communications over lossy networks. One such algorithm is Distributed Video Application (DIVA) [2], which aims at creating two balanced descriptions each of which is an H.264/AVC compliant bit stream. It uses the concepts of primary and redundant slices, creating high-quality primary slices and low-quality redundant slices. Primary and Redundant slices are interleaved to form two balanced descriptions. At the decoder, if both descriptions of a given slice are received, only the primary representations are retained and decoded. The integration of the DIVA method into the Scalable Video Coding (SVC) base layer, leads to so-called S-DIVA technique (DIVA for SVC) [3]. In S-DIVA results are determined through rate distortion optimization, taking into account both the rate repartition among layers and the redundancy allocated to MDC.

Usual concealment techniques such as Recursive Optimal per-Pixel Estimate (ROPE) [4] algorithm allows the encoder to estimate the pixel-by-pixel predictable distortion of the decoded video series appropriate to channel failure. This algorithm needs an input for approximating the packet deficit rate and this information is used for error concealment, employed at the decoder to decode frames. A Slepian-Wolf based inter frame transcoding (SWIFT) algorithm [5] transcodes an inter-coded P-blocks to a new type of block called X-block for the purpose of error robustness. The X-block can be lossless transcoded back to the exact original P-block when there is no transmission error, and can also be decoded robustly like an I-block when there are errors in the predicted block. Wavelet based Wyner-Ziv [6] video codec uses intra-frame encoding and inter-frame decoding. This type of codec is useful for systems which require simple encoders but have to handle more complex decoders. The encoder is composed of a scalar quantifier and a rate compatible turbo encoder. The decoder performs turbo decoding using an interpolated frame as side information. In this paper, we propose a new MDC technique called HIS-MDC and use it for video streaming in P2P wireless networks. Based on the multiple state video coding, two streams of lower-resolution pictures are added to enhance the error resilience and in order to improve the video quality when a description stream is missing spatial temporal hybrid interpolation scheme is exploited. This feature allows the decoder to conceal the corrupted frames more effectively and provide better video quality.

3 HIS-MDC Architecture

Figure 1 shows the architecture for the proposed multiple description coding technique. The architecture contains encoder and a decoder part. A video clip is fragmented into even and odd frames, even frames are down sampled to one-eighth and a weight is calculated using the two frames in encoder and transmitted. Down sampling is done, so as to increase the transfer speed among peers by decreasing the size of the frame. $E_{oneeighth}$ and O_{full} are transmitted independently. In decoder, the proposed MDC technique HIS-MDC is used to improve the video quality when some portions of frame are corrupted, by using both spatial and temporal information. HIS-MDC uses the concept of multiple state video coding, where the original video sequence is separated into even and odd streams, each encoded and transmitted independently.

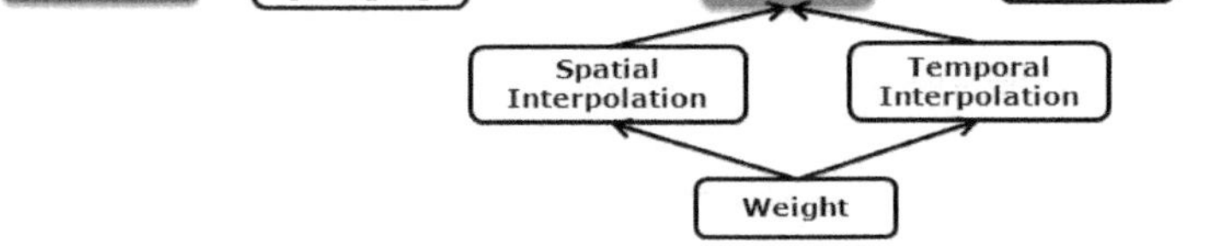

Fig. 1. HIS-MDC Architecture

Basically each description can reproduce the video at a half frame rate. When both descriptions are received, a full rate video can be played.

3.1 Encoder and Decoder

In the encoder (Figure 2a) odd frames are encoded normally while even frames are down-sampled to one-eighth size before encoding i.e. pre-processing stage and a weight is calculated using combination of down sampled even and full odd frames. In the decoder (Figure 2b), a better video quality is obtained, if both descriptions are received and there is no need to construct interpolated frames. If some portions of frame are corrupted then we apply interpolation techniques to conceal it.

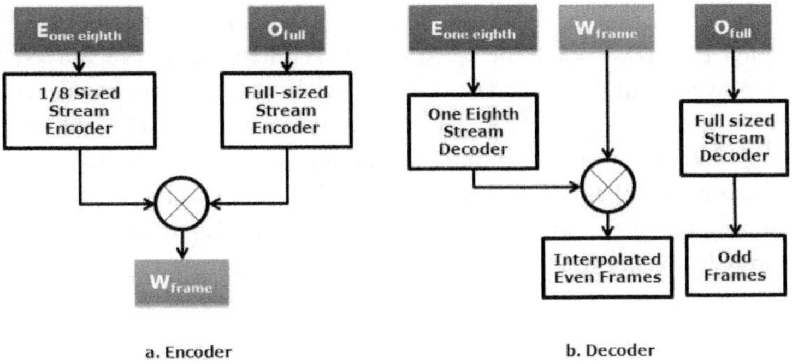

Fig. 2. Encode and Decoders

3.2 Spatial and Temporal Interpolation

Spatial interpolation is a process of approximating the rate properties at a sampled position inside the area enclosed by existing observations. In spatial interpolation, missing pixels are predicted as of surrounding pixels of the identical frame i.e. using the neighboring pixels. Temporal Interpolation is based on the resulting areas of the reference frames. Steps involved in temporal interpolation are i) first, the even frames are down sampled to one-eighth size, ii) then, by using the down-sampled even frame as the reference frame, motion vectors are obtained each 8x8 block in the one-eighth sized even frame and iii) finally, by passing the corresponding 64x64 blocks of the odd frame pointed to by motion vectors, concealed temporally interpolated frame is obtained. Figure 3, depicts the operation of hybrid interpolation where both spatial and temporal interpolations are used to conceal the corrupted portions of a frame. The decoded O_{full} and $E_{oneeighth}$ are used to produce the temporal and spatial interpolated frame, and the missing frame is replaced by a linear combination of the two interpolated frames with the weight determined at the encoder.

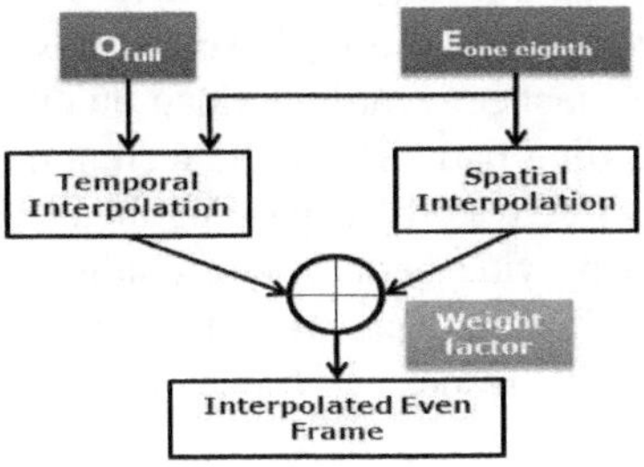

Fig. 3. Hybrid Interpolation

4 System Scenario

Figure 4, illustrates a possible P2P networking for multiple streaming using the proposed framework. The number of circles indicates roughly the amount of resources available in a network. In turn, N_2, N_3, N_6 shares video file (V_1) and displays it to all nodes in a network. For example, if N_8 wants to download or stream the video file, then N_2 sends $E_{oneeighth} + O_{full}$ due to most powerful machine, N3 may send $E_{oneeighth}$ due to less powerful machine and N_6 may send all. N_8 can receive all video streams from N_2, N_3, N_6 i.e. $E_{oneeighth} + O_{full}$. If a received frame is corrupted, then it is concealed using the proposed interpolation technique, a better quality of video is obtained. Compared to other MDC designs, HIS-MDC offers superior flexibility for P2P video streaming as all the streams are independently encoded and a node may deliver suitable combinations according to the peer's capabilities.

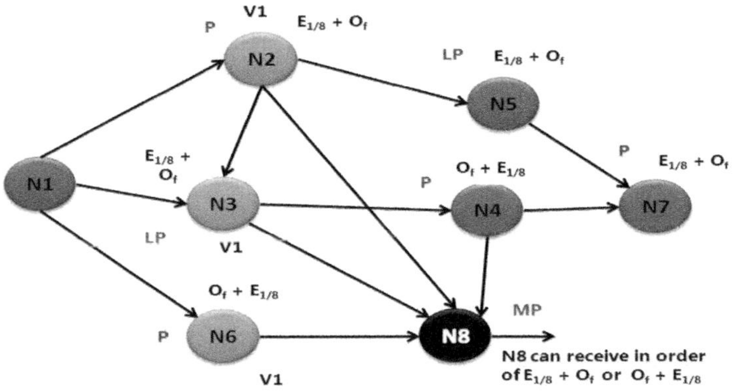

Fig. 4. System Scenario

4.1 Experimental Results

The parameters used for analyzing the proposed framework are: i) PSNR vs Frame rate, ii) Computational complexity, iii) Video Quality. DIVA algorithm offers scalability, but does not adopts any error concealment methods, ROPE exhibits high computational costs and long encoding time, whereas SWIFT method improves the error resilience capability for off-line compressed videos, and Wyner-Ziv coding does not reach the desired performance regarding video quality for inter-frame but outperforms in intra-frames encoding part. HIS-MDC technique allows the decoder to conceal the lost frames more effectively by exploiting the combination of both spatial and temporal information and provide better video quality. The Figure 5 compares the PSNR value obtained for HIS-MDC and with other various MDC methods. There is a linear increase in PSNR with respect to frame rate for HIS-MDC technique when compared to other techniques.

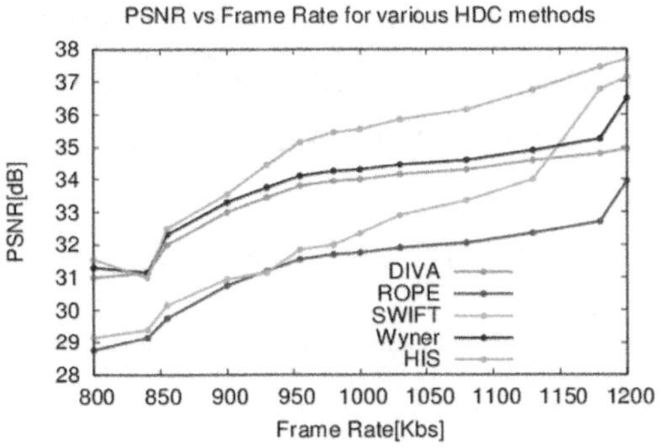

Fig. 5. Comparison between HIS-MDC and Other methods

5 Conclusion

In this paper, we proposed a new MDC technique called HIS-MDC which can be helpful in peer-to-peer video streaming. HIS-MDC makes the system resilient to data loss and adaptive to network heterogeneity. The experimental results indicate that video streaming in wireless P2P network with new multiple description coding technique is a disputing approach to adaptive provisioning of video broadcast on heterogeneous networks and performance improved when compared with previous multiple description coding.

References

1. Goyal, V.K.: Multiple description coding: compression meets the network. IEEE Signal Processing Magazine 18(5), 74–93 (2001), doi:10.1109/79.952806
2. Zezza, S., Magli, E., Olmo, G., Grangetto, M.: Seacast, A protocol for peer-to-peer video streaming supporting multiple description coding. In: Multimedia and Expo. ICME 2009, pp. 1586–1587 (2009), doi:10.1109/ICME.2009.5202819
3. `http://www.ist-sea.eu/Public/SEAD3.4.1POLFF20090515.pdf`
4. Liao, Y., Gibson, J.D.: Enhanced error resilience of video communications for burst losses using an extended ROPE algorithm. In: IEEE International Conference on Acoustics, Speech and Signal Processing, pp. 1853–1856 (2009)
5. Fan, X., et al.: On improving the robustness of compressed video by Slepian-Wolf based lossless transcoding. In: IEEE International Symposium on Circuits and Systems, ISCAS 2009, pp. 884–887 (2009)
6. Li, Z., Fang, S., Huang, G., Bai, L.: A New Framework of Scalable Video Compression with High Error Resilience Based on Wavelet Using Wyner-Ziv Technique. IEEE Multimedia Information Networking and Security, 218–221 (2009), doi:10.1109/MINES.2009.71

A Novel Framework for Target Tracking and Data Fusion in Wireless Sensor Networks Using Kernel Based Learning Algorithm

B. Kalpana and R. Sangeetha

Department of Computer Science, Avinashilingam University for Women, Coimbatore
kalpanabsekar@yahoo.com, sangeethadj@gmail.com

Abstract. In recent years, a number of powerful kernel based learning algorithms like Support Vector machines, Kernel Fisher Discriminant Analysis, Kernel Principal Component Analysis have reported their success in various domains like Signal Processing, Image Processing and Pattern Classification. In a nutshell, a kernel-based algorithm is a nonlinear version of a linear algorithm where the data has been nonlinearly transformed to a higher dimensional space, in which we need to compute the inner products via a kernel function as in [1]. The attractiveness of such algorithm stems from their elegant treatment of nonlinear problems and their efficacy in high-dimensional space. In this paper, we illustrate a comprehensive review of kernel methods and its usefulness in signal processing problems like target detection and target tracking.

Keywords: Support Vector Machine, Kernel Function, Target Tracking, Target Detection.

1 Introduction

Wireless sensor network is being composed of a large number of nodes which are deployed compactly in close proximity to the phenomenon which is to be monitored. Each of these nodes collects the data, process it, stores and transfers information from one node to another. Sensor nodes are fitted with an on-board processor. Instead of sending the raw data to the nodes responsible for the fusion, sensor nodes use their processing abilities to locally carry out simple computations and transmit only the required and partially processed data. Factors influencing [2] sensor network are Fault Tolerance, Transmission Media, Environment, Scalability, Power Consumption etc.

The rest of this paper is organized as follows. Section II discusses the Target Detection, Tracking and Data Fusion. Section III explicates the proposed framework using kernel based target tracking, and finally conclusion in Section IV.

2 Target Detection, Tracking and Data Fusion

A fundamental problem addressed by the sensor network applications is target detection in the field of interest. This problem has two primary levels of difficulties (i)

V.V. Das, J. Stephen, and Y. Chaba (Eds.): CNC 2011, CCIS 142, pp. 170–174, 2011.

Single target detection and (ii) ***Multiple target detection***. The single target detection problem can be solved using statistical signal processing methods and the multiple target detection problems are very difficult to solve. The detector effectively computes a running average of the signal power over a window of pre-specified length. For target localization, the detector output at any desired instant within the offset and onset times may also be communicated. All nodes in sensor network execute in same logic as ***initialization, track information, parameter estimation, detections, threshold, future track, track update*** and lastly ***obtaining the result***.

In Target tracking, selecting only a proper group of sensors from the available set to perform tracking is of research value [3, 4]. Target tracking is often complicated by the measurement noise. The noise must be "filtered" out in order to predict the true path of a moving target. Data-filtering target trackers are utilized in many situations to provide an estimate of the position and velocity of a target at the time of measurement.Kalman Filter [5] and its derivatives give better result in target tracking.

Data fusion in [6] seeks to combine information, about a state x, from a variety of sources. The mechanism for combining prior information with observed and communicated information is Bayes theorem i.e.

$$P(x/z) = \frac{P(z/x)P(x)}{P(z)} \tag{1}$$

The value of this theorem lies in the interpretation of the probability density functions P(x / z), P (z / x) and P(x).

3 Proposed Framework Using Kernel Based Learning Algorithm

Figure 1 presents a proposed framework for tracking and fusion system along with its relation to the central sensor system. There are some steps involved in the proposed framework .They are as follows

1. Signal received in the waveform from the sensor node
2. Wave passes through the detector device and data rate reduction takes place
3. Signal processing unit provides estimates of signal parameters
4. Sensor Performance, Target characteristics and Target environment of the signal are received from the sensing hardware (e.g. radar antennas).
5. From these preprocessed estimates, sensor reports are formed, i.e. measured quantities possibly related to the target of interest
6. Then measured quantities possibly related to the target of interest are the input information for the tracking system.
7. In the tracking system itself we have all sensor data, in which the relevant data can be associated to the existing tracks and used for track maintenance (prediction, filtering and retrodiction).
8. The remaining non-associated data are processed in order to establish new tentative tracks (track initiation, multiple frame track extraction).
9. Evidently, a priori knowledge in terms of statistical models of the sensor performance, Target characteristics (including their dynamical behaviour), and the Target environment is prerequisite to both track maintenance and track initiation.

10. Track confirmation/termination, Target classification/identification, and fusion of tracks representing identical information is performed in the track processing unit.
11. The Framework of a tracking and fusion system is completed by a interface with displaying and interaction functions.
12. The available information on the sensor, the Target of interest, and the environment can be specified, updated, or corrected by direct human interaction as well as by the track processor itself, e.g. as a consequence of a successful Target classification.

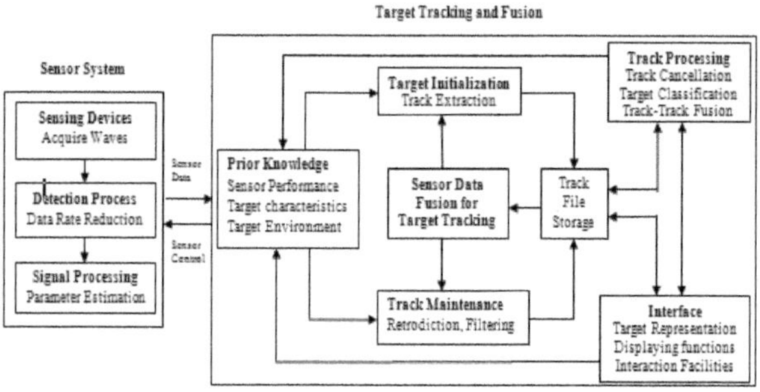

Fig. 1. Novel Framework for target tracking and Fusion

3.1 Kernel Based Target Tracking [9, 10]

Generally kernel functions are classified as ***Translation invariant*** and ***Rotation Invariant***. Our aim is to make the kernel generalized for every domain. In paper [7, 8], authors explained about the usage of hybrid kernel in target classification. It works as an efficient classifier while comparing with standard kernel classifier and gives high accuracy for classification.

Kernel based Posterior Estimation

The state and observation value of target is represented respectively as x_t and z_t at discrete time t where, xt $\in \Re nx$, zt $\in \Re n_y$. Let $z_t = \{z_0 \ldots\ldots z_t\}$ be the history of observations up to time t. In recursive Bayesian estimation, the Posterior Probability Density Function is estimated through propagating the PDF as time:

$$P(x_t / Z_t)\alpha \in P(z_t \mid x_t)P(x_t \mid Z_{t-1}) \tag{2}$$

where
$$P(x_t / Z_{t-1}) = \int P(x_t \mid x_{t-1})P(x_{t-1} \mid Z_{t-1})dx_{t-1} \tag{3}$$

A particle set $S_t = \{S_t^n\}_{n=1}^N$ is given at time *t* and weights $\{w_t^n\}_{n=1}^N$ is associated. Then, the kernel density estimation of the posterior with kernel K can be drawn up as:

$$\hat{P}(x_t / Z_t) = \sum_{n=1}^{N} K_\lambda (x_t - s_t^n) w_t^n \tag{4}$$

where λ is the kernel width : $K_\lambda(x_t - s_t^n) = \frac{1}{\lambda^{n_x}} K(\frac{x_t - s_t^n}{\lambda})$ (5)

Kernel functions and window size are determined by the rules i.e. mean square error is minimum, which combined the posterior probability density distribution function with the approximate nuclear density estimation.

Posterior Gradient Estimation

Posterior estimation is given in equation (4). The author in [9] estimates its gradient and move particles to the modes of the posterior along the gradient direction. It can be achieved using the mean shift procedure. In this procedure, each particle is moved to its sample mean determined by

$$m(s_t^n) = \frac{\sum_{l=1}^{N} G_\lambda(s_t^n - s_t^l) w_t^l s_t^l}{\sum_{l=1}^{N} G_\lambda(s_t^n - s_t^l) w_t^l} \quad (6)$$

It is shown that the mean shift vector m(x) -x using kernel G would be in the gradient direction of equation (9) if the kernel profiles satisfy g(λ) = −ck′(r) for all r Є [0,∞) and some c Є [0,∞)

Particle Re-weighting

The mean shift from [9] can be applied repeatedly to a particle set. A problem arises when particles change their positions: the new particles do not follow the posterior distribution anymore. This is compensated in KPF by re-weighting the particles. Denote the particle set after the i-th mean shift procedure at time t as $\{S_{t,i}^{(.)}\}$. After each mean shift procedure, the weight is re-computed as the posterior density evaluated at the new particle positions augmented with a particle density balancing factor

$$w_{t,i}^n = \frac{p(s_{t,i}^n \mid Z_t)}{q_{t,i}(s_{t,i}^n)} \quad (7)$$

where the denominator is the new proposal density that captures the non-uniformity of the new particle set and the posterior density is given by (2):

$$q_{t,i}(x_t) = \sum_{l=1}^{N} K_\lambda(x_t - s_{t,i}^n) \quad (8)$$

$$P(s_t^n / Z_t) \alpha \in P(z_t \mid s_t^n) . \sum_{l=1}^{N} p(s_t^n \mid s_{t-1}^l) w_{t-1}^l \quad (9)$$

The second item on the right side of (9) is a sample based approximation of the prior density. When mean shift is executed on St, the next generation of particles will focus more on the approximated density modes, but the weight will contain a factor that compensates the effect. In kernel-based detection algorithms the data is implicitly mapped into a high dimensional kernel feature space by a nonlinear mapping which is associated with a kernel function. The detection algorithm is then derived in the feature space which is *kernelized* in terms of the kernel functions in order to avoid explicit computation in the high dimensional feature space.

4 Conclusion

In this paper, we have presented a kernel-based learning approach to the Target detection in wireless sensor networks. The flexibility, fault tolerance, high sensing fidelity, low-cost and rapid deployment characteristics of sensor networks create many new and exciting application areas for remote sensing.. Discrete kernel functions exploit the ordinal nature of local sensor decisions and compute appropriate smoothing parameters, this approach resulted in Bayes risk consistent fusion rules which provide optimal detection performance in the asymptotic case. In Future, the proposed methodology can be applied to the real time applications using hybridized kernels. Also, it can be used for multiple target tracking and classification.

References

1. Müller, K.-R., Mika, S., Rätsch, G., Tsuda, K., Schölkopf, B.: An Introduction to Kernel-Based Learning Algorithms. IEEE Transaction on Neural Networks (2001)
2. Papageorgiou, P.: Literature Survey on Wireless Sensor Networks (2003)
3. Tharmarasa, R., Kirubarajan, T., Hernandez, M.L.: Large-Scale Optimal Sensor Array Management for Multitarget Tracking. IEEE Trans Systems, Man and Cybernetics (2007)
4. Soe, K.T.: An energy-efficient target tracking system in wireless sensor networks. In: ICCA 2008, Yangon, Myanmar, Februarysa 14-15 (2008)
5. Smets, P., Ristic, B.: Kalman Filter and Joint Tracking and Classification based on Belief Functions in the TBM Framework (2005)
6. Zhu, A.F., Jing, Z.R.: Multisensor information fusion for target tracking. In: Jisuanji Gongcheng yu Yingyong (Computer Engineering and Applications) (2009)
7. Sangeetha, R., Kalpana, B.: A comparative study and choice of an appropriate kernel for support vector machines. In: Das, V.V., Vijaykumar, R. (eds.) ICT 2010. CCIS, vol. 101, pp. 549–553. Springer, Heidelberg (2010)
8. Sangeetha, R., Kalpana, B.: Optimizing the Kernel Selection for Support Vector Machines using Performance Measures. In: A2CWiC 2010 (2010), ISBN: 978-1-4503-0194-7
9. Chu, H., Wang, K.: Research of Kernel Particle Filtering Target Tracking Algorithm Based on Multi-feature Fusion. In: World Congress on Intelligent Control and Automation (2010)
10. Xuena, Q., Shirong, L., Fei, L.: Kernel-based Target Tracking with Multiple Features Fusion. In: IEEE Conference on Decision and Control and 28th Chinese Control Conference (2009)

Low Power Lapped Bi-orthogonal Transform (LBT) for FPGA's

P. Deepa[1] and C. Vasanthanayaki[2]

[1] Research Scholar
Department of Electronics and Communication Engineering,
Government College of Technology, Coimbatore, Tamil Nadu, India
[2] Associate Professor
Department of Electronics and Communication Engineering,
Government College of Technology, Coimbatore, Tamil Nadu, India
deepap05@gmail.com, vasanthi@gct.ac.in

Abstract. The power reduction is the crucial part in hardware architecture especially in the case of battery driven applications. In this paper, a low power and low area Lapped Biorthogonal Transform (LBT) for Field Programmable Gate Arrays (FPGAs) has been proposed based on flow graph algorithm (FGA) which significantly reduces the power and area. The proposed LBT is implemented in Cylcone III Nios II embedded evaluation kit using Quartus II tool. The simulation and synthesis result reveals that the proposed LBT significantly exhibits an improved performance. Hence it is suitable for low power, low area and high speed applications.

Keywords: Low power, Lapped Biorthogonal Transform, Discrete Cosine Transform, Area.

1 Introduction

With the exponential growth of the performance and power of integrated circuits, power consumption has become one of the most crucial restraining factors in the Integrated Circuit (IC) design flow. It is generally accepted that most power savings are achieved at the algorithmic and architectural levels since there are wider degrees of design freedom [1]. Applying low power techniques at the logic, circuit and physical levels also gains importance but in general depend on the technology available with which the optimized architecture is to be implemented. Therefore initial work on one-dimensional (1D) LBT has its focus on optimizing the number of basic operations such as architectural/algorithmic level additions/ subtractions, multiplications and processing stages.

The first useful LT (Lapped Transform) with a well defined factorization was the Lapped Orthogonal Transform (LOT). Later LBT has been proposed as an improvement over LOT with increase in peak signal to noise ratio (PSNR) [2]. FGA is the most popular way of realization in consideration of very large scale integration

V.V. Das, J. Stephen, and Y. Chaba (Eds.): CNC 2011, CCIS 142, pp. 175–181, 2011.

(VLSI) implementation [3]. To implement LBT for FPGAs binDCT has been considered which approximates multiplications with add and shift operations [4]. Although binDCT reduces the computational complexity significantly, PSNR is reduced about 3 dB as compared to other DCTs like Loeffler DCT [5]. However, the multiplications consume about 40% of the power and almost 45% of the total area [5]. Hence the low complexity based FGA has been considered for implementing LBT which basically requires 11 multiplications and 29 addition operations [6]. The low power and high performance LBT has been developed and it also focuses on architectural level approaches.

The rest of the paper is organised as follows: Section 2 provides an overview of LBT and lift based LBT. In Section 3 the proposed LBT is presented. Results and discussions are presented in section 4. Finally in section 5 the paper is concluded.

2 Lifting-Based Lapped BiOrthogonal Transform

LBT can be viewed as a combination of the common block based Discrete Cosine Transform (DCT)/Inverse DCT with a simple time domain pre- and post- filtering. Both classes of algorithms share one common goal: to eliminate or reduce the severity of coding artifacts in the reconstructed signal. By analyzing the structures of these two classes of algorithms, time-domain pre-processing of DCT called LOT is used in this paper which is attractive for distributed implementation. The LOT is a larger improvement over the DCT for image compression and it reduces the blocking artifacts. But the blocking artifacts are not completely eliminated because of the format of the low frequency bases of LOT. To resolve this problem Malvar later proposed to modify the LOT, creating LBT [2].

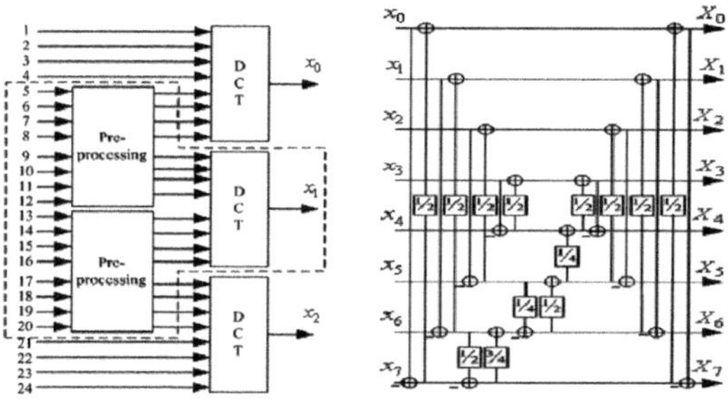

(a) LT (b) Pre-processing

Fig. 1. Implementation flow of 1D LT

DCT has more number of computations and consumes majority of power in the hardware architecture. The power reduction is a crucial part in the hardware architecture design. Hence it is considered that the LBT hardware architecture is compatible for power reduction. The structure depicted in Fig. 1a is the LT as pre-processor of DCT. Fig. 1b shows the structure of the binary LBT time domain pre-processing. The analysis polyphase matrix of type-II fast lapped orthogonal transform (LOT) can be written as:

$$E^{II}(z) = D_M C_M^{II} \begin{bmatrix} I & 0 \\ 0 & zI \end{bmatrix} \begin{bmatrix} 0 & I \\ I & 0 \end{bmatrix} P, \quad (1)$$

where

$$P = \frac{1}{2} \begin{bmatrix} I & J \\ J & -I \end{bmatrix} \begin{bmatrix} I & 0 \\ 0 & V \end{bmatrix} \begin{bmatrix} I & J \\ J & -I \end{bmatrix}$$

$$V = J (C_{M/2}^{II})^T C_{M/2}^{IV} J^{-1}$$

$$D_M = diag\{1, -1, 1, -1, ...\}$$

The matrix V controls pre-filtering and contains all of the degrees of freedom in the structure. By inserting an invertible diagonal matrix S between $(C_{M/2}^{II})^T$ and $C_{M/2}^{IV}$, V can be represented in the single-value decomposition form:

$$V = J (C_{M/2}^{II})^T S C_{M/2}^{IV} J^{-1}$$

S is limited to $diag\{s, 1, ..., 1\}$, where s is a scaling factor. I is $N \times N$ identity matrix. The DCT data sequence denoted as $x(n)$, $n = 0, 1, ..., N-1$, by $X(k), k = 0, 1, ..., N-1$. Equation 2 gives DCT expression,

$$X(k) = e(k) \sum_{n=0}^{N-1} x(n) \cos[\frac{(2n+1)\pi k}{2N}], k = 0, 1, ..., N-1 \quad (2)$$

where

$$e(k) = \begin{cases} \frac{1}{\sqrt{2}}, k = 0 \\ 1, otherwise \end{cases}$$

The direct implementation of DCT or IDCT requires N(N-1) multiplication operations, i.e., $O(N^2)$, which is very hardware expensive. Therefore, an algorithmic architecture has been proposed for LBT with less number of computations for low power.

3 Proposed LBT

The LBT structures considered in our case uses the pre-processor as in Fig. 3b. The efficient LBT implementation can be achieved by considering the FGA. For an 8-point LBT without pre-processing generally requires 56 multiplications which can be reduced to 22 multiplications for basic LBT to 13 multiplications for fast LBT and to 11 multiplications for the proposed LBT for power reduction.

3.1 Basic LBT

Consider an 8-point DCT and it can be written in the matrix form and for an 8-point DCT the algorithmic architecture mapping can be done in three steps.

Step1: Apply trigonometric properties, and form FGA with 22 multiplications [7]. Fig. 2(a) shows the equivalent hardware architecture design for the LBT. The addition, subtraction, multiplication, D-flip flop and look up table (LUT) in architecture perform its own operation based on the clock input. The block B (acts as a buffer), the three inputs are named as A, B, C and the clock assigned to each inputs is CLKA, CLKB, CLKC and an extra clock for processing B is CLKP and Fig. 2(b) depicts the structure of block B. In the structure 2(b), when the input A is possible the CLKA is enabled, similarly for inputs B and C, the CLKB and CLKC are enabled. When any of the inputs are feasible to the unit, clock (CLKP) will be in enable condition for passing the data from input to output. Thus the power consumed by this unit will be saved by local clock.

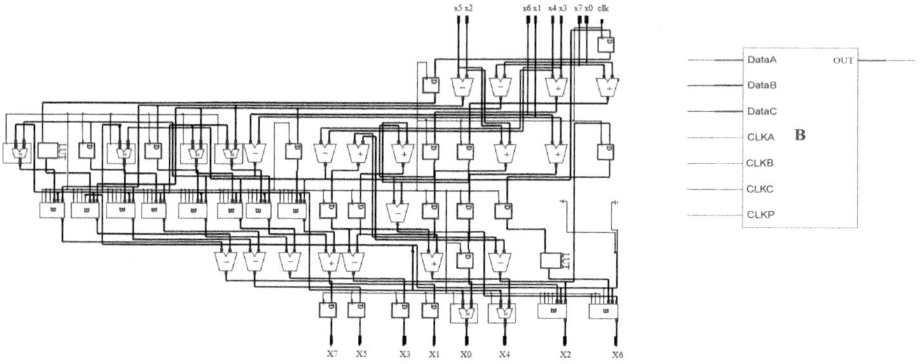

(a) (b)

Fig. 2. First step 8-point LBT (a) Equivalent Hardware architecture for basic LBT (b) Structure of Block B

Step 2: The LBT structure from Fig. 2 is grouped into different functional units represented by blocks and then the whole LBT structure is transformed into a block diagram.
Step 3: Reduced complexity implementations of the various blocks are exploited which leads to a fast 8-point LBT.

3.2 Fast LBT (Decimation in Frequency) and Proposed LBT

Fast 8-point LBT with decimation in frequency approach has been considered for the implementation of LBT. This algorithm reduces the number of multiplications to about $(N/2)\log_2 N$ by power-of-2 decomposition. Fig. 3(a) illustrates the 8-point fast LBT hardware architecture. This structure requires only 13 multiplications which by design reduces the power consumed by the architecture.

The proposed LBT has been considered for implementation which reduces the number of computations. It is proved the number of multiplications has been reduced from 13 to 11 [6]. Fig. 3b shows the hardware architecture with reduced number of multiplications and also the block B in it saves the power for FPGAs. Table 1 shows the comparison of computational complexity for three LBT in terms of multiplication. Since the number of multiplications for the proposed LBT is 80.36% less compared to the LBT of equation 2 without pre-processing hence, the power consumed by the architecture will also be reduced. The power analysis of all the three LBT structures has been discussed in the next section

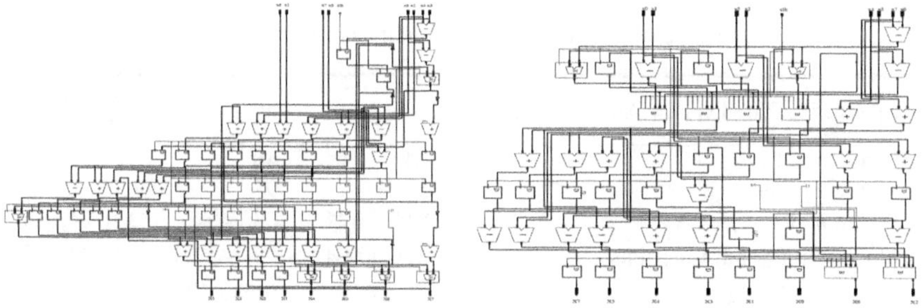

(a) (b)

Fig. 3. Eight-point LBT Equivalent hardware architecture (a) Fast LBT (b) proposed LBT

Table 1. Comparison of computational complexity for the proposed LBT

Type\Operation	Order of Multiplication	No. of Multiplications	% of reduction
Basic LBT	$\mathcal{O}[N^2/2-(N+2)]$	22	60.71 %
Fast LBT	$\mathcal{O}[(N/2)\log_2 N]$	13	76.76%
Proposed LBT	$\mathcal{O}[((N/2)\log_2 N)-1]$	11	80.36%

4 Results and Discussions

The three architectures i) Basic LBT ii) Fast LBT iii) Proposed LBT are implemented using Intel core i3 processor, with a processor speed of 4GB RAM using Altera Quartus II 9.1 in Nios II Embedded Evaluation kit of Cyclone III edition. The operating conditions set for all LBT are VCCINT voltage of 1.2V, junction temperature of 15^0 Celsius and the Toggle rate used for input signals is 80% and default toggle rate of 20% for the remaining signals.

4.1 Area, Power and Time Utilization

After synthesis of three LBT, the proposed LBT put aside 16.67% and 19.7% resources (on an average) as with others. Also the PowerPlay suite of power analysis tool shows that the power dissipation for the proposed LBT is 0.02% and 0.03% low as compared to the basic LBT and fast LBT for N=8. Since the number of computations in proposed LBT is less, the dynamic and static power in the design gets reduced. Thus the implementation of the proposed LBT for FPGAs consumes lesser power. Table 2 consolidates the resource and power (mW) utilization for all the three LBT architectures.

The Quartus II Classic Timing Analyzer uses Altera-specific Setup time (t_{SU}), Clock-to-output time (t_{CO}), and Hold time (t_H) timing constraints. Table 3 shows the quantitative results of the slack and the actual time. The time complexity between the proposed LBT, fast LBT and basic LBT are first quantitatively compared. In the proposed LBT the slack is relatively reduced by 2.36% and 1.25% as compared to basic and fast LBT for the setup time. Thus results depicted in Table 3 justifies that the proposed LBT structure is best suitable for implementing with a high speed.

Table 2. Resource utilization and Power analysis for Three Architectures

Device utilization\ Type of Architecture	Basic LBT	Fast LBT	Proposed LBT
Estimated total Les	424	248	229
Device resources	557	393	358
Logic elements	381/24,624	208/24,624	201/24,624
Total Combinational functions	381/24,624	202/24,624	186/24,624
Dedicated logic registers	64/24,624	60/24,624	58/24,624
Total registers	128	98	88
Total Pins	129/216	129/216	129/216
Average fan-out	1.98	1.65	1.59
Parameters/Hardware structure	Basic LBT	Fast LBT	Proposed LBT
Thermal power Dissipation	103.12	102.95	102.92
Core Dynamic Thermal Power Dissipation	0.05	0.03	0.01
Core Static Thermal power Dissipation	80.98	80.98	80.97
I/O Thermal Power Dissipation	22.09	21.99	21.94
Total thermal power estimate for the design	103.12	102.94	102.92
Average toggle rate (millions of transitions/s)	0.100	0.018	0.018

Table 3. Timing for Three LBT Structures

Type	Parameters	Worst case t_{SU}	Worst case t_{CO}	Worst case t_H	Worst case Minimum t_{CO}
Basic LBT	Slack	92.92	19.03	19.11	4.87
	Required Time (ns)	100	25	20	1
	Actual Time (ns)	7.08	7.22	0.15	5.87
FAST LBT	Slack (ns)	91.88	12.77	12.1	4.87
	Required Time (ns)	100	25	20	1
	Actual Time (ns)	8.12	7.23	0.90	5.87
Proposed LBT	Slack (ns)	90.73	12.78	19.41	4.87
	Required Time (ns)	100	20	20	1
	Actual Time(ns)	7.27	7.22	0.59	5.87

4.2 Simulation Results

The result has been taken for number of standard test images. Only a sample of two test images with PSNR and compression ratio (CR) is shown in Table 5. Fig. 7 depicts the reconstructed lena image by LBT for bits per pixel (bpp) of 0.25. The proposed LBT algorithm also reduces the blocking artifacts in reconstructed images at low bit

Table 5. Performance of the proposed LBT

Bit rate in Bpp	Lena image (512x512)		Barbara image (512x512)	
	CR	PSNR (dB)	CR	PSNR (dB)
1	8:1	39.98	8:1	36.54
0.5	16:1	36.71	16:1	31.95
0.25	32:1	33.83	32:1	29.01

PSNR: 33.83 dB

(a) (b)

Fig. 7. (a) Input Lena image (b) Reconstructed image using LBT at bpp=0.25

rates which can be observed from Fig. 7. This significant reduction is because of the lapped property of the transform.

5 Conclusion

In this paper, synthesis of the proposed LBT for FPGAs has been analysed in benchmark to basic and fast LBT. It is observed that the proposed LBT consumes low power and area. The reduction in power is mainly because of the significant reduction in the number of multiplications. Thus the proposed low power LBT can also be extended for battery driven applications.

Acknowledgment

The authors would like to be grateful for the financial support provided by the Ministry of Science and Technology, under Women Scientist Scheme-A India, for this research.(Ref. File No.: SR/WOS-A/ET-25/2009).

References

1. Ruby, W.: Low Power To The People. EDAVision Magazine (2002)
2. Malvar, H.S.: Biorthogonal and nonuniform lapped transforms for transform coding with reduced blocking and ringing artifacts. IEEE Trans. on Signal Processing 46, 1043–1053 (1998)
3. Chen, W.H., Smith, C., Fralick, S.: A Fast Computational Algorithm for the Discrete Cosine Transform. IEEE Trans. Commun. 25, 1004–1009 (1977)
4. Liang, J., Trac, T.D.: Fast multiplierless approximations of the DCT with the lifting scheme. IEEE Trans. Acoust., Speech, Signal Processing 49, 3032–3044 (2001)
5. Sung, C.C., Ruan, S.J., Lin, B.Y., Shie, M.C.: Quality and Power Efficient Architecture for the Discrete Cosine Transform. IEICE Transactions on Fundamentals Special Section on VLSI Design and CAD Algorithms (2005)
6. Loeffler, C., Lightenberg, A., Moschytz, G.S.: Practical fast 1-D DCT algorithms with 11-multiplications. In: Proc. ICASSP, vol. 2, pp. 988–991 (1989)
7. Parhi, K.K.: VLSI Digital Signal Processing Systems. A Wiley-Interscience Publication, John Wiley & Sons Inc. (1999)

CASE-JigDFS: Context-Aware, Smart and Efficient JigDFS Network

Anand Gupta, Ankit Kumar, and Hemant Malik

Department of Instrumentation and Control,
Netaji Subhas Institute of Technology,
Delhi University, New Delhi, India
omaranand@nsitonline.in, ankit5990@gmail.com, malikhem@gmail.com

Abstract. Jigsaw Distributed File System (JigDFS) [3] is a secure P2P distributed file system. However, observations during implementation conveyed that JigDFS [3] has certain limitations. It [3] deals with all the files in a preprogrammed manner (which is rather obsolete) and takes impractically high file encoding time. This paper introduces "CASE-JigDFS", an improved JigDFS [3] network that is smart enough to distinguish between file attributes so as to dynamically manipulate data encryption and IDA parameters to intelligently utilize system's resources. The optimized network is best suited for mobile network devices viz. Smart phones and PDAs. The results have been arrived at, by comparing processing time vs. file size plots for different file types for 'JigDFS'[3] and 'CASE-JigDFS' networks.

Keywords: IDA (Information Dispersal Algorithm), JXTA, Buffer Slicing, CAFF (Context-Aware File Filter), Heap memory.

1 Introduction

Jigsaw Distributed File System (JigDFS) is a secure P2P Distributed File System (DFS). In P2P Distributed File Systems (DFS), the file is sliced into several pieces and stored on the network. It is implemented on JXTA and employs DHT to index resources and peers. It offers features like Data security, User Anonymity, Plausible Deniability and Fault Tolerance. To achieve these (features), techniques like Hashed key chain encryption, Information Dispersal Algorithm (IDA), and Cauchy Reed Solomon coding are used. Some related work on the use of the previously mentioned features is discussed below.

In the field of Peer-to-Peer File Distribution Systems earliest works involved CFS (2001) (Co-operative File Systems) [1]. Then, Pastis (2005) [2] was proposed, but it was prone to hacking. Colony FS (2009) [4] used Fragment Redundancy Scattering (FRS) and optimization algorithms but did not pay much attention to the security aspect. JigDFS (2009) [3] incorporated a high degree of security to p2p networks, but it was unable to customize its procedures according to altering file attributes to achieve an optimum performance. This feature prevented its [3] use in mobile devices. This served as a motivation for invoking some degree of smartness in the network.

V.V. Das, J. Stephen, and Y. Chaba (Eds.): CNC 2011, CCIS 142, pp. 182–186, 2011.

In order to describe the "CASE-JigDFS" in an appropriate manner, the rest of the paper is divided into three sections: Section 2 describes CASE-JigDFS as a whole, along with the additional features introduced in original JigDFS [3]. Section 3 demonstrates various experiments performed to testify an increase in performance offered by CASE-JigDFS over JigDFS [3]. Finally, Section 4 concludes the paper with possible future work.

2 CASE-JigDFS Network

JigDFS [3] was designed specifically for dealing with the computer networks and is not suited for mobile computing. To address this limitation a restructured version of JigDFS [3], CASE-JigDFS is introduced.

CASE-JigDFS being an improved version of JigDFS [3] network inherited all its [3] modules. The modules 'BufferSlicing' and 'Context-Aware File filter (CAFF)' were added to CASE-JigDFS, giving it an edge over the existing system [3].

Algorithm 1: BufferSlicing( )

Input: file Size
Output: number of buffer parts

If file size <limit-A
  Then number of buffer parts $\rightarrow$ 1
else If file size <= limit-A and >limit-B
  Then number of buffer parts $\rightarrow$10
else If file size >limit-B
  Then number of buffer parts $\rightarrow$ 20

Algorithm 2: CAFF( )

Input: file Object, file Size
Output: number of file slices, securitythreshold, IDAchunksize, encryption parameter

1. Determine file type
2. Compare file size with standard ranges
3. Set value for number of file slices
   Set value for security threshold
   Set value for IDA chunksize
   Set value for encryption parameter

The 'BufferSlicing module' utilizes the 'BufferSlicing' property which actually refers to the process of loading a file to the RAM (Random Access Memory) in a piecewise manner rather than as a single chunk (as in JigDFS [3]).This module can be explained by algorithm 1 above.

The other module namely 'Context-Aware File Filter (CAFF)' can be described as in algorithm 2 shown above. The file, before being uploaded to the network is first sent over to the CAFF which determines the file type to find out if the file is predominantly a document or a multimedia type.

Buffer Slicing and CAFF modules will be linked with the file distribution implementation of the original JigDFS [3] in the ensuing section.

CASE-JigDFS Network Implementation

CASE-JigDFS implements its procedures and protocols in a manner very similar to that of JigDFS [3]. The only difference lays in the fact that it employs two additional modules viz. 'BufferSlicing' and 'CAFF'.

The process of file saving in CASE-JigDFS can be explained as follows:

A block diagram for the complete file saving operation on the network is shown in Fig. 1. The shaded blocks are the ones that are added to the existing system [3] to optimize it.

The file being shared by the user is first loaded into the system memory using "Buffer Slicing" procedure. A framework of this is shown in Fig. 1(a). 'Buffer Slicing' enables the system to encode a file of any size without the memory and processing constraint hindrances. The file is next passed on to the 'Context-Aware File Filter' (CAFF). A framework for the CAFF is shown in Fig. 1(b). The CAFF( ) returns the optimal combination of IDA parameters for the file uploaded on the CASE-JigDFS network. A log file is also maintained on the uploading node which contains a list of files that are uploaded on the network. This list is used in generating corresponding IDA and decryption parameters when the file is retrieved from the network.

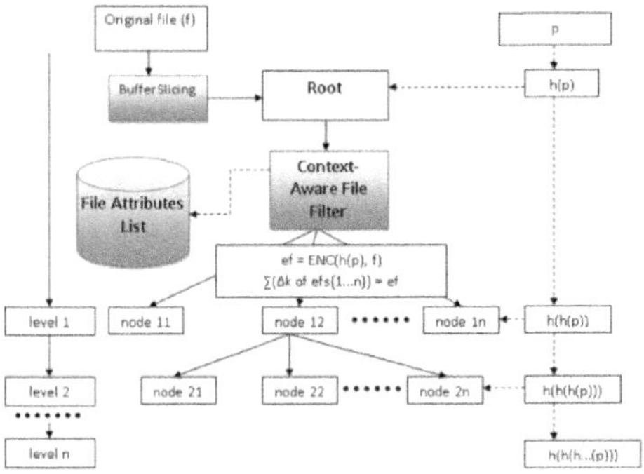

Fig. 1. File saving on CASE-JigDFS

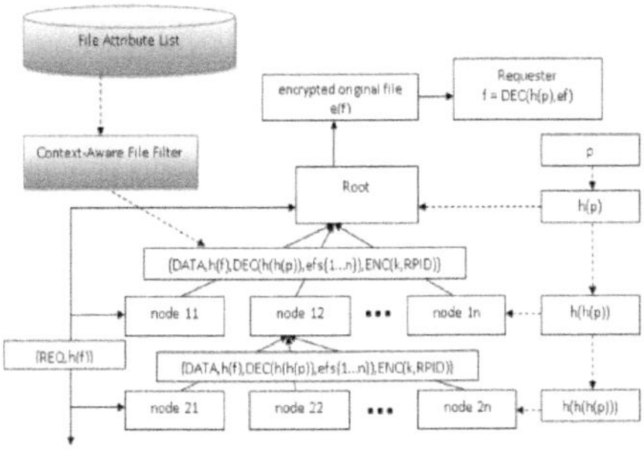

Fig. 2. File retrieval on CASE-JigDFS

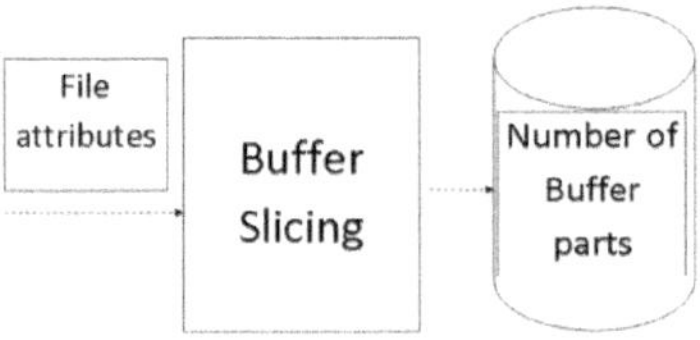

Fig. 1(a). BufferSlicing block

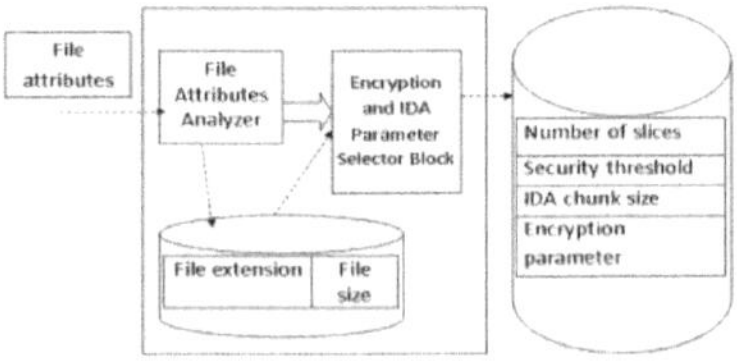

Fig. 2(b). Context-Aware File Filter

Rest of the procedures involving data encryption and distribution of the file slices in CASE-JigDFS is same as that of JigDFS [3] with only difference being that instead of using hardcoded IDA and Encryption parameters, the system now uses the parameters provided by the CAFF block (already explained in the previous section).

The file indexing in the CASE-JigDFS is being done exactly in the same manner as of the JigDFS [3] because it can be conveniently inherited in mobile computing too.

A framework for file retrieval in CASE-JigDFS is shown in Fig. 2. The shaded blocks are the ones that are added to existing system [3] to optimize it. The process of retrieving a file in the CASE-JigDFS network requires the user to know only the password for the file uploaded, similar to the original system [3], from which a corresponding hash value is generated. This hash value is checked with the list maintained by the system to determine the file name and its size. These attributes are hereafter used to compute the dynamically determined Encryption and IDA parameters that are used during file saving. This procedure ensures that even though the encoding of the file is accomplished dynamically to suit system performance, the same changes are reflected in the subsequent file retrieving too.

The rest of the decoding process in CASE-JigDFS network goes on in accordance with the JigDFS [3] network.

The next section provides experimental results that show a clear difference between performances of JigDFS [3] and CASE-JigDFS.

3 Test Results and Discussions

Both JigDFS [3] and CASE-JigDFS were tested on a system with the configuration:- Installed Memory (Ram): 4.00 GB DDR3, System Type: 64-bit Operating System, Processor: Intel®(R) Core™ i7 CPU Q720@ 1.60 GHz, Java: 1.6.0_11; JavaHotSpot(TM) Client VM 11.0-b16.

For testing, the code was run on NetBeansIDE 6.9.

JigDFS [3] and CASE-JigDFS systems are tested with document and media files with the file sizes varying from a few kilobytes to a few gigabytes, in order to analyze the total time taken by the two systems to process a particular file.

A Detailed comparative data for different Document File Sizes is depicted in Fig. 3. It can easily be discerned that, in comparison to the JigDFS [3], CASE-JigDFS takes considerably less encoding time for the file sizes less than 100 MB. Also, while the original code could measure only up to 100 MB(for the current testing system), the 'Buffer Slicing' block introduced in CASE-JigDFS allowed it to read files of much higher sizes (in Gigabytes too) theoretically.

The data comparison for media files is portrayed in Fig.4. After a detailed study of the graph we can infer that, for the media files too, the CASE-JigDFS is much more efficient as compared to the original system [3]. In Fig. 4, the encoding time has been measured for file sizes up to 1.02 GB (for indicative purposes) and it is clearly evident from the graph that CASE-JigDFS takes lesser encoding time for processing files of same size when compared with JigDFS.

The comparative response of the CASE-JigDFS network for document and media files is shown in Fig. 5.

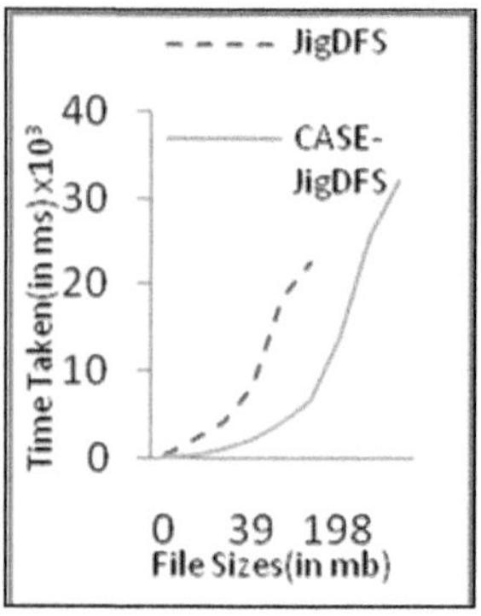

Fig. 3. Encoding time comparison for document files

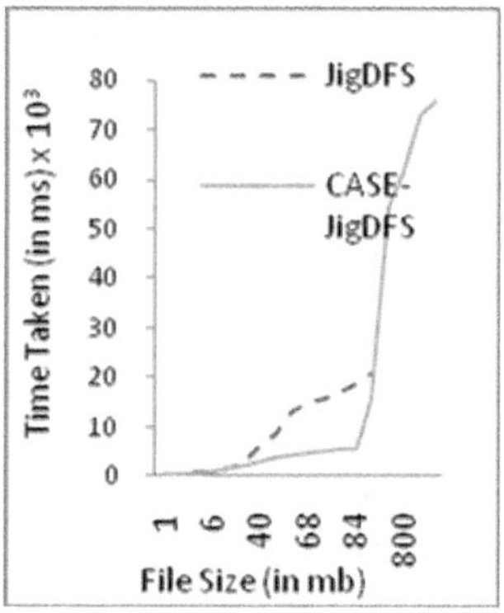

Fig. 4. Encoding time comparison for media files.

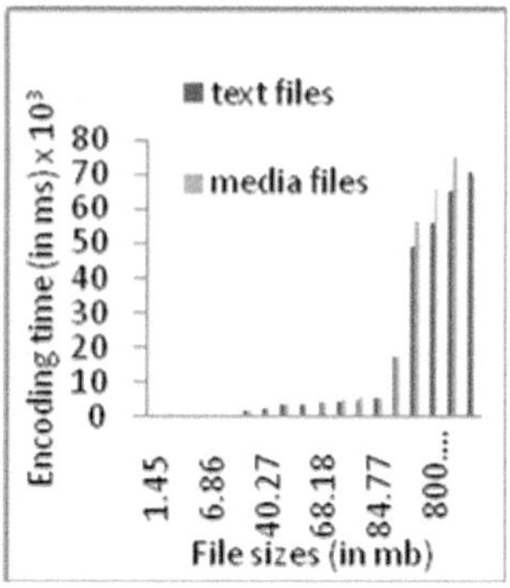

Fig. 5. Comparison for document and media files

CASE-JigDFS, unlike JigDFS[3], deals with the files of different sizes and types in a dynamic mannner so as to improve encoding time for each file in accordance with its attributes. It obtains optimum values for the given file. Hence as is evident from Fig. 5,the processing time for same sized media and text files is different.

4 Conclusion and Future Work

The CASE-JigDFS proposed here is intelligent enough to manipulate its procedures according to varying input in terms of file types and sizes, in order to strike a balance between system resource usage and output performance. The tests mentioned in the paper substantiate the claims and present a detailed comparison between the existing [3] and the improved framework. As a result, it can be concluded that CASE-JigDFS is more suitable in mobile computing as compared to JigDFS [3].

While working on the JigDFS [3] code we have observed that energy-efficient error-control requires a detailed contemplation for future work.

References

1. Dabek, F.: A Co-operative File System. Master's Thesis. In the Department of Electrical Engineering and Computer Science, M.I.T, USA (September 2001)
2. Busca, J.-M., Picconi, F., Sens, P.: Pastis: a Highly-Scalable Multi-User Peer-to-Peer File Systems. In: Cunha, J.C., Medeiros, P.D. (eds.) Euro-Par 2005. LNCS, vol. 3648, pp. 1173–1182. Springer, Heidelberg (2005)
3. Bian, J., Seker, R.: Jigdfs: A secure distributed file system. In: Proceedings of Symposium on Computational Intelligence in Cyber Security (CICS 2009), Nashville, USA, March 30-April 2, pp. 76–82 (2009)
4. Nolan, C.: ColonyFS: A Peer-to-Peer Filesystem, `https://launchpad.net/colonyfs`

Novel Stock Market Prediction Using a Hybrid Model of Adptive Linear Combiner and Differential Evolution

Minakhi Rout[1], Babita Majhi[1], Ritanjali Majhi[3], and G. Panda[4]

[1] Dept. of CSE/IT, ITER, Siksha O Anusandhan University, Bhubaneswar, India
minakhi.rout@gmail.com, babita.majhi@gmail.com
[2] School of Management, Naitional Institute of Technology Warangal, India
ritanjalimajhi@gmail.com
[3] School of Electrical Sciences, Indian Institute of Technology Bhubaneswar, India
ganapati.panda@gmail.com

Abstract. The paper proposes a novel forecasting model for efficient prediction of small and long range predictions of stock indices particularly the DJIA and S&P500. The model employs an adaptive structure containing a linear combiner with adjustable weights implemented using differential evolution. The learning algorithm using DE is dealt in details. The key features of known stock time series are extracted and used as inputs to the model for training its parameters. Exhaustive simulation study indicates that the performance of the proposed model with test input is quite satisfactory and superior to those provided by previously reported GA and PSO based forecasting models.

Keywords: Stock market prediction, hybrid model, adaptive linear combiner, differential evolution.

1 Introduction

In the modern business world, the forecasting of correct stock prices is a complex task as many factors influence the behavior of the stock market. Many statistical models based on autoregressive (AR)[1], autoregressive moving average (ARMA)[2] and autoregressive integrated moving average (ARIMA)[2] methods have been employed for this purpose. But these methods being linear, and the stock time series are inherently nonlinear the prediction performance of these methods is poor. On the other hand, soft computing based approaches such as artificial neural network(ANN) and fuzzy logic (FL) have found extensive applications in diverse fields including finance [3]. The inherent nonlinearity associated with ANN makes it more suitable to handle nonlinear data and hence is a potential candidate for forecasting nonlinear stock time series.

Recently a hybrid model using modular morphological neural network (MMNN) combined with a particle swarm optimizer (PSO) is proposed for efficient forecasting of stock prices [4]. In a recent work the forecasting of the stock market has been carried out by integrating genetic algorithm and neural network [5]. Another hybrid

V.V. Das, J. Stephen, and Y. Chaba (Eds.): CNC 2011, CCIS 142, pp. 187–191, 2011.

model using rough set theory is suggested for accurate prediction of stock markets[6]. Very recently an integrated approach based on genetic, fuzzy systems and artificial neural networks is reported [7] to develop an efficient stock price prediction system.

In this paper we propose a simple but efficient prediction model using an adaptive linear combiner (ALC) as a basic model and differential evolution (DE) as a learning algorithm to predict the movements of prices in the DJIA and S&P500 stock indices. The paper is organized as follows :

In Section 2 the development of forecasting model is discussed in which the learning of the weights of the adaptive linear combiner is performed using DE based optimization rule. The simulation of the proposed model for training and testing are presented in section 3. It also deals with the results of the simulation and comparative performance evaluation. Finally the conclusion is presented in Section 4.

2 Development of a Hybrid ALC-DE Based Prediction Model

The hybrid model consists of an adaptive linear combiner (ALC) shown in Fig.1 which is essentially an adaptive finite impulse response (FIR) filter having its number of inputs corresponding to the number of features in the input pattern. The weights of the ALC are updated by the principle of DE[8]. In this case each member of the population in DE represents one possible weight vector of the ALC which are updated following mutation and crossover operations so that the mean square error (MSE) of the model progressively minimized. The development of the prediction model proceeds as follows :

(i) The initial weights of the ALC model are taken as a member of the initial target population having P individuals. Each member of the population constitutes D number of parameters (number of weights of ALC) and each parameter represents a weight of the linear combiner.

(ii) K number of feature vectors each containing D features or technical indicators are obtained from the previous stock indices.

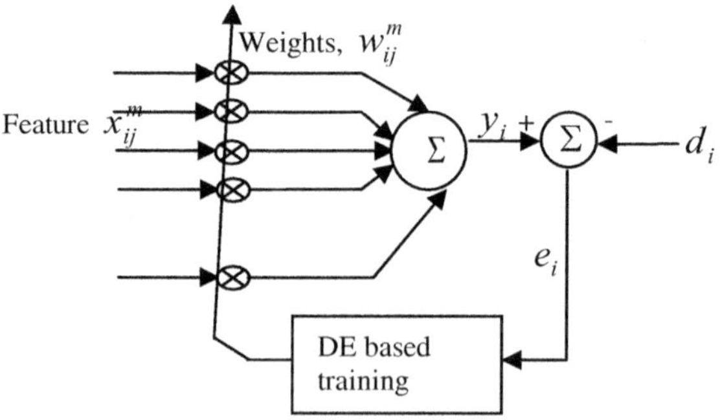

Fig. 1. Hybrid ALC-DE prediction model

(iii) Out of K set of features L sets (80% of K approximately) are used for training the hybrid model and the remaining $(K-L)$ sets are used for testing the performance of the model.

(iv) All K patterns are sequentially applied to the linear combiner and the output of the model is estimated as

$$y_i = \sum_{n=1}^{N} w_n x_n \tag{1}$$

(v) The output of the linear combiner, y_i is then compared with corresponding normalized desired value, d_i to produce the error, e_i. At the completion of all patterns K errors are produced. The mean square error (MSE) for a set of parameters (corresponding to ith target vector) is determined by using the relation.

$$MSE(i) = \frac{\sum_{k=1}^{K} e^2(k)}{K} \tag{2}$$

This is repeated for P times.

(vi) The perturbs vector is then calculated and the mutant population is obtained.

(vii) Then the crossover operation is done and the trial population is created.

(viii) The selection of members of the target population for the next generation is done and the whole process is repeated for certain generations.

(v) In each generation the minimum MSE (MMSE) is obtained and plotted against generation to show the learning characteristics of the model

(vi) The learning process is stopped when MMSE reaches the minimum possible floor level.

(vii) The prediction capability of the model so developed is tested with known stock value.

3 Simulation Study and Discussion on Results

In this section simulation study of the proposed model has been carried out using DE based training of its parameters. The practical data used for training and testing of the models are from 3rd January 1994 to 23rd October 2006. In total 3228 data are used from which 3168 patterns are generated. Each pattern consists of ten technical features using the formulae given in [3]. Out of 3168 known patterns 2510 patterns are used for training the model and the remaining 658 patterns are used for testing purpose. The model is developed for prediction of two stock indices (DJIA and S&P 500) for one day, three and seven days' ahead. Figs 2(a)-(b) represent the convergence characteristics of one day ahead prediction of S&P 500 and one day ahead prediction of DJIA stock price respectively. Similarly Figs 3(a)-(b) show the comparison of actual and predicted S&P 500 and DJIA stock indices during testing of the models for one day ahead prediction. The mean absolute percentage of error (MAPE) defined as

$$MAPE(k) = \frac{1}{M}\sum_{m=1}^{M}\frac{|y(m) - \hat{y}(m)|}{y(m)} \times 100 \quad (3)$$

is used as a performance measure of the prediction model where $y(m)$, $\hat{y}(m)$ and M represent actual stock price, predicted stock price and number test samples used respectively. Table 1 demonstrates the MAPE values obtained from the three different models for various days (1 to 7) ahead prediction of two stock prices. The different training characteristics displayed in Fig. 2 clearly indicate that DE offers fastest and least mean square error convergence compared to those obtained from the PSO and GA based models. Out of these three the GA offers the slowest and least accurate prediction capability. Further the comparison of matching performance between the true and predicted one during testing shown in Figs. 3(a)-(b) is quite excellent for both the stock indices. In general it is noticed that the prediction performance of the DE model is superior to that of the other two. This observation is also evident from the MAPE values listed in Table 1.

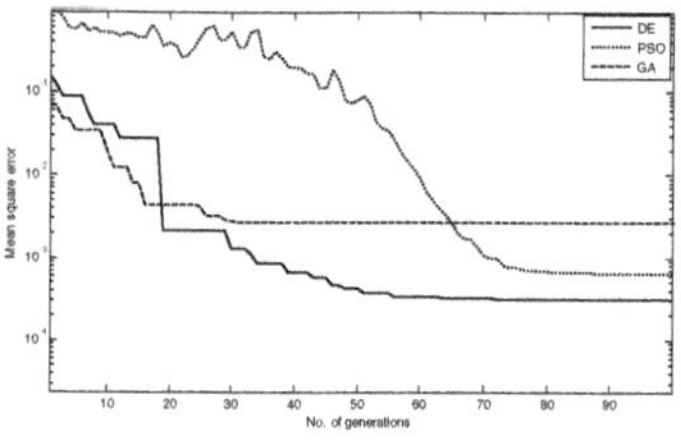

Fig. 2(a). Comparison of convergence characteristics for 1day ahead prediction using S&P 500 Indices

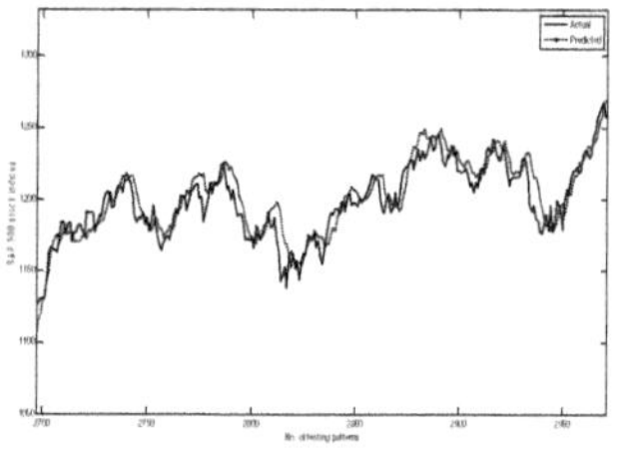

Fig. 3(a). Comparison of actual and predicted value during testing of S&P 500 indices for 1day ahead prediction using DE

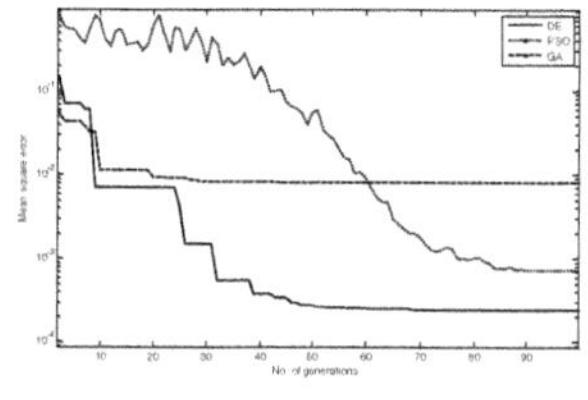

Fig. 2(b). Comparison of convergence characteristics for 1day ahead prediction using DJIA Indices

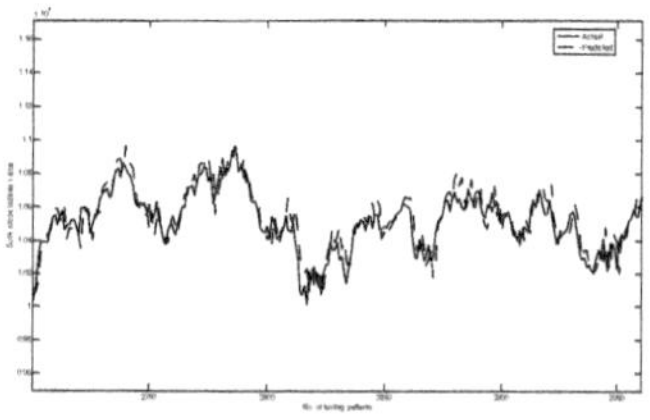

Fig. 3(b). Comparison of actual and predicted value during testing of DJIA indices for 1day ahead prediction using DE

Table 1. Comparison of MAPE of different algorithms for DJIA and S&P 500 stock indices

Days ahead	DE		PSO		GA	
	DJIA	S&P 500	DJIA	S&P 500	DJIA	S & P 500
1	0.6494	0.7765	0.8070	0.8634	1.8465	1.5985
3	0.9420	0.9897	1.0770	0.9846	1.9800	1.4489
7	1.3984	1.3953	1.4024	1.4136	1.5278	1.5547

4 Conclusions

The paper has developed an efficient DE based stock market prediction models using several features extracted from past stock data. The validity and potentiality have been tested using known stock indices and comparing the results obtained using PSO and GA based models. The comparison of results with PSO and GA models demonstrates superior performance and faster training of the proposed model.

References

1. Champernowne, D.G.: Sampling theory applied to autoregressive schemes. Journal of the Royal Statistical Society: Series B 10, 204–231 (1948)
2. Box, G.E.P., Jenkins, G.M.: Time series analysis: Forecasting and control, 3rd edn. Prentice Hall, Englewood Cliffs (1994)
3. Majhi, R., Panda, G., Sahoo, G.: Development and performance evaluation of FLANN based model for forecasting of stock markets. 36(3), 6800–6808 (April 2009)
4. Araújo, R.d.A.: Swarm-based translation-invariant morphological prediction method for financial time series forecasting. Information Sciences 180, 4784–4805 (2010)
5. Armano, G., Marchesi, M., Murru, A.: A hybrid genetic-neural architecture for stock indexes forecasting. Information Sciences 170, 3–33 (2005)
6. Teoh, H.J., Chen, T.-L., Cheng, C.-H., Chu, H.-H.: A hybrid multi-order fuzzy time series for forecasting stock markets. Expert Systems with Applications 36, 7888–7897 (2009)
7. Hadavandi, E., Shavandi, H., Ghanbari, A.: Integration of genetic fuzzy systems and artificial neural networks for stock price forecasting. Knowledge-Based Systems 23, 800–808 (2010)
8. Price, K.V., Storn, R.M., Lampinan, J.A.: Differential Evolution – A practical approach to global optimization. Springer, Heidelberg (2005)

Secure Self Organization of Wireless Sensor Network: A Need Based Approach Using Certificate Less Public Key Infrastructure

Hemanta Kumar Kalita[1] and Avijit Kar[2]

[1] TCS Innovation Lab Performance Engineering
TATA Consultancy Services, Akruti Business Port,
MIDC, Andheri(E) Mumbai
hemanta91@yahoo.co.in
[2] Dept. of Computer Science & Engg.
Jadavpur University, Kolkata, India
Dr.avijit.kar@gmail.com

Abstract. Wireless Sensor Network (WSN) is envisioned as a static ad hoc network. Sensor Nodes in this type of network needs to self organize themselves. During self organization sensor nodes initially discovers themselves and then joins the network. Security plays an important role when a new node joins the WSN. Only the authorized nodes should be allowed to join. However, security comes with a cost and same level of security may not be required always. Hence authors propose an approach to secure self-organization of WSN with need based security in this paper.

Keywords: WSN, PKI, Self Organization, Ad hoc Network, SNEP, MAC, ECC, SSL, IP, RSA, ZigBee, WMN.

1 Introduction

In many applications of WSN, wireless sensor nodes (henceforth nodes) will be deployed in an ad-hoc fashion, without careful planning. In these cases, the nodes must organize themselves to form a multi hop, wireless communication network. While self-organizing, it is essential that only the authorized nodes are allowed to join the network. This is not trivial. Authentication of new nodes can be applied starting from simple 'Mac id verification in plain text' up to 'multi-level mutual authentication in cipher text'. Stricter the security more will be the cost of the network in terms of energy. Therefore, authors propose three levels of security for self-organization of WSN. They are – low, medium and high.

Remainder of this paper is divided into four sections. In section 2 authors discuss background or related work. In section 3 proposed need-based securities for self organization of WSN are discussed. Section 4 is for results analysis and finally section 5 concludes the paper.

V.V. Das, J. Stephen, and Y. Chaba (Eds.): CNC 2011, CCIS 142, pp. 192–196, 2011.

2 Related Work

[1] presents SPINS: Security Protocol for Sensor Networks. SPINS has two building blocks: SNEP and μTESLA. [3] probes into various security requirements with regard to WSN and analyze status quo of the security in WSN. [4] analyzes three complementary WSN applications with the goal of developing a modular toolbox to support an integrated security and reliability architecture for medium and large-scale WSNs. [5] makes a critical analysis of WSN security and gives a layered-based classification. [6] presents ECC based lightweight SSL for IP based WSN. [7] analyzes public key cryptography for WSN security and suggests that RSA is not well-suited for WSN. [8] proposes an asymmetric key based architecture for WSN. [9] gives an analysis of WSN security management based on ZigBee. [10] presents secFleck, a Trusted Platform Module (TPM)-based public key framework for sensor network. [11] proposes WMNSec, an adaptation of IEEE 802.11i for Wireless Mesh Network with limited CPU power, node mobility and interruption free connectivity.

3 Proposed Need Based Security

Proposed need based security for self organization of WSN has three levels where degree of security varies from low to high.

3.1 Low Security

Low Security mode advocates only device authentication in plain text. A new node needs to send joining request to the base station in plain text. Any intermediate node including a new node's nearest neighbor is liable to forward the joining request to the base station. Base station verifies its unique id (Mac id) and accepts or rejects the joining request. This mode has many security loop holes. For example, device id or Mac ID can be spoofed by the adversary and any third party can sniff and read the data.

3.2 Medium Security

In this case a new node is required to authenticate itself with the help of a Hook key pre-distributed to it. Hook key is a device specific key. In Fig.1 the medium security

Step 1: A new node sends the 'Join Request' message by encrypting using its hook key, HKK and authenticating it using $K_{mac.}$
Step 2: Any intermediate node (including the new node's nearest neighbor) verifies the integrity of the message and forwards it to the base station.
Step 3:
a) Base station verifies integrity of the message.
b) Base station matches the HKK with the list available to it against the Mac id. If the HKK is found in the list, then base station can decrypt the 'Join Request' message and hence allows a new node to join the network.

Fig. 1. Modified SNEP Algorithm

based scheme of node joining called Modified SNEP is outlined. In this scheme a sensor node needs to possess two keys pre-distributed to it. They are–master key (from where the keys K_{encr} and K_{mac} are derived) and hook key, HKK. The 'master key' leakage, which is a symmetric key, is a major drawback in this scheme.

3.3 High Security

This is an asymmetric key based scheme where public keys are certificate less. Therefore, 'public key' exchange is done only after 'mutual authentication', according to this scheme. 'Mutual authentication' is carried out using symmetric keys: one hop key (network specific) and hook key (device specific). In Fig. 2 'high security' scheme called 'HySecNJoining' is outlined. The scheme has two phases and a trusted neighbor plays the role of a 'broker' to send initial 'Join Request' to the base station. Since there are two levels of authentications for any new node to join, therefore 'flooding attack' is eliminated using this scheme. Moreover, usage of 'asymmetric key' gives the advantage of public key infrastructure (PKI).

Phase 1: Neighbor authentication.

a) New node, α sends its 'public key' encrypting using one hop key and authenticating using its 'private key' to the nearest neighbor, δ. δ is already a trusted member of the network.

b) δ upon successful verification supplies its 'public key' to α.

Phase 2: Authentication by the base station.

a) α sends 'Join Request' (encrypted with α's Hook Key) to δ by encrypting with δ's public key and authenticating using α's private key.

b) Based upon successful verification, δ 'forwards' the 'Join Request' to base station, β by encrypting using β's public key and authenticating using δ's private key.

c) β verifies 'Join Request' by matching the Hook Key of α against the Mac Id and request δ to give β's public key and 'multicast key' to α.

d) β multicasts α's public key to the trusted members of the network which falls in the 'key path'.

Fig. 2. HySecNJoining Algorithm

Self organization of WSN is divided into three phases: node discovery, node joining and data communication. In this section three schemes for 'node joining' are proposed: low, medium and high security. Depending upon the need of security vs. performance any one option for node joining will be chosen. Discussion of 'node discovery' and 'data communication' is out of scope of this paper.

4 Analysis

In this section the proposed algorithms are analyzed with the help of a simulator. For simulation authors modify and enhance the WSN simulator v1.1 designed by [2]. For comparison 'total residual energy' of WSN is taken as the metric. Starting with same number of nodes in same configuration what is the total residual energy of a WSN after certain time gives us an idea about overall energy efficiency or performance of the algorithms.

Algorithms Modified SNEP and HySecNJoining are implemented using C# and simulated using the WSN Simulator for comparing 'total residual energy' of the WSN. In Fig. 3, 'total residual energy' after certain time gap for both the algorithms is compared. This shows that Algorithm Modified SNEP is better than HySecNJoining in terms of 'energy conservation'. This validates authors' assumption that stricter the security more will be the cost of WSN in terms of energy consumption.

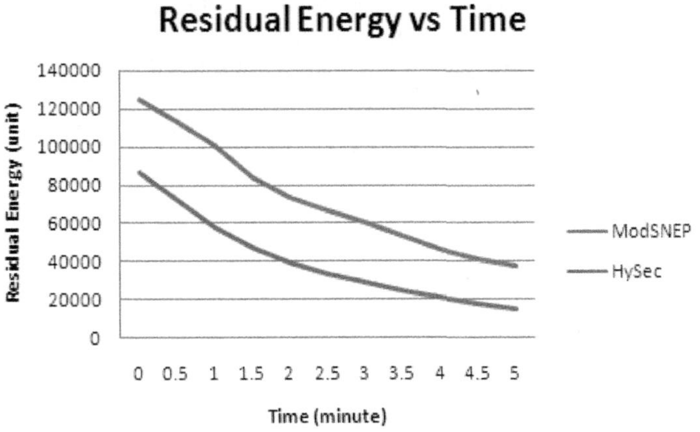

Fig. 3. Comparison of Energy

5 Conclusion

In this paper, a need based secure joining method for self-organization of WSN is proposed. Once node discovery is over which can be triggered through the use of a preset timer (or any other method deemed suitable), part 2 of the self-organization starts. In part 2 also called secure joining, nodes are allowed to be part of the network with three modes of security: low, medium and high. In low security mode any node with a valid Mac id are allowed to join the network. In medium security mode any node with valid hook key and Mac id can join the network provided it has the master key. Medium security mode of secure joining (Modified SNEP Algorithm) is based on symmetric master key and vulnerable to flooding attacks and has single point of failure, i.e. the master key. If the master key is compromised then the whole network is compromised. However, it is light as compared to High security mode of secure joining. High security mode of secure joining eliminates flooding attack and has the benefits of PKI. HySecNJoining Algorithm proposed in this case is based on certificate less PKI and requires no third party certification authority as public keys are exchanged between nodes only after mutual authentication. HySecNJoining is a complex, hybrid secure joining algorithm. Since the cost of using HySecNJoining is very high it is suggested for use in a situation where security is high priority for the WSN.

References

[1] Perrig, A., Szewczyk, R., Wen, V., Culler, D., Tygar, J.: SPINS: Security Protocols for Sensor Networks Mobile Computing and Networking (2001)
[2] Stein, D.J., Esq: Wireless Sensor Network Simulator v1.1 (2005)
[3] Weimin, L., Zongkai, Y., Wenqing, C., Yunmeng, T.: Research on the Security in Wireless Sensor Network. Asian Journal of Information Technology 5(3), 339–345 (2006)
[4] Westoff, D., Girao, J., Sarma, J.: Security Solutions for Wireless Sensor Networks Special Issue: Dependable IT and Network Technology (2006)
[5] Saxena, M.: Security in Wireless Sensor Networks- A Layer based Classification Purdue University, West Lafayette, IN 47907-2086 (2007)
[6] Jung, W., Hong, S., Ha, M., Kim, Y.-J., Kim, D.: SSL-based Lightweight Security of IP-based Wireless Sensor Networks (2007)
[7] Amin, F., Jahangir, A.H., Rasifard, H.: Analysis of Public-Key Cryptography for Wireless Sensor Networks Security World Academy of Science. Engineering and Technology 41 (2008)
[8] Haque, M.M., Pathan, A.-S.K., Hong, C.S., Huh, E.-N.: An Asymmetric Key-Based Security Architecture for Wireless Sensor Networks (2008)
[9] Soderman, P.: An Analysis of WSN Security Managemant School of Computer Science and Engineering. Royal Institute of Technology (2008)
[10] Hu, W., Corke, P., Shih, W.C., Overs, L.: secFleck: A Public Key Technology Platform for Wireless Sensor Networks (2009)
[11] Lukas, G., Fackroth, C.: WMNSec Security for Wireless Mesh Networks. In: IWCMC 2009, Leipzig, Germany (2009)

Design of High Resolution, Fast Locking, Low Power Multiphase-Output Delay Locked Loop

Anu Gupta, Nirmalya Sanyal, Suman Kumar Panda,
Satish Sankaralingam, and Deepak Kumar Gupta

Birla Institute of Technology and Science,
Pilani, 333031 Rajasthan, India

Abstract. This paper presents a fast locking, multiphase-output Delay Locked Loop(DLL). We propose a novel method to reduce locking time using a circuit which determines the input frequency thereby enabling the DLL to start output clock closest to reference clock(ref_clk). The DLL is designed in TSMC 0.18um technology. It has a frequency range from 105 MHz to 183 MHz with worst case resolution less than 350 ps. The DC power consumption of the DLL is approx. 2.8 mW at 1.8 V.

1 Introduction

To meet the rising performance and complexity demands, high frequency multi phase clock is required, which is generated using DLL and Phase Locked Loops (PLL). Here we propose a novel way for reduction of locking time of the DLL using a control circuit. Section II describes the architecture of the DLL. Section III discusses the calculation and results. Section IV discusses the novel control circuit for fast locking. Section V concludes the work.

2 Multiphse Output DLL

A Multiphase Output DLL(Fig. 1) consists of a *PFD, charge-pump (CP), loop filter(LF) and VCDL*. The ref_clk propagates through the VCDL. The output signal (out_clk) at the end of the delay line is compared with ref_clk. The PFD generates UP and DOWN signals depending on leading or lagging phase of out_clk w.r.t ref_clk. The CP generates a voltage depending on PFD output to control VCDL. The DLL also consists of a start controlled digital circuit. The period of the ref_clk should satisfy the following condition to avoid false or harmonic locking problem [1]:

$$Max\left(T_{VCDL,min}, \frac{2}{3}T_{VCDL,max}\right) < T_{clk} < Min\left(T_{VCDL,min}, 2*T_{VCDL,min}\right) \quad (1)$$

V.V. Das, J. Stephen, and Y. Chaba (Eds.): CNC 2011, CCIS 142, pp. 197–201, 2011.

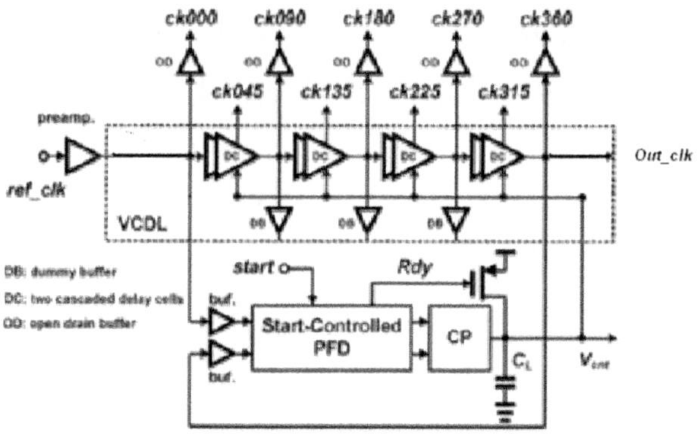

Fig. 1. Architechture of multiphase-output DLL [1]

3 Calculation and Results

3.1 Locking Time and Locking Range

8 delay elements are used which resulted in a $D_{VCDL,min}$ = 4.75ns and $D_{VCDL,max}$ = 8.21ns with a mean delay $D_{VCDL,mean}$ = 6.48ns. The locking range of the VCDL is 5.47ns to 9.5ns. The frequency range of the DLL is 105MHz to 183MHz.

3.2 Charge Pump with Loop Filter

Loop bandwidth : $$\omega_{loop} = \frac{2\pi}{T + D1 + D2} = 1.31 * 10^{9}\, rads/sec \tag{2}$$

PFD Gain: $$\frac{V_{dd}}{2\pi} = \frac{1.8}{2\pi}\, rads/sec. \tag{3}$$

VCDL gain: $$K_{VCDL} = \frac{Delay}{V_{VCNTL}} \cong 4ns \tag{4}$$

Loop Filter Gain: $$\frac{I_{cp}}{sC} \tag{5}$$

Transfer Function: $$\frac{\varphi_{out}}{\varphi_{in}} = \frac{1}{1 + \dfrac{2\pi s C}{V_{dd} I_{cp} K_{vcdl} 2\pi f}} \tag{6}$$

Band Width: $$\omega_{dll} = \frac{V_{dd} I_{sc} K_{VCDL}}{C.T} \tag{7}$$

Hence to minimize jitter the required charge pump[3] must have:

$$\frac{I_{cp}}{C} \leq \frac{T * 2\pi}{10*(T+D1+D2)*V_{dd} * K_{VCDL}} \tag{8}$$

Hence,for I_{cp} = 52.6 uA the $C_{LOOPFILTER}$ =1.5 pf. If we choose a higher value, the response would be over damped, hence increasing locking time.

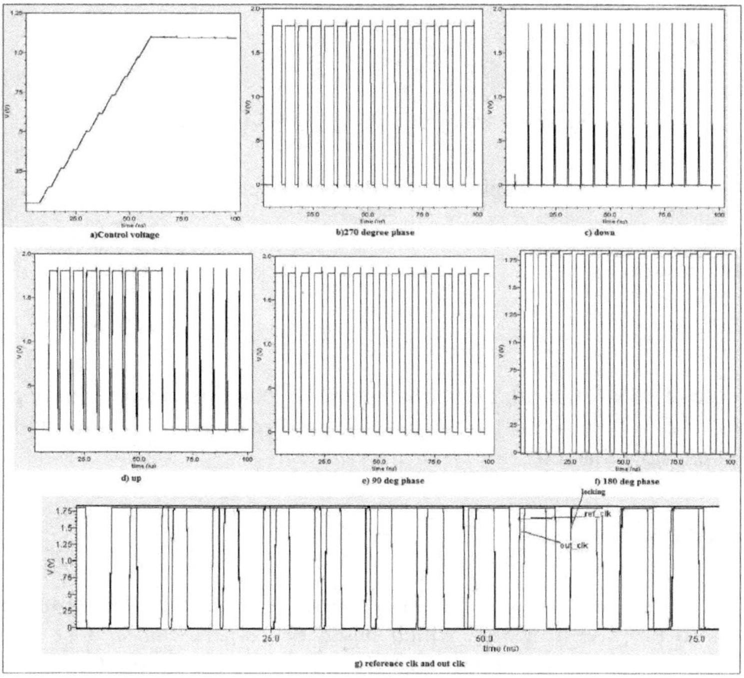

Fig. 2. Plots of simulated waveforms of the DLL

Table 1. Phase Error

PhaseerrorinLockedState	*Duration(ps)*
Relative(between clock phase)	0
Absolute(w.r.t ref_clk)	47

Table 2. Characterization of the DLL at 125 MHz

Characterstics(125Mhz)	TT(1.8V 27°C)	SS(1.62V 0°C)	SF(1.8V 27°C)	FS(1.8V 27°C)
Cycle to Cycle jitter(ps)	66	20	118	54.
Locking Time(cycles)	23	21	19	24
Dynamic Power(mW)	2.02	1.543	2.05	1.96
Resolution(ps)	160	180	150	350

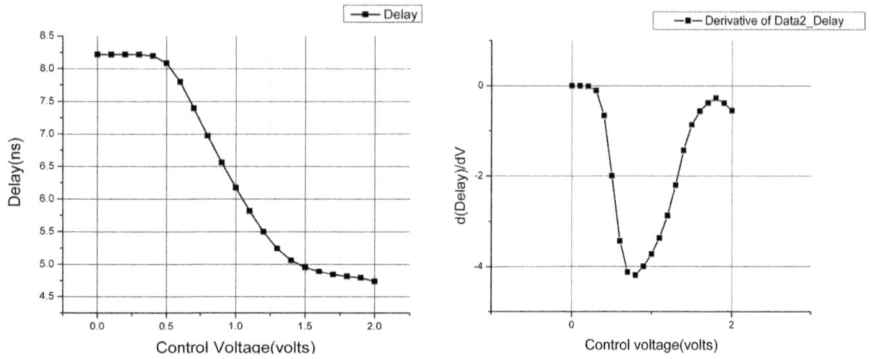

Fig. 3. a) Delay vs V_{CNTRL} b) K_{VCDL} vs V_{CNTRL}

Table 3. Static Power Consumption

Block	*PowerConsumption*
Delay Elements(8)	2.304 mW
Charge Pump	0.403 mW
PFD(Leakage)	25.52 pW

In FF case (1.98V, 100) the DLL was unable to lock due to small delay generated . So with no. of delay elements increased to 14, locking time of 18 cycles with a resolution of 100 ps was achieved.

4 Control Circuit

The current design proposed by Chang [1] always starts with the minimum delay. This led us to propose a design, which based on ref_clk can start at minimum or maximum delay. Thus, if the input is >154 MHz the control circuit will generate appropriate UP and DOWN signals to start with minimum delay. Otherwise the DLL starts at maximum delay. The ref_clk being within the locking range eliminates any chances of harmonic or false locking. $D_{VCDL,mean}$ being constant, the count corresponding to $D_{VCDL,mean}$ required for comparison is known. An external clock faster than the maximum ref_clk frequency is available for counting (here 2 ns). The external clock is at least twice the ref_clk frequency. Initially, FF's are reset.

Because of initial reset, final count is always available after 2 ref_clk cycles.

Algorithm:
If $D_{VCDL,mean}$ count < ref_clk count
Out = 0, UP(CP) = ! (DN), DN(CP) = UP, start from maximum delay.
else if $D_{VCDL,mean}$ count > ref_clk count
out = 1, UP(CP) = ! (UP), DN(CP) = DN, start from minimum delay.

The no. of locking cycles was reduced from 23 to 10 (plus 2 overhead) at 125 MHz when starting from maximum delay meaning a reduction of 50% in the locking time.

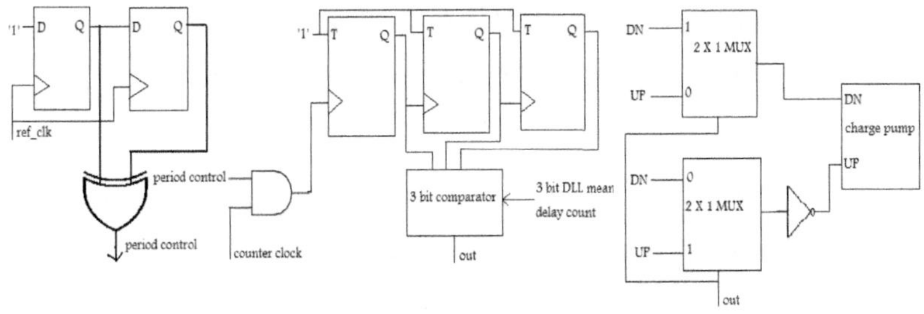

Fig. 4. Proposed Control Circuit

5 Conclusion

The paper presents a multiphase-output DLL. Also, a novel control circuit to dynamically adjust the initial delay is proposed to reduce locking time. The locking frequency range of the DLL is 105 to 183 MHz. The results show that the DLL can lock on to the range of frequencies with a resolution of less than 350 ps.

References

[1] Huang, P.-J., Chen, H.-M., Chang, R.C.: A novel start-controlled phase/frequency detector for multiphase-output delay-locked loops. In: Proceedings of 2004 IEEE Asia-Pacific Conference on Advanced System Integrated Circuits, August 4-5, pp. 68–71 (2004), doi:10.1109/APASIC.2004.1349408

[2] Cheng, J.: A delay-locked loop for multiple clock phases/delays generation, 005, 94 pages. Georgia Institute of Technology, 3198555

[3] Booth, E.R.: Wide range, low jitter delay-locked loop using a graduated digital delay line and phase interpolator (November 2006)

Dynamic Materialized View Selection Approach for Improving Query Performance

A.N.M. Bazlur Rashid[1], M.S. Islam[2], and A.S.M. Latiful Hoque[3]

[1] Department of Computer Science & Engineering,
Comilla University, Comilla, Bangladesh
bazlur.rashid@yahoo.com
[2] Institute of Information & Communication Technology,
Bangladesh University of Engineering & Technology, Dhaka, Bangladesh
mdsaifulislam@iict.buet.ac.bd
[3] Department of Computer Science & Engineering,
Bangladesh University of Engineering & Technology, Dhaka, Bangladesh
asmlatifulhoque@cse.buet.ac.bd

Abstract. Because of the query intensive nature of data warehousing or online analytical processing applications, materialized view is quite promising in efficiently processing the queries and for improving the query performance. It is costly to rematerialize the view each time a change is made to the base tables that might affect it. So, it is desirable to propagate the changes incrementally. Hence, all of the views cannot be materialized due to the view maintenance cost. In this paper, we have developed a dynamic cost model based on threshold level incorporating the factors like view complexity, query access frequency, execution time and update frequency of the base table to select a subset of views from a large set of views to be materialized. A number of algorithms and mathematical equations have been designed and developed to define the dynamic threshold level.

Keywords: Materialized view, View materialization, Dynamic selection, Query processing cost, View maintenance cost.

1 Introduction

A view is defined as a function from a set of base tables to a derived table and the function is recomputed every time the view is referenced. On the other hand, a materialized view is like a cache *i.e.*, a copy of data that can be accessed quickly. Utilizing materialized views that incorporate not just traditional simple SELECT-PROJECT-JOIN operators but also complex online analytical processing (OLAP) operators (*i.e.*, PIVOT and UNPIVOT) play crucial role to improve the OLAP query performance. Materialized views are useful in applications such as data warehousing, replication servers, data recording systems, data visualization and mobile systems [1-3].

V.V. Das, J. Stephen, and Y. Chaba (Eds.): CNC 2011, CCIS 142, pp. 202–211, 2011.

The materialization of views is the most important tribulation in data warehousing. Materialization of all possible views is impossible as large computation and storage space is necessitated. Consequently, the primary concern in data warehousing is the "*view selection*" that deals with the selection of suitable set of views to materialize that strikes stability among computational cost and increased query performance [4]. The number of times a query is executed or requested at a particular time period is the access frequency of that query. A view or a query can be more complex if it contains more algebraic operators, joining of two or more tables etc. In a dynamic environment, the selection of appropriate set of views to materialize necessitate consideration of additional factors like query access frequency, execution time and complexity; hence, it is a demanding task. The selection of materialized views is affected by numerous factors. Thus the process of selecting the suitable views to materialize in data warehouse implementation is a critical issue. Several methods have been proposed in the literature to address the materialized view selection problem such as [5 – 7].

In this paper, we have designed dynamic cost model to select a subset of views from a large set of views to be materialized in order to achieve improved query response in low time based on the high access frequency or those queries are complex and have higher execution costs. All the cost metrics associated with the materialized views like query access frequency, execution time, complexity, base table update frequency are used in the model to calculate the query processing cost and view maintenance cost. Mathematical equations have been developed to define the dynamic threshold level. Experimental performance results have been carried out for the designed model for view selection.

The rest of the paper is structured as follows: Section 2 presents a brief review of related works in materialized view selection. Section 3 illustrates the dynamic cost model of the selection of views to materialize. Section 4 presents the experimental performance results. Section 5 discusses the comparisons with the existing related works. Finally, Section 6 concludes the research work.

2 Related Works

The problem of finding views to materialize in order to efficiently answering queries has traditionally been studied under the name of view selection.

Gupta *et al.* [5] developed algorithms to select a set of views to materialize in a data warehouse in order to minimize the total query response time under the constraint of a given total view maintenance time. Zhang *et al.* [6] explored the use of a hybrid evolutionary algorithm for materialized view selection based on multiple global processing plans for queries. An efficient solution has been proposed by Lee *et al.* [7] to the maintenance-cost view selection problem using a genetic algorithm for computing a near optimal set of views used to search for a near optimal solution.

Yu *et al.* [8] presented a new constrained evolutionary algorithm for the maintenance-cost view-selection problem where the constraints were incorporated through a stochastic ranking procedure. Wang *et al.* [9] proposed a modified genetic algorithm for the selection of a set of views for materialization. An optimized framework has been designed by Ashadevi and Balasubramanian [10] for the selection of views to materialize for a given storage space constraints to achieve the best combination good query response, low query processing cost and low view maintenance cost. The proposed framework considered the query execution frequencies, query access costs, view maintenance costs and system's storage space constraints for materialized view selection.

From the above mentioned works, it is found that the research works have been provided different approaches for the selection of views to materialize considering view maintenance cost and storage space. The appropriate selection of views to be materialized based on dynamic threshold level has not been explored so far.

3 Dynamic Selection of Views

The costs and factors that have been considered for selection of views are shown in Table 1. The first column in the table represents the SQL statements (Q_n) that are called queries or views.

Table 1. Costs and factors associated with queries

Q_n	f_n	w_n	e_n	t_n	j_n	a_n	c_n	QP_n	MT_1	MT_2	..	MT_i	MC_n	TC_n
Q_1	f_1	w_1	e_1	t_1	j_1	a_1	c_1	QP_1	*cost*	*0*	..	*cost*	MC_1	TC_1
Q_2	f_2	w_2	e_2	t_2	j_2	a_2	c_2	QP_2	*cost*	*cost*	..	*0*	MC_2	TC_2
…	…	…	…	…	…	…	…	…	…	…	..	…	…	…
Q_n	f_n	w_n	e_n	t_n	j_n	a_n	c_n	QP_n	*0*	*cost*	..	*cost*	MC_n	TC_n

All the queries used in the table are unique SQL statements in a particular time. The second column is the access frequency (f_n) count of each query at that particular time period. A weighting factor (w_n) is assigned to reflect the importance of the query considering that the higher the frequency of the query in the system, the higher the weight. The weight is calculated as,

$$Weight(w_n) = \log_k (f_n + 1) \tag{1}$$

The value of k is either 2 or 10. If the difference between the maximum and the minimum frequency is using high, k is 10; otherwise it is 2.

The fourth column presents the query execution time and it is called the response time of that query. The fifth, sixth and seventh columns are used to calculate the structural complexity of the query and it is calculated as, Complexity (c_n) = No. of tables involved (t_n) + No. of joining occurred (j_n) + No. of aggregate operator used (a_n). The query processing cost (QP_n) is calculated using query access frequency, weighting factor, execution time, selectivity and complexity.

The columns MT_1 to MT_j are used to calculate the maintenance cost (MC_n) of the view by summing up the total table maintenance costs for those tables that are involved in the query. For the maintenance cost of the view, corresponding involved table maintenance cost is placed in the columns.

If a table maintenance cost is not associated with the query then a "0" is placed in that column. For example, if table t_1 and table t_2 are involved in the query Q_1, then the maintenance cost of the table t_1 and t_2 are placed in the columns for the view maintenance cost. Remaining table maintenance costs are set as "0". The total cost (TC_n) is calculated by summing up the query processing cost and the maintenance cost. After filling up the table, a (n, 11 + m) matrix is formed where n is the total number queries and m is the total number tables in the schema.

3.1 Query Processing Cost

To compute the total query processing cost (QP_n), we have considered the query access frequency, weighting factor, query execution time, selectivity, and complexity of the query. In data mining, we can find the large item set for the mining process as specified in the data mining algorithms like in [11]. After completion of the (n, 7 + m) cost associated table matrix, by applying the modified data mining algorithm as used in [12] for finding the large total cost associated with the queries, the query processing cost of each query can be calculated as,

Query processing cost of query Q_n is

$$(QP_n) = w_n e_n c_n \tag{2}$$

3.2 View Maintenance Cost

View maintenance is the process of updating pre-computed views when the base fact table is updated. The maintenance cost for the materialized view is the cost used for refreshing this view whenever a change is made to the base table. We have calculated the maintenance cost using the update frequency of the base table and the weighting factor to reflect the importance of the base table update frequency.

The associated update frequencies, corresponding weighting factors and maintenance cost of the tables are depicted in the following Table 2.

Table 2. Table maintenance costs

Table (t_j)	Update Frequency (u_j)	Weighting Factor (w_j)	Table Maintenance Cost (MT_j)
t_1	u_1	w_1	MT_1
t_2	u_2	w_2	MT_2
...	...	...	...
t_j	u_j	w_j	MT_j

The update frequency (u_j) of the table (t_j) is the counting of the total number of updates done on the base tables at a particular time period. The updates to the base tables can be insertion of new rows, modification of existing rows and deletion of rows from the tables. We have calculated the update frequency of the base tables as,

$$Update\ \ frequency\ \ (u_j) = (i_j + m_j + d_j)/3 \tag{3}$$

where i_j is the no. of insertions m_j is the no. of modifications and d_j is the no. of deletions. As like the weighting factor of query access frequency, the weighting factor for the table is defined as,

$$Weight\left(w_j\right) = \log_k\left(u_j + 1\right) \tag{4}$$

The maintenance cost of a table is defined as,

$$Maintenance\ cost\ of\ a\ table_j, MT_j = w_j \tag{5}$$

The total maintenance cost associated with a query is the summation of the maintenance costs of the tables that are involved with the query and is defined as,

$$Maintenance\ Cost\ of\ Query\ Q_n, MC_n = MT_1 + MT_2 + MT_3 + \ldots + MT_j \tag{6}$$

3.3 Total Query Cost

The total cost of the query and the associated maintenance cost can be calculated by summing up the query processing cost and the view maintenance cost.

$$\text{Total}\ Cost\ of\ Query\ Q_n, TC_n = QP_n + MC_n \tag{7}$$

$$\text{Therefore, the } AVG\ (TC) \text{ or threshold level} = \frac{\sum_{x=1}^{n} TC_x}{n} \tag{8}$$

3.4 Dynamic View Selection Algorithm

The following algorithm *DynamicViewMaterializationSelection* in Fig. 1 selects the views dynamically for materialization to improve the query performance based on the query processing cost and the view maintenance cost. The algorithm first calculates the query processing cost and the maintenance cost associated with the queries and the total cost for materialized view maintenance. Then, it finds the minimum of the total cost. Finally, the algorithm selects the queries with higher total processing cost

```
Algorithm DynamicViewMaterializationSelection{
Given: Input: A set of queries or views;
        Output: Query cost (n × 4) table matrix;
Begin
        For i = 1 to total number of queries (n) {
        Assign: Q_i = getNextQuery ( );
        Calculate the query processing cost (QP_i);
        Calculate the view maintenance cost (MC_i);
        TC_i = Calculate the total cost (QP_i + MC_i)
        } End For;
        Find the minimum of the total cost Min (TC);
        For i = 1 to total number of queries (n) {
                If TC_i > Min (TC) Then {
                Select Q_i for materialization;
                Insertlist (Q_i, QP_i, MC_i, TC_i);
                } End If;
        }
        End For;
End;
}
```

Fig. 1. Algorithm for dynamic selection of views to be materialized

than the minimum total processing cost, *AVG* (*TC*) where the *AVG* (*TC*) is the dynamic threshold level.

4 Experimental Results

The experiments reported here have been carried out under the following hardware and software environment: Processor: Intel® Core™ 2 Duo, 2.00 GHz; L2 Cache: 2 MB; RAM: 3 GB and Hard disk: 150 GB. The system software used in the experiments is Microsoft Windows XP Professional Service Pack 3. The database used in the experiments is Oracle 11*g* Release 1 (11.1.0.6.0) Enterprise Edition. The popular sales history schema has been used in the experiments by generating appropriate set of data to meet the research experimental goal.

First, we have calculated the maintenance cost associated with the tables using the update frequencies of the tables and the weighting factors reflecting the importance of the tables. Now the different cost associated with the queries are calculated in Table 3.

The maintenance cost calculation of the tables is shown in the following Table 4 where the weighting factor of the table is calculated using

$$Weight(w_j) = \log_2(u_j + 1) \tag{9}$$

Table 3. Query associated costs calculation

Q_n	f_n	w_n	e_n	t_n	j_n	a_n	c_n	QP_n	MT_1	MT_2	MT_3	MT_4	MT_5	MT_6	MT_7	MT_8	MC_n	TC_n
Q_1	12	3.70	77.4	3	2	0	5	1431.90	0	0	0	4.39	0	0	0.41	6.08	10.88	1442.78
Q_2	13	3.81	32.31	6	5	6	17	2092.72	0	0	0	0	0	0	0.41	6.08	6.49	2099.21
Q_3	11	3.58	9.6	5	4	3	12	412.42	0	0.00	0	4.39	2.11	0	0.41	6.08	8.60	421.02
Q_4	18	4.25	69.34	3	2	0	5	1473.48	0	0	0	0	2.11	0	0.41	6.08	4.21	1477.69
Q_5	20	4.39	9.62	4	3	1	8	337.85	0	0	0	4.39	2.11	0	0.41	6.08	12.99	350.84
Q_6	11	3.58	40.7	4	3	1	8	1165.65	0	0	0	4.39	2.11	0	0.41	6.08	12.99	1178.54
Q_7	11	3.58	30.09	4	2	1	7	754.06	0	0	0	4.39	2.11	0	0.41	6.08	12.99	767.05
Q_8	20	4.39	8.53	5	4	1	10	374.47	1.42	0.00	0	4.39	0	0	0.41	6.08	12.30	386.77
Q_9	17	4.17	7.54	5	4	1	10	314.42	1.42	0.00	0	4.39	0	0	0.41	6.08	12.30	326.72
Q_{10}	7	3.00	36.18	5	4	1	10	1085.40	1.42	0.00	0	4.39	0	0	0.41	6.08	12.30	1097.70
Q_{11}	4	2.32	125.59	5	3	1	9	2622.32	1.42	0.00	0	4.39	0	0	0.41	6.08	12.30	2634.62
Q_{12}	13	3.81	111.06	5	3	1	9	3808.25	1.42	0.00	0	4.39	0	0	0.41	6.08	12.30	3820.55
Q_{13}	19	4.32	118.46	5	3	1	9	4605.73	1.42	0.00	0	4.39	0	0	0.41	6.08	12.30	4618.03
Q_{14}	5	2.58	41.51	5	4	1	10	1070.96	1.42	0.00	0	4.39	0	0	0.41	6.08	12.30	1083.26
Q_{15}	14	3.91	47.17	5	4	1	10	1844.35	1.42	0.00	0	4.39	0	0	0.41	6.08	12.30	1856.65
Q_{16}	11	3.58	22.28	6	5	1	12	957.15	1.42	0.00	0	4.39	2.11	0	0.41	6.08	14.41	971.56
Q_{17}	13	3.81	10.37	5	7	2	14	553.14	0	0.00	0	4.39	2.11	0	0.41	6.08	12.99	566.13

Table 4. Table maintenance costs calculation

SL	t_i	i_i	m_i	d_i	u_i	w_i	MT_i
1	*channels*	5	0	0	1.67	1.42	1.42
2	*countries*	0	0	0	0.00	0.00	0.00
3	*costs*	10	10	5	8.33	3.22	3.22
4	*customers*	50	0	10	20.00	4.39	4.39
5	*products*	10	0	0	3.33	2.11	2.11
6	*promotions*	5	2	0	2.33	1.74	1.74
7	*times*	1	0	0	0.33	0.41	0.41
8	*sales*	200	0	0	66.67	6.08	6.08

The minimum of the total cost or threshold level, *AVG* (*TC*) = 25099.12/17 = 1476.42. So, according to the *DynamicViewMaterializationSelection* algorithm, the following queries in Table 5 are selected to be materialized.

Table 5. Candidate queries for view materialization

Query	Query Processing Cost	Maintenance Cost	Total Cost
Q_2	2092.72	6.49	2099.21
Q_4	1473.48	4.21	1477.69
Q_{11}	2622.32	12.30	2634.62
Q_{12}	3808.25	12.30	3820.55
Q_{13}	4605.73	12.30	4618.03
Q_{15}	1844.35	12.30	1856.65

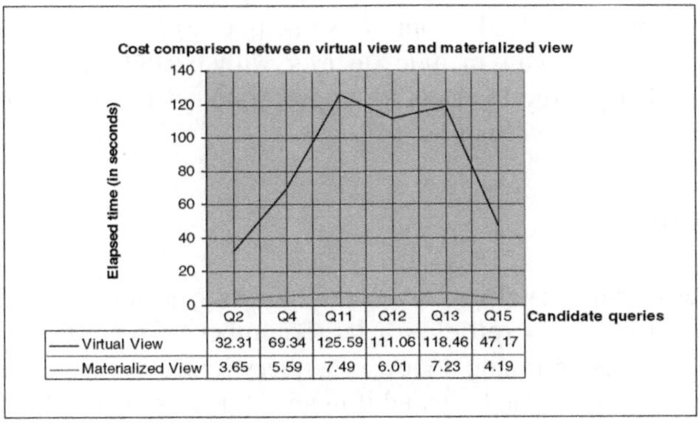

Fig. 2. Query answering cost comparison

From the selected queries for materialization, it is seen that the dynamic view selection approach selects queries not having only higher access frequency but also considers a combination of higher execution time, more complex queries. The graph in Fig. 2 shows the cost comparison of the selected queries for answering through virtual view and materialized view. From the Fig. 2, it is seen that after the candidate queries are materialized, query response times have reduced with significant amount of time.

5 Comparison with Related Work

The factors considered here in comparison with the work done by Ashadevi *et al.* [10] are listed in Table 6.

Table 6. Comparison of factors with [10]

SL	Factors	Existing Method [10]	Proposed Method
1.	Query access frequency	Yes	Yes
2.	Query weight	No	Yes
3.	Query execution time	No	Yes
4.	View complexity	No	Yes
5.	Base table update frequency	Yes	Yes
6.	Table weight	No	Yes
7.	Priority of table	Yes	No
8.	Threshold level	Arbitrary	Dynamic

Assigning a priority to the base table is more difficult than defining the importance of the table based on the update frequency of the table. Also, arbitrary selection of threshold level is difficult to choose for different level rather selection of threshold level based the calculated data that is called dynamic threshold and it is easy to define. Finally, in [10], queries with higher access frequencies are selected first for further processing which may lead to opt out some queries having higher execution time or more complexity but can include queries with execution time in considerable amount of time. Our method selects an appropriate set of views from the large set where the above scenario will not be occurred.

6 Conclusions

Faster query performance is always expected by a user or in a decision support system environment where a query retrieves a large number of records, aggregates data or joins lots of tables. Materializing a view or a query improves these query performance by storing the result in the database and it incurs storage space and maintenance issue. So, we need to select an appropriate set of queries from the number of queries requested by the user at a particular snapshot. Our proposed method selects an appropriate set of queries with a combination of higher access frequency, complexity of higher execution time.

References

1. Chaudhuri, S., Dayal, U.: An Overview of Data Warehousing and OLAP Technology. SIGMOD Record 26(1), 65–74 (1997)
2. Chen, S., Rundensteiner, E.A.: GPIVOT: Efficient Incremental Maintenance of Complex ROLAP Views. In: Proceedings of the 21st International Conference on Data Engineering (ICDE 2005), pp. 552–563 (2005)
3. Rashid, A.N.M.B., Islam, M.S.: Role of Materialized View Maintenance with PIVOT and UNPIVOT Operators. In: Proceedings of the IEEE International Advance Computing Conference (IACC 2009), Patiala, India, pp. 951–955 (2009)
4. Zhuge, Y., Molina, H.G., Hammer, J., Widom, J.: View Maintenance in a Warehousing Environment. In: Proceedings of the ACM SIGMOD Conference, San Jose, California, pp. 316–327 (1995)
5. Gupta, H., Mumick, I.S.: Selection of Views to Materialize Under a Maintenance Cost Constraint. In: Beeri, C., Bruneman, P. (eds.) ICDT 1999. LNCS, vol. 1540, pp. 453–470. Springer, Heidelberg (1998)
6. Zhang, C., Yao, X., Yang, J.: An Evolutionary Approach to Materialized View Selection in a Data Warehouse Environment. IEEE Transactions on Systems, Man and Cybernetics 31(3), 282–293 (2001)
7. Lee, M., Hammer, J.: Speeding up Materialized View Selection in Data Warehouses using a Randomized Algorithm. International Journal of Cooperative Information Systems 10(3), 327–353 (2001)
8. Yu, J.X., Yao, X., Choi, C., Gou, G.: Materialized View Selection as Constrained Evolutionary Optimization. IEEE Transactions on Systems, Man and Cybernetics, part c 33(4), 458–467 (2003)
9. Wang, Z., Zhang, D.: Optimal Genetic View Selection Algorithm under Space Constraint. International Journal of Information Technology 11(5), 44–51 (2005)
10. Ashadevi, B., Balasubramanian, R.: Optimized Cost Effective Approach for Selection of Materialized Views in Data Warehousing. Journal of Computer Science and Technology 9(1), 21–26 (2009)
11. Agrawal, R., Srikant, R.: Fast Algorithms for Mining Association Rules in Large Databases. In: Proceedings of the 20th International Conference on VLDB, Santiago, Chili, pp. 487–499 (1994)
12. Shafey, M.A.L.: Performance Evaluation of Database Design Approaches for Object Relational Data Management. M. Sc. Engg. Thesis. Institute of Information and Communication Technology, Bangladesh University of Engineering and Technology, Dhaka (2008)

Approximate Pattern Matching for DNA Sequence Data

Nagamma Patil, Durga Toshniwal, and Kumkum Garg

Department of Electronics and Computer Engineering,
Indian Institute of Technology Roorkee, Roorkee, India
parildec@iitr.ernet.in, durgafec@iitr.ernet.in,
kumkum.garg@manipal.edu

Abstract. In real world biological applications, most relevant sequences are "similar" instead of exactly the same. It is therefore useful to search patterns using soft computing techniques. Our proposed work uses the fuzzy approach for approximate pattern matching. Given a database of sequences, our aim is to find total number of candidate patterns that will approximately match to the query pattern with specified fault tolerance. We show the complete analysis of approximate matching patterns by fuzzy and exact matching approach. We use DNA sequences of bacteria downloaded from National Center for Biotechnology Information (NCBI).

Keywords: Soft computing, fuzzy sets, approximate matching, exact matching, DNA sequence.

1 Introduction

Soft computing is a collection of techniques in artificial intelligence, which can be used to handle uncertainty, imprecision and partial truth. Fuzzy sets, which constitute the oldest component of soft computing, can provide approximate solutions faster [1].

Mining of sequence data has many real world applications [2]. Transaction history of a bank customer, product order history of a company, performance of the stock market [3] and biological DNA data [4] are all sequence data, where sequence data mining techniques are applied. For sequential data, pattern matching can be considered as either (1) exact matching or (2) approximate matching [5]. Approximate matching is the finding of the most similar match of a particular pattern within a sequence. Quite often in real world data mining applications, exact patterns do not exist therefore, an approximate matching algorithm is required.

Bioinformatics is text rich field [6]. The analysis of protein and DNA sequence data has been one of the most active research areas in the field of computational molecular biology [7]. In the present work, we propose a methodology based on fuzzy Levenshtein distance to extract approximately matching patterns from the DNA sequences of bacteria downloaded from National Center for Biotechnology Information (NCBI) [8]. We show the complete analysis of the approximate matching patterns by varying candidate length and distance. The results are compared with exact distance method and it was observed that the fuzzy approach resulted in more number of patterns that are approximately matching to given query.

V.V. Das, J. Stephen, and Y. Chaba (Eds.): CNC 2011, CCIS 142, pp. 212–218, 2010.

Rest of the paper is organized as follows. In section 2, we discuss related work. Section 3 includes problem statement and proposed pattern extraction algorithm. In section 4, we show the experimental results obtained by using data set of DNA sequences. In section 5, we draw the conclusions and future work.

2 Related Work

In exact sequence pattern matching problems, we aim to find a substring in text *T* that is exactly the same as the searching pattern *P*. There exist, efficient algorithms that solve this problem. Boyer–Moore [9] published a exact pattern matching algorithm that performed the character comparison in reverse order to the pattern being searched for and had a method that did not require the full pattern to be searched if a mismatch was found. The Knuth–Morris–Pratt [10] string searching algorithm searches for occurrences of a "word" within a main "text string" by employing the observation that when a mismatch occurs, the word itself embodies sufficient information to determine where the next match could begin, thus bypassing re-examination of previously matched characters. These existing exact methods are useless for the biological data, because, in real world biological applications, most of the sequences are "similar" instead of exactly the same. This gave a motivation to "search allowing errors" or approximate match [11].

The Smith-Waterman algorithm [12] is a representative one for DNA subsequence matching, that basically assumes the sequential scan for accessing DNA sequences. It determines the optimal local alignment between two sequences *S* and *Q* so that their similarity score gets maximized. However, it requires a long processing time. BLAST [13] is a heuristic algorithm that resolves the performance problem of the Smith-Waterman algorithm. alignment. BLAST searches for high scoring sequence alignments between the query sequence and sequences in the database using a heuristic approach. Thus, these algorithms are based on sequence alignment technique.

Current approximate searching algorithm (such as Prosite Scan [14]) is rigid in a way that its definition of "similarity" is fixed. Chang et al. [15] proposed approximate pattern matching algorithm based on a fuzzy inference technique in which, the user is able to define the meaning of "similarity" by adjusting the membership functions. There are four main steps in this algorithm. Firstly, a searching pattern (string), *P* is decomposed to obtain events and event intervals. Then, the obtained events and event intervals are fuzzified in the sequence fuzzification step. A Sequence Inference step follows to determine the sequences that are "similar" to the searching pattern *P*. Finally, a sequence search is conducted to determine the "similarity" between a text *T* and the pattern *P*. The algorithm is quite complicated involving pattern decomposition, fuzzification and sequence inference step.

In the present work, we propose a methodology based on fuzzy Levenshtein distance to extract approximately matching patterns from the DNA sequence data. In our method similarity is measured by percentage of tolerance i.e. we mainly work on fuzzy distance between candidate and query pattern.

3 Proposed Approach

In this section, first we discuss algorithm for finding fuzzy distance between two patterns. We also discuss the use of fuzzy distance to find the total number of approximately matching patterns.

3.1 Algorithm for Finding Fuzzy Levenshtein Distance

Let *s* be the candidate pattern and *t* be the query pattern. The fuzzy Levenshtein distance is the number of deletions, insertions, or substitutions required to transform *s* into *t*.

1. Set n to be the length of s. Set m to be the length of t. Construct a matrix with 0..m rows and 0..n columns.
2. Initialize the first row to 0..n. Initialize the first column to 0..m.
3. Examine each character of s (i from 1 to n).
4. Examine each character of t (j from 1 to m).
5. If s[i] equals t[j], the cost is 0. If s[i] doesn't equal t[j], the cost is 1.
6. Set cell d[i,j] of the matrix equal to the minimum of (d[i-1,j] + 1, d[i,j-1] + 1, d[i-1,j-1] + cost).
7. After the iteration steps (3, 4, 5, 6) are complete, the distance is found in cell d[n,m].

For example, Levenshtein distance (LD) between candidate pattern (s) and the query pattern (t) is:

- If s is "levenshtein" and t is "meilenstein", then LD(s,t) =4, because two substitutions (change "l" to "m" and change "v" to "l"), one insertion(insert “i”) and one deletion (delete “h”) is sufficient to transform s into t.

The greater the Levenshtein distance, the more different the patterns are.

3.2 Algorithm for Finding Total Number of Fuzzy Patterns Using Fuzzy Levens-htein Distance

Given a sequence of length *l*, query pattern of length *p*, candidate length *n* and percentage of fault tolerance allowed *k*. We want to find total number of candidate patterns that will approximately match to the query pattern with at most *k* tolerance.

1. Generate candidates from the given sequence.
2. Count=0.
3. Find the fuzzy Levenshtein distance between candidate pattern and query pattern.
4. If distance <= *k*, then increment count.
5. Repeat steps 3 and 4 for all generated candidates.
6. Output count as total number of fuzzy patterns.

Example: Given a sequence “TTTATC”, candidate length 4, query pattern TCAT and allowed tolerance 50%. Then to find total number of patterns obtained by fuzzy distance and exact distance, first we have to generate candidates of length 4 from the sequence: TTTA, TTAT, TATC. Fuzzy distance between query and generated candidates are 3, 1, 2 respectively. Similarly exact distances are 3, 1, 3 respectively. Total number of candidates that are matching to query within given tolerance by fuzzy distance and exact distance are 2 and 1 respectively.

We compare the results obtained using fuzzy distance to exact distance. To find the exact distance between candidate and query pattern, we count number of places in which the two patterns differ, i.e., have different characters.

4 Results and Discussion

We consider three DNA sequences of bacteria (bacillus-subtilis) for our experiment. Table 1 shows details of each sequence length and the number of candidates generated with different candidate length.

To find approximately matching patterns we used a query of length 5 and allowed fault tolerance of 40%. Candidate length is varied from 3 to 7, to allow both positive and negative tolerance i.e., when the query length is 5, candidate length should be 3 for -40% tolerance and it should be 7 for +40% tolerance. Table 2 show cumulative result i.e. total number of approximate patterns obtained for all 3 sequences using fuzzy and exact distances. We can observe that total numbers of patterns obtained by using fuzzy approach are grater than exact approach because, patterns rarely match the sequences exactly. The detailed analysis is explained in the following cases.

Table 1. Sequence length and number of candidates generated

SN	*SL*	*Number of Candidates Generated*				
		CL=3	CL =4	CL =5	CL =6	CL =7
1	1554	1552	1551	1550	1549	1548
2	2928	2926	2925	2924	2923	2922
3	1904	1902	1901	1900	1899	1898
Total	6386	6380	6377	6374	6371	6368

Legend : SN : Sequence Number; SL : Sequence Length; CL : Candidate Length.

Case 1. In this case candidate length is fixed and the distance between candidate and query pattern is varied. For example, if the candidate length is 4, then for 40% tolerance the distance can be either 1 or 2 with query length 5.

Table 2. Total number of patterns obtained for 3 sequences

Approach Used	Patterns Obtained for different CL				
	3	4	5	6	8
Fuzzy Approach	493	897	917	453	147
Exact Approach	117	406	783	159	22

Fig. 1 shows that when candidate length is 3 and query length is 5, then for 40% fault tolerance all 3 characters in the candidate should match to the query (D=2). Total number of patterns obtained using fuzzy distance and exact distance are 493 and 117 respectively.

Fig. 2 shows that when candidate length is four and query length is five, then for 40% fault tolerance all four characters in the candidate can match (D=1) or three characters can match (D=2) to query. We can observe that as distance decreases, the number of obtained patterns will also decrease.

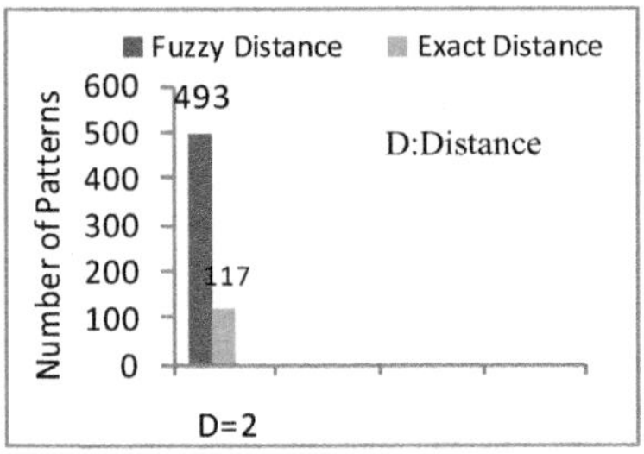

Fig. 1. Candidate length 3

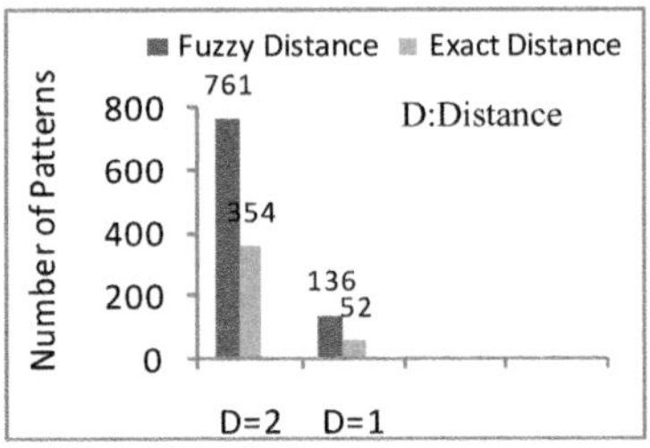

Fig. 2. Candidate length 4

Table 3 summarizes similar results for candidate length 5, 6 and 7. When candidate length is 5, then for 40% tolerance either 3 (D=2), 4 (D=1) or all 5 (D=0) characters in the candidate can match to query and for each distance, number of patterns obtained are as tabulated. Distance equal to zero indicates exact match of candidate and query. Therefore number of patterns obtained by using fuzzy distance and exact distance are equal. For our dataset, when distance is 1, once again the total patterns obtained by using both fuzzy and exact distance are equal. When distance is 2, any 3 characters out of 5 characters can match then the number of patterns obtained are more in case of fuzzy distance compared to exact distance.

When the candidate length is 6, then for 40% tolerance either 5 (D=1) or 4 (D=2) characters can match to query of length 5 . Similarly when candidate length is 7, any 5 (D=2) characters can match to query. In all these cases the number of patterns obtained are more in case of fuzzy distance compared to exact distance and as distance decreases, the number of obtained patterns will also decrease. A zero value for number of patterns indicates that, there are no approximately matching patterns with given fault tolerance.

Case 2. In this case candidate length is varied and the distance between candidate and query pattern is kept fixed with query length 5 and tolerance of 40%.

If the distance is zero, candidate length must be 5 for a query of length 5 and the number of patterns obtained using both fuzzy and exact distance are equal i.e. 22.

Table 3. Total number of patterns obtained with query length 5 and 40% fault tolerance

CL	*Number of Patterns*					
	D=2		D=1		D=0	
	FA	EA	FA	EA	FA	EA
5	760	624	137	137	22	22
6	388	137	65	22	0	0
7	147	22	0	0	0	0

Legend: CL : Candidate Length; D: Distance; FA: Fuzzy Approach; EA: Exact Approach.

Fig. 3 shows that if the distance is 1, candidate length can be 4, 5 and 6 and the number of character matches must be all 4, any 4 and any 5 respectively.

Fig. 4 shows that if the distance is 2, candidate length can be 3, 4, 5, 6 and 7 and the number of character matches must be all 3, any 3, any 3, any 4 and any 5 respectively.

With all these distances, as the candidate length increases, total number of approximately matching patterns will increase on the negative side (candidate length from 3 to 5) and will decrease on the positive side (candidate length 6 and 7). We can observe that the total number of patterns obtained in case of fuzzy distance are more compared to exact distance.

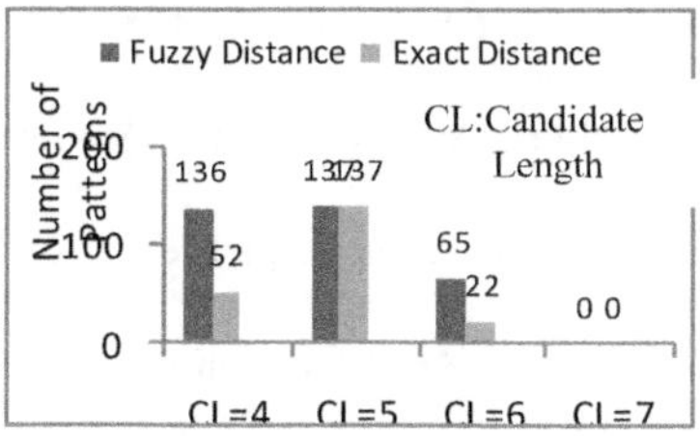

Fig. 3. Distance 1

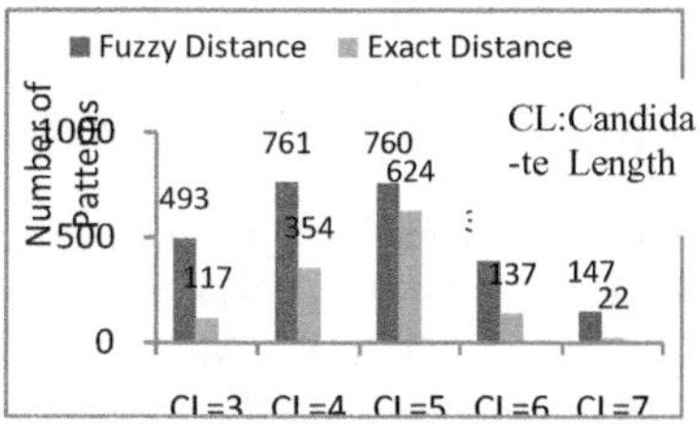

Fig. 4. Distance 2

5 Conclusions and Future Work

In real world biological applications, most of the sequences are "similar" instead of exactly the same. Patterns rarely match the sequences exactly. Fuzzy sets, which are a component of soft computing, are suitable for handling the issues related to understandability of patterns, incomplete/noisy data, and can provide approximate solutions faster. We have proposed a novel methodology based on fuzzy Levenshtein distance to extract approximately matching patterns from the DNA sequences data of bacteria obtained from National Center for Biotechnology Information (NCBI). We have also shown the complete analysis of the approximate matching patterns (fuzzy patterns) by varying candidate length and distance. The results are compared to pattern matching with exact distance method and it was observed that the fuzzy approach resulted in more number of patterns that are approximately matching to given query. We are also working on DNA sequence classification by using approximately matching patterns with fault tolerance.

References

1. Mitra, S., Pal, S.K., Mitra, P.: Data Mining in Soft Computing Framework: A Survey. IEEE Trans. on Neural Networks 13(1), 3–14 (2002)
2. Han, J., Kamber, M.: Data Mining: Concepts and Techniques. Academic Press, New York (2001)
3. Gately, E.: Neural Network for Financial Forecasting. John Wiley & Sons, Chichester (1996)
4. Misener, S., Krawetz, S.A.: Bioinformatics: Methods and Protocols. Human Press Inc. (2000)
5. Gusfield, D.: Algorithms on Strings, Trees, and Sequences. Cambridge University Press, Cambridge (1997)
6. Luscombe, N.M., Greenbaum, D., Gerstein, M.: What is bioinformatics? A proposed definition and overview of the field. Methods Inf. Med. 40, 346–358 (2001)
7. Salzberg, S., Searls, D., Kasif, S.: Computational Methods in Molecular Biology. Elsevier, Amsterdam (1998)
8. National Center for Biotechnology Information, `ftp://ftp.ncbi.nlm.nih.gov/genbank/genomes/bacteria/bacillus-subtilis/AL009126.frn`
9. Boyer, R.S., Moore, J.S.: A fast string searching algorithm. Communications of the ACM 20(10), 762–772 (1977)
10. Knuth, D.E., Morris Jr., J.H., Pratt, V.R.: Fast pattern matching in strings. SIAM J. of Computing 6(2), 323–350 (1977)
11. Navarro, G.: A guided tour to approximate string matching. ACM Computing Surveys 33, 31–88 (2001)
12. Smith, T., Waterman, M.: Identification of common molecular subsequences. J. of Mol. Biol. 147, 195–197 (1981)
13. Altschul, S., Gish, W., Miller, W., Myers, E., Lipman, D.: Basic local alignment search tool. J. of Mol. Biol. 215, 403–410 (1990)
14. Prosite Scan, `http://npsa-bil.ibcp.fr/cgi-bin/npsa_automat.pl?page=npsa_prosite.html`
15. Chang, B.C.H., Halgamuge, S.K.: Approximate Symbolic Pattern Matching for Protein Sequence Data. Int. J. of Approximate Reasoning 32(2-3), 171–186 (2003)

DFIG Based Wind Energy System (WES)

M. Arun Bhaskar, B. Vidya, V. Meenatchi, and R. Dhanduyathabani

Department of EEE, Velammal Engineering College
Chennai, Tamil Nadu
m.arunbhaskar@gmail.com, viya.bhaskaran@gmail.com

Abstract. This paper presents the DFIG wind energy system. The modelling of the system has been discussed and also the system parameters are plotted. MATLAB Simulink has been used as the tool to evaluate the system. The grid parameter variations are also discussed.

Index Terms: Doubly-fed induction generator (DFIG), Wind energy conversion system (WECS), distributed generation (DG).

1 Introduction

In recent years, the average rate of world primary power consumption has increased to about 16TW (or) 16000GW. Moreover, the environmental impact on usage of the conventional sources has been disintegrative with the environmental issues such as pollution, global warming, excessive greenhouse effect etc. Because of these problems and our dwindling supply of petroleum, finding sustainable alternatives is becoming increasingly urgent. Perhaps, the greatest challenge is in devising a sustainable future, which relies on integration and control of renewable energy sources in grid distributed generation. Generation of power locally at distribution voltage level by using non-conventional (or) renewable energy sources like, solar photo voltaic cells, wind power, biogas, and fuel cell is known as distributed generation (DG).

Hence for the present situation utilization of renewable energy has become a vital strategy. Hence wind energy conversion has been discussed in this paper.

2 Modelling of DFIG Based WECS

A commonly used model for induction generator converting power from the wind to serve the electric grid is shown Fig.3 The stator of the wound rotor induction machine is connected to the low voltage balanced three-phase grid and the rotor side is fed via the back-to-back PWM voltage-source inverters with a common DC link. Grid side converter controls the power flow between the DC bus and the AC side and allows the system to be operated in sub-synchronous and super synchronous speed. The proper rotor excitation is provided by the machine side power converter and also it provides active and reactive power control on stator and rotor sides respectively by employing

V.V. Das, J. Stephen, and Y. Chaba (Eds.): CNC 2011, CCIS 142, pp. 219–223, 2011.

vector control. DFIG can be operated as a generator as well as a motor in both sub-Synchronous and super synchronous speeds, thus giving four possible operating modes. Only the two generating modes at sub-synchronous and super synchronous speeds are of interest for wind power generation. To exploit the advantages of variable speed operation, the tracking of optimum torque-speed curve is essential. Speed can be adjusted to the desired value by controlling torque. So, an approach of using active power set point from the instantaneous value of rotor speed and controlling the rotor current *i*ry in stator flux-oriented reference frame to get the desired active power will result in obtaining the desired values of speed and torque according to the optimum torque speed curve. The reactive power set point can also be calculated from active power set point using a desired power factor.

In the stator flux-oriented reference frame, reactive power can be controlled by controlling the *d*-axis rotor current. In stator flux-oriented control, both stator and rotor quantities are transformed to a special reference frame that rotates at an angular frequency identical to the stator flux linkage space phasor with the real axis (*x*-axis) of the reference frame aligned to the stator flux vector. At steady state, the reference frame speed equals the synchronous speed. This model is called dynamic vectorized model[3].

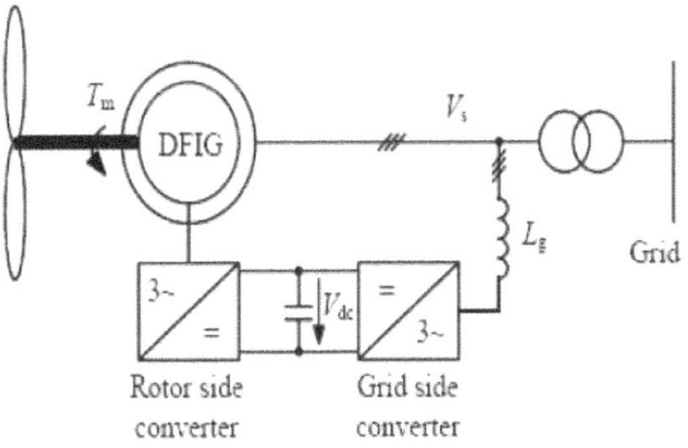

Fig. 1. Wind turbine driven DFIG

The main variables of the machine in rotating frame are flux linkages φ_{qs}, φ_{ds} φ'_{qr} φ'_{dr} in state space form. Substituting the conditions $\omega = \omega_r$ and $V_{qr}=V_{dr}=0$ in the flux linkage equation, we get

$$\varphi_{qs} = \omega_b \int v_{qs} - \omega_r/\omega_b \varphi_{ds} + r_s/x_{ls} (\varphi_{mq} - \varphi'_{qs}) \quad (1)$$

$$\varphi_{ds} = \omega_b \int v_{ds} - \omega_r/\omega_b \varphi_{qs} + r_s/x_{ls} (\varphi_{md} - \varphi'_{ds}) \quad (2)$$

$$\varphi'_{qr} = \omega_b \int r'r/xl_r(\varphi_{mq} - \varphi'_{qr}) \quad (3)$$

$$\varphi'_{dr} = \omega_b \int r'r/xl_r(\varphi_{md} - \varphi'_{dr}) \quad (4)$$

The currents can now be calculated

$$I_{qs}=(\varphi_{qs}-\varphi_{mq})/x_{ls} \quad (5)$$

$$I_{ds}=(\varphi_{ds}-\varphi_{md})/x_{ls} \quad (6)$$

$$I'_{qr}=(\varphi'_{qr}-\varphi_{mq})/x'_{lr} \quad (7)$$

$$I'_{dr}=(\varphi'_{dr}-\varphi_{md})/x'_{lr} \quad (8)$$

Solving equations (4-8), the φ_{mq}, φ_{md} are obtained as

$$\varphi_{mq}=x_m(\varphi_{qs}/x_{ls}+\varphi'_{qr}/x'_{lr}) \quad (9)$$

For maintaining proper flow of variables and for convenience of simulating, the above equations are separated into the q-axis, the d-axis and the rotor circuits. In the q-axis circuit, the Equations (2), (4), (6), (8) and (10) are used to calculate, φ_{qs} , φ'_{qr},i_{qs} and i'_{qr} respectively and φ_{qs}, i_{qs} arc used in the calculation of electromagnetic torque. In the q-axis circuit, the Equations (3), (5), (7), (9) and (11) are used to calculate, φ_{ds}, φ'_{dr}, i_{ds} and i'_{dr} respectively and φ_{ds}, i_{qs} arc used in the calculation of electromagnetic torque. The rotor circuit makes use of the φ_{qs}, i_{qs} obtained from the q-axis circuit and φ_{ds}, i_{qs} obtained from the d-axis circuit and calculates the electromagnetic torque using equation (12).The rotor circuit also takes the input mechanical torque values supplied to it and computes W_r/W_b from equation (12).

$$\Phi_{md}=x_m(\varphi_{ds}/x_{ls}+\varphi'_{dr}/x'_{lr}) \quad (10)$$

Where $x_m=1(1/x_m+1/x_{ls}+1/x_{lr})$

Now that we know φ_{qs}, i_{qs} and φ_{ds}, i_{ds}, the electromagnetic torque can be calculated by

$$T_{em}=(3p/4\omega_b)(\varphi_{ds}i_{qs}-\varphi_{qs}i_{ds}) \quad (11)$$

The equation that governs the motion of rotor is obtained by equating the inertia torque to the accelerating torque

$$J(d\omega_m/dt)=T_{em}+T_{mech}-T_{damp} \quad (12)$$

Expressed in per unit values, equation (9) becomes:

$$2H\, d(\omega_r/\omega_b)/dt=T_{em}+T_{mech}-T_{damp} \quad (13)$$

3 Simulation Study and Results

In fig4 wind farm consists of six 1.5 MW wind turbines connected to a 25 kV distribution system exporting power to a 120 kV grid through a 30 km 25 kV feeder. A 2300V, 2 MVA plant consisting of a motor load (1.68 MW induction motor at 0.93 PF) and of a 200 kW resistive load is connected on the same feeder at bus B25. A 500 kW load is also connected on the 575 V bus of the wind farm.

The wind turbine model is a phasor model that allows transient stability type studies with long simulation times. In this case study, the system is observed during 50 s. The 6-wind-turbine farm is simulated by a single wind-turbine block by multiplying the following three parameters by six, as follows:

The nominal wind turbine mechanical output power: 6*1.5e6 watts, specified in the Turbine data menu. The generator rated power: 6*1.5/0.9 MVA (6*1.5 MW at

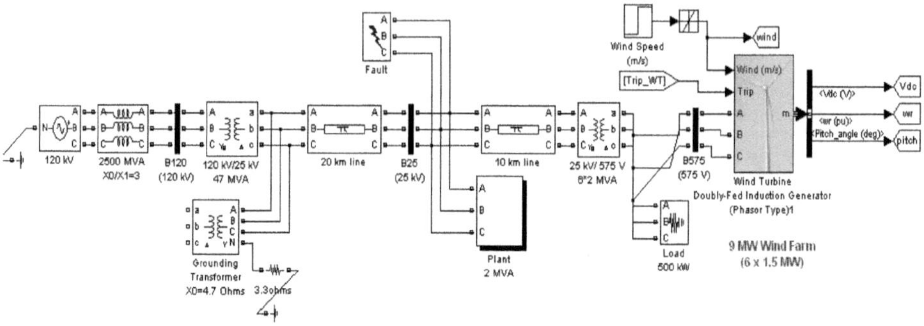

Fig. 2. Simulink model of WECS

0.9 PF), specified in the Generator data menu. The nominal DC bus capacitor: 6*10000 microfarads, specified in the Converters data menu.

Initially, in fig 5 wind speed is set at 8 m/s, and then at t=5s, wind speed increases suddenly at 14 m/s. At t=5 s, in fig 6 the generated active power starts increasing smoothly (together with the turbine speed) to reach its rated value of 9MW in approximately 15s.

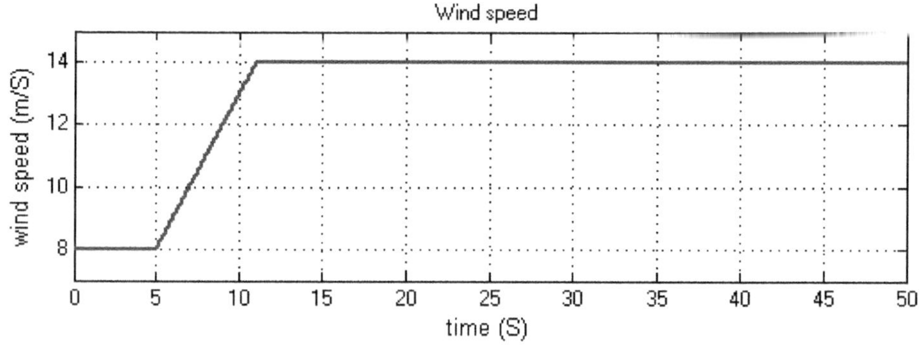

Fig. 3. Wind speed

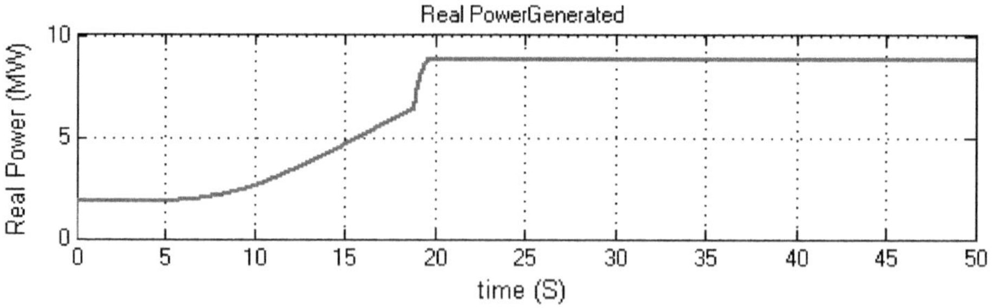

Fig. 4. Real power

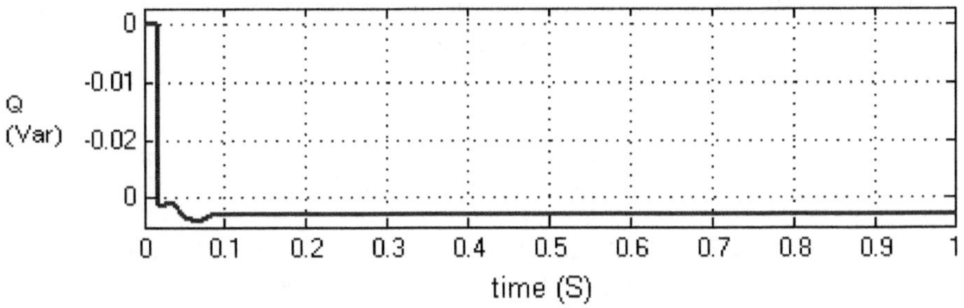

Fig. 5. Reactive power

4 Conclusion

A novel concept of utilization of wind energy for grid connection has been proposed and validated through MATLAB/Simulink simulations. The wind energy conversion system has been modelled. This novel strategy utilizes renewable energy ie wind energy and also basic electrical drive. Hence there is reduction in cost and also environmental welfare as wind energy poses to be a pollution free system.

References

1. Chitti Babu, B., Mohanty, K.B.: Doubly-Fed Induction Generator for Variable Speed Wind Energy Conversion Systems- Modelling & Simulation. International Journal of Computer and Electrical Engineering 2(1), 141–147 (2010)
2. Yazdani, A., Dash, P.P.: A Control Methodology and Characterization of Dynamics for a Photovoltaic (PV) System Interfaced With a Distribution Network. IEEE Transactions on Power Delivery 24(3), 1538–1551 (2009)
3. Patel, M.R.: wind and solar power systems: design, analysis and operation. Wiley/IEEE Press, NJ (2000)
4. Petersson, A., Thiringer, T., Harnefors, L., Petru, T.: Modeling and Experimental Verification of GridInteraction of a DFIG Wind Turbine. IEEE Transactions on Energy Conversion 20(4), 878–886 (2005)

Modelling and Voltage Stability Enhancement Using "Sen" Transformer

M. Arun Bhaskar[1], A. Venkatesh[1], S.S. Dash[2], C. Subramani[2], and M. Jagadeesh Kumar[1]

[1] Department of EEE, Velammal Engineering College, Chennai-600066
[2] Department of EEE, SRM University, Chennai
m.arunbhaskar@gmail.com

Abstract. This paper covers the modeling of a 2 bus system using the Matlab package. The basic system is tested under large and small disturbances to study the dynamic behavior of the system and the stability margins associated with the different configurations of the system. As a suggested solution to increase stability margins of the system, Sen transformer is added, modeled and tested to show the effect of the controller on the different stability margins under both large and small disturbances.

Keywords: Voltage Stability, "Sen" Transformer, Simulation.

1 Introduction

A. Voltage Stability

Voltage stability refers to ''the ability of a power system to maintain steady voltages at all buses in the system after being subjected to a disturbance from a given initial operating condition'' (IEEE-CIGRE, 2004). If voltage stability exists, the voltage and power of the system will be controllable at all times. In general, the inability of the system to supply the required demand leads to voltage instability (voltage collapse).

The nature of voltage instability phenomena can be either fast (short-term, with voltage collapse in the order of fractions of a second to a few seconds) or slow (long-term, with voltage collapse in minutes to hours) (IEEE-CIGRE, 2004). Short-term voltage stability problems are usually associated with the rapid response of voltage controllers (e.g., generators' automatic voltage regulator [AVR]) and power electronic converters, such as those encountered in flexible AC transmission system or FACTS controllers and high voltage DC (HVDC) links. In the case of voltage regulators, voltage instability is usually related to inappropriate tuning of the system controllers. Voltage stability in converters, on the other hand, is associated with commutation issues in the electronic switches that make up the converters, particularly when these converters are connected to ''weak'' AC systems, i.e., systems with poor reactive power support.

V.V. Das, J. Stephen, and Y. Chaba (Eds.): CNC 2011, CCIS 142, pp. 224–228, 2011.

2 Modelling of Sen Transformer

A digital computer simulation model of the ST has been developed using MATLAB. The model consists of two subsystems: the electrical subsystem and the tap-selection algorithm subsystem.

A. Electrical System

The electrical system is comprised of two ac systems connected by a three phase transmission line. The ST is connected at the sending end of the transmission line. Table I gives the parameters for both the ST and the network.

1) Electrical Network Model: The ac sources at both sending and receiving ends are modeled as infinite sources with the same magnitude but at a phase difference of 20 (the receiving end voltage lags the sending end voltage). The transmission line is modeled as a lumped series impedance.
2) ST Model: The ST is a specially designed transformer with multiple windings having multiple tap positions in the secondary. The model for such a transformer is not available in matlab. Therefore, nine single-phase transformers, each having on-load tap changing capability have been used to model the ST. By using single-phase transformers, inter-phase. mutual flux linkage and thus mutual inductance has not been considered, which may cause some discrepancies in the results.These nine single-phase transformers are modeled with a small resistance and leakage reactance as shown in Table I. Output voltages of three transformers (contributing from phase , and) are added in series and then fed to one phase of the transmission line. The nine outputs (aa, ab, ac, ba, bb, bc, ca, cb, and cc) from the tap-selection algorithm supply the value of tap setting to all nine transformer Tap terminals. Should these outputs undergo any changes, the transformers readjust their tap positions and produce the required compensating voltages.
3) Tap Changer Model: In a practical transformer, tap changing is performed through a tap selector, where a resistor or inductor is used in parallel with the tap positions to limit the current through a shorting winding segment between two consecutive taps. In Fig. 1, an example of tap-changing operation has been shown along with the equivalent matlab model for each position of the tap selector. Although, practical transformers with on load tap changers (OLTC), such as the ones from Reinhausen use taps with voltage difference in the range of 0.02 p.u. to 0.067 p.u., in this model a voltage difference of 0.1 p.u. between taps has been assumed for theclarity of simulation. The time required to move the tap selector between adjacent tap positions is 2 s. In order to move the tap selector from its initial position, terminal E (0.2 p.u.) to its final position, terminal C (0.1 p.u.), the following four steps of approximately 0.5 s each are performed.

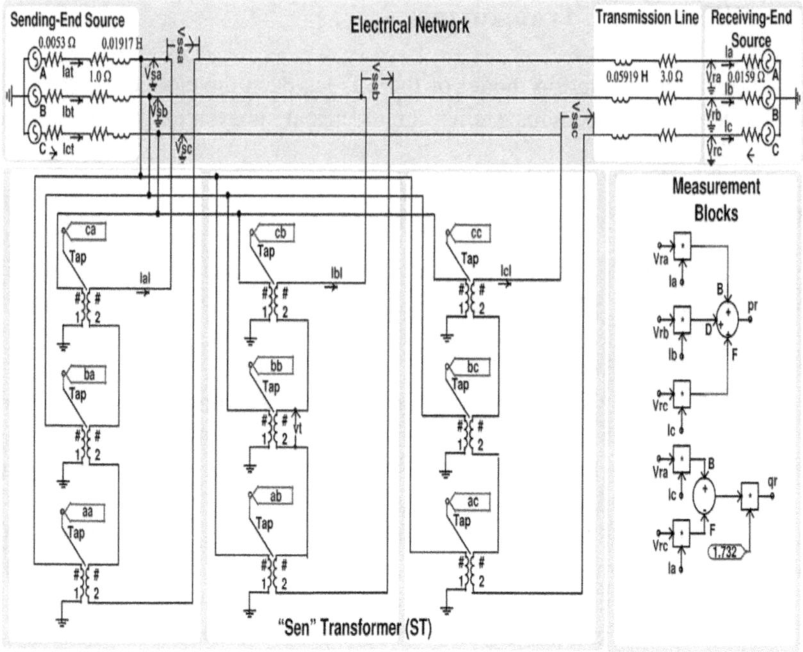

Fig. 1. Sen Transformer modelling

3 Testing

A. Modelling of 2 bus system

Fig 2 shows the simulink model of a simple 2 bus system connected through a transmission line which is represented by a three phase RLC branch connected in series.

The experiment is carried out for both cases, with and without "Sen" transformer. A three phase L-L-L-G fault is induced by connecting a three phase circuit breaker in the system. The simulation is run from 0 to 1 sec and the circuit breaker is opened from 0.5 to 1 sec. The line voltages and the line currents are calculated during fault condition. The line voltage will become zero from 0.5 to 1 sec due to the short circuit. And the fault current in all the three phases will increase above rated current. The Sen Transformer is connected in the system to compensate for this drop in voltage from 0.5 to 1 sec and to reduce the fault current.

There is no specific model for Sen Transformer available in simulink. So it is modeled using nine single phase transformers, three for each phase. Tap changers are used to control the operation of these nine transformers. In case of a sag or dip in the line voltage, the tap changers operate so that the required transformers are activated in order to compensate the voltage dip.

The transformers used are single phase multi winding transformers which have a fixed winding on the primary side and the secondary windings are split into a number of taps. By operating the taps, the required number of windings can be selected. For tap changing

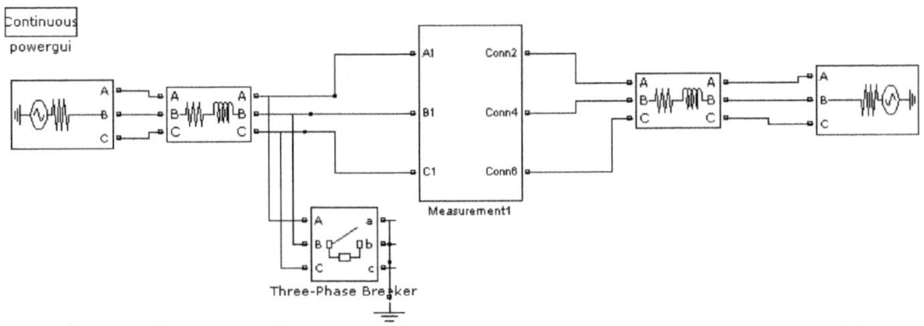

Fig. 2. 2 bus system

operation, circuit breakers are connected to each of the tap and they are operating by giving a specific transition by which they can be opened and closed accordingly.

4 Simulation and Results

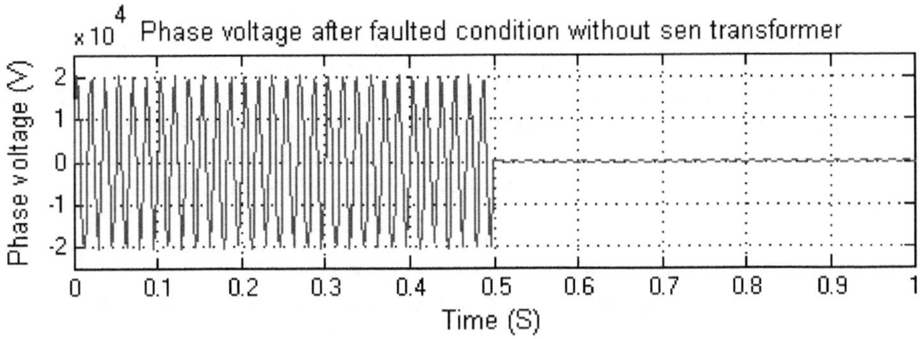

Fig. 3. Phase Voltage after faulted condition without Sen Transformer

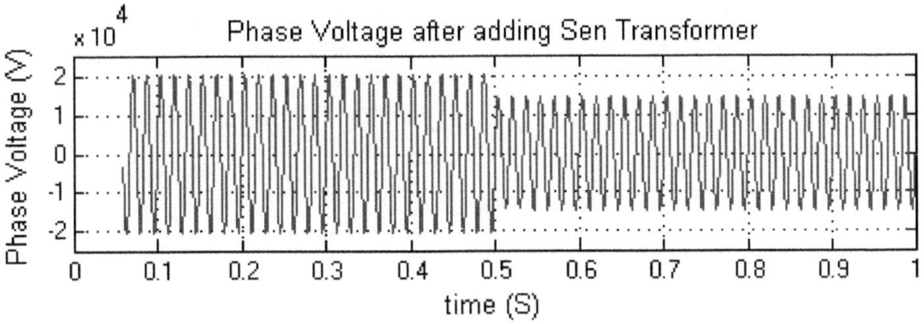

Fig. 4. Phase Voltage after adding Sen Transformer

Table 1.

Parameter	Normal	Faulted condition Without SEN transformer	With SEN transformer
Phase Voltage	20 kV	0 Kv	15 Kv
Fault Current	0 A	830 A	260 A

5 Conclusion

This paper discusses about the improvement in voltage stability using a Sen Transformer. A simple 2 bus system is used for this purpose. A three phase L-L-L-G fault is created in the system and the line voltages and the fault currents are calculated without connecting Sen Transformer and then after connecting Sen transformer. It is seen that the voltage stability is improved after connecting Sen Transformer.

References

1. Schulz, R.P.: Synchronous Machine Modeling. Symposium on Adequency and Philosophy of Modeling System Dynamic Performance. IEEE Pub. 75 CH O970-PWR (1975)
2. Sen, K.K., Sen, M.L.: Comparison of the 'Sen' Transformer with the Unified Power Flow Controller. IEEE Trans. Power Delivery 18(3), 1523–1533 (2003)
3. Sen, K.K., Sen, M.L.: Introducing the Family of 'Sen' Transformers: A set of Power Flow Controlling Transformers. IEEE Trans. Power Delivery 18(1), 149–157 (2003)
4. Sen, K.K., Keri, A.J.: Comparison of Field Results and Digital Simulation Results of Voltage-Sourced Converter-Based FACTS Controllers. IEEE Trans. Power Delivery 18(1), 300–306 (2003)
5. Sen, K.K., Stacey, E.J.: UPFC-unified power flow controller: Theory, modeling and applications. IEEE Trans. Power Del. 13(4), 1453–1460 (1998)
6. Load Tap Changer, Type RMV-A, Reinhausen Manufacturing, instruction manual, TL 8001.01

DBAM: Novel User Based Bandwidth Allocation Mechanism in WiMAX

Niharika Kumar[1], K.N. Balasubramanya Murthy[1], and Amitkeerti M. Lagare[2]

[1] PESIT, Bangalore, India
niharika.kumar@gmail.com, principal@pes.edu,
[2] IIIT-Bangalore, India
amitkeerti@rediffmail.com

Abstract. IEEE 802.16e classifies packets into 5 different service classes, namely, UGS, rtPS, ertPS, nrtPS and BE. Bandwidth is allocated to SS based on the need of the service class. In this paper we propose a user based bandwidth allocation mechanism called Differentiated Bandwidth Allocation Mechanism (DBAM). WiMAX users shall be divided into 3 different priority levels as high-priority users, regular users and low-priority users. Bandwidth shall be allocated to the service flows of these service class based on the priority of the users. This method of bandwidth allocation provides a differentiated QoS to users for the same service class. Simulation results revel that when the number of Subscriber Stations (SS) is 10, high-priority and Regular users are able to transmit 25% more ertPS data compared to low-priority users and when the number of SS is 14, high-priority user transmits 30% more data compared to regular user.

Keywords: QoS, WiMAX, Bandwidth Allocaiton, MAC, DBAM.

1 Introduction

IEEE 802.16e provides differentiated Quality-of-Service (QoS) by providing five different service classes, namely, Unsolicited Grant Services (UGS), extended Real Time Polling Services (ertPS), Real Time Polling Services (rtPS), non Real Time Polling Services (nrtPS), Best Effort(BE) [1][2][3].

WiMAX supports demand based bandwidth allocation mechanism where each service flow at the Subscriber Station (SS)/Mobile Stations (MS) requests uplink bandwidth based on the amount of data present in the respective queues. The Base Station (BS) will allocate bandwidth based on the type of service flow and availability of bandwidth. The bandwidth can be allocated on Grant Per Connection (GPC) or Grant Per Subscriber Station (GPSS). GPC mechanism allocates bandwidth on a per connection basis where as GPSS allocates aggregate bandwidth on a per SS basis.

In order to allocate bandwidth, BS will apply a bandwidth allocation scheme. Various bandwidth allocation schemes have been proposed by researchers. In [4]

V.V. Das, J. Stephen, and Y. Chaba (Eds.): CNC 2011, CCIS 142, pp. 229–236, 2011.

authors propose aggregated bandwidth request mechanism. In [5] authors propose service class based bandwidth allocation. In [6] adaptive bandwidth request scheme is proposed where in contention-free bandwidth request mechanism is incorporated within the contention based opportunities. In [7] a predictive bandwidth allocation mechanism which predicts the future data flow based on current data rates is proposed. In [8] a channel aware bandwidth allocation for downlink is proposed.

In this paper we propose a bandwidth allocation scheme called Differentiated Bandwidth Allocation Mechanism (DBAM). This allocation scheme divides the WiMAX users into three different Priority levels, namely high-priority users, regular users and low-priority users (defined in section 3). For each service class the high priority users shall be allocated bandwidth first followed by regular users and finally the low-priority users in a round robin fashion. To the best of our knowledge, according to available literature, user based bandwidth allocation within the framework of service based bandwidth allocation is yet to be studied in detail for 4G broadband wireless access network.

2 Existing Bandwidth Allocation Algorithm

WiMAX supports demand based bandwidth allocation mechanism. If a particular service flow has data to be transmitted on the uplink, a bandwidth request is made either by using a Bandwidth request header or piggybacking the bandwidth request with uplink data.

Upon receiving the bandwidth request from the service flows of all SS in the network, the BS shall apply a bandwidth allocation algorithm to distribute the uplink bandwidth among the SSs that have requested for bandwidth. While allocating the bandwidth the BS has to ensure that it takes care of the QoS constraints for the services flows that have requested for bandwidth. One of the most commonly used bandwidth allocation algorithm is the strict priority algorithm. A high level view of the algorithm is presented in Fig. 1 (a). Within each service class, the service flows can be allotted bandwidth in a round robin fashion.

3 Differentiated Bandwidth Allocation Mechanism

We would like to propose a user-based bandwidth allocation mechanism called DBAM. There shall be 3 different types of WiMAX users called as: High-Priority users, Regular Users and Low-Priority Users. High-Priority users could be those users who are ready to pay more to receive higher priority for their traffic. Regular users are those users who do not wish to be associated with any priority. Low-Priority users could be those users who are ready pay less for wireless broadband service and settle with lesser quality of service. A high level view of DBAM is provided in Fig. 1(b).

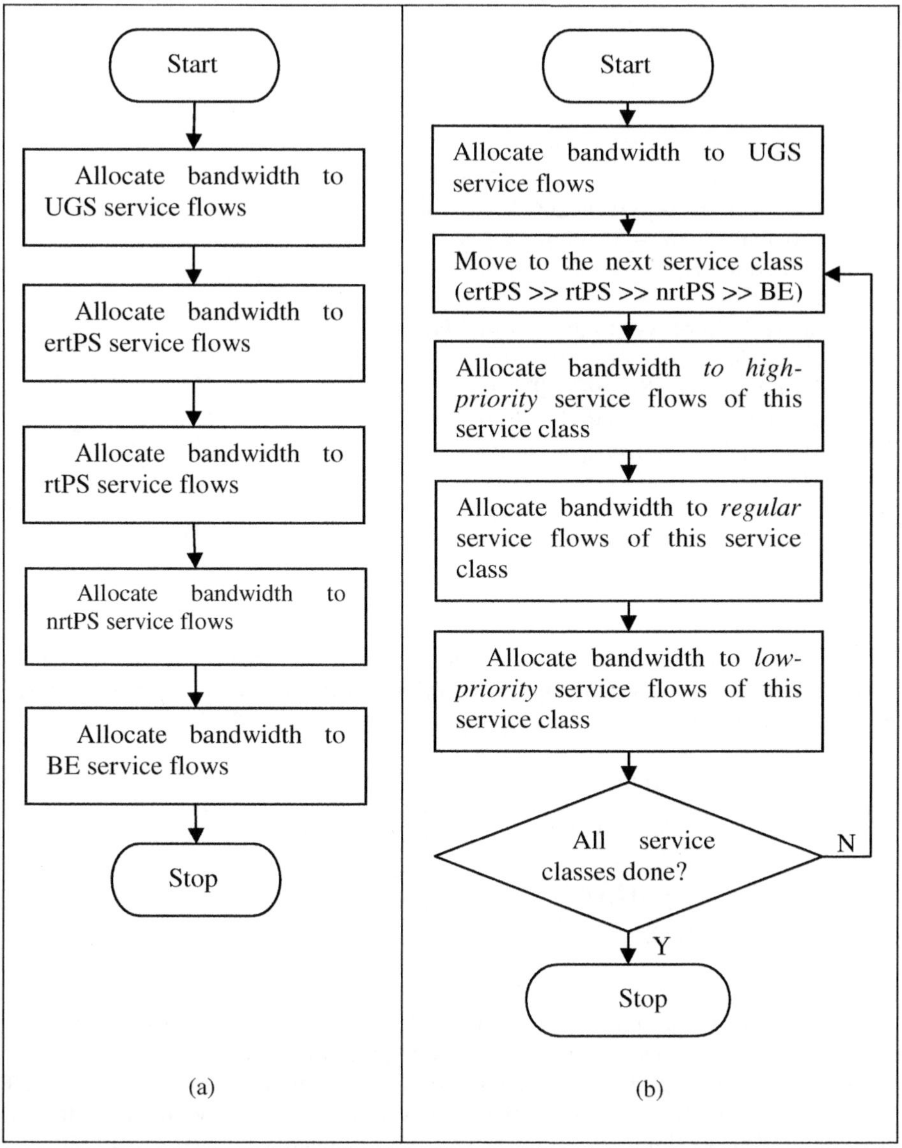

Fig. 1. Strict Priority Bandwidth Allocation Algorithm[5] **v/s** Differentiated Bandwidth Allocation Mechanism

DBAM algorithm is described in detail below:

```
differentiatedBandwidthAllocation() Begin
    int available_bw := calc_total_bw( );
   //Step 1: Allot bandwidth to ertPS
   //Step 1.1: Allot BW to priority ertPS
    for(i:=0; i<no_high_pri_ertps_sf; i++) Begin
```

```
            int allotted := allocate_ertPS_bw(fl[i]);
           available_bw :=  available_bw – allotted;
           if(available_bw EQ 0)   return;
        End
        //Step 1.2: Allot BW to regular ertPS
        for(i:=0; i<no_reg_ertps_sf; i++) Begin
            int allotted := allocate_ertPS_bw(fl[i]);
            available_bw := available_bw - allotted
            if(available_bw EQ 0)  return;
        End
        //Step 1.3: Allot BW to low-priority ertPS
        for(i:=0; i<no_low_pri_ertps_sf; i++) Begin
          int allotted := allocate_ertPS_bw(fl[i]);
          available_bw :=  available_bw – allotted;
          if(available_bw EQ 0)  return;
        End
    End
```

4 Implementation

Authentication, Authorization and Accounting (AAA) server shall maintain a table of MAC address of the SS and the associated priority value of the SS. A sample state of the Table could be as given in Table-1.

Table 1. Sample Table State at AAA Server

MAC Address	Priority Value
01:67:43:82:98:ba	1
04:56:84:ba:12:97	0
56:23:85:56:37:86	2
………………………	…………………

Priority value 0 is assigned to regular IEEE-802.16e users. Priority value 1 is for high-priority user and priority value 2 is for low-priority users.

When the SS initiates the initial ranging process by sending a range request (RNG-REQ) to the BS, the BS shall obtain the priority value associated with the SS from the AAA server. On obtaining the priority values associated with the SS/MS, BS shall allocate bandwidth for each service flow of SS based on the priority of SS and the type of service flow.

5 Analytical Modeling

In this section we will describe a model for the system. For the purpose of brevity we will consider only the ertPS service flow. The system of equations would be similar for the other service classes. Table-2 shows the modeling parameters.

Table 2. Notations Used for System Modelling

Symbol	Description
ertps_pri_bw_req(p)	Bandwidth needs of pth ertPS service flow of priority SS.
ertps_reg_bw_req(p)	Bandwidth needs of pth ertPS service flow of regular SS.
ertps_npr_bw_req(p)	Bandwidth needs of pth ertPS service flow of low-priority SS
ertps_pri_bw_allot(p)	Bandwidth allotted to the pth ertPS service flow of priority SS.
ertps_reg_bw_allot(p)	Bandwidth allotted to the pth ertPS service flow of regular SS.
ertps_npr_bw_allot(p)	Bandwidth allotted to the pth ertPS service flow of low-priority SS.
tot_bw	Total bandwidth available on the uplink for the current frame
tr	Minimum Reserved traffic rate
avl_bw	Amount of unallocated bandwidth available in the frame.
m	Number of high-priority ertPS service flows
n	Number of regular ertPS service flows
o	Number of low-priority ertPS service flows

Bandwidth is allocated to high-priority ertPS service flows as per (1)

$$ertps\_pri\_bw\_allot(p) = \begin{cases} ertps\_pri\_bw\_req(p) \\ \quad \text{if } ertps\_pri\_bw\_req(p) < tr \\ \quad \text{and } ertps\_pri\_bw\_req(p) < avl\_bw \\ tr \quad \text{if } tr \le ertps\_pri\_bw\_req(p) \\ \quad \text{and } ertps\_pri\_bw\_req(p) < avl\_bw \\ avl\_bw \quad \text{if } avl\_bw \le ertps\_pri\_bw\_req(p) \\ \quad \text{and } avl\_bw \le tr \\ tr \quad \text{otherwise} \end{cases} \tag{1}$$

After allotting bandwidth to the pth priority ertPS service flow, the leftover bandwidth is calculated as:

$$avl\_bw = tot\_bw - \left(\sum_{j=1}^{x} ertps\_pri\_bw\_allot(j) \right) \tag{2}$$
$$x \le m,$$

After allotting bandwidth to the priority ertPS service flows, bandwidth is allotted to the regular ertPS service flows as:

$$ertps\_reg\_bw\_allot(p) = \begin{cases} ertps\_reg\_bw\_req(p) \\ \quad \text{if } ertps\_reg\_bw\_req(p) < tr \\ \quad \text{and } ertps\_reg\_bw\_req(p) < avl\_bw \\ tr \quad \text{if } tr \le ertps\_reg\_bw\_req(p) \\ \quad \text{and } ertps\_reg\_bw\_req(p) < avl\_bw \\ avl\_bw \quad \text{if } avl\_bw \le ertps\_reg\_bw\_req(p) \\ \quad \text{and } avl\_bw \le tr \\ tr \quad \text{otherwise} \end{cases} \tag{3}$$

Leftover bandwidth after allotting bandwidth to regular ertPS is calculated as:

$$avl\_bw = avl\_bw - \left(\sum_{j=1}^{x} ertps\_reg\_bw\_allot(j) \right) \tag{4}$$

$$x \le n,$$

After allotting bandwidth to the regular ertPS service flows, bandwidth is allotted to the low-priority ertPS service flows as given in (5):

$$ertps\_npr\_bw\_allot(p) = \begin{cases} ertps\_npr\_bw\_req(p) & \\ \quad \text{if } ertps\_npr\_bw\_req(p) < tr & \\ \quad \text{and } ertps\_npr\_bw\_req(p) < avl\_bw & \\ tr \quad \text{if } tr \le ertps\_npr\_bw\_req(p) & \\ \quad \text{and } ertps\_npr\_bw\_req(p) < avl\_bw & \\ avl\_bw \quad \text{if } avl\_bw \le ertps\_npr\_bw\_req(p) & \\ \quad \text{and } avl\_bw \le tr & \\ tr \quad \text{otherwise} & \end{cases} \tag{5}$$

Leftover bandwidth is calculated as:

$$avl\_bw = avl\_bw - \left(\sum_{j=1}^{x} ertps\_npr\_bw\_allot(j) \right) \tag{6}$$

$$x \le o,$$

Bandwidth allocation for service flows of rtPS, nrtPS, BE service classes for high-priority, regular and low-priority users shall happen in the same fashion as given in eqn (1) – (6).

6 Results and Discussion

Simulations were carried out to analyze the proposed DBAM. Simulations were carried out using the network simulator NS 2.29[10]. Light WiMAX Module (LWX)[9] was used on top of the NS-2.29 to simulate WiMAX environment. We modified the strict priority bandwidth allocation algorithm of LWX to implement DBAM. Simulations were carried out using the parameters of LWX as per Table-3.

Table 3. Simulation Parameters

Parameter	Value
Uplink data rate	10 Mbps
OFDMA Frame Duration	5 ms
OFDMA symbol time	100.94 μs
eRTPS data arrival rate	1 Mbps

We setup the simulation network in such a way that at any given time 1/3rd of SS in the network are high-priority SS, another 1/3rd of the SS are regular SS and the remaining 1/3rd of the SS in the network are low-priority SS. Along with the uplink traffic from these SS, we also introduced downlink ftp traffic at the rate of 1Mbps.

Simulation was carried out to analyze the throughput of the SS for priority user, regular user and low-priority user. Simulation results are as shown in Fig. 2(a). Theoretical analysis was also done using the parameters of Table-3 results are shown in fig. 2 (b). The theoretical results closely match the simulation results.

When the number of SS in the system is less than 9, there is enough bandwidth support all SS. But when it exceeds 9, high-priority SS and Regular-SS are able to transmit their data at the expense of low-priority SS. Beyond 13 SS, high-priority SS enjoy higher throughput compared to Regular-SS and low-priority SS.

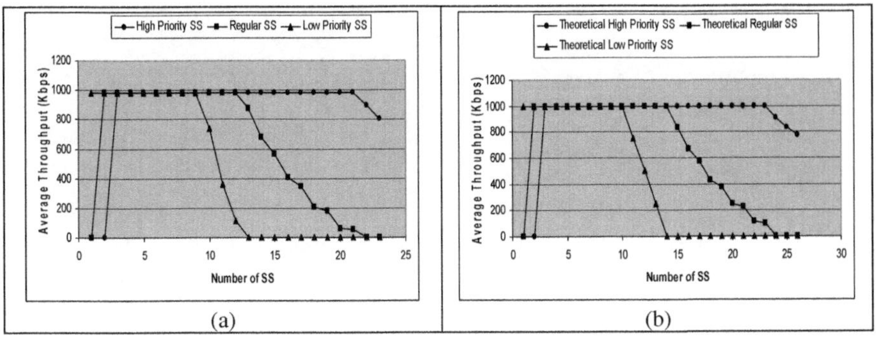

Fig. 2. Simulation results for ertPS traffic v/s Theoretical results for ertPS traffic

7 Conclusion

In this paper we presented a novel user based bandwidth allocation algorithm, DBAM. Simulation results show marked improvement in throughput for high-priority and regular users under high load conditions. DBAM is a win win for both users and operators.

References

1. IEEE std 802.16-2004, Air Interface for Fixed Broadband Wireless Access Systems
2. IEEE std 802.16e-2005, Air Interface for Fixed and Mobile Broadband Wireless Access Systems
3. Vaughan-Nichols, S.J.: Achieving Wireless Broadband with WiMAX. IEEE Comp. 37(6), 10–13 (2004)
4. Tao, S.Z., Gani, A.: Intelligent Uplink Bandwidth Allocation Based on PMP Mode for WiMAX. In: Proceedings of the 2009 International Conference on Computer Technology and Development, Malaysia, pp. 86–90 (2009)
5. Wee, K.K., Lee, S.W.: Priority based bandwidth allocation scheme for WIMAX systems. In: Proceedings of 2nd IEEE International Conference on Broadband Network & Multimedia Technology, pp. 15–18 (2009)
6. Liu, C.Y., Chen, Y.C.: An Adaptive Bandwidth Request Scheme for QoS Support in WiMAX Polling Services. In: Proceedings of The 28th International Conference on Distributed Computing Systems Workshops, pp. 60–65 (2008)

7. Peng, Z., Guangxi, Z., Haibin, S., Hongzhi, L.: A Novel Bandwidth Scheduling Strategy for IEEE 802.16 Broadband Wireless Networks. In: Proceedings of International Conference on Wireless Communications, Networking and Mobile Computing, pp. 2000–2003 (2007)
8. Lin, Y.N., Wu, C.W., Lin, Y.D., Lai, Y.C.: A Latency and Modulation Aware Bandwidth Allocation Algorithm for WiMAX Base Stations. In: Wireless Communications and Networking Conference, pp. 1408–1413 (2008)
9. Light WiMAX Module, `http://code.google.com/p/lwx/`
10. ns2, `http://www.isi.edu/nsnam/ns/`

Applying a Predictive Approach for QoS Monitoring in Web Service

Mahmoud Hossein Zadeh and Mir Ali Seyyedi

No. 23, 22 All, Javad Aleme St.,
Qom, Iran
Mh_zadeh@hotmail.com, maseyyedi@azad.ac.ir

Abstract. This present paper discuss a new approach aimed a monitoring the Quality of Service(QoS). This paper focuses on a predictive approach (PA) for QoS. QoS monitoring measures quality parameters and present true judgment about quality parameters. By using PA, components decisions would be more intelligent and efficiency, because decisions is taken according to current and future values of quality parameter. By applying PA, some period of time measuring of quality, isn't required, then monitoring cost is decreasing and consequently Service Oriented Architecture (SOA) overhead. For PA, Neural Network (NN) has been applied. We believe that applying a PA will result optimized, more efficiency an intelligent for SOA.

Keywords: SOA, Web Service, QoS Monitoring, PA, NN.

1 Introduction

SOA is a successful approach in IT and computing. SOA gives the best solution for distributed application problems, and have finally gained acceptance from both industry and research. Web services represent the most common realization of SOA.

With the popularity increase of developing web services with similar operational functions, QoS parameters plays a key role in selecting, detecting and combining web services. QoS monitoring methods receiving update the real values of quality parameters.

The monitoring may be performed differently at three levels, namely: service level, communication level and orchestration level. The service level monitoring considers the basic monitoring approach. The communication level monitoring intercepts exchanged messages between web service providers and requesters. The orchestration level monitoring supervises orchestrated services as BPEL using handlers provided by the orchestration engine such as Active BPEL. Besides, the monitoring is done at three places: consumer, server and third person like a BUS.

All methods shared in one point that is QoS monitoring at every time of web service calling. In the related works, much attention has been paid to the design QoS model and QoS monitoring with three levels.

Whether we can accomplish QoS monitoring with less overhead, more efficiency and more intelligent? To answer this question, we started this research.

V.V. Das, J. Stephen, and Y. Chaba (Eds.): CNC 2011, CCIS 142, pp. 237–241, 2010.

2 Approach

QoS monitoring to achieve less overhead, more efficient and intelligent by using of the PA. In PA, having the current values of QoS parameters, future values of QoS parameters is obtained based on its past behavior. The current values of QoS parameters making time series (as shown in Fig1.). Time Series Forecasting (TSF) takes an existing series of data $x_{t-n},\ldots,x_{t-2},x_{t-1},x_t$ and forecasts the $x_{t+1},x_{t+2},\ldots$ data values. TSF methods for the analysis of the present values of time series can use a variety of techniques.

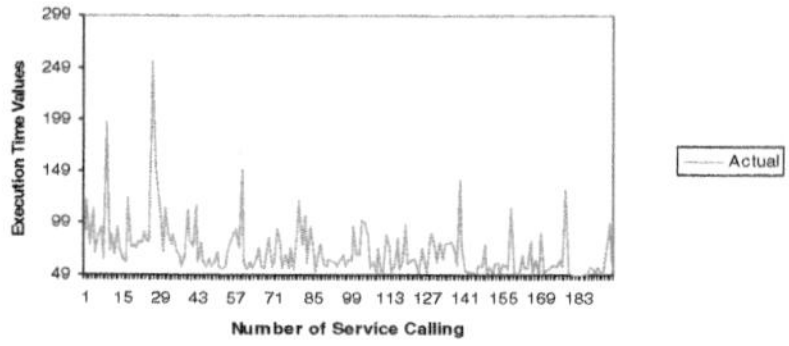

Fig. 1. QoS time series

In below, we describe more efficiency and intelligent and less overhead that is result of using PA in QoS monitoring, then we describe NN.

With using PA in Qos monitoring, future values of quality parameters (i.e. execution time) send to consumer, server and QoS management. In this condition, decisions would be more intelligent and efficiency, because decisions is taken according to current and future values of quality parameter. In this condition, components can to prevent of the failures in system. In this context, design self-healing architecture based on PA in QoS monitoring is our future work.

In the web service, QoS monitoring at every time of service calling, imposes some overhead on the SOA, then overhead is causing adverse effects on the service quality parameters such as Response Time, as shown in Fig2.[1].

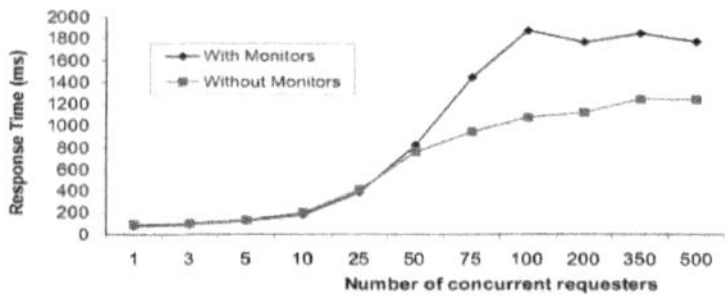

Fig. 2. Monitoring overhead

In order to estimate the monitor overhead, we conduct a large scale experiments under the gird5000 to measure the response time of web services while varying the requester's number from 1 to 500. They obtained the two curves shown in Figure 1. In the first, the monitoring is achieved using the ReqSideMon and the ProvSideMon (monitoring components).In the second, the measurement is done in the client code and without using monitors [1]. By applying PA, some period of time measuring of quality, isn't required, then monitoring cost is decreasing and consequently Service Oriented Architecture (SOA) overhead. Neural networks as a powerful technique is introduced to predict. Neural networks are shown their ability to solve complex problems. To implement the method we use the neural network. Given a training set of data, the NN can learn the data with a learning algorithm. Back propagation is used the most common algorithm. Through back propagation, the NN forms a mapping between inputs and desired outputs from the training set by altering weighted connections within the network.

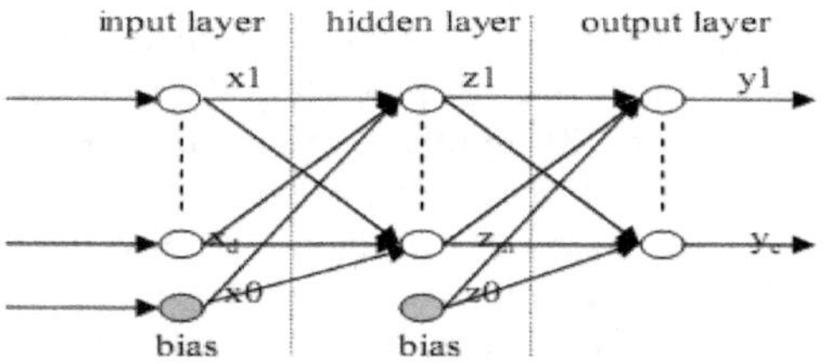

Fig. 3. 3-layer BPNN structure

(1) Forward pass: the outputs are calculated and the error at the output units calculated; and (2) Backward passes: The output unit error is used to alter weights on the output units. Then the error at the hidden nodes is calculated (by back-propagating the error at the output units through the weights), and the weights on the hidden nodes altered using these values. For each data pair to be learned, a forward pass and backwards pass is performed. This procedure is repeated over and over again until the error is at a low enough level. [2]

3 Implementation and Experiments

For implement PA in QoS monitoring in web services, we suggest three components: Data Gathering, TSF by NN, Core Monitoring.

Data gathering: Data gathering component plays the role of monitoring through collecting the information of the quality parameters. It mesure the present values of

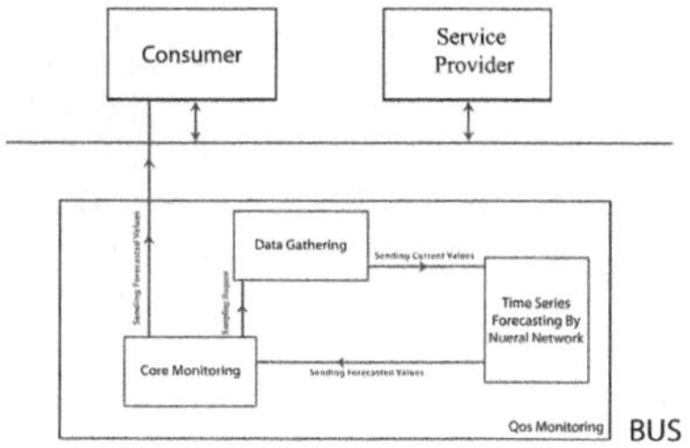

Fig. 4. QoS Monitoring by Time Series Forecasting using Neural Network

quality parameters. When an order for stop was sent by the Monitoring Core, this component sends the present values to the forecasting component. Forecasting component attempts to forecast the future values of quality parameters based on the present values through TSF with NN. Monitoring core: Monitoring core component plays the role of managing the method. Monitoring Core is sending request to Data Gathering for gathering the current values of quality parameters. Then, the values are sent forecasting component to fulfil the forecast. After receiving the future values, the Monitoring Core requests for monitoring to be stopped (the request is delivered to data gathering component) and it receives the values from the TSF with NN component. We hope that designing an intelligent monitoring core will be done by the future works and researches. TSF by NN: This component to implement the forecasting part. The validation of time series forecasting with NN at the execution time has been done through a reliable and quick device for predicting the time series by NN, which has named as Alyuda XL software [3]. In this paper, the method for training a network is to first partition the data series into two disjoint sets: the training set, and the test set. The network is trained (e.g., with Back propagation) directly on the training set, and its ability to forecast is measured on the test set. 70 percent has been applied for the network training and the rest, 30 percent, has been applied for the network training.

Fig. 5. measured times for QoS monitoring

The Execution Time is a implemented quality parameter for QoS monitoring that's defined as the time elapsed for processing a request; Execution time = t3 – t2 that's shown in Fig.5.

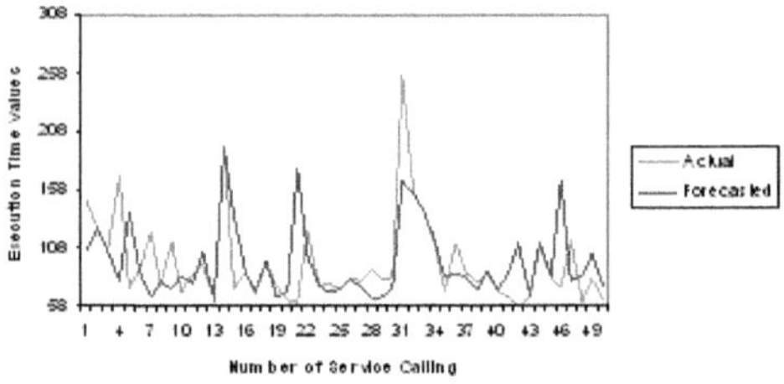

	Training set	Test set
# of rows:	35	15
Average AE:	15.1333 81	25.9769 21
Average MSE:	988.676 46	940.339 44
Tolerance type:	Relative	Relative
Tolerance:	10%	30%
# of Good forecasts:	24 (69%)	11 (73%)
# of Bad forecasts:	11 (31%)	4 (27%)
Good Trend Forecasting = 75%		

Fig. 6. Execution Time with one requester (Experience duration=5 min) and requests 1-50[1]

[1] http://homepages.laas.fr/khalil/TOOLS/QoS-4-SHWS/GridDBtxt/ResTime1C_5m.sql

Fig.6 shows forecasting values prove reasonably accuracy .For validating the NN performance, error analysis methods can be used. Mean Square Error, Mean Absolute Error and Good Trend Forecasting are used in the paper. Mean Square Error (MSE) the MSE of an estimator ô with respect to the estimated parameter o is defined as MSE = $[(\hat{o} - o)^2]$ Absolute Error (AE) given some value v and its approximation v_{approx}, the absolute error is AE = $| v - v_{approx} |$. Good Trend Forecasting means that the network can duly forecast the next values decrease or increase in the context of the previous obtained values. In the words, when the chances of forecasting are in alignment with the real changes, they should be considered as the correct forecasting.

$exr_{t+1} > exr_t \rightarrow e\check{x}r_{t+1} > e\check{x}r_t$ & $exr_{t+1} = exr_t \rightarrow e\check{x}r_{t+1} = e\check{x}r_t$ & $exr_{t+1} < exr_t \rightarrow e\check{x}r_{t+1} < e\check{x}r_t$

exr_t :Desired value at t time & $e\check{x}r_t$:Estimator value at t time

The percentage of Good Trend Forecasting = Good Trend Forecasting / the Total Number of Forecasting.

Good trend forecasting is the most important factor. If the good trend forecasting have a reasonable accuracy, then customer appears to correctly judge based on these data about QoS level of web service.

4 Conclusion and Future Work

In this paper, we demonstrated the PA is method for lessening the overhead of the SOA and increasing efficiency and intelligent. This paper proposes to use the TSF technique for estimating the quality of service parameters in SOA. NN technique is used on the proposed approach and simulation data are provided. The presented method has been evaluated, and the forecasting is reasonable accurate. We will also working on the monitoring core component, for design an intelligent tuning. The intelligent tuning delineates the number of forecasting values by NN. In this context, design self-healing architecture based on PA in QoS monitoring is our future work.

References

[1] Halima, R., Guennoun, K., Drira, K., Jmaiel, M.: Non-instrusive QoS Monitoring and Analysis for Self-Healing Web Services, pp. 549–554. IEEE Computer Society, Los Alamitos (2008)

[2] Gao, Z., Wu, G.: Combining_QoSbased_service_selection_with_performance_prediction. In: IEEE ICEBE 2005 (2005)

[3] Alyuda, http://www.alyuda.com/

An Effective Performance of Fuzzy Hierarchical Clustering Using Time Series Data Streams

V. Kavitha[1] and M. Punithavalli[2]

[1] Research Scholar, Dept of Computer Science, Karpagam University, Coimbatore, Tamilnadu, India
Kavithaanand11@gmail.com
[2] Director and Head, Dr.SNS Rajalakshmi College of Arts and Science, Coimbatore, Tamilnadu, India
mpunitha_srcw@yahoo.co.in

Abstract. Mining Time Series data has a remarkable development of concentration in today's humanity. Clustering time series is a trouble that has applications in a wide variety of fields and has recently fascinated a huge amount of researchers. Time series data are frequently large and may surround outliers. In addition, time series are a special type of data set where elements have a chronological ordering. Therefore clustering of such data stream is an important issue in the data mining process. The clustering algorithms and its effectiveness on various applications are compared to extend an innovative method to solve the existing problem. This paper presents a new method called Fuzzy Hierarchical clustering algorithm is implemented. Experimental results on ECG data evaluate the processing qualities of the system is more accurate than the previous system.

Keywords: Time Series Data Streams, Hierarchical Clustering, Fuzzy Logic.

1 Introduction

Time Series Data has a remarkable augmentation of concentration in today's humanity. Clustering time series is a trouble that has applications in a wide variety of fields and has recently fascinated a huge amount of researchers. A Time series is a progression of real numbers, each number representing a value at a time point. For example, the sequence could represent stock or commodity prices, sales, exchange rates, weather data, biomedical measurements etc. Recently, stream time series data management has become a hot research topic due to its wide application practice. All these applications require continuously observing time series data stream. A data stream is an structured series of points $x_1, , , , , , , x_n$ that must be accessed in order and that can be read only once or a small number of time. Conventional models cannot acclimatize to the high speed arrival of new data set. This way the innovative algorithms have been developed, which endeavor at processing data in real time. These algorithms should be capable of processing each example in constant time and memory while consistently providing a compact data report at each given moment. Clustering methods developed for handling various data into five major categories:

V.V. Das, J. Stephen, and Y. Chaba (Eds.): CNC 2011, CCIS 142, pp. 242–248, 2011.

Partitioning methods, Hierarchical methods, Density Based methods, Grid Based methods and Model Based Method.

In the next section, Section 2 Discuss about the Hierarchical Clustering Algorithm. Section 3, Fuzzy Logic. Section 4, Experimental Results is gathered. Section 5, where concluding remarks.

2 Methodology

A clustering is a group of data objects that are similar to one another within the same cluster and are dissimilar to the objects in other clusters. Clustering is the assignment of a set of observations into subsets (called cluster) so that observations in the same cluster are similar. The clustering problem has been addressed in many contexts and by researchers in many disciplines. This reflects its broad appeal and usefulness as one of the steps in exploratory data analysis. However, clustering is a difficult problem combinatorial, and differences in assumptions and contexts in different communities have made the transfer of useful generic concepts and methodologies slow to occur.

2.1 Hierarchical Clustering

A hierarchical method generates a hierarchical decomposition of the certain set of data objects. A hierarchical can be classified as being either agglomerative or divisive. The Agglomerative also describe the bottom up approach begins with each object forming a separate group. It successively merges the objects or groups that are close to one another, until all of the groups are merged into one hierarchy. The divisive approach is also called the top down approach, starts with the entire object in the same cluster. In each iteration, a cluster is split up into smaller clusters, until ultimately each object is in one cluster, or until a termination condition holds.

Hierarchical does not require us to pre specify the number of clusters and most hierarchical algorithms that have been used in IR are deterministic. Advantage of Hierarchical comes at the cost of lower efficiency.

2.2 Clustering Time Series Data Stream

Data streams typically consist of variables generating examples continuously over time. Let $X = \{x^t_1, x^t_2, \ldots\ldots\ldots\ldots x^t_n\}$ be the example containing the observations of all streams x_i at a specific time t. The objective of a clustering system for multiple time series is to discover a partition P of streams, where streams in the same cluster tend to be more comparable than streams in different clusters.

2.3 Online Divisible Agglomerative Clustering

The ODAC (Online Divisible Agglomerative Clustering) system is a variable clustering algorithm that creates a hierarchical tree shaped structure of clusters using top down strategy.

The leaves are the resulting clusters with each leaf grouping a set of variables. The unification of all leaves is the complete set of variables. The Intersection of any two leaves is the empty set. The System encircles an incremental distance measure and performs procedures for expansion and aggregation of the tree based structure based on the diameters of the clusters.

The system constantly monitors existing clusters diameters over time. The diameter of a cluster is the utmost distance among variables of that cluster. For each existing cluster, the system discovers the two variables defining the diameter, the system divides the cluster and consigns each of the elected variables to one of the new clusters, becoming the pivot variable for that cluster. Subsequently, all outstanding variables on the old cluster are assigned to the new cluster, which has the closest pivot. New leaves start new statistics, assuming that only forthcoming information will be useful for deciding whether or not this cluster should be split. The test is executed to verify if the formerly determined split still represents the structure of variables. On stationary data streams, the overall intra cluster dissimilarity should reduces with each split. This way, if a cluster is split into two child leaves, the diameter of the new cluster should be less than or equal to the diameter of the parent node. If the diameter of the leaf is greater then its parent's diameter, then the previously taken decision no longer reflects the structure of data. The System reaggregates on the clusters parent, restarting statistics. Overview of the system.

2.4 Fuzzy Clustering

Hierarchical clustering methods are more flexible than their partitioning counterparts, in that they do not need the number of clusters as an input. The disadvantages of Hierarchical clustering algorithm are,

They only create crisp clustering , i.e. they do not support degrees of membership in their output.

They only create partitions, i.e. they do not allow for overlapping among the detected clusters.

In order to overcome the above disadvantage the fuzzy clustering is to be introduced. Traditional clustering approaches create partitions; each pattern belongs to one and only one cluster. Consequently, the clusters in a hard clustering are displace. Fuzzy clustering extends this perception to correlate each pattern with every cluster using a membership function. The output of such algorithms is a clustering, but not a partition. In hard clustering (non fuzzy clustering), data is divided into distinct clusters, where each data element belongs to exactly one cluster. In soft clustering (fuzzy clustering), data elements can belong to more than one cluster, and associated with each element is a set of membership levels. These indicate the strength of the association between that data element and a particular cluster. Fuzzy clustering is a process of assigning these membership levels, and then using them to assign data elements to one or more clusters. Fuzzy data is a kind of data that is inaccurate or with some source of uncertainty. This data type has been readily used in natural language, social science, knowledge representation, etc. an extension for hierarchical clustering algorithm that can work on fuzzy data too. The main difference between the traditional hard clustering and fuzzy clustering is, in hard clustering an entity belongs only to one cluster, in fuzzy clustering entities are allowed to belong to many clusters with different degrees of membership.

2.5 Fuzzy Hierarchical Clustering Algorithm

Input: The similarity threshold value α, the difference threshold value λ, the full-threshold value δ of the degree of effect, and the semi-threshold value γ of the degree of effect, where the values of α, λ, δ and γ are determined by the user, $\alpha[0, 1]$, $\lambda[0, 1]$, $\delta[0, 1]$ and $\gamma[0, 1]$.

Output: Data clusters.

Step 1: let the membership degree of each attributes belonging to its data be equal to 1.

Step 2: Find a pair of data clusters Ci and Cj among the set of attributes which have the largest degree of similarity, where the degree of similarity between document clusters Ci and Cj is ψ, and $\psi[0, 1]$.

Step 3: Find a pair of data clusters Ck and Cl among the set of attribute clusters which have the second largest degree of similarity and have an identical document cluster with the document cluster pair obtained in step2(i.e., Ci = Ck).

Step 4: If the difference between ψ and ω is smaller than the threshold value λ (i.e., $\psi - \omega < \lambda$) and both ψ and ω are not less than α, then perform Step 4.1 to Step 4.5. Otherwise (i.e., the difference between ψ and ω is larger than or equal to the threshold value λ), go to Step 5.

Step 4.1: Make a copy of data cluster from Ci.

Step 4.2: Merge cluster Ci with cluster Cj into a new cluster. Assume that cluster Ci contains n documents. If the membership degree of each data belonging to cluster Ci is μis, where $\mu is[0, 1]$ and $1 \leq s \leq n$, then the membership degree of data belonging to the newly generated cluster is equal to $\mu is \times w \, \psi + \omega$. Assume that cluster Cj contains m documents. If the membership degree of each data belonging to cluster Ct is μit, where $\mu it \in [0, 1]$ and $1 \leq t \leq m$, then the membership degree of data belonging to the newly generated cluster is equal to μit.

Step 4.3: Determine the type of combination between clusters Ci and Cj and generate the dynamic cluster center.

Step 4.4: Merge the cluster copied from Ci with document cluster Ck into a new cluster. Assume that the cluster Ci contains n documents, and if the membership degree of each data belonging to cluster Ci is μis, where $\mu is \in [0, 1]$ and $1 \leq s \leq n$, then the membership degree of data belonging to the newly generated cluster is equal to $\mu is \times w \, \psi + \omega$. Assume that cluster Ck contains g data, and if the membership degree of each data belonging to cluster Cu is μku, where $\mu ku \in [0, 1]$ and $1 \leq u \leq g$, then the membership degree of data belonging to the newly generated cluster is equal to μku.

Step 4.5: Determine the type of combination of clusters Ci and Ck and generate the dynamic cluster centers.

Step 5: Merge cluster Ci with cluster Cj into a new document cluster. Assume that cluster Ci contains n data. If the membership degree of each data belonging to cluster Ci is μis, where $\mu is \in [0, 1]$ and $1 \leq s \leq n$, then the membership degree of data belonging to the newly generated cluster is equal to μis. Assume that cluster Cj contains m data. If the membership degree of each data belonging to cluster Ct is μit, where μit

∈ [0, 1] and $1 \leq t \leq m$, then the membership degree of data belonging to the newly generated cluster is equal to μit.

Step 6: Recalculate the degree of similarity between each pair of clusters.

Step 7: If the degree of similarity between any two clusters is smaller than the threshold value α, where α[0, 1], then Stop. Otherwise, go to Step 2.

3 Experimental Results

To evaluate clustering methods concentrate on the quality of hierarchy. There are three approaches to investigate clusters the quality known as external, internal and relative criteria. External criteria make an evaluation based on a pre specified structure, which reflects our prior knowledge.

3.1 Clustering Validity

External criteria include the folks and mallows Index and Hubert's T statistics. Internal criteria usually stands for an evaluation based on measures involving the data members themselves. Usually used the divisive coefficient or the cophenetic correlation coefficient CPCC. Third criteria relative approach, which compares the resulting cluster with other clustering results for the same data set. Relative approach was the Dunn's Index and the Modified Hubert's T Index.

3.2 Hierarchical Quality

ODAC system permits the analysis of another quality measure, the CPCC, is used measures quality in hierarchical structures. The CPCC is defined as

$$CPCC = \frac{\sum_{i=1}^{N-1}\sum_{j=1+1}^{N} C_{ij}P_{ij} - \mu_P\mu_C}{\sqrt{\sum_{i=1}^{N-1}\sum_{j=1+1}^{N} P_{ij}^2 - \mu_P^2 \sum_{i=1}^{N-1}\sum_{j=1+1}^{N} C_{ij}^2 - \mu_C^2}}, \quad . \tag{1}$$

Where C is the cophenetic proximity matrix, with each c_{ij} being the proximity level at which the two objects I and j appeared together in the same cluster. The closer the value of this is to 1, the better the match, and the hierarchy fits the data. High values of any of these indices indicate the presence of a good clustering structure.

3.3 System Performance on ECG Data

Time series of ECG data are one of the most broadly studied sets of data. This paper, the Hierarchical clustering, ODAC Clustering and fuzzy hierarchical clustering algorithms are compared using ECG Data sets. In this comparison, we have tried finding the right number of clusters for each user with more than 1,800 examples, which is possibly the same for all. We should note that the best result is defined using the quality measure. The results are presented in Table 1.

Table 1. Fuzzy Hierarchical clustering and Fuzzy ODAC Clustering with ECG Data set

No of Clusters	Fuzzy Hierarchical Clustering	Fuzzy ODAC Clustering
2	0.5784	0.9941
3	0.5784	0.9954
4	0.5784	0.9972
8	0.5784	0.9999

3.4 Scalability Evaluation

A different set of data was used to conclude the scalability characteristics of the system. The following graph shows that the performance of Fuzzy Hierarchical Clustering algorithm and Fuzzy ODAC Clustering algorithm.

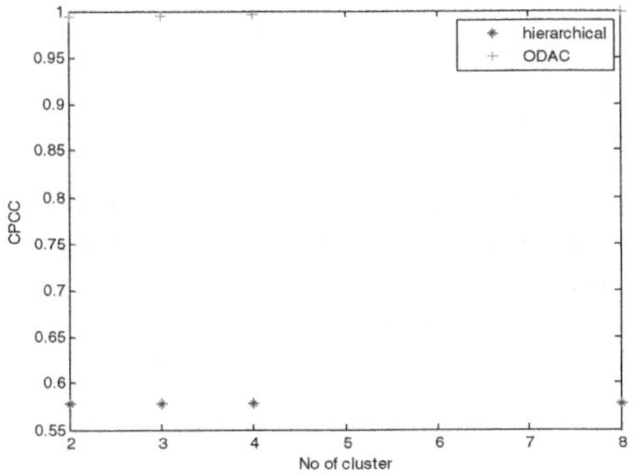

This graph shows that the Fuzzy Hierarchical and ODAC Clustering algorithms performance is more accurate than the previous system on Hierarchical Quality. Experimental results indicate that the performance is competitive for clustering time series, showing that this system is adaptable in the presence of the evolving concept.

4 Conclusion

Mining Time Series data has an incredible development of attention in today's world. Clustering time series is a trouble that has applications in an extensive assortment of fields and has recently attracted a large amount of researchers. Time series data are frequently large and may contain outliers. This paper presents the performance of Fuzzy Hierarchical Clustering algorithm using ECG data sets. Experimental results indicate that the performance is competitive for clustering time series, showing that this system is adaptable in the presence of the evolving concept. Using Fuzzy Hierarchical clustering algorithm, they reveal good performance on finding the cluster quality.

References

1. Dingi, H., Trajcevski, G., Scheuestern, P., Wang, X., Keogh, E.: Querying and Mining of Time Series Data: Experimental Comparison of Representations and Distance Measures. ACM Proceedings of the VLDB Endowment 1(2), 1542–1552 (2008)
2. Talbot, L.M., Talbot, B.G., Peterson, R.E., Tolley, H.D., Mecham, H.D.: Application of fuzzy grade-of membership clustering to analysis of remote sensing data. Journal on Climate 12, 200–219 (1999)
3. Rodriguess, P.P., Pedroso, J.P.: Hierarchical Clustering of Time Series Data Streams. IEEE Transactions on Knowledge and Data Engineering 20(5), 615–627 (2008)
4. Hautamaki, V., Nykanen, P., Franti, P.: Time Series Clustering by Approximate Prototypes. In: IEEE 2008 (2008)
5. Bagnall, A.J., Janacek, G.J.: Clustering time series from ARMA models with Clipped data. In: ACM Proceedings of the Tenth ACM SIGKDD International Conference on Knowledge Discovery and Data Mining, pp. 49–58 (2004)
6. Guha, S., Meyerson, A., Mishra, N., Motwani, R.: Clustering Data Streams: Theory and practice. IEEE Transactions on Knowledge and Data Engineering 15(3), 515–528 (2003)
7. Yin, J., Zhou, D., Xie, Q.-Q.: A Clustering Algorithm for Time series Data. In: Seventh International Conference on Parallel and Distributed Computing, Applications and Technologies (PDCAT 2006), pp. 119–122 (2006)
8. Sato, M., Sato, Y.: Fuzzy clustering model for fuzzy data. IEEE Transactions on Fuzzy Systems 4, 2123–2128 (1995)
9. Rodrigues, P.P., Gama, J.: A semi Fuzzy Approach for Online Divisive Agglomerative Clustering. In: EPIA 2007 Proceeding of the Artificial Intelligence (2007) ISBN : 3-540-77000-3 978-3-540-77000-8
10. Sato, M., Sato, Y.: Fuzzy clustering model for fuzzy data. IEEE transactions on Fuzzy Systems 4, 2123–2128 (1995)

Credit Scoring Using PCA-SVM Hybrid Model

M.A.H. Farquad[1,2], V. Ravi[1,*], Sriramjee[3], and G. Praveen[4]

[1] Institute for Development and Research in Banking Technology, Castle Hills Road #1, Masab Tank, Hyderabad – 500 057 (A P) India
Tel.: +91-40-2353 4981
farquadonline@gmail.com, rav_padma@yahoo.com
[2] Department of Computer & Information Sciences, University of Hyderabad, Hyderabad – 500 046 (A P) India
[3] Indian Institute of Technology, Department of Mathematics, Kharagpur-721302, India
sriramjee17@gmail.com
[4] Indian Institute of Technology, Department of Mathematics, Hauz Khas, New Delhi-110016, India
gautam.praveen01@gmail.com

Abstract. A new approach for credit scoring using principal component analysis (PCA) and support vector machine (SVM) in tandem is proposed in this paper. The proposed credit scoring hybrid algorithm consists of two basic steps. In the first step, PCA is employed for dimension reduction and in the second, SVM is employed for classification purpose, resulting in PCA-SVM hybrid model. The effectiveness of PCA-SVM model is evaluated using German and UK credit data sets. It is observed that PCA-SVM outperforms stand-alone SVM and PCA-Logistic Regression (LR) hybrid. However, in terms of sensitivity alone, LR outperformed PCA-SVM hybrid.

Keywords: Credit Scoring, SVM, PCA, Logistic Regression, dimension reduction.

1 Introduction

In the financial industry, consumers regularly request credit to make purchases. The risk for financial institutions to extend the requested credit depends on how well they distinguish the good credit applicants from the bad ones. One widely adopted technique for solving this problem is "Credit Scoring". These credit scoring models decide who will get credit, how much credit they should get, and what operational strategies will enhance the profitability of the lenders and it also helps assess the risk in lending. Typical application areas in the consumer market include: credit cards, auto loans, home mortgages, home equity loans, mail catalogue orders, and a wide variety of personal loan products. Specifically, the objective of credit scoring is to label credit applicants as either good customers or bad customers. Therefore it turns out to be a classification problem.

* Corresponding author.

V.V. Das, J. Stephen, and Y. Chaba (Eds.): CNC 2011, CCIS 142, pp. 249–253, 2011.

2 Literature Review

Research in credit scoring models proliferated during the past 20 years with the application of linear discriminant analysis, logistic regression (LR), decision tree, Bayes network, linear programming, backpropagation trained neural network (BPNN), support vector machines (SVM) etc. these models is the need to increase the scoring accuracy of the credit decision. An improvement in accuracy of even a fraction of a percent translates into significant future savings.

Linear discriminant analysis (LDA) [1] was one of the first credit scoring models. The appropriateness of LDA for credit scoring has been questioned because of the categorical nature of the credit data unequal covariance structure of the good and bad credit classes and non-normality. Later, Rosenberg and Gleit reported a survey that included the use of discriminant analysis (DA), classification trees, and expert systems for static decisions, and dynamic programming, linear programming, and Markov chains for dynamic decision models [2]. Then, Henley explored a LR model for credit scoring applications [3]. Later, West investigated the credit scoring using BPNN, mixture-of-experts, radial basis function, learning vector quantization, and fuzzy adaptive resonance theory (ART) [4]. He concluded that the both the mixture-of-experts and radial basis function neural network models were the best. LR is found to be the most accurate of the traditional methods.

Paolo reported the use of Bayesian methods coupled with Markov Chain Monte Carlo (MCMC) techniques for credit scoring [5]. Then, Lee et al., reported that MLP - LDA hybrid outperformed LDA and LR [6]. Later, Baesens et al. found that both the Least Square-SVM (LS-SVM) and NN classifiers yielded a very good performance, while LR and LDA did not lag far behind [7]. A new fuzzy support vector machine is proposed to discriminate good creditors from bad ones [8]. Recently, a recursive algorithm for extracting classification rules from feedforward neural networks (NNs) is proposed for credit scoring [9]. They concluded that the algorithm generated accurate rules. Later, Zhang et al. developed a hybrid credit scoring model (HCSM) by incorporating GP and SVM [10]. They reported that HCSM outperformed SVM, GP, Decision Tree, LR, and MLP. More recently, a credit scoring model using the Simulated Annealing and decision tree hybrid is proposed [11]. Later, Zhang et al. [12] reported that SVM performed better than BPNN. Further, neighbourhood rough set is used to select input features for SVM and grid search to optimize RBF kernel parameters for solving the credit scoring problem [13]. Most recently, an improved BP algorithm by adding hierarchical learning rates is proposed to simulate personal credit scoring instead of using an ensemble [14]. Then, several SVM ensemble models that aggregate LS-SVM classifiers are proposed for credit scoring problem [15]. In this paper, we propose a hybrid consists of two steps: (i) PCA is employed for dimension reduction and (ii) SVM and LR classifiers are employed for classification separately.

3 Proposed Approach

Now we briefly present the constituents of the hybrid viz., PCA and SVM as follows:

3.1 Support Vector Machine (SVM)

SVM developed by Vapnik [16] found numerous real world applications. For classification problems SVM finds a maximal margin hyperplane that separates two classes. The main objective of SVM is to find an optimal separating hyperplane that correctly classifies data points as much as possible by minimizing the risk of misclassifying the training samples and unseen test samples. To deal with non-linear problems, SVM first projects data into higher dimensional feature space and tries to find the linear margin in the new feature space.

3.2 Logistic Regression

Logistic regression, a member of generalized linear models the dependence of a binary response variable on a vector of explanatory variables [17]. It is too well-known to be described here.

4 Dataset Description

UK credit data and German credit data are analysed in this study. UK credit data is obtained from a financial services company of England [18]. There are 14 explanatory variables and 1 outcome variable. The dataset has information of 1225 applicants, of which 323 are bad and the rest 903 are good. German credit data is publicly available in UCI Machine Learning Repository (http://kdd.ics.uci.edu/).

5 Results and Discussions

We performed the principal component analysis on dataset, only the PCs that together accounted for 80% variance. 10 fold cross validation test method is followed in this study and the average results obtained are discussed here. The effectiveness of the proposed approach is evaluated using Accuracy, Sensitivity, Specificity and AUC. The measures used to evaluate the performance of the hybrid are defined as follows:

$$Sensitivity = \frac{BB}{BB+BG}, \quad Specificity = \frac{GG}{GG+GB}$$

$$Accuracy = \frac{GG+BB}{GG+GB+BB+BG}$$

Where BB = Bad predicted as Bad, BG = Bad predicted as Good, GG = Good predicted as Good, GB = Good predicted as Bad.

Since sensitivity is paramount important from business perspective, we base our discussion on sensitivity alone. It is always better to identify bad customer and avoid lending money. Table 1 presents the results obtained using UK credit data. Using UK

Table 1. Average Results Obtained Using PCA-SVM for UK Credit Data

Classifier	Sensitivity	Specificity	Accuracy	AUC
SVM	48.6	58.6	62.1	5360
PCASVM	*73.02*	43.45	51.18	5823.5
LR	82.69	7.13	27.18	4499.85
PCALR	52.85	46.36	47.89	4960.3
Zhou et al, [15]	68.25	46.98	52.66	-

credit data, it is observed that the proposed classifier PCA-SVM yielded best average sensitivity of 73.02%, average specificity of 43.45%, average accuracy of 51.18% with average AUC of 5823.5, whereas standalone SVM yielded only 48.6% average sensitivity. It is observed that LR yielded the best average sensitivity of 82.69% and very low average specificity of 7.13%.

Recently, Zhou et al., [15] proposed LSSVM based credit scoring model. While discussing the results they mixed up sensitivity and specificity and in the process reported erroneously high sensitivities. In this paper, we compare our results with their best results that were corrected by us. It is observed that using the same UK credit data, Zhou et al., [15] reported highest sensitivity of 68.25% with the individual classifier quadratic discriminant analysis (QDA). It is also observed that, our proposed hybrid outperforms QDA [15]. Table 2 presents the average results obtained using the German credit data. Using German credit data, it is observed that our proposed PCA-SVM hybrid outperforms the standalone SVM. On German credit data PCA-SVM yielded the average sensitivity of 70%, average specificity of 47.29%, average accuracy of 54.1% and average AUC of 5864.3 while standalone SVM yielded an average sensitivity i.e. 60%. However, LR yielded the highest average sensitivity of 93.33%, but with an average specificity of 7.22%. It is observed that using PCA for dimension reduction increases prediction rate of the SVM classifier. Even though LR yielded high sensitivity, it failed miserably on average specificity. Further, it is observed that PCA-SVM yielded high average sensitivity and average specificity and hence, would be preferred as a hybrid model for credit scoring.

Table 2. Average Results Obtained Using PCA-SVM for German Credit Dataset

Classifier	Sensitivity	Specificity	Accuracy	AUC
SVM	60	51.57	54.1	5578.5
PCASVM	*70*	47.29	54.1	5864.3
LR	93.33	7.322	33.05	5032.75
PCALR	28.67	74.87	61.1	5176.85
Zhou et al., [15]	68.75	73.34	72.07	-

6 Conclusions

In this study, we proposed a hybrid model for credit scoring, where first dimension reduction is carried out using PCA followed by SVM and LR separately as classifiers. Since sensitivity is paramount from business perspective, we based our discussion on sensitivity alone. It is observed that using reduced features, SVM yielded better sensitivity than using full feature data. It is also observed that LR stand-alone yielded the best sensitivity but the specificity yielded by LR is very low.

References

1. Durand, D.: Risk Elements in Consumer Installment Financing. National Bureau of Economic Research (1941)
2. Rosenberg, E., Gleit, A.: Quantitative methods in credit management: a survey. Operations Research 42(4), 589–613 (1994)
3. Henley, W.E.: Statistical aspects of credit scoring. Dissertation, The Open University, Milton Keynes, UK (1995)
4. West, D.: Neural network credit scoring models. Computers and Operations Research 27, 1131–1152 (2000)
5. Paolo, G.: Bayesian data mining with application to benchmarking and credit scoring. Applied Stochastic Models in Business and Society 17, 69–81 (2001)
6. Lee, T.S., Chiu, C.C., Lu, C.J., Chen, I.F.: Credit scoring using the hybrid neural discriminant technique. Expert Systems with Applications 23(3), 245–254 (2002)
7. Baesens, B., Van Gestel, T., Viaene, S., Stepanova, M., Suykens, J., Vanthienen, J.: Benchmarking state-of-the-art classification algorithms for credit scoring. Journal of the Operational Research Society 54(6), 627–635 (2003)
8. Wang, Y., Wang, S., Lai, K.K.: A New Fuzzy Support Vector Machine to Evaluate Credit Risk. IEEE Transactions on Fuzzy Systems 13(6) (2005)
9. Setiono, R., Baesens, B., Mues, M.: Recursive Neural Network Rule Extraction for Data with Mixed Attributes. IEEE Transactions on Neural Networks 19(2) (2008)
10. Zhang, D., Hifi, M., Chen, Q., Ye, W.: A Hybrid Credit Scoring Model Based on Genetic Programming and Support Vector Machines. In: Fourth International Conference on Natural Computation, pp. 8–12 (2008)
11. Jiang, Y.: Credit Scoring Model Based on the Decision Tree and the Simulated Annealing Algorithm. In: World Congress on Computer Science and Information Engineering, pp. 18–21 (2009)
12. Zhang, L.-L., Hui, X.-F., Wang, L.: Application of Adaptive Support Vector Machines Method in Credit Scoring. In: 16th International Conference on Management Science & Engineering, Moscow, Russia, September 14-16, pp. 1411–1415 (2009)
13. Yao, P.: Hybrid Classifier Using Neighborhood Rough Set and SVM for Credit Scoring. In: International Conference on Business Intelligence and Financial Engineering, pp. 138–142 (2009)
14. Qin, R., Liu, L.L., Xie, J.: An Application of Improved BP Neural Network in Personal Credit Scoring. In: Second International Conference on Computer Modeling and Simulation, pp. 238–241 (2010)
15. Zhou, L., Lai, K.K., Yu, L.: Least squares support vector machines ensemble models for credit scoring. Expert Systems with Applications 37, 127–133 (2010)
16. Vapnik, V.N.: The Nature of Statistical Learning Theory. Springer, New York (1995)
17. Strauss, D.: The many faces of Logistic Regression. The American statistician 46(4), 321–326 (1992)
18. Thomas, L.C., Edelman, D.B., Crook, J.N.: Credit scoring and its applications. SIAM, Philadelphia (2002)

Energy Efficient Data Aggregation and Routing in Wireless Sensor Networks

A.V. Sutagundar[1], S.S. Manvi[2], and B.S. Halakarnimath[3]

[1] Department of Electronics and Communication Engineering,
Basaveshwar Engineering College, Bagalkot-587102, India
[2] Department of Electronics and Communication Engineering,
REVA Institute of Technology and Management, Bengaluru, India
[3] Department of Computer Science and Engineering,
S.G. Balekundri Institute of Technology, Belgaum - 590010, India
ashok_ec@yahoo.com, agentsun2002@yahoo.com,
basaprabhu97@gmail.com

Abstract. The objective of this work is to maximize the network lifetime by utilizing data aggregation and in-network processing techniques. Here, data aggregation and data routing is addressed as joint problem to achieve the objective. This paper considers the problem of correlated data gathering in WSNs with the objective of minimizing the total transmission cost in terms of power consumption. To minimize the energy consumption, two-level aggregation in fixed virtual grid is used. The first level aggregation is within the cluster by using cluster head and second level aggregation is from the set of cluster heads to group aggregators. In this paper mainly focused on the problem of finding the set of aggregation points that satisfy the objective. Fixed virtual wireless backbone that is built on top of the physical topology is used to reduce clustering and routing overhead. Numerical results show that the proposed work provides substantial energy savings.

Keywords: ILP, LBA, Aggregators.

1 Introduction

Wireless sensor networks are composed of sensor nodes that must cooperate in performing specific functions. WSNs are well suited to perform event detection, which is clearly a prominent application of wireless sensor networks. For military applications sensor nodes battery is not replenished during operation. Energy is consumed both in radio transmission and reception. This imposes a number of energy constraints, one for each sensor. Thus, energy is critical resource that must be conserved throughout the lifetime of sensor network. Since sensor nodes are densely deployed in sensor networks, the data collected by nearby nodes are either redundant or correlated. Data aggregation reduces the communication overhead in the network, thus saving the sensor scare energy resources and reduces channel contention and packet collisions.

As discussed in [1], data aggregation/fusion is categorized based on relationships among the input data. The recent researches have significantly highlighted the

V.V. Das, J. Stephen, and Y. Chaba (Eds.): CNC 2011, CCIS 142, pp. 254–259, 2011.

importance of in-network aggregation in sensor networks, as they require intercepting data at network layer itself. Different routing protocols are designed to fulfill the shortcomings of the resource constraint nature of the WSNs [2]. A number of studies that compared aggregation scheme, e.g., [3], [4], [5] concluded that enhanced network throughput and more potential energy savings are highly possible using data aggregation and in-network processing in WSNs. In [3], proposed a Directed Diffusion (DD) where aggregation is used to reduce communication costs. In [5], a hierarchical clustering algorithm for WSNs, called low energy adaptive clustering hierarchy (LEACH) is discussed. Linear programming formulation is introduced in [6], to solve the optimal routing problem in WSNs. The work in [7] introduces the PEGASIS algorithm that uses the 'energy X delay' metric over the routing tree. This algorithm finds chains of nodes instead of clusters. A family of adaptive protocols called sensor protocols for information via negotiation (SPIN) [8] are designed to conserve less energy by sending a descriptive version of data instead of sending all the data and by using meta-data negotiations before any data is transmitted to the base-station. In [9], the problem of correlated data gathering in WSNs with the objective of minimizing the total transmission cost in terms of power consumption is addressed. In order to achieve, an exact solution using an integer linear program (ILP) formulation and a near optimal, but simple and efficient heuristic, called load balancing with aggregation (LBA) strategies were used. In this paper, we propose, clustered network, a 2-level hierarchy.

In this paper, section 2 introduces proposed work along with contributions. Simulation model and performance parameters are given section 3. In section 4, results are discussed and this work is concluded in section 5.

2 Proposed Work

We assume energy constrained, homogeneous sensor nodes are randomly deployed in sensor field within bounded region. Each sensor has sense range as same as transmission range **tr** in meters. We assume sensor nodes are fixed, while sensing object is mobile. We focus on energy related to the sensor node which is limited resource. We assume that initially the total energy within a zone is same. It is assumed that the different levels of energy for transmission and reception is required as sensor nodes, during a sensing interval generates packets are of variable lengths. We use clustering infrastructure, which will allow nodes that are closed to each other to share data before sending it to the base station (BS). In this work, the clustering approach presented in [10] is used to create a fixed rectilinear virtual topology, called Virtual Grid Architecture (VGA), on top of the physical topology. To provide fairness and to avoid single node failure cluster heads are chosen at periodic intervals. VGA divided into, non-overlapping, G groups. MAC layer solves the contention between sensors.

2.1 Contributions

The algorithms which are discussed in [9] are used here and following contributions are added to improve the lifetime of network.

Formation of sensor group
Sensors monitor the same phenomenon involves in the formation of group. As we have considered virtual grid to form network infrastructure we believe that, the nodes belongs to particular zone are monitoring the same phenomenon/object. Each group $\mathbf{g_i}$ must have master aggregator $\mathbf{M_i}$ to perform second level aggregation. Here, the problem is how to form a group which monitors the same phenomenon. We have addressed this problem by computing the groups in advance based on the M, master aggregator which is input parameter.

Routing in group
To construct the path the following algorithm *GroupRouting* is used. GroupRouting invokes the Route algorithm recursively.

Algorithm: GroupRouting(Adj[][], G,SS[])

```
begin
// I/p Adj[][]: Adjacent table consists of adjacent zones // of each zone in group, SS: super
sensor
// O/p: R[][]: Routing table
  for (i=1 to G) do
     SS[i] = get MA for the group i from LBA
     Route(SS[i],Adj[i][])
  end
end
```

Algorithm: Route(SS, Adj[i][])

```
begin
// Input SS: super sensor,
// input Adj[i][]: Adjacent node table of group i
  while (all LAs in group i)
    if ( LA[j] adjacent to MA) then
      R[i][] = LA[j]
      Route(SS,Adj[LA[j]][]
    end
  end
end
```

Once the paths are set up within group, local aggregators can send the data to master aggregator. Then master aggregator of each group directly communicates to base station. The location of base station is known to all local aggregators.

Mobility of sensor nodes
In previous section we discussed with related to the static nodes deployed in virtual grid. Mobility **m** is another issue where we can think during routing and data aggregation.

3 Simulation Model and Performance Metrics

We have carried out the simulation using Turbo C language on windows platform. Simulation environment comprises of six models namely Network Model, Channel Model, Propagation model, Battery model, Data Aggregation Model, and Mobility Model.

Simulation scenario: Each experiment corresponds to a random placement of sensors in a fixed network area. We assume a single base-station attempting to gather information from a number of data sources in the network area. The Base-Station (BS) is located at middle of the grid i.e., at (m, n) where m=n. We randomly place sensor nodes in square field. The sensor field is divided into the appropriate number of zones i.e. NumberofZones = (size X size) / 100.

Performance parameters considered in this simulation are

- **Energy:** Each node consumes some energy for receiving/ transmitting, computation and for its transmitting amplifier unit.
- **Delay:** Time to transmit the data
- **Mobility:** Mobility of percentage of **m** of deployed nodes is selected for computing the effectiveness of energy and delay.

Simulation Parameters: For our simulation we have considered 30 – 80 nodes, 4-6 MAs, 36 zones, TrRcE as 50μJoules, mobility 30-50%, delay 5-15 seconds, transmission range 10 meters and grid size is 60X60.

4 Results and Discussions

In this section, we present the results which are simulated using C programming language and discuss the experimental issues regarding the results given.

4.1 Node's Battery Usage

The fig. 2 presents Nodes Vs Energy for NA, aggregation and with varying mobility for 4 MAs, and for exact solution i.e. ILP. The thick line in graph indicates no aggregation which consumes maximum energy than the other methods. The dotted line of mobility indicates that the energy consumption is more if mobility is increased. We observe similar kind of results for 6 MAs and for exact solution which is shown in fig. 3.

Energy for Master Aggregators: In second level, aggregation and routing is performed by identifying the master aggregator and finding the route from LA to MA and MA to BS.

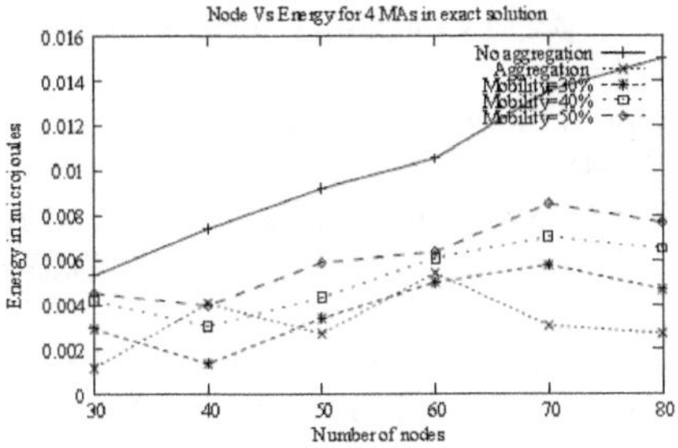

Fig. 2. The energy consumption for 4MAs, exact solution with mobility

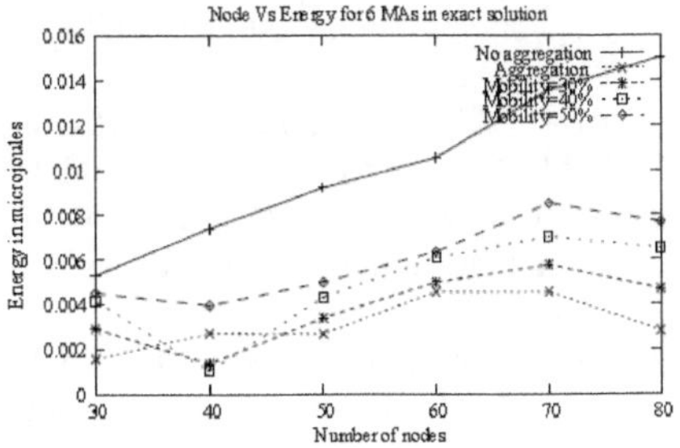

Fig. 3. The energy consumption for 6MAs, exact solution with mobility

4.2 Delay vs Nodes

In this work, the delay is the time taken to transmit the data from one node to another. The fig. 4 and fig. 5 shows the delay for 4MAs and 6MAs taken for exact solution. We observe that the dotted lines shown for mobility would increase delay than the aggregation.

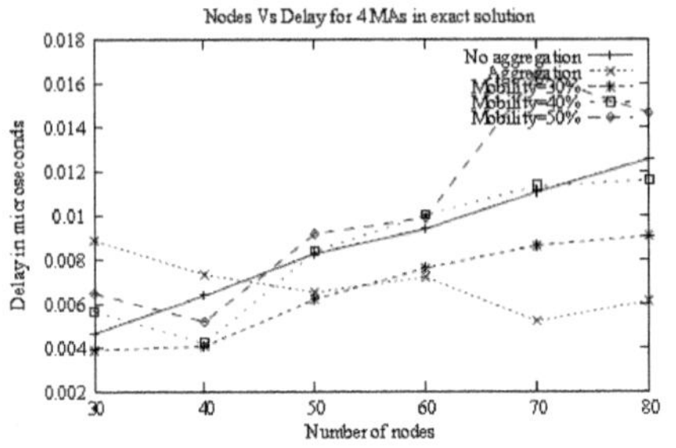

Fig. 4. The delay required for 4MAs, exact solution with mobility

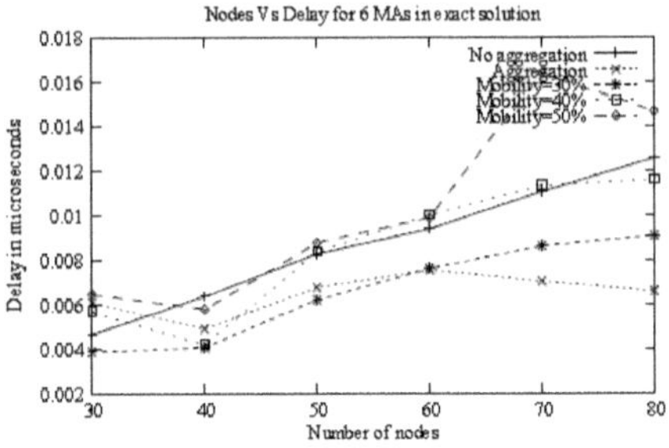

Fig. 5. The delay required for 4MAs, exact solution with mobility

5 Conclusions

In this work, we discussed the approaches, which increase the lifetime of sensor network. This approach uses a fixed virtual backbone to perform energy-efficient routing. We compared the exact results with the heuristic results. Further we have tested our work by adding mobility feature to nodes and results shows that, LBA approach is would take less energy and less delay than the ILP approach. We also noticed that, if mobility is more, then the delay would be more. Finally we would like to conclude our discussion by saying aggregation with different techniques would be energy efficient for sensor network.

References

1. Nakamura, E.F., Loureiro, A.A.F., Frery, A.C.: Information Fusion for Wireless Sensor Networks: Methods, Models, and Classifications. ACM Computing Surveys 39(3), Article 9 (August 2007)
2. Al-Karaki, J.N., Kamal, A.E.: Routing techniques in wireless sensor networks: a survey. IEEE Wireless Communications 11(6), 628 (2004)
3. Intanagonwiwat, C., Govindan, R., Estrin, D.: Directed diffusion: a scalable and robust communication paradigm for sensor networks. In: Proc. of ACM Mobi-Com 2000, Boston, MA, pp. 56–67 (2000)
4. Krishnamachari, B., Estrin, D., Wicker, S.: Impact of data aggregation in wireless sensor networks. In: International Workshop on Distributed Event-Based Systems, Vienna, Austria (July 2002)
5. Heinzelman, W.R., Chandrakasan, A., Balakrishnan, H.: Energy efficient communication protocol for wireless micro sensor networks. In: Proc. of HICS 2000 (January 2000)

6. Chang, J.H., Tassiulas, L.: Maximum Lifetime Routing in Wireless Sensor Networks. IEEE/ACM Trans. On Networking 12(4), 609–619 (2004)
7. Lindsey, S., Raghavendra, C.S., Sivalingam, K.: Data Gathering in Sensor Networks using the Energy*Delay Metric. In: Proc. of IPDPS Workshop (2001)
8. Heinzelman, W.R., Kulik, J., Balakrishnan, H.: Adaptive protocols for Information Dissemination in Wireless Sensor Networks. In: MOBICOM 1999, Seattle, WA, pp. 174–185 (1999)
9. Al-Karaki, J.N., Kamal, A.E.: On the Correlated Data Gathering Problem in Wireless Sensor Networks (2004)
10. Al-Karaki, J.N., Kamal, A.E.: Efficient Virtual-Backbone Routing in Mobile Ad Hoc Networks. Elsevier Computer Networks 52(2), 327–350 (2008)

Design of High Speed Optimized Flash ADC

Kapil Chaudhary[1], B.K. Kaushik[2], and Kirat Pal[3]

[1] Deptt. of Computer Science & Engg., Roorkee Institute of Technology, Roorkee, India
kapil.cse@gmail.com
[2] Deptt. of Electronics & Computer Engg., Indian Institute of Technology, Roorkee, India
brajesh_k_k@yahoo.com
[3] Department of Earthquake Engg., Indian Institute of Technology, Roorkee-247667, India
kirat.pal@gmail.com

Abstract. This paper presents the design methodology of high speed flash ADC by individually optimizing its various components so that the overall performance of the resulting flash ADC is improved over traditional flash ADCs. The design parameters chosen are speed, sampling frequency, power consumption, supply voltage and chip area. With high speed as a parameter, components are designed so that they operate with sampling frequency as high as 70-75 MHz. The power consumption is reduced and the ADC operates at power supply voltage down to 2.5V compatible with low power digital portion of the design. The designed ADC occupies less chip area too. All the components are designed using the 0.35μm CMOS technology.

Keywords: Comparator, Residue Amplifier, DAC.

1 Introduction

An analog-to -digital converter (ADC) is a device which converts continuous signals to discrete digital numbers [1]. The analog signal is continuous in time and it is necessary to convert this to a flow of digital values. It is therefore required to define the rate at which new digital values are sampled from the analog signal. The rate of new values is called the sampling rate or sampling frequency of the converter.

At present, there exists a variety of ADCs with different architectures, resolutions, sampling rates, power consumptions, and temperature ranges. Since the performance - sampling rate, resolution, and power consumption- of an ADC is basically determined by its architecture, one single ADC type cannot cover all applications. For instance, flash (parallel) ADCs can be used in high speed and low resolution applications [2].The highly parallel flash architectures lend it to easily implemented low voltage circuit blocks. They have high bandwidth, but the pipeline architecture results in exponential growth in number of comparators for every bit of increased resolution [2, 3]. This paper uses flash ADCs because they are the fastest way to convert an analog signal to digital signal. They are suitable for systems requiring very large bandwidths. However, flash converters consume lot of power, have relatively low resolution, and can be quite expensive. This limits them to high frequency applications. Two step approach is the preferred design as it reduces area as well as power [4].

V.V. Das, J. Stephen, and Y. Chaba (Eds.): CNC 2011, CCIS 142, pp. 260–263, 2011.

2 Flash ADC Architecture with Two Step Approach

The potential of two-step flash architectures for realizing fast, high resolution ADCs are demonstrated in number of designs [5,6,7]. With the conversion rates approaching half those of fully parallel (flash ADC) these architectures provide relatively small input capacitances together with the low power dissipation and can be used to achieve resolutions in the range of 10-14 bits which is well above that obtained in the single stage flash designs. The basic structure of the two step converter is shown in Fig. 1.

The first converter generates a rough estimate of the value of the input, and the second converter performs a fine conversion. The advantage of this architecture is that the number of comparators is greatly reduced from that of the flash converter. The proposed architecture reduces the number of comparators from 2^N-1 to $2(2^{N/2}-1)$. For example, an 8 bit flash converter requires 255 comparators, while the step requires only 30. The tradeoff is that the conversion process takes two steps instead of one, with the speed limited by bandwidth and settling time required by the residue amplifier and the summer.

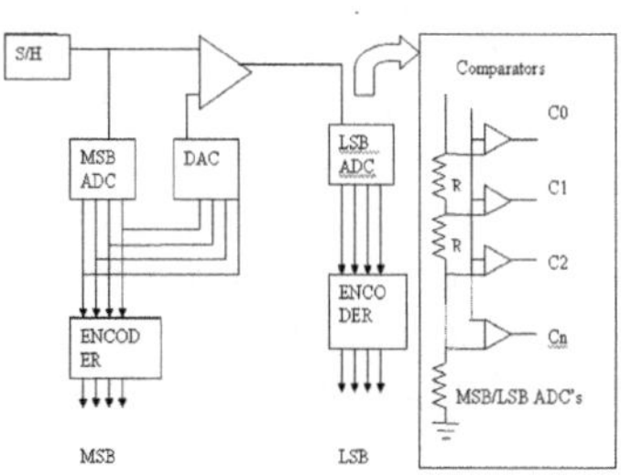

Fig. 1. Two-Step Flash ADC Architecture

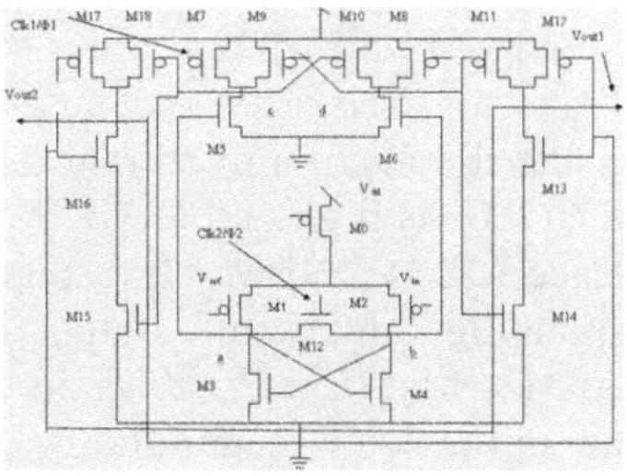

Fig. 2. Comparator Circuit

3 ADC Components Optimization

Comparator Design: In high speed ADCs, comparator design crucially influences overall performance. Converter architecture that incorporates large number of comparators in parallel to obtain a high throughput rate imposes stringent constraint on resolution, power dissipation, and chip area. Limitations are also imposed on the input voltage ranges that accompany the integration of comparator circuits in low-voltage scaled VLSI technologies, which in fact severely compromises the precision. Depending on the architecture, comparators can be classified into open loop and regenerative comparators. The open loop comparators are operational amplifiers without compensation. The comparator design implemented for the A/D application is based on references [5, 8, 9], which is shown in Fig. 2. Operating analysis of the comparator is given by [7, 5]. Based on the analysis carried out by [8, 9] a small signal model is developed and analyzed in the reset stage. The relation obtained is

$$\frac{W_{12}}{L}K_{pn}\left[(V_{dd}-V_b-V_{Tn12})(V_a-V_b)-0.5(V_a-V_b)^2\right]>\frac{W_5}{L}K_{pn}\left[(V_a-V_{ss}-V_{Tn5})(V_b-V_{ss})-0.5(V_b-V_{ss})^2\right]$$

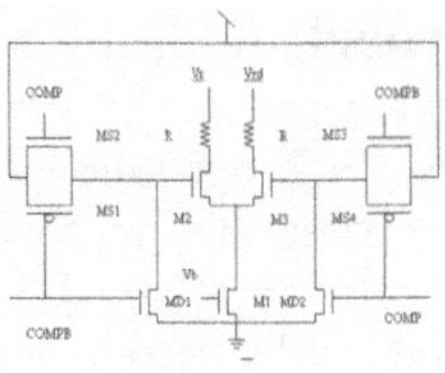

Fig. 3. Single Cell Circuit Diagram

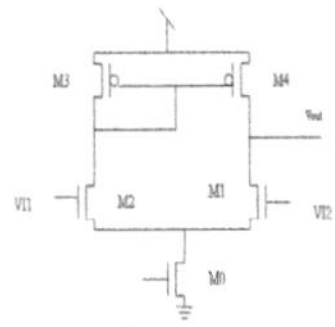

Fig. 4. Differential Amplifier

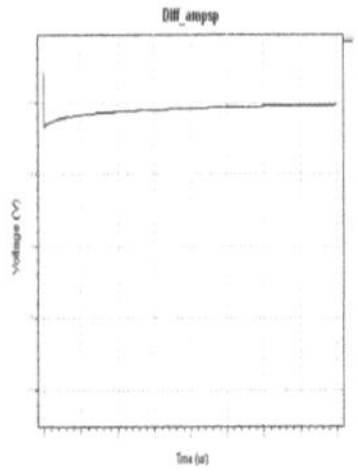

Fig. 5. Output of D .A

Single Cell DAC Circuit Design: There are number of means of converting a digital signal into an analog signal representation. The approaches differ in speed, chip area, power efficiency and achievable accuracy etc. The Single cell of our DAC has the structure as given in Fig.3. Single cell of the DAC corresponds to 1-bit DAC.

Working of the Single Current Cell: Here the transistors MS1, MS2, MS3 and MS4 act as switch whose control is governed by the output of the comparator. The transistor MD1 and MD2 are used for the fast charging and discharging of M2 and M3. Transistor M1 is used as a constant current source, which is biased such that it can handle the maximum current during all operational stages. When the comparator output *COMP* is high and *COMPB* is low, switch made by the MS1 and MS2 is ON and hence M2 is ON, which results in some current flow through resistor *R* and the corresponding value of the output voltage will be generated. The same operation is true for M3, when the *COMP* is low and *COMPB* is high. Current supplied by all the cells are of the equal magnitude.

Differential Amplifier/Residue Amplifier Design: Simple circuit of differential amplifier is used as shown in Fig. 4. This residue amplifier enhances the difference between the two input signals to such a desired level so that it can be easily interpreted by the input devices.

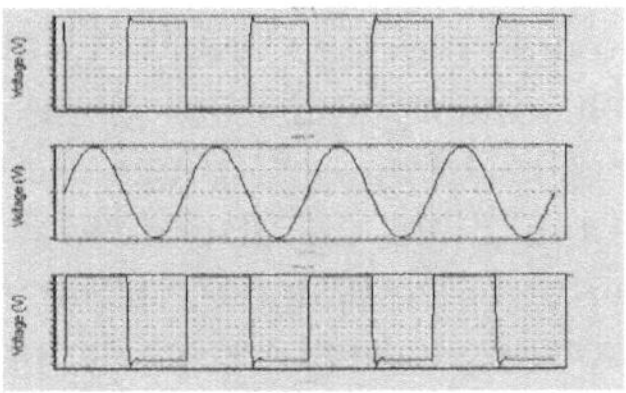

Fig. 6. Comparator output (Vref=0.125V)

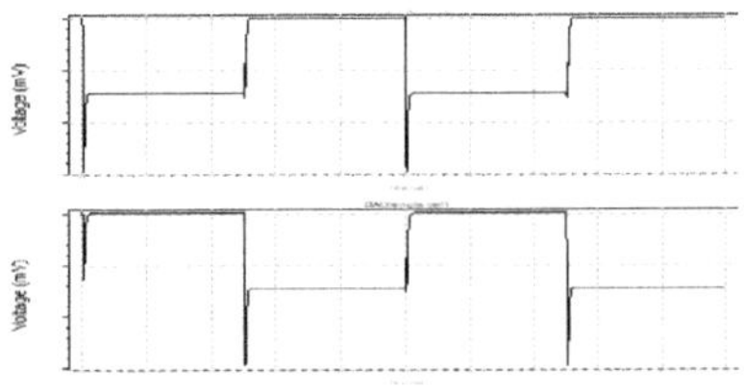

Fig. 7. Output of Single Cell

4 Conclusion

In this paper we designed and analyzed the performance characteristics of the two-step flash ADC components. The goals of high speed and low power are accomplished. The design architecture is developed for individual components of the ADC. Based on the scheme developed, W/L ratios for all the transistors are calculated using the model parameters of the 0.35μm CMOS technology. Individual components are implemented independently and later on cascaded together. The high performance of circuit is ensured by using preferred technology for individual components such as comparator, residue amplifier and single cell DAC. The scaling of operating voltage down to 2.5V for 0.35μm CMOS technology ensured reduced power dissipation. The same technique and architecture can be further used for deep submicron technology for achieving ultra low power dissipation and extremely higher speeds. The designed components are best suited for a complete flash ADC.

Table 1. Operational Parameters of Designed ADC

	Specification	Unit	Target Value	Measured Value
Static	Resolution	Bits	8	8
	INL / DNL	LSB	+0.6/1.7/ +0.62/-0.64	+0.6/1.7/ +0.62/-0.64
Dynamic	SFDR	dBc	40.3	40.3
	SNDR	dBc	32.5	32.5
	Sampling Frequency	MHz	70-75	67-75
Environmental	Conversion Rate	Samples/s	2.7 G	2.1 G
	Power Supply	V	2.5	2.5
Optimization	Power	μW	250	260
	Area	mm^2	1.30	1.30

References

1. Gustavsson, M., Wikner, J.J., Tan, N.: CMOS Data Converters for Communications. Kluwer Academic Publishers, Dordrecht (2000)
2. Uyttenove, K., Steyaert, M.S.J.: A 1.8V 6-Bit 1.3 GHz Flash ADC in 0.25μm CMOS. IEEE J. Solid-State Circuits 38(7), 1115–1122 (2003)
3. Taft, R.C., Tursi, M.R.: A 100-MS/s 8b CMOS Subranging ADC with Sustained Parametric Performance from 3.8V Down to 2.2V. IEEE J. Solid-State Circuits 36(3), 331–338 (2001)
4. Kerth, D.A., Sooch, N.S., Swanson, E.J.: A 12-bit 1-MHz Two-Step Flash ADC. IEEE J. Solid-State Circuits 24(2), 250–255 (1989)
5. Razavi, B., Wooley, B.A.: A12-b 5-Msample/s Two Step CMOS A/D Converter. IEEE J. Solid-State Circuits 27(12), 1667–1678 (1992)
6. Brianti, F., Manstretta, A., Torelli, G.: High Speed Autozeroed CMOS Comparator for multistep A/D Conversion. Microelectronics J. 29, 845–853 (1998)
7. Van Der Ploeg, H., Remmers, R.: A 3.3-V, 10-b, 25-MSamples/s Two-Step ADC in 0.35-μm CMOS. IEEE J. Solid-State Circuits 34(12), 1803–1811 (1999)
8. Cho, T.B., Cline, D.W., Conroy, C.S.G., Gray, P.R.: Design Considerations for High-Speed Low-Power Low-Voltage CMOS Analog-to-Digital Converters. In: Proc. Third Workshop on Advances in Analogue Circuit Design, Delft, The Netherlands (1994)
9. Yin, G.M., Eynde, F.O., Sansen, W.: A High Speed CMOS Comparator with 8-b Resolution. IEEE J. Solid-State Circuits 27(2), 208–211 (1992)

Alleviating the Data Loss in Congested Sensor Networks Using a Differentiated Routing Methodology

K. Saranya, Jayapreetha Lovely, J. Thangakumar, and M. Roberts Masillamani

School of Computer Science & Engineering, Hindustan Institute of Technology & Science, Padur, Chennai, Tamilnadu, India
{saran0575,preethaME86,thang.kumar,deancs}@gmail.com

Abstract. The congestion in the sensor networks leads to the poor delivery of the data. This causes the heavy loss of data in the sensor network. In order to alleviate this problem, a combination of two algorithms is used in this paper. The algorithms are Active Congestion-less Routing (ACR) and Medium Access Control Enhanced Active Congestion-less Routing (MACR). Both these algorithms used a differentiated routing for High Precedence data and Low precedence data. ACR lessen the congestion by routing only the high precedence data, where as the MACR lessen the congestion in mobile network by routing High precedence data as well as Less precedence data. These algorithms perform this by the nodes in the network.

Keywords: Sensor networks, congestion control, High precedence, Active Congestion-less Routing, Medium Access control Enhanced Active Congestion-less Routing.

1 Introduction

Large deployment sizes of the sensor network causes congestion, which may lead to indiscriminate dropping of data (i.e., high-precedence packets may be dropped while Less-precedence packets are delivered). It also results in an increase in energy consumption to route packets. In this paper, we examine data delivery issues in the presence of congestion. The use of data prioritization and a differentiated routing protocol and/or a prioritized medium access scheme to mitigate its effects on high priority data traffics is proposed in this paper. The solution will accommodates both less precedence and high precedence data traffic.

2 Active Congestion-Less Routing

In the presence of sensor network all the ACR data packets are classified into High precedence or less precedence data by the data sources, and nodes within a congested zone forward the high precedence data packets. Less precedence data traffic is routed out of and/or around the congested zone. ACR performs the slow start mechanism in case if the acknowledgement is not received.

V.V. Das, J. Stephen, and Y. Chaba (Eds.): CNC 2011, CCIS 142, pp. 264–266, 2011.

High-Precedence data Routing Network Formation
After the nodes in the sensor network has been deployed, the High Precedence data destination (the sink) starts building the high precedence routing network (HPNet).

Identification of congested zones and active differentiated routing
After building the HPNet, the next task is to dynamically discover the congested zone. The congested zone is formed when one area is generating high precedence data. This area is referred as the critical area.

Slow Start mechanism
ACR handles the slow start mechanism, if the acknowledgement is not received by the source. In this mechanism the source always calculates a congestion window for a receiver. If the acknowledgement arrives the source increases the congestion window by one segment which makes one round trip time (RTT).

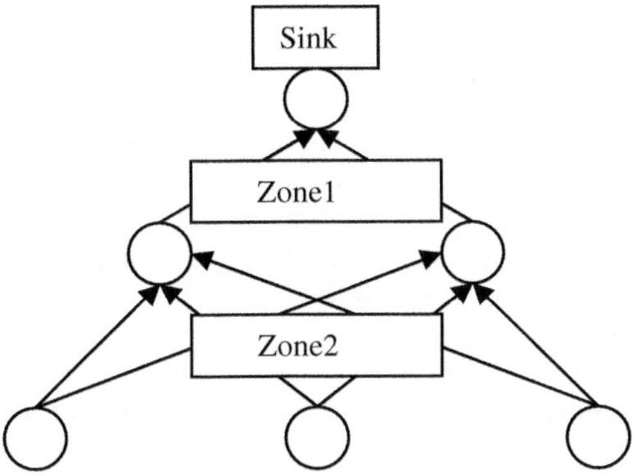

Fig. 1. Differentiated routing in different zones

3 MAC-Enhanced Active Congestion-Less Routing

3.1 Machine for MACR

The node state machine used by MACR to support differentiated routing is based on 5 MAC-layer enhancements and the following modes are used to establish the MACR mechanism.

Less Precedence mode. In this mode, nodes forward less precedence data. Upon receiving or overhearing a less precedence packet, nodes remain in the less precedence mode. If a node in the less precedence mode overhears a high precedence packet, it transitions to the suppression mode.

High Precedence mode. Nodes in the path of high precedence data are in the high precedence mode. Upon transitioning to this state, the node sets two timers: a received timer and an overhearing timer.

Suppression mode. Nodes in this state are within the communication range of high precedence data traffic but not on a forwarding path. Nodes in this state suppress less precedence data traffic, thus preventing it from interfering with high precedence traffic in the network.

4 Conclusion

This paper addresses the active differentiated routing of data in wireless sensor network. ACR protocol assigns precedence and routes the data according to that precedence. ACR increases the fraction of High Precedence data delivery and decrease delay and jitter for such delivery while using energy more uniformly in the deployment along with slow start mechanism. ACR also routes less precedence data in the presence of congestion. This additionally shows that MACR maintains high precedence data delivery rates in the presence of mobility

References

[1] Hull, B., Jamieson, K., Balakrishnan, H.: Mitigating Congestion in Wireless Sensor Networks. In: Proc. Second ACM Conf. Embedded Networked Sensor Systems (SenSys), pp. 184–203 (2004)

[2] Draft Supplement to Part 11: Wireless Medium Access Control (MAC) and Physical Layer (PHY) Specifications: Medium Access Control (MAC) Enhancements for Quality of Service (QoS). IEEE 802.11e/ D4.0

[3] Ahn, G.-S., Hong, S.G., Miluzzo, E., Campbell, A.T., Cuomo, F.: Funneling-MAC: A Localized, Sink-Oriented MAC for Boosting Fidelity in Sensor Networks. In: Proc. Fourth ACM Conf. Embe.

Interoperable Dynamic Composite Web Service

N. Sasikaladevi[1] and L. Arockiam[2]

[1] Research Scholar, Department of Computer Science, St. Joseph's College, Trichy, Tamilnadu, India
sasikalade@yahoo.com
[2] Associate Professor, Department of Computer Science, St. Joseph's College, Trichy, Tamilnadu, India
larockiam@yahoo.co.in

Abstract. As Windows and Java enterprise solutions continue to grow, companies increasingly emphasize interoperability to extend existing software infrastructure and application investments. For application architects, the challenge is to find strategies to integrate and extend existing deployments without 'rip and replace' options or costly application re-writes. The proposed ideas will be an effective and widely supported middleware technology enabling interoperability between Sun's J2EE and Microsoft's COM & .NET applications. This architecture can be used by companies to effectively integrate and harness the most powerful features of Java and Microsoft technologies.

Keywords: Web Service, XML, SOAP, WSDL, BalckBox Tesing.

1 Introduction

To build a usable web services system, there is more than meets the eye. The web services must be reliable, highly available, fault-tolerant, scalable, and must perform at acceptable levels. These needs are no different than the needs of any other enterprise application. J2EE and .NET are evolutions of existing application server technology used to build such enterprise applications. The earlier versions of these technologies have historically not been used to build web services. Now that web services have arrived, both camps are repositioning the solutions as platforms that used to build web services. The shared vision between both J2EE and .NET is that there is an incredible amount of 'plumbing' that goes into building web services, such as XML interoperability, load-balancing, and transactions.

2 Dynamic Composite Web Service

Consider the task of taking out a loan on the Web. In the absence of automation, the user invests considerable resources visiting numerous sites, determining appropriate service providers, entering personal information and preferences repeatedly, integrating information, and waiting for responses. The user sends a single request to a loan finding service containing personal information, the type of loan desired, and some provider preferences. The loan finder distributes its work among two partner

V.V. Das, J. Stephen, and Y. Chaba (Eds.): CNC 2011, CCIS 142, pp. 267–269, 2011.

services: a credit assessor, which consumes the user's personal information and provides a credit history report, and a lender service, which consumes a credit report and a loan request and returns a rejection or a loan offer and its terms. The loan finder first invokes a credit assessor to generate a credit report for the user, which it then passes to the lender service along with the user's personal information. The lender service generates a result, which the loan finder reports to the user. It is no longer required that the user enter information multiple times, determine which services are appropriate, or standby to bridge output from one service to another as input. These responsibilities have been offloaded to the loan finding service and its service provider—the party responsible for the form and function of the loan finding service.

3 Interoperability Issue in Data Exchange

At the most basic level, implementing interoperability between application platforms involves the exchange of data. When implementing a .NET and J2EE interoperability project, we confront three main data exchange challenges. Primitive data type mappings, Non-existent data types and Complex data types.

4 Proposed Model for Interoperable Dynamic Composite Web Service with XML Serialization

A framework model is proposed in Figure 1. A client request for composite web service is received. WSDL of the selected atomic web service are given to data type inconsistency checker. This module checks weather the atomic web service uses the incompatible data type or not. If it uses the incompatible data type, then it is passed to the XML serializer.

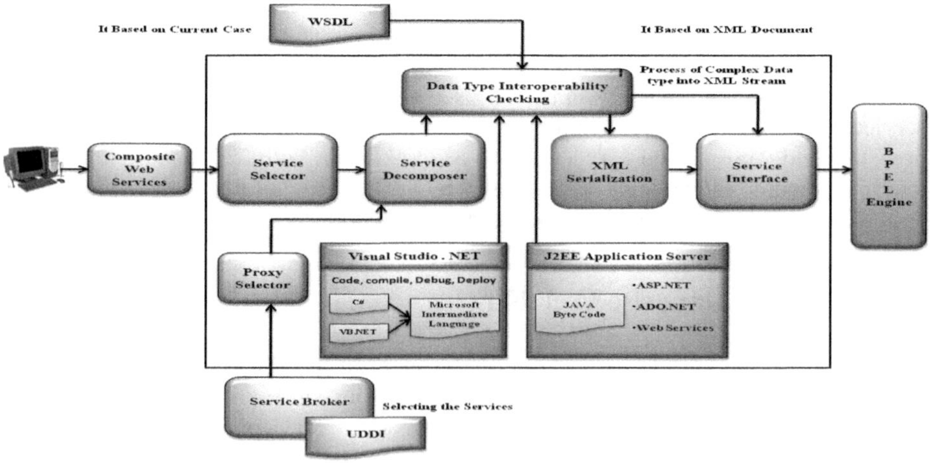

Fig. 1. Interoperable Dynamic Composite WS with XML Serialization

XML serialization is the process of taking a complex data type (or object) and encoding it into an XML stream. Then this XML stream is converted to a persistent state in the form of an XML document, transport it, and then later decode (de-serialize) it back into the original complex data type (or object). If it is possible to serialize an object or data type from one platform into an XML document, it should be relatively easy to read, understand, and de-serialize this document back into an object or data type on the other platform. Unfortunately, this is not always the case, but XML serialization does provide an interoperability route in most situations. The.NET Framework provides numerous classes and APIs for XML serialization. The main one of these is the *System.Xml.Serialization.XmlSerializer* class. This class allows the serialization of .NET Framework types to an XML document and back again. XML serialization converts (serializes) only the public fields and properties of an object into an XML stream.

5 Conclusion

Success of the web service relies on interoperability, because it is available in the internet. Different frameworks are available for designing the web service. When we compose the composite web service, each atomic web service involved in the composite web service is designed by different frameworks. Mapping the data type is the tedious task today. This problem can be solved by using the XML serialization. This framework model will improve the degree of the language interoperability in the dynamic composite web service.

References

1. Fisher, M., Lai, R., Sharma, S., Moroney, L.: Java EE and .Net Interoperability: Integration Strategies, Patterns, and Best Practices. Prentice Hall, Englewood Cliffs (2006)
2. Hochgurtel, B.: Cross-Platform Web Services Using C# and Java. Charles River Media, Hingham (2003)
3. Laudati, P., et al.: Application Interoperability: Micosoft.NET and J2EE Patterns and practices. Microsoft press, Redmond (2003)
4. Bukovics, B.: .NET 2.0 Interoperability Recipes: A problem solution approach. Apress (2006)
5. McMillan, R.: IDC: Web Services to Enable $4.3B Hardware Market by 2007. Computerworld (May 23, 2003)
6. Langdon, C.S.: The State of Web Services. IEEE Computer, 93–94 (July 2003)
7. Naedele, M.: Standards for XML and Web Services Security. IEEE Computer, 96–98 (April 2003)
8. Cardoso, J., Miller, J., Sheth, A., Arnold, J.: Modeling Quality of Service for Workflows and Web Service Processes. Technical Report #02-002, LSDIS Lab, Computer Science, University of Georgia, International Symposium on Service-Oriented System Engineering (2002)

A Message Passing System with Faulty Node Identification for Mobile Agents

K. Lakshman and C.B. Reshma

DOEACC Centre Calicut
Calicut, India
lakshman.korra@gmail.com

Abstract. Mobile agents are programs that can migrate from host to host in a network, at times and to places decided by the agent. These loosely-coupled decentralized systems widely use broadcast as an important communication mechanism to pass control signals, share resources, notify urgent messages etc. The main problem with flooding is the redundant broadcast messages, which constitute great network load, thereby decreasing the speed of delivery of messages and the entire network as such. In the context of mobile agents, fault tolerance is crucial to enable the integration of mobile agent technology into today's business applications. This paper proposes a broadcast technique based on the broadcasting algorithm by Peter Merz, for passing messages between the mobile agents and also detects faulty agents and adapts accordingly, based on the algorithm put forth by JayaDev Misra, the results are proved by simulating the model.

Keywords: Mobile agents, broadcasting, fault detection, message-passing, distributed applications, flooding.

1 Introduction

Mobile agents are processes that can transport its state and data from one environment to another, and perform appropriately in the new environment. Mobile agents decide when and where to move next. A mobile agent accomplishes this move through data duplication. When a mobile agent decides to move, it saves its own state and transports this saved state to next host and resumes execution from the saved state. It is very essential for mobile agents to send and receive messages and information in order to achieve the said function. This paper aims at creating a message passing system.

2 The Broadcast Procedure

The agents form a decentralized system. Every agent maintains an (incomplete) database about the rest of the network. This database contains entries (e.g. addresses) on some other agents (neighbors) together with timestamps of the last successful contact to that neighbor. Every agent stores in its neighbor list, addresses of some

V.V. Das, J. Stephen, and Y. Chaba (Eds.): CNC 2011, CCIS 142, pp. 270–272, 2011.

other agents near it. The agents use regular epidemic contacts to randomly chosen neighbors from the neighbor list to refresh timestamps and to find better neighbors with smaller distances to stated 'ideal' positions. To add a new peer to the network it is necessary to add only one living address to the network. At the beginning, the new agent s_n starts a lookup for an optimal position in the network and sends the lookup-query to its direct neighbor. The model uses two kinds of signals: A-signals or Activity signals, which correspond to the signals and B signals to inform them about the waiting status of the neighbors.

The broadcast can be initiated by every node in the network. In general, the broadcast message contains the information to be broadcasted and a *limit* argument. This *limit* is used to restrict the forwarding space of a receiving node. When a node s_a receives the broadcast message, it picks from its neighbor list the neighbor s_b nearest to the middle between its own address and *limit*: $middle = adr(s_a)+0.5 \cdot dist(s_a, limit)$ and sends to s_b the broadcast message with *limit* argument (in other words, s_b then must forward the broadcast between its address and *limit*). The node s_a itself continues the broadcast distribution within an interval between its address and $adr(s_b)$-1. When an agent is sending information, both data and control signals it uses both A & B signals to inform the system that it is sending/ receiving information. An analysis of the nodes provides information on the faulty agent.

3 The Experiment

The performance of the algorithm was tested using a prototype. The experiment focused on speed of message delivery and message complexity of different broadcast scenarios. The prototype was tested with multiple scenarios of agents

- constant-sized stable networks: new agents join the network at the simulation start and remain alive until the simulation ends;
- steady-state networks: agents randomly join and leave the network. The rates of joining/leaving are equal, the average network size remains constant;

The simulated network architecture had the following properties:

- Number of nodes in the network: 120; Number of agents: 50,100,150; interval between messages sent: 500 time units; length of neighbor lists: 10 neighbors; limit was set to 25;

4 Result

The experiment was tested 100 times for each scenario and the average results have been used to display the graphs obtained. The test was carried out by a Java program and the experiment was carried out by broadcasting to agents adapting the epidemic algorithm and results obtained are as in Figure 1. The program used a regular epidemic algorithm and the time taken to pass messages was noted. The time taken to pass messages using the adapted algorithm was also noted. The graph shows that the time taken by the adapted algorithm is less then the regular epidemic algorithm. The traffic generated in the system was also considerably less than the regualr epidemic one.

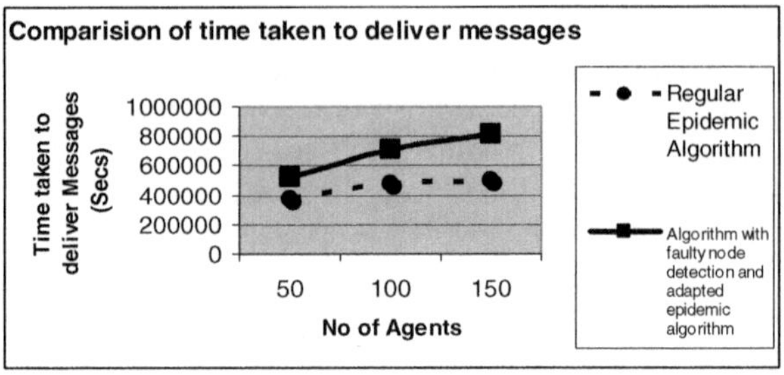

Fig. 1. Delivery time using broadcasting in the adapted algorithms

References

1. De Capitani di Vimercati, S., Ferrero, A., Lazzaroni, M.: A Mobile Agent Platform For Remote Measurements. In: IMTC 2005 – Instrumentation and Measurement Technology Conference Ottawa, vol. 1, pp. 570–575 (2005)
2. Merz, P., Gorunova, K.: Efficient Broadcast in P2P Grids. In: Proceedings of the IEEE/ACM International Symposium on Cluster Computing and the Grid (CCGrid 2005), Cardiff, UK (2005) (online paper)
3. Misra, J., Chandy, K.M.: Termination Detection of Diffusing Computations in Communicating Sequential Processes. ACM Transactions on Programming Languages and Systems (TOPLAS) 4, 37–43 (1982)
4. Murphy, A.L., et al.: Tracking Mobile Units for Dependable Message Delivery. IEEE Transactions on Software Engineering, 433–448 (2002)
5. Murphy, A.L., Roman, G.-C., Varghese, G.: Algorithms for Message Delivery in a Micromobility Environment. Technical Report WUCS98-02, Washington, Univ. St. Louis, Mo. (February 1998)
6. Dijkstra, E.W., Scholten, C.: Termination Detection for Diffusing Computations. Information Processing Letters 11(1) (1980)

Automatic Traffic Signal Controller for Roads by Exploiting Fuzzy Logic

Parveen Jain

Delhi Technological University (Formerly Delhi College of Engineering)
New Delhi, India
Parveenjaincta@gmail.com

Abstract. As the number of road users constantly increases and resources provided limited, intelligent traffic signal controller is a very major requirement. Therefore the need arises for simulating and optimizing traffic control algorithms to better accommodate this increasing demand. However, some limitations to the usage of intelligent traffic control exist. Avoiding traffic jams for example is beneficial to both environment and economy like a decrease in each of delay, number of stops, fuel consumption, pollutant emissions, etc., but improved traffic-flow may also lead to an increase in demand Optimization of traffic light switching increases road capacity and traffic flow, and can prevent traffic congestions. The fuzzy controller will regularly query the traffic conditions in order to decide whether to extend or terminate a current phase. The work shows that replacing the conventional traffic control system by fuzzy based traffic control system can considerably reduce traffic congestion delay.

Keywords: Traffic, Controller, Optimization, Delay, Time.

1 Introduction

The traffic signals affect the life of nearly everyone everyday. As the number of traffic over road is ever increasing, it has been a major concern of city authorities to facilitate effective control of traffic flows. It is important to improve traffic congestion in urban traffic. The degree of the traffic congestion depends on performance of the signal control and urban planning. Especially in certain rush hours, even a small interval of poor control at traffic signals may result in a long time traffic jam causing a chain of delays in traffic flows or many unexpected results. People accept that traffic signals ensure safety and mobility. One long standing as well as challenging problem is to apply traffic signal control so as to

1. Maximize the efficiency of existing traffic systems without new road construction,
2. Reduce the vehicle delay or equivalently queue length.
3. Minimize the air and noise pollution.

The primary function of any traffic signal is to assign right-of-way to conflicting movements of traffic at an intersection. Permitting conflicting streams of traffic to

V.V. Das, J. Stephen, and Y. Chaba (Eds.): CNC 2011, CCIS 142, pp. 273–277, 2011.

share the same intersection by means of time separation does this. A controller continuously (or at regular intervals) gathers information and evaluates the status of each approach and takes the most appropriate option. As a good traffic controller not only reduce delays or optimize traffic flow but also improve the environmental conditions. There are two different types of traffic signal control:

1. Time Control: Preset time signal control is based on the historical traffic data, assuming traffic conditions are unchanged in the time periods.
2. Signal Control: As an improvement over preset time control, this control scheme uses the online data that are collected from sensors and detectors installed at an intersection.

2 Fuzzy Based Traffic Signal Controller

While Using the conventional traffic signal controller (fixed time controller), where time for each signal was fixed, whether there is any number of traffic. Fuzzy logic based controllers are designed to capture the key factors for controlling a process without requiring many detailed mathematical formulas. Due to this fact, they have many advantages in real time applications. The controllers have a simple computational structure, since they do not require many numerical calculations.

The main goal of traffic signal control is to ensure safety at signalized intersections by keeping conflict traffic flows apart. The entire knowledge of the system designer about the process, traffic signal control in this case, to be controlled is stored as rules in the knowledge base. Thus the rules have a basic influence on the closed-loop behavior of the system and should therefore be acquired thoroughly.

3 Design Procedure of Fuzzy Controller

The In order to design a fuzzy controller, first road condition is to be identified. For this purpose, two variables are considered.

1. Distance of Road Occupied
2. Number of Vehicles on the road

As there are two input variable i.e. distance covered and number of vehicle on the road and output variable is time of the signal that also has four values. Both variables are having four values as below:

Distance	**Number of vehicle**	**Time of Signal**
Very small	Very Less	Very less
Small	Less	Less
Far	Large	Medium
Very far	Very Large	Large

So totally three variables are there to design a fuzzy controller. These are the real life terms used for the traffic signal controller.

3.1 Membership Functions of Input Variables

Membership functions of the input and output variables are either Triangular or trapezoidal. The membership's functions are simulated using Fuzzy toolbox in Matlab.

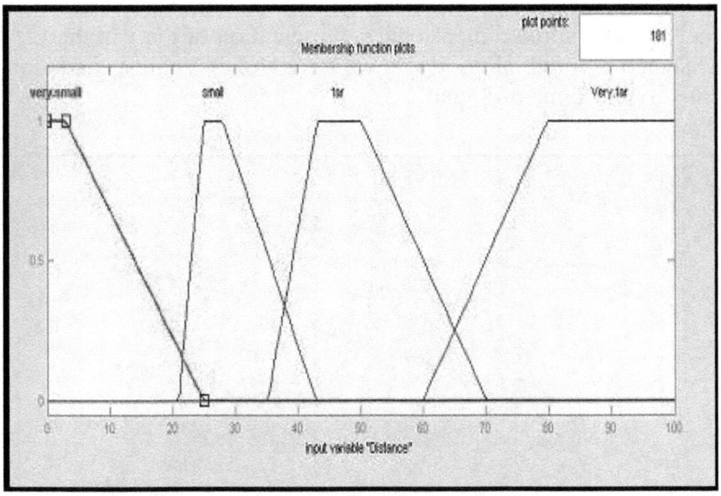

Fig. 1. Membership Function of Distance

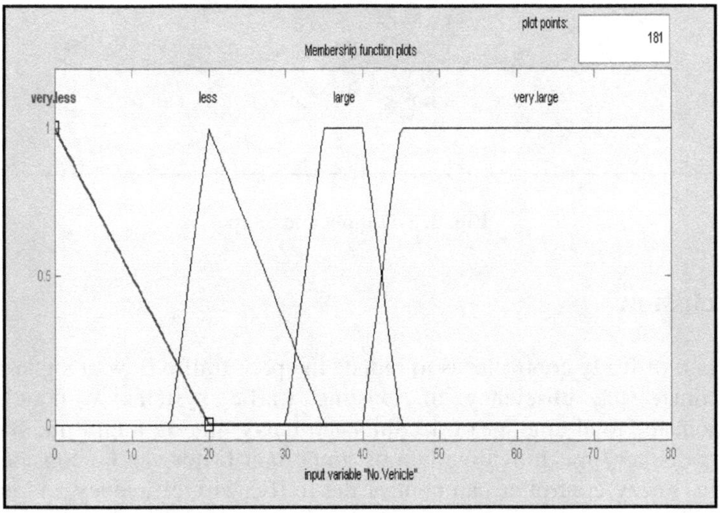

Fig. 2. Membership Function of No. of Vehicle

As few variable has trapezoidal shape others having triangular shape depend upon how their value changes.In both of the figure above, Four membership functions are defined fore both variable. Similar we can have for the output variable i.e. Time of Signal.

3.2 Mapping of Crisp Output Variable vs. Input Variable

Diagram below is the three dimensional representation of the variables. Along x-axis we have Distance covered, along y axis we have No. of Vehicles and along z axis we have output variable Time of Signal.

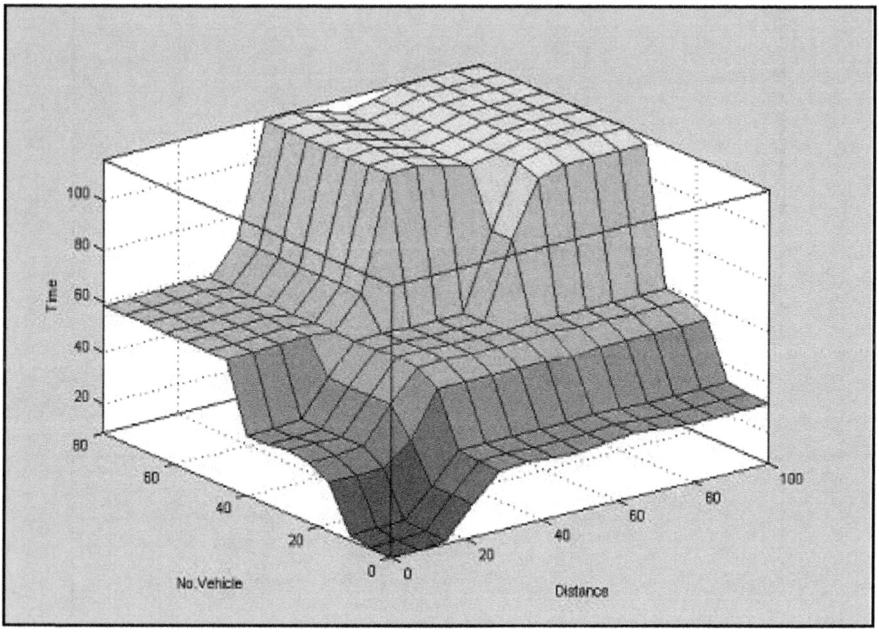

Fig. 3. 3-D Input and Output

4 Conclusion

As main aim of fuzzy controller is to reduce the poor traffic flow at signals or we can say Maximize the efficiency of existing traffic systems without new road construction, the total amount of accumulated delay time in a city due to waiting at signal stops is very much as time is most important factor can be considered in cost calculation. Fuzzy controller can control the traffic with efficiency and ease. Future work considers the pattern recognition to modify the rules, so that the controller can give better exact time for the signals.

Acknowledgments. The author wishes to acknowledge Dr.(Mrs.) Daya Gupta for her support and valuable guidance.

References

1. Chiu, S.L.: Fuzzy logic traffic signal control system Document Type and Number: United States Patent 5357436, http://www.freepatentonline.com/5357436.html
2. Sankar, G., Kumar Saravana, S.: Fuzzy Logic Based Automatic Braking System in Trains. In: IEEE Region 10 Conference on TENCON 2006, pp. 1–4 (2006)
3. Hellmann M.: Fuzzy Logic Introduction. University of Rennes, http://epsilon.nought.de/tutorials/fuzzy/fuzzy.pdf
4. Robert, H., Ulrich, J.: Fuzzy Control of Traffic Lights. In: IEEE International Conference on Fuzzy Systems, vol. 3, pp. 1526–1531 (2007)

Multimedia Conferencing for MANET in Cluster Based Signaling

R. Ganeshan, Amritha Ramachandran, J. Thangakumar, and M. Roberts Masillamaní

School of Computer Science and Engineering,
Hindustan Institute of Technology and Science, Chennai, India
(ganeshramasamy111,amritha.ramachandran4,thang.kumar,
rmasillamani)@gmail.com

Abstract. Mobile adhoc networks (MANETs) are infrastructure-less and can be set up anywhere, anytime. They can host a wide range of applications in rescue operations, military, private, and commercial settings. Some examples are audio/video conferencing, multiplayer games, and online public debating. Signaling is the nerve center of multimedia conferences it establishes, modifies, and tears down conferences. This paper focuses on signaling for multimedia conferences in MANETs. We review the state of the art and propose a novel architecture based on application-level clusters. Our validation employed SIP as the implementation technology and OPNET as our simulation tool. Our clusters are constructed dynamically and the nodes that act as cluster heads are elected based on their capabilities. The capabilities are published and discovered using a simple application-level protocol.

Keywords: Mobile ad hoc networks, conferencing, signaling, clustering, SIP.

1 Introduction

Mobile Ad hoc Networks can be defined as transient networks formed dynamically by a collection of arbitrarily located wireless mobile nodes, without the use of existing network infrastructure or centralized administration [1]. They rely on wireless technologies, such as IEEE 802.11 and Bluetooth, and are either stand alone or connected to other networks. Many of these applications involve multimedia conferencing. This paper is devoted to signaling for multimedia conferencing in stand-alone MANETs. The cluster heads are elected based on their capabilities. Our clusters are significantly different from routing layer clusters, such as those discussed in [2]. Routing level clusters are formed using the physical or geographical location of nodes, and they have no relation to whether or not there is a conference at the application layer. Our signaling clusters are formed only when there is a conference, and thus, they are completely independent of the routing layer clusters. We have selected Session Initiation Protocol (SIP) [3] as the implementation technology.

2 Existing System

Signaling for Conferencing in Infrastructure less Networks Mark and Kelley [4] describe an SIP application for a fully distributed conference model. The approach is

V.V. Das, J. Stephen, and Y. Chaba (Eds.): CNC 2011, CCIS 142, pp. 278–280, 2011.

of special interest to us because it only involves an SIP end system (UA) and no server is required. In addition, the approach is lightweight because there is no extra control message added to the baseline SIP but the session related information is carried by the basic SIP messages. However, this approach has several drawbacks, one of which is how the session related information is propagated. There is a problem when two (or more) parties are invited to join an ongoing conference at the same time. There is no general solution to ensure that each of the invited parties is made aware of the other invited parties. The problem is identified in [4] as the "coincident join," and no solution is provided. Another drawback is that unintentional departures are not considered and are not handled gracefully.

3 Proposed System

3.1 Cluster Based Signaling Architecture

Clusters enable scalability without centralization and we believe that they can aid in meeting all of the requirements. The overall architecture principles is given below.

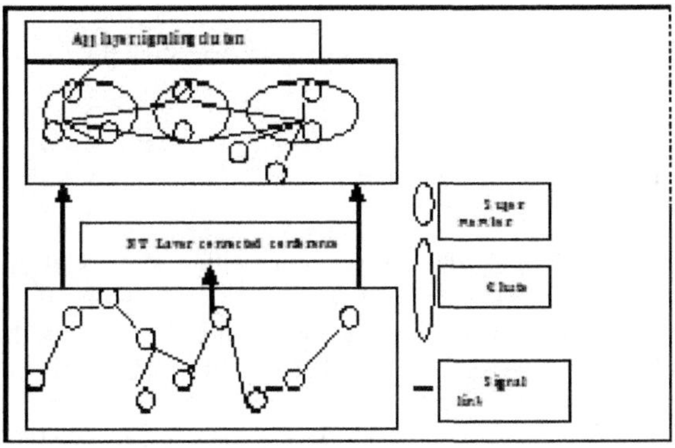

Fig. 1. Overall signaling architecture for conferencing in MANETs. Overall Architecture and Principles.

Figure 1 gives an overall view of the proposed cluster-based architecture. The only functional entity is the signaling agent (SA). There is one per party, or more generally, one per node in a stand alone MANET. All super members have direct links to the super members of the neighboring clusters. There are two general parameters of a cluster: Split value (Sv) and Merge value (Mv).

3.2 Operational Procedures of Clusters

In our architecture, clusters are dynamically created and deleted when conferencing. The signaling system is responsible for maintaining the state of the conference and the clusters. Each signaling cluster has a life cycle.

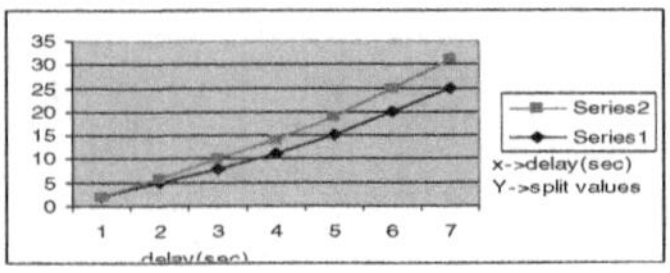

Fig. 2. Performance Analysis Graph

4 Result

Conferencing enables a set of attractive applications in MANETs. In this paper, we addressed an important conferencing issue signaling. We have presented the challenges, reviewed the state of the art signaling protocols, and proposed a cluster based signaling scheme. The performance of the signaling scheme has been evaluated through simulation. For the simulation experiments, we have developed all of the signaling entities that we defined. The results show that our signaling scheme outperforms the existing conferencing framework for ad hoc networks. It can support different scales of conferences and it can run on different lower layer routing protocols. In this work, we have evaluated conference scales up to 100 nodes. It would be very interesting to know the behaviors of the signaling scheme with hundreds and even thousands of nodes. However, the current experimental environment has a limited scope.

References

1. Liu, J., Chlamtac, I.: Mobile Ad-Hoc Networking with a View of 4G Wireless: Imperatives and Challenges. In: Mobile Ad Hoc Networking, vol. ch. 1. Wiley-IEEE Press (2004)
2. Fu, C., Glitho, R.H., Dssouli, R.: A Novel Signaling System for Multiparty Sessions in Peer-to-Peer Ad Hoc Networks. In: Proc. IEEE Wireless Comm. and Networking Conf., WCNC 2005 (2005)
3. Yu, J.Y., Chong, P.H.J.: A Survey of Clustering Schemes for Mobile Ad Hoc Networks. IEEE Comm. Surveys 7(1), 32–48 (2005)
4. Schooler, E.M., Casner, S.L.: An Architecture for Multimedia Connection Management. In: Proc. IEEE Fourth Comm. Soc (ComSoc) Int'l Workshop Multimedia Comm. (MM 1992), pp. 271–274 (1992)

Specification Based Test Case Generation Using Classification Tree Method

J. Viswanath, M. Parthiban, J. Thangakumar, and M. Roberts Masillamani

School of Computer Science and Engineering,
Hindustan Institute of Technology and Science, Padur, Chennai, India
{viswaj20,rsmparthiban,thang.kumar,rmasillamani}@gmail.com

Abstract. The systematic test is an inevitable part of the verification and validation process for software. The most important prerequisite for a thorough software test is the design of relevant test case, since they determine the kind and scope. Hence the quality of the test the classification-tree method and the graphical editor CTE support the systematic design of black box test cases. The classification-tree method is an approach to partition testing which uses a descriptive tree-like notation and which is especially suited for automation.

Keywords: CT-Classification tree, CTE-Classification tree Editor.

1 Introduction

Specifications are the primary source of information for deriving test cases. Specification based test case generation is becoming more and more important together due to the increase in the size of software systems and the widely use of components in software development. The classification tree method is one of the approaches for deriving test cases. A method is proposed for deriving classification trees from specifications. Furthermore, not all information from specifications is captured in the classification trees, and hence there is room for improvement. Although proposed structuring classification tree in the problems cannot be solved completely. We noticed that the CT method was able to express fully the information that is obtained from specifications.

1.1 Existing System

Human decisions are required. Categories and choices are identified. The test cases are generated from a formal specification. The lack of methods and tools for test case design using a black-box approach. The existing system finds greater difficulties as the decision making depends on skill based workers. This also leads to lack of Customization and scalability.

V.V. Das, J. Stephen, and Y. Chaba (Eds.): CNC 2011, CCIS 142, pp. 281–283, 2011.

Disadvantages

- No systematic techniques
- No test case development from the functional specification.
- Involves much manual work

1.2 Proposed System

The proposed system enables decision making process more easy. The System also provides an interface which is highly customizable and scalable. The Automated communication between the roles and the process speeds up service delivery. The classification tree processing service act key enables Improvement by Management Decision Making and Reporting. The advantages are

- Supports black box test cases
- Uses a descriptive tree like notation
- Test cases can be generated automatically using functional specification
- Test cases are based on classification tree

1.3 Problem Definition

Three steps were found to be most important when using the classification-tree method for real world applications:

- Selecting test objects,
- Designing a classification tree,
- Combining classes to form test cases.

A large, real-world system cannot be tested reasonably with a single classification-tree, as such a tree would become too large to handle. Therefore the functionality of the system under test has to be divided into several separate test objects.

2 System Implementation

The following procedure is a brief outline of the method of generating test cases using classification trees.

- Identify all the classifications and their classes from the specifications.
- Construct the CT with the classifications and classes.
- Construct the combination table from the classification tree.

Select all the feasible combinations of classes from the combination table. Thus the classification tree shown in the fig. 1.

The lower part shows three possible combinations of classes, i.e., three rows of the combination table. Each row is obtained by selecting one class under each top level classification.

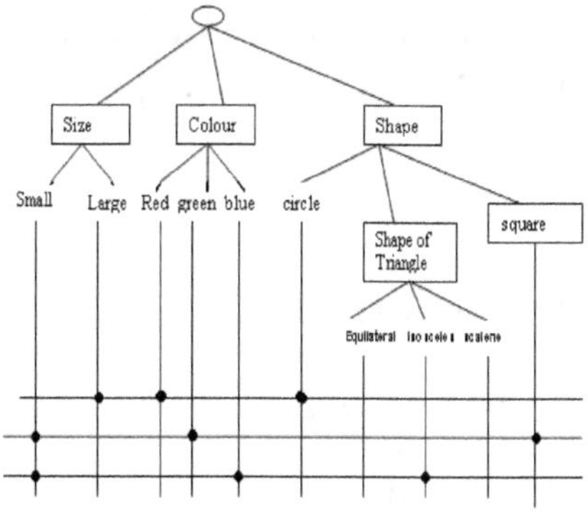

Fig. 1. Classification tree

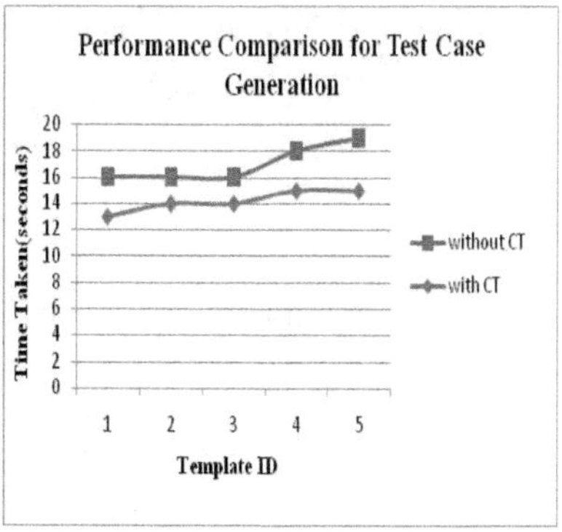

Fig. 2. Performance comparison

3 Result

The paper is implemented using the methodologies described in this document and also performed experimental evaluations to determine the performance measures corresponding to the project which shown in the fig 2. The project is developed to deliver a novel method to classify, analyze, and generate test cases. The proposed technique is versatile and customizable and gives significant data representation power to the user – domain experts like Software testing analyst have the ability to adjust parameters and testing mechanisms to fine tune testing and increase productivity through quality of service.

References

1. Weiser, M.: Program Slicing. IEEE Trans. on Software Engineering SE-10(4), 352–357 (1984)
2. Huang, H., Tsai, W.T., Subramanian, S.: Generalized Program Slicing for Software Maintenance. In: Proc. of Software Engineering and Knowledge Engineering, pp. 261–268 (1996)
3. Clarke, L.A., Richardson, D.J.: Validation by Symbolic Evaluation or Symbolic Execution. In: Hausen, H.L. (ed.) Software Validation. North-Holland, Amsterdam (1984)
4. Boehm, B.: Verifying and Validating Software Requirements and Design Specifications. IEEE Software 1(1), 75–88 (1984)

Mammogram Problem Solving Approach: Building CBR Classifier for Classification of Masses in Mammogram

Valliappan Raman[1,*], Patrick Then[1], and Putra Sumari[2]

[1] Swinburne University of Technology Sarawak,
93350 Kuching, Sarawak, Malaysia
{vraman,pthen}@swinburne.edu.my
[2] University Sains Malaysia
11800, Penang, Malaysia
putracs@usm.my

Abstract. Breast cancer continues to be a significant public health problem in the world. Early detection is the key for improving breast cancer prognosis. The aim of the research presented here is in twofold. First stage of research involves machine learning techniques, which segments and extracts features from the mass of digital mammograms. Second level is on problem solving approach which includes classification of mass by performance based case base classifier. We study the application of these algorithms as a classification system in order to differentiate benign from malignant mass, obtained from MIAS database, and we compare the classifier with other classification techniques to test the accuracy.

Keywords: Mammography, Segmentation, CaseBase, Classification.

1 Introduction

Early detection of breast cancer is considered as a major public health issue. Breast cancer incidence is the highest among female cancers and the second cause of mortality in Europe. To address this problem, it is necessary to create the adequate conditions allowing for the installation of mass detection campaigns, i.e. involving the maximum number of women at risk. However, an early detection of breast cancer depends on the adequacy of the mammogram from which the diagnosis is made an important research question addressed within this paper is performance evaluation. Since segmentation evaluation is essential for providing a scientific basis for image analysis in general and for medical image analysis in particular, this question is as important as the development of the segmentation algorithms themselves.

* Corresponding author.

V.V. Das, J. Stephen, and Y. Chaba (Eds.): CNC 2011, CCIS 142, pp. 284–289, 2011.

A Case-Based Reasoning algorithm is used for classifying these cases into benign or malignant cases. We have to be aware that Case-based Reasoning means using previous experience in form of cases to understand and solve new problems. A case-based reasoner remembers former cases similar to the current problem and attempts to modify their solutions to fit the current case. The underlying idea is the assumption that similar cases have similar solutions. Though this assumption is not always true, it holds for many practical domains. However, in this domain, reliable hypotheses are required because a single wrong prediction could be fatal. Reliability and accuracy are the main goal to be achieved in this research.

A number of CAD techniques have been developed for the detection and classification of microcalcifications [1] [2] [3] [5], in which several image processing techniques are applied ranging from gray-level image analysis [1] [5]) to morphological methods [5], as well as a great number of classifiers ranging from Bayesian classifiers to neural networks [4] [5].

Bottigli et al. presented a comparison of some classification system for massive lesion classification. An algorithm based on morphological lesion differences was used to extract the features. The two classes (pathological or healthy ROIs) were differentiated by utilizing the features. A supervised neural network was employed to check the discriminating performances of the algorithm against other classifiers and the ROC curve was used to present the results. In comparison with the other recent studies; the results of the new representation applied are comparable or better, owing to its better ability to distinguish pathological ROIs from the healthy ones. Vibha L proposes a method for detection of tumor using Watershed Algorithm, and further classifies it as benign or malignant using Watershed Decision Classifier (WDC).Experimental results show that this method performs well with the classification accuracy reaching nearly 88.38%.

In this paper we focus on building performance based Case Base Classifier to classify tumor from benign to malignant. Section 2 explains the research objective. Section 3 explains the Segmentation of mass by region growing in mammogram. Section 4 describes the performance base Case base classifier for classification of masses. Section 5 provides the experimental results of preliminary classification outputs. Finally conclusion is made at Section 6.

2 Mass Segmentation

First, the X-ray mammograms are digitized with an image resolution of 100 × 100 $\mu m2$ and 12 bits per pixel by a laser film digitizer. To detect mass & microcalcifications on the mammogram, the X-ray film is digitized with a high resolution. Because small masses are usually larger than 3mm in diameter, the digitized mammograms are decimated with a resolution of 400 × 400 mm 2 by averaging 4 × 4 pixels into one pixel in order to save the computation time. Fig 1 illustrates the mammogram segmentation and classification.

$$B_t = I_{\max} + 2.5\sigma_{bg} \tag{1}$$

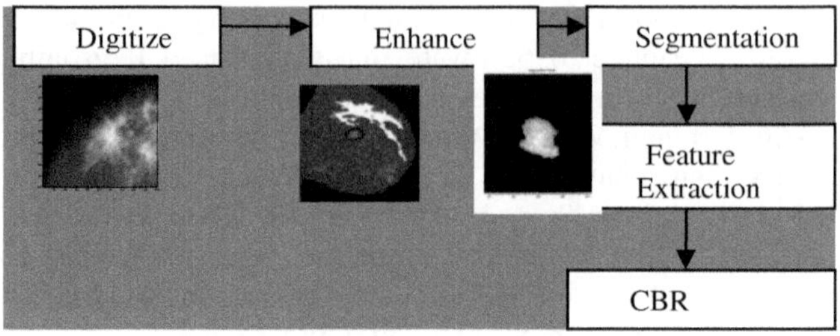

Fig. 1. Illustrates the mammogram segmentation and classification

Where I_{max} is the pixel intensity at the maximum peak count in the histogram of the decimated image and σ_{bg} is a standard deviation of all pixel values less than I_{max} under the assumption that the histogram of the background has Gaussian distribution centered at I_{max}. After breast region extraction, the extracted breast region *Br* is divided into three partitioned regions, the fat region, the glandular region, and the dense region. Then, automatic seed selection and segmentation by region growing are performed for each partitioned region. After the detected mass candidates are selected by the region growing, feature extraction and CBR classification are applied to the detected mass candidates [3].

Feature extraction and selection which includes area, perimeter, skewness, entropy, circularity and standard deviation (i.e.12 local and global features) chosen for next stage of classification.

3 CBR Classifier

Classifiers play an important role in the implementation of computer-aided diagnosis of mammography. The features or a subset of these features are employed by classifiers to classify mass into benign and malignant. Case Based Reasoning (CBR) seems as a potential method in making computerized screening decision. Case-Based Reasoning (CBR) integrates in one system two different characteristics: machine learning capabilities and problem solving capabilities. CBR uses a similar philosophy to that which humans sometimes use: it tries to solve new cases (examples) of a problem by using old previously solved cases. The process of solving new cases contributes with new information and new knowledge to the system. This new information can be used for solving other future cases. The basic method can be easily described in terms of its four phases. The first phase retrieves old solved cases similar to the new one. In the second phase, the system tries to reuse the solutions of the previously retrieved cases for solving the new case. The third phase revises the proposed solution. Finally, the fourth phase retains the useful information obtained when solving the new case. In a Case-Based Classifier System, it is possible to simplify the reuse phase. Classifying the new case with the same class as the most similar retrieved case can do reuse.

Performance based Case Base Classifier is a classifier where the re-use phase can be simplified. The kernel in a Case-Based Reasoning system is the retrieval phase (phase 1). Phase 1 retrieves the most similar case or cases to the new case. Obviously, the meaning of most similar will be a key concept in the whole system. Similarity between two cases is computed using different similarity functions. For our purpose in this paper, we use the similarity functions based on the distance concept. The most used similarity function is the Nearest Neighbor algorithm, which computes the similarity between two cases using a global similarity measure. The future practical implementation (used in our system) of this function is based on the Minkowski's metric.

Minkowski's metric is defined as:

$$Similarity\ (Case\_x, Case\_y) = \sqrt[r]{\sum_{i-1}^{F} W_i \times |x_i - y_i|^r} \quad (2)$$

Where *Case_x*, *Case_y* are two cases, whose similarity is computed; F is the number of features that describes the case; $x_i\, y_i$ represent the value of the ith feature of case *Case_x* and *Case_y* respectively; and w_i is the weight of the ith feature. In this study we test the Minkowsky's metric for three different values of r: Hamming distance (r = 1), Euclidean distance (r = 2), and Cubic distance (r = 3). This similarity function needs to compute the feature relevance (w_i) for each problem to be solved. Assuming an accurate weight setting, a case-based reasoning system can increase their prediction accuracy rate. We use also the Clark's and the Cosine distance, both are based on distance concept and also use weighting features. Sometimes human experts can not adjust the feature relevance, automatic method can solve this limitation.

3.1 Building Perform Case Base Classifier

To improve the performance and accuracy of the system, Perform CB classifier is build with 10 data sets, therefore data sets will undergoes for a formal converter to extract the features and select the features for classification. Therefore comparison is made between the cases based on similarity measures which help to increase the accuracy and reliability.

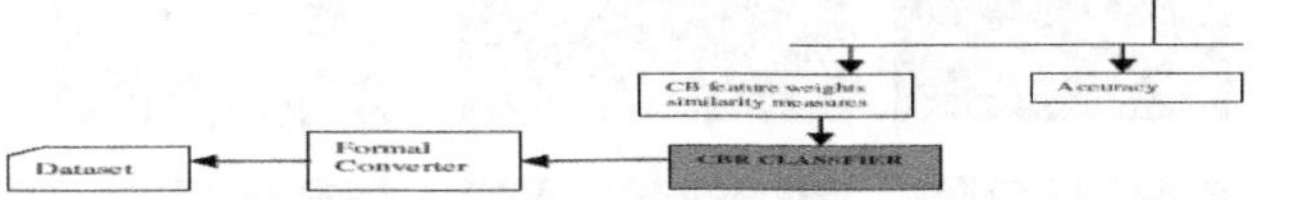

Fig. 2. Illustrates the Performance Based CB classifier

3.2 Improving the Accuracy of Perform CB Classifier

The combination of classifiers has long been proposed as a method to improve the accuracy achieved in isolation by a single classifier. Stacking allows us to combine different decisions of each classifier to obtain a definitive prediction. There are several ways that stacking can be used for variety of task. For example, if the two classifiers that perform best classify the same, then we take their decision ignoring the other classifiers. Bagging and boosting have been used to improve the accuracy of classification over a single classifier learned on the entire data. The Bagging technique uses repeated random samples of the dataset whereas in boosting the samples are weighted. These methods can be applied to CBR Classifiers and we can use them for improving the performance of these classifiers. Currently the work is ongoing of building the classifier with more data set.

4 Experimental Results

A total of 25 mammograms were considered for this study and experiment is at final stage. Mammogram data's have been taken from the MIAS database of mammograms containing cancerous masses. The mammograms are scanned from X-rays with a maximum resolution of 512x512 pixels. Mammographic image is reduced to an m × n matrix. This matrix contains as many rows, m, as the number of Mass present in the image, and as many columns (n=12) as the number of features that describe one mass. Next, this m × 12 matrix is transformed into a vector. This transformation computes the average value for each column (feature) across all the rows (mass in the image). Finally, the computed vector is labeled using the class (benign or malign) obtained from the diagnosis done by surgical biopsy. We performed two kinds of experiments in order to compare the performance of the different algorithms. First, we maintained the proportion of original images - now, a set of features for each image- as training and test sets proposed by human experts. Thus, we compared the results obtained by other classifiers with those achieved by human experts, and the statistical modeling terms of classification accuracy. We also included in this comparison the true positive (malignant cases) rate of classified examples (sensitivity) and the true negative rate of classified examples (specificity).

4.1 Results

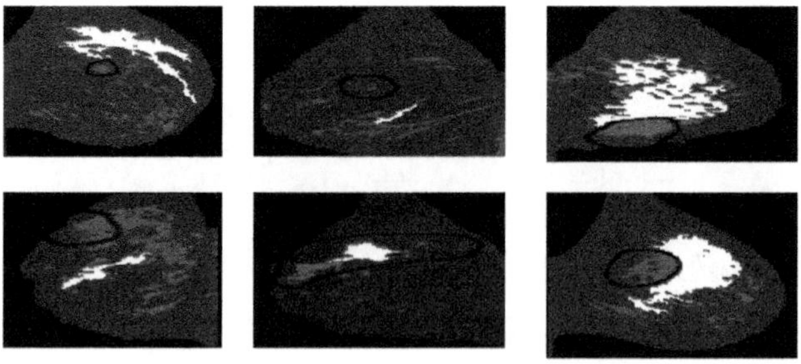

Fig. 4. Illustrates the Determination of ROI and Segmentation

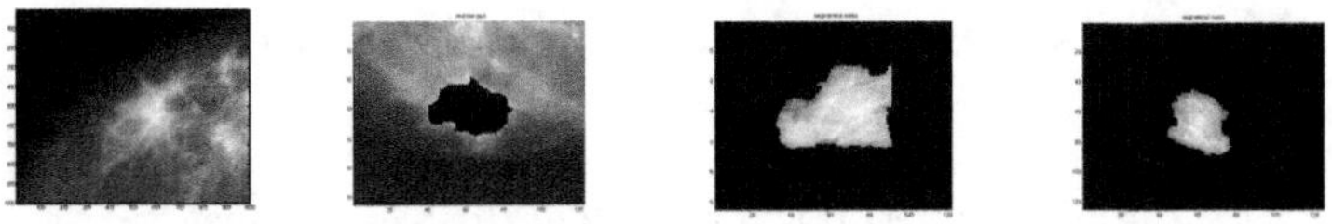

Fig. 5. Illustrates (i) orginal image (ii) identying the regions & segmenting the regions (iii) maligant tumor (iv) benign tumor after classification

Table 1. Illustrates the Classifier Results

Variant	Not Classified	SENSITIVITY	SPECIFICITY	ACCURACY
HAMMING	0.00	82.7	70	72.5%
CUBIC	0.00	70.5	72.7	71%
EUCLIDEAN	0.00	71.6	75.9	72%
CLARK	0.00	69.8	80.4	76%

The current initial results show that accuracy in our classifier schemes is higher than the accuracy obtained by the human experts and the statistical model.

5 Conclusion

We propose a performance based CBR classifier framework based on region growing segmentation. This integrated technique leads us to satisfactory knowledge base construction. The formulated knowledge base is in the structure of a concept lattice. Its implications provide classification rules efficiently where the time and space complexities are reduced. The CBR classifier will be improved by adding bagging and stacking to improve the reliability and accuracy. In future, algorithms will be constructed for stacking and bagging rules automatically.

References

1. Zhang, W., et al.: An improved shift-invariant artificial neural network for computerized detection of clustered microcalcifications in digital mammograms. Medical Physics 23, 595–601 (1998)
2. Lai, S.M., Li, X., Bischof, W.F.: On techniques for detecting circumscribed masses in mammograms. IEEE Transactions on Medical Imaging 8, 377–386 (1989)
3. Otsu, N.: A threshold selection method from gray-level histograms. IEEE Trans. System Man Cabernet SMC 9, 62–66 (1979)
4. Rabottino, G., Mencattini, A., Salmeri, M., Caselli, F., Lojacono, R.: Mass Contour Extraction in Mammographic Images for Breast Cancer Identification. In: 16th IMEKO TC4 Symposium, Exploring New Frontiers of Instrumentation and Methods for Electrical and Electronic Measurements, Florence, Italy (2008)
5. Valliappan, R., Sumari, P.: Digital Mammogram Segmentation: An Initial Stage. In: 4th IASTED International Conference on Advanced Computing Science and Technology, Langawi, Malaysia (2008)

New Mutants Generation for Testing Java Programs

Kapil Kumar, P.K. Gupta, and Roshan Parjapat

Department of Computer Science, Jaypee University of Information Technology, Solan, H.P, India
{kapil.cs89,pradeep.juit,roshanprajapat89}@gmail.com

Abstract. Mutation testing is a different approach of testing that tests the test cases by introducing bugs in the code. These bugs are the small modification in the program codes and that produces new versions of code called mutants. The correctness of mutation testing depends heavily on the types of faults that the mutation operators are designed to represent. Many mutation operators have previously been presented, and many researches have been done to show the usefulness of the operators. In this paper a set of mutant operator is proposed for the Java programming language. In this paper some new mutant operators are briefly described for testing java exception handling and string handling programs.

Keywords: mutation testing, mutation operators, class-level operators, web specific operators, mujava.

1 Introduction

Testing is a crucial activity in the software lifecycle. Mutation testing method [1] is a fault based test technique that measures the effectiveness of test suite [1, 2]. It uses a set of "rules" called mutant operators to create programs slightly different from the program under test. The goal of mutation testing is the generation of a test set that distinguishes the behavior of the mutants from the original program. The ratio of distinguished mutants (also called dead or killed mutants), over the total number of mutants, measures the adequacy of the test set. A test case is adequate if it is useful in detecting faults in a program [3]. If the original program and all mutant programs generate the same output, the test case is inadequate.

Many mutation tools have been developed and distributed for procedural programs. There are many mutation tools for java also like mujava [6], jester [7], and jumble [8], Judy [9]. Mujava is a publicly available Java mutation system. MuJava supports both generation and execution of mutants to help testers and researchers carry out mutation testing. But in mujava no mutant operators are there for exception handling and string handling. So our research interest in this paper is to present the new mutation operators for these topics of java programming. Because they are also important and frequently used techniques in java programming.

2 Related Work

Mutation testing has a huge research history[5]. It mainly focused on only four kind of research categories related to mutation testing: 1) Proposing mutation operators, 2)

V.V. Das, J. Stephen, and Y. Chaba (Eds.): CNC 2011, CCIS 142, pp. 290–294, 2011.

Applying these operators (Experimentation), 3) Developing Mutation tools, and 4) Reducing the cost of mutation analysis. These are the only four research activities related to mutation testing. The first one involves defining new mutation operators for different languages and types of testing [3, 6, 11]. The second research activity is experimentation with mutations [12, 13]. The third kind of activities in mutation testing research is developing mutation tools. Mothra[5] and Proteum[4] were developed for FORTRAN and C, respectively. Jester [7], Jumble[8], Judy[9] MuJava [6] for Java. Fourth one activities involves research activities is investigating ways to reduce the cost of mutation analysis.

3 Proposed Mutation Operators for Java

In mujava, there are no operators for exception handling and string handling. These topics are also very crucial like others, so there should be some mutation operators for them. In this paper some operators are proposed, explanation is given as follows.

3.1 Mutation Operators for Exception Handling

In Java exception handling is managed via five key words: try, catch, throw, throws and finally [16]. We handle an exception i.e., catch the exception by declaring a try-catch block or propagate it by declaring it to throw in a throws statement of method declaration. There are many common mistakes that are frequently done by the author of the code, these mistakes result an output difference. New mutation operators for exception handling are given below.

Catch Clauses Deletion (CCD). In some case, more than one exception could be raised by a single piece of code. To handle this type of situation, two or more catch clauses can specify, each catch block can catch a different type of exception.

The CCD removes catch clauses one by one when there is more than one catch clauses(no effect whether or not the finally clause exist in the code). CCD working is given by the following codes.

The original code-

1. class ccd{
2. try {.........}
3. catch(ArithmeticException e){......}
4. catch(ArrrayIndexOutOfBoundsException e){......}
5. catch(ArrayStoreException e){.........}
6. catch(IllegalArgumentException e){...} }

CCD mutant:

1. class ccd{
2. try{......}
3. catch(ArithmeticException e){......}
4. catch(ArrrayIndexOutOfBoundsException e){.........}
5. catch(ArrayStoreException e){...}}

Throw Statement Deletion (TSD). In java, exceptions are thrown by the Java run-time system. However, it is possible to throw exception explicitly, using the throw statement.

TSD operator deletes the throw statement that is shown by the following codes with the different outputs –

The original code-

```
class tsd{
try{
System.out.println("1st time exception throw");
Throw new NullPointerException("demo");
}catch(NullPointerException e){
System.out.println("exception throw");
throw e; } } }
```

TSD mutant -

```
class tsd{
try{
System.out.println("1st time exception throw");
}catch(NullPointerException e){
System.out.println("exception throw");
throw e; } } }
```

Exception Name Change (ENC). In the throw statement ENC change the exception name with the other exception name that throw different exception at the run time.

The original code:

```
class enc{
try{
throw new NullPointerException ("demo");
}catch(NullPointerException e)
{System.out.println("cought");
throw e;}}}
```

ENC mutant:

```
class enc{
try{
throw new ArithmeticException ("demo");
}catch(NullPointerException e){
System.out.println("cought");
throw e;}}}
```

Finally Clause Deletion (FCD). The finally block execute whether or not an exception is thrown. It is useful for closing file handles and freeing up any other resources that might have been allocated at the beginning of a method with the intent of disposing of them before returning.

The FCD operator delete this finally clause and produce an output difference.

The original code:

```
class fcd{
try{..................
...........{catch(......){
...........}catch(......){...}
...}finally{..............}
```

FCD mutants:

```
class fcd{
try{..................
...........{catch(........){
...........}catch(......){...}
```

Exception Handling Modification (EHM). The EHM operator change the way of handling an exception in the code. It modify an exception handling statement to an exception propagate statement, and an exception propogation statement to an exception handling statement.

The original code:

```
1. class ehm{try{.....
2. ....}catch(ArrayIndex
OutOfBounds ar){.....}
```

EHM mutants:

```
1. class ehm{
2. throwsArrayInadexOu
tOfBounds {....}
```

3.2 Mutation Operators for String Handling

Unlike most of the other languages, Java treats a string as a single value rather than as an array of characters. There are many constructors for handling the string object.

ValueChangeOperator(VCO). There are many String methods(charAt(), getChars(), getbyte(), substring(), replace(), deleteCharAt() and many more) for different purpose. By using these methods and by passing some specified values to these methods, they returns desired output (may be any character or any substring etc). VCO change the parameter values that are passed in these methods by plus 1.

Table 1. Value change operators

Mutation Operator	String Methods	Original code	Mutants
	charAt()	s.charAt(4);	s.charAt(5);
	setCharAt()	s.setCharAt(6);	s.setCharAt(7);
VCO	getChars()	getChars(10,14,str,0)	getChars(11,15,str,1
	substring()	substring(5,6);	substring(6,7);
	replace()	replace(5,7,"new");	replace(6,8,"new");
	deleteCharAt()	deleteCharAt(3);	deleteCharAt(4);

MethodSwapOperator(MSO). The String class includes several methods that compare strings or substrings within string.There are pair of methods of same type such as equals() and equalsIgnoreCase() method.Both methods compare two string and return true or false,but there are small difference between these methods that is equals() is a case sensitive comparison but other method ignore the case.In Java String handling there are many mehods of such type i.e."startWith()" and "endWith()", "equals() and "==" ,"IndexOf()" and lastIndexOf".

Table 2. Method swap operators

Mutation Operator	Original code	Mutants
	s1.equals(s2);	s1.equalsIgnoreCase(s2);
	s1.startWith("prefix");	s1.endWith("prefix");
MSO	s1.equals(s2);	(s1 == s2);
	s.indexOf('t');	s.lastIndexOf('t');
	s1.equalsIgnoreCase(s2);	s1.equals(s2);
	s1.endWith("prefix");	s1.startWith("prefix");
	(s1 == s2);	s1.equals(s2);
	s.lastIndexOf('t');	s.indexOf('t');

4 Conclusion

Mutation testing has been shown to be an effective way of testing software in terms of fault revealing capacity and it has been widely explored .In this paper we described new mutation operators for Java mutation testing. We have presented practical approach also so that these new mutation operators can be applying in Java mutation tools. Seven new mutation operators are defined in this paper with the results of applying faults in the codes.

References

1. Demillo, R.A., Lipton, R.J., Sayward, F.G.: Hints on test data selection: help for the practicing programmer. IEEE Comput. 11(4), 34–41 (1978)
2. Hamlet, R.G.: Testing programs with the aid of a compiler. IEEE Trans. Softw. Eng. 3(4), 279–290 (1977)
3. Ma, Y.S., Kwon, Y.R., Offutt, J.: Inter-class mutation operators for java. In: Proc. 13th Int. Symp. Software Reliability Engineering, pp. 352–363 (November 2002)
4. Delamaro, M.E., Maldonado, J.C.: Proteum – A tool for the assessment of test adequacy for C programs. In: Proceedings of the Conference on Performability in Computing Systems, pp. 75–95 (July1996)
5. DeMillo, R.A., Guindi, D.S., King, K.N., McCracken, W.M., Offutt, A.J.: An extended overview ofthe Mothra software testing environment. In: Proceedings of the Second Workshop on Software Testing, Verification, and Analysis, pp. 142–151 (July1988)
6. Ma, Y.S., Offutt, J., Kwon, Y.R.: MuJava: a mutation system for Java. In: Proc. 28th Int. Conf. Software Engineering, pp. 827–830 (May 2006)
7. Moore, I.: Jester – a JUnit test tester. In: Proc. Second Int. Conf. Extreme Programming and Flexible Processes in Software Engineering, pp. 84–87 (May 2001)
8. Irvine, S.A., Tin, P., Trigg, L., Cleary, J.G., Inglis, S.,Uttingm: Jumble Java byte code to measure the effectiveness of unit tests. In: Proc. Testing: Academic and Industrial Conf. Practice and Research Techniques, pp. 169–175 (September 2007)
9. Madeyski, L., Radyk, N.: Judy – a mutation testing tool for Java. IET Softw. 4(1), 32–42 (2010)
10. Offutt, A.J., Untch, R.H.: Mutation 2000: uniting the orthogonal. In: Wong, W.E. (ed.) Mutation Testing for the New Century, 1st edn., pp. 34–44. Kluwer Academic Publishers, Dordrecht (2001)
11. Offutt, J., Ma, Y.S., Kwon, Y.R.: The class-level mutants of MuJava. In: Proc. Int. Workshop Automation of Software Test, pp. 78–84 (May 2006)
12. Offutt, A.J., Lee, A., Rothermel, G., Untch, R.H., Zapf, C.: An experimental determination of sufficient mutant operators. ACM Trans. Softw. Eng. Meth. 5(2), 99–118 (1996)
13. Andrews, J.H., Briand, L.C., Labiche, Y.: Is mutation an appropriate tool for testing experiments?. In: Proc. 27th Int Conf. Software Engineering, pp. 402–411 (May 2005)
14. Schildt, H.: 'The Complete Reference Java 2 Fifth edition' Book

A Novel Approach to Hindi Text Steganography

Mayank Srivastava[1], Mohd. Qasim Rafiq[2], and Rajesh Kumar Tiwari[1]

[1] Dept. of Computer Science, G.L.A Univeristy, Mathura (U.P.), India
[2] Dept. of Computer Engg., Z.H.C.E.T., A.M.U., Aligarh (U.P.), India
mayank_sri2001@rediffmail.com, mqrafiq@hotmail.com, rajeshkrtiwari@yahoo.com

Abstract. The digital information revolution has brought important changes in our society and our life. Now a days, large amount of data are transmitted over the network. If the data that is being transmitted is important, one should use secure technique like steganography to transmit it. Steganography is a method of hiding a secret message in a cover media such as image, text, audio, video etc in way that hides the existence of the secret data. This paper introduces three new methods for Hindi text steganography which is based on the use of directory/dictionary structure.

Keywords: Steganography, Data Hiding, Hindi, Sanskrit, Synonyms, Punctuation Marks.

1 Introduction

The growing use of the Internet has led to a continuous increase in the amount of data that is being exchanged and stored in various digital media. Security and authentication techniques are used to enhance the various security features and to preserve the intellectual property of digital data. Steganographic techniques are one of the most successful techniques which support the hiding of critical data in a way that prevent the detection of hidden messages [4]. Steganographic jobs have been performed on images, video, sound and text. However, text steganography is difficult to achieve because text files lack the redundant information and the structure of text documents is identical with what we observe.

In this paper three new methods for Hindi text steganography are presented. The techniques makes use of the directory/dictionary, which will be created depending on the punctuation marks, synonym of Hindi words and the use of Sanskrit words in Hindi. The rest of the paper is arranged as follows: section 2 deals with the related work, section 3 briefs the proposed methods and section 4 gives the conclusion.

2 Related Work

Few works have been done in Hindi text steganography. In syntactic structure based method, the secret data was stored in the Hindi text by using the encoding scheme of the Hindi diacritics [1]. Another method creates numerical directory of the Hindi vowel and consonants to hide the secret data [1].The next method in the same

V.V. Das, J. Stephen, and Y. Chaba (Eds.): CNC 2011, CCIS 142, pp. 295–298, 2011.

category makes the use of Hindi words modifiers and characteristics of the Hindi characters like bar, no bar etc to hide the secret data. [2].

In feature based method, representation of one of the matra of Hindi text is changed to store the secret data [3].

3 Proposed Algorithm

The proposed secret data embedding algorithms are as follows.

3.1 Punctuation Marks Based

The method we are proposing here will make use of a directory of the punctuation marks of the Hindi text to hide the secret data. We encode all the punctuation marks with a specific numerical code having a range between 0 and 3 (Table 1). At the sending site, input message is converted to its binary equivalent and then Hindi sentence is created that makes use of the punctuation marks according to input bit stream. At the receiving end, the receiver receives the stego text and extracts the binary data by examining the punctuation marks in it. After the rearranging the bits, the original message is displayed.

Table 1. Encoding table for Punctuation Marks

0	,	“ ”	“	………………
1	।	॥	‘ ’	( )
2	;	—	:-	०
3	–	?	!	^

We are showing here the result of encoding the secret message **“se”**, which is represented in binary as **01 11 00 11 01 10 01 01**.

01 11 00 11 01
भारत एक विशाल देश है। यहाँ नागरिको में धर्म – निरपेक्षता , कर्तव्य – परायणता की भावना है।
10 01
भारत की आजादी के लिए एक व्यक्ति का योगदान सबसे महत्वपूर्ण है — महात्मा गाँधी । भारत कई
01
सभ्यताओं एवं संस्कृतियों का मेल है।

3.2 Synonym Based

The goal of this method is to hide secret data in Hindi text by substituting the words which have synonym in Hindi text. For this, we prepare a four level dictionary of Hindi words where the basic Hindi word is represented as a level zero and the next three levels contain the synonyms of the basic Hindi word (Table 2). At the sending site, input message is converted into its binary equivalent and then Hindi sentence is created that will use the words from the dictionary according to input bit stream. Now at the receiving end, the program extracts the bit level data from the stegotext by examining the words of different levels of the dictionary used in it. After rearranging the bits, the original message is displayed.

Table 2. Encoding table for Synonym based

0	1	2	3
सफल	समृद्ध	धन्य	सौभाग्यशाली
संघ	मातृभूमि	राष्ट्र	देश
आजादी	स्वतंत्र	स्वाधीन	मुक्त
अंशदान	भागदान	सहयोग	योगदान
असरदार	महत्वपूर्ण	आवश्यक	प्रभावशाली
बहुत	बहुतेरे	अनेक	कई
अध्यापन	शिक्षा	सीख	उपदेश
प्रबन्ध	संगठन	संघटन	व्यवस्थापन

Here, for illustration we are showing the encoding result for secret text "**se**" which is represented in binary as **01 11 00 11 01 10 01 01**.

01 11 00 11 01
भारत एक समृद्ध देश है। भारत की आजादी के लिए एक व्यक्ति का योगदान सबसे महत्वपूर्ण है —
10 01 01
महात्मा गाँधी । भारत में अनेक , शिक्षा प्रदान करने वाले संगठन है।

3.3 Sanskrit Classification Based

The method we are proposing here will make use of a dictionary of the words to hide the secret data. Here the dictionary will consist of two levels. Level zero corresponds to tatbhav words and level one corresponds to tatsam words. We encode the level zero words with bit 0 and level one words with bit 1 (Table 3). To implement the algorithm, firstly the secret message is converted to its equivalent binary data. Now, we create multiple Hindi sentences that use multiple words of the dictionary to represent the secret binary data. At the receiving end, the receiver extracts the bit level data from the stego Hindi text by examining the words of two level dictionary used in it. After rearranging the bits, the original message is displayed.

Table 3. Encoding table for Sanskrit based

0	1
सावन	श्रावण
छाता	छत्र
मोर	मयूर
नाच	नृत्य
रात	रात्रि
साँप	सर्प
आग	अग्नि
सूरज	सूर्य

Here, for illustration we are showing the encoding result for secret text “**s**” which is represented in binary as **01 11 00 11**.

0 1
सावन में समान्यतः अच्छी वर्षा होती है। लोग छत्र लेकर घरों से निकलते है।
1 1 0 0 1
मयूर नृत्य करने लगते है। रात में साँप इत्यादि निकल आते है। अग्नि की भांति
1
सूर्य लुप्त हो जाता है।

4 Conclusion

Security in secret data transmission has become a very critical aspect of modern internet and intranet systems. In this paper suitability of steganography as a tool to conceal highly sensitive information has been discussed by using three different methods which are based on directory/dictionary structure. Here directory/dictionary is created depending upon the use of punctuation marks, synonyms and Sanskrit words in Hindi. As we know Hindi is one of the most popular language, the stegotext generated by the above methods can be transmitted to anybody anywhere across the globe in a complete secured form that we generally expect. It is worth mentioning that all the proposed methods have used the same directory/dictionary at both the ends.

References

1. Alla, K., Siva Rama Prasad, R.: An Evolution of Hindi Text Steganography. In: Proceedings of the Conference in IEEE Computer Socienty Digital Library, 6th ITNG 2009, pp. 1577–1579 (2009), doi:IEEE.computersociety.org/10.1109
2. Alla, k., Siva Rama Prasad, R.: A new approach to hindi text steganography using Maraye, Core classification and HHK Scheme. In: Proceedings of the Conference in IEEE 7th ITNG 2010 (2010)
3. Chandger, S., Debnath, N.C., Ghosh, D.: A New approach to hindi text steganography by shifting matra. In: 2009 International Conference on Advances in Recent Technologies in Communication and Computing (2009)
4. Johnson, N.C., Jajodia, S.: Exploring steganography: seeing the unseen. IEEE Computer, 26–34 (February 1998)

Object Detection and Tracking in Videos Using Snake and Optical Flow Approach

E. Jayabalan[1] and A. Krishnan[2]

[1] Assistant Professor of Computer Science, Government Arts College (Autonomous), Salem-7, Tamilnadu, India
[2] Dean (Academic), KSR College of Engineering, Tiruchengode, Tamilnadu, India
ejksrcas@rediffmail.com, ejksrcas@gmail.com,
ejksrcas@yahoo.co.in

Abstract. In real time situations, non rigid object tracking is a challenging and important problem. Due to the non rigid nature of objects in most of the tracking applications, deformable models are appealing in tracking tasks because of their capability and flexibility. The proposed approach uses an observation model based on optical flow information used to know the displacement of the objects present in the scene. After finding the moving regions in the initial frame, we are applying active contour model (ACM) to track the moving objects in the further frames dynamically. These models have been used as a natural means of incorporating flow information into the tracking. The formulation of the Active Contour Model by incorporating an additional force driven optical flow field improves the tracking speed. This algorithm efficiently works to track for low contrast videos like Aerial videos.

Keywords: Object, Tracking, Video, frames, Contour, Optical flow, non rigid.

1 Introduction

In real time situations, non rigid object tracking is a challenging and important problem. Due to the non rigid nature of objects in most of the tracking applications, deformable models are appealing in tracking tasks because of their capability and flexibility. There are two general types of active contour models in the literature today: parametric active contours and geometric active contours. This work focuses on parametric active contours called Snakes, which synthesize parametric curves within an image domain that can move under the influence of internal forces within the curve itself and external forces derived from the image data[1][2]. There are two key difficulties with active contour algorithms. First, the initial contour is close to the true boundary or else it will likely converge to the wrong result. The second problem is that active contours have difficulties progressing into concave boundary regions. The object tracking is done using Optical Flow approach [3] in the proposed work.

2 Proposed System

In the proposed approach we extended the formulation of active contour model using information about the optical flow to make the contour in less number of iterations.

V.V. Das, J. Stephen, and Y. Chaba (Eds.): CNC 2011, CCIS 142, pp. 299–301, 2011.

With out user intervention, by using optical flow automatically initializes a contour in the first frame and then use the snake propagation method (Traditional Snakes/GVF Snakes) to find the final contour, thus detecting the object of interest in the first frame. The initializations of the contour for frame ***i+1*** is done by using the final contour obtained from frame ***i*** by using motion prediction. Thus we can achieve fast real time object tracking using this method. In order to achieve the tracking for the moving objects, Automatic initialization is achieved by optical flow method and the deformation is achieved by both snake and optical forces.

3 Experimental Results

3.1 Tracking by Active Contour Models

This approach is implemented in both MATLAB 6.5 and JDK 1.5 with JMF 2.0 and JAI 1.0.1. Fig.1 shows the person tracking by using active contour models (GVF Snake) in MATLAB 6.5. In this simulation the contour is manually initialized by the program, then the contour is easily tracked by using the parameters α =0.08, β =0.02 and δ =1. But the problem is as shown in the figure.

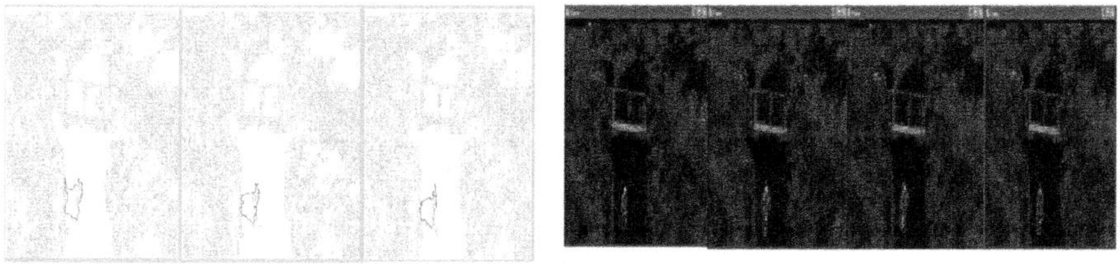

(a) (b)

Fig. 1. (a) and (b) Shows the tracking nature of the boundary of a person using MATLAB 6.5 and Java

3.2 Trcking by Optical Flow Method

The limitations of human intervention for initialization of the contour eliminated by using this method. Fig. 2 shows a test video where the person is moving from left to right and the flow detection of the person in shown in first frame. Further the tracking of flow vectors is shown in Fig. 2.

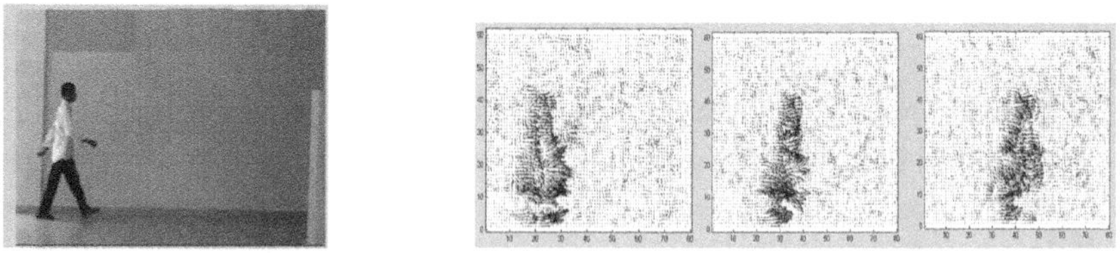

(a) (b)

Fig. 2. (a) and (b) Optical flow detection of a moving person for the shown frame

4 Conclusion

In Uncompressed videos within the scene the proposed algorithm describes motion of the object by using both deformation forces and optical flow forces. In active contour propagation, tracking of a person is simulated with both the traditional snake and GVF snakes. By using a new external force model for snakes called gradient vector flow solves the problem of convergence to the boundary concavities. But the major problem with active contour model is its necessary to initialize the contour for ROI. The manual initialization was eliminated by Optical flow approach. Optical flow information is used to know the displacement of the objects present in the scene. The Optical flow estimation is one of the simplest estimation even though this approach provides dense estimation.

References

1. Xu, C., Prince, J.L.: Gradient Vector Flow: A New External Force for Snakes. In: IEEE Conference on Computer Vision and Pattern Recognition, pp. 66–71 (1977)
2. Kass, M., Witkin, A., Terzopoulos, D.: Sanakes: Active Contour Models. J. International Journal of Computer Vision, 321–331 (1987)
3. Barroon, J.L., Fleet, D.J., Beauchemin, S.S.: Performance of optical flow techniques. J. International Journal of Computer Vision 12, 43–77 (1994)
4. Yokoyama, M., Poggio, T.: A Contour-Based Moving Object Detection and Tracking. In: 2nd Joint IEEE International Workshop on Visual Surveillance and Performance Evalution of Tracking and Surveillance, pp. 271–276 (2005)
5. Takaya, K.: Tracking a Video Object with the Active Contour (Snake) Predicted by the Optical Flow. In: IEEE Canadian Conference on Electrical and Computer Engineering, pp. 369–372 (2008)

An Improved Technique for Handling Information Regarding Location of Moving Object

E. Jayabalan[1] and A. Krishnan[2]

[1] Assistant Professor of Computer Science, Government Arts College (Autonomous), Salem-7, Tamilnadu, India
[2] Dean (Academic), KSR College of Engineering, Tiruchengode, Tamilnadu, India
ejksrcas@rediffmail.com, ejksrcas@gmail.com, ejksrcas@yahoo.co.in

Abstract. A technique is proposed to drastically reduce the update cost. Specifically, we propose to model the current position of a moving object as the distance from its starting position, along a given route. The distance continuously increases as a function of time, without being updated. In the context of position attributes that change continuously with time, the main issues that are addressed are position-update policies and imprecision. To accomplish tracking with certain accuracy, each wireless device monitors its real position (its TRAVEL DATABASE position) and compares this with a local copy of the position that the central database assumes. When needed in order to maintain the required accuracy in the database, the wireless device issues an update to the server.

Keywords: Tracking, Video, Database, Wireless, Travel-Database, Update, Moving Object.

1 Introduction

In this age of significant telecommunications competition, mobile network operators continuously seek new and innovative ways to create differentiation and increase profits. One of the best ways to do accomplish this is through the delivery of highly personalized services. One of the most powerful ways to personalize mobile services is based on location. I will discuss Location Based Services (LBS) [1][2][3], but we will first discuss the basis of LBS - location technology. IN step with the emergence of an infrastructure for mobile, online location-based services (LBSs) for general consumers, such services are attracting increasing attention in industry and academia. An LBS is a service that provides location-based information to mobile users. The main idea is to provide the service user with a service that is dependent on positional information associated with the user, most importantly, the user's current location. The service may also be dependent on other factors, such as personal preferences and interests of the user. A service might inform its users about traffic jams and weather situations that are expected to be of relevance to each user. A friend monitor may inform each user about the current whereabouts of friends. Other services may track

V.V. Das, J. Stephen, and Y. Chaba (Eds.): CNC 2011, CCIS 142, pp. 302–304, 2011.

the positions of emergency vehicles, police cars, security personnel, hazardous materials, or public transport.

2 Proposed System

Point-Based Tracking Using this approach, the server represents a moving object's future positions as the most recently reported position. An update is issued by a moving object when its distance to the previously reported position deviates from its current TRAVEL DATABASE position by the specified threshold. In vector tracking, the future positions of a moving object are given by a linear function of time, i.e., by a start position and a velocity vector. Point tracking corresponds to the special case where the velocity vector is the zero-vector. A TRAVEL DATABASE receiver computes both speed and heading for the object it is associated with—the velocity vector used in this representation is computed from these two.

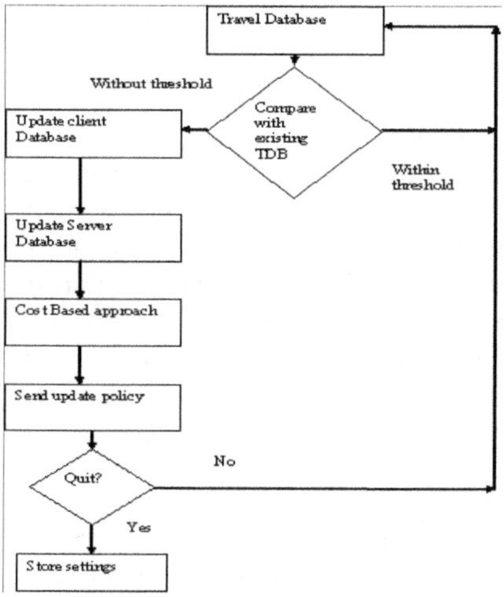

3 Experimental Results

The road networks based tracking object is based on the localization based services. The data set is transferred to destination in form the form of packets it can travel through any nodes if it travel through a single node one or more times it will be considered as a default node then the packets will be transferred through that node only. Then the dataset will be stored .it can travel through any nodes but it should reach the destination. Default node will be stored separately.

3.1 Implementation

The main idea is to provide the service user with a service that is dependent on positional information associated with the user, most importantly, the user's current location. The service may also be dependent on other factors, such as personal preferences and interests of the user. With the continued advances in wireless communications, geo-positioning, and consumer electronics, an infrastructure is emerging that enables location-based services that rely on the tracking of the continuously changing positions of entire populations of service users, termed moving objects. This scenario is characterized by large volumes of updates, for which reason location update technologies become important. In this paper we address position update policies and impression. Namely, assuming that the actual position of a moving object m derivates from the position computed by the DBMS, when should m update its position in the database in order to eliminate the deviation? Furthermore, how can the DBMS provide a bound on the error when it replies to a query: what is the current position of m? We propose cost based approach to update policies that answers both questions.

4 Conclusion

In this paper we considered databases that model the location of objects moving on routes. We formulated the database-position update problem as a mathematical cost optimization problem using the concept of a position-update policy, i.e. a quintuple (deviation cost function, update cost, estimator function, fitting method, predicted speed). Such a policy is adaptive and predictive in the sense that the update time-points depend on the current and predicted behavior of the deviation (of the actual position from the database position) as a function of time. Then we devised and analyzed three position update policies. We showed that the DBMS is able to bound the uncertainty at any point in time. Actually, for two of the policies (the immediate ones) the position-uncertainty decreases as time-since-the-last- update increases.

References

1. Ilarri, S., Mena, E., Illarramendi, A.: Location-Dependent Query Processing: Where We are and Where We Are Heading. J. ACM Computing Surveys 42(3), Article 12 (2010)
2. Wolfson, O., Chamberlain, S., Dao, S., Jiang, L., Mendez, G.: Cost and Imprecision in Modeling the Position of Moving Objects. In: 14th International Conference on Data Engineering (ICDE 1998), Florida, pp. 588–596 (1998)
3. Wolfson, O., Sistla, A.P., Camberlain, S., Yesha, Y.: Updating and Querying Databases that Track Mobile Units. J. Distributed and Parallel Databases 7(3), 257–287 (1999)
4. Lam, K., Ulusoy, O., Lee, T.S.H., Chan, C., Li, G.: An Efficient Method for Generating Location Updates for Processing of Location-Dependent Continuous Queries. J. Database Systems for Advanced Applications, 218–225 (2001)

Advanced Bioinformatics Approach in Machine Learning for Analyzing Genome Wide Expression Profiles and Proteomic Data Sets

Archana Dash[1], Tripti Swarnkar[2], and Mamata Nayak[3]

[1] ITER (SOA) University,
Jagamohan Nagar, Bhubaneswar, India
dash.sweetstar@gmail.com
[2] Department of Computer Applications, S.O.A University, Jagamohan Nagar, Bhubaneswar
tswarnakar@iter.ac.in
[3] Department of Computer Applications, S.O.A University, Jagamohan Nagar, Bhubaneswar
mamata2005@gmail.com

Abstract. Biological research is becoming increasingly database driven, motivated, in part, by the advent of large-scale functional genomics and proteomics experiments such as those comprehensively measuring gene expression. Consequently, a challenge in bioinformatics is integrating databases to connect this disparate information as well as performing large-scale studies to collectively analyze many different data sets. These composite data sets are conducive to extensive computational analysis and present new opportunities for data mining. *Both supervised and unsupervised* approaches can often be used to analyze the same kinds of data, depending on the desired result and the range of features available. Large-scale experiments, such as those performed with microarrays, yield large homogenous data sets that are well suited for computational analysis.

Keywords: Machine Learning, Genome, Protein Chips, DNA arrays, SOM.

1 Introduction

Biological research is becoming increasingly database driven, motivated, in part, by the advent of large-scale functional genomics and proteomic experiments such as those comprehensively measuring gene expression. These provide a wealth of information on each of the thousands of proteins encoded by a genome. This approach represents a paradigm shift away from traditional single-gene biology, and it often involves statistical analyses focusing on the occurrence of particular features (e.g., folds, functions, interactions, pseudo genes, or localization) in a large population of proteins. However, the information provided by different views may not always be easily reconciled, due to the incomplete nature of the data, or the fact that some patterns will be present in one view but not in another.

V.V. Das, J. Stephen, and Y. Chaba (Eds.): CNC 2011, CCIS 142, pp. 305–307, 2011.

2 Types and Methods

2.1 Biological Research Is Database Oriented

Databases have defined the information structure of molecular biology for over a decade, archiving thousands of protein and nucleotide sequences and three-dimensional (3-D) structures. As large-scale genomics and proteomics move to the forefront of biological research, the role of databases has become more significant than ever. The current landscape of biological databases includes large public archives, such as GenBank, DDBJ, and EMBL for nucleic acid sequences [1] give dynamic reports on the occurrence of protein families in various genomes. Databases have also been developed to provide comprehensive access to sequence, expression, and functional data for all the known genes of specific model organisms.

2.2 Machine Learning Approaches to Genomic Data Analysis

A general problem in data analysis is how to structure information into meaningful taxonomies or categories. This issue is of great importance when trying to infer relationships in diverse biological data sets. Statistical methods for finding trends and patterns in experimental results have played a large role in their interpretation. Principal component analysis (PCA) can be an effective method of identifying the most discriminating features in a data set. This technique usually involves finding two or three linear combinations of the original features that best summarize the types of variation in the data. The *k*-means algorithm is a popular instance-based method of cluster analysis. The self-organizing feature map (SOM)[2] consists of a neural network whose nodes move in relation to category membership.

2.3 Supervised Learning and Classification

Analysis of large data sets that contain diverse information often involves the explicit application of supervised learning. Supervised learning is conducted in two phases: training and testing the classifier model. Using this strategy, the data set is divided into two mutually exclusive sets. The first set is used to train the model, where correct classifications/responses of the input examples are known a priori. This information is used to improve the performance of the model and reduce the classification error rate, a process that incrementally adjusts an *n*-dimensional hyperplane that serves to partition the data set into categories. Afterward, unseen instances in the test set are classified according to the partitioning established during the training phase.

2.4 Classifying Gene Function with Support Vector Machines

An example of a supervised learning method applied to functional genomics data is the use of support vector machines (SVMs) to classify a set of objects in to two separate classes. The SVM maps an n-dimensional input space onto a higher-dimensional feature space, simultaneously transforming a nonlinear class boundary into a simple hyperplane.

The *maximum margin hyperplane* is defined as the hyperplane that is maximally distant from both convex hulls; this will bisect the shortest line connecting them. The data points closest to the maximum margin hyperplane are called *support vectors* and uniquely define the maximum margin hyperplane (the minimum number of support vectors from each class is one).

3 Conclusion

As with biological sciences in general, the emphasis of computational biology has changed in recent years from the characterization of individual molecules to the analysis of genome-wide expression profiles and proteomics data sets. An essential part of this work is the integration of a wide variety of experimental data to assemble a larger picture of biological function and frame pertinent information within a meaningful context. This capability can allow investigators to interpret data classifications and potentially elucidate functional properties. To facilitate this, it will be necessary to establish consistent formats for database interoperation, which will add greatly to the utility of experimental results in terms of potential data mining application.

References

1. Berman, H.M., Westbrook, J., Feng, Z., Gilliland, G., Bhat, T.N., Weissig, H., Shindyalov, I.N., Bourne, P.E.: The Protein Data Bank. Nucleic Acids Res. 28, 235–242 (2000)
2. Kohonen, T.: Self-organization and Associative Memory. Springer, Berlin (1984)
3. Baldi, P., Brunak, S.: Bioinformatics: the Machine Learning Approach, 2nd edn. MIT Press, Cambridge (1998)
4. Lones, D.T.: Gent READER: an efficient and reliable protein fold recognition method for genomic sequences. J. Mol. Biol. 287, 797–815 (1999)
5. Ding, C.H.Q., Dubchak, I.: Multi-class protein fold recognition using support vector machines and neural networks. Bioinformatics 17(4), 349–358 (2001)

Simulation and Performance Analysis of OLSR under Identity Spoofing Attack for Mobile Ad-Hoc Networks

Srishti Shaw, Kanika Orea, P. Venkateswaran, and R. Nandi

Dept. of Electronics & Tele-Communication Engineering
Jadavpur University, Kolkata – 700 032, India
srishtishaw@yahoo.com, kanikaorea@yahoo.in,
pvwn@ieee.org, robnon@ieee.org

Abstract. Mobile ad hoc networks (MANETs) consist of a collection of wireless mobile nodes, which dynamically exchange data without fixed base station. But, in an unfriendly and uncontrolled terrain conditions, the performance of the routing protocol may deteriorate due to intrusion. Thus, secured routing is an essential requirement for MANETs, to provide protected communication between mobile nodes. In this paper, we examine security issues related to the Optimized Link State Routing (OLSR) protocol, a proactive routing protocol for MANETs. In particular, we concentrate on the identity spoofing attack in OLSR and the network performance under this attack. We enumerate a number of possible attacks against the integrity of the OLSR routing infrastructure, due to identity spoofing. The simulation in OPNET shows that the performance of the networks degrades with increase in number of malicious nodes.

Keywords: MANETs, OLSR, Routing protocol, Security, Identity Spoofing.

1 Introduction

The inherent flexibility offered by MANETs, originally conceived for military and rescue operations only, is now employed in commercial applications like, convention meetings, electronic classroom, law enforcement, PAN. In this paper, we describe the identity spoofing attack and its effect on OLSR routing protocol [1, 2], a proactive, table driven routing protocol, specifically designed for MANETs.

2 The Optimized Link State Routing Protocol

The OLSR protocol[1, 2] exchanges routing information periodically and has routes immediately available when needed. OLSR provides the following features:

• *minimization of flooding* by using only a set of selected nodes, called *multipoint relays* (MPRs), to diffuse its messages to the network; this mechanism is based on having each node select a subset of its neighbours such that this subset ensures connectivity to every two-hop neighbours.

• reduction of the size of control packets by declaring only a subset of links with its neighbours who are its multipoint relay selectors (MPR selectors).

V.V. Das, J. Stephen, and Y. Chaba (Eds.): CNC 2011, CCIS 142, pp. 308–310, 2011.

3 OLSR Vulnerabilities

Existing OLSR [1] does not specify validation procedures and security mechanisms. Thus, numerous opportunities exist for intruding OLSR nodes to launch attacks. The OLSR basically employs two types of control traffic: HELLO messages and TC messages. Incorrect Control Message Generation and Incorrect Control Message relaying are a serious setback for the routing in OLSR.

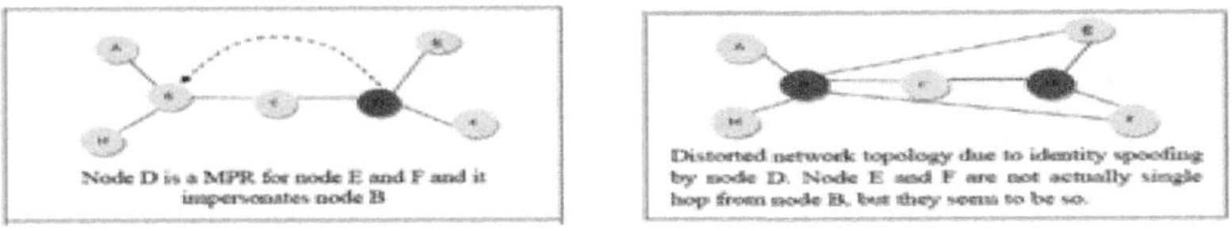

Fig. 1. Identity Spoofing attack in OLSR

This paper focuses on the identity spoofing attack in OLSR MANET. In order to attract traffic to itself in a network using proactive routing like OLSR, an attacker impersonates a well-behaved node and spoofs the MANET node's IP address as the source address to send HELLO message in the intruder's neighbourhood. Fig. 1 shows how the network topology is distorted due IP impersonation. Node D is the MPR for node E and Node F. Node D impersonates node B. Thus, when sending data packets to node E and node F, data will be sent to node B, which is believed to be their MPR. The data packets will, thus be lost.

4 Simulation Results and Analysis

We performed the network simulation using OPNET Modeler 14.0[3]. In evaluating the performance of the OLSR protocol under security attack, we used the following performance metrics: network throughput, average end to end delay, and packet delivery ratio. The simulation settings for the experiment are listed in Table 1.

Table 1. Parameter settings for the simulation

Simulation Parameters	Parameters Value
No. Of nodes	20,30 or 50
Simulation time	1000 seconds
Movement Area	1000mX1000m
Mobility Model	Random way-point mobility
Packet size	1024 bytes
Traffic	Constant Bit rate traffic
No. of malicious nodes	20%,25%,30% or 40% of the total nodes in a given scenario

Fig. 2a shows the relative throughput of the network. The network throughput is normalised with respect to the throughput value of the network with no malicious nodes. As evident, the throughput falls rapidly with the increase in concentration of malicious nodes. This is expected, as with malicious nodes the probability of data packet drops increases.

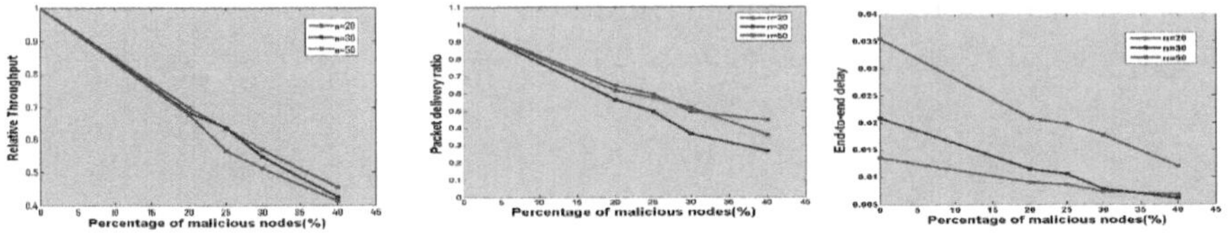

(a) (b) (c)

Fig. 2. Performance metric curves ('n' indicates the number of nodes in the scenario)

Fig. 2b shows the packet delivery ratio of the simulation. As expected, the packet delivery ratio falls with the increase in the number of intruders because under active attack, OLSR fails to find the true destination and data packet intercepted by a malicious node, does not reaches the intended destination (or intermediate) node and gets dropped. With the increase in concentration of attacker, the chances of falling victim to such an attack increases, and hence the packet drop increases considerably. We observed that, with increase in node density, the end to end delay increases (Fig. 2c). This is due to high level of network congestion and multiple access interferences at certain regions of ad hoc network.

5 Conclusion

Mobile ad hoc routing and forwarding are vulnerable to misbehaviour. In this paper, we have done a simulation based study of the impact of identity spoofing attack on OLSR. The study shows that the performance of the network degrades by nearly 30% even with a small number of attackers .Thus, network administrators can conclude of an impersonation attack if the performance metrics show such a trend.

References

1. Adjih, C., Clausen, T., Jacquet, P., Laouiti, A., Minet, P., Mühlethaler, P., Qayyum, A., Viennot, L.: Optimized link-state routing protocol, March 3 (2003) (work in progress) `http://draft-ietf-manet-olsr-08.txt`
2. Jacquet, P., Mühlethaler, P., Clausen, T., Laouiti, A., Qayyum, A., Viennot, L.: Optimised link state routing protocol for ad hoc networks. In: Proceedings of the 5th IEEE Multi Topic Conference (INMIC 2001), Lahore, Pakistan, December 28-30, pp. 1–7. IEEE Press, Los Alamitos (2001)
3. OPNET Documentation, `http://www.opnet.com`

Bayesian Belief Networks – Based Product Prediction for E-Commerce Recommendation

S.S. Thakur[1,*], Anirban Kundu[2], and J.K. Sing[3]

[1] M.C.K.V. Institute of Engineering, Liluah, Howrah 711204, India
subroto_thakur@yahoo.com
[2] Netaji Subhash Engineering College, Garia, Kolkata 700152, India
[3] Department of Computer Science & Engineering, Jadavpur University, Kolkata

Abstract. Prediction systems apply knowledge discovery techniques to the problem of making personalized product recommendations. New recommender system technologies are needed that can quickly produce quality recommendations, even for very large-scale problems. This paper presents a new and efficient approach that works using Bayesian belief networks (BBN) and that calculate the probabilities of inter-dependent events by giving each parent event a weighting (Expert systems). To get best result for the sales prediction, different weights have been applied on the proposed algorithm. Finally we got results for prediction, for a given product, using our proposed algorithm and final recommendation [1] has been made to the customers based on preference similarity.

Keywords: Bayesian Belief Networks (BBN), Expert Systems, E-Commerce.

1 Introduction

In recent years, E-commerce has been growing fast. More and more companies are providing services on the Web. Recommendation systems goals are to provide customers with what they want or need without requiring them to ask for it explicitly [2, 3, 4]. In order to achieve its goals, understanding customer's behaviours deeply is necessary. Recommender systems [5, 6] work by analyzing databases of customer interaction with Web sites. This data is often in the form of purchase information, the contents of shopping baskets and explicit rating data (i.e., data gathered via some form of questionnaire, generally indicating the degree of like or dislike for certain products).

This paper proposes a recommender system based on Bayesian Belief Network (BBN) and product preference and taste similarity. It calculates the probabilities of inter-dependent events, then calculate initialized probability and finally revised probability of the events. This paper also outlines the implementation of Bayesian Belief Network (BBN) by applying weights [7, 8, 9, 10] to increase the product sales.

---

* Corresponding author.

V.V. Das, J. Stephen, and Y. Chaba (Eds.): CNC 2011, CCIS 142, pp. 311–314, 2011.

2 Proposed Approach

Our prediction and recommendation system emphasizes four key areas. These are

- Introducing Bayesian Belief Network
- Calculating Initialised probabilities [2]
- Product Prediction [2]
- Product Recommendation which is dealt in section 3.

3 Product Recommendation

In E-commerce, personalization is mainly about recommendation system, which can be used for recommending products to customers according to their preferences Customer profiles are the most important component of learning-based recommendation system. An example of customer profile is as follow:

$$\{<obj_1, w_1>, <obj_2, w_2>, \ldots <obj_n, w_n>\} \quad (1)$$

where w_i is the weight of Web object obj, which can be measured in terms of several methods, for example, by the time consumption of frequencies of objects.

Customer Data: Customer data can be obtained from customer feed back forms or Web server logs. A static profile is stable or changes very slowly and can be constructed based on customer basic data, which includes customer's name, sex, age, job, salary etc. With respect to static profile, dynamic profile can be built and updated in real time, which is on the basis of customer behavior data. Dynamic profile has two parts: preference profile and rule profile.

Product hierarchy: Product hierarchy is an instance of concept hierarchy. A concept hierarchy defines a map sequence, which maps lower level concepts to higher level concepts. In order to learn a customer's preference on products, we use product hierarchy. Each node is a product category, each product is a leaf node and the root of the tree is "Product". Customer behavior data can tell us which type of product a customer likes, while product hierarchy gives us a detailed or global view of a specific product. So we combine customer behavior data with product hierarchy, to learn customer preferences and construct his (or her) profile. For each node, we can obtain the view of a preference of a customer. Because the leaf nodes are specific products, we can extract customer preferences about their parents from customer purchase histories. We define customer preferences in terms of products (or product category) attributes. So the customer preference has the same presentation with products. That is,

$$\text{Customer preference} = (<T_1,w_1>,<T_2,w_2>, \ldots <T_n,w_n>) \quad (2)$$

where T_i is the taste of the preference. We define a taste as the sub preference of a customer preference, which has the more particular description about a customer. For example "Nokia Mobile Phone" is a customer preference, while "Price", "Make", "Model No" etc, are all tastes of "Mobile Phone", w_i is the weight of T_i, which is specified by experts. For numeric product attributes, we extract tastes by using clustering techniques. Thus the attributes are grouped into clusters. When a customer has new purchase records, his (or her) profile can be updated.

The goal of the recommendation system is to generate a "recommendation set" for active customers, of which the products most closely match active customer's profile. In this method, both customer's profiles and products are represented as the same model. So it is fairly convenient to compute the similarity between a customer profile and a product.

$$\text{Preference}= (<T_1,tw_1>,<T_2,tw_2>, \ldots <T_n,tw_n>)$$
$$\text{Product}= <A_1,aw_1, A_2,aw_2,\ldots,A_n, aw_n>$$

So, the similarity of preference and product is Preference Similarity

$$=\sum_{i=1}^{n} (\textit{Taste Similarity x taste weight}) \quad (3)$$

4 Experimental Results

One year data i.e. from 2009 July – June 2010, customer's data collected from Spencer's, Mumbai.

Table 1. Comparison Data

Sr. No	Mobile (Make) - Preference	Features (Taste)	Approx. Price (taste)	Discount offered (Purchase Y/N?)
1	NOKIA	Camera, battery back up, memory etc.	Rs. 2500 – 15,000/-	Yes – 124 customers
2	SAMSUNG/ Motorola	-------	Rs. 2600 – 15,000/-	Yes – 35 customers
3	Sony Erickson/LG	-------	Rs. 1500 – 15,000/-	Yes -17 customers

We have taken feedback from approx. 1565 persons in Liluah, Kolkata and accordingly calculate preference and taste similarity based on available data, and found that 176 persons preference and taste are matching for a particular product i.e. mobile phones. If discounts offered are from 30% to 50%, it has been observed that 70% customer's preference is to purchase of Nokia Mobile phones, 20% customers prefer Samsung/Motorola, and 10% prefer LG/Sony Erickson Mobile phones.

5 Conclusion and Further Work

In this paper, we used Bayesian Belief Networks (BBN) in e-commerce area. Sales Prediction has been proposed for customers, who are using either cash and debit card or Credit cards for purchase of items. We apply weights (i.e. discounts) on sales data, and it has been observed that sales will increase or at least remains constant for the set of data available. Then final recommendation has been done to the Customers, based on preference similarity. The next few years will be an exciting time in the evolution of recommender systems, both in E-commerce and in traditional commerce applications.

Acknowledgments

The authors are thankful to M/s Spencers, Mumbai for providing us the data for research work. The authors are also thankful to Ms. Shruti Pansari, 7th Semester CSE student of MCKV Institute of Engineering, Liluah, for helping in data collection from local market.

References

1. Basu, C., Hirsh, H., Cohen, W.W.: Using social and Content-based Information in Recommendation. In: Proceedings of the AAAI 1998. AAAI Press, Menlo Park (1998)
2. Thakur, S.S., Kundu, A., Sing, J.K.: A Novel Approach – using Bayesian Belief Networks in Product Recommendation. Accepted for publication in EAIT, India (2011)
3. Niu, L., Yan, X.W., Zhang, C.Q., Zhang, S.C.: Product Hierarchy-Based Customer Profiles for Electronic Commerce Recommendation. In: 1st International Conference on Machine Learning and Cybernatics, Beijing (November 2002)
4. Mittal, B., Lassar, W.: The Role of Personalization in Service Encounters. Journal of Retailing 72(1), 95–125 (1996)
5. Mulvenma, M.D., Annual, S.S., Buchner, A.G.: Personalization on the Net using Web Mining. Communication of the ACM 43(8), 122–125 (2000)
6. Hohen, W.W.: Fast Effective Rule Induction. In: Proceedings of the Twelfth International Conference on Machine Learning. AAAI Press, Lake Tahoe (1995)
7. Prasad, B.: A knowledge-based product recommendation system for e-commerce. International Journal of Intelligent Information and Database Systems 1(1) (2007); ISSN: 1751-5858
8. Stahl, A.: Learning feature weights from case order feedback. In: Aha, D.W., Watson, I. (eds.) ICCBR 2001. LNCS (LNAI), vol. 2080, pp. 78–84. Springer, Heidelberg (2001)
9. Wettschereck, D., Aha, D.W.: Weighting features. In: Aamodt, A., Veloso, M.M. (eds.) ICCBR 1995. LNCS, vol. 1010, Springer, Heidelberg (1995)
10. Han, J., Kamber, M.: Data Mining Concepts and Techniques. Elsevier, India (2005)

Yield Prediction Using Artificial Neural Networks

Seshadri Baral[1], Asis Kumar Tripathy[2], and Pritiranjan Bijayasingh[3]

[1] Dept. of Computer Science & Engg., International Institute of Information Technology, Bhubaneswar, Odisha, India
[2] Dept. of Computer Science & Engg, NM Institute of Engg. & Technology, Bhubaneswar, Odisha, India
[3] Dept. of Computer Science & Engg, Balasore College of Engg. & Technology, Balasore, Odisha, India
{seshadri.baral,asistripathy,pritiprbs}@gmail.com

Abstract. Artificial Neural Network (ANN's) technology with PSO as optimization technique was used for the approximation and prediction of paddy yield at 3 different districts in different climatic zones based on 10 years of historical data sets of yields of paddy ,daily temperature(mean and maximum) and precipitation(rainfall).

Keywords: ANN, PSO, Yield Prediction of Paddy, Models for yield prediction.

1 Introduction

Crop production in general mostly affected by the prevailing weather condition of a region. Most climatic change impact studies use expected changes of mean values of climatic variable, neglecting the inter-annual (year to year) or intra-annual (month to month) changes of climatic variables and their effect on the agricultural productivity.

2 Related Work

Several types of supervised Feed-forward neural networks were investigated in an attempt to identify methods which were able to relate soil properties and grain yields on a point by point basis within ten individual site years [3]. Investigation were done to show that if artificial neural network (ANN) models could effectively predict Maryland corn and soybean yields for typical climatic conditions and thus prediction capabilities of models were compared at state, regional, and local levels. ANN model performance relative to variations of developmental parameters; and the effectiveness of multiple linear regression models to ANN models were done [2].

3 Methodologies

The data Obtained for the model is processed for missing values using neighboring mean method in which the missing value is the mean of its adjacent values. If T_{max}^{n-1},

V.V. Das, J. Stephen, and Y. Chaba (Eds.): CNC 2011, CCIS 142, pp. 315–317, 2011.

T^{n+1}_{max} were two known values for maximum temperature and T^{n}_{max} was the unknown temperature the

$$T^{n}_{max} = \frac{T^{n}_{max}+T^{n}_{max}}{2} \tag{1}$$

The resulting data was then divided into two data sets, to be used for the model. One data set was prepared using 3 day average (D_3) and other was prepared using 5 day average (D_5).If $T^{n}_{max}, T^{n+1}_{max}, T^{n+2}_{max}, T^{n+3}_{max}, T^{n+4}_{max}$ were the consecutive maximum temperatures of a month then

$$D^{n}_{3} = \frac{T^{n}_{max}+T^{n+1}_{max}+T^{n+2}_{max}}{3} \tag{2}$$

$$D^{n}_{5} = \frac{T^{n}_{max}+T^{n+1}_{max}+T^{n+2}_{max}+T^{n+3}_{max}+T^{n+4}_{max}}{5} \tag{3}$$

The datasets thus prepared were normalized using min-max normalization method. The normalized data was then used in the model.

The model processed data according to Algorithm 1.The fitness function used was a single layer standard feed forward network. The feed forward network was used for error calculation and the error thus obtained was optimized using PSO. The best weights obtained after maximum iterations were stored for testing. For each random weight generated the model was executed for 5 runs and the standard deviation of errors (optimized error in each run) was recorded.

Algorithm 1. Algorithm for getting optimized error from the model

```
Initialize number of hidden neurons, population size,
maximum iterations and model parameters(c1, c2,m),
Read weather and yield data from file
generate random population and biases
 1. for each population i in 1, . . . , populationsize do
 2.   calculate fitness
 3. end for
 4. calculate Population best(pbest) and Global
    Best(gbest)
 5. for i in 1, . . . , maxIterations do
 6.   for i in 1, . . . , populationsize do
 7.         for j in 1, . . . , dimensions do
 8.                     Calculate velocity
 9.                     Calculate Position
 10.        end for
 11.  end for
 12.  Calculate fitness value
 13.  for i in 1, . . . , populationsize do
 14.        Calculate population best(pbest)
 15.  end for
 16.  Calculate global best (gbest)
 17.  end for

 Display final Weights and Biases
```

Algoritm 2. Testing of the model with optimized parameters obtained from training Input weather data, yield data, optimized weight and number of neuron,

```
for j=1:no_of_hidden_neurons
  for k=1:no_of_input_neurons
       Calculate the output of first layer using the
     input test data of two years
  end for
End for
for n=1:no_of_hidden_neurons
 Calculate the output using the input of fist layer
End
Calculate Error for the output
```

4 Results and Conclusion

ANN with PSO model produced consistently lower error rates with variation of 8.24 ± 0.6 % for 5-day dataset and 8.28 ± 0.08 % for 3-day dataset. Testing of the prepared model gives an error rate of 8.3 ± 0.1 % on an overall basis. For 5 day average dataset prepared from the test data the error rate was 8.53 % and for 3 day dataset the error rate was 8.44%.The overall error rate for the model is 8 % which is in the allowable error range. The model can be given extra input parameters like soil types, nutrient contents , variety of crop to get a more accurate results with reduced error rates

References

[1] Heinzow, T., Tol, S.J.: Prediction of Crop yields across four climatic zones in Germany: An Artificial Neural Network Approach. Research unit Sustainability and Global Change, Hamburg University (September 2003)

[2] Kaul, M., Hill, L.R., Walthall, C.: Artificial neural networks for corn and soybean yield prediction. Agricultural Systems 85, 1–18 (2005)

[3] Drummond, S.T., Sudduth, K.A., Joshi, A., Birrell, S.J., Kitchen, N.R.: Statistical and neural methods for site-specific yield prediction. Transactions of the ASAE 46(1), 5–14 (2003)

A Novel Approach for Matrix Chain Multiplication Using Greedy Technique for Packet Processing

Hitesh Nimbark[1], Shobhen Gohel[1], and Nishant Doshi[2]

[1] B.H. Gardi College of Engineering and Technology, Rajkot, Gujarat, India
{hitesh.nimbark,shobhen.ce}@gmail.com
[2] Sardar Vallabhbhai National Institute of Technology, Surat, Gujarat, India
doshinikki2004@gmail.com

Abstract. Matrix chain multiplication is one of well known application of optimization problem. There are N matrices of different size were given and we have to multiply them based on their row and column size. The order in which matrices are multiplied may change the number of computations. This shows the use of greedy technique to solve this problem in O(NlogN) time which we are applying for packet processing in routing.

Keywords: Greedy, Matrix, Matrix chain multiplication.

1 Introduction

Matrix chain multiplication [1] is well known problem in which there were more than 1 matrix given with different row and column size and goal is to find their multiplication with minimum number of computations required. There is vast varicosity of literatures on this problem. The book [2] which is talking about this approach. In [11,12] the first (NlogN) time algorithm proposed based on the convex polygon approach. [3,4,7,10] gives the applications and discuss the variations of matrix multiplication approach. In [5] the lower bound for matrix multiplication was given and proved. In [2,13-15] the matrix chain multiplication was discussed on more than one processor and tries to reduce it up to logN processor and discuss various application of it. In [8] the theorems for matrix ordering chain multiplication were discussed. In [6,8,9] there are all based on some approximations like inputs are in increasing order or decreasing order, all have same size etc. but in none of the above literature they had try to solve this using greedy strategy which were useful for many techniques like knapsack problem, Minimum spanning tree etc. in greedy technique we have to select optimal criteria which was true at every stage and it will lead to optimal solution. Matrix chain multiplication is one of the applications which shows that if we sole problem using dynamic approach it require $O(N^3)$ time and if we solve using greedy approach it require O(NlogN) time complexity. This section cover introduction and literature survey in section 2 we proposed the greedy approach using an example. In section 3 we had given algorithm and proof for greedy technique.

V.V. Das, J. Stephen, and Y. Chaba (Eds.): CNC 2011, CCIS 142, pp. 318–321, 2011.

2 Greedy Approach

Let assume that we have matrices a1, a2,...,an and the greedy approach is given as follow. Make list 1 contain in the column major order and list 2 as row major order. Then start from last node of list1 and match the row of that matrix with the column of matrix from list 2 start from last node. After the match of matrices say a_k from list 1 and a_j from list 2. Then do multiply a_k and a_j. Then delete the corresponding matrix from list 1 and list 2 and add the multiply matrix in list 1. To select a_j for a_k is the greedy strategy. This will yield the optimal sub solution for the problem solution.

Example 1

Consider the problem of a chain (a1, a2, a3) of three matrices. Suppose that the dimensions of the matrices are 10 × 100, 100 × 5 and 5 × 50. So execution of the greedy strategy is as follow.

List 1 contains a1, a3 and a2. List 2 contains a2, a1 and a3. Now we match a2 with matrix in list2 and the matching matrix is a1 as we scan from last matrix in list2. So we multiply a2 × a1 and resulting matrix say ak is added in list1. Now we delete matrix a1 from list1 and matrix a2 from list2. So list1 contains a3 and ak and list2 contains a3. So we multiply ak × a3 so the final sequence is (a1 ×a2) ×a3=7500 scalar multiplications.

Example 2

Consider the problem of a chain (a1, a2, a3, a4, a5, a6) of six matrices. Suppose that the dimensions of the matrices are 30 × 35, 35 × 15, 15 × 5, 5 ×10, 10 × 20 and 20 × 25. So execution of the greedy strategy is as follow.

So the list 1 contains a1, a6, a5, a2, a4 and a3. List 2 contains a2, a1, a6, a3, a5 and a4.

Table 1. Iteration Table

Itera'n	List
Initially	Make list 1 as column wise sorted in nonincreasing order and list 2 as row wise sorted in nonincreasing order. Sum=0
	List1- a1(30 × 35), a6(20 × 25), a5(10 × 20), a2(35 × 15),a4(5 ×10),a3(15 × 5)
	List2- a2(35 × 15),a1(30 × 35),a6(20 × 25),a3(15 × 5),a5(10 × 20),a4(5 ×10)
1	Multiply ak=a3×a2. Add ak in list 1. Delete a2 and a3 from both lists. Sum=35×15×5=2625.
	List1- a1(30 × 35), a6(20 × 25), a5(10 × 20),a4(5 ×10),ak(35 × 5)
	List2- a1(30 × 35),a6(20 × 25),a5(10 × 20),a4(5 ×10)
2	Multiply aj=ak×a1. Add aj in list 1. Delete a1 and ak from list 1 and delete a1 from list 2. Sum=2625+30×35×5=7875.
	List1- a6(20 × 25), a5(10 × 20),a4(5 ×10),aj(30 × 5)
	List2- a1(30 × 35),a6(20 × 25),a5(10 × 20),a4(5 ×10)

Table 1. (*continued*)

3	No matching for aj so we come to a4 and continue. Multiply ak=a4×a5. Add ak in list 1. Delete a4 and a5 from list 1 and list 2. Sum=7875+5×10×20=8875. List1- a6(20 × 25), ak(5 ×20),aj(30 × 5) List2- a6(20 × 25)
4	No matching for aj so we come to ak and continue. Multiply ah=ak×a6. Add ah in list 1. Delete ak and a6 from list 1 and a6 from list 2. Sum=8875+5×20×25=11375. List1- ah(5 ×25),aj(30 × 5) List2- NULL
5	As list2 is null. Multiply the remaining matrices in list1. Multiply ak=ah×aj. Delete ah and aj from list 1 and add ak in list 1. Sum=11375+30×5×25=15125. So ak=((a1(a2a3))((a4a5)a6)

3 Algorithm and Proof

Let assume that we have matrices a1, a2,...,an and the greedy approach. Here we have to create two lists and for each of the element of list1 we scan through list 2 to find the matching and then we add and delete that particular matrix. So we follow the algorithm stepwise.

1. Sort the given matrixes according to column wise and row wise and assign them into list 1 and list 2 respectively. Make sum=0 initially.
2. Repeat the steps 3-4 till list 1 is not empty.
3. Take the first element in list 2 and find the corresponding element in list 1.
4. Multiply them both. Increase the value of sum by adding the number of multiplications. Remove both matrices from both list and add the resultant matrix in place of them.
5. Return the sum as optimal answer.

So the step 1 and 2 takes O(NlogN) time as it is lower bound on the sorting[1]. Step 3-5 require to scan list linearly so it takes O(N) time. So total time is O(N+NlogN) is equal to O(NlogN).Now we will discuss about the proof.

Assume there is random oracle which return optimal sequence O=O1,O2,…,ON. And let assume that greedy technique return G=G1,G2,…,GN. Assume that for some k,0<k<N we assume that Op=Gp for 0<p<=k. so that Ok+1 and Gk+1 are different. So if columns(Ok+1)< columns(Gk+1) than by selecting Ok+1 we get loss as if we select matrix with more column values reduce the total comparisons. So we can replace the Ok+1 by Gk+1. So by this we can efficient transform the sequence G to sequence O. this proves the optimality criteria for our approach.

4 Conclusion

The approach given in [11-12] gives the same bound as our approach. Which show that if we apply our approach to network packet processing problem then it gives better performance in terms of comparisons, efficiency etc. in future we can make analysis for better understanding of the greedy approach. But approach proves that in matrix chain multiplication the dynamic approach is no better than the greedy approach.

References

1. Cormen, T.H., Leiserson, C.E., Rivest, R.L., Stein, C.: Introduction to algorithms, 2nd edn. McGraw-Hill publication, New York (2002)
2. Thomson Leighton, F.: Introduction to parallel algorithms and architectures: array, trees, hypercubes. Morgan Kaufmann Publishers Inc., San Francisco (1991)
3. Huang, X., Pan, V.Y.: Fast rectangular matrix multiplication and applications. Journal of Complexity 14(2), 257–299 (1998)
4. Raz, R.: On the Complexity of Matrix Product. SIAM Journal on Computing 32(5), 1356–1369 (2003)
5. Shpilka, A.: Lower Bounds for Matrix Product. SIAM Journal on Computing 32(5), 1185–1200 (2003)
6. Ezouaoui, S., Ben Charrada, F., Mahjoub, Z.: On instances of the matrix chain product problem solved in linear time. In: ROADEF 2009 (2009)
7. Godbole, S.: An Efficient Computation of Matrix Chain Products. IEEE Trans. Computers 22(9), 864–866 (1973)
8. Chandra, A.: Computing Matrix Chain Products in Near-Optimal Time. technical report, IBM T.J. Watson Research Center, Yorktown Heights, N.Y., IBM Research Report RC 5625(#24393) (1975)
9. Santoro, N.: Chain mulitplication of matrices of approximately or exactly the same size. Communications of the ACM 27(2), 152–156 (1984)
10. Yuster, R., Zwick, U.: Fast sparse matrix multiplication. ACM Transactions on Algorithms (TALG) 1(1), 2–13 (2005)
11. Hu, T.C., Shing, M.T.: Computation of matrix chain products. Part II. SIAM Journal on Computing 13(2), 228–251 (1984)
12. Hu, T., Shing, M.: Computation of matrix chain products I. SIAM J. Comput. 11(2), 362–373 (1982)
13. Bradford, P.G., Rawlins, G.J.E., Shannon, G.E.: Efficient matrix chain ordering in polylog time. In: Eighth International Proceedings Parallel Processing Symposium, April 26-29, pp. 234–241 (1994)
14. Atallah, M.J., Kosaraju, S.R.: An Efficient Parallel Algorithm for the Row Minima of a Totally Monotone Matrix. In: SODA 1991, pp. 394–403 (1991); And to appear in the Journal of Algorithms
15. Bradford, P.G., Rawlins, G.J.E., Shannon, G.E.: Matrix Chain Ordering in Polylog Time with n/lgn Processors (October 1992) (full version)

Width of a General Tree for Packet Routing

Hitesh Nimbark[1], Shobhen Gohel[1], and Nishant Doshi[2]

[1] B.H. Gardi College of Engineering and Technology, Rajkot, Gujarat, India
{hitesh.nimbark,shobhen.ce}@gmail.com
[2] Sardar Vallabhbhai National Institute of Technology, Surat, Gujarat, India
doshinikki2004@gmail.com

Abstract. Till date in all the books they talks about height of a general tree like height balanced tree. In this paper we talk about width of a general tree. Width of a general tree is defined as number of nodes at each level and maximum of them. The recursive algorithm for width has been given. Later in the paper the iterative version using the stack is given.

Keywords: Algorithm, Height, Stack, Tree, General tree, Width.

1 Introduction

General tree is a tree in which every internal node contains at least zero nodes and at most any number of nodes. We can use the m-ary term for the general tree. The root node at level 0 and so on. Leaf nodes are at last level with no children. The width of binary tree will be defined based on the number of nodes at each level and find the level contains the maximum nodes [2]. Till now in literature they had discussed the height balanced trees [1]. In [2] authors proposed the width for a binary tree. the limitation of the paper was it is only for binary tree. In this paper we had extend the notion to the general tree. The concept we can utilized in computer network packet routing. In Figure 1(a) we had example of general binary tree.

2 Recursive Approach

Here we can the same concept as in [5] with little modification.

```
Algorithm Findwidth (TREEPTR *t, l)
{
If(t = NULL)
Then
   Return
Level[l]=Level[l] +1
For each child X of t
Do
```

V.V. Das, J. Stephen, and Y. Chaba (Eds.): CNC 2011, CCIS 142, pp. 322–324, 2011.

```
    Findwidth(X,l+1)
Done
Return
}
```

We had illustrated one example in figure 1(b) for number of nodes at each level.

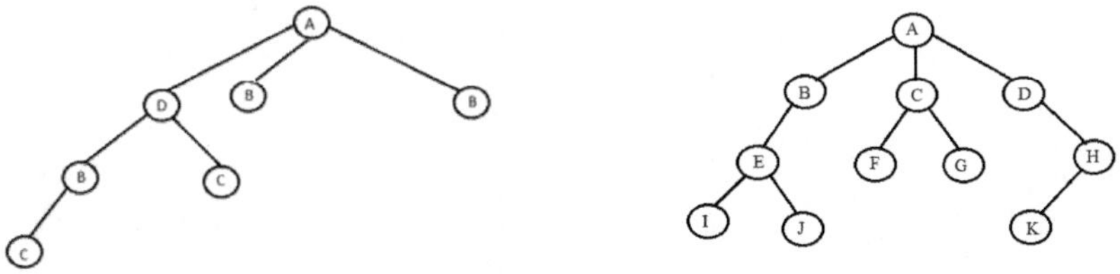

(a) General Binary Tree Example (b) General Tree

Fig. 1. (b) showing that at level 0 number of node is 1, at level 1 number of nodes are 3 and so on

3 Iterative Approach

Now we will use the same concept of stack as given in [10].

```
Algorithm Findwidth (Root r, Level [ ])
{
Stack S
PUSH(S,r,0)
While (NotEmpty(S))
Do
    Curr_level=GetTopLevel(S)
    Level[Curr_level]= Level[Curr_level] + 1
    Node = POP(S)
    For each child X of Node
    Do
          PUSH(S, X, Curr_level+1)
    Done
Done
Return  FindMax(Level)
```

Here we have to store each level's node in array Level which passed as argument and in last step we had to find the maximum element and return which shows the width of a general tree. Now let's take example of Figure 1(b) and see the step by step evaluation of stack in Table 1.

Table 1. Execution for width algorithm

Iteration	Stack Content		Level Array			
0	0	A	0	0	0	0
1	1	D	1	0	0	0
	1	C				
	1	B				
2	2	H	1	1	0	0
	1	C				
	1	B				
3	3	K	1	1	1	0
	1	C				
	1	B				
4	1	C	1	1	1	1
	1	B				
5	2	G	1	2	1	1
	2	F				
	1	B				
6	2	F	1	2	2	1
	1	B				
7	1	B	1	2	3	1
8	2	E	1	3	3	1
9	3	I	1	3	4	1
	3	J				
10	3	J	1	3	4	2
11	EMPTY		1	3	4	3

4 Conclusion

In this paper we proposed the width for the general tree for routing the packet in computer network. Previous literatures were about binary tree and height balanced tree. Currently we had to use array to store value for each level in array so we require more space complexity in iterative compare to recursive approach and in future may be we can find the application of the width parameter.

References

1. Cormen, T.H., Leiserson, C.E., Rivest, R.L., Stein, C.: Introduction to algorithms, 2nd edn. McGraw-Hill Publication, New York (2002)
2. Doshi, N., Sureja, T., Akbari, B., Savaliya, H., Daxini, V.: Article: "Width of a Binary Tree". International Journal of Computer Applications 9(2), 41–43 (2010)

Unified Knapsack Problem

Umesh Chandra Jaiswal, Ankit Singh, Omkar Maurya, and Anil Kumar

Department of Computer Science and Engineering,
M.M.M. Engineering College, Gorakhpur, U.P. India - 273010
ucj_jaiswal@yahoo.com,
{ankit.singh77,omkar0312,159anil}@gmail.com

Abstract. Knapsack problem is a classical optimization problem. It is widely used in resource allocation problems. We are aware of 0-1 knapsack and fractional knapsack problem. Here, we will discuss about a newer version, unified knapsack problem, which is a combination of above two versions. In 0-1 knapsack problem each item is either be taken or left behind but in fractional knapsack problem a fraction of any item can be taken, while in unified knapsack problem some items can be fractioned and some cannot.

Keywords: Resource allocation, fractional knapsack problem, unified knapsack problem, 0-1 knapsack problem.

1 Introduction

The basic aim of knapsack problem is to find a combination of different objects that a thief chooses for his knapsack such that total value of all objects of his/her choice is maximized. This situation is common in complexity theory, cryptography, business, combinatorics, and applied mathematics. Owing to its vast applications, knapsack problem has been rigorously studied by both theorists and experimentalists [1, 2, 3]. This problem has also been modeled into many industrial situations, such as cargo loading and capital budgeting [4, 5, 6, 7].

In unified knapsack problem, divisibility of items plays an important role. It decides whether an item can be fractioned or not. We will use a character variable which shows the divisibility of an item. In fractional and 0-1 knapsack problem, there is no need of such variable because all items are of same type.

2 Types of Knapsack Problem

A typical knapsack problem is defined as: there are finite number of items n and a knapsack. Item j has a weight w_j and the knapsack has a weight limit W. If item j is placed into the knapsack, we will obtain a benefit b_j. The problem is to maximize the total profit under the constraint that the total weight of all chosen items is at most W [8,9]. Two types of knapsack problems are described below:

V.V. Das, J. Stephen, and Y. Chaba (Eds.): CNC 2011, CCIS 142, pp. 325–330, 2011.

a) 0-1 Knapsack Problem: The 0-1 knapsack problem can be formulated as:

$$\max f(x) = \sum_{j=1}^{n} b_{j*}x_j$$

$$\text{subject to } \sum_{j=1}^{n} w_{j*}x_j \le W, \textbf{ where } x_j \in \{0, 1\} \ (j = 1,2,3.....n).$$

If there are n items and weight limit of knapsack is W then dynamic solution of 0-1 knapsack problem will be of order $O(n*W)$.

b) Fractional Knapsack Problem: It is stated mathematically as follows:

$$\max f(x) = \sum_{j=1}^{n} b_{j*}x_j$$

$$\text{subject to } \sum_{j=1}^{n} w_{j*}x_j \le W, \textbf{ where } x_j \in [0,1] \ (j = 1,2,3.....n).$$

If there are n items and weight limit of knapsack is W then greedy solution of fractional knapsack problem will be of order $O(n*lg(n))$.

3 Unified Knapsack Problem

Here we are going to introduce a new term '*Divisibility*', which decides whether items in unified knapsack problem, can be fractioned or not. The divisibility of item j is either "Y" or "N"(Y- divisible, N - indivisible). According to divisibility, if a fraction x_j of item j is placed in the knapsack, then a profit, $g_{j*}x_j$, is earned. Then knapsack problem is stated mathematically as follows:

$$\max f(x) = \sum_{j=1}^{n} b_{j*}x_j$$

$$\text{subject to } \sum_{j=1}^{n} w_{j*}x_j \le W$$

$x_j \in \{0, 1\}$ if item can't be fractioned,

$x_j \in [0,1]$, if item can be fractioned, **where** $j = 1,2,3.....n$.

3.1 Terms Used in Algorithm

Following are the terms used in algorithm: *items* is a 2-D array of string of items. Format of entry in *items* array will be as "Weight Cost Divisibility". W is the capacity of knapsack, n is the number of items. *maximize_cost* represents obtained maximum benefit. *temp_w* is a variable used to represent current used weight. *w*[n], *d*[n], *cost*[n], *cpw*[n] are the arrays of weights, divisibility, cost and cost per unit weight of items respectively, in sorted order according non-increasing cost per unit weight. *used*[*n*] is an array list of indexes of items those have been selected for filling the knapsack. *top*, *top1* are the pointers to point the last selected item for filling the knapsack. *cost*, *cost1* are the values to represent the current cost of items in knapsack. *flag*[*n*] is a boolean variable to check whether an item has been used to fill the knapsack or not.

unified_knapsack (n, items, W)

```
▶ check the divisibility of each item
If each item is indivisible   ▶ 0-1 Knapsack problem
then solve using dynamic solution
else
  do
    calculate cost per unit weight for each item
    sort items list according to non-increasing cost per unit weight.
    i=0, temp_w = 0,top = -1
   while (i<n)
   do begin
     If (temp_w < W)
     then If (W- temp_w>=w[i])
            then top=top+1
                 used[top]= i
                 temp_w = temp_w + w[i]
                 cost=cost + cost[i]
           else
              If (d[i]='Y')
              then top=top+1
                   used[top]=i
                   cost=cost + (W-temp_w)*cpw[i]
                   temp_w= W
     i=i+1
   end
  top1=top
  cost1=cost
  If (temp_w<W)
  then
    while (top!= -1)
    do begin
        If (d[used[top]='Y')
       then temp_w=temp_w-w[used[top]]
              cost1=cost1- cost[used[top]]
              for k=0 to n
                   flag[k]=0
                   for j=0 to top1
                        if (k=used[j])
                        then flag[k]=1
                             break
              for i=0 to n
                 if (d[i]='N' & w[i]<=W-temp_w & flag[i]=0)
                 then temp_w=temp_w + w[i]
                        cost1 += cpw[used[top]]*(W-temp_w)
                        break
              break
          top=top-1
     end
```

```
49 If (cost1>cost)
50 then maximize_cost=cost1
51 else
52     maximize_cost=cost
```

4 Explanation of Algorithm

Initially the algorithm checks the type of problem by checking the divisibility attribute of each item. If any item has divisibility value '*Y*' problem falls into unified category otherwise problem will be of 0-1 knapsack type. In unified solution, first we calculate cost per unit weight for each item. We put the values of weight, cost, divisibility and cost per unit weight of items in arrays w[], cost[], d[] and cpw[] respectively. Then we sort the entries in arrays according non-increasing cost per unit weight. The variables temp_w and *cost* are initialized to *zero*. *top* points to the current visiting item in the list and is initialized to *-1*. Until all items are not visited for filling the knapsack or knapsack is not full following procedure continues:

Select an item from the list which is next to '*top*' pointer, check weight of the item, if weight of the item is less than empty space in knapsack, it will be selected to fill the knapsack, increment the temp_w and cost variables by adding w[i] and cost[i] respectively. '*i*' represents the index of current item. If weight of the item is greater than empty space in knapsack then we check whether the item is divisible or not, if it is divisible then it can be selected by adding appropriate fraction of that item otherwise it can't be selected. When an item is selected, it is added in used[] list. '*top*' is a pointer to the used[] list, while '*i*' points to the sorted item's list (or array).

In this way items can be selected to give maximize cost, but it is not complete solution of the problem. We will give complete explanation of the solution of problem after giving some examples which are as following…

E.g-1. Let us consider following items with a knapsack of capacity 150.
1. "100 1000 Y" 2. "100 100 N" 3. "100 100 N".

According to our prescribed explanation output will be 1000 and it is correct. Items are in sorted order. Control of program selects item on top of the list and increase temp_w and cost variables by 100 and 1000 respectively. On second iteration of while loop, control checks whether knapsack have sufficient capacity to select next item and finds result "no". Then it checks whether next item can be fractioned or not and finds result "no". So the item can't be selected. Same process is repeated for the third item. Now control comes to the *line* 26 of psuedocode and saves *cost* and *top* values in *cost1* and *top1* variables respectively. If there is still some space left in knapsack, control removes the latest added divisible item from knapsack and searches those items in the *items* array which have not been used and having higher cost per unit weight value and adds that item in the knapsack. If there is still some capacity remains in the knapsack then control adds appropriate fraction of that item which was removed from knapsack. On applying this procedure to above example value of *cost* variable becomes 600, which is less than that of previous value, so finally output comes as 1000.

Now a question arises, "Why we need all the later procedure?". To give the answer of above question we will take another example.

E.g.-2 Let us consider following items with a knapsack capacity 150.
1. "100 199 Y" 2. "100 100 N" 3. "100 100 N".

There are two options for the answer 1) 199 and 2) 199.5. Correct answer is 199.5 and we are going to explain it. After execution of first while loop, value of *cost* and *temp_w* variables become 199 and 100 respectively. After that, control finds that there is extra capacity left in knapsack. It removes latest used divisible item (100 199 Y) from knapsack and adds "100 100 N" (second item) to the knapsack, after that control adds 50% weight of (100 199 Y, which was removed from knapsack).By this way cost becomes 199.5 which is maximum and is the output.

5 Simulation and Performance of Algorithm

For simulation of above algorithm, we have implemented the above algorithm in *C programming language*, run this program for different inputs and we have found satisfactory results. Screenshot of result for some inputs is as follows:

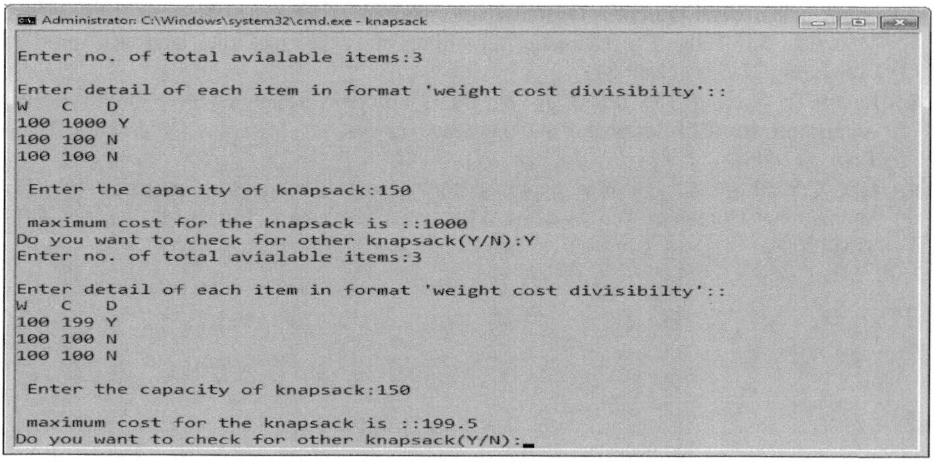

Fig. 1. A screenshot showing result for some inputs

Performance: In worst case, running time of algorithm is of order O (n^2). Worst case occurs when in first while loop, all items have been selected and knapsack still has some capacity to bear load. In average case running time is of order O ($n*p$), where p lies between 1 and n.

6 Conclusion and Future Scope

With this purview, we are introducing a new algorithm named as **unified knapsack.** Initially we developed this algorithm for our interest but the need, usage and its efficacy encouraged us to present it before you so that it can be made available to all algorithm users and designers.

Right now, **unified knapsack** solves all types of knapsack problems, but it can be enhanced for better running time. Also we intend to work on this algorithm to make it simpler.

References

1. Cormen, T.H., Leiserson, C.E., Rivest, R.L., Stein, C.: Greedy Algorithms. Introduction to Algorithms, 2nd edn., pp. 382–383. PHI (2009)
2. Martello, S., Toth, P.: Knapsack problems: Algorithms and Computer Implementations. John Wiley & Sons Ltd., Chichester (1990)
3. Dynamic programming 0-1 Knapsack Problem, `http://www.cse.unl.edu/~goddard/Courses/CSCE310J`
4. Martello, S., Pisinger, D., Toth, P.: New Trend in exact algorithms for the 0-1 knapsack problem. European Journal of Operational Research 123, 325–332 (2000)
5. Gavish, J., Pirkul, H.: Efficient algorithms for solving multiconstraint zero-one knapsack problems to optimality. Mathematical Programming 31, 78–105 (1985)
6. Magazine, M., Oguz, O.: A heuristic algorithm for the multidimensional 0-1 knapsack problem. European Journal of Operational Research 16, 319–326 (1984)
7. Martello, S., Toth, P.: Heuristic algorithms for the multiple knapsack problem. Computing 27, 93–112 (1981)
8. Lin, F.T.: Solving The Imprecise Weight Coefficients Knapsack Problem by Genetic Algorithms. In: IEEE International Conference on Systems, Man, and Cybernetics, Taipei, Taiwan (2006)
9. Lee, C.Y., Lee, Z.J.: A new approach for solving 0-1 knapsack problem. In: IEEE International Conference on Systems, Man and Cybernetics, Taiwan, China, pp. 3138–3143 (2006)

Autonomous System Dealings in the Internet

Mayur M. Jadhav and N.B. Pokale

Research Scholar - Department Information Technology,
Asst. Professor - Department Information Technology
Sinhgad Academy of Engineering, Pune, India
jadhavmayurm@gmail.com, nbpokale@gmail.com

Abstract. Within the Internet, an autonomous system is a collection of connected Internet Protocol routing prefixes under the control of one or more network operators that presents a common, clearly defined routing policy to the Internet[5] Even though there are multiple Autonomous Systems supported by the ISP, the Internet only sees the routing policy of the ISP. That the ISP must have an officially registered Autonomous System Number (ASN)[10]. To the best of our knowledge, there has been no publicly available information about inter-AS relationships and this is the first attempt in understanding and inferring AS relationships in the Internet.

Keywords: Automomous System,BGP,Inter domin routing protocols.

1 Introduction

The Internet has experienced a tremendous growth in its size and complexity since its commercialization. The Internet connects thousands of Autonomous Systems (ASes)[6] operated by many different administrative domains such as Internet Service Providers (ISPs), companies and universities. Since two ISPs might merge into one and each administrative domain can possess several ASes, an administrative domain can operate one or several ASes. Routing within an AS is controlled by intra domain routing protocols such as static routing, OSPF[7], IS-IS[8], and RIP[9]. Border Gateway Protocol (BGP). The *Border Gateway Protocol (BGP)* is an inter-autonomous system routing protocol. An autonomous system is a network or group of networks under a common administration and with common routing policies. BGP is used to exchange routing information for the Internet and is the protocol used between Internet service providers (ISP). Customer networks, such as universities and corporations, usually employ an Interior Gateway Protocol (IGP) such as RIP or OSPF for the exchange of routing information within their networks. Customers connect to ISPs, and ISPs use BGP to exchange customer and ISP routes. When BGP is used between autonomous systems (AS), the protocol is referred to as External BGP (EBGP). If a service provider is using BGP to exchange routes within an AS, then the protocol is referred to as Interior BGP (IBGP). Figure 1 illustrates this distinction.

V.V. Das, J. Stephen, and Y. Chaba (Eds.): CNC 2011, CCIS 142, pp. 331–336, 2011.

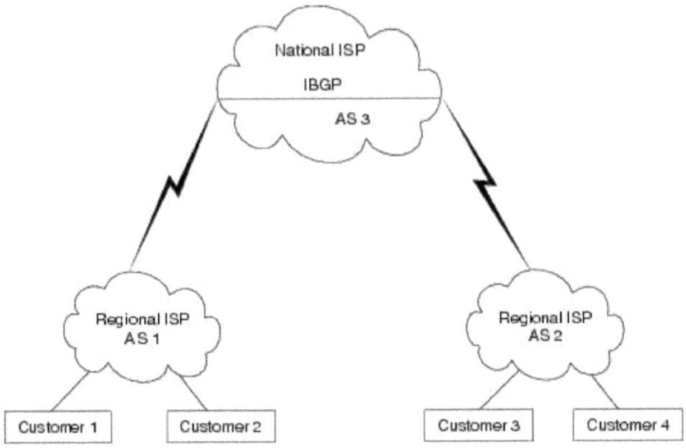

Fig. 1. External and Interior BGP

BGP is a very robust and scalable routing protocol, as evidenced by the fact that BGP is the routing protocol employed on the Internet. At the time when this was wrriten, the Internet BGP routing tables number more than 90,000 routes. To achieve scalability at this level, BGP uses many route parameters, called attributes, to define routing policies and maintain a stable routing environment. The commercial agreements between pairs of administrative domains can be classified into customer-provider, peering, and mutual-transit agreements 121. A customer pays its provider for connectivity to the rest of the Internet. Therefore, a provider does transit traffic for its customers. However, a customer does not transit traffic between two of its providers. A pair of peers agree to exchange traffic between their respective customers free of charge. A mutual-transit agreement allows a pair of administrative domains to provide connectivity to the rest of the Internet for each other. This mutual transit relationship is typically between two administrative domains such as small ISPs or universities who are located close to each other and who cannot afford additional Internet services for better connectivity. These contractual commercial agreements between administrative domains play a crucial role in shaping the structure of the Internet and the end-to-end performance characteristics. Previous work on the Internet topology has been focused on the interconnection structure at either AS or router level. Since routing between ASes is controlled by BGP, a policy-based routing protocol, connectivity does not imply reachability. For example, national ISPs A and B are connected to their customer a regional ISP C respectively. Although ISPs A and B are connected through ISP C, ISP A can not reach ISP B via ISP C since C as a customer does not provide transit services between its providers. Even if ISPs A and B can reach each other via other ISPs, the end-to-end performance characteristics between A and B can not be inferred from that of between A and C and between C and B. For example, the delay between A and B is independent of the total delay between A and C and between C and B. This has been observed by several measurement studies. Therefore, the connectivity (at either the AS or the router level) alone does not fully characterize the structural properties of the Internet.

2 Related Work

The increasing importance and complexity of the Internet routing infrastructure has sparked interest in understanding Internet topology and its effect on the end-to end performance. Previous work consists of discovering the Internet topology, constructing the Internet distance map, and identifying inherent structural properties of the Internet. Several studies, present heuristics on discovering the router adjacencies by effectively using the traceroute tool. Motivated by important problems such as the mirror site placement, Jamin and Theilmann et al study the construction of the distance map by estimating the end-to-end distance using strategically placed measurement servers. Faloutsos et al identify the power-law properties of the Internet connectivity at both router and AS levels. In, inter-AS connectivity is characterized by a hierarchy of ASes, where ASes are classified into four levels by the degree of the AS.

With the exception of, all of the aforementioned work do not have explicit notion of AS hierarchy. To the best of our knowledge, all studies have assumed that the connectivity is equivalent to the reachability and there is no explicit notion of AS relationships in the topology characterization. Our paper is the first study that explores AS relationships, which is an inherent aspect of the policy-based Internet routing structure. The information about AS relationships is crucial in fully understanding the structural properties of the Internet. Further, AS relationships can help to effectively place measurement servers and better approximate the end-to-end distance.

3 Internet Architecture

The Internet consists of a large collection of hosts interconnected by networks of links and routers. The Internet is divided into thousands of distinct regions of administrative domain, each of which possesses one or several autonomous systems.

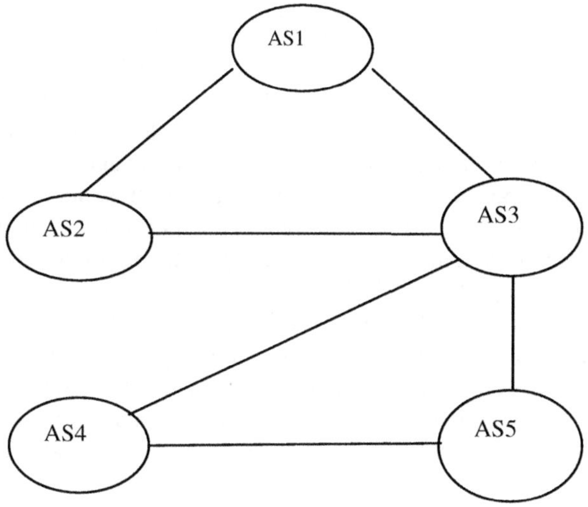

Fig. 2. An AS graph example

Since the commercialization of the Internet in 1995, the Inter- net has experienced tremendous growth in both size and complexity. The interconnection between ASes are dynamically evolving since ISPs can add or remove connections to other ASes and companies can change their Internet service providers. Furthermore, the contractual agreements between ASes can change due to ISP merging and restructuring. There are several registration services for the administration and registration of IP and AS numbers.

Each AS has responsibility for carrying traffic to and from a set of customer IP addresses. The scalability of the Internet routing infrastructure depends on the aggregation of IP addresses in contiguous blocks, called prefixes, each consisting of a 32-bit IP address and a mask length (e.g., 1.2.3.0/24). An AS employs an intradomain routing protocol (such as OSPF or IS-IS) to determine how to reach each customer prefix, and employs an interdomain routing protocol (BGP) to advertise the reachability of these prefixes to neighboring ASes.

3.1 Routing Policies

What is a routing policy? A routing policy allows an administrative authority to control the routing behavior of an autonomous system. A routing policy can be applied to all external and internal routes, and can control the propagation of routes, vastly improving the stability of the network. A well-planned routing policy will help the administrator control network behavior automatically, reducing the administrative overhead associated with maintaining the network.

Routing policies are created by applying a series of route filters and route maps. Route filters control which routes are advertised and received. Route maps control the metrics on those routes to determine which routes should be received and used normally, and which routes should be administratively adjusted to change traffic flow and routing behavior.

BGP is a distance-vector protocol that constructs paths by successively propagating updates between pairs of BGP speaking routers that establish BGP peering sessions The simplest distance-vector protocol would employ shortest-path routing, where each AS selects a route with the shortest AS path. However, BGP allows a much wider range of routing policies so as to honor contractual agreements that control the exchange of traffic. Upon receiving an update, a router must decide whether or not to use this path according to import policies and, if the path is chosen, whether or not to propagate the update to neighboring ASes according to export policies Each AS sends only its best route for a prefix to a neighbor.

Export policies allow an AS to determine whether to send its best route to a neighbor and, if it does send the route, what hint it should send to its neighbor on using the route Each BGP speaking router keeps a BGP routing table, which stores a set of candidate routes for the router. We refer to a candidate route as a routing table entry, which contains a destination prefix, next-hop, med, local preference and AS path of the route.

4 Routing Table and Topology Table

A basic routing table includes the following information:

- Destination: The IP address of the packet's final destination
- Next hop: The IP address to which the packet is forwarded
- Interface: The outgoing network interface the device should use when forwarding the packet to the next hop or final destination
- Metric: Assigns a cost to each available route so that the most cost-effective path can be chosen
- Routes: Includes directly-attached subnets, indirect subnets that are not attached to the device but can be accessed through one or more hops, and default routes to use for certain types of traffic or when information is lacking.

Routing tables can be maintained manually or dynamically. Tables for static network devices do not change unless a network administrator manually changes them. In dynamic routing, devices build and maintain their routing tables automatically by using routing protocols to exchange information about the surrounding network topology. Dynamic routing tables allow devices to "listen" to the network and respond to occurrences like device failures and network congestion.

A routing Table is a table which contains the best paths, that a router uses to forward packets. You can also say it's a decision making table, what ever the routing protocol or mechanism is being used by the routers they decide where to forward the packet on the basis of routing table. Some times it's also called a forwarding table. While topology table is a table that can be defined as a road map of any network topology, which contains all the possible routing information and all the paths, good or bad but all the paths(routes/networks). But Topology table is not used by some protocols. In the case of EIGRP and OSPF, yes!! They both use it!! Topology Table is used to find the best paths, and every protocol run their algorithms, on the Topology table, but the topology table is never used for the final decision it's just used to build the best paths and routes according to specific protocols, which ever the algorithm they use.

In EIGRP Topology Table is called Topology Table, and it contains all the possible information about all the networks/paths/routes, that a router has learnt from it's neighbors, gus or ba, but all paths. EIGRP Topology Table contains back up paths also called Feasible Successors(Remain in Topology Table, Making EiGRP Fastest!!) and Successors (best paths, copy of it is included inside Routing Table).

In ospf Topology table is also named as LSDB, or Link state Database. it has the same function, keeps all the information of all the possible routes inside an OSPF Area. in eigrp there is no area concept, but OSPF uses Areas concept. And an Area is a group of routers having same LSDB!!

In OSPF there is no back up path concept, so OSFP LSDB doesnt contain any backup paths. That's why ospf convergence time is slower than EIGRP!!

5 Conclusion

Interdomain routing policies are constrained by the commercial contractual relationships between administrative domains. As a result, AS relationships are an inherent aspect of the Internet routing structure.

References

1. Gao, L.: On Inferring Autonomous System Relationships in the Internet. IEEE, Los Alamitos (2000)
2. http://www.cisco.com/en/US/docs/internetworking/technology/handbook/bgp.html
3. Stewart, J.W.: BGP4: Inter-Domain Routing in the Internet. Addison-Wesley, Reading (1999)
4. Zhang, R., Bartell, M.: BGP Design and Implementation. Cisco Press
5. http://tools.ietf.org/html/rfc1930
6. http://en.wikipedia.org/wiki/Autonomous_system_(Internet)
7. http://www.ietf.org/rfc/rfc2328.txt
8. http://www.cisco.com/en/US/products/ps6599/products_white_paper09186a00800a3e6f.shtml#wp39303
9. http://www.faqs.org/rfcs/rfc1058.html
10. http://www.faqs.org/rfcs/rfc5398.html

Federated Query Processing Service in Service Oriented Business Intelligence

M.S. Hema[1] and S. Chandramathi[2]

[1] Senior Lecturer, Department of Computer Science and Engineering, Kumaraguru College of Technology, Coimbatore, Tamil Nadu, India
[2] Professor, Sri Krishna College of Engineering and Technology, Coimbatore, Tamil Nadu, India
ghema_shri@yahoo.co.in

Abstract. In current business scenario, most of the software are offered as services which are exposed using contracts by the service providers. The client or user pays the service provider for the duration of usage. In process of business intelligence data integration is inevitable. Data federation is a form of data integration which is implemented as a servicer.

The federated query service gets the input query, decomposes into sub queries, post the sub queries to different data sources, extract the required data from sources and integrate into a single virtual view. This paper proposes the algorithms for federated query processing service and uses distributed XML database for data federation. The experimental results are provided.

Keywords: Business Intelligence, Service Oriented Architecture, Data Federation, Analytical Applications.

1 Introduction

Business Intelligence is a broad category of applications and technologies to gather, store, analyze and provide access to data to help enterprise users make better business decisions. The above functionalities are provided as services in service oriented business intelligence framework. Data integration techniques are used to gather the data from distributed heterogeneous data stores.

Data Integration provides a unified view of the business data that is scattered throughout an organization. Data Integration facilitates query over autonomous and heterogeneous data sources through a common and uniform schema (global schema).

1.1 Data Integration Techniques

There are three main techniques used in data integration a. Data consolidation, b. Data propagation and c. Data federation.

Data federation provides a single virtual view of data from multiple heterogeneous sources without permanently moving or replicating to a new location and enables applications to see the dispersed data as though it resided in a single data source. Data federation is an approach to real-time data integration which is easier and cost

V.V. Das, J. Stephen, and Y. Chaba (Eds.): CNC 2011, CCIS 142, pp. 337–340, 2011.

effective than data consolidation for certain type of application. Enterprise Information integration (EII) is the technology that supports data federation. The source data remains where they are and results persist in the server only as needed for caching.

The data federation is implemented as service in service oriented business intelligence framework. A Service-Oriented Architecture is an application architecture where the functionalities are defined as independent services, with well defined interfaces, which can be called in built-in sequences to build business processes [1].

This architecture provides a framework that allows heterogeneity, integration and reusability of the participant components in a flexible environment. In this context, it is interesting to analyze how a Federated System [2] can be designed within the ideas proposed by the Service-Oriented Architecture.

2 Related Work

Integration of heterogeneous distributed data source can be addressed either by replicating the data in a consolidated data store or by creating integrated view on-demand. The later method is more appropriate for dynamically changing business environment.

A few data federation techniques already available in literature are Integration Brokers for Heterogeneous Information Source [3], Dynamic data Integration using Web Service [4], Service Oriented Architecture for Federated Database System [5], A Multi layered Service Oriented Architecture for Business Intelligence [6], and Three tier Architecture for Service Oriented Business Intelligence [7].

3 Proposed Query Decomposition Algorithm

Query decomposition service accepts xpath query built based on global schema and outputs the sub-query that conforms to the local schema for which the sub query need to be generated. The xpath query based on global schema of the form /p1/p2/p3/…../pn-1/pn is evaluated from right to left as in xpath query. It is the actual result which the user wants to get from federated system. The local schema tree is traversed from leaf to root level.

If pn does not exist in local schema, it can be concluded that there is no sub-query for this schema. Otherwise if pn is found at a node in the local schema tree, the subsequent searches for node pi for (i = (n-1)…1) will be performed from the ancestor nodes of the matched node. Instead of searching the whole tree only the ancestor nodes of last matched node need to be searched. This can significantly reduce the search time. If node pi is found, then pi is concatenated with pn to form query.

Data access service accepts the sub-query from query decomposition service. This xpath sub-query conforms to local schema of data source in which data access service runs as opposed to user query based on global schema before decomposition. This service executes the xpath query and outputs a dataset that contains the requested data found in this data source. Each data source has an instance of data access service running at a port receiving the xpath sub-query, executes the query and provides the dataset.

The algorithm for query decomposition is shown below

Algorithm:

Input : userquery Q_{global} based on global schema, local schema S

Output: sub-query for S

Service splitter(userquery, local schema)

1. Initialize the variables 'query' and 'subquery' to NULL.
2. Initialize the arrays parts[], condition[] and value[] to NULL.
3. Split Q_{global} (global schema) and store it in an array parts[].
4. For each parts[i] repeat
   If parts[i] has condition and value then
   i. Split them and store in condition[i] and value[i] arrays respectively.
   ii. Create subquery based on local schema with condition
   Else
   i. Create subquery based on local schema
5. Return subquery.

The algorithm for sub query execution and data set generation is as follows

Algorithm:

Input : xpath subquery

Output :Dataset containing requested data from respective XML data source

Service Dataaccess (subquery)

1. Create a navigator XNav for XML Document
2. Evaluate subquery using XNav
3. Create an node iterator XNodeIter for nodes selected with Xnav
4. While there exist nodes in XNodeIter
5. If node has attributes
6. While there exist node.attribute
7. add the value to the Dataset
8. End While
9. If node has child elements
10. While node.child elements
11. add the value to the Dataset
12. End While
13. **return** Dataset

Algorithm:

1. Input : Dataset DSA,DSB....DSn
2. Output : Merged Dataset DSM
3. Service ResultIntegrator (DSA,DSB....DSn)
4. Dataset DSM := null
5. DSM :=Merge(DSA,DSB....DSn)
6. DSM :=Remove_duplicate_records(DSM)
7. DSM :=Sort(DSM,sortcolumnname,ASC/DESC)
8. return DSM

Fig. 1. Algorithm for query decomposition, Algorithm for sub query execution and data set generation and Algorithm for Result Integrator

Result Integrator Service accepts the dataset from different data access service containing the data from multiple data sources. It combines the datasets eliminating any duplicates, orders the data for consistent display of records and sorts the records if required. The combined dataset is displayed to the user.

4 Experimental Results

For experimental study a database containing 10,000 records is used. The databases which serve as data sources are geographically distributed and run in heterogeneous platform with structural and semantic heterogeneity. The data are used are pertinent to car sales and resale of an organization which have number of branches. All these data sources are initially converted to XML databases to suite xpath query decomposition. The local schemas are used for source identification and federated schema generation. The time taken to execute same query increased proportionally.

The web services were placed in a remote server. And the above process was repeated. The time taken to execute query is not proportional with number of records. The factors like server load, bandwidth availability affects the execution time.

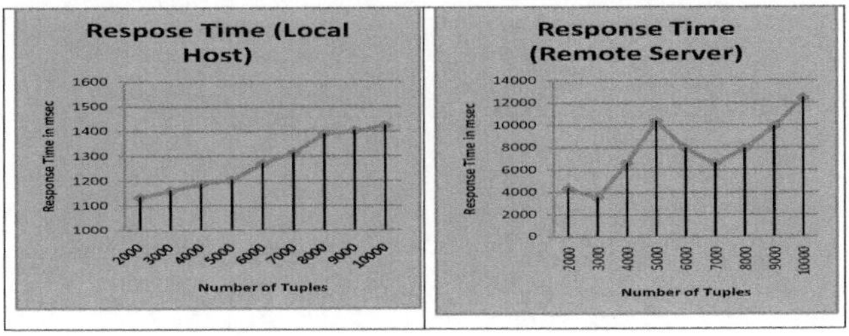

Fig. 2. Performance Chart

5 Conclusions and Future Enhancement

The data federation service in service oriented business intelligence was implemented. All services used to perform the operation are coarse grained and belongs to one service provider. Since the process is implemented using services they are independent of underlying software and hardware. The proposed algorithm efficiently integrates the distributed data efficiently. The algorithm is scalable and reliable.

The services like ad-hoc query processing can be designed for effective decision making. A common mapping layer can be proposed for business intelligence operation to have service composed of simple services of different service providers. An efficient service composition mechanism is required for federated query service in service oriented business intelligence.

References

1. Erl, T.: Service Oriented Architecture: Concept, Technology and Design. Pearson Education, London (2009)
2. Sheth, A., Larson, J.: Federated database system for managing distributed, heterogeneous and autonomous databases. ACM Computing Surveys 22(3), 183–236 (1990)
3. Kotsiopoulos, I., Keane, J., Turner, M., Layzell, P.J., Zhu, F.: IBHIS: Integration Brokers for Heterogeneous Information Sources. In: Proceedings of COMSAC 2004. IEEE Computer Society Press, Los Alamitos (2004)
4. Zhu, F., Turner, M., Kotsiopoulos, I., Bennett, K., Russell, M., Budgen, D., Brereton, P., Keane, J., Layzell, P.J., Rigby, M., Xu, J.: Dynamic Data Integration using Web Services. In: Proceedings of IEEE International Conference on Web Services, ICWS 2004 (2004)
5. Calegari, D., Viera, M., Motz, R.: Design of a Service Oriented Architecture for Federated System. Journal of JCST 5(4), 167–172 (2005)
6. Wu, L., Barash, G., Bartolini, C.: A Service Oriented Architecture for Business Intelligence. In: IEEE International Conference on Service Oriented Computing and Applications, SOCA 2007, 19-20, pp. 279–285 (2007)
7. Nageswara Guptha, M., Rajan, P.T., Chitra, A.: Data Sourcing using Federated Database System in Service Oriented Architecture. CSI Communications 33(1), 12–18 (2009)

Comparative Study of P2P and Cloud Computing Paradigm Usage in Research Purposes

Vineet Sinha, Aditi Gupta, and Guneet Singh Kohli

Computer Science Engineering, Maharaja Surajmal Institute of Technology, C-4 Janakpuri, New Delhi – 110058
vinesinha@gmail.com, aditi_gpc4@gmail.com, guneet_singh89@yahoo.com

Abstract. Cloud computing has emerged as an exciting new computing paradigm which offers business organizations the flexibility of scaling their computing resource usage [1], but the amount of effort and money being spent for developing cloud platforms for research and development is very little as compared to the huge amount of money being spent on developing cloud computing platforms for different commercial and business oriented uses. Peer to peer on the other hand is a popular concept where each participatory node can act as a "producer" as well as a "consumer" but again, the application of P2P on research platforms is minimal. This paper focuses on application and comparative study of these paradigms on research based projects and applications.

1 Introduction

The Internet has provided a boost to modern education and research technology. Internet being the core of both P2P and Cloud Computing paradigm has the potential not to only to improve current research practices but also provide researchers with resources required to carry out research varying from a small scale like a college based research to a large scale industry based research.

There is an ever growing need for faster processing and more storage hence to share remote resources for conducting research for example working on processing intensive softwares like Matlab and Mathematica can be made much more efficient if implemented on a distributed network.

This paper is divided into the following Sections, Section 2) The P2P paradigm, Section 3) The Cloud Computing paradigm, Section 4) Application of P2P paradigm for research purposes, Section 5) Application of Cloud Computing paradigm for research purposes, Section 6) A comparative study of P2P and Cloud Computing paradigms in research and Section 7) Conclusion of the paper.

2 The P2P Paradigm

Peer to peer (P2P) beginning with the 1990s has emerged as the dominant usage component of internet bandwidth [2]. A P2P network relies on dynamic and distributed

V.V. Das, J. Stephen, and Y. Chaba (Eds.): CNC 2011, CCIS 142, pp. 341–347, 2011.

addition of nodes to the network architecture. The participatory nodes make a portion of their resources directly available to other nodes in the network to better utilize bandwidth using individual station's computing power and storage capacity.

The technology allows computers, along with their users, to tap idle resources that would otherwise remain unused on individual workstations. The evolution of p2p clients from Napster to KaZaA, Gnutella, Morpheus and others has dramatically increased the amount of data transferred across service provider networks. [2]

2.1 Underlying Architecture

In a P2P network, when a user wants a file, installed P2P software locates any copies of the file within the P2P network and then the user is allowed to create multiple connections with several sources that have all or part of the requested file. As parts of the file are received, they are also uploaded to other users that are requesting that file. This protocol of matching several sources to a request makes for an efficient download scheme. [3]

P2P frameworks can be centralized which utilizes the client server architecture where the central server acts as a "traffic cop" maintaining directories of shared files stored on each node. The central server cross references the client requests with its database and finds and displays any matches to the requesting client. In contrast to the traditional client server architecture the actual file is never stored on the central server.

Contrasting to a Client-Server design the nodes of the network act as both the client and server themselves thus eliminating the need for a central server. Peers thus, as such, can join or leave the system without any intervention from any centralized server, which facilitates seamless integration of any number of new nodes (peers) to existing systems. [4]

3 The Cloud Computing Paradigm

A cloud is a collection of computers, usually owned by a single party, connected together such that users can lease access to a share of their combined power. [5] Cloud computing is a recent trend in IT that moves computing and data away from desktop and portable PCs into large data centers.[6]

Cloud computing involves the delivery of applications as services over the Internet as well as to the cloud infrastructure. The steady decline in data storage costs and the omnipresence of wireless networking paved the way for this computing paradigm which was only supported by progressive improvements in Internet computing software.

3.1 Underlying Architecture

The Cloud Computing system can be thought to be divided into two sections – front end and a back end. The front end constitutes the client side and consists of what the user "sees". On the other hand the back end comprises of the "Cloud".

The front end includes the various applications with which the users access the cloud and the terminal which the user has currently employed. The back end has the components which create the actual "cloud" including servers, and data storage systems. Each application on the client has a server committed to it. Also a central server ensures and maintains the smooth running of the entire system which includes monitoring of the entire system which includes traffic and demands of the clients.

Cloud computing has a clear implementation for business applications and services. An implementation in an academic environment can be achieved with a research cloud which can be implemented as scalable resources provided as XaaS where X can be a software (SaaS), hardware (HaaS), platform (PaaS), Infrastructure (IaaS) etc. [7] The cloud attributes and technologies are "coupled" in order to serve various services as mentioned above.

4 Application of P2P Paradigm for Research Purposes

To work efficiently on any research oriented software the two major requirements are (1) high processing speed and (2) enough storage space.

While the P2P paradigm might not have any influence on the storage part but it can play a very important role for the processing power required for computation, as the processing power increases with the increase in the number of systems in the network.

If we consider that our P2P network is simulated on Java 2 Platform SE and in particular, uses its multi-thread technology to model peers in the network, then each peer in the simulated network can be thought as an initiation of a single thread in the Java runtime environment. All threads run concurrently to emulate simultaneous peer execution in a real P2P network. Threads interact with each other by invoking methods of other threads through referencing, thus emulating interactions among peers. [8]

As shown in the Figure 1, a sender terminal can send a program which can be divided into a number of threads. The server keeps track of the all the systems attached to it which have the capability of executing that particular type of program. When all

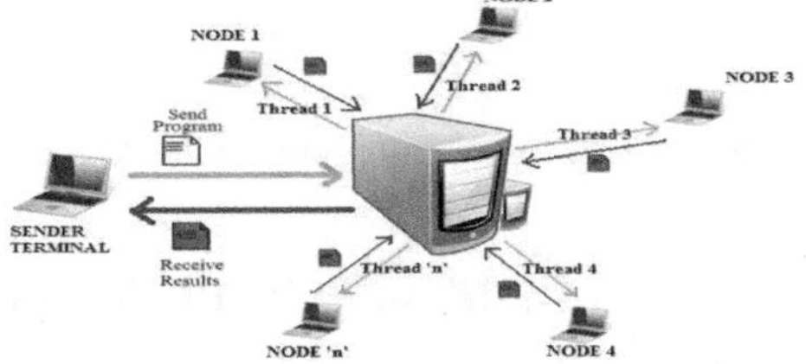

Fig. 1. Execution of a program on a centralized P2P network

the systems have been found out then the server can deliver the threads to different computers for execution. Once the execution is complete then the result can be delivered back to the sender terminal.

5 Application of the Cloud Computing Paradigm for Research Purposes

As explained in the topic above the efficiency of implementation of any research software basically depends upon the processing power and the storage space. Cloud computing being a paradigm which provides both of these on a pay per use basis (utility computing) is an excellent approach towards providing researchers with endless storage and processing speed according to the amount of money they or their organization is ready to pay.

Cloud computing services can be collectively called XaaS (X being software, infrastructure, platform etc.) hence all the resources required for implementing research oriented softwares can be delivered on the cloud by the service provides as explained in the subsections below.

5.1 Software (SaaS)

If we take Matlab as an example, the latest Release 2010a recommends 2048 MB RAM and 3-4 GB disk space for a typical installation [9] hence it can be seen that it requires a considerable amount of processing speed and might not be suitable for executing research based programs on computers that do not meet these standards. Some of the large scale projects even require mainframe computers and supercomputers (like space projects and research based projects for the army) but with the use of 'cloud computing' even a low end computer can have access to high speed processing and huge storage as shown in the Figure 2.

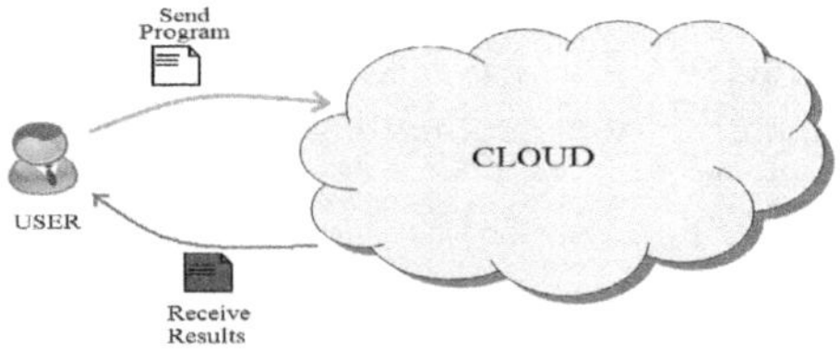

Fig. 2. Execution of a program on the "cloud"

6 A Comparative Study of P2P and Cloud Computing Paradigm for Research Purposes

Table 1 is a comparative study of P2P and the cloud computing paradigm when used in research according to some very important aspects.

Table 1. Comparison of P2P and Cloud Computing

S. No.	Aspects	Peer to Peer	Cloud Computing
1.	Storage	Considerable storage space required in peers in the network.	Storage services are available on the cloud itself.
2.	Processing	P2P places every node on an equal footing enabling them to be both the provider and end user of information. Nodes with significant processing capacities are joined together electronically to integrate their resources. Thus an ample number of nodes can create very significant processing capabilities. However incase of large amount of data being shared congestion arises.	Processing in the cloud computing paradigm is dependent on the resources bought from the cloud provider. Hence processing on bulky data is the key technology in the cloud computing infrastructure.
3.	Environ-mental Impact	P2P does not affect the environment as such but has significant economical impact because of piracy issues.	Although cloud com-puting has a positive environmental impact through effective hardware utilization, the boom in the cloud computing industry will leave a significant carbon footprint.[1]
4.	Piracy	The current software piracy rate in India is 68% and these pirated softwares including research softwares are accessible through p2p networks. [10]	Cloud computing curbs piracy as the resources are directly obtained from the cloud provider in the form of XaaS.
5.	Privacy	Maintenance of privacy in a P2P network is user dependent as how to much data a particular user has demarcated for sharing on the P2P network. Hence for research purposes confidential data and results can be kept private.	Privacy can be an issue of concern in cloud computing as the data resides on hardware provided by a third party system. Hence it can be vulnerable to security breaches. Leak of confidential research data and results is undesirable and under-mines privacy required for research and research based systems.
6.	Reliability	In case of a failure or disruption in a P2P network there is no authority to regulate and allocate measures for improvement and correction.	Cloud computing paradigm here is par-ticularly reliable as any discrepancies in opera-tion can be reasoned with the cloud provider.

Table 1. (*continued*)

7.	Perform-ance	Performance is dependent on the number of active nodes in the network. Hence performance for a research oriented system requires the presence an optimum significant number of nodes in the network.	Cloud computing paradigm provides consistent performance "anytime anywhere" as it is not peer dependent.
8.	Software Variety	In a P2P network if the software is not available on all the peers then the processing power or computation speed is affected proportionately.	If the software is available on the cloud the processing is dependent only the amount of resources rented or hired by the user.
9.	Setting Up Data Centers	Setting up of exclusive data centers is not required in P2P.	Setting up of data centers is an expensive and tedious job.
10.	Network Sizes	P2P is suitable for small to middle sized network as the size of the network increases congestion creeps in.	Cloud computing paradigm is suitable for any network size but is not cost effective for small networks.

7 Conclusion

Cloud Computing as compared to P2P is still an emergent paradigm. Its application in research purposes is still in its initial stages and under development and debate. Cloud computing has some distinct advantages over P2P such as consistency of performance, but P2P is preferable for small networks such as college or university networks whereas Cloud Computing proves to be more beneficial for larger scale projects. Privacy is a key issue while dealing with public clouds whereas P2P based systems offer user controlled privacy. With the advancements in Cloud Computing, the clouds will grow in size as soon as the available bandwidths and complementary service models mature, paving the way for revolutionary changes in research practices.

References

1. Gupta, A., Awasthi, L.K.: Peer Enterprises: A viable alternative to cloud computing? In: IEEE International Conference on IMSAA (2009)
2. Peer to Peer file sharing, the impact of file sharing on service provider networks, `http://www.sandwine.com`
3. Chao, G.: Study on Privacy Protection and Anonymous Communication in Peer-to-Peer Networks, mines. In: Proceedings of International Conference on Multimedia Information Networking and Security, MINES 2009, vol. 2, pp. 522–525 (2009)
4. Mondal, A., Kitsuregawa, M.: Privacy, Security and Trust in P2P environments: A Perspective. IEEE DEXA (2006)
5. Grids and clouds: what's in a name?, `http://www.gridcafe.org/grids-and-clouds.html`

6. Dikaiakos, M.D., Katsaros, D., Mehra, P., Pallis, G., Vakali, A.: Cloud Computing, Distributed Internet Computing for IT and Scientific Research. IEEE Internet Computing 13(5) (2009)
7. Rimal, B.P., Choi, E., Lumb, I.: A Taxonomy and Survey of Cloud Computing Systems. In: 5th International Joint Conference on INC, IMS and IDC (2009)
8. Rong, L.: Multimedia Resource Replication Strategy for a Pervasive Peer-to-Peer Environment. Journal of Computers 3(4) (April 13, 2008)
9. "Mathworks – Matlab R2010a system requirements", http://www.mathworks.com/support/sysreq/current_release/index.html
10. Piracy rates in India, http://www.bsa.org/country.aspx

Cyclic Prefix Based Modulation Selection Using Comparative Analysis of EVM and RCE for WiMAX Network

Budhaditya Bhattacharyya, Iti Saha Misra, and Salil Kumar Sanyal

Department of Electronics and Telecommunication Engineering
Jadavpur University, Kolkata-700032, India
bhattacharyya.budhaditya@gmail.com, itisahamisra@yahoo.co.in,
s_sanyal@ieee.org

Abstract. The study of any individual performance measuring attribute does not necessarily confirm the selection of modulation, or the coding to be used for a specific communication system. In this paper a Simulink based WiMAX 802.16e Physical Layer model has been simulated using Vector Signal Analyzer (VSA) as a real time display scope to evaluate EVM and RCE with minimum complexity. The focus is maintained on the comparative analysis associated with these performance metrics to arrive at a proposition regarding the selection of modulation in a WiMAX Physical Layer (PHY) under the influence of varying Cyclic Prefix using proper simulations results.

Keywords: EVM; RCE; OFDM; Cyclic Prefix; Modulation; VSA.

1 Introduction

Physical layer attributes like EVM, IQ Offset, RCE, Frequency Error etc. play a crucial and significant role for improving the overall efficiency of the communication system [1], [2]. In this paper, we provide a comprehensive comparative analysis of two of the most important measuring attribute, Error Vector Magnitude(EVM) and Relative Constellation Error(RCE) in order to achieve the best possible inference regarding the selection of a WiMAX specified modulation scheme under a given channel condition. WiMAX specifies BPSK, QPSK, 16-QAM and 64-QAM as the standard modulation techniques used as backend modulations for subcarriers. In OFDM technology however, a cyclical redundancy is offered to the subcarriers in form of Cyclic Prefix (1/4, 1/8, 1/16 and 1/32) as guard interval to combat Inter Symbol Interference (ISI). Although modulated, the unpredictability in the nature of combination of energy component makes a phase/amplitude distorted received signal.

The extent of deviation of the constellations is greatly affected by the introduction of channel noise. Error vector being *e (k)*, formed by the difference between reference vector (*s(k)*) and received symbol vector (*r(k)*) while, RCE being the RMS-averaged magnitude error experienced by multiple constellation point. In this paper, performance metrics like EVM, RCE have been rigorously calculated using Simulink-VSA based WiMAX model for afore mentioned modulations with a unique addition of the

V.V. Das, J. Stephen, and Y. Chaba (Eds.): CNC 2011, CCIS 142, pp. 348–350, 2011.

effect of changing Cyclic Prefix (CP). Due to absence of recent study in this domain, we take this opportunity to highlight the cause and effect relationship amongst EVM, RCE, and Modulations.

2 Simulation Background and Modeling Environment

Introduction of CP reduces the receiver design complexity as the cyclically extended symbols make the channel effect to be multiplicative. Keeping these basic concepts in mind we have designed a WiMAX 802.16e Physical Layer in Simulink. To improve the proficiency of the system, integration with VSA has been successfully established as in Fig. 1. Thus VSA becomes the virtual display scope for categorical analysis of EVM and RCE.

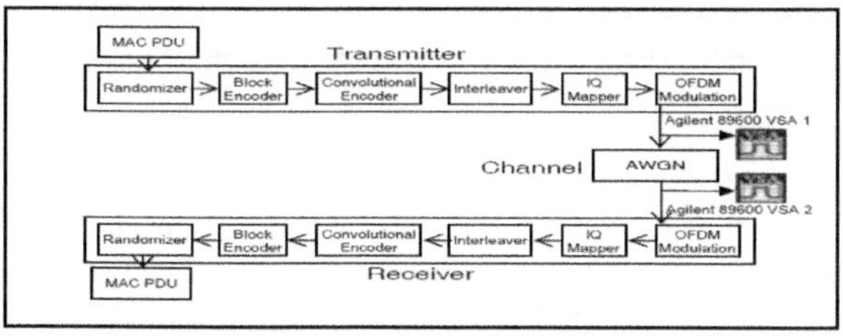

Fig. 1. The Complete Simulation Model including the interface with VSA

3 Modulation Selection Algorithm and Simulation Results

The Simulink-VSA based WiMAX model follows the following algorithm in selecting the proper modulation scheme for a given CP and channel condition,

1. For each of the modulation schemes, each and every CP is chosen, while Channel SNR varies from 10 dB (worse) to 120dB (best) like a virtual varying wireless medium.
2. The modulation to be selected would depend on minimum EVM and RCE obtained by the network at a given CP value for a specific channel SNR.
3. In case either EVM or RCE has a higher value the specific selection will defer till minimum value is achieved for that specific metrics.
4. If EVM and RCE do not reduce at all, then the average error (which should be close to zero) would decide on the selection of specific modulation at a selected CP.

Selection of modulations based on CP is illustrated in Table1.that strictly follow the aforementioned algorithm. Our discussion is well supported by Fig.2 that shows EVM and RCE summary occurring at CP=1/4. Undoubtedly at CP=1/4, the repetition fraction is largest compared to other CP, hence introduction of error will be minimum at low value of channel SNR. Fig. 2(b) shows a small value of RCE for QPSK, compared to the observed high EVM value in Fig.2 (a). Hence as suggested by our algorithm, QPSK can't give a conclusive verbatim to the selection at CP=1/4. However, for 16 QAM (high data rate) both EVM and RCE show a near zero proximity for changing SNR. The results as presented in this paper show that 16-QAM is preferred for both CP=1/4 and 1/32, whereas QPSK is preferred for CP=1/8 and 64-QAM for CP=1/16, depending on the smallest possible average error introduced in the system.

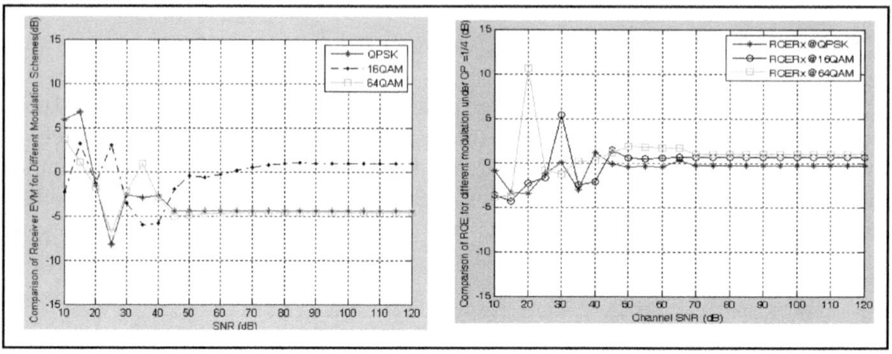

(a) (b)

Fig. 2. (a) EVM and (b) RCE observation at CP=1/4 for QPSK, 16-QAM and 64-QAM

Table 1. Comparative analysis of EVM and RCE A: EVM B: RCE C: Average Error 1: Very Poor, 2: Poor, 3: Average, 4: Good, 5: Very Good

Different Coding Schemes	CP=1/4			CP=1/8			CP=1/16			CP=1/32		
	A	B	C	A	B	C	A	B	C	A	B	C
QPSK	1	5	3	4	4	4	3	1	2	2	2	2
16-QAM	4	4	4	1	3	2	1	3	2	3	3	3
64-QAM	1	3	2	3	3	3	4	4	4	3	2	2

Acknowledgement

The authors deeply acknowledge the support from DST, Govt. of India for this work in the form of FIST 2007 Project on "Broadband Wireless Communications" and in the form of PURSE 2009 Project on "Cognitive Radio" in the Department of ETCE, Jadavpur University. Equal acknowledgement is also accorded to TEQIP, Jadavpur University for providing the necessary software infrastructure behind this work.

References

1. Angrisani, L., Napolitano, A.: Modulation Quality Measurement in WiMAX Systems Through a Fully Digital Signal Processing Approach. IEEE Transactions on Instrumentation and Measurement 59(9), 2286–2302 (2010)
2. McKinley, M.D., Remley, K.A., Kenney, J.S., Myslinski, M., Koepke, G., Camell, D., Johnk, R.T.: Comparison of QPSK and 64QAM EVM for an OFDM WLAN signal at 2.4 GHz and 4.9 GHz in a multipath environment. In: Antennas and Propagation Society International Symposium, APSURSI 2009, June 1-5, pp. 1–4. IEEE, Los Alamitos (2009)

Experimental Study of Rank 1 Chaos in Chua's Oscillator with Cubic Nonlinearity

K. Gopakumar[1,*], K.G. Gopchandran[2], and B. Premlet[3]

[1] Department of Electronics, TKM College of Engineering, Kollam, Kerala - 691 005
Tel.: 091 9544510625
gopaan@asianetindia.com
[2] Department of Optoelectronics, University of Kerala,
Trivandrum, Kerala
Tel.: 091 9447862247
[3] Department of Physics, TKM College of Engineering, Kollam, Kerala
Tel.: 091 9447333285

Abstract. Chua's oscillator is one of the simplest electronic circuits that are capable of producing chaos. It can exhibit a wide array of behaviour including a great variety of attractors, bifurcations, and routes to chaos. In this paper we report an experimental investigation focused on the dynamics of two-dimensional parameter spaces of a three parameter Chua system, where the nonlinearity of the Chua diode is represented by a cubic polynomial. We investigated the creation of strange attractors in the circuit. The design and use of this circuit is motivated by a recent mathematical theory of rank one attractors developed by Wang and Young. Strange attractors are created by periodically kicking a weakly stable limit cycle emerging from the centre of a supercritical Hopf bifurcation. The periodic pulses are applied directly as an input. For this scheme of creating rank-one attractors to work, the applied periodic pulses must have short pulse widths and long relaxation periods.

Keywords: Strange attractors; switch-controlled circuits; Hopf bifurcation; chaos; cubic nonlinearity.

1 Introduction

Nonlinear systems may exhibit many types of complex behaviour such as chaos, and this complexity has attracted many scientists from various fields (e.g. physics, mathematics, engineering, etc.) to study such systems. Since it has been shown that simple electronic circuits may exhibit chaotic behaviour, the study of chaos in nonlinear electronic circuits received a great deal of attention in the last decade. An electronic circuit that has come to be known as Chua's oscillator [1] is to be perceived among such endeavors. Such a circuit has undoubtedly become a standard benchmark in the study of nonlinear dynamics and chaos.

This paper aims at introducing a new chaos theory, namely the theory of rank one maps [2,3]. The studies on rank1 chaos are based on a recent mathematical theory of

* Corresponding author.

V.V. Das, J. Stephen, and Y. Chaba (Eds.): CNC 2011, CCIS 142, pp. 351–355, 2011.

rank-one maps [4, 5] originated from the principles of the logistic family ($f_a = 1-ax^2$) and the H´enon family ($x_1 = 1- ax^2 + y$, $y_1 = bx$). The important mathematical breakthroughs along the way are (i) Benedicks' and Carleson's theory on the H´enon maps [6] and (ii) the proof of Benedicks and Young on the existence of Sinai–Ruelle–Bowen (SRB) measure for good H´enon maps [7]. To fill the gap between the theory and its applications to physical systems, it became necessary to verify certain assumptions such as the positivity of Lyapunov exponents or the existence of SRB measures [8, 9]. The theory of rank one maps helped to circumvent the above issues. The Hopf limit cycle required for the smooth case required in this investigation is created by employing Chua's oscillator with a nonlinear resistor of cubic nonlinearity [10]. After creating a limit cycle, periodic pulses were added to the circuit so as to modulate its state variables, namely, the capacitor voltages and the inductor current. The periodic pulses were applied in intervals controlled by analog switches to generate the kicking effect suggested by Wang and Young [4, 11] for creating rank 1 chaos. Consequently, strange attractors appeared supporting the theory of rank one maps. The implementation of switch-controlled Chua's oscillator, experimental results obtained and the concluding remarks are presented in subsequent sections.

2 Cubic-Polynomial Chua's Diode and Switch-Controlled Chua's Oscillator

For the investigation, we specifically chose the well-known Chua's [1] oscillator shown in Fig. 1. This circuit contains four linear elements (two capacitors, one inductor, and one resistor) and a nonlinear resistor N_R called Chua's diode. Chua's diode is described by a piecewise-linear function $g(V_R) = m_0 V_R + 0.5 (m_1 - m_0) [| V_R + B_P | - | V_R - B_P |]$. But all features of a real circuit are not captured correctly by the simple piecewise circuit[10]. Therefore we need to realise a smooth nonlinearity by a cubic polynomial defined as

$$g(V_R) = a_o + aV_R + bV^2_R + cV^3_R \tag{1}$$

Since the desired V-I characteristics of the nonlinear resistor N_R in Chua's oscillator is an odd symmetric function with respect to origin, we use the cubic polynomial of Eqn. 1 with the coefficients

$$ao = 0,\ a < 0,\ b = 0,\ \text{and } c > 0.\ \text{i.e., } i_R = g(V_R) = aV_R + cV^3_R \tag{2}$$

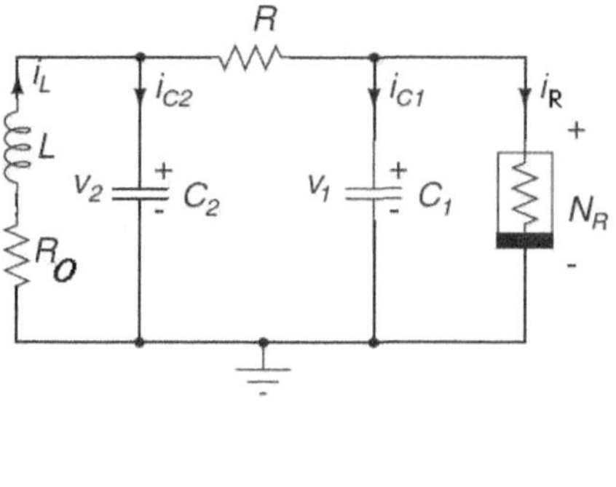

Fig. 1. Chua's oscillator

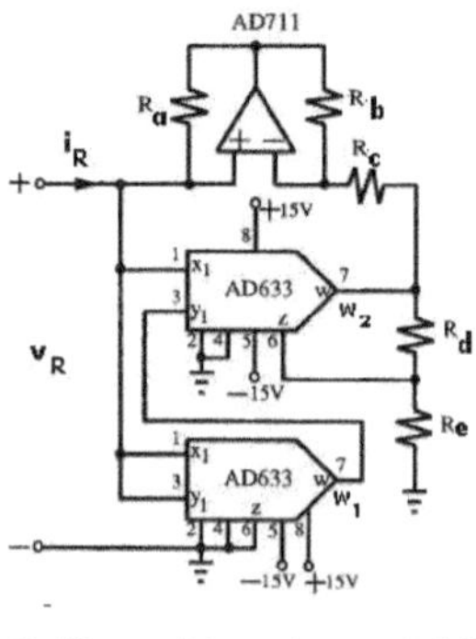

Fig. 2. The cubic polynomial Chua's diode

The basic circuit to realise the cubic polynomial of Eqn. 2 is a multiplier with a feedback loop and the complete circuit is shown in Fig. 2. For the circuit

$$a = \frac{1}{R_c},\quad c = \frac{R_d + R_e}{R_d . R_c} \cdot \frac{1}{10V} \cdot \frac{1}{10V},\ \text{and gain factor} = \frac{R_d + R_e}{R_d} \tag{3}$$

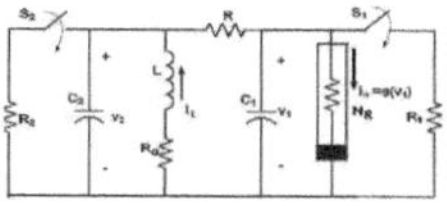

Fig. 3. Switch- controlled Chua's oscillator

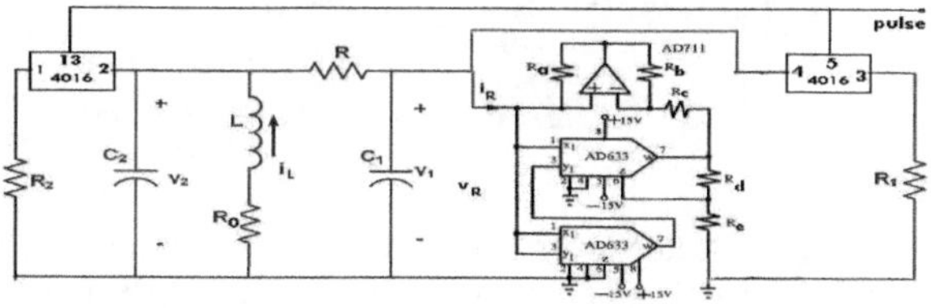

Fig. 4. The complete circuit

To obtain rank 1 chaos in Chua's oscillator, we add to it an extra forcing term $P_{T,p}(t)$, a periodic pulse train of period T, and pulse width p. To modulate the system state variables, we can connect controlled switches as shown in Fig. 3. For the switch controlled Chua's oscillator shown in Fig. 3, we obtain the following dimensionless set of equations:

$$\overset{o}{x} = -\alpha h(x) + \alpha y - \varepsilon_1 x P_{T,p}(t)$$

$$\overset{o}{y} = \gamma x - \gamma y + \gamma \eta z - \varepsilon_2 y P_{T,p}(t) \tag{4}$$ where

$$\overset{o}{z} = -\beta y - \beta \xi z$$

$$P_{T,p}(t) = \frac{1}{P}\sum_{-\infty}^{\infty} F_n(t),\quad p = p_0\omega_n,\quad T = T_0\omega_n,\quad h(x) = b_1 x + b_3 n x^3$$

$$b_1 = 1 + aR,\quad b_3 = cRV_0^2,\quad \alpha = \frac{1}{RC_1\omega_n},\quad \beta = \frac{R_n}{L\omega_n},\quad \gamma = \frac{1}{RC_2\omega_n} = 1 \tag{5}$$

$$\varepsilon_1 = \frac{\alpha R p}{R_1},\quad \varepsilon_2 = \frac{\gamma R p}{R_2},\quad \eta = \frac{R}{R_n},\quad R_n = \frac{V_0}{I_0},\ and\quad \xi = \frac{R_0}{R_n}$$

3 Experimental Results

The complete circuit shown in Fig. 4 was constructed on a breadboard with parameter values

$\alpha = 2.0$, $\beta = 1.04$, $\gamma = 1.0$, $p_0 = 1.23$, $b_1 = 0.35$, and $b_3 = -1.0$.The component values are then obtained as R2 = 2.5kΩ pot, R1 = 4.7kΩ pot, R0 = 15Ω, R = 1.5kΩ, Ra = 2kΩ, Rb =2kΩ, Rc =1.8kΩ, Rd = 3.3kΩ, Re =7.8kΩ, C1 = 2.2nF, C2 = 4.4nF and L = 10mH. AD711 BiFET operational amplifier and AD633 analog multipliers were used with supply voltage ±15V. CD 4016 quad CMOS switch was used to provide the required kick. With these component values, and with the CMOS switch 4016

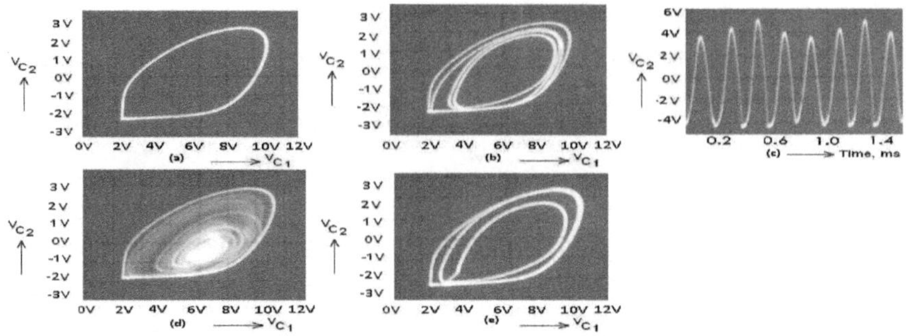

Fig. 5. Strange attractors from smooth Chua's oscillator (a) a Hopf limit cycle with switches S1 and S2 opened (b) with S2 closed T = 40 (c) time response plot with T = 40 (d) with T = 85 (e) with T = 110

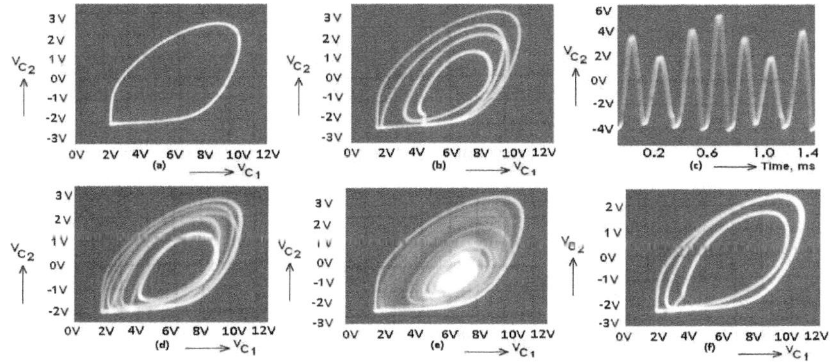

Fig. 6. Strange attractors from smooth Chua's oscillator (a)Hopf limit cycle with switches S1 and S2 opened (b) with S1 closed T = 40 (c) time response plot with T = 40 (d) with T = 75 (e) with T = 93 (f) with T = 115

opened, the Hopf limit cycle obtained experimentally is depicted in Fig. 5(a). This limit cycle was then kicked periodically by closing the switch connected to R2 and by keeping the one connected to R1 opened. This was achieved by connecting rectangular pulses of long duration to pin number 13 of CD4016. The results obtained for different duty cycles are shown in Figs. 5(b) to 5(e). The experiment was also carried out for the single switch case for resistor R1 by connecting the pulse to pin number 5 of CD4016 and by keeping the other switch opened. The various attractors obtained are depicted in Figs. 6(a) to 6(f).

Let us conclude this section by explaining how strange attractors are created by the application of pulses controlled by external switches. In the absence of external forcing, the phase portrait appears like a circle. When the switch is closed, this closed loop is deformed to become the solid curve and two factors are put into action, the shearing and the attraction. The shearing exaggerates the initial deformation brought in by the kicking force and the attraction acts against this deformation. If the shearing is weak, and the stability of the original limit cycle is strong, then attraction overcomes the initial deformation. In this case, all points nearby are attracted to a simple closed curve. On the other hand, if the shearing is strong, and the stability of the original limit cycle is weak, then the deformation overcomes the attraction. Now, the attracting sets get disintegrated into a finite collection of periodic saddles and sinks. As shearing gets further strengthened, the deformation is exaggerated further into the dominance of sinks as the only observable set of attraction. Simultaneously, strange attractors with complicated fractal structure will occur and eventually, chaotic attractors will dominate.

4 Conclusions

In this paper, we have provided the experimental proof of rank 1 chaos by using a switch-controlled Chua's oscillator. Initially, a smooth limit cycle was generated and then by subjecting it to periodic kicks, rank 1 attractors were obtained. The important advantage of this circuit is the availability of a recipe-like procedure to generate rank one chaos and the major disadvantage being the complexity of the resulting circuit. We find that the experimental results are in perfect agreement with the predictions of the theory.

References

1. Chua, L.O.: Chua's circuit: an overview ten years later. J. Circuits, Syst. Computers 4, 117–159 (1994)
2. Wang, Q., Oksasoglu, A.: Rank one chaos: Theory and applications. International Journal of Bifurcation and Chaos 18, 1261–1319 (2008)
3. Chen, F., Han, M.: Rank one chaos in a class of planar systems with heteroclinic cycle. Chaos. 19(043122), 5 pages (2009)
4. Wang, Q., Oksasoglu, A.: Strange attractors in periodically kicked chua's circuit. International Journal of Bifurcation and Chaos 15, 83–98 (2005)
5. Oksasoglu, A., Wang, Q.: A new class of chaotic attractors in Murali–Lakshmanan–Chua circuit. International Journal of Bifurcation and Chaos 16, 2659–2670 (2006)
6. Benedicks, M., Carleson, L.: The dynamics of the H´enon map. Ann. Math. 133, 73–169 (1991)
7. Benedicks, M., Young, L.S.: Sinai–Bowen–Ruelle measure for certain H´enon maps. Invent. Math. 112, 541–576 (1993)
8. Young, L.-S.: What Are SRB Measures, and Which Dynamical Systems Have Them? Journal of Statistical Physics 108(5/6) (2002)
9. Alves, J.F., Bonatti, C., Viana, M.: SRB measures for partially hyperbolic systems whose central direction is mostly Expanding. Invent. Math. 140, 351–398 (2000)
10. Zhong, G.-Q.: Implementation of Chua's circuit with a cubic nonlinearity. IEEE Trans Circuits Syst.- 1: Fund. Theories and appl. 41, 934–941 (1994)
11. Wang, Q., Young, L.-S.: From Invariant Curves to Strange Attractors. Commun. Math. Phys. 225, 275–304 (2002)

A Novel Algorithm for the Implementation of PSO-Based SVM

Shravan Karadkal[1], Sunil N. Putty[1], and K. Manikantan[2]

[1] Final Year, Department of Electronics and Communication,
M.S. Ramaiah Institute of Technology, Bangalore
{shravan.667,sunil.putty2008}@gmail.com
[2] Assistant Professor, Department of Electronics and Communication,
M.S. Ramaiah Institute of Technology, Bangalore
kmanikantan2009@gmail.com

Abstract. This paper proposes a new algorithm for the implementation of Support Vector Machines using Particle Swarm Optimization (PSO), called *Intelligently Initialized Particle Swarm Classifier (I2PSC)*. Initialization of the particles is done close to the optimal solution by using basic concepts of Geometry and Statistics, so that the rate of convergence of the swarm towards the optimal solution is better as compared to a randomly initialized swarm. The fitness function is designed in such a way that the trade-off between the distance between the bounding planes and the number of misclassifications is greatly reduced compared to traditional SVMs.

Keywords: Support Vector Machines, Particle Swarm Optimization, I^2PSC.

1 Introduction

Support vector machines (SVMs) have been extensively used for classification of data sets of two different classes [1]. Since its inception, it has undergone major changes so that the Minimum Generalization Error is greatly reduced. But in certain situations, wherein datasets are closely spaced, optimization is required so that the Minimum Generalization Error is reduced.

PSO [2] and its variants have been successfully used to generate optimum classifiers. Classifiers are generated by adopting a chaotic search method with improved PSO parameters [3]. Parameters involved in SVM are tuned and optimized using PSO since the performance of SVM is based on proper selection of its parameters [4, 5]. Optimal SVM models can be selected in a dynamic fashion combing the power of the swarm intelligence theory with the conventional grid-search method using dynamically updated training datasets [6]. PSO can also be used to generate the classifier directly, while the parameters of PSO are selected using fuzzy systems [7]. Here we propose a novel algorithm for generating classifiers using PSO with intelligent initialization of its swarm. By using a well defined fitness function we eliminate the need to tune any parameters of either PSO or SVM, while generating classifiers of greater

V.V. Das, J. Stephen, and Y. Chaba (Eds.): CNC 2011, CCIS 142, pp. 356–361, 2011.

efficiency in lesser iterations. The proposed software architecture is very simple compared to other PSO–SVM hybrid models.

1.1 SVM

SVM is a class of Machine Learning algorithm, in which a set of inputs is given which has to be classified based on the label associated with each input. SVMs classify the input data by introducing a hyperplane between the two datasets belonging to two different classes. The points which are used to train the SVMs are known as the training set. Bounding planes are drawn parallel to the hyperplane which decide the margin between the two datasets. The distance between the bounding planes should be maximum in order to reduce the Minimum Generalization Error. The distance between the bounding planes is given by $2/\|W\|$. Where, W is the vector containing the coefficients of the hyperplane.

1.2 PSO

Particle Swarm Optimization (PSO) is swarm intelligence technique inspired by the social behavior of bird flocking. In PSO, potential solutions called as particles, move around solution space of the problem. These particles search for the optimal solution of the problem in the predefined solution space. The direction of motion of these particles is influenced by three factors namely: Inertial influence, Social influence, Personal influence

The particle swarm optimizer tracks the best solution (fitness value) achieved so far for a given particle; this is called the personal-best (*pbest*) value of the given particle. Apart from this, the optimizer tracks the best fitness value achieved so far by any particle in the entire flock; this is called the global-best (*gbest*). In PSO, with every iteration, the velocity of each particle is updated based on the pbest value of that particle and the gbest of the swarm. Hence the definition of the fitness function plays a vital role in performance of the PSO algorithm for solving a given problem.

The rate of convergence of the swarm towards the optimal solution also depends on the initialization of the swarm. Initializing in a random manner may result in a better solution but will take more number of iterations. Hence there is always a scope for initializing the particles in an efficient way.

2 Intelligent Initialization of Swarm

The swarm used in the PSO algorithm has a population size of 20.For the working of PSO the swarm has to be initialized first. Random initialization will lead to significantly higher number of iterations to reach the optimal solution or the swarm might get stuck at certain regions of the search space leading to solutions which are not perfectly optimized (Premature Convergence). In order to generate optimal classifiers in a relatively lower number of iterations, the swarm is initialized in an intelligent way. Hence, the classifier generated is called Intelligently Initialized Particle Swarm Classifier (I^2PSC).

In order to initialize the particles of the swarm, an approximate classifier is generated using basic concepts of statistics and geometry as given:

- A best fit line l_1 is constructed passing through both the datasets .
- A line, l_2 is constructed perpendicular to l_1 and passing through the mean of both the datasets.
- The coefficients of l_2 are divided by half the distance between the means of each dataset to get an approximate classifier C. Let the equation of the classifier C be

$$W_i' X_i - \gamma' = 0 \; i \geq 1,2,3 \ldots\ldots\ldots N$$

Where N is the dimension of the classifier.

2.1 Manner in Which Particles Are Initialized

- First N dimensions of any two particles in the swarm are initialized to W_i' for i= 0, 1, 2......N and the $(N+1)^{th}$ dimension is initialized to value γ'.
- Any ten particles in the swarm are initialized as in above step, but with ±30% variation in their value compared to that of the values assigned in above step.
- Remaining eight particles are initialized randomly in the search space.

3 Fitness Function

The way in which the fitness function of the classifier is designed is very important to generate an efficient classifier. Here, the fitness function is designed in such a way that the classifier with the highest fitness value will require no further optimization. The fitness of a classifier is based on two factors:

- Distance between the bounding planes.
- Number of misclassified points.

The designed fitness function has two components - distance component and misclassification component. The distance component tries to maximize the distance between the bounding planes and obeys the following conditions:

- It varies from -1 to +1.
- The maximum value is attained when the distance between the bounding planes is infinity.
- The minimum value is attained when the distance between the bounding planes is zero.
- The mid value (0) is attained when the distance between the bounding planes is half the distance between the means of each dataset.

The misclassification component tries to minimize the misclassification and obeys the following conditions:

- It varies from -1 to +1.
- The maximum value is attained when the number of misclassifications is zero.
- The minimum value is attained when the number of misclassifications is equal to half of the total number of data points.
- The mid value (0) is attained when the number of misclassifications is one fourth of the total number of data points.

Based on the above conditions, the following fitness function is designed:

$$\text{Fitness\_value} = \left(1 - 4^{(0.5 - ndis)}\right) + \left(1 - (4 * nmis)\right)$$

where,

$$\text{ndis} = \frac{\text{Distance between Bounding Planes}}{\text{Distance between m1 and m2}}$$

$$\text{nmis} = \frac{\text{Number of misclassified points}}{\text{Total number of data points}}$$

4 Software Architecture

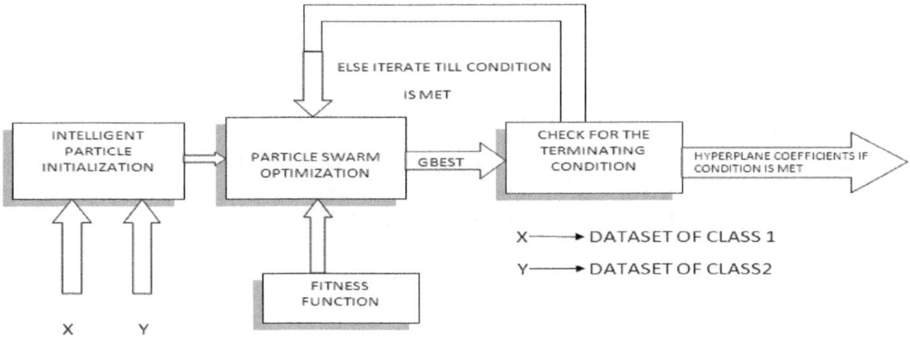

5 Results

The graphs and tabulated results shown below have been obtained through simulations in MATLAB.

The data sets used have the following parameters:

1. Number of points in each dataset=50
2. Variance of the datasets=2
3. Distance between the means of each dataset=5 (for Fig.1 & Fig.3)

PSO parameters are set according to Trelea type 1model as in [8].

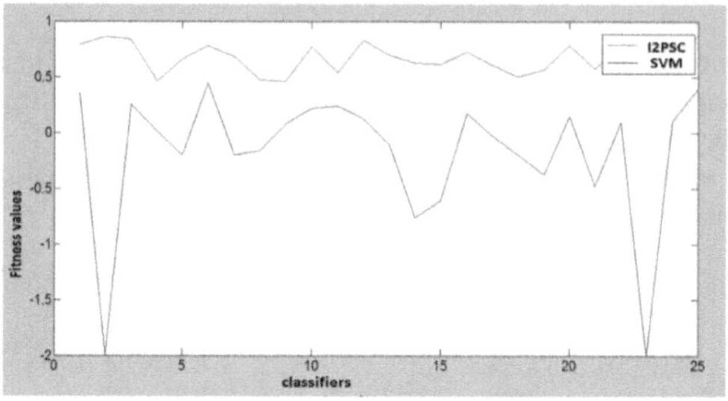

Fig. 1. Shows the comparison between the classifiers generated by the proposed method and traditional SVM

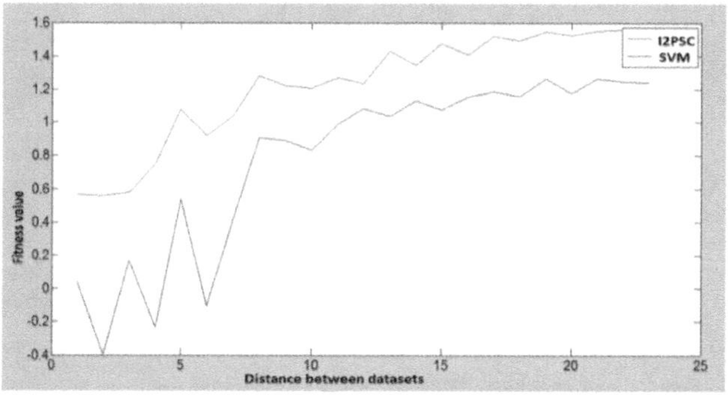

Fig. 2. Shows fitness values v/s distance between the means of the data sets with other parameters remaining the same. This shows how the fitness of the classifier varies with the distance between the datasets in I^2PSC based SVM and traditional SVM.

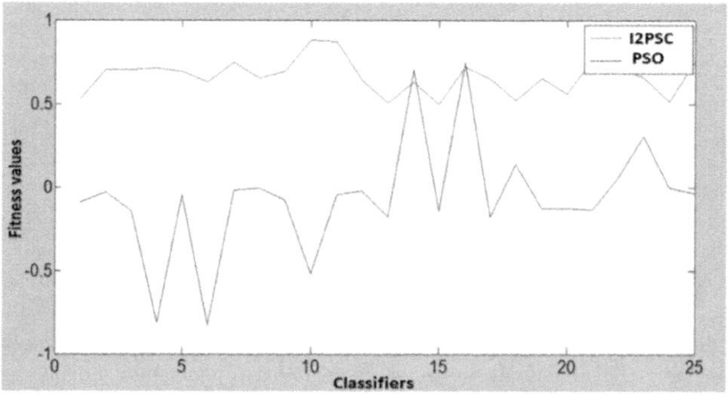

Fig. 3. Shows the comparison between fitness values of I^2PSC and classifiers generated by randomly initialized PSO. Under the terminating conditions mentioned, I^2PSC has an average fitness value of 0.55.

Table 1. Shows the comparison between I^2PSC and classifiers generated using randomly initialized PSO for a 2-Dimensional data. It is evident from the above table that I^2PSC gives classifiers with fitness values almost same as those of PSO, but the number of iterations required is less.

Distance b/w the datasets	I^2PSC		PSO	
	Fitness value	Iterations taken	Fitness value	Iterations taken
5	.5305	29	.4986	154
6	.8521	21	.8834	131
7	.8230	31	.8539	174
8	.9796	35	1.000	150
9	1.0331	35	1.0207	135
10	1.0900	21	1.0971	82

6 Conclusions

This paper proposes an efficient algorithm for classifying datasets. This method can be extensively used for pattern recognition. The MATLAB simulation results show that the proposed algorithm is more accurate as compared to SVM classifiers. Also, the convergence rate of I^2PSC is found to be better than randomly initialized PSO. Further, the software architecture proposed is reasonably easy to implement on a hardware.

References

1. Vapnik, V.N.: Statistical learning theory. John Wiley & Sons, New York (1998)
2. Kennedy, J., Eberhart, R.C.: Particle swarm optimization. In: Proceedings of IEEE International Conference on Neural Networks, Piscataway, NJ, pp. 1942–1948 (1995)
3. Li, Y.-b., Zhang, N., Li, C.-b.: Support Vector Machine forecasting method by chaotic particle swarm optimization and its application. Journal of South University of Technology volume 13 (2009)
4. Sivakumari, S., Praveena Priyadarsini, R., Amudha, P.: Performance evaluation of svm kernels using hybrid pso-svm. ICGST-AIML Journal (2009)
5. Lin, S.-W., Ying, K.-C., Chen, S.-C., Lee, Z.J.: Particle swarm optimization for parameter determination and feature selection of support vector machines. Expert systems with applications 35 (2008)
6. Kapp, M.N., Sabourin, R., Maupin, P.: Proceedings of the 11th Annual Conference on Genetic and Evolutionary Computation (2009)
7. Zahiri, S.-H., Seyedin, S.-A.: Swarm Intelligent based classifiers. Journal of Franklin University (2007)
8. CristiaTrelea, I.: The particle swarm optimization algorithm: convergence analysis and parameter selection. Inf. Process. Lett. 85(6), 317–325 (2003)

Omega Loops of Proteins in *Homo Sapiens*: Role in Diseases

Kuchi Srikeerthana and Patrick De Causmaecker

Department of Computer Science, CODeS Group,
Interdisciplinary Research on Technology Education and Communication (iTec),
Katholieke Universiteit Leuven Campus Kortrijk, Kortrijk, Belgium
{Srikeerthana.Kuchi,Patrick.DeCausmaecker}@kuleuven-kortrijk.be

Abstract. Omega(Ω-) loops are non-regular secondary structures formed by poly-peptide chains, that are characterized by loop-shaped structures in three-dimensional space. Omega loops have been found to involve in protein stability, folding and function. The study on the structural geometry and prediction of Ω-loops have been carried out by many researchers [1], [2], [3], however, analysis on the involvement of these loops in disease-related interactions of *Homo sapiens* is significantly rare. In this study, Ω-loops that are involved in the interactions, causing potential diseases in *Homo sapiens* have been analyzed, at the level of their structures, functions and their involvement in diseases. The information on the understanding of the disease-causing interactions and other properties of Ω-loops could be used in potential drug design against these diseases.

Keywords: Omega loops, non-regular secondary structures, *Homo sapiens*, sequence, structure, function, diseases, drug design.

1 Introduction

Secondary structures of proteins have been classified into regular and non-regular secondary structures; α-helix, β-strand constitutes the former, while loops, random coils which are included in the latter, lack repeating backbone dihedral angles, however have repeating motifs [1]. Omega (Ω-) loops are compact, where the motif's backbone forms a loop(Ω)-shaped structure in three-dimensional space and hence the name.

Ω-loops are often found clustered on the surface of a protein and consists usually of 6-16 amino acids. Residues correlated with reverse turns, glycine and proline and hydrophilic residue, asparagine are found in Ω-loops often. Charged residues, except aspartic acid occur with average frequency, however, hydrophobic residues, such as, phenylalanine, alanine, methionine, leucine, isoleucine and valine are rarely found [1]. Ω-loops play roles in protein functions requiring certain flexibility, such as serving as substrate, inhibitor or as ligand-binding sites. Ω-loops, being a structure, having very few hydrogen bonds, still play an important role in the protein structure and stability, because single and multiple mutations of these motifs have drastic effects on

V.V. Das, J. Stephen, and Y. Chaba (Eds.): CNC 2011, CCIS 142, pp. 362–367, 2011.

overall stability of the protein [1], thus authenticating the role of these loops in stabilizing and destabilizing protein structures. It has also been suggested that these loops play an important role in protein folding.

Previous studies have been focused on the various structural and functional features of Ω-loops, however, there are not enough studies, that have been devised to analyze, the involvement of these loops in disease-causing interactions in humans. Thus, this study focuses on the involvement of Ω-loops in human diseases and provides an initiator step for the detailed study of the disease causing interactions of the Ω-loops in *Homo sapiens*, providing clues for drug design against the diseases caused by mutations of Ω-loops.

2 Methods

The database of all known structures, the PDB [4] was searched using the PDBelite [5] database, for the x-ray crystal structures of Human proteins, that constitutes "Omega loop" and the search yielded a total of 4 non-redundant hits. Structural homologues of these 4 proteins have been identified using the FUGUE algorithm [6] and the secondary structure elements (SSE) of the four proteins and their structural homologues have been aligned and compared using the secondary structure matching server (SSM) [7]. The results have been analyzed for the similarities and uniqueness of Ω-loop of the 4 proteins, with respect to their structural homologues.

3 Results

The keyword search "Omega loop" in PDBelite yielded a list of 4 non-redundant hits that included, Phosphatidylcholine Transfer Protein (1LN1) [8], Human factor IX Gla domain (1NL0) [9], Human neuroserpin (3FGQ) [10] and Peroxisome proliferator activated receptor (3PRG) [11]. The crystal structure literature of the 3 out of 4 proteins confirmed the presence of Ω-loops in their three-dimensional structures and are connected to disease-related mechanisms in *Homo sapiens*. Thus, these proteins have been preferred for further analysis.

Comparison of the domains of a protein, with its structural homologues provides insights into the unique structural patterns of the protein and leads to the identification of the probable regulatory binding sites and helps in drug-design against the disease-causing motifs as targets. In this analysis, an eminent algorithm, FUGUE [6] has been employed for protein sequence-structure homology recognition, to correlate the proteins that have functional Ω-loops with their distant homologues by sequence-structure comparison. To identify similar and to distinguish the unique secondary structure patterns, the 4 proteins have been compared with the secondary structure elements of their corresponding structural homologues using the SSM [7] server.

3.1 1LN1

Phosphatidylcholine Transfer Protein (PC-TP) transfers Phosphatidylcholine (PtdCho) between membranes and belong to START (Steroidogenic Acute Regulatory protein related transfer) domain superfamily of ligand-binding proteins. The crystal structure of PC-TP with its ligand comprises of 2 Ω-loops [8], which are inserted between β5- β6 (Ω1) and β7- β8 (Ω2). Phosphatidylcholine is a major constituent of cell membranes and plays a role in membrane-mediated cell signaling.

```
1ln1:A   hSSSHhsSS-SssH
2pso:A   hSSSHs-SShSssH

1nl0:L   -SShSSSShSsSshssss hss
1hxm:A   sSShSSSS-S-Ssssssss-

3fgq:B   HsHHhSHSHhSHSSSShHSSHHShsss
2ach:A   H-HHhSHSH-SHSSSShHSSHHSs---

3prg:A   HhShHhHSShhHHhH
1skx:A   HsS-HhHSSh-HHhH
```

Fig. 1. Secondary structure matching of the 4 proteins with their structural homologues as aligned by SSM server

FUGUE yielded Human StarD13 (DLC2) lipid transfer and protein localization domain (2PSO) as a profile hit with a Z-score of 26.07. The alignment in SSM shows that the SSE of PC-TP shares a high similarity with the SSE of the DLC2 and hence the reason for their structural homology (Fig. 1). Manual inspection of the superimposed structures shows that, they in fact share similarity in the regions of the Ω-loops (Fig 2a). The presence of Ω-loops in the lipid transfer protein is just a hypothesis in this study, since the crystal paper is yet to get published. DLC2, the lipid transfer protein has been implicated to be a tumor suppressor protein.

3.2 1NL0

Human factor IX, a serine protease of the coagulation system, belongs to peptidase family S1. Deficiency of this protein causes hemophilia B, an inherited disorder in which one of the proteins needed to form blood clots is either missing or reduced. Binding of factor IX to membranes during blood coagulation is mediated by N-terminal Gla (γ-carboxyglutamic acid-rich) domain [9]. Factor IX antibodies are directed at the Ca^{2+} stabilized Gla domain to interfere in the Factor IX-membrane interaction. Such antibody, 10C12 forms a hydrophobic pocket, to accommodate the Gla domain and is directed at the membrane binding site in the ω-loop of Factor IX, that is important for cell surface anchoring and blocks its interaction with membranes [9].

FUGUE yielded gamma-delta t-cell receptor (1HXM) [14] as a profile hit with a Z-score of 33.12. The relative similarity of their secondary structures has been determined using the SSM server (Fig 1), which showed that, they have similarity at their regular secondary structures, however, have considerable differences at loops. The 3D structural alignment of both the proteins showed that the t-cell receptor lacks the ω-loop at the corresponding site of 1NL0, which shows the uniqueness of ω-loop of 1NL0. However, they both share a similar fold and retain Immunoglobulin v-set and c1-set domains, revealed by a Pfam search.

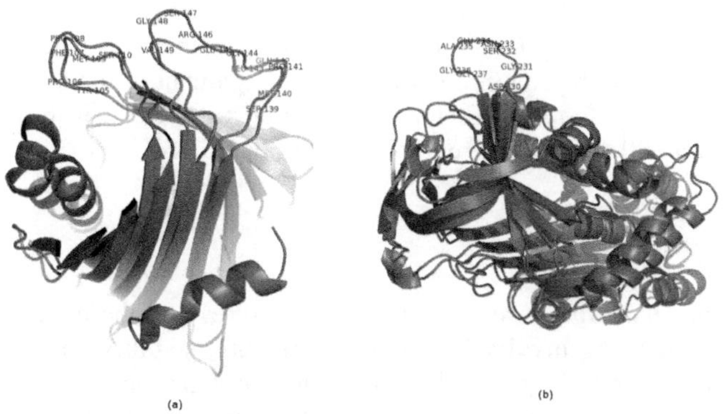

Fig. 2. **(a)** Superimposition of 1LN1 (Shown in Pink) and 2PSO (Shown in Green). The residues that correspond to the formation of Ω-loops have been numbered. Numbering corresponds to 1LN1. **(b)** Superimposition of 3FGQ (Shown in Pink) and 2ACH (Shown in Green). The residues that correspond to the formation of Ω-loop in Human neuroserpin have been numbered. The images have been created with PyMOL [18].

3.3 3FGQ

Human neuroserpin is a selective inhibitor of tissue-type plasminogen activator (tPA) and plays a role in neuronal plasticity, memory and learning. The crystal structure of the human neuroserpin shows an Ω-loop between strands 1B and 2B [10], which contributes to the inhibition of tPA. An Ω-loop deleted mutant of neuroserpin, formed a less covalent complex and had a lower K_a (association constant) with tPA, suggesting the importance of an Ω-loop, for the interaction between neuroserpin and tPA. Deficiency of neuroserpin results in Familial encephalopathy with neuroserpin inclusion bodies (FENIB), dementia and serpinopathy, confirming the role of neuroserpin Ω-loop in human diseases.

FUGUE yielded Human α1-antichymotrypsin (2ACH) [15] as the topmost hit, since they both retain the Serpin domain in a Pfam search. The Z-score was 61.0, and the SSE of 2ACH has been aligned with that of Human neuroserpin, which showed the complete absence of Ω-loop in the α1-antichymotrypsin (Fig 1). This along with the manual inspection of the superimposed structures confirms that, the Ω-loop is a unique structural pattern of Human neuroserpin, which is absent in all the other members of the serpin superfamily (Fig 2b).

3.4 3PRG

Peroxisome proliferator activated receptor (PPAR) are members of the nuclear receptor supergene family and in contrast to other nuclear receptor ligand binding domains (LBDs), the ω-loop is extended in 3PRG and stands between helix 1 and 3 in PPARγ LBD [11]. Researchers have recently found the nature of abnormal lung maturation in the absence of airway epithelial cell PPARγ and identified a putative genetic determinant of dysanapsis [16]. Thus, PPARγ seem to have a disease-related functional omega loop.

FUGUE yielded 1SKX (orphan nuclear receptor pxr) [17] as a profile hit with a Z-score of 28.44, since they both retain the LBD of nuclear hormone receptor domain, revealed by Pfam search. The three dimensional alignment of the superimposed structures show that the ω-loop in PPARγ has the largest deviation from the fold and has an insertion of a +20 amino acid region, with respect to 1SKX.

4 Discussion

Previous studies have discussed the structural features and the geometry of Ω-loops, which are proved to be involved in protein structure, stability and in folding. However, the analysis of Ω-loops in disease-related proteins have not been carried out thoroughly. The comparison of structural features of such proteins with their structural homologues, results in the identification of the similarities and also helps to distinguish the unique structural patterns of the disease related Ω-loops in proteins of *Homo sapiens*. This study, focuses on the involvement of Ω-loops in human diseases and thus provides insights for drug design against these diseases, caused by undesirable interactions and mutations of Ω-loops.

References

1. Fetrow, J.S.: Omega loops: nonregular secondary structures significant in protein function and stability. FASEBJ 9, 708–717 (1995)
2. Maity, H., Rumbley, J.N., Englander, S.W.: Functional Role of a Protein Foldon—An Ω-Loop Foldon Controls the Alkaline Transition in Ferricytochrome c. Proteins: Structure, Function and Bioinformatics 63, 349–355 (2006)
3. Fetrow, J.S., Horner, S.R., Oehrl, W., Schaak, D.L., Boose, T.L., Burton, R.E.: Analysis of the structure and stability of omega loop A replacements in yeast iso- 1 -cytochrome c. Protein Sci., 6197–6210 (1997)
4. Berman, H.M., Westbrook, J., Feng, Z., Gilliland, G., Bhat, T.N., Weissig, H., Shindyalov, I.N., Bourne, P.E.: The Protein Data Bank. Nucl. Aci. Res. 28, 235–242 (2000)
5. Velankar, S., Best, C., Beuth, B., Boutselakis, C.H., Cobley, N., Sousa Da Silva, A.W., Dimitropoulos, D., Golovin, A., Hirshberg, M., John, M., Krissinel, E.B., Newman, R., Oldfield, T., Pajon, A., Penkett, C.J., Pineda-Castillo, J., Sahni, G., Sen, S., Slowley, R., Suarez-Uruena, A., Swaminathan, J., van Ginkel, G., Vranken, W.F., Henrick, K., Kleywegt, G.J.: PDBe: Protein Data Bank in Europe. Nucl. Aci. Res. 38, D308–D317 (2010)
6. Shi, J., Blundell, T.L., Mizuguchi, K.: Fugue: sequence-structure homology recognition using environment-specific substitution tables and structure-dependent gap penalties. J. Mol. Biol. 310, 243–257 (2001)
7. Krissinel, E., Henrick, K.: Secondary-structure matching (SSM), a new tool for fast protein structure alignment in three dimensions. Acta Cryst. 60, 2256–2268 (2004)
8. Roderick, S.L., Chan, W.W., Agate, D.S., Olsen, L.R., Vetting, M.W., Rajashankar, K.R., Cohen, D.E.: Structure of human phosphatidylcholine transfer protein in complex with its ligand. Nat. Struct. Biol. 9, 507–511 (2002)
9. Huang, M., Furie, B.C., Furie, B.: Crystal structure of the calcium-stabilized human factor IX Gla domain bound to a conformation-specific anti-factor IX antibody. J. Biol. Chem. 279, 14338–14346 (2004)

10. Takehara, S., Onda, M., Zhang, J., Nishiyama, M., Yang, X., Mikami, B., Lomas, D.A.: The 2.1-Å crystal structure of native Neuroserpin reveals unique structural elements that contribute to conformational instability. J. Mol. Biol. 388, 11–20 (2009)
11. Uppenberg, J., Svensson, C., Jaki, M., Bertilsson, G., Jendeberg, L., Berkenstam, A.: Crystal structure of the ligand binding domain of the human nuclear receptor PPARgamma. J. Biol. Chem. 273, 31108–31112 (1998)
12. Lehtio, L., Busam, R., Arrowsmith, C.H., Collins, R., Dahlgren, L.G., Edwards, A., Flodin, S., Flores, A., Graslund, S., Hammarstrom, M., Hallberg, B.M., Holmberg-Schiavone, L., Johansson, I., Kallas, A., Karlberg, T., Kotenyova, T., Moche, M., Nordlund, P., Nyman, T., Ogg, D., Sagemark, J., Stenmark, P., Sundstrom, M., Thorsell, A.G., Van den Berg, S., Weigelt, J., Welin, M., Persson, C.: The crystal structure of human StarD13 (DLC2) lipid transfer and protein localization domain
13. Finn, R.D., Mistry, J., Tate, J., Coggill, P., Heger, A., Pollington, J.E., Gavin, O.L., Gunesekaran, P., Ceric, G., Forslund, K., Holm, L., Sonnhammer, E.L., Eddy, S.R., Bateman, A.: The Pfam protein families database. Nucl. Aci. Res. 38, D211–D222 (2010)
14. Allison, T.J., Winter, C.C., Fournié, J.J., Bonneville, M., Garboczi, D.N.: Structure of a human gammadelta T-cell antigen receptor. Nature 411, 820–824 (2001)
15. Baumann, U., Huber, R., Bode, W., Grosse, D., Lesjak, M., Laurell, C.B.: Crystal structure of cleaved human alpha 1-antichymotrypsin at 2.7 A resolution and its comparison with other serpins. J. Mol. Biol. 218, 595–606 (1991)
16. Simon, D.M., Tsai, L.W., Ingenito, E.P., Starcher, B.C., Mariani, T.J.: PPARγ deficiency results in reduced lung elastic recoil and abnormalities in airspace distribution. Respiratory Research 11, 69 (2010)
17. Chrencik, J.E., Orans, J.O., Moore, L.B., Xue, Y., Peng, L., Collins, J.L., Wisely, G.B., Lambert, M.H., Kliewer, S.A., Redinbo, M.R.: Structural disorder in the complex of human pregnane X receptor and the macrolide antibiotic rifampicin. Mol. Endocrinol. 19, 1125–1134 (2005)
18. DeLano, W.L.: The PyMOL Molecular Graphics System DeLano Scientific, Palo Alto, CA, USA, `http://www.pymol.org`

Voice Record on RISE (Radio Broadcasting Integrated System) on Air Application

Meidy Fangelene De Jong[1] and Agus Pratondo[2]

[1] Department of Information Technology
Telkom Polytechnic
Bandung, Indonesia
meidy.fangelene@gmail.com
[2] Department of Information Technology
Telkom Polytechnic
Bandung, Indonesia
agus@politekniktelkom.ac.id

Abstract. The real effect of telecommunication and information technology development in broadcasting, such as radio communication sector. Normally, radio able to serve all kind of information for all ages without need a lot of money to enjoy it. Nowadays many radio stations are developing an integrated system with one management control. Hopefully this technique can increase RADEX (radio cost commercial) that record RADEX and ADEX (media cost commercial) in Indonesia does not balance. As long as the system that was applicable better, there are many domestics station radio join to be the member of this integrated system. Many advertisement owner trusts to publish the product on radio. Every important event in broadcast such as talk shows or conferences will be record as data for the station radio and hope it can be reusable (i.e. broadcasted again) in other opportunity, of course with the permission from the advertisement owner. That is an important work system for all radio station is broadcasting time. For this reason, come along an idea to change the record process. Need a tool to record all data for station achieve like voice record.

Keywords: Technology, Broadcast, Radio, Voice record.

1 Introduction

RISE (Radio Broadcasting Integrated System) is an integrated application technology for broadcasting based on Internet. RISE adopts some popular radio technology like Google Radio Automation and Pandora that was service for radio and music on-line in United State. RISE (Radio Broadcasting Integrated System) is an innovation technology and business solution for radio stations. Radio reach, time spent listening, podcasting, operational & broadcast management, advertise monitoring, and broadcast

V.V. Das, J. Stephen, and Y. Chaba (Eds.): CNC 2011, CCIS 142, pp. 368–372, 2011.

monitoring will be fixed systematically used coexistence and convergence on RISE technology.

Just as for call centers or call with banks, station radio events need to be recorded in order to give the possibility to listen again at later time [1]. Broadcast activity recording is one of the constraint for some radio station today, broadcaster just can dating all broadcast activity by hand or manually. Voice record is a choose idea to fill this miss, coming in RISE (Radio Broadcasting Integrated System) on air application system to saving all broadcast recording. It means that RISE (Radio Broadcasting Integrated System) on air application need to support broadcaster work flow. Voice Record on RISE (Radio Broadcasting Integrated System) On Air Application is related to the Rise (Radio Broadcasting Integrated System), the overall large scale, so that the developers are divided into several modules. In this module, further discussion will lead to the development of innovative tools for the application of voice record RISE.

2 Design and Implementation

A. Design

User is a broadcaster that responsible to manage this application. When user start the application all voice will be listening by speaker on desktop computer as analog signal. The signal transform into digital signal and saving automatically on internal hard disk as mp3 format and the data recording will be saving in central database per table. That's the main work flow of voice record on RISE (Radio Broadcasting Integrated System) on air application.

Additional feature on RISE (Radio Broadcasting Integrated System) on air application is "setting". In "setting", the broadcaster can change the sound quality, (e.g. bit rate, record's title, record's description, etc).

1) System Design

UML (Unified Modeling Language) is a standard notation for the modeling of real-world objects as a first step in developing an object-oriented design methodology [2]. There are four diagrams that used to describe system design of voice record on RISE (Radio Broadcasting Integrated System) application on air. As followed:

a) Use Case Diagram

A use case diagram is "a diagram that shows the relationships among actors and use cases within a system." Use case diagrams are often used to:

- Provide an overview of all or part of the usage requirements for a system or organization in the form of an essential model or a business model
- Communicate the scope of a development project
- Model your analysis of your usage requirements in the form of a system use case model [3].

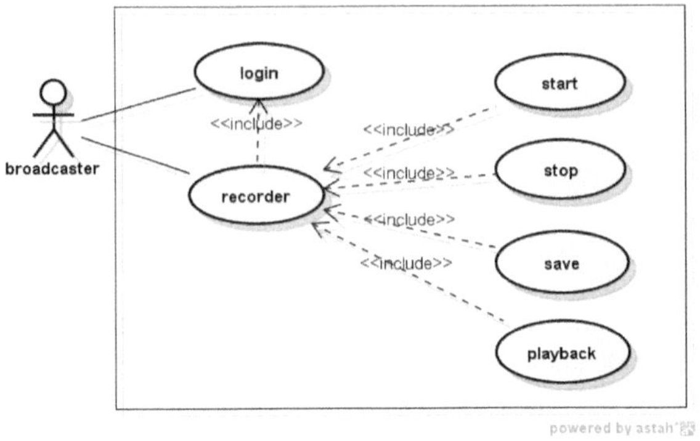

Fig. 1. Use case diagram for voice record application

b) Activity Diagram

Activity diagrams describe the workflow behavior of a system. Activity diagrams are similar to state diagrams because activities are the state of doing something. The diagrams describe the state of activities by showing the sequence of activities performed. Activity diagrams can show activities that are conditional or parallel [4].

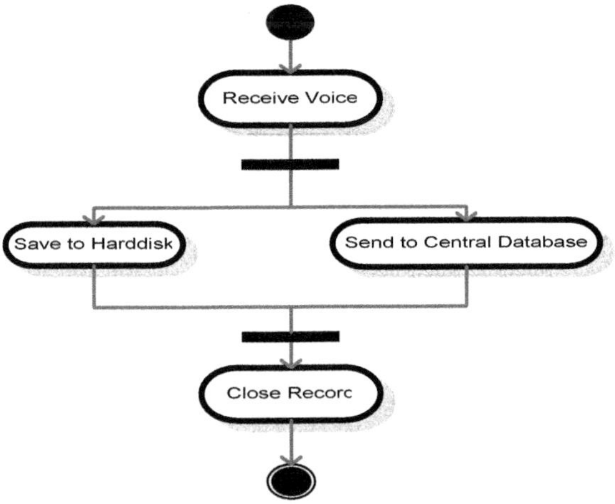

Fig. 2. Activity diagram for voice record application

c) Sequence Diagram

Sequence diagrams model the flow of logic within your system in a visual manner, enabling you both to document and validate your logic, and are commonly used for both analysis and design purposes. Sequence diagrams focuses on identifying the behavior within your system [5].

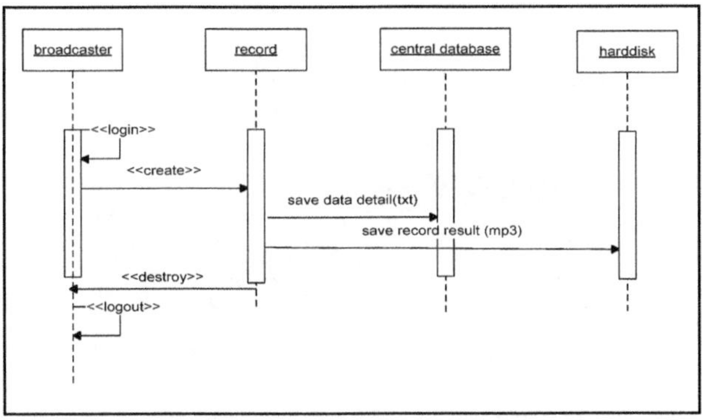

Fig. 3. Sequence diagram for voice record application

First, the broadcaster as actor need login to create a record a show. All record data detail will be saving in text format in central database. Beside, mp3 format save in internal computer harddisk. To exit all operation the actor need to logout. That the way to read sequence diagram for voice record application.

d) Class Diagram

Class diagram is a diagram showing a collection of classes and interfaces, along with the collaborations and relationships among classes and interfaces. A class diagram consists of a group of classes and interfaces reflecting important entities of the business domain of the system being modeled, and the relationships between these classes and interfaces [6].

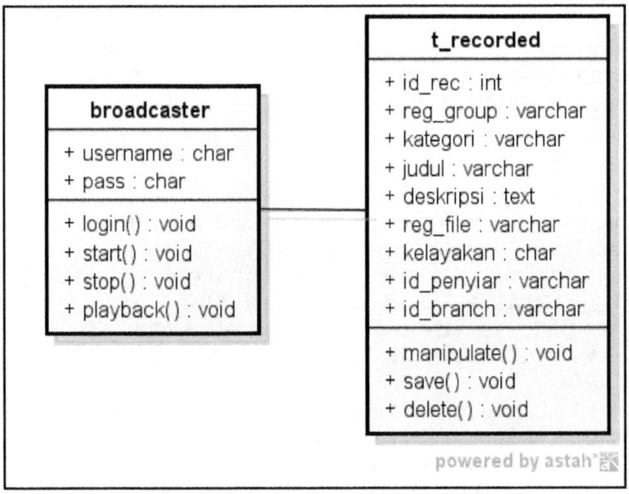

Fig. 4. Class diagram for voice record application

2) Implementation

Generally, this application can be implemented for some society for recording activity like, 1) Station radio: can record events as achieves; 2) Organization: can record

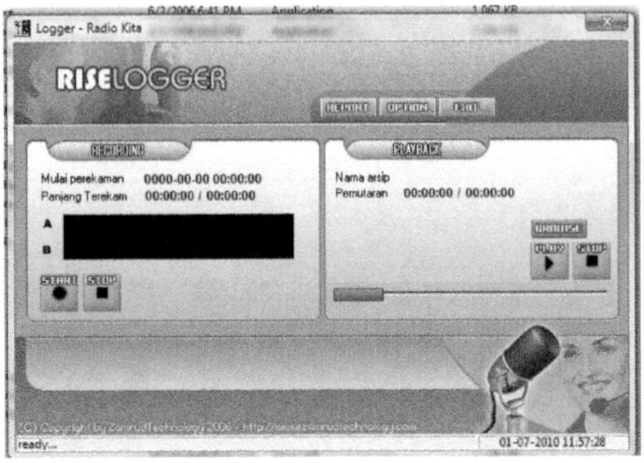

Fig. 5. Example of voice record application

meeting result to help the writer; 3) Personal: to help people record all important things as data in internal computer harddisk.

3 Closing

3.1 Conclusion

1) One integrated system management in industry radio can be the way out to increase Indonesia revenues back.
2) Voice record can be useful to record all data and it's possible to reuse the data again.

3.2 Suggestion

1) For safety reason, better to back up the data every six month to avoid data crash.
2) Next development can use another programming language to knowledge explore special about sound manipulate.

References

[1] Siemens enterprise communications.Voice recording with SIP end devices (2007)
[2] Shnitman, A.: Unified modeling language (2007), `http://searchsoftwarequality.techtarget.com/sDefinition/0,,sid92_gci214158,00.html` (August 9, 2010)
[3] Ambler, S.W.: UML 2 use case diagramming guidelines (2002-2009), `http://www.agilemodeling.com/style/useCaseDiagram.htm` (August 9, 2010)
[4] Braun, D., Sivils, J., Shapiro, A., Versteegh, J.: Activity diagrams (2000), `http://atlas.kennesaw.edu/~dbraun/csis4650/A&D/UML_tutorial/activity.htm` (August 9, 2010)
[5] Ambler, S.W.: UML 2 sequence diagrams (2002-2009), `http://www.agilemodeling.com/artifacts/sequenceDiagram.htm` (August 9, 2010)
[6] Ravikitran: The UML class diagram: part 1 (2009), `http://www.developer.com/design/article.php/2206791/The-UML-Class-Diagram-Part-1.htm` (August 10, 2010)
[7] Malik, J.J.: Tip & trik unik Delphi lanjutan. Andi Yokyakarta (2006)

Design of Hybrid Billboard with Scrolling Static Image and Remotely Updatable Matrix Display

Arko Djajadi and Gary Gregorius Gunarman

Mechatronics Department, Faculty of Engineering
Swiss German University
EduTown BSD City, Tangerang 15321, Indonesia
arko@sgu.ac.id

Abstract. This paper is concerned with design of a hybrid billboard with scrolling printed images and remotely updatable light emitting diodes matrix display for advertisement purposes. The images showing the advertisements are printed on a backlit sheet that is rolled on both left and right ends of the sheet. Thus, rolling a sheet on one direction allows translational moves of the image that is being shown to the viewer. The system is combined with the LED matrix display that shows the message of the products that is currently being displayed in the rotating image part. This message could be modified wirelessly anytime in a very short time by using a user interface program that is operated from distance in the office of the advertisement company.

Keywords: billboard, hybrid, wireless, updatable.

1 Introduction

Every trading process needs advertising as a way for a company to introduce and promote their products or services. There are many kinds of advertisements in various media, and one is billboard advertisement. However, billboard currently exists in various display systems, for example traditional vinyl sheet, tri-vision mechanical billboard, scrolling billboard, and various digital billboards that diverge from Light Emitting Diode to LCD display. Each has its own advantage and disadvantage compared to the others. Digital billboards could be updated in a short time but it is very expensive to be made. On the other hand, traditional and scrolling billboards are relatively cheaper but they are not easily updateable. In this paper, the advantages of both billboard with scrolling images and digital display with wirelessly updatable running messages are combined into one flexible advertisement system. Overall design and implementation of such a system is fundamentally a combination of mechanical, electronic, and computer system including telecommunication. Such approach is commonly known as Mechatronical approach [1].

2 System Design

The overall design is a combination of two available billboard types, scrolling billboard and remotely updatable LED matrix display. The scrolling billboard shows a

V.V. Das, J. Stephen, and Y. Chaba (Eds.): CNC 2011, CCIS 142, pp. 373–379, 2011.

static image and its static messages of a product or service. The image displayed in the billboard is frozen for several seconds and then scrolled into other image from other product or service. In total, it could show up to eight different images displayed in one advertisement system. Figure 1 illustrates the system.

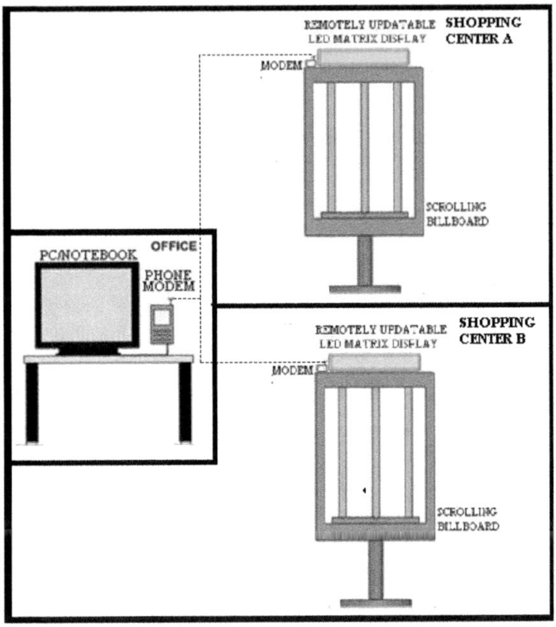

Fig. 1. System Design Overview

In addition, the running text in the remotely updatable LED matrix display is synchronized with the display in the scrolling billboard. It means that when the scrolling billboard advertises product A, the LED matrix display will change the text to run the promotional words of product A. The text in the LED matrix display can be changed easily and immediately from a long distance, i.e from the office of the advertising company or from a mobile console through a user interface software.

The same information can be distributed to similar billboard located anywhere along the high traffic roads or in and around shopping centers. All of them can be easily managed from a management center. Overall, the efficiency in term of time, money, and energy is improved.

2.1 Scrolling Billboard Mechanism

Principally the sheet would be rolled in two sides of the billboard, left side and right side. Each side consists of two shafts; shaft a from Figure 2 is connected to a motor which will rotate the shaft, and the other shaft, shaft b, follows the rotation of the sheets to make the sheet stretches and therefore resulting in a non-wavy display. Each shaft end is also connected to a bearing to hold the position of the shaft and at the same time allows the shaft to rotate freely. The shaft is made from steel and was

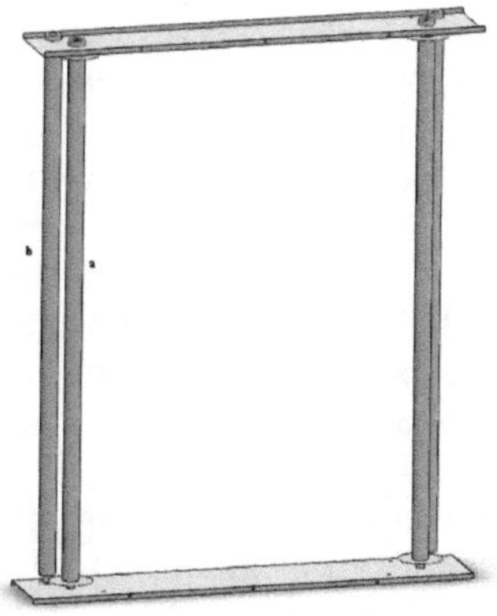

Fig. 2. Design of scrolling billboard mechanism, rear view

covered with a soft surface material pipe to protect the sheet form scratches. There are also four plastic disks to hold the roll, as shown in Figure 2.

2.2 Electrical System Block Diagram

Figure 3 is the overall electrical system and its interface with the remote computer and its local scrolling mechanism. Initially, the user interface software in the laptop will send an AT command that instructs a phone modem to send a message. The message will be sent as a Short Messaging Service (SMS) similar with GSM-to-GSM SMS sending. It will be received by the other modem in the billboard system. The received message then will be forwarded to the motor microcontrollers via serial connection. Afterwards, the motor microcontroller will translate the message and save it inside the RAM of the microcontroller. Up to 8 messages can be stored in this RAM, one for each scrolling image.

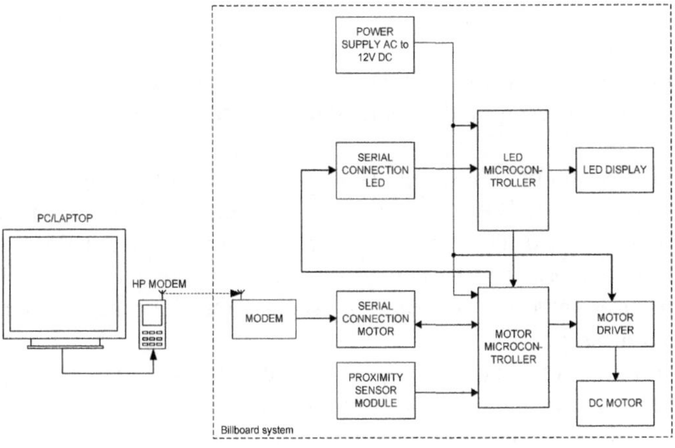

Fig. 3. Electrical system block diagram

Assume that static image of product A is displayed on the billboard for a few seconds. During this time, the LED matrix display will also show the message of product A. After a predetermined time, the scrolling billboard will display the image of next product B. At the same time, message B is retrieved from the memory and then used to update the message on the LED matrix display accordingly to match the image on the scrolling billboard. The rest of the process is similar.

For the motor mechanism for the scrolling billboard, the motor driver will turn the motor on or off depending on the built-in sensor for image border colored black. The motor will keep turning until the proximity sensor detects a black bar. If so the motor will stop for a certain predetermined time.

AC to DC power supply is connected to microcontrollers and motor driver while other devices take power supply from microcontrollers which have internal power supply regulator that convert 12V DC to 5V DC.

2.3 Programming Design

The programming design of the billboard consists of three main parts: (1) the user interface software programming within a laptop, (2) the motor microcontroller programming [5], and (3) the LED display microcontroller programming.

2.3.1 User Interface Software Programming Design

The parameters to be considered are for example the location of billboard, its address, customer name, customer product and message, customer product position in the scrolling billboard and other control parameters. The user interface program is the software that is used by the operators of the advertising company to manage the display of the advertisement billboard especially the LED Matrix display part. From the software, the operators can add a new customer or product, change the message of a customer, change the billboard locations of a customer, add a new location, change location's GSM number, and change the arrangement of an advertisement board. It consists of six menus for each of those tasks; add new customer, edit message, customer's location, add location, edit location, and position arrangement.

The user interface software was written using four different programming languages. The first programming language is ActionScript 2.0[3] that is used in Adobe Flash software. By using Flash software, the user interface software becomes more attractive, live, and colorful, adding a modern and professional look for the software itself.

The second programming language is Visual Basic [2] language. In this user interface software, the Visual Basic programming language is like the bridge for all the languages, the center and the brain of the program itself. Visual Basic is usually intended to modify database and Visual Basic is easily connected to many hardware and software. This user interface software is used to edit the database and instruct the modem to send message.

The third programming language is Structured Query Language (SQL)[4] that is used to add, edit, and delete an item of the database and gather information from the database. The SQL is executed from the Visual Basic in this case.

The last programming language is the Attention (AT) Command language that is used by the GSM modem to do task like calling or sending SMS. In this software, the AT Command is used to send a message and it is also executed from the Visual Basic.

2.3.2 Motor Microcontroller Programming Design

The motor microcontroller programming principally controls the scrolling displays sheet movement by turning either motor on or off. When the sensor gives input to the microcontroller that a black bar is detected, the program will instruct the running motor to stop. Then, it will delay to let the display freeze for several seconds and afterwards check whether the black bar is short or long. If the black bar is short, the same motor will be turned on, resulting on the same direction of scrolling displays sheet movement. Inversely, if the black bar is long, which indicating the end of the sheet either the most right or most left display, the other motor will be turned on so the sheet will move in the reversed direction.

When the display is stopped, the program in this motor microcontroller will check if a new message is available in the modem or not by sending a read SMS AT command to the modem. Then, it will wait for an interrupt from the modem as a reply of the read SMS command. If a message is available, then the content of the message will be stored temporarily to a register and then copied to an EEPROM depending on the display order number of the message. For instance, if product A is in the third order of the display, the message from product A will be saved in the third page of EEPROM. Afterwards, the message will be deleted from the modem by sending an erase message AT command to the modem. The message needs to be erased to keep the modem away from full memory error. At last, whether a new message is available or not, the message of the current display will be sent from EEPROM to the serial connection from the motor microcontroller to the LED display system.

2.3.3 LED Display Programming Design

The LED display programming controls the message of the current display. When an interrupt occurs from the motor microcontroller serial connection to the LED microcontroller, the old message that is stored in the EEPROM of the LED Matrix Display system is replaced with the new message just received. Every character in the EEPROM is then translated into six bytes data and the results are copied to the RAM to make the reading faster due to simpler instruction set to read from RAM rather than to read from EEPROM. The six bytes represent the five columns available for each character that will be displayed in 7x5 LEDs matrix and one byte represents a space. From the RAM, the bytes are inserted one by one to the column shift register until all the columns are filled. Then, the row to be displayed is turned on to let the LEDs emerge light. The process will be repeated several times for different row and RAM pointer, which points to the starting byte to be displayed, to display a running text display.

3 System Test

The whole system test is performed by sending a new message of a particular customer, Test 1, from a notebook as shown in figure 4.

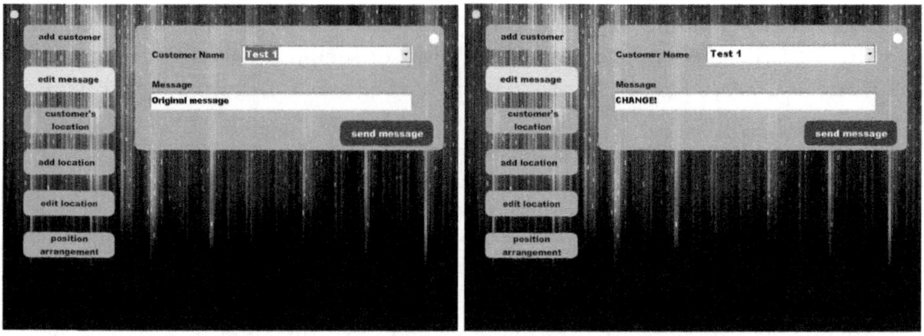

Fig. 4. A new message is changed from a notebook

As a result, the billboard is then showing the new message for the first customer as shown in figure 5.

Fig. 5. New message (right) replacing the old message (left) in the billboard after it is changed from a notebook

4 Results

The system is working properly and run as it was design to. When the operators add a customer in the user interface program, the new customer will be added to the database. If a location is assigned by the new customer, the LED display will directly be modified by adding the message of the new customer in the last position of the LED display's texts. When a message of a specified customer is changed by the operators, the new message will replace the old message in the database and the LED display's message of the customer will be modified. If a customer is deleted, the record of the customer will be deleted in the database and if a location contains the deleted customer, the LED display will remove the message of the customer and rearrange the text. If a location is added to or deleted from a customer, the database will be changed and the LED display will either add or remove the message of the customer. If a location is added, it will be stored in the database and the LED display of the new location will be filled with the messages of the assigned customers. If a location GSM number is modified, the old number in the database will be replaced with the new one. When a location is being deleted, the records in the database that contains the deleted location will be deleted. Last, when the position of a certain location is rearranged, the new arrangement will be saved in the database and the LED display will adapt the changes.

5 Future Development

The physical appearance could be varied to various forms, for example could be placed on the wall or could be resized to meet the customer's requirements. From the electrical site, another buttons could be added so the viewers could change the advertisement manually. For example, if they want to look through all the advertisements without waiting all the delay times, they can hit the next button until the last advertisement or previous button until the first advertisement. For the LED display, it could be modified to tricolor LED so it can display multicolor text. The LED display could also be enriched to perform symbols that are not listed yet. Meanwhile, the user interface software could be modified by adding shortcuts to make the process faster and menu bar could also be added as an option. Another improvement could be a feedback from the modem if a message has successfully been received. Therefore, there will be no missing message. For the real application, there will be more than one computer. Hence, the database should be made accessible from multicomputer. Larger bulletin-sized billboard could also be made by first doing a research to prevent failures due to temperature, rain, and wind factors. Other modification from the programming site also could be made by controlling the stop time for each advertisement. A billboard that has more customers of course would run faster than the billboard that has fewer customers. Therefore, it would be very nice if the stop time could be controlled. Other system features like restarting the microcontroller could also be added to make the system easier to be controlled. The passkey for the SMS that carries the new message, which in this case 'Z' character, could also be modified to more than one character in order to reduce the chance that the message is changed by an unknown party. Longer message and more customers could also be added. Other improvement also could be a feedback that is sent by the microcontroller through the modem when an error occurs. Therefore, the operators could know if the system is having a trouble.

References

[1] Alciatore, D.G., Histand, M.B.: Introduction to Mechatronics and Measurement Systems. McGraw Hill, Colorado (2007)

[2] Brown, S.: Visual Basic 6 Complete. Sybex, California (1999)

[3] Francis, C., Haan, J.D., Dickson, R.L., Rahim, S.: ActionScript 2.0 Language Reference for Macromedia Flash 8. Macromedia Press, California (2005)

[4] Du Bois, P.: MySQL, 4th edn. Pearson Education, New Jersey (2009)

[5] Mazidi, M.A., GillispieMazidi, J., Mckinlay, R.D.: The 8051 Microcontroller and Embedded Systems Using Assembly and C, 2nd edn. Pearson Education, New Jersey (2006)

Determining Factors of Online Auction Prices Analysis with Data Mining Algorithms on the Basis of Decision Trees

Richard Lackes, Chris Börgermann, and Erik Frank

Department of Business Information Management
Technische Universität Dortmund
Dortmund, Germany
richard.lackes@udo.edu

Abstract. The impressive scale of online auctions sales and the economic ecosystem of private and commercial providers, made eBay & Co. interesting for science. As part of this work determinants will be identified which affect the outcome of online auctions to a large extent. We observed that for identical (new) products often different prices were charged and that the price volatility is bound to the auction characteristics. By evaluating a large number of online auctions using the decision tree method we derive significant rules that determine the auction success. This gives the market participants, but also the platform operator, additional insights they can take into account with regard to their market activities.

Keywords: online auction, price determinants, internet auction, data mining, price prediction, decision tree.

1 Introduction

Online auctions have a revenue share of 25% of global e-commerce [1]. eBay, with over 200 million members is the largest auction site before Yahoo! Auctions and Amazon Market Place [2]. The annual sales volume of goods on the eBay platform is 50 billion [3]. In particular, the many professional traders, who are responsible for about 50% of sales, have contributed to the rapid growth of eBay. The attractiveness of the platform for such Powersellers is due to (a) the simplified offering of goods on the Internet, (2) the secured payment through PayPal, and finally (3) the enormous customer potential. While independent online shops need to consider the technical infrastructure, marketing and management in order to sell products, for eBay Powersellers it is only crucial to make the auctions attractive to buyers and maximize the revenue. The principles identified by us will support sellers with this task. Yet, the identified interrelations of the auction-relevant data and the final auction-price can also be used by buyers to identify interesting products at an early stage and then adjust the bids.

V.V. Das, J. Stephen, and Y. Chaba (Eds.): CNC 2011, CCIS 142, pp. 380–390, 2011.

2 Online Auctions and Auction Success

Following the example of an eBay auction, the general auction process can be modeled as follows. A supplier provides a product to be auctioned with a textual and visual product description. The duration of the auction and the auction end as well as the starting price $p_{start} \geq 1$ and any shipping costs are setup by the seller. Details of the advertiser, such as his nickname, his address and his reputation can be derived from previous market activity. This information is accessible by all market participants. After the auction started interested parties can submit bids for the product. The current auction price at the time t $p_t$which would be achieved if the auction would end immediately (i.e. no more bids would be taken into account), is continuously published. Another bid will be accepted if its maximum price $p_H \geq p_t + \Delta p$. The minimum increment Δp is stipulated by the platform operator and amounts to 0.5 for lower and 1 for higher prices. The auction procedure corresponds to an English auction, for which the final price is calculated based as the maximum price of the second highest bidder plus Δp [4]. Immediately after an accepted bid during the auction period the current price $p_{t+1}$is updated. It has to be taken into account that although the current price of the product is not disclosed, the maximum price of the current highest bidder p_m is not published. Thus we obtain as a new price $p_{t+1} = p_H + \min\{\Delta p, p_H - p_M\}$, if the new bid $p_H$is higher than the previous p_M. Otherwise, $p_{t+1} = p_H + \min\{\Delta p, p_H - p_M\}$

Empirically, it can be concluded that most of the activities or bids happen immediately before the auction ends, often not until the last seconds. The reason is that this behavior conceals the bidder's maximum price because of the limited time remaining, so that no competing bidders can successively approximate this value. This phenomenon suggests that the definition of the auction end could be relevant for the auction success. A successful auction, requires the existence of potentially interested parties and that the respective offer is found by all of these. In this respect, any measures to improve the search result listings will be significant. In this regard we have to distinguish between the crowd of buyers with the product in the results of their search queries and those that actually visit the auction website itself. In case the offer is sufficiently attractive people that are interested will add it to the watch list or provide a bid already. This is certainly dependant of the current auction conditions, but also of the tactical considerations of the bidders (e.g. placing bids just before the auction ends). The products in the watch list can be found later without a new search. In addition an imminent end of an observed auction will be automatically indicated as well, in order for the potentially interested parties to compass the current conditions and place a bid shortly before the auction ends (see figure 1).

Therefore, in regard to the auction success the following points are essential:

- The auction should be found by as many potential buyers as possible,
- and the actual auction website should then be visited.
- This website should be as attractive as possible, so that the auction will be added to the favorites or even receive a bid.
- Furthermore, one should ensure that as many last minute bidders participate (i.e. those bidders that place a bid within the last few minutes of an auction). This is determined by the ending time of the auction.

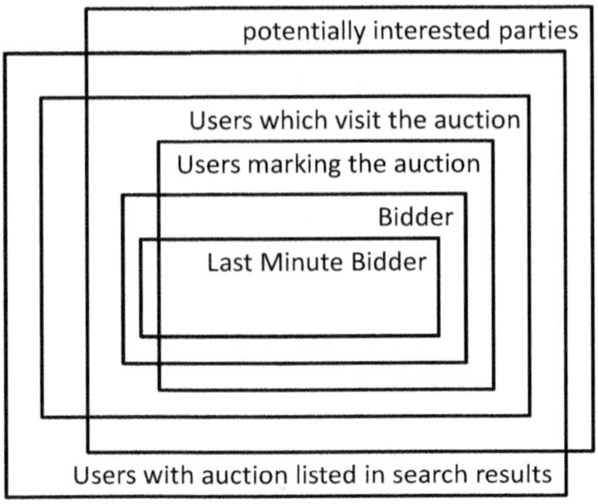

Fig. 1. eBay user classes

Considering the design of auctions, it is crucial that an auction be found in the 1086 categories and in competition with the 800.000 other auctions per day by other eBay members [5]. If a potential buyer uses the search function to find articles, it is important to choose the auction title wisely, so that the auction gets listed in as many search queries as possible. And both in the search results and in the category view, it is in turn important to set apart the auction from others. In most cases, an appealing picture of product to be auctioned in combination with the headline is essential. However, such a picture has to be paid for. The seller can pay for further extra options in order to market an auction e.g. by a more prominent placement in search results regardless of the activated categories and search filter. In this work we will be limited to an examination of the usual design options. After an auction has gained the attention of the bidders, only the direct auction characteristics largely determine the final bid price. In this regard, eBay gives the following advice to sellers [6]:

- A good description factually informs, is short but complete in regard to the goods to be auctioned. It is important that the key search terms are included, used by potential buyers interested in such goods.
- The end of the auction must suit the time, at which the relevant target group visits eBay. Auctions which are aimed at men e.g. should preferably end in the early evening hours. In contrast, items that appeal to femals, are usually successful in the morning and on weekdays.
- The starting price of an auction should be determined depending on the individual price range and the expected demand. If an article starts at one euro, then the auction for the bidder seems attractive. Yet, under certain circumstances, the necessary revenue can not be achieved. Starting prices that are too high can result in an auction, that ends entirely without any bid, because no one had placed a bid. Still in general, products with a high demand should be sold at a low starting price. On average, the revenue is higher than sold at a fixed price or a higher starting price [7].

In regard to the product image, the auction duration, shipping costs and the seller's reputation eBay does not provide any tips. However, it is obvious that these factors also influence the final price of an auction. In particular, the reputation of a seller determines whether a buyer is willing to bid and at which price [8]. However, the reputation has an asymmetric influence on the auction. Thus, a positive rating, depending on the individual preferences of the buyer is hardly promotional. Conversely, negative feedback has a much bigger effect on the perception of an auction [9]. However, the influence of a negative reputation will be diminished by a detailed product description. The latter will strengthen the confidence of the buyer in the supplier [10]. For our research we will consider the following influencing factors:

- Product-specific factors (P)
- Presentation-specific factors (T)
- Seller-specific factors (V)
- Auction procedure-specific factors (A).

3 Dataset

In this study we have evaluated 994 eBay auctions in the period from 11/2009 to 6 / 2010 in order to identify the auction characteristics that influence the revenue. Based upon these data interrelations between the attribute values and the auction revenue have been detected. In order to limit the influence of product-specific factors, which are mostly irrelevant in regard to auction design, we have decided to take a look at a common standard product. A product of such kind is the Levi's 501. The jeans are manufactured without change since 1872, and accordingly the price over a longer period of several months is more or less stable [11]. In Contrast i.e. electronic products are subject to a rapid technological progress, so that a significant fall in prices can be observed over a short period. Furthermore, our target, the identification of possible universal rules for the design of an auction, requires the choice of a product without a limited target group with specific habits or preferences. A circumstance which has not been taken into account in other works i.e. with the choice of silver and gold coins or Pokémon toys [12] [13]. Another problematic product-specific attribute is the quality, which is changed by aging and usage. Although a differentiation of "new" and "used" would have been possible, the class of "used" products is too heterogeneous (ranging from "worn once" to "completely worn out") in order to derive good results. Therefore, we considered only those auctions, which contained new goods (a new product with label). For the specific data mining analysis the following attributes as determinants of the output attribute "price class" have been collected (in brackets the assignment to the categories that have been defined above):

- Modernity as the only product-specific attribute (P)
- Number of pictures (T)
- Number of ratings (V)
- Ratio of positive ratings (V)
- Duration of the auction (A)
- Auction end day(A)
- Auction end time (A)
- Shipping costs (A)
- Starting price (A)

The price has been used as the classification attribute. The evaluated data were pre-coded. In regard to the modernity it was necessary to classify the article with respect to fashionable aspects. Based on the color, the popular colors "black" and "blue" were classified as "fashionable, the others as "not fashionable". For the presentation-specific attribute "number of pictures" three classes were formed, namely (1) no image, (2) exactly one image, and (3) multiple images with the following breakdown.

Amount of photos	Distribution
None	0.91 %
One	46.27 %
Some	52.81 %

Fig. 1. Distribution of the photos

The distribution of ratings within the three classes (1) few, (2) average and (3) many has been carried out by appointing the 0.33-quantile and 0.66-quantile as respective class boundaries.

Ratings	From	Until	Distribution
Few	0	81	33.33 %
Average	82	253	33.33 %
Many	254	5520	33.33 %

Fig. 2. Distribution of rating amount

For the seller rating the same classes were formed, but with different class boundaries. Here, the 0.05-quantile and the 0.12-quantile were defined as class boundaries.

Reputation	From	Until	Distribution
Bad	0	97.8	5.00 %
Average	97.9	99.0	12.00 %
Good	99.0	100	83.00 %

Fig. 3. Distribution of reputation

Duration	From	Until	Distribution
Short	1	5	21.43 %
Long	7	10	78.57 %

Fig. 4. Distribution of duration

The auction duration has been divided in (1) short and (2) long and each auction over seven days has been rated as long.

For both the ending day and the ending time of an auction four classes were formed: (1) weekday, (2) Friday, (3) Saturday and (4) Sunday or holiday, and (1) in the morning, (2) at noon, (3) in the evening and (4) at night. Accordingly, the following distributions subject to the defined boundaries resulted.

Auction end day	Distribution
Weekday (Monday - Thursday)	28.03 %
Friday	3.16 %
Saturday	7.21 %
Sunday[a]	48.79%

a. includes holidays

Fig. 5. Distribution of the auction end day

Auction end time	From	Until	Distribution
In the morning	08701	12:00	3.82 %
At noon	12:01	17:00	26.76 %
In the evening	17:01	23:29	66.19 %
At night	23:31	07:00	3.23 %

Fig. 6. Distribution of the auction end time

At last, the shipping costs were divided into two classes, depending on the common national shipping fees of 4.90 €.

Shipping costs	From	Until	Distribution
Suitable	0.00 €	4.90 €	65.00 %
High	4.91 €	12:50 €	35.00 %

Fig. 7. Distribution of shipping costs

In regard to the classification attribute "price" we distinguished between a below-average or above-average price. The average price for a 501 jeans based on the collected eBay auction data was 31.81 €. The auction end price ranged from 1.00 € to 68.70 €.

Auction end price	From	Until	Distribution
Under average	1.00 €	31.81 €	46.28 %
Above average	31.82 €	68.70 €	53.72 %

Fig. 8. Distribution of the auction end price

Auction end price	Distribution
1 Euro	46.28 %
Other	53.72 %

Fig. 9. Distribution of the starting price

The colleted data of the 1000 ebay auctions was then analyzed with the freely available data mining software RapidMiner[1], by developing a corresponding decision tree.

4 Decision Trees

Starting point of the decision tree method is an object set X with an associated attribute set M. The objects $x \in X$ are represented accordingly as feature vectors with

(1) $$x = [x_1, \dots, x_n]$$

Furthermore, a characteristic K for classification exists in order to determine the classes of X with $K(x)$ as the class of object x. The construction of a decision tree $T(D)$ from a subset of training data $D \subseteq X$ with

(2) $$d = [x, K(x)]$$

is processed according to the following procedure: if for all $d = [x, K(x)] \in D$

(3) $$K(x) = K_i$$

is valid as well as

(4) $$T(D) = \{K_i\}$$

a leaf of the decision tree $T(D)$ is found which corresponds to class K_i. Otherwise, an attribute M_n has to be chosen in order to split the decision tree $T(D)$ into sub-trees $T(D_i)$

[1] http://www.rapidminer.com/download

(5) $T(D) = \{T(D_1), \ldots, T(D_s)\}$

and

(6) $D_i = \{[x_1, \ldots, x_n], K(X) \subseteq D\}$.

The choice of the attribute M_i, which is most suitable for further branching of the decision tree, is chosen in regard to the information gain. The information gain is based on the entropy as a measure of order. The latter is defined as

(7) $H(D) = \sum_{i=1}^{k} -p_i \, log_2 \, p_i$

with

(8) $p_i = p(K_i|D) = \frac{|D \cap K_i|}{|D|}$

as the probability of $d \in D$ belonging to the class K_i. The resulting information gain due to a split in regard to attribute M_i with corresponding sub-trees $T(D_i)$ is calculated as:

(9) $\Delta\, H(D_i) = H(D) - \sum_{i=1}^{s} p(D_i|D) H(D_i)$

with D_i according to (6) and with

(10) $p(D_i|D) = \frac{|D_i|}{|D|}$

as the probability of $d \in D$ belonging to D_i. For branching the attribute M_i is chosen so that the information gain is maximum in the corresponding node, i.e. the entropy $H(D_i)$ and thus the order D_i has the lowest value.

The method of inductive learning of decision trees (ID3) generally constructs compact trees and branches in consideration of a high information gain close to the root [14]. The fact, that only a local optimization may be found and a globally optimal solution may be missed, is accepted deliberately.

Taking into account the basic finding that a variety of no need must be limited to the simplest and that simple models in the presence of statistical noise will lead to more accurate predictions [15], the ID3 method for developing decision trees has been modified and supplemented with a pre-and post-pruning. In terms of pre-pruning a further split-up into sub-trees $T(D_n)$ in regard to attribute M_i is overruled, because no significant information gain can be achieved. Instead, a leaf node is directly formed i.e. a classification K_i as a function of the dominant class in D_i is carried out. However, in case of post-pruning, at first the complete decision tree is developed and then transformed into a set of rules. These rules are then ranked in regard to the confidence and support and eventually evaluated. Rules that apply only with a low probability or are rarely used, are not considered and thus ignored. The modified ID3 method is also known as C4.5 and is used consecutively for the analysis of eBay auctions [16].

5 Results

The used mining software provided a decision tree for the auction data in accordance with the procedure described above. Each path of the decision tree from the root to a leaf represents a conjunctive rule. The inner nodes represent the attributes of the premise. The attribute values are noted on the edges of the tree. The leaf contains the implicit class and in our case is determined by the majority-rule, i.e. whether a price is higher or a below average.

Since our goal was not to classify all possible attribute values, which is typical for decision trees, but to identify a "interesting" interrelations in the form of rules, the overall assessment of the decision tree does not matter. The decisive question is whether the tree has a sufficiently large number of good rules. The quality of a rule or a path in the decision tree is determined primarily by the confidence and the support.

The confidence is calculated as the ratio of the correctly classified records and all records described by the rule's premise and correspond to the "correctness" of a rule. The support reflects the applicability of a rule by the number of records described by the rule's premise in relation to all available records. Good rules must have a certain minimum confidence and a specified minimum support. Too many exceptions or correctly classified data sets are not interesting for further consideration. However, the minimum support per rule, which expresses the frequency of its use should not be overestimated, since not only a single rule but a rule set has been determined. In our study, we set a minimum confidence of 60% and a minimum support of 2% for each rule. Under these conditions, the following rules resulted:

(R1) Unfashionable articles achieve below-average rates (confidence: 75.6% and support: 4.13%).

(R2) If an auction ends on Saturday and the seller has many ratings, it achieves an above-average price (confidence: 75.8% and support: 2.92%)

(R3) If an auction ends at noon, it achieves above-average prices, except on Saturday at noon (confidence: 66.41% and support: 12.8%)

(R4) If an auction ends between Monday and Friday and the seller does not have a good reputation it will achieve below-average prices (confidence: 68.5% and support: 5.43%)

(R5) If an auction only has a short duration and ends in the evening, it achieves below-average prices, unless the seller has a lot of ratings (confidence: 60% and support: 8.05%)

(R6) If the seller's reputation is low, the auction achieves under-average prices (confidence: 60% and support: 7.65%)

More highly confident rules with less application were:

(R7) Without images mostly below-average prices are obtained (confidence: 85%, support: 0.71%)

(R8) If an auction ends at night, it still obtains above-average prices, if it had a long duration (approximately 92.3% confidence, support: 1.31%)

Furthermore, it shows in comparison that auctions which end in the morning tend to achieve smaller prices than others.

6 Conclusion

Because in contrast to other work, we analyzed data non dependent on a specific target audience, it seems that our analysis of price determinants and the corresponding results are of general validity. Based on the available data a number of interesting correlations have been identified. These interrelations can be an evidence to deduce how the auction process data and product presentation should be designed. However, we have to recognize, that the experience and the reputation of a seller is the most important factor that needs to be administered by the seller. The found results can lead to a significant increase in auction revenue and are interesting primarily for Powersellers.

References

[1] Ghani, R., Simmons, H.: Predicting the End-Price of Online Auctions. In: Proceedings of International Workshop on Data Mining and Adaptive Modelling Methods for Economics and Management, Pisa, Italy (2004)

[2] Pugh, R.: The eBay Business Handbook – How Anyone Can Build a Business and Make Money on eBay, p. 27. Harriman House, Petersfield (2009)

[3] Capon, N.: Managing Marketing in the 21st Century, p. 286. Wessex Publishing, Gillingham (2008)

[4] Hirshleifer, J., Glazer, A., Hirshleifer, D.: Price Theory and Applications. Decisions, Markets and Information, p. 445. Cambridge University Press, Cambridge (2005)

[5] Lujanac, M.P., Blacharski, D.W.: How and Where to Locate the Merchandise to Sell on eBay: Insider Information You Need to Know from the Experts Who Do It Every Day, p. 15. Atlantic Publishing, Ocala (2007)

[6] eBay, eBay Sales Manual, p. 36, http://pics.ebaystatic.com/aw/pics/de/pdf/ebay_verkaeuferhandbuch.pdf (accessed: June 29, 2010],

[7] Ku, G., Galinsky, A.D., Murninghan, J.K.: Starting Low but Ending High: A Reversal of the Anchoring Effect in Auctions. Journal of Personality and Social Psychology 90(6), 975–986 (2006)

[8] Houser, D., Wooders, J.: Reputation in Auctions: Theory and Evidence From eBay. Journal of Economics & Management Strategy 15(2), 353–369 (2006)

[9] Standifird, S.S.: Reputation and E-Commerce: eBay Auctions and the Asymmetrical Impact of Positive and Negative Ratings. Journal of Management 27, 279–295 (2001)

[10] Melnik, M.I., Alm, J.: Does a Seller's eCommerce Reputation Matter? Evidence from eBay Auctions. Journal of Industrial Economics 50(3), 337–349 (2002)

[11] Witzel, M.: The Encyclopedia of the History of American Management, p. 483. Thoemmes Continuum, Bristol (2005)

[12] Reiley, D.H., Bryan, D., Prasad, N., Reeves, D.: Pennies from Ebay: The Determinants of Price in Online Auctions. Journal of Industrial Economics 55(2), 223–233 (2007)

[13] Katkar, R., Reiley, D.H.: Public Versus Secret Reserve Prices in eBay Auctions: Results from a Pokémon Field Experiment. The B. E. Journal of Economic Analysis and Policy 6(2) (2007), http://www.bepress.com/bejeap/advances/vol6/iss2/art7

[14] Quinlan, J.R.: Induction of Decision Trees. Machine Learning 1(1), 81–106 (1986)

[15] Akaike, H.: Information Theory and an Extension of the Maximum Likelihood Principle. In: Petrov, B.N., Csaki, F. (eds.) The Second International Symposium on Information Theory, Budapest, pp. 267–281 (1973)

[16] Quinlan, J.R.: C4.5: Programms for Machine Learning, p. 267. Morgan Kaufmann, San Francisco (1993)

A Framework for Efficient Information Retrieval Using NLP Techniques

R. Subhashini[1] and V. Jawahar Senthil Kumar[2]

[1] Research Scholar, Sathyabama University, Chennai-119, India
subhaagopi@gmail.com
[2] Assistant Professor, Anna University, Chennai, India
veerajawahar@yahoo.com

Abstract. In the academic area, the Internet is used as a scientific resource. However, finding appropriate information on the Web remains difficult. To simplify this process, we designed the Information retrieval framework to retrieve information from the Web Using NLP Techniques. Many information retrieval systems are based on vector space model (VSM) that represents a document as a vector of index terms. To remedy this problem, documents and queries are optimized using NLP. In this paper, the architecture of the proposed tool is presented and it proposes a new approach over the traditional VSM by considering only the nouns and verbs of the documents extracted from NLP as the constituting terms for the VSM instead of the traditional term "word". Such a mechanism may raise the effectiveness of the Information Retrieval by increasing the evaluation metrics values.

Keywords: NLP, Text Mining, Information Retrieval, Vector Space Model.

1 Introduction

Many Natural Language Processing (NLP) techniques [5], including stemming, part of-speech tagging, compound recognition, de-compounding, chunking, word sense disambiguation and others, have been used in Information Retrieval (IR). The core IR task we are investigating here is document retrieval. Several other IR tasks use very similar techniques, e.g. document clustering [2], filtering, new event detection, and link detection, and they can be combined with NLP in a way similar to document retrieval. Today, the amount of documents published on the internet grows dramatically, especially in education and e-learning activity and if the same information retrieval methods is used, then the time to find relevant document will also increase the same way. One possible solution is to develop software framework, which would improve the quality of current search engines and decrease time dedicated to searching process.

2 System Architecture

Our proposed system sends the optimized query to the search engine API [4]. This search engine API is a web service and it collects the top n documents from the search

V.V. Das, J. Stephen, and Y. Chaba (Eds.): CNC 2011, CCIS 142, pp. 391–393, 2011.

engine like yahoo. The optimized query obtained by NLP is compared with the optimized retrieved documents by cosine similarity measure [3] and it finds relevant documents to that query and displays them as a ranked list according to their similarity score. The Fig. 1 shows the architecture of the proposed system. The uniqueness of our proposed method to the existing method is that it calculates the similarity measure for only the nouns and verbs of the query and documents optimized by NLP by using sharp NLP [1]. We used Sharp NLP parser to parse the English sentences in to subject, verb and object. The noun phrases and verb phrases are extracted from the parser and they are used for calculating the similarity measures.

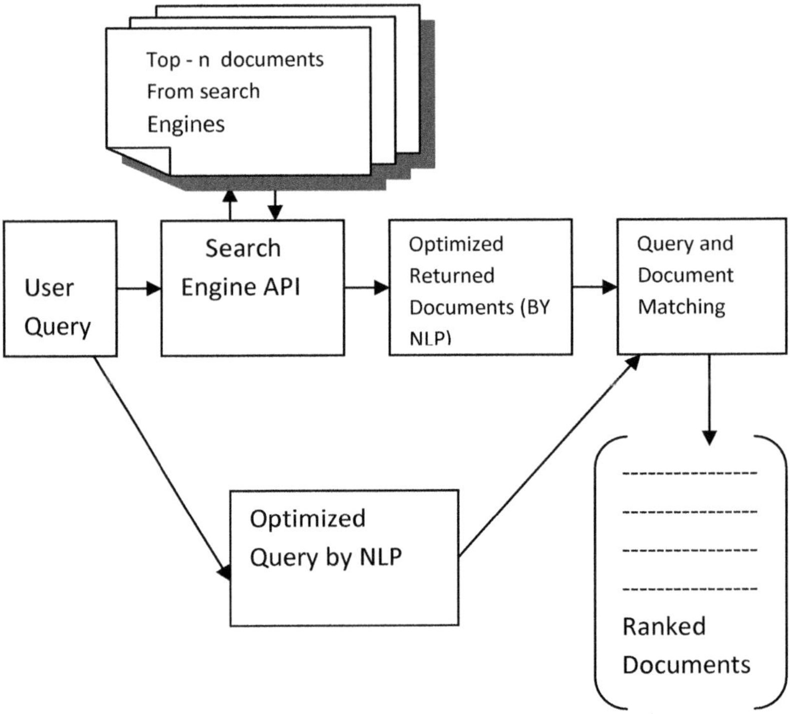

Fig. 1. Architecture of the proposed System

3 Experimental Setup and Evaluation Measures

Traditional methods used text datasets but in this system, web documents are collected using yahoo API and similarity measures are implemented on it. BOSS (Build your Own Search Service) is Yahoo open search web services platform. The application was written in c#. The query is given to yahoo API and it returns the top n documents i.e., the xml documents and it contains the URL's. Here, the value of n is set to 10 i.e., it represents the first 10 documents retrieved by the search engine. With that URL's the web pages are retrieved, optimized and then compared with the optimized query to get the relevant output. The relevant documents were ranked based on their similarity score. The system is tested with 50 queries which is represented as 5 sets such as computer, graphics, multimedia, Internet and Games. Each set contains the related terms of ten queries. The Fig. 2 Compares the precision for these 5 query sets between the traditional VSM and NLP based VSM. It shows that the precision value of the proposed system is comparatively higher for all query sets.

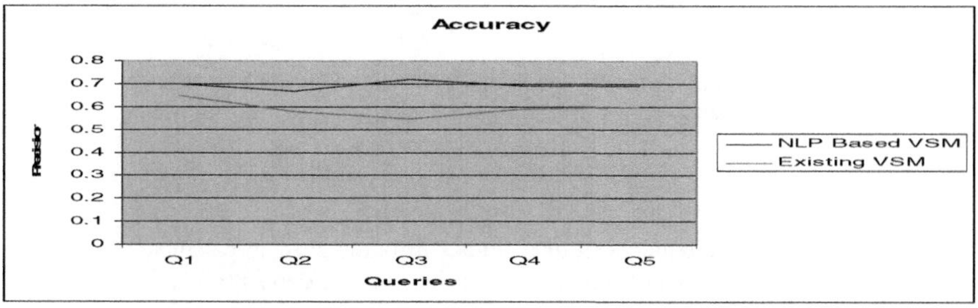

Fig. 2. Accuracy of Retrieval Performance

4 Conclusion

Natural Language Processing (NLP) techniques have been used in Information Retrieval mainly for presenting the user the documents that best satisfy their information needs. NLP is used to retrieve the relevant documents and hence it improves the performance. The Future work is extended to use the NLP for longer queries and also it should be combined with Word net for calculating the semantic similarity measures.

References

1. http://www.codeplex.com/sharpnlp
2. Supreethi, K.P., Prasad, E.V.: Web Document Clustering Technique Using Case Grammar Structure. In: IEEE International Conference on Computational Intelligence and Multimedia Applications 2007 (2007)
3. Salton, G., Wong, A., Yang, C.S.: A Vector Space Model For Automatic Indexing. Cornell University, Ithica
4. http://developer.yahoo.com/search/boss/
5. Liddy, E.D.: Enhanced Text Retrieval Using Natural Language Processing. Bulletin of the American Society for Information Science 24(4) (1998)

Modified Graph Spectra Approach for Music Information Retrieval

Madhur Sarin

Vipasyana Research and Analytics Solutions
No-9,3rd Cross, Nehru Road, Kammanahalli, Bangalore-84, India
madhur.sarin@vipasyana.org

Abstract. For large collections of melodies, it is too slow to calculate the distance of the query object from all the test objects. A reasonably fast method is needed to coarse sieve and identify a set of closest tunes .One way to achieve this is by using nearest neighbour methods on indexed database. Graph Spectra [1] Indexing method, which index tunes using their Laplacian spectra have proved successful in finding nearest neighbours of a query melody. In this paper we propose a modified version of Graph Spectra indexing method called Ratio Graph Spectra. A comparison of commonly used geometric methods such as Earth Movers Distance (EMD) and Indexing methods specifically Graph Spectra with Ratio Graph Spectra is conducted. We observed that R Graph Spectra indexing method outperformed other methods giving up to 80 % accuracy in retrieving nearest neighbour for a query tune.

Keywords: Graph Spectra, EMD methods, Music Information Retrieval (MIR), Query by Humming, Laplacian Spectra.

1 Introduction and Background

Music Information Retrieval (MIR) is an interdisciplinary area of science concerned with developing theory as well as applications for effectively accessing and retrieval of musical information from a database. Presently music databases are arranged in a way similar to a normal database with metadata this has severe limitation with respect to retrieval. MIR systems are now being developed to look for variations of songs, catch music plagiarism, tune matching etc. This paper deals with only a small part of this research area and improves upon the existing graph based melody pattern matching technique known as the Graph Spectra Indexing [1] method. It is found that melodic contour is important to the perception of melody [2]; this contour is given by the pitch interval between two successive notes. When melody is transposed to another key, even then the melodic contour remains the same. It is this property of the melody that we wish to utilize.

The laplacian spectra are commonly used for image feature characterization [1].This property of the laplacian spectra was utilized for characterizing tune's features [1]. A melody generally uses a subset of all possible pitches and different melodies have different frequency of intervals. A graph for a melody is a time independent structure.

V.V. Das, J. Stephen, and Y. Chaba (Eds.): CNC 2011, CCIS 142, pp. 394–396, 2011.

Let M be a melody i.e. a set of intervals of pitches $\{p_j\}$ $j \in I, \{I = 1, \ldots, m\}$, where m is the length of the tune. The graph G is then defined with vertex set V and edges such as a_j such that

$$a_j : \begin{cases} p_j \rightarrow p_{j+1} & \text{for every couple } (p_j, p_{j+1}) \subseteq M \\ p_m \rightarrow p_1 & \text{for the couple } (p_m, p_1) \subseteq M \end{cases}$$

The graph gives the geometry of a melody, but in order to allow computations in metric space, it is converted into an algebraic form; the most frequently used algebraic structure associated with graph is the Laplacian matrix L(G),which is given as the difference between degree matrix D(G) and the adjacency matrix A(G). Given a melody of length m =|M|, consisting of pitches, a Graph spectra index is created by computing the modulus of eigenvalues of the Laplacian matrix L (G). Eigenvalues are sorted in order of their magnitude to get the index. An example of a graph for the song in Fig1.shown below is depicted in Fig 2.

Fig. 1. Song name - Kiss me fast

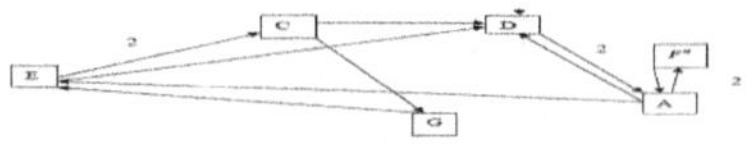

Fig. 2. Directed Graph for "Kiss me fast"

The first 16 notes of the song above have the pitch E, D, A, , A, E, C, G, E, C, D, A, , A, D, A.

2 Modification and Experiments

To be successful in matching tunes, the Graph Spectra method must produce similar indices for melodies that share topological similarities. However two similar tunes with varying length may end up having different spectra, unless the database objects as well as query tune are of same length, the Graph Spectra method will not yield the best results. This means having a fixed length of query and database tune. We propose a modified form of Graph Spectra indexing method called Ratio Graph Spectra, which is query/database tune length (number of notes) independent. The ratio of the length of the database tune p to the length of the query tune q is denoted by R. The sorted eigenvalues for a database object are divided by the ratio R to obtain Ratio Graph Spectra of the object. R is given as

$$R = \frac{|M_p|}{|M_q|}$$

The Athole collection of Scottish music is used as the database for conducting tests.It should be noted that the Athole collection contain songs with varying length. This implies that variants of the same song may or may not have the same length. In this experiment the proposed method will be tested for accuracy and comparison is made with commonly used geometric methods [2] .First, In order to test the genuineness of the proposed modification, a test sample of tunes is selected in such a manner that the sample tunes' exact match and as well as its variants are available in the Athole collection. The classification of melody as a variant of a sample is based on human understanding of melodic similarity. Euclidian distance is used to measure distances between tunes' spectres. K-nearest neighbours to query tune are analysed to check, if any variants of query are present in them or not. Table 1, gives the results for comparison between Ratio Graph Spectra with Graph Spectra and other commonly used geometric methods. Poot [3] implement these methods for the Athole collection.

Table 1. Accuracy comparison of ratio graph spectra with EMD and its variants

Size(k)	Percentage Accuracy				
	EMD	PTD	EMDα	Graph Spectra	R Graph Spectra
100	64	68	56	56	80
50	52	60	52	44	76
10	36	32	44	24	44

Table 2. Index comparison between graph spectra and ratio graph spectra

	q	p-Graph Spectra	p-Ratio Graph Spectra
Tune	Dumfries House (#169)	Dumfries House	Dumfries House
Length	170	107	107
λ_1	41.78	27.76	44.1
λ_2	30.61	19.69	31.28
λ_3	23.22	15.27	24.26
λ_4	20.59	12.72	20.21
λ_5	12	7.98	12.68
λ_6	12	7.54	11.98
λ_7	1.29E-15	4.64E-15	7.37E-15
λ_8	1.29E-15	1.46E-15	2.32E-15
λ_9	7.15E-16	1.46E-15	2.32E-15
λ_{10}	1.35E-16	1.30E-15	2.06E-15
λ_{11}	7.00E-33	2.24E-31	3.56E-31
λ_{12}	0	0	0

In table 2, it can be observed that even the laplacian for exactly similar tunes of different length will be different. Comparison between eigenvalues of expected match (p) using Graph Spectra and Ratio Graph Spectra is shown for a query (q) tune.

3 Conclusions

The results from experiments show that the proposed R Graph Spectra method outperformed commonly used methods for tune similarity matching. Good results obtained during nearest neighbour retrieval with tunes of arbitrary length justify modification to Graph Spectra method. Ratio Graph Spectra allows for variable tune length, both for query as well as database tunes, which is generally the case.

References

1. Pinto, Van Leuken, R.H., Faith Demirci, M.: Indexing Music Collection through Graph Spectra. In: Proc. ISMIR, Vienna, pp. 23–27 (September 2000)
2. Weber, R., Schek, H.J., Blott, S.: A Quantitative Analysis and performance study for Similarity search methods in high dimensional spaces. In: Proc. 24th International Conference on Very large Databases, VLDB, pp. 194–205 (1998)
3. Poot, T.: Measuring Melodic Similarity, MSc. thesis, School of Mathematics, University of Edinburgh. Edinburgh, Scotland, UK (2008)

Comparative Analysis of CK Metrics across Object Oriented Languages

N. Kayarvizhy[1] and S. Kanmani[2]

[1] AMC Engineering College, Bangalore, India
[2] Pondicherry Engineering College, Puducherry, India

Abstract. Measurement is fundamental to any engineering discipline. There is considerable evidence that object oriented design metrics can be used to make quality management decisions. This paper focuses on an empirical evaluation of object oriented metrics proposed by Chidamber and Kemerer across the two languages. Three projects in Java and two projects in C++ have been considered as input for the study. The resulting values have been analyzed to provide significant insight about the object oriented characteristics of the projects.

Keywords: CK Metrics, Java, C++.

1 Introduction

Object oriented systems continue to share a major portion of software development in the twenty first century. As customer base for these systems is on the rise, quality takes precedence. Lots of time, money and effort is spent in ensuring that the system is of the highest quality. One approach towards ensuring quality is by evaluating key attributes of the software through the use of metrics. The introduction and subsequent use of metrics as a means to evaluate the software quality has had deep and useful impact on the overall system. But the success of software quality assessment through metrics is hampered by the need for constant validation to ensure the accuracy of such predictions. There are several metrics proposed in the literature to capture the quality of Object Oriented design. Several empirical validation studies have been attempted for various object oriented metric sets in [3], [4], [5], [6], [7], [8], [9]. Chidamber and Kemmerer's suite of object oriented metrics is widely used and the most referred. They proposed six metrics – weighted method per class (WMC), coupling between objects (CBO), response for a class (RFC), depth of inheritance tree (DIT), number of children (NOC) and lack of cohesion in methods (LCOM). Amjan Shaik et al.[1] have done statistical analysis on CK metric suite by validating the data collected from student projects. Subramanyam et al [2] conducted an empirical analysis on a subset of CK metrics suite in determining software defects. The general conclusion is that these metrics are important indicators of external quality factors. Cartwright and Shepperd [10] studied the inheritance measures from the CK suite (DIT, NOC) and found that both these measures were associated with defect density of classes.

V.V. Das, J. Stephen, and Y. Chaba (Eds.): CNC 2011, CCIS 142, pp. 397–399, 2011.

2 Research Objective and Evaluation Results

Five projects have been considered for the study – three from Java and two from C++. The first project in Java is a standard library available as part of the Servlet package. This consists of the interface classes required to realize a web application using Java. The next project in Java is JForum which is open source software for a Java based discussion forum. It is being widely used among online communities to implement their discussion board. The final project in Java is 'Online Ticket Reservation System' which was a student project. For C++, two projects have been considered for the study and both were student projects. The first project is "Library Management System" for a college. The requirement specification was obtained from the existing Library management software and was used for this project. The next project is a "Graphical Scene Creator" which can be used to define and draw a scene comprised of basic shapes.

The table below tabulates the CK metrics that were obtained for the projects. First the CK Metrics were calculated for all the classes of a project. Then their mean was found. Finally the values were normalized considering on the highest value for a metric among the projects and assigning a unity value to it. All other values for the same metric in other projects were then given a number relative to this maximum.

Table 1. CK Metric values for the various projects considered for this study

CK Metric	Java #1	Java #2	Java #3	C++ #1	C++ #2
DIT	0.98	1	0	.90	1
NOC	0.25	0.36	0	0.89	1
WMC	1	0.82	0.22	0.29	0.29
CBO	0	1	0.32	0.58	0.70
RFC	1	0.95	0.98	0.27	0.45
LCOM	1	0.98	0	0	0.02

The CK metrics clearly identifies a lot of characteristics of the underlying project. For example the Servlet program is seen having the highest WMC and LCOM. This is because being a library of classes; the application interface is given importance resulting in a high WMC. Also most public methods of a library will work upon only a subset of the class attributes resulting in a high LCOM as well.

JForum software also has high CK Metric values. This is due to the fact that JForum is an inherently complex application with a lot of class interaction enabling it

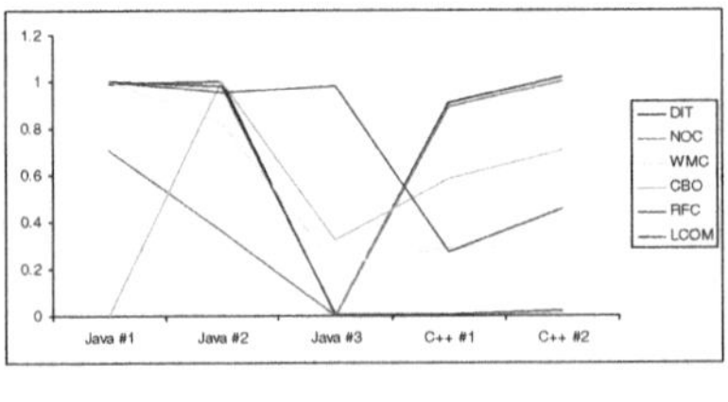

Fig. 1. This shows a figure consisting of all CK metrics for the different projects

to have many features supported with limited code. The three student projects (two in C++ and one in Java) have comparable CK Metric values. An interesting observation is that the DIT values are comparable between the Servlet package, the JForum and the student projects in C++ reflecting the fact that similar level of inheritance has been used. The student project in Java has not used inheritance which is evident from a value of 0 for NOC and DIT.

3 Conclusion

In this study, CK metrics have been measured for five projects in two object oriented languages and the results have been analyzed. This provides an idea of how the metrics are distributed among different types of systems in different object oriented languages. The metrics have then been used to interpret certain characteristics of the underlying system and make valuable assessment that can help in identifying problematic areas during the design and development of a system. This clearly indicates the usefulness of object oriented metrics like CK metrics in predicting the quality of an object oriented system.

References

1. Amjan, S., Reddy, C.R.K., Damodaran, A.: Statistical Analysis for Object Oriented Design Software security metrics. International Journal of Engineering Science and Technology 2(5), 1136–1142 (2010)
2. Subramanyam, R., Krishnan, M.S.: Empirical Analysis of CK metrics for Object Oriented Design Complexity - Implication of Software defects. IEEE Transactions on Software Engineering 29(4) (2003)
3. Abreu, F.B., Walcelio, M.: Evaluating the Impact of Object-Oriented Design on Software Quality. In: Proceedings of the 3rd International Symposium on Software Metrics, p. 90 (1996)
4. Audun, F.: Dynamic coupling measurement for object-oriented software. IEEE Transaction on Software Engineering, 491–506 (2004)
5. Prechelt, L., Barbara, U., Michael, P., Walter, T.: A controlled experiment on inheritance depth as a cost factor for code maintenance. Journal of Systems and Software 65, 115–126 (2003)
6. Chidamber, K.R., David Darcy, P., Chris Kemerer, F.: Managerial use of metrics for object-oriented software. IEEE Transaction on Software Engineering, 629–639 (1998)
7. Lavazza, L., Denaro, G., Pezze, M.: An empirical evaluation of object oriented metrics in industrial setting. In: 5th CaberNet Plenary Workshop (2003)
8. Khaled, E., Walcelio, M., Javam, M.C.: The prediction of faulty classes using object-oriented design metrics. Journal of Systems and Software, 3–75 (2001)
9. Victor, B.R., Lionel, B.C., Walcelio, M.L.: A validation of object-oriented design metrics as quality indicators. IEEE Transaction on Software. Engineering, 751–761 (1996)
10. Cartwright, M., Sheppered, M.: An Empirical Investigation of an Object-Oriented Software System. IEEE Transactions of Software Engineering (2009)

Integration of Spatial and Transform Domain in LSB Steganography

K.B. Shiva Kumar[1], K.B. Raja[2], R.K. Chhotaray[3], and Sabyasachi Pattnaik[4]

[1] Department of TC, Sri Siddhartha Institute of Technology, Tumkur, Karnataka, India
[2] Department of ECE, University Visvesvaraya College of Engineering, Bangalore University, Bangalore, India
[3] Department of CSE, Seemanta Engineering College, Mayurbhanj, Orissa, India
[4] Department of Computer Science, FM University, Balasore, Orissa, India
kbsssit@gmail.com, raja_kb@yahoo.com

Abstract. In this paper, we propose an integration of spatial and transform domains in LSB steganography (ISTDLS) where both the cover image as well as payload are divided into two cells each and RGB components of cell-1 are separated. The extracted RGB components of cover image cell-1 are individually transformed from spatial to transform domain using FFT/DCT/DWT and embedded differently, the components of cell-2 are retained in spatial domain itself.

Keywords: Steganography, Cover Image, Payload, Stego Image, DCT, DWT, LSB.

1 Introduction

The desire to communicate between two entities in secrecy has existed since the beginning of human communication which can be achieved through steganography in recent data embedding techniques. The steganography is an information embedding technique in which the secret information to be communicated is hidden in a data such as image, audio, video and text files [1]. The secret information is embedded into a data to derive stego object in such a way that the existence of it can't be detected by a hacker and retrieved with a minimum distortion at the destination by an authorized person. In modern steganography, normally the images are used as carrier of secret information since images have highly redundant insignificant data that can be replaced by the secret data by retaining the perceptual quality of stego object. Mohan Kumar and Shunmuganathan [2] described a multilayered architecture for hiding executable files in 3D images. Samir Kumar Bandyopadhyay and Indra Kanta Maitra [3] proposed an alternative approach of steganography using reference image. Four bit image is converted into four shaded Gray Scale image which acts as reference image to hide the text. Moazzam Hossain et al., [4] proposed a variable rate Steganographic technique for gray scale digital image using neighborhood pixel information and psycho visual redundancy.

V.V. Das, J. Stephen, and Y. Chaba (Eds.): CNC 2011, CCIS 142, pp. 400–402, 2011.

2 Algorithm

Problem definition: The algorithm is given in Tale. 1 and the objectives are:

1. To embed the payload into the cover image to derive stego image.
2. High robustness with reasonable PSNR to be achieved.

Table 1. ISTDLS Embedding Algorithm

- Input: Cover image , Payload
- Output: Stego image

(i) Read the cover image and payload images
(ii) Divide the cover image and payload images into two cells each of desired dimension
(iii)RGB components are separated from both cover image and payload image
(iv)The cover image RGB components of cell 1 are transformed from spatial domain to transform domain, where as the cell 2 components are retained in spatial domain itself.
(v) Four MSB bits of payload cell 1 and cell 2 are embedded in the second and fourth LSB positions of cover image cell 1 and cell 2 respectively to improve security.
(vi)Components in cover image cell 1 are transformed back to spatial domain and added with cell 2 components to generate stego image.

3 Performance Analysis

The PSNR values between cover image and stego image for different transform domains are compared in the Table 2. The ISTDLS with DWT has better PSNR compared to ISTDLS with DCT as well as individual transform domain techniques.

Table 2. PSNR for different Techniques

Images C1(512 x 512) PL(128 x 128)	PSNR				
	FFT	DCT	DWT	ISTDLS (DCT)	ISTDLS (DWT)
CI:peppers PL: old image	39.30	34.30	35.87	41.21	42.8
CI:peppers PL: araras	38.25	34.45	35.07	41.58	42.3
CI:peppers PL: audrey	37.00	34.31	35.22	41.3	42.7

4 Conclusions

In this paper both the cover image and payload are divided into two cells each and RGB components of only the cover image cell 1 are transformed into frequency domain using FFT/DWT/DCT and retaining Cell 2 in spatial domain. The four MSB bits of payload cell 1 and cell 2 are embedded in the second and fourth LSB positions of cover image cell 1 frequency coefficients and cell 2 spatial pixel values respectively.

References

1. Venkat Narayan Rao, T., Govardhan, A., Badashah, S.J.: Improved Lossless Embedding and Extraction-A Data Hiding Mechanism. International Journal of Computer Science and Information Technology 2(2), 75–86 (2010)
2. Mohan Kumar, P., Shunmuganathan, K.L.: A Multilayered Architecture for Hiding Executable Files in 3D Images. Indian Journal of Science and Technology 3(4), 402–407 (2010)
3. Bandyopadhyay, S.K., Maitra, I.K.: An Alternative Approach of Steganography using Reference Image. International Journal of Advancements in Technology 1(1), 95–102 (2010)
4. Hosain, M., Haque, S.a., sharmin, F.: Variable Rate Steganography in Gray Scale Digital Images Using Neighbourhood Pixel Information. The International Arab Journal of Information Technology 7(1), 34–38 (2010)

Rail Road Strategy for SJF Starvation Problem

Shashank Jain, Vivek Mishra, Raghvendra Kumar, and Umesh Chandra Jaiswal

Department of Computer Science and Engineering,
M. M. M. Engineering College, Gorakhpur, U. P. India - 273010
{shashank.neocfc,vivekmishra89,raghvendra3}@gmail.com,
ucj_jaiswal@yahoo.com

Abstract. Problem statement: Providing an ageing strategy for Shortest Job First, keeping the waiting time as efficient as possible better than any other algorithm of similar kind. Approach: Relative waiting time and burst time of each process are compared to get an effective scheduling policy in a constant scheduling overhead. Result: Based on experiments and calculations, our algorithm have provided better waiting time than Highest Response Ratio Next (HRRN) in some cases, equal waiting time in most cases, very negligible more waiting time in some cases and a constant scheduling overhead. HRRN was the best known improvement on Shortest Job First till date. Conclusion: Our algorithm can prove very effective for batch processing systems and situations where burst time can be easily predicted.

Keywords: CPU scheduling, shortest job first, highest response ratio next, starvation, scheduling overhead.

1 Introduction

There exist a number of proportional share scheduling algorithms [1], [2], [3], [4]. They have used the concept of assigning weights for fair distribution of CPU time among resources. We have used similar means for developing our algorithm.

In this paper, we have discussed an improvement on SJF solving its starvation problem. SJF leads to minimum possible waiting time. Using it to our advantage, we have used the relative comparison of burst time & waiting time as an ageing strategy.

We have divided our paper into following sections: Section 2 gives the introduction about related work done, Highest Response Ratio Next. Section 3 talks about materials and methods. Section 4 depicts about our algorithm to solve the problem of starvation. Section 5 tells that how our algorithm is implemented. Section 6 compares the performance of our algorithm with HRRN and at last section 7 is conclusion.

2 Related Work

One of the best known improvements of SJF is Highest Response Ratio Next [5]. It uses the following equation (1):

V.V. Das, J. Stephen, and Y. Chaba (Eds.): CNC 2011, CCIS 142, pp. 403–408, 2011.

$$\text{Priority}=1 + (\text{Waiting time}/\text{burst time}) \tag{1}$$

Process with the higher priority as a result of above equation is allocated by scheduler first. HRRN though solving starvation problem does not fully utilizes the efficient waiting time offered by SJF as would be shown by us later.

There have been numerous modifications done on HRRN to suit to a particular application need [6], [7], [8]. Our algorithm being pre-emptive & comparatively less overhead can replace HRRN quite efficiently in those applications.

3 Materials and Method

We have used the following equation (2) to decide which process is to be run next:

$$P1b+P2w < P2b+P1w \tag{2}$$

Where P1b, P1w are burst time and waiting time of a process respectively.P2b, P2w are burst and waiting time of another process.. Process whose burst time is added to wait time of another process to achieve the smaller value of the two should be given preference in scheduling order. If both the sides are equal then the process with shorter burst time is allocated to CPU. Starvation is solved because as waiting time of process having longer burst time increases, their chance to run next on the CPU increases.

```
first=initially_sorted_array()
second=empty
while(true){
        if (interrupt){
           if(new_process_runnable)
               add_process_to_second()
      put_interrupted_process_on_first()
               scheduler()
        }
        if(process terminates){
            If (first is empty){
               first=second
               second=initialize_new()
              scheduler()
            }
        }
}
```

```
scheduler(){
        If(second is empty) then
            top element of first p1 runs
        If(p1b+p2w<p2b+p1w) then
            top element of first p1 runs
        else if(p1b+p2w>p2b+p1w)
            top element of second p2
            runs
        else{
          if(p1b<p2b) then
              top element of first p1 runs
         else
          top element of second p2
          runs
        }
}
```

Fig. 1. Algorithm

4 Algorithm

We sort all the process at a time t according to their burst time and place them in a *first* array. These processes being sorted in ascending order as per the equation are given the entirety of their burst time to run when scheduled. When a new process arrives it is placed into a *second* array which is also sorted in ascending order with the help of above equation. Interrupted process would always be at the top of the *first* array. So, the top element of both the arrays contains the best two processes which can be scheduled next to achieve minimum waiting time. Therefore, we use the equation to compare the top element of both arrays and choose whichever gives less value. Let p1 be the process at top of first and p2 at top of second. Fig.1 describes it.

5 Implementation

We have used an array named *first* to contain the initially sorted processes. We initialize another array named *second* to contain the processes arriving dynamically at any time t second is initially empty. Following terms are further defined.

- Gbw which is abbreviated as global waiting time
- P1wf is waiting field time of a process
- Gbw=total amount of time ongoing schedule of processes have been allowed to run
- Waiting field time of the newly runnable process which is inserted into second array is minus of gbw.
- Waiting field time of currently executing interrupted process is equal to current waiting field time of interrupted process+(minus of time process have executed in its current run).
- Reduce the burst time of interrupted process accordingly depending on the time left un-allocated.

See, in the equation (3) to calculate the waiting time of a process we add the global waiting time to the process individual waiting field time. This sum is then used in the equation as a process waiting time.

$$P1w=P1wf+gbw \tag{3}$$

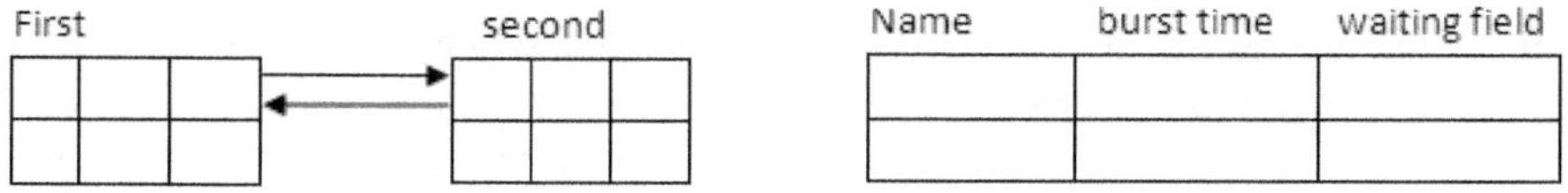

Fig. 2. Data Structure for Rail Road

In the figure drawn below: I1 with burst-time 5 arrives at t=3, I2 with burst time 9 arrives at t=9. We've shown some of the steps in the implementation till t=9.

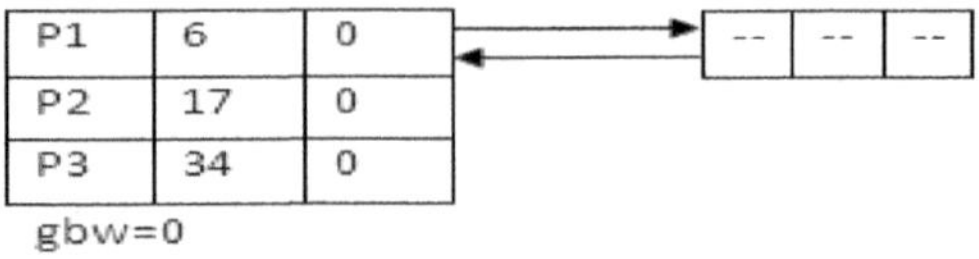

Fig. 3. At time t=0

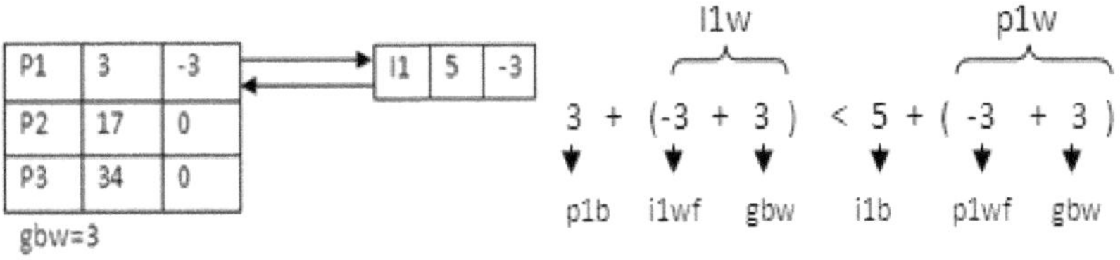

Fig. 4. At time t=3

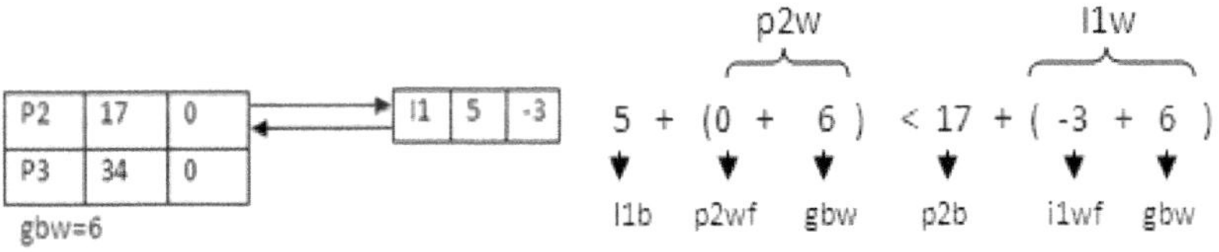

Fig. 5. At time t=6

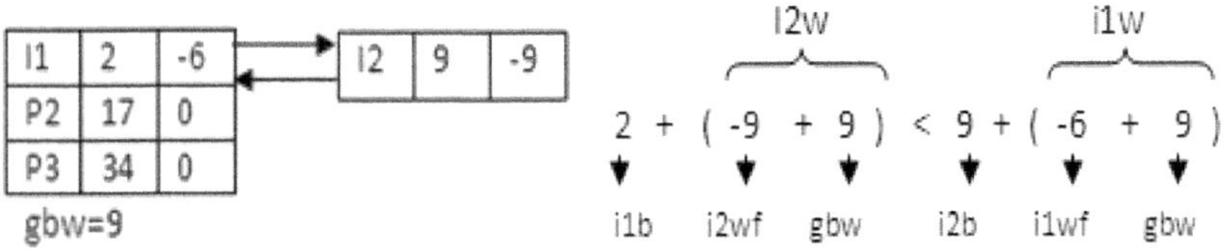

Fig. 6. At time t=9

6 Comparison and Performance

Initially the processes can be sorted using insertion sort- O(n),RB-tree-O(log h).This over-head is incurred in any scheduling algorithm which aim at a specific criteria like less waiting time, fair share allocation. Whenever a new process comes in, it also needs to be inserted at a proper place. This overhead is also same in our and other scheduling algorithms. Our scheduling time overhead is O(1) because scheduler just needs to compare the top two elements of both arrays to decide which one to run next. Whereas, in HRRN scheduling time overhead is O(n) on each scheduler invocation.

Of course HRRN sorts the processes when a previous running process have finished, while our strategy inserts the process as soon as it is brought in. In our case, we have to insert the newly arrived process with the help of insertion sort. While in case of HRRN no of iterations performed would be the number of processes present. In our strategy only in the worst case n iterations would be performed, while in HRRN it happens on each scheduler invocation. Also, the expensiveness of division operation performed in each of n iterations of HRRN decreases its efficiency all the more [8].

Look at table 1, initially process priority order to run is P4>P2>P3>P1. After P4 executes, priority order changes to P1>P2>P3. This happens because we are considering ratio to find the priority of process. This ratio would keep on changing depending on increase in waiting time. In our strategy once a process is inserted in ordered manner it won't change its order of execution. Therefore, no of process remaining same, HRRN would have to sort them again to determine the one with highest priority [8].

Table 1. Comparison between Rail Road and HRRN

Name	Waiting/burst time	Burst time	Waiting time
P1	0.333	30	10
P2	0.375	40	15
P3	0.34	50	17
P4	0.383	60	23

HRRN is non-preemptive in nature [7] but our algorithm is pre-emptive in nature. If the incoming process would satisfy the equation to run next, then it would be next given processor time. Also longer process wouldn't starve as after a while the equation would not satisfy for incoming shorter process. Thereby, we offer better responsiveness than HRRN.

We've written a simulator in C to calculate the effect of our strategy & HRRN when no of processes are large. We've considered a initial set of process, with new process interrupt arriving at different time of execution. We've drawn the graph for up to 50 processes. For varying distribution of processes we can clearly see as number of process increases our strategy offers much better waiting time than HRRN [Figure. 7]. Also, the number of iterations in our algorithm is significantly less as compared to HRRN [Figure. 7]. This is a major advantage as so many complex division operations are avoided.

7 Conclusion

Our algorithm is more suitable for batch processing of programs where burst time can be easily predicted. It is also useful among users who happen to use the same share of daily programs. We have provided an improvement of waiting time and scheduling overhead on HRRN which is the best known modification of SJF in a constant scheduling over-head of O(1). This algorithm is also applicable in all real life scheduling situations with its less overhead.

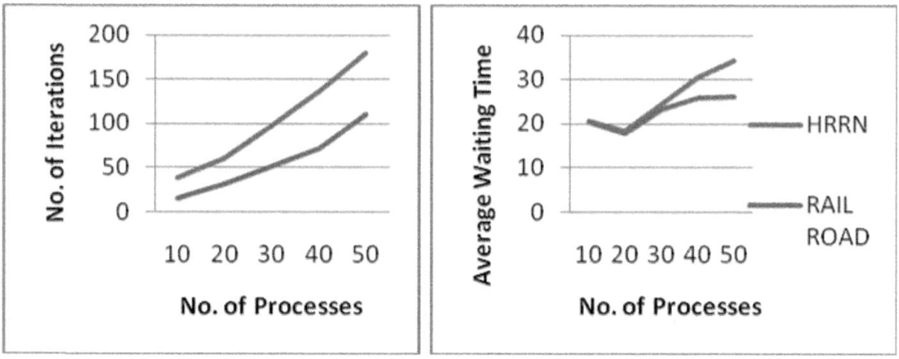

Fig. 7. Comparison between HRRN and Rail Road

References

1. Caprita, B., Chan, W.C., Stein, J.N.C., Zheng, H.: Group Ratio Round-Robin: O(1) Proportional Share Scheduling for Uniprocessor and Multiprocessor Systems. Columbia University, Technical Report CUCS-028-04
2. Waldspurger, C.A., Weihl, W.E.: Lottery and Stride Scheduling: Flexible Proportional-Share Resource Management. In: Proceedings of the 1st USENIX conference on Operating Systems Design and Implementation, OSDI 1994 (1994)
3. Bennett, J.C.R., Zhang, H.: WF^2Q: Worst-Case Fair Weighted Fair Queueing. In: INFOCOM 1996, vol. 1, pp. 120–128 (1996)
4. Neih, J., Vaill, C., Zhong, H.: Virtual-Time Round Robin: An O(1) Proportional Share Scheduler. In: Proceedings of the General Track: USENIX, Annual technical conference (UAT 2001), pp. 245–260 (2001)
5. Tanenbaum, A.S.: Modern Operating Systems, 3rd edn., p. 1104. Prentice Hall, Englewood Cliffs, ISBN: 13:9780136006633
6. Hong, L., Ningyi, X., Zucheng, Z., Jihu, P.: N New HRRN in the Bus Arbitration of SoC Design. In: Proceedings of 5th International Conference on ASIC (2003)
7. Aburas, A.A., Miho, V.: Fuzzy Logic based Algorithm for Uniprocessor Scheduling. In: Proceedings of International Conference on Computer & Communication (2008)
8. Saboori, E., Mohammadi, S., Parsazad, S.: A New Scheduling Algorithm for Server Farms Load Balancing. In: 2nd International Conference on Industrial and Information Systems (IIS) (2010)

A Descriptive Approach towards Data Warehouse and OLAP Technology: An Overview

Ardhendu Tripathy[1] and Kaberi Das[2]

[1] I.T.E.R (S.O.A University),
Jagamohan Nagar, Bhubaneswar
ardhendutripathy2007@gmail.com
[2] Department of Computer Applications, S.O.A University, Jagamohan Nagar, Bhubaneswar
kaberidas@iter.ac.in

Abstract. Data warehousing and on-line analytical processing (OLAP) are essential elements of decision support, which has increasingly become a focus of the database industry. Data warehouses support business decisions by collecting, consolidating, and organizing data for reporting and analysis with tools such as online analytical processing (OLAP) and data mining. In this paper address approaches and choices to be considered when designing and implementing a data warehouse. The paper begins by contrasting data warehouse databases with OLTP databases, introducing OLAP and then adds information about design issues.

Keywords: Data Warehouse, OLAP, Schema, Dimension.

1 Introduction

Data warehousing is a collection of ***decision support*** technologies, aimed at enabling the ***knowledge worker*** (executive, manager and analyst) to make better and faster decisions. A data warehouse is a "subject-oriented, integrated, time varying, non-volatile collection of data that is used primarily in organizational decision making." The data warehouse supports on-line analytical processing (OLAP), the functional and performance requirements of which are quite different from those of the on-line transaction processing (OLTP) applications traditionally supported by the operational databases.

2 Data Warehouse, OLTP and OLAP

A relational database is designed for a specific purpose [1, 2, 3]. Data Warehouse Database is designed for analysis of business measures by categories and attributes. OLAP Database is designed for real-time business operations.

2.1 Data Warehouse Support for OLTP Systems

A data warehouse supports an OLTP system by providing a place for the OLTP database to offload data as it accumulates, and by providing services.

V.V. Das, J. Stephen, and Y. Chaba (Eds.): CNC 2011, CCIS 142, pp. 409–411, 2011.

2.2 OLAP as a Data Warehouse Tool

Online analytical processing (OLAP) is a technology designed to provide superior performance for ad hoc business intelligence queries.

3 Data Warehouse Designing: Prerequisite

On designing of a data warehouse, it is imperative that the architectural goals of the data warehouse are to be clear and well understood.

3.1 Data Warehouse Architecture Goals

A data warehouse exists to serve its users such as analysts and decision makers. A data warehouse must be designed to satisfy the following requirements:

a. Deliver a great user experience: user acceptance is the measure of success.
b. Provides a central repository of consistent data.
d. Answer complex queries quickly.

4 Developing the Data Warehouse: Details

The phase of a data warehouse project generally starts with identifying requirements and ends with deploying the system.

4.1 Identify and Gather Requirements

Identify Sponsors: A successful data warehouse project needs a sponsor in the business organization.

4.2 Designing the Dimensional Model

The model design must result in a relational database that supports OLAP cubes to provide "instantaneous" query results for analysts [4, 5, 6]. A typical dimensional model uses a star or snowflake design that is easy to understand and relate to business needs.

4.3 Designing the OLAP Cubes

Typical OLAP queries **slice and dice** different parts of the cube comparing aggregations from one level to aggregation from another level.

5 Developing the Data Maintenance Application and Analysis Application

The data maintenance applications, including extraction, transformation, and loading processes, must be automated. OLAP cubes and data mining models are constructed using Analysis Services tools.

6 Testing and Deploying the System

It is important to involve users in the testing phase. After initial testing by development and test groups, users should load the system with queries.

7 Conclusion

Data warehousing approaches and techniques are well established, widely adopted, successful, and not controversial. The key to data warehousing is data design. The remaining tasks flow naturally from a well-designed model: extracting, transforming, and loading the data into the data warehouse.

References

1. Adamson, C., Venerable, M.: Data Warehouse Design Solutions. J. Wiley & Sons, Inc., Chichester (1998)
2. Agrawal, R., Gupta, A., Sarawagi, S.: Modeling Multidimensional Databases. In: ICDE (1997)
3. Ballard, C.: Data Modeling Techniques for Data Warehousing. SG24-2238-00. IBM Red Book (1998), ISBN: 0738402451
4. Kortnik, M.A.R., Moody, L.: From Entities to Stars, Snowflakes, Clusters, Constellations and Galaxies: A Methodology for Data Warehouse Design. In: Akoka, J., Bouzeghoub, M., Comyn-Wattiau, I., Métais, E. (eds.) ER 1999. LNCS, vol. 1728. Springer, Heidelberg (1999)
5. Chaudhuri, S., Dayal, U.: An overview of Data Warehousing and OLAP Technology. SIGMOD Record 26(1) (1997)
6. Hacid, M., Sattler, U.: An Object-Centered Multi-dimensional Data Model with Hierarchically Structured Dimensions. In: Proc. of the IEEE Knowledge and Data Engineering Workshop (1997)

Energy Consumption Analysis of Multicast Routing Protocols in Wireless Ad Hoc Network Environment

M.P.A. Mala @ Aarthy[1], M. Shanmugaraj[1], and V.R. Sarma Dhulipala[2]

[1] M.E. Pervasive Computing Technologies/CCT
[2] Assistant Professor (Physics)/CCT
Anna University of Technology, Tiruchirappalli,
TamilNadu, India
{aarthy.mala,innoraj,dvrsarma}@gmail.com

Abstract. Ad hoc Network is a wireless network with a group of wireless hosts. It does not depend on any existing network infrastructure such as base stations or routers for its assistance. Energy consumption of the nodes is an important issue while designing Ad hoc network. This paper presents an analysis of selected multicast routing protocols for a selected scenario under wireless Ad hoc network environment. The results from our analysis showed the suitability of PIM and DVMRP in large scale scenarios.

Keywords: Ad hoc Networks, Energy Consumption, Multicast Routing, On-Demand Multicast Routing Protocol (ODMRP), Distance Vector Multicast Routing Protocol (DVMRP) and Protocol Independent Multicast routing protocol (PIM).

1 Introduction

Ad hoc Networks provide rapidly deployable mobile infrastructures which introduce several constraints while designing protocol. Nodes within Ad hoc network must utilize limited bandwidth and conserve battery power. Energy conservation is one of the important factors to be considered while designing Ad hoc networks [1]. Multicasting is used in group-oriented application where the need for one-to-many data dissemination is quite frequent. It reduces the communication costs by minimizing the link bandwidth and delivery delay [2]. Energy consumption varies based on number of nodes and protocol implemented [3].

2 Related Works

The paper [4] presents an architectural analysis of multicast routing protocols for wireless Ad Hoc networks. In which he has discussed about the advantage and disadvantage of tree-based and mesh-based multicast routing protocol. The author of paper [5] presents a design, implementation and performance evaluation of ODMRP in a multihop wireless networks. The paper [6] presents a power aware on-demand multicast routing protocol for maximizing the network life time. In [7] author has

V.V. Das, J. Stephen, and Y. Chaba (Eds.): CNC 2011, CCIS 142, pp. 412–414, 2011.

discussed about the proactive and reactive multicast routing protocols over a wireless mesh network.

3 Multicasting Protocols

Different routing protocols have been proposed to reduce the energy consumed by the node. Multicast routing can be used in path, tree and mesh based network. ODMRP, DVMRP and PIM are some of the multicast routing protocol for Ad hoc network [8].

DVMRP is based on distance vector routing algorithm. In this protocol, each router maintains a multicast routing table by exchanging distance vector information among routers [1]. Once a tree is created, it broadcasts the datagram packet from source to destination. If a member wants to join the group a graft message is sent and a new branch is created. Prune message is sent to truncate the branch. ODMRP is on demand and mesh based protocol that forwards data packets from source to receivers by creating mesh [9]. It applies on-demand procedures to dynamically built routes and maintains multicast group membership. When a multicast source has packets to send, it floods a member advertising packet with data payload attached. PIM is a multicast routing protocol that runs over existing unicast infrastructure. PIM provides one to many and many to many distributions of data over networks. PIM consist of two separate protocols, namely PIM Dense Mode (PIM-DM) and PIM Sparse Mode (PIM-SM). Main difference with DVMRP and PIM is its independence from underlying unicast routing operation.

4 Experimental Results

Based on various multicast routing protocol the consumption of energy are to be considered here. Simulation is carried out by varying the number of nodes and also by using various multicast routing protocols such ODMRP, DVMRP and PIM. The energy consumed by the network by using ODMRP has a minute variation with the increase in the node density.

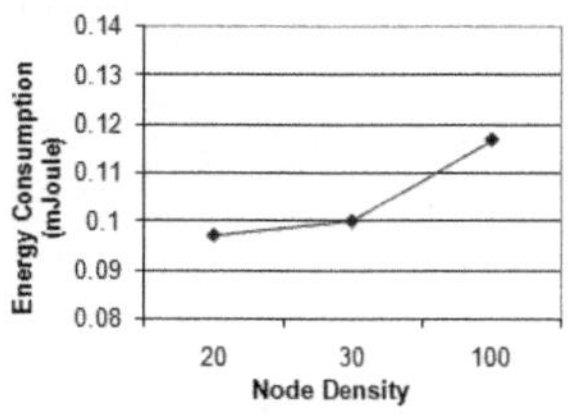

Fig. 1. Energy Consumption Using DVMRP

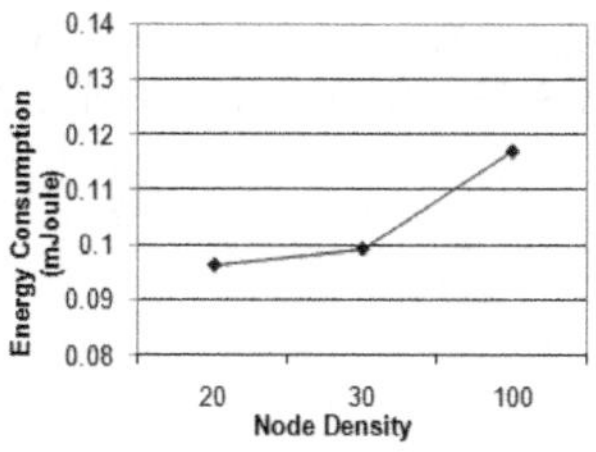

Fig. 2. Energy Consumption Using PIM

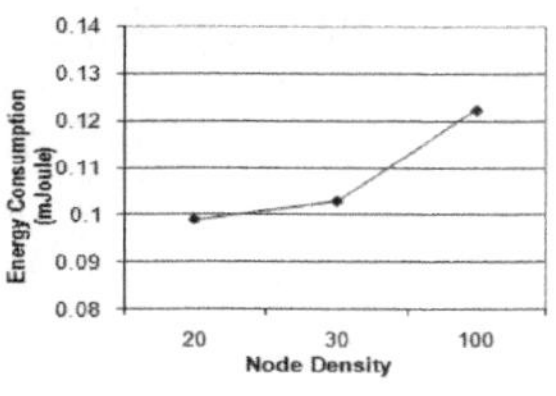

Fig. 3. Energy Consumption Using ODMRP

Fig 1 and Fig 2 shows that the energy consumed by the network using the protocols DVMRP and PIM are similar as the node density increases. Fig 3 shows that the energy consumed by the network by using ODMRP has a minute variation with the increase in the node density.

5 Conclusion

Our study of this work started with an idea of studying about the various multicasting routing protocols and their use in wireless sensor networking environment. We succeeded in uderstanding the situational and scalability utilization of the protocols. But this experiment is performed with a selected number of protocols. In future we are concentrating on the energy model based framework useful for military application. As a future work we are trying to enhance the study by considering more protocols.

References

1. Bae, S.H., Lee, S.-J., Gerla, M.: Multicast Protocol implementation and validation in an Ad hoc network testbed, `http://www.cs.ucla.edu/NRL/wireless`
2. Junhai, L., Liu, X., Danxia, Y.: Research on multicast routing protocol for mobile ad-hoc networks, `http://www.sciencedirect.com`
3. Sivaraman, R., Sarma Dhulipala, V.R., Aarthy, V., Kavitha, K.: Energy Comparison and analyisis for cluster based Environment for Wireless sensor Networks. In: IJRTE, vol. 2(4) (November 2009)
4. Ahmed, D.T., Shirmohammadi, S.: Architectural Analysis of Multicast Routing Protocols for Wireless Ad Hoc Networks. In: Proc. IEEE ICN 2007, Issue, pp. 21–21 (April 2007)
5. Bae, S.H., Lee, S.-J., Su, W., Gerla, M.: The Design, Implementation, and Performance Evaluation of on-demand multicast routing protocol in multihop wireless networks. Proc. IEEE Network 14(1), 70–71
6. Venkatalakshmi, B., Manjula, S.: Power Aware On-Demand Multicast Routing Protocol. In: CoMPC 2008 (2008)
7. SaifulAzad, M., Anwar, F., Arafatur Rahman, M., Abdulla, A.H.: Performance Comparison of Proactive and Reactive Multicast Routing Protocols over Wireless Mesh Networks. In: IJCSNS, vol. 9(6), pp. 55–62
8. Vasiliou, A., Economides, A.A.: Evaluation of Multicasting Algorithms in Manets. Proceedings of World Academy of Science, Engineering and Technology 5 (April 2005), ISSN 1307-6884
9. Begdillo, S.J., Asadi, M., Haghighat, A.T.: Improving Packet Delievery Ratio in ODMRP with Route Diversity. In: IJCSNS, vol. 7(12) (December 2007)

Fuzzy Logic Based Computer Vision System for Classification of Whole Cashew Kernel

Mayur Thakkar, Malay Bhatt, and C.K. Bhensdadia

Dept. of Comp. Engg., Dharmsinh Desai University, Nadiad, Gujarat – 387001
mayurthakkar.er@gmail.com, malaybhatt202@yahoo.com,
ckbhensdadia@ddu.ac.in

Abstract. This paper proposed an intelligent fuzzy logic based computer vision system for classification of whole cashew kernels by assigning the grades. Proposed approach is divided into various phases. Image acquisition phase captures images of the cashew kernel under investigation. Pre-processing phase smoothes acquired image to handle distortion. The cashew part (region of interest) is differentiated from the background part of the image in segmentation phase. In this system color along with other morphological features such as length, width and thickness are considered as essential features, and extracted in the feature extraction phase. Obtained quantitative information is fuzzified and taken as input to the fuzzy inference system (FIS) during the fuzzy classification phase. Final grade of the cashew kernel is decided based on the result of the classification.

Keywords: Cashew kernel, Computer vision system, Fuzzy inference system, Grade, Quality.

1 Introduction

Cashew is one of the most popular and important commercial crop in India. India is second largest producer of raw cashew nut and the largest exporter of processed cashew nut. International markets for the cashew nut are still expanding, at the same time quality requirements and standards are increasingly applied to suppliers. To ascertain the quality, the grade standard has been designed by considering the color and the size of the cashew kernel as important characteristic as given in Table 1 and Table 2.

With exception of few mechanical methods, cashew kernels are mostly graded manually by skilled labor, employed only for grading, but the quality decisions may vary among the graders and are inconsistent. Grading systems based on mechanical instruments, optical devices and computer vision principles are in existence. The Mechanical grading system is not appropriate for grading because of direct contact which can cause the damage to the cashew kernel. On the other hand, optical grading devices utilize light with certain wavelength reflected from the object, to assess quality. This kind of optical devices cannot be used for the size or shape based grading.

Inspection and grading of the whole cashew kernels using computer vision provides an alternative to manual, mechanical and optical methods because it offers automated,

V.V. Das, J. Stephen, and Y. Chaba (Eds.): CNC 2011, CCIS 142, pp. 415–420, 2011.

high speed, non-destructive and cost-effective technique [3]. In [2], computer vision based classification of pistachio nut using neural network has been introduced, in which morphological features and fourier descriptor of the boundary has been considered as important feature to classify the pistachio nuts. Similarly, it is also possible to develop the automatic grading system for the whole cashew kernels using computer vision. In [1], the physical properties of the raw cashew nut and cashew kernel have been evaluated, which plays vital role in designing the cashew kernel grading system. Designing such equipment without taking the physical properties of cashew kernel into consideration may yield poor results. Neural network and the fuzzy logic have been successfully implemented for decision making in control system [3]. As the grading standard for cashew kernels are defined, fuzzy logic is appropriate because of the presence of the imprecision.

Table 1. Color characteristic of whole cashew kernels

Cashew Kernel Type	Color Characteristic
White Whole (W)	Cashew kernels shall be white and free from damage.
Scorched Whole (SW)	Cashew kernels shall be light brown and free from damage.
Dessert Whole (DW)	Cashew kernels shall be dark brown, it may show deep black spot and free from damage.

Table 2. Weight characteristic of whole cashew kernels

White Whole Grades	Number of Kernels Per 454 gms.	Scorched Whole Grades	Number of Kernels Per 454 gms.	Dessert Whole Grades	Number of Kernels Per 454 gms.
W180	170-180	SW180	170-180	DW	No sepcification
W210	200-210	SW210	200-210		
W240	220-240	SW240	220-240		
W280	260-280	SW280	260-280		
W320	300-320	SW320	300-320		
W400	350-400	SW400	350-400		
W450	400-450	SW450	400-450		
W500	450-450	SW500	450-450		

2 Methods and Material

The samples of whole cashews of different grades, used in this study were collected from Orbitta Exports, one of the cashew production companies of Gujarat. Initially the different samples of the each grade are taken and weight of each cashew kernel is measured individually with accuracy of 0.001 gm.

Cashew classification system consists of two digital cameras placed in front and top of cashew sample under investigation at distance of 15 cm from the sample position as well as perpendicular to each other, an image capturing box, fluorescent lamp and computer system. Image processing toolbox in the MATLAB is used as image analysis and image processing software. Fig. 1(a) shows the system architecture of the cashew classification system.

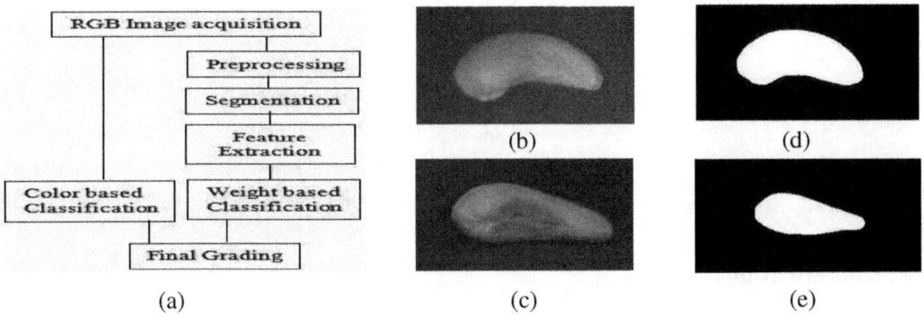

Fig. 1. (a) System Architecture (b) RGB top view (c) RGB front view (d) Binary top view (e) Binary front view of cashew kernel

2.1 RGB Image Acquisition

Image acquisition involves capturing of RGB front and top view images of each cashew kernel under study. These RGB images are as shown in fig. 1(b) and 1(c).

2.2 Preprocessing and Segmentation

3x3 average filter is used to smooth the image during Preprocessing phase. Because of black background, obtained histogram is always bimodal as shown in fig. 2(b). Threshold segmentation differentiates the cashew kernel region from background and converts the gray scale image into the binary image as shown in fig. 1(d) and 1(e).

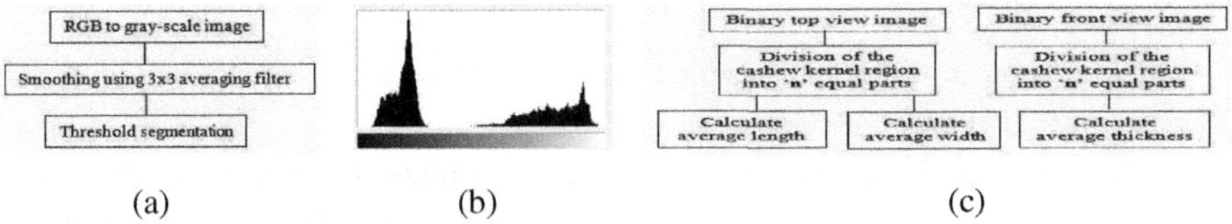

Fig. 2. (a) Preprocessing and Segmentation (b) Bi-modal Histogram of cashew kernel image (c) Morphological feature extraction

2.3 Feature Extraction

To estimate the weight of the cashew kernel, quantitative information of the morphological features like Length (L), Width (W) and Thickness (T) are extracted by dividing the cashew kernel region into '**n**' samples as shown in fig. 3. As the shape of cashew kernel is irregular, for better accuracy, averaging of these samples using equations (1), (2) and (3) is calculated.

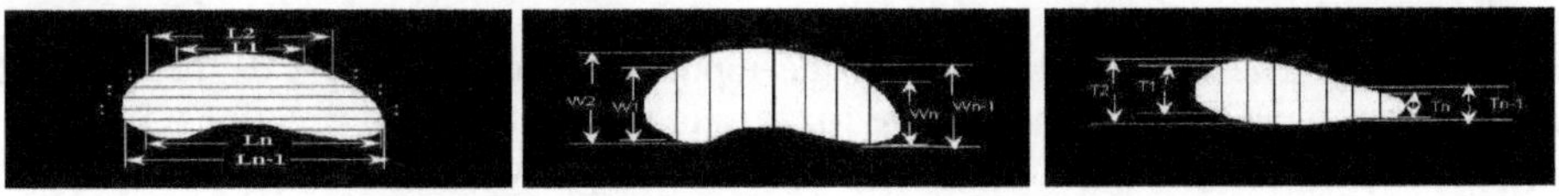

Fig. 3. Division of cashew kernel region into 'n' samples

$$\text{Length (L)} = \frac{L_1+L_2+ \ldots +L_{n-1}+L_n}{n} \quad (1)$$

$$\text{Width (W)} = \frac{W_1+W_2+ \ldots +W_{n-1}+W_n}{n} \quad (2)$$

$$\text{Thickness (T)} = \frac{T_1+T_2+ \ldots +T_{n-1}+T_n}{n} \quad (3)$$

2.4 Classification

In order to classify the whole cashew kernel, in this method two level of classification has been employed. In the first level, cashew kernel is classified based on its color characteristic and in the second level it is classified on the basis of its weight.

Color based Classification. At the first level of classification, cashew kernel is classified as whether it is white whole, scorched whole or dessert whole. Initially RGB image is converted to gray-scale image, then the average intensity of the pixels that belongs to the cashew region in the gray-scale image is determined and this intensity value is used for color based classification of the cashew kernel.

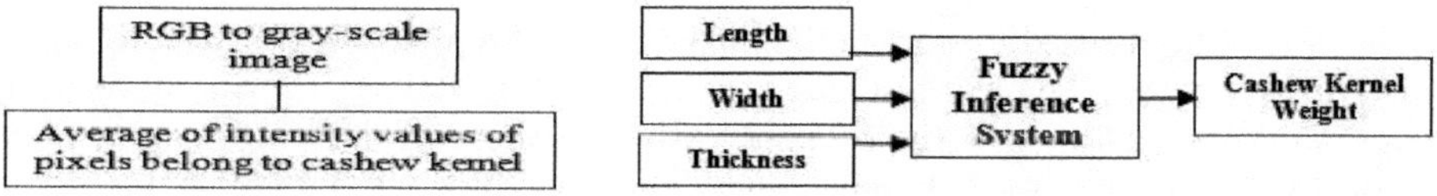

(a) (b)

Fig. 4. (a) Color information extraction (b) Fuzzy inference system

Weight based Classification. In second level, Fuzzy Inference System (FIS) is designed to estimate the weight of the cashew kernel. Fuzzy inference is the process of formulating the mapping from given input to an output using fuzzy logic [4]. In the FIS, extracted value of the length, width, and thickness are interpreted as linguistic input variables and the cashew weight is considered as linguistic output variable.

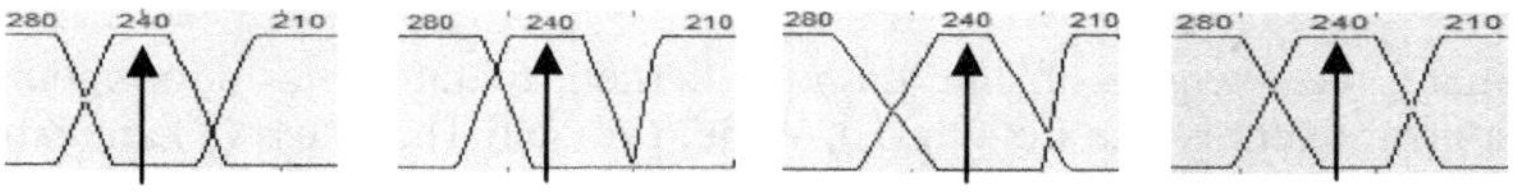

If length=240 AND width=240 AND thickness=240 then Weight = 240

Fig. 5. Application of AND operator in rules

Fuzzification. The first step is to take the inputs and determine the degree to which they belong to each of the appropriate fuzzy sets via membership functions. Linguistic input variables 'length, width and thickness' and output variable 'weight' are fuzzified using trapezoid membership function because for more than one value of input, it is possible that membership function gives value 1, as shown in fig. 5.

Applying AND/OR operator. There are eight fuzzy sets have been defined for each input variable, therefore at most 512 rules are possible with AND operator. The fuzzy operator AND is applied on three inputs to obtain the result of the antecedent for that rule. This resultant will be applied to the output function. This procedure is shown in fig. 5.

Rule aggregation. Rule Aggregation is the process by which the fuzzy sets that represent the outputs of each rule are combined into a single fuzzy set.

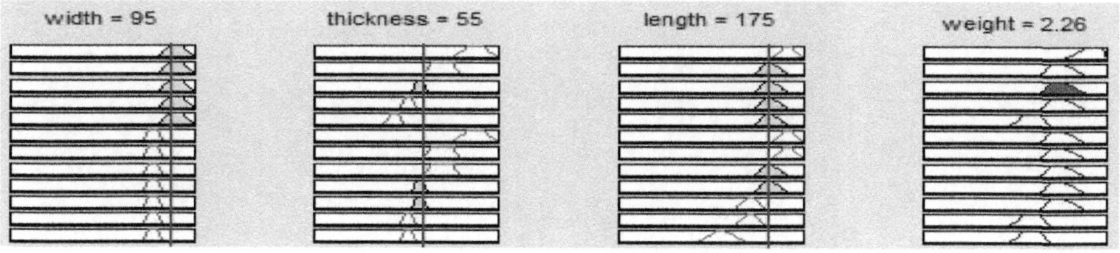

Fig. 6. Rule viewer

Defuzzification. The final desired output for each variable is generally a single number. Therefore, the aggregation of fuzzy sets must be defuzzified in order to resolve a single output value that indicates the weight of the cashew kernel.

2.5 Final Grading

The grade of the cashew is decided based on the result of these two classifications. If at first level, cashew is classified as white based on the estimated color and at second level as 240 based on the weight, then the cashew is graded as W240.

3 Conclusion

More than 100 samples of each grade are collected which results into a total of more than 1700 samples and form the data set. The sample data set is partitioned into training set which consists of 66% of total samples and remaining 34% forms the testing set. Training set is used to trim down fuzzy rules from 512 to 61 and testing set is used to validate the final rules. Implementation of Fuzzy logic based computer vision system for classification of whole cashew kernel is an effective and efficient technique that provides automated, non-destructive, high speed solution with classifier accuracy of 89%.for cashew kernel grading.

References

1. Balasubramanian, D.: Physical properties of raw cashew nut. Journal of agricultural Engineering Research 78(3), 291–297 (2001)
2. Ghazanfari, A., Irudayaraj, J., Kusalik, A., Romaniuk, M.: Machine vision grading of pistachio nuts using fourier descriptors. Journal of agricultural Engineering Research 68, 247–252

3. Brosnan, T., Sun, D.W.: Improving quality inspection of food products by computer vision. Journal of Food Engineering 61, 3–16 (2004)
4. Yen, J.: Fuzzy Logic, Intelligence, Control and Information. India Pearson Education, London (2006)
5. Gonzalez, R.C., Woods, R.E.: Digital Image Processing. Prentice-Hall International, Englewood Cliffs (2002)
6. MATLAB documentation on Image Processing Toolbox, `http://www.mathworks.com`

Efficiently-Enabled Inclusive Approach Preventing SQL Injection Attacks

Abhijot Singh Mann[1] and Sheela Jain[2]

[1] Department of Computer Science, UCOE, Punjabi University, Patiala, India
[2] International Institute of Informatics and Management, Mansarovar, Jaipur, India
jot.abhi@gmail.com, sheela.iiim@gmail.com

Abstract. Commercial, scientific, and social activities are increasingly becoming dependent on Web database applications. Worldwide implementation of database applications demands the need of new tools that allow testing web applications against security issues along with perfect handling the unique features of these systems. Structured query language (SQL) injection has become one of the most serious security threats for web based applications. There are automated ways to test the applications written in imperative and structured languages. However, it is still a challenge for the engineers as the sources of SQL injection vulnerabilities vary widely. Moreover the proposed testing approaches still lack the issue of generating adequate test data sets in order to detect SQL injection vulnerabilities. In the present work we propose a new efficient comprehensive approach to test these applications for SQL vulnerabilities. The approach is novel and experimental result claims its efficiency.

Keywords: SQL injections, Database Testing, Information security.

1 Introduction

The world wide use of internet has urged a wide range of web services such as online stores, social network services, Business network, e-commerce applications etc. Database (DBs) plays a central role in all these above mentioned operations. A lot of effort has been put in to ensure both the work efficiency and protecting the integrity of data stored, via the algorithms and data structures used by DBMSs. On the other hand little attention has been given to the security issues related to these database systems. Today's organizations are highly dependent on applications that heavily interact with databases by generating and issuing queries to database systems to perform necessary tasks. DBs application programs are mostly written in semi declarative language, such as SQL, or a combination of imperative and declarative languages. Nonetheless the programs written in imperative languages can be tested using white box or black box, depending on the tests. However the existing testing techniques are designed explicitly for imperative languages, and therefore are not directly applicable to the DBs application programs. SQL injection technique is maliciously used to obtain unrestricted data access to DB by inserting maliciously crafted strings to SQL queries via a web application. In Literature, different approaches were proposed in different times for the detection and testing of SQL injection attacks including SQLunitGen, Sania, SQLrand,

V.V. Das, J. Stephen, and Y. Chaba (Eds.): CNC 2011, CCIS 142, pp. 421–423, 2011.

SQL-IDS, SQLCheck, SQL mutation etc, where some of them includes white box testing part too [1]. In this paper we propose an inclusive approach which is Injection Sensitive for testing DBs applications. It is based on input validation test and diversified validation rules based on exhausted rule library. As an example, In case of online banking there are two fields: User name and the corresponding password in the user table. When the credential are submitted, they are inserted in a query template which is

Statement = "SELECT * FROM users WHERE login = "' + $user_id + "' AND Password = password "' + $pwd + "';"

Instead of a valid user name the attacker can use ' or 1=1; --then

WHERE login = '' or 1=1; --'' AND Password = password "any text'';"

The single quotes balance the quotes in the predefined query, and the double hyphen out the remainder of the query. In this case the password becomes irrelevant and may be set to any character string.

2 Proposed Approach

The proposed approach in this paper consists of mainly four steps.

Input Validation Scheme

Secure Input: Input validation is a secure way to verify the user input. A negative security model (black list) allows more abundant input data than a positive security model (white list). Another way to sanitize content is to encode the process. It is helpful when the range of input that is allowed cannot guarantee that the input is safe. This can be anything from stripping a null from the end of a user supplied string to escape out values so they are treated as literals [2].

Input validation test: Besides this, in many cases individual fields require specific validation. The implementation of one rule to constrain all inputs may lead errors in decision process. Choosing proper rules is a challenging issue and the initial burden of the solution falls on the application developers. To avoid this we use the automatic validation scheme, which can create diversified validation rules produced by an automatic mechanism [3]. The backbone of the mechanism is an exhausted rule library where an automatic mechanism chooses proper rules for validating input data accurately and implements a system which is designed to protect websites from attack.

Web service instrumentation: Aspect-oriented programming (AOP) programming paradigm is used for web service instrumentation which isolates secondary or supporting functions from the main program's business logic. In practice, here AOP is used to intercept key web service execution points and introduce the vulnerabilities detection mechanisms. It detects vulnerabilities by automatically identifying all the locations in the web service code where SQL commands are executed [4].

Crafting Test Cases: In the third phase we need to generate attack input test cases based on the rules mentioned in the previous phases. These test cases will include parameter values that attempt to inject the SQL queries. This attempt will expose all the loopholes in the web application. A set of malicious values are inserted by applying the rules in the validation rule library.

3 Implementation and Results

This new approach was implemented in a Java-based web application with a list of attack codes in XML. One of the applications used to test the approach is a web portal to share paid research articles. The new approach is used to discover the SQL injection vulnerabilities. In order to compare the results, a freeware tool "Paros" has been used. Our technique was able to identify 17 vulnerabilities where as Paros can identify only 6.

Table 1. Analysis of False Positives

Tool	No. of false positives		Description
Injection Sensitive Approach	12	6	Length of code too long
		3	Backslash mistaken
		1	Authentication failed
		2	Failed to delete inserted SQL query
Paros	39	16	Content of the page was changed after editing
		23	Including rest all

Our approach injects the malicious code based on the proper validation rules that are determined runtime. This stands helpful in exposing context based vulnerabilities by targeting pinpointed attacks. This approach also reports false positives results.

References

1. Shahriar, H., Zulkernine, M.: MUSIC: Mutation-based SQL Injection Vulnerability Checking. In: 8th International Conference on Quality Software, pp. 77–86 (2008)
2. Meier, J.D., Mackman, A., Dunner, M., Vasireddy, S., Escamilla, R., Murukan, A.: Microsoft Corporation: Improving Web Application Security. In: Threats and Countermeasures (2003)
3. Lin, J., Chen, J., Liu, C.: An Automatic Mechanism for Sanitizing Malicious Injection. In: The Proceedings of the 9th International Conference for Young Computer Scientists, pp. 1470–1475 (2008)
4. Kiczales, G., Lamping, J., Mendhekar, A., Maeda, C., Lopes, C., Loingtier, J.-M., Irwin, J.: Aspect-Oriented Programming. In: Liu, Y., Auletta, V. (eds.) ECOOP 1997. LNCS, vol. 1241. Springer, Heidelberg (1997)

Simulation of Ultrasonic Phased Array Sensors for Time of Flight Measurements

Ashwini Naik[1] and M.S. Panse[2]

[1] Research scholar, Dept. of Electrical Engineering, V.J.T.I.
ashwinivyas@yahoo.com
[2] Professor, Dept. of Electrical Engineering,
V.J.T.I., Matunga, Mumbai-400019, India
mspanse@vjti.org.in

Abstract. Most of the ultrasonic distance measurements are based on the determination of the Time of flight. Digital signal processing techniques for obtaining high accuracy in ultrasonic distance measurements are presented in this paper. The proposed method employs a Wavelet Transform or Short Time Fourier Transform to extract the envelope of the reflected pulse echo, together with a Cross-correlation pulse detection algorithm for Time of flight estimation. The simulated system consists of five elements transducer array as both transmitter and receiver. The transducer array is formed by aligning the transducers with minimum spacing between elements of 2 wavelengths. The results show the accuracy with the theoretical value. The simulation for the above said application has been implemented using LabVIEW (Laboratory virtual instrument engineering workbench) developed by National Instrument, is a graphical Programming environment suited for high-level or system level design. Advantage of this approach is the flexibility and very rapid development time offered by this graphical programming software.

Keywords: LabVIEW- Time of flight – ultrasonic phased array, beam steering, ranging – DSP.

I Introduction

Ultrasonic sensors are widely used to measure object distance in air. The operating principle is based on the measurement of the Time of flight (T.o.F.), that is, the time necessary for an ultrasonic wave to travel from the transmitter to the receiver after being reflected from the target. The object distance from the transducer **R** is given [1][3][4]

$$\mathbf{R=Cx(T.o.F.)/2} \tag{1}$$

where **c** is the velocity of sound (air medium 340m/s). The main motivation for this problem of Time of flight estimation stems from aiming to design a prototype, of Sensory aid for the visually impaired. For this application Ultrasonic phased array sensors are used to transmit/ receive sound waves in air. Ultrasonic sensors are widely regarded as simple and inexpensive devices to sense the proximity of objects. Their

V.V. Das, J. Stephen, and Y. Chaba (Eds.): CNC 2011, CCIS 142, pp. 424–430, 2011.

ability to provide relatively accurate geometrical information concerning the existence, the localization, and the nature of insonified objects. This simulation uses Pulse-echo technique to generate ultrasonic waves. With this method a short train of waves is generated, enabling the same transducer to be used both as a transmitter and as receiver.

In most basic sense, a phased array system utilizes the wave physics principle of phasing [9], varying the time between a series of outgoing ultrasonic pulses in such a way that the individual wave fronts generated by each element in the array combine with each other to add or cancel energy in predictable ways that effectively steer and shape the sound beam. This is accomplished by pulsing the individual probe elements at slightly different times. Frequently the elements will be pulsed in groups to improve effective sensitivity. The wave fronts in turn combine constructively and destructively in to single beam that travel in a particular direction. If there is an object in the path of this pulse, part or all of the pulse will be reflected back to the transmitter as an echo and can be detected through the receiver path. The beam can be dynamically steered through various angles, focal distances in such a way that a single probe assembly is capable of scanning across a range of different perspectives. So that a scan from multiple angles or with multiple focal depths can be performed in a small fraction of second. The returning echoes are received by the various elements or groups of elements, amplified, and time shifted as necessary to compensate for varying delays and then summed. Then by using Digital Processing Algorithms Time of Flight of the sensor is calculated.

In this paper ultrasonic transducer array of 5 elements arranged in column is considered. Each element is excited with a delay for steering angle θ [9] is given by

$$\mathbf{\Delta t_n = nd/c\ (\sin \theta) + t} \tag{2}$$

where **d** is the distance between the elements. The reflected signals are summed and denoised using Wavelet Transform. Envelope is detected using Hilbert transform. The transmitted and echo pulses are digitized, and their cross-correlation is computed in digital form. The time index of the peak of the correlation function measures the delay between the two signals that is the Time of flight.

2 Principle of Operation

Measurement of the Time of flight can be considered in the general framework of time delay estimation. Each element of the transducer array is excited by the time sequence pulse that propagates at any desired azimuth angle in the transmission medium and reflects back to the sensors after a delay. The Time of flight measurement system has to determine this time interval. The propagating medium introduces attenuation that increases with frequency, and can distort the reflected wave.

The simplest and most common way to detect the echo signal is the threshold method, where detection occurs when the signal crosses a given amplitude level [1]. Although it has proved to be simple and cheap, the technique may suffer from poor resolution, particularly when the echo pulse has been greatly attenuated. An improvement consists in the adoption of an adjustable amplitude threshold, in which case detection does not depend on the magnitude of the signal but only upon its shape [3]. In the presence of noise due to air turbulence, and/or mechanical vibrations close to the transducer operating frequency, both methods are known to achieve a repeatability of some tenths of the ultrasonic wavelength. This corresponds to different distance values, In fact, the received echoes reach the threshold level some time after the exact beginning making the target to appear slightly farther away than it actually is. The cross-correlation method is generally used as signal processing algorithms can provide improved resolution at least comparable accuracy.

To Calculate the Time of flight a suitable mathematical model [12] of the transmitted ultrasonic signal is considered:

S (t) = A $\mathbf{t^m e^{-t/u}}$ **cos (2πft+Φ),** Where A, f and Φ are the pulse amplitude, frequency and phase respectively, while m and u are parameters that depend on the type of ultrasonic transducer and model the echo waveform envelope. Actually, the parameter **m** models the initial finite slope of the pulse, while the parameter **u** determines the final one. It must be noted that actual transducers always produce pulses with both **m > O** and **u > O** [5].

The method by which a linear array may be excited to produce a wave front which propagates at any desired azimuth angle is illustrated in Fig. 1(a) [11]. In this figure, the time sequence of excitation pulses to each of the elements is indicated on the left side of the array. First, the top element is excited which produces a circular wave front which propagates towards the right into the air medium at any desired steering angle. A short time later, the second element is excited which also propagates in to the air medium. This process continues until the last element has been pulsed.

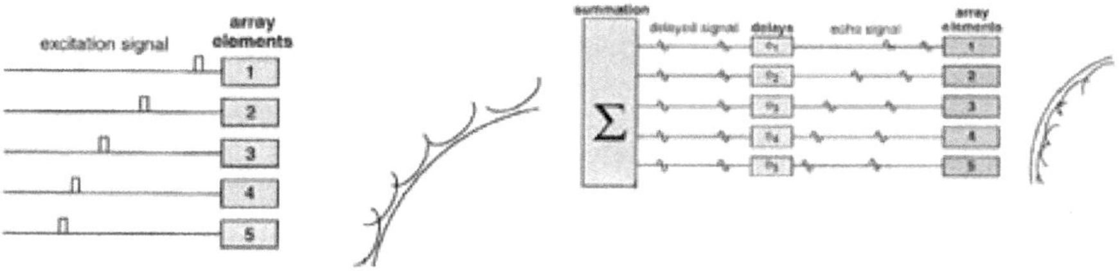

Fig. 1(a). **Fig. 1(b).**

Fig. 1. Steering of ultrasonic beam using 5 element phased array (a) in the Transmit mode (b) in the receive mode

Figure 1(b) [11] shows block diagram for receiving an echo. The echo is incident on each array element at a different time interval. The received signals are electronically delayed so that the delayed signals add in phase for an echo originating at that steering angle. For echoes originating elsewhere, at least some of the delayed signals will add out of phase.

3 Modelling of Ultrasonic Signals

The Ultrasonic signal generated can be represented mathematically [12] as

$$\mathbf{S\,(t) = A\, t^m\, e^{-t/u} \cos(2\pi ft+\Phi)} \qquad (3)$$

The transmitted signal of ultrasonic phased array is as shown in the figure 2(a). The signal is generated using f = 40 kHz, m = 3.6 E-4 and u = 2. The time sequence of the

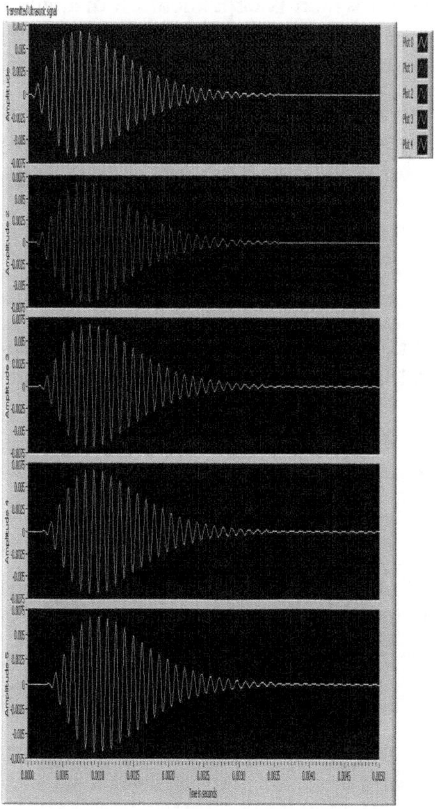

Fig. 2(a).

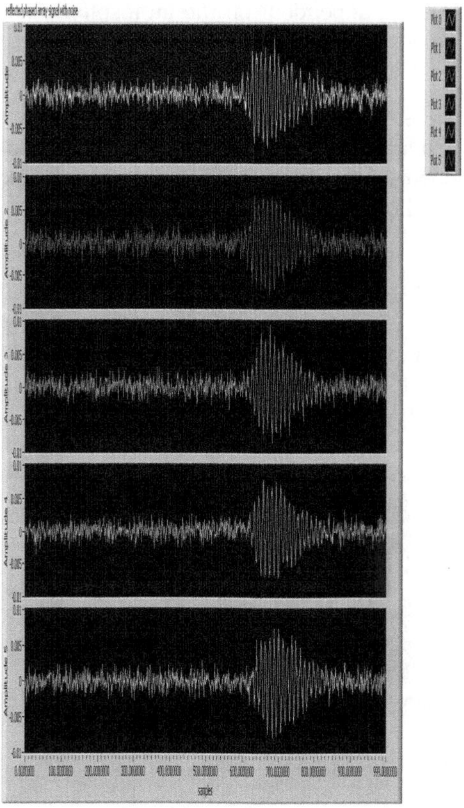

Fig. 2(b).

Fig. 2. Front panel showing the ultrasonic wave for steering angle of 10^0 (a) Transmitted signal (b) Reflected signal with delay and noise

excitation signal is calculated using equation (2). The interelement distance is considered to be twice the wavelength [9][10].

The transmitted signal is a noise free signal, while the received echo signal is an attenuated and delayed version of S(**t**). The source code is constructed in LabVIEW's Graphical programming language. A LabVIEW program called as virtual instrument has two main parts, the front panel and the block diagram. The front panel is used for user interactions and display of results. The block diagram is used for designing the source code. Figure 2(a) is the front panel of Virtual instrument in transmit mode. The Time of flight measurement is performed by considering the digital received echo signal and can be represented as

$$\mathbf{R(nT)=A(nT\text{-}T.O.F)^{m}e^{-(nT-T.O.F)/u}os(2\pi f(nT\text{-}T.O.F)+\Phi)+W(nT)} \quad (4)$$

Where T is the sampling period, T.O.F is the time of flight, A is the amplitude of the received signal W(t) is the white Gaussian noise with zero mean. The Reflected signal of the phased array is as shown in figure 2(b). The amplitude of the reflected pulse signal is dependent on the atmospheric attenuation which is dependent on distance d is given as [3]

$$A(d)=A\exp(-\acute{\alpha} d)/d \quad (5)$$

where ά is the c oefficient of attenuation in air (1.6E-11 f^2dbsg^2/mm). The echoes at each array elements are amplified and then delayed so that they can be simultaneously added by the summer. The envelope of the echo is then detected for Time of flight calculation.

4 Time of Flight Calculation

An important and difficult step in Time of flight estimation is to recover the envelope from the noisy reflected signal. The distorted reflected signal at the output of summer is as shown in figure 2(c).

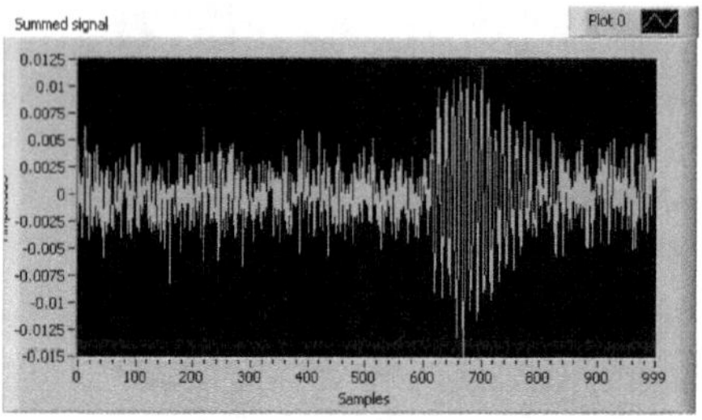

Fig. 2(c). Summed output of Phased array

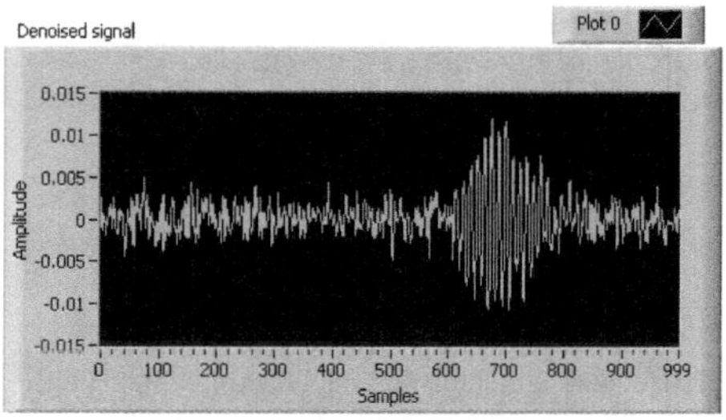

Fig. 2(d). Noise Rejected Using biorthogonal wavelet

Next step is to remove the distortion from the reflected signal. The aim of the present approach is to produce a better resolution and a good noise-rejection performance. Discrete Wavelet Transform is used to get better noise rejection. Figure 2(d) shows the noise rejection using biorthogonal type of wavelet.

The Cross-Correlation method of Time of flight measurement compares the echo envelope with the transmitted signal envelope. For this Echo envelope must be detected. Analog rectifiers followed by low pass filters are traditionally used for this operation, but these devices are less accurate compared to signal-processing algorithm [6]. Therefore Hilbert transform is used to detect the magnitude of the received signal. The detected envelope shown in figure 2(e) is then correlated with the transmitted envelope.

The cross correlation virtual instrument tool built in LabVIEW software has been used to compute the cross correlation between the template envelope, and received echo envelope. Correlation is used to reveal the degree of similarity between one sequence of data and the other as a function of time shift between them. The measurement system acquires the two digital sequences **S(nT)** and **R(nT)** representing the

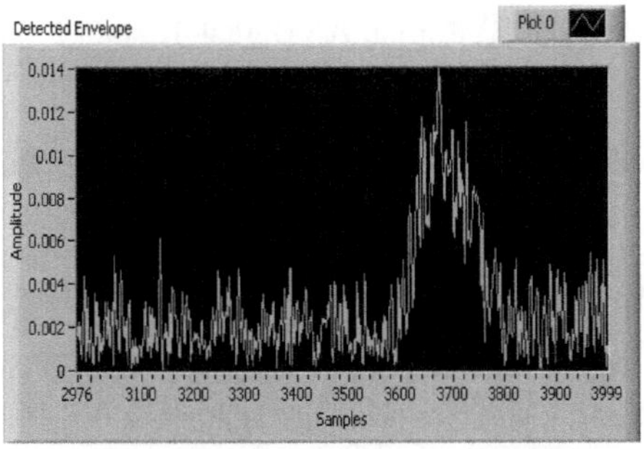

Fig. 2(e). Detected envelope

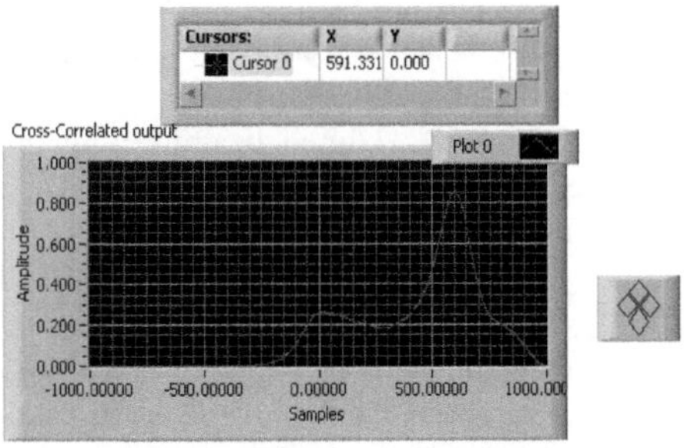

Fig. 2(f). Cross-correlated signal

transmitted and echo signals respectively, where T is Sampling period, considered here is 100Khz . The cross-correlation of the two sequences is given by:

$$\mathbf{C(kT) = \sum_{n=-\infty}^{n=+\infty} S\,(nT)\,R(nT + kT)} \tag{6}$$

k = 0, 1, 2, 3 …, and kT is the shifted sampling points in respect to the S (nT). The delay can be estimated by finding the peak of the cross-correlation function in equation (2), since the amplitude of each sample in the cross-correlation signal is a measure of how much the received signal resembles the template, at that location. In other words, the value of the cross-correlation is maximized when the target signal is aligned with the same features in the received signal is as shown in the figure 2(f)

5 Conclusion

The main program starts with the measure of range in turn Time of flight measurement. Here the Ultrasonic signal of 40 kHz is generated. Figure 2(a) shows the graph of the generated ultrasonic phased array signal. The signal is transmitted for defined distance and delay time is calculated. The received signal is processed by wavelet transform and echo is detected by Hilbert transform. The detected envelope is shown in figure 2(e).To calculate Time of Flight, Cross correlation technique is used. Correlation increases the accuracy of Time of flight measurements as compared to other detection methods. From figure 2(f) the time delay between the two waveforms is 5.9 ms. Then by using equation (1), distance is calculated and compared with the assumed distance. The defined distance was 1m and the output calculated is also 1 metre. Hence Time of flight calculation of ultrasonic phased array signal can be implemented for different ranges. This Time of Flight Simulation is been implemented using LabVIEW. This simulation is now being implemented using hardware component of ultrasonic phased array of five elements.

References

1. Hammad, A., Hafez, A., Elewa, M.T.: A LabVIEW Based Experimental Platform for Ultrasonic Range Measurements. DSP Journal 6(2), 1–8 (2007)
2. Wooh, S.-C., Shi, Y.: A Simulation Study of the Beam Steering Characteristics for Linear Phased Arrays. Journal of Non-destructive Evaluation 18(2) (1999)

3. Marioli, D., Narduzzi, C., Offelli, Petri, D., Sardini, E., Taroni, A.: Digital Time of Flight Measurement of Ultrasonic Sensor. IEEE Transactions on Instrument and Measurements 41(1), 93–97 (1992)
4. Steinberg, D.: Digital beamforming in ultrasound. IEEE Transactions, Ferroelectronics, Frequency Control 39(6), 716–721 (1992)
5. Andria, G., Attivissimo, F., Lanzolla, A.: Digital Measuring Techniques for High Accuracy Ultrasonic Sensor Application. In: IEEE Instrumentation and Measurement Technology Conference, pp. 1056–1061 (1998)
6. Andria, G., Attivissimo, F., Giaquinto, N.: Digital Methods for very accurate Ultrasonic sensor measurements. IEEE Instrumentation and Measurement Technology, 1687–1692 (1999)
7. Li, C., Zhang, L., Hu, X.: Virtual Instrumentation. IEEE Engineering in Medicine And Biology (2005)
8. Parrila, M., Anaya, J.J., Fritsch, C.: Digital Signal Processing Techniques for High Accuracy ultrasonic Range Measurements. IEEE Transactions on Instruments and Measurements 40(4) (1991)
9. von Ramm, O.T., Smith, S.W.: Beam steering with linear arrays: IEEE Transactions on Biomedical Engineering, vol. BME 30(8), 438–452 (1983)
10. Harput, S., Bozkurt, A., Yamaner, F.Y.: Ultrasonic Phased Array Device for Real- Time Acoustic Imaging in Air. In: IEEE International Ultrasonic Symposium Proceedings, pp. 619–621 (2008)
11. Goldberg, R., smith, S., Mottley, J.g.: The Biomedical Engineering Handbook, ch. 65, 2nd edn., pp. 1–17. CRC Press LLC
12. Haykin, S.: Handbook on Array Processing and Sensor Networks, ch. 1, 2, pp. 5–30. Wiley, Chichester

Verifying and Enhancing the Correctness of the Dynamic Composite Web Service

N. Sasikaladevi[1] and L. Arockiam[2]

[1] Research Scholar, Department of Computer Science, St. Joseph's College, Trichy, Tamilnadu, India
sasikalade@yahoo.com
[2] Associate Professor, Department of Computer Science, St. Joseph's College, Trichy, Tamilnadu, India
larockiam@yahoo.co.in

Abstract. Organizational workflows are now automated by incorporating the dynamic composite web services. As more and more web service is evolving now, enforcing the Quality of service (QoS) factors in web service is in need today. Among the various QoS factor, reliability play a vital role. Reliability is defined by using the functional and non functional metrics. Functional metrics are correctness, fault tolerance and testability of the web service. Non functional metrics are timeliness and Interoperability. In this paper, a model is proposed to verify and enhance the correctness of the web service in the composite web service environment.

Keywords: Web Service, XML, SOAP, WSDL, BalckBox Tesing.

1 Introduction

A composite WS, denoted as W0, is a WS whose operations are interfaces to operations provided by other WSs, also known as atomic WSs. The each atomic WS may provide only a subset of operations as required by the composite WS In general, when a client application invokes operations in a composite WS, the composite WS does not execute these operations itself, but serves as a broker that delegates the operations to some atomic WSs instead. One of the more comprehensive characterizations of web services reliability is provided by Zhang and Zhang [1] where both functional and non-functional components are considered, and reliability is defined as a composite of correctness, fault tolerance, testability, availability, performance and interoperability. Correctness here implies that the Web service generates reasonable output from input derived from the problem domain. In this paper, Correctness of the web service is considered.

2 Related Work

Web services testing is a fast moving research area as the complexity of web services increases. The prevalent approach in this area is model-based web service testing.

V.V. Das, J. Stephen, and Y. Chaba (Eds.): CNC 2011, CCIS 142, pp. 431–433, 2011.

Sinha et al [2] proposed a model-based web service test framework where a WSDL operation is augmented with Preconditions and Effects using OWL ontology to specify its functional behavior. Wang et al [3] proposed a web services test generation approach by using Petri net translated from OWL-S process descriptions. Again, the "control flows" have been prescribed instead of being discovered as in our approach. Carroll [4] described a way to match RDF graphs based on graph isomorphism theory. It shows that RDF equality and equivalence can be resolved by testing graph isomorphism. Li Li et. all [5] presents a service oriented architecture for testing Web Services. Narayanan et al [6] adopted a DAML-S based ontological representation of web services and proposed a Petri net formalism for encoding the service semantics for simulation and verification. WSAT [1] is a formal analysis tool for web services composition developed by Fu et al. This tool takes some preexisting "control flows" in BPEL or Conversation Protocol and translated them into GFSA (Guarded Finite State Automata) which is subject to a series of formal analyses.

3 Proposed Model for Testing the Web Service

This section will describe an approach for testing the correctness of the dynamic composite web service. Figure 1 describes the overall architecture of the Dynamic composite Web Services correctness testing approach proposed in this paper.

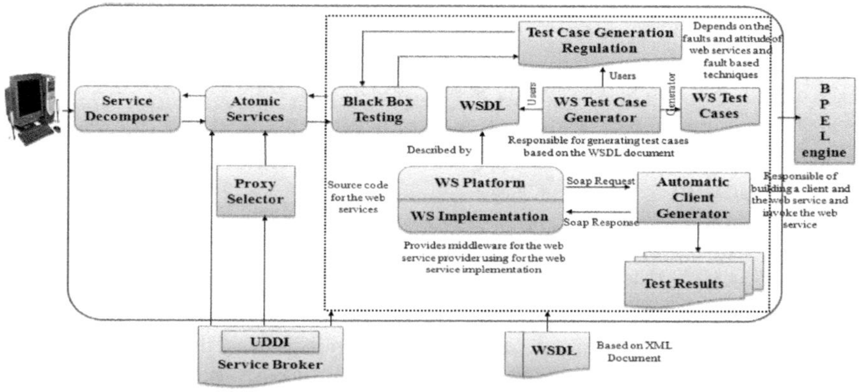

Fig. 1. Web service testing

The components of the architecture in Figure 1 are describes as follows: Service decomposer selects the atomic services that are involved in composite web service. They could measure every node of this business process. For the operation relationship, they adopt Jorge Cardoso's Stochastic Workflow Reduction (SWR) algorithm. Service Broker is used to find out the web services that are involved in a dynamic composite service. Proxy selector is used to select alternative service from the service broker if the primary service fails in the correctness testing. Black box tester is used to perform functional testing on the selected web service. Web Service platform is the platform or middleware that the Web Service Provider is using for his Web Service implementation. Test case generation rules are the rules that is proposed by this paper for test case generation, these rules depends on the faults that affects the robustness quality attribute of Web Services and the fault-based testing techniques. Web Service implementation is the source code of the Web Service that is written by the Web Service Provider. Automatic Client Generator is the component that is responsible of building a client to the Web service under test and invoking the Web service under test using the test data provided by the Web Service test case generator component. After defining the different components of the overall architecture in Figure 1, the interaction between those components will be described: Client request the composite web service. Primary services involved in the composite web services are selected from Service broker. WSDL file is given to the Black box tester to perform the functional testing. Black box tester receives the test cases from the Test case generator. If the selected primary service fails in the functional test, then alternative proxy service is selected and it will undergo functional test. Functionally correct web services are given to BPEL engine to composition.

4 Conclusion

A model for verifying the correctness of the dynamic composite web service is proposed. Black box testing is used to perform the functional testing on the web service. In future, other functional and nonfunctional metrics will be considered to enhance the reliability of the dynamic composite web service. This Frame work is not implemented but in future it makes a growth in web service environment.

References

1. Fu, X., Bultan, T., Su, J.: WSAT: A Tool for Formal Analysis of Web Services. In: Alur, R., Peled, D.A. (eds.) CAV 2004. LNCS, vol. 3114, pp. 510–514. Springer, Heidelberg (2004)
2. Sinha, A., Paradkar, A.: Model Based Functional Conformance Testing of Web Services Operating on Persistent Data. In: Proceedings of the 2006 Workshop on Testing, Analaysis, and Verification of Web Services and Applications TAV-WEB 2006, Portland, Maine, July 17, pp. 17–22 (2006)
3. Wang, Y., Bai, X., Li, J., Huang, R.: Ontology-Based Test Case Generation for Testing Web Services. In: Proceedings of 8th IEEE International Symposium on Autonomous Decentralized Systems (ISADS 2007) (2007)
4. Carroll, J.J.: Matching RDF graphs. In: Horrocks, I., Hendler, J. (eds.) ISWC 2002. LNCS, vol. 2342, pp. 5–15. Springer, Heidelberg (2002)
5. Li, L., Chou, W., Guo, W.: Control Flow Analysis and coverage driven testing for web service. In: Proceedings of the IEEE International Conference on Web Service, pp. 473–481 (2008)
6. Narayanan, S., McIlraith, S.A.: Simulation, Verification and Automated Composition of Web Services. In: Proceedings of WWW 2002, Honolulu, Hawaii, USA, May 7-11, pp. 77–88 (2002)
7. Cardoso, J., Miller, J., Sheth, A., Arnold, J.: Modeling Quality of Service for Workflows and Web Service Processes. Technical Report #02-002, LSDIS Lab, Computer Science, University of Georgia (2002); International Symposium on Service-Oriented System Engineering (2008)

Script Identification for a Tri-lingual Document

Prakash K. Aithal[1], G. Rajesh, Dinesh U. Acharya,
M. Krishnamoorthi, and N.V. Subbareddy

[1] Manipal Institute of Technology, Manipal University, Manipal, India
prakash.aithal@manipal.edu

Abstract. India is a multilingual multi-script country. States of India follow a three language formula. The document may be printed in English, Hindi and other state official language. For Optical Character Recognition (OCR) of such a multilingual document, it is necessary to identify the script before feeding the text lines to the OCRs of individual scripts. In this paper, a simple and efficient technique of script identification for Tamil, Hindi and English text lines from a printed document is presented. The proposed system uses horizontal projection profile to distinguish the three scripts. The feature extraction is done based on the horizontal projection profile of each text line. The knowledge base of the system is developed based on 20 different document images containing about 600 text lines. The proposed system is tested on 20 different document images containing about 200 text lines of each script and an overall classification rate of 100% is achieved.

Keywords: Multiscript, Horizontal projection profile, Rule based classifier.

1 Introduction

In multi-lingual document analysis, it is important to automatically identify the scripts before feeding each text line of the document to the respective OCR system. Quite a few results have already been reported in the literature, identifying the scripts in a multi-lingual and multi-script document dealing with Roman and other Oriental scripts such as Chinese, Korean, Japanese, Arabic and Hindi. The Classifiers used include statistical analysis, linear discriminate analysis, cluster analysis and template matching based on the features like texture, upward Concavities, optical densities and characteristic shapes or symbols [5].

In India, a multi-lingual multi-script country, 18 official languages are written in 12 different scripts. Under three-language formula, the documents for a region are printed in English, Hindi (Devnagari) and the regional official language. Our aim is to find approaches of separating all script triplets formed in this way. In this context there is a potential requirement of script identification in tri-script documents. The proposed system addresses the script identification pertaining to Tamilnadu, a state in south India [1, 3, 13].

In the context of Indian language document analysis, major literature is due to Pal and Choudhari. The automatic separation of text lines from multi-script documents by extracting the features from profiles, water reservoir concepts, contour tracing [1, 2].

V.V. Das, J. Stephen, and Y. Chaba (Eds.): CNC 2011, CCIS 142, pp. 434–439, 2011.

Santanu Choudhury, Gaurav Harit, Shekar Madnani and R. B. Shet has proposed a method for identification of Indian languages by combining Gabor filter based technique and direction distance histogram classifier considering Hindi, English, Malayalam, Bengali, Telugu and Urdu [4]. Chanda and Pal have proposed an automatic technique for word wise identification of Devnagari, English and Urdu scripts from a single document [6]. Gopal Datt Joshi et al. has proposed script Identification from Indian Documents [7]. Word level script identification in bilingual documents through discriminating features has been developed by B V Dhandra, et al. [8]. Neural network based system for script identification (Kannada, Hindi and English) of Indian documents is proposed by Basavaraj Patil and N. V. SubbaReddy [9]. Lijun Zhou Yue Lu and Chew Lim Tan have developed a method for Bangla and English script identification based on the analysis of connected component profiles [10]. Vijaya and Padma has developed methods for English, Hindi and Kannada script identification using discriminating features and top and bottom profile based features [11,15]. Ahmed M Elgammal et al. have done work on identification of English and Arabic languages they used the horizontal projection profile based feature[12].D Dhanya et al. have proposed a method to identify Tamil and English script using character density and gabor filters[14]. Prakash K. Aithal et al. have proposed a technique for separating English, Hindi and Kannada Scripts [16]. Vijaya and Padma have worked on wavelet based texture feature based method to classify ten Indian Languages[17].

This paper deals with line-wise script identification for Tamil, Hindi and English script pertaining documents from Tamilnadu. The proposed system has used horizontal projection profile based feature extraction technique to classify the scripts. The first and second peak points from the horizontal projection profile of the text line are extracted for identifying Hindi script from others. The number of histograms crossing the mean between first and second peak points is used to classify the English and Tamil.

The rest of the paper is organized as follows: Section 2 gives the segmentation method used. The feature extraction method is discussed in Section 3. Section 4 presents the classification based on rule based classifier. Finally, result analysis is done in Section 5.

2 Segmentation

White space between text lines is used to segment the text lines. The line segmentation is carried out by calculating the horizontal projection profile of the whole document. The horizontal projection profile is the histogram of number of ON (black) pixels along every row of the image. The projection profile exhibits valleys of zero height corresponding to white space between the text lines. Line segmentation is done at these points. Fig. 1 shows the horizontal profile for a sample document.

3 Feature Extraction

The proposed system has used horizontal projection profile based feature extraction technique to classify the scripts. The first and second peak points from the horizontal projection profile of the text line are extracted for identifying Hindi script from others. The number of histograms crossing the mean between first and second peak points is used to classify the English and Tamil.

4 Rule-Based Classifier

The Proposed system uses a rule-based classifier for script identification. The rules are as follows

1) For each text line calculate the first (V1) and second (V2) maxima of the horizontal projection profile.
2) If first maximum greater than 1.5 times the second maximum then the script is identified as Hindi.
3) The unidentified lines are classified by the following steps
   a) Calculate the mean (M) of the projection profile.
   b) Find out number of histograms crossing the mean (M) which are between first and second maxima excluding both.
   c) If the number of histograms crossing the mean is 5 or 2 then the line is classified as English. If the number of histograms crossing the mean is 4 or 1 or 0 then the line is Tamil.

The proposed system uses minimum distance classifier to classify English and Tamil. (Minimum distance classifier works as follows suppose that each pattern class, w_j, is characterized by the mean vector m_j and we have set of patterns arranged as the rows of a Matrix Z then selecting the smallest distance is equivalent to evaluating the function $d_j(Z)=Z^T m_j -(1/2)\ m_j{}^T m_j$ and decision boundary between classes w_i and w_j will be $d_{ij}(Z)=d_i(Z)-d_j(Z)$)The minimum distance query matrix will be [4, 5, 1, 2, 0, 3]. If the minimum distance index is 1, 3, 5 then script is Tamil. If the minimum distance index is 2, 4, 6 then script is English. The minimum distance index for a sample document is shown in Fig.2.

5 Experimental Results

The dataset includes 1200 text lines from 40 different document images. The document images are downloaded from e-news papers (Times of India, Dinamalar and Navbharat times respectively for English Tamil and Hindi). The knowledge base (rules) of the system is developed based on 20 document images containing about 600 text lines of all three scripts. The proposed system is tested on 20 different document images containing about 200 text lines of each script and an overall classification rate of 100% is achieved.Fig.3 shows the relation between the first and second maxima of the different script text lines which used as a discriminating feature to identify Hindi scripts from others.

Table 1 shows the range of V2/V1 for 600 text lines Hindi and non Hindi scripts and Table 2 gives the minimum distance index for the Tamil and English text lines. Table 3 give the confusion matrix for the proposed script identification. The comparison of the proposed work with contemporary works is shown in table 4.

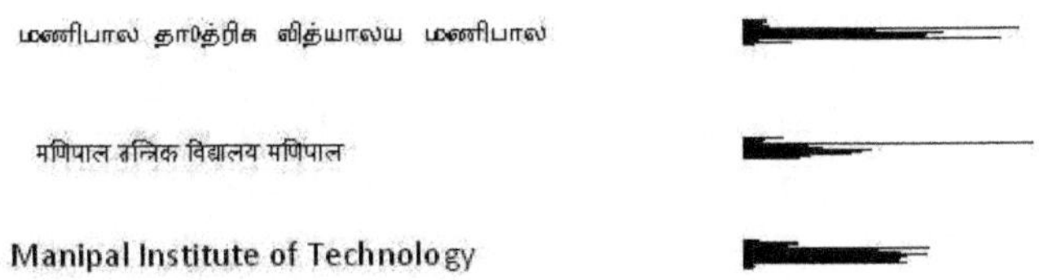

Fig. 1. Horizontal Projection Profile of a sample document with all 3 scripts

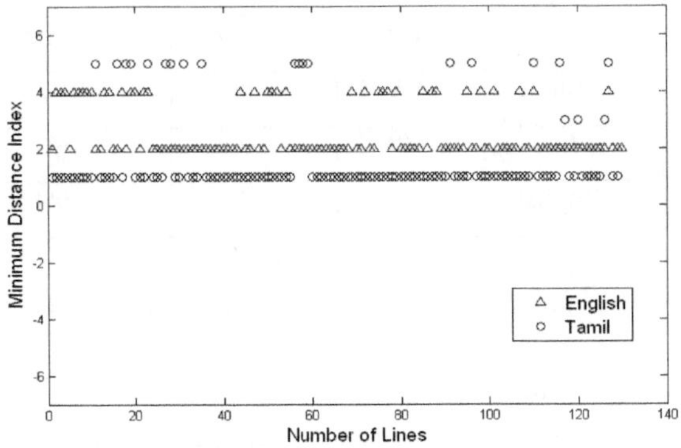

Fig. 2. Minimum distance index obtained from minimum distance classifier

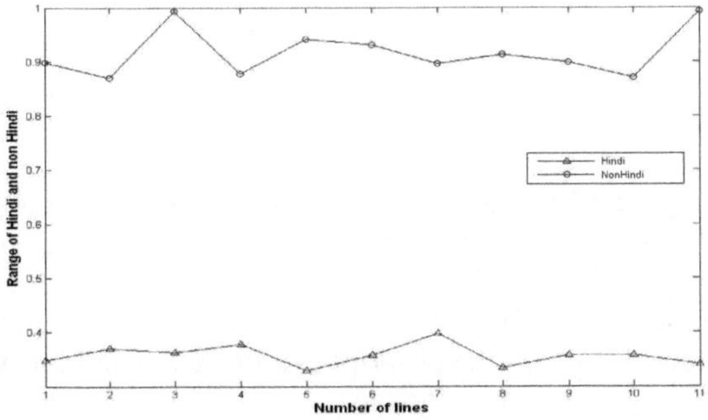

Fig. 3. Range V2/V1 for Hindi and Non Hindi scripts

Table 1. Font sizes Range of V2/V1 for Hindi and Non-Hindi text lines

Language	Range
Hindi	0.01 to 0.58
Non Hindi	0.69 to 0.99

Table 2. Minimum distance index for Tamil and English text lines

Language	Minimum Distance Index
Tamil	1 or 3 or 5
English	2 or 4 or 6

Table 3. Confusion matrix of script identification

	Tamil	Hindi	English
Tamil	100%	0%	0%
Hindi	0%	100%	0%
English	0%	0%	100%

Table 4. Comparison of proposed work with contemporary works

Methodology	Dataset	Overall accuracy
Separation scheme using script characteristics and shape based features [1]	500 (approx.)	97.7%
Proposed work	600	100%

6 Conclusion

In this paper, a simple and efficient algorithm for script identification of Tamil, Hindi and English text lines from printed documents is proposed. The approach is based on the analysis of horizontal projection profile and does not require training or any character or word segmentation. The system exhibits an overall accuracy of 100%. The work could be extended to word level script identification and for other Indian scripts.

References

1. Pal, U., Choudhuri, B.B.: Script line separation from Indian multi-Script documents. In: Fifth International Conference on Document Analysis and Recognition, pp. 406–409. IEEE Computer Society Press, Los Alamitos (1999)
2. Pal, U., Sinha, S., Choudhuri, B.B.: Multi-Script line identification from Indian documents. In: Seventh International Conference on Document Analysis and Recognition, ICDAR, pp. 880–884 (2003)
3. Padma, M.C., Nagabhushan, P.: Identification and separation of text words of Kannada, Hindi and English languages through discriminating features. In: Second National Conference on Document Analysis and Recognition, Karnataka,India, pp. 252–260 (2003)
4. Choudhury, S., Harit, G., Madnani, S., Shet, R.B.: Identification of Scripts of Indian Languages by Combining Trainable Classifiers. In: ICVGIP, Bangalore, India (2000)

5. Tan, T.N.: Rotation Invariant Texture Features and their use in Automatic Script Identification. IEEE Trans. on Pattern Analysis and Machine Intelligence 20(7), 751–756 (1998)
6. Chanda, S., Pal, U.: English, Devanagari and Urdu Text Identification. In: International conference on Document Analysis and Recognition, pp. 538–545 (2005)
7. Joshi, G.D., Garg, S., Sivaswamy, J.: Script Identification from Indian Documents. In: Bunke, H., Spitz, A.L. (eds.) DAS 2006. LNCS, vol. 3872, pp. 255–267. Springer, Heidelberg (2006)
8. Dhandra, B.V., Hangarge, M., Hegadi, R.: Word Level Script Identification in Bilingual Documents through Discriminating Features. In: IEEE - ICSCN 2007, Chennai, India, pp. 630–635 (2007)
9. Patil, B., SubbaReddy, N.V.: Neural network based system for script identification in Indian documents. In: Sadhana, India, vol. 27(part 1), pp. 83–97 (2002)
10. Zhou, L., Lu, Y., Tan, C.-L.: Bangla/English ScriptIdentification Based on Analysis of Connected Component Profiles. In: Bunke, H., Spitz, A.L. (eds.) DAS 2006. LNCS, vol. 3872, pp. 243–254. Springer, Heidelberg (2006)
11. Vijaya, P.A., Padma, M.C.: Text line identification from a multilingual document. In: International Conference on Document Image Processing (ICDIP-2009), Bangkok, pp. 302–305 (2009)
12. Ahmed Elgammal, M., Mohamed Ismail, A.: Techniques for language identification for hybrid Arabic-English document images. In: Sixth International Conference on Document Analysis and Recognition, Seattle, pp. 1100–1104 (2001)
13. Kasturi, R., Lawrence Gorman, O., Raju, V.G.: Document image analysis- a primer. In: Sadhana, India, vol. 27(part 1), pp. 3–22 (2002)
14. Dhanya, D., Ramakrishnan, A.G., Basapati, P.: Script Identification in printed bilingual documents. In: Sadhana, India, vol. 27(part 1), pp. 73–82 (2002)
15. Padma, M.C., Vijaya, P.A.: Identification and separation of Text words of Kannada, Telugu, Tamil, Hindi and English languages through visual discriminating features. In: International Conference on Advances in Computer Vision and Information Technology (ACVIT 2007), Aurangabad, India, pp. 1283–1291 (2007)
16. Prakash, Aithal, K., Rajesh, G., Dinesh Acharya, U., Krishnamoorthi, M., Subbareddy, N.V.: Text Line Script Identification for a Tri-lingual Document. In: International Conference on Computing Communication and Networking Technologies, Karur, India, pp. 1–3 (2010)
17. Padma, M.C., Vijaya, P.A.: Wavelet Packet Based Texture Features for Automatic Script Identification. International Journal of Image Processing 4(1) (2010)

Analysis of Different Key Distribution Schemes for Distributed Sensor Networks

Peeyush Jain and Zia Saquib

Centre for Development of Advanced Computing,
Mumbai, India
{peeyushj,saquib}@cdacmumbai.in

Abstract. DSN is a special network which can provide a useful interface to the real world with their data acquisition and processing capabilities. However, DSN is extremely vulnerable against any type of internal or external attacks, due to resource constraints, lack of tamper-resistant packages, and the nature of its communication channels. Therefore, it is necessary to develop useful security mechanisms which are suitable for DSN. With the recent acceleration in research into secure key distribution, we compare and contrast different key distribution approaches. We aim to identify the distinguish features and weighing up the benefits and drawbacks of each.

Keywords: Key Predistribution, Distributed Sensor Network, Network Security.

1 Introduction

In general, DSN consist of battery-operated sensor devices with computing, data processing, and communicating components. In most application domains, the sensors are used to collect a specific type of data from particular target areas, and the collected data are often considered secret and are not intended for public disclosure. DSN is extremely vulnerable against any kind of internal or external attacks than traditional network. Hence, efficient and secure mechanisms are needed to transmit acquired data securely to the appropriate recipients. All security mechanisms require a certain amount of resources for the implementation such as: data memory, code space and energy to power the sensor. These resources, however, are very limited like Limited Memory and Storage Space, Limited bandwidth and Power Limitation. The limited computation and power resources of sensor nodes often makes it undesirable to use public-key algorithms. Sensor nodes may be deployed in public or hostile locations in many applications. Furthermore, the large number of nodes that are deployed implies that each sensor node must be low-cost, which makes it difficult for manufacturers to make them tamper-resistant. This exposes sensor nodes to physical attacks by an adversary and it may be able to undetectably take control of a sensor node and compromise the cryptographic keys. Sensor networks are mostly deployed via random scattering, the sensor network protocols cannot know beforehand which nodes will be within communication range of each other after deployment. Hence, a security

V.V. Das, J. Stephen, and Y. Chaba (Eds.): CNC 2011, CCIS 142, pp. 440–445, 2011.

protocol should not assume any prior knowledge. Sensor Networks are vulnerable to many attacks. It should have the following general security goals:

1. Confidentiality: protecting secret information from unauthorized entities
2. Integrity: ensuring message has not been altered by malicious nodes
3. Data Origin Authentication: authenticating the source of message
4. Access Control: restricting access to resources to privileged entities
5. Availability: ensure desired service may be available whenever required
6. Survivability: providing service in the presence of failures or attacks
7. Freshness: ensure data is recent and no adversary can replay old messages
8. Scalability: supporting a great number of nodes
9. Efficiency: storage, processing and communication limitations.

2 Key Distribution Schemes

To implement a fundamental security service, pair-wise key establishment should be used, enabling secure communications among the sensor nodes. There are three types of general key agreement schemes: trusted-server scheme, self-enforcing scheme, and key predistribution scheme. The trusted-server scheme depends on a trusted server for key agreement between nodes e.g. Kerberos[7]. This scheme is not suitable for sensor networks because there is no trusted infrastructure in sensor networks. The self-enforcing scheme depends on asymmetric cryptography, such as key agreement using public key certificates. However, limited computation and energy resources of sensor nodes often make it undesirable to use public key algorithms, such as Diffie-Hellman key agreement[8] or RSA[9]. So, it is necessary to use new algorithms based on more optimal approaches, such as Elliptic Curve Cryptography (ECC). The third type of key agreement scheme is key predistribution, where key information is distributed among all sensor nodes prior to deployment. However, most sensor network deployments are random; thus, such a priori knowledge does not exist. There exist number of key predistribution schemes which do not rely on a priori knowledge.

2.1 The Eschenauer-Gligor (EG) Scheme

Eschenauer and Gligor first proposed a random key pre-distribution scheme[1] which consists of three phases: key predistribution, shared-key discovery, and path-key establishment. Before sensor nodes are deployed, an initialization phase is performed. In the key predistribution phase, the basic scheme picks a random pool of keys S out of the total possible key space. After the sensor nodes are deployed, a key-setup phase is performed. The nodes first perform key-discovery to find out with which of their neighbors they share a key. The shared key then becomes the key for that link. After key-setup is complete, a connected graph of secure links is formed. Nodes can then set up path keys with nodes in their vicinity whom they did not happen to share keys with in their key rings. If the graph is connected, a path can be found from a source node to its neighbor. The source node can then generate a path key and send it securely via the path to the target node. Consider a random graph G(n, pl) of n nodes for which the probability that a link exists between any two nodes is pl. Calculate the necessary expected node degree d in terms of the size of the network n as:

$$d= ((n-1)/n)(\ln(n) - \ln(-\ln(c)))$$

From the formula, d = O(log n). For a given density of sensor network deployment, let n' be the expected number of neighbors within communication range of a node. Since the expected node degree must be at least d as calculated, the required probability p of successfully performing key-setup with some neighbor is: p = d/n'. Since the models of connectivity are probabilistic, there is always the chance that the graph may not be fully connected. This chance is increased if the deployment pattern is irregular or has unpredictable physical obstacles to communication.

2.2 q-Composite Random Key Pre-distribution Scheme

This scheme[2] propose a modification to the basic scheme where q common keys (q > 1) are needed, instead of just one. By increasing the amount of key overlap required for key-setup, it increase the resilience of the network against node capture. As the key overlap increases, it becomes exponentially harder for an attacker with a given key set to break a link. However, to preserve the given probability p of two nodes sharing sufficient keys to establish a secure link, it is necessary to reduce the size of the key pool |S|. This allows the attacker to gain a larger sample of S by breaking fewer nodes. The interplay of these two opposing factors results in an optimal amount of key overlap to pose the greatest obstacle to an attacker for some desired probability of eavesdropping on a link.

In the initialization phase, a set S of random keys out of the total key space are picked. For each node, m random keys are selected from S and store them into the node's key ring. In the key-setup phase, each node must discover all common keys it possesses with each of its neighbors. This can be accomplished with a simple local broadcast of all key identifiers that a node possesses. After key discovery, each node can identify every neighbor node with which it shares at least q keys. A new communication link key K is generated as the hash of all shared keys, e.g., K= hash(k1||k2|| . . ||kq). The keys are hashed in some canonical order, for example, based on the order they occur in the original key pool S. Key-setup is not performed between nodes that share fewer than q keys. q-composite key scheme strengthens the network's resilience when the number of nodes captured is low.

2.3 Peer Intermediaries for Key Establishment (PIKE)

PIKE, involves using one or more sensor nodes as a trusted intermediary to facilitate key establishment [6]. It shows that both the communication and memory overheads of PIKE scale sub-linearly (O(pn)) with the number of nodes in the network yet achieving higher security against node compromise than others. It is designed to address the lack of scalability of existing schemes as these schemes incur linearly increasing costs (O(n)) in either communications per node or memory per node. PIKE achieves a trade-off by achieving $O(n^{1/2})$ overheads in both. It establishes keys between any two nodes regardless of network topology or node density which makes it applicable to a wider range of deployment scenarios. PIKE is more resilient than previous approaches against sensor node compromise.

The basic idea is to use sensor nodes as trusted intermediaries to establish shared keys between nodes. Each node shares a unique pairwise key with each of O(n1/2) other nodes in the network. The keys are deployed such that for any two nodes A and B, it is possible to find some node C in the network that shares a unique pairwise key with both A and B. A can then securely route the key establishment message through C to B. Since we only predistribute unique pairwise keys that are shared between exactly two nodes, the established key is secure if C has not been compromised.

2.4 Blom's Key Pre-distribution Scheme

Blom proposed a key pre-distribution scheme to make any pair of users of a network find a pairwise secret key[1]. The construction of the scheme starts with creating a (L + l)xN matrix H over a finite field GF(q), where the N is the size of the network. This H will be known by all users in the network. Then the key generating authority create a random symmetric matrix D over GF(q), and D must be kept secret. The keys to be used by the user pairs are then given by $K = (DH)^T H$. K is symmetric and hence if user i knows row i of K and user j knows row j of K, then have a common key $K_{i,j} = K_{j,i}$. For example, if node i and j need to find the pairwise key between them, node i should store the ith row of matrix $(DH)^T$ and the ith column of matrix H and node j should store the jth row of matrix $(DH)^T$ and the jth column of matrix H before they are deployed, and then they exchange their columns of H. So, they can compute $K_{i,j}$ and $K_{j,i}$, respectively, using their private rows of $(DH)^T$. This scheme is L secure in that as long as no more than L nodes are compromised, the network is secure[11].

2.5 The Du-Deng-Han-Varshney (DDHV) Scheme

Some slight modifications to the Blom scheme in order to make it more suitable for sensor networks have been made, but the essential features remain unchanged. It is agreed-upon (L+1)*N matrix G over a finite field GF(q), where N is the size of the network and q>N. G is public information and may be shared by different systems. In key generation phase, the base station creates a random (L+1)*(L+1) symmetric matrix D over GF(q), and computes an N*(L+1) matrix $A=(D.G)^T$, where $(D.G)^T$ is the transpose of (D.G). Matrix D must be kept secret (although, one row of $(D.G)^T$ will be disclosed to each sensor node). Because D is symmetric, it is easy to see that

$$A.G = (D.G)^T.\ G = G^T.D^T.G = G^T.D.G = (A.G)^T$$

A.G is a symmetric matrix. let K = A.G, and $K_{i,j} = K_{j,i}$, where $K_{i,j}$ is the element in the ith row and jth column of K. To carry out the above computation, nodes i and j should be able to compute $K_{i,j}$ and $K_{j,i}$, respectively. This can be easily achieved using the following key predistribution scheme, for k =1, . . . , N:

1. store the kth row of matrix A at node k, and
2. store the kth column of matrix G at node k.

When nodes i and j need to establish their pairwise key, they first exchange their columns of G and then compute $K_{i,j}$ and $K_{j,i}$, respectively, using their private rows of A. This L-secure property guarantees that no coalition of up to L nodes (not including i and j) have any information about $K_{i,j}$ or $K_{j,i}$.

3 Analysis

Eschenauer and Gligor proposed a random key predistribution scheme[1]: before deployment, each sensor receives a random subset of keys from a large key pool. Two neighboring nodes find one common key within their subsets and use that key as their shared secret key. It cannot guarantee the connectivity and the exclusive key pairing between any pair of nodes in a network. Because each node is randomly supplied with some secret keys from a large key pool, and then it tries to find some direct or indirect common keys with its neighbors in certain probability. Based on the EG scheme, Chan et al. proposed a q-composite random key predistribution scheme. The major difference between these schemes is that q common keys (q1), instead of just a single one, are needed to establish secure communications between a pair of nodes. E-G scheme is suitable for large wireless sensor networks which allows addition and deletion of sensors, resilient to sensor node capture. Data and Information Spoofing attacks could be deterred. Chan et al. scheme provide resilience of the network, protect the network even if part of the network is compromised. The limitation of these approaches is that a small number of compromised sensor nodes may affect the secure communication between a large number of non-compromised sensor nodes. In addition to improving Blom's scheme in [2], Du et. al. proposed two schemes for key predistribution[10,3]. In the early one they introduced a deployment knowledge based scheme that improves Blom's[2] by avoiding the unnecessary memory, communica tion, and computation with reasonable connectivity[10]. Motivated by the random key predistribution schemes [2], Du et al. developed an improved key pre-distribution scheme (DDHV scheme) using multiple key spaces [3]. The DDHV scheme first constructs w Spaces using Blom's scheme and then has each sensor node carry key information from t (2 <= t < w) randomly selected key spaces. The resilience of this scheme is better than E-G scheme. However, it costs too much computational expense. A similar method was also developed by Liu and Ning, using pre deployment knowledge. The grid-based idea was first proposed in Liu and Ning [2003a] to arrange the secrets in sensor networks based on a logical grid. A similar idea was later used in PIKE [Chan and Perrig 2005]. PIKE involve lower memory storage requirements than random key distribution while requiring comparable communication overheads. PIKE is currently the only symmetric key predistribution scheme which scales sub-linearly in both communication overhead per node and memory overhead per node while being resilient to an adversary capable of undetected node compromise. PIKE enjoys a uniform communication pattern for key establishment, which is hard to disturb for an attacker. The distributed nature of PIKE also does not provide a single point of failure to attack, providing resilience against targeted attacks. In contrast to the currently popular random-key predistribution mechanisms, PIKE has the advantage that key establishment is not probabilistic, so any two nodes are guaranteed to be able to establish a key.

A number of key establishment protocols based on predistribution are explored, but they do not scale effectively to large sensor networks. For a given level of security each protocol incurs a linearly increasing overhead in either communication cost per node or memory per node. These symmetric key based schemes are computationally efficient, the trade-off has to be paid for complicated key predistribution and key management. In particular, the public key cryptography, symmetric key encryption and the addition of SKE-based key management will likely make strong security a more realistic expectation in the future. We also expect that the hardware of sensor node will be improved so that it can be suitable for the application of cryptography.

References

1. Eschenauer, L., Gligor, V.D.: A key management scheme for distributed sensor networks. In: Proceedings of the 9th ACM Conference on Computer and Communication Security, pp. 41–47 (November 2002)
2. Chan, H., Perrig, A., Song, D.: Random key predistribution schemes for sensor networks. In: IEEE Symposium on Security and Privacy, Berkeley, California, pp. 197–213 (May 2003)
3. Du, W., Deng, J., Han, Y.S., Varshney, P.K.: A pairwise key pre-distribution scheme for wireless sensor networks. In: Proc. of the 10th ACM conference on Computer and communications security, pp. 42–51 (2003)
4. Liu, D., Ning, P.: Establishing pairwise keys in distributed sensor networks. In: Proc. of the 10th ACM Conference on Computer and Communications Security, CCS 2003 (2003a)
5. Liu, D., Ning, P.: Improving key predistribution with deployment knowledge in static sensor networks. ACM Trans. Sensor Netw. 1(2), 204–239 (2005)
6. Chan, H., Perrig, A.: PIKE: Peer Intermediaries for Key Establishment in Sensor Networks. In: IEEE INFOCOM (2005)
7. Clifford Neuman, B., Ts'o, T.: Kerberos: An Authentication Service for Computer Networks. IEEE Communications 32(9), 33–38 (1994)
8. Diffie, W., Hellman, M.E.: New directions in cryptography. IEEE Transactions on Information Theory 22, 644–654 (1976)
9. Rivest, R.L., Shamir, A., Adleman, L.M.: A method for obtaining digital signatures and public-key cryptosystems. Communications of the ACM 21(2), 120–126 (1997)
10. Du, W., Deng, J., Shan, Y., Chen, S., Varshney, P.K.: A Key Management Scheme for Wireless Sensor Networks Using Deployment Knowledge. In: INFOCOM 2004, vol. 1, pp. 7–11. IEEE Computer and Communications Societies (2004)
11. Du, W., et al.: A pairwise key pre-distribution scheme for wireless sensor networks. In: Proc. 10th ACM conf. on CCS, Washington, DC, pp. 42–51 (2003)
12. Perrig, A., Szewczyk, R., et al.: SPINS: Security protocols for sensor networks. ACM Wireless Networks (5), 521–534 (2001)
13. Blom, R.: An optimal class of symmetric key generation systems. In: Beth, T., Cot, N., Ingemarsson, I. (eds.) EUROCRYPT 1984. LNCS, vol. 209, pp. 335–338. Springer, Heidelberg (1985)

Audio Database Watermarking for Tamper Detection

V. Prasannakumari and V. Balu

Department of computer science, SCSVMV University, Enathur, Kanchipuram, TN, India
vprasannakumari@gmail.com, balukanchi@gmail.com

Abstract. The piracy of digital assets such as software, images, video, audio and text has long been a concern for owners of these assets. Protection of these assets is usually based upon the insertion of digital watermarks into the data. Proving ownership rights, authenticity and origin on outsourced relational databases is a crucial issue in internet-based application environments and in many content distribution applications. This paper is an approach towards watermarking audio databases for tamper detection.

Keywords: Audio Databases, Watermarking, Tamper detection, MMDBMS.

1 Introduction

The increasing use of databases in applications beyond "behind-the-firewalls data processing" is creating a similar need for watermarking databases. In the last years watermarking techniques have emerged as an important building block which plays a crucial role in addressing the ownership problem. Such techniques allow the owner of the data to embed an imperceptible watermark into the data.

A watermark describes information that can be used to prove the ownership of data, such as the owner, source, or recipient of the content. Secure embedding requires that the embedded watermark must not be easily tampered with, forged, or removed from the watermarked data. Imperceptible embedding means that the presence of the watermark is unnoticeable in the data. Furthermore, the watermark detection is blinded, that is, it neither requires the knowledge of the original data nor the watermark.

2 Audio Database Watermarking

There have been numerous watermarking techniques [1] proposed and implemented for audio files treating them as domain of frequency,time, compressed or wavelet. Similarly, there are a number of watermarking techniques [2..6] implemented for relational databases, securing them against tampering and ownership rights.

Our study is on audio database where audio files are organized and accessed as a relational databaseTheir applications typically access the audio file that is stored in the database with other metadata such as author, casting, musician, lead instruments, album, subject, lyrics title etc.

V.V. Das, J. Stephen, and Y. Chaba (Eds.): CNC 2011, CCIS 142, pp. 446–448, 2011.

3 Proposed Watermarking Model for Audio in DB

3.1 Encoding

For representation, Base64 holds good for the audio file also. It is used to watermark the audio content when appended with the generated string for all the attributes in the tuple. This serves as a strong check point to compare and detect any suspicious tampering of data in its original audio file or at any of the attributes in the table.

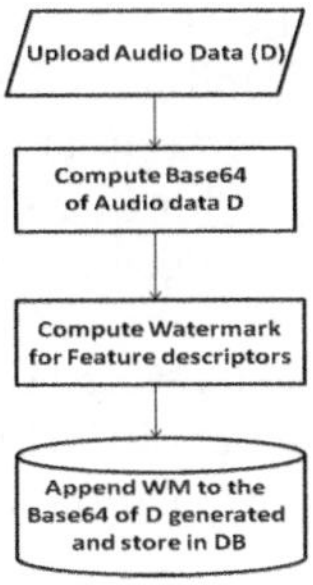

Fig. 1. Encoding of Audio content and its feature descriptors

3.2 Verification for Tampering

The audio content represented in its Base64 form in the database is verified for authenticity before it is provided to the user. User is given a choice to choose from top 3 results that match his search criteria and the file selected by him is verified against the entry in the database by converting it to its base64 form. If they both happen to be same, the audio file is certified to be tamper-proof or authentic. Otherwise the user is let know that there has been an attempted tampering on it.

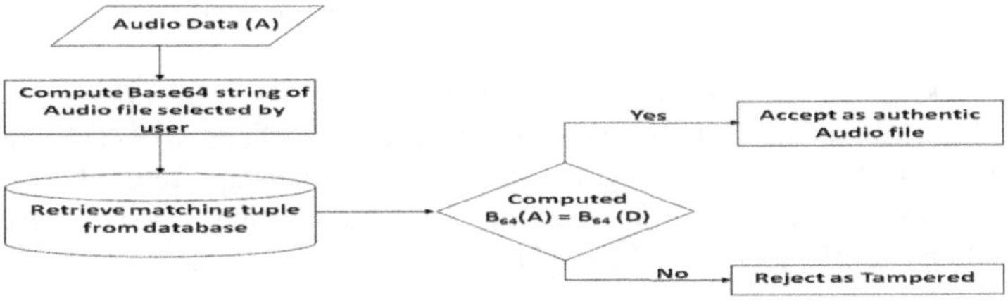

Fig. 2. Verification of audio file to be authentic

3.3 Search Capability

Our watermarking system also provides a search capability by any of the contextual attributes or clippings of audio. We have used stemming as a method to trim all related words to a root word which will enable more number of relevant data to be retrieved for a query. Learning by feedback concept is used to update the stemming database to improve the precision after every search with a new search word.

4 Attach Model

The proposed model of representing and storing the audio files for constructing an audio database is examined to be resilient to the following common attacks possible.

Insert attack: If the attack is to insert an audio file into the location where all files are located, since there is no corresponding tuple in the database, that will never come to the point of user selection or retrieval.

Modify/alter attack: If the music file is altered or overwritten with the same name, the base64 that has been computed and stored in the database stands unique, thus differentiating the tampered. If the alteration is done on the tuple level either on the feature descriptors or on link to the audio file, it will be discovered during the comparison done before authenticating the selected content by the user.

Delete attack: If an audio file has been deleted from the stored location or folder, the tuple becomes orphaned. Though the data is lost because of this type of attack, it can be reconstructed from the base64 representation that is available in the database.

References

1. Alsalami, M.A.T., Al-Akaid, M.M.: Digital Audio Watermarking Survey, `http://ducati.doc.ntu.ac.uk/uksim/ESM2003/Papers/Track-General/CS-16/paper%20CR.pdf`
2. Agarwal, R., Kiernan, J.: Watermarking Relational Databases. In: Proceedings of 28th VLDB Conference, Hong Kong, pp. 155–166 (2002), `http://www.cse.ust.hk/vldb2002/VLDB2002-proceedings/papers/S05P03.pdf`
3. Shehab, M., Bertino, E., Ghafoor, A.: Watermarking Relational Databases using Optimization Based Techniques. IEEE Transactions on Knowledge and Data Engineering 20(1), 116–129 (2008)
4. Sion, S., Atallah, M., Prabhakar, S.: Relational data rights protection through watermarking. IEEE Trans. Knowl. Data Eng. 16, 912–926 (2004)
5. Al-Haj, A., Odeh, A.: Watermarking relational database systems. In: Conf. on Applications of Digital Information and Web Technologies, ICADIWT 2008, pp. 270–274 (2008)
6. Bertino, E., Ooi, B., Yang, Y., Deng, R.: Privacy and ownership preserving of outsourced medical data. In: Proc. of the 4th Intl. Conf. on Data Engineering, CA, USA, April 5-8, pp. 521–532 (2005), doi:10.1109/ICDE.2005.111

Agent Based Information Aggregation and Routing in WSN

Prashant Sangulagi[1], A.V. Sutagundar[1], and S.S. Manvi[2]

[1] Department of Electronics and Communication Engineering
Basaveshwar Engineering College, Bagalkot-587102, India
[2] Department of information Science and Engineering
REVA institute of Technology and Management Bangalore, India
psangulgi@gmail.com, ashok_ec@yahoo.com

Abstract. This position paper uses multi-agent agent system for the aggregation of the information and routing it to the sink node. Mobile agent is responsible for aggregating information from different sensor nodes that makes energy efficient routing to consume less bandwidth and static agents for estimation of cost between the nodes which makes the routing table to be efficient for routing. Hence this position paper proposes agent based information aggregation and routing in WSN.

Keywords: information aggregation, mobile agent, routing, static agent, WSN.

1 Introduction

The sensor no de's radio is usually the most power hungry component t[1] hence reducing transmission and reception is very desirable. Instead of sending the raw data to the nodes responsible for the fusion, nodes use their processing ability to locally carryout simple computations and transmit only the required and partially processed data. Power supply of each individual sensor being provided by a battery so both communication and computation activities must be optimized. To that effect, information aggregation has been put forward as an essential technique to achieve power efficiency by reducing redundancy and minimizing bandwidth usage. Information aggregation in WSN is notably different from traditional computer network [6]. Aggregation can be done in every node of the sensor network. The aggregation may be of maxima, minima or average of data.

An agent is a piece of software that can achieve a special task in an autonomous way, in other words agents are autonomous program situated within an environment which sense the environment and act upon it to achieve the goals. The agents have possibility of co-coordinating, communication and co-operating with system or to other agents. Some of the special characteristics of agents include autonomy, social ability, reactivity and pro-activeness [2]. Agents have the special properties such as mandatory property and orthogonal property [3].The MA based computing model has the following features that can address the unique challenges posed by sensor networks[4] are scalability, reliability, exibility, maintainability and task adaptively[5].

V.V. Das, J. Stephen, and Y. Chaba (Eds.): CNC 2011, CCIS 142, pp. 449–451, 2011.

The mobile agent has a technique to aggregate the information by using only its code and transfer final aggregated information to the sink node. Our contribution in this position paper includes 1) using static agent and mobile agent to perform several tasks that aid information gathering, processing, aggregation, routing etc. 2) Aggregation of information is done using Aggregator agent when event occurred information strength is more than predefined value. 3) Shortest path Routing is done by using mobile agent. Further sections 2 provide the proposed work and 3 conclude the paper.

2 Proposed Work

Every sensor node of WSN has predetermined value of the signal strength. Once the information is sensed by sensor no de, it compares the information with predetermined value, if it is greater, a message is sent to sink node by using static agent and mobile agent. Static agent which finds the path between the nodes and create the routing table and mobile agent is responsible for moving information from one node to another node and make the aggregation to make high quality information.

2.1 Node Agency (NA)

Node agency will process the event occurred information. The Node Agency comprises a Node Black Board (NBB) and Node Manager Agent (NMA) [3], which is generally used to compare the event occurred information with the predefined information and if any change occurs NMA creates two agents, routing agent (RA) and aggregator agent (AA) both are mobile agents.

Node Black Board (NBB): This is knowledge base or data base and updated by agents, it is responsible for synchronizing the actions of the agents within themselves and outside as well. This black board creates routing table by gathering the information from routing agent. Routing table consist of event occurred nodes identity, its location, status, battery in volts, signal strength and power in watts.

Node Manager Agent (NMA): It is a static agent who compares the sensed information with the predefined information and creates the NBB as well updates the NBB. NMA makes the node to operate in sleeping mo de and active node in case of sleeping mode the node will not transmit any information, NMA generates two agents, RA and AA which are helpful for aggregation and routing in WSN.

2.2 Routing Agent (RA)

Routing agent is a mobile agent generated by static agent called NMA. When the event sensed by the node first NMA compares the event occurred information with the predefined information, if any changes occur it updates NBB and creates a mobile agent called Routing Agent (RA), this RA routes in the network and find event occurred nodes, and stores information about them and comes back to the node from where it was generated and updates NBB.

2.3 Aggregation Agent (AA)

Aggregator agent is a mobile agent which is equipped with event information codes that migrate from one to another node depending upon the routing information provided by NBB. Aggregator agents are generated by NMA. AA moves to nearest no de by having information from routing table with only co de of the information, and aggregates the information of the neighbor node and moves to sink node with high quality information about the event by consuming less energy and bandwidth. Routing method used in this paper is shortest path routing which is very efficient routing method and consumes very less amount of energy to route the information from source node to sink node. Only coded information is used instead of whole information for aggregation and routing purpose.

3 Conclusion

This position paper proposes energy efficient Routing and Aggregation in wireless sensor network by using multi-agents, static agent and mobile agent. Aggregation and routing is done by using mobile agents and static agents for calculating the path between the nodes and generating Routing table. By using this concept we can overcome the problems occurring in sensor network and can be used in large applications especially in military applications.

References

[1] Hill, J.: System architecture for wireless sensor network. PhD thesis, UC Berkely (2003)
[2] Wooldridge, M., Jenning, N.R.: Agent theories, architectures and languages: a survey, pp. 1–22. Springer, Woolridge (1995)
[3] Sutagundar, A.V., Manvi, S.S.: Agent based Information Fusion in Wireless Sensor Networks. In: Proc. IEEE TENCON, Hydrabad, India, pp. 1–6 (2008)
[4] Qi, H., Iyengar, S., Chakrabarty, K.: Multiresolution data integration using mobile agents in distributed sensor networks. IEEE Transaction on Systems, Man and Cybernetics, 383–391 (2001)
[5] Sutagundar, A.V., Manvi, S.S., Birje, M.N.: Agent based Location Aware Services in Wireless Mobile Networks. Proc. IEEE, AICT, Mauritius (2007)
[6] Krishnachari, B., Estrin, D., Wicker, S.: The impact on data aggregation in WSN. In: Proc. of the 22nd Int. Conf. on Distributed Computing Systems (2002)

Dynamic IP Address Auto-configuration in MANETs

S. Zahoor Ul Huq[1], D. Kavitha[1], K.E. Sreenivas Murthy[2], and B. Satyanarayana[3]

[1] Associate Professor, Dept. of Computer Science & Engineering, G. Pulla Reddy Engineering College, Kurnool, Andhra Pradesh, India
{s_zahoor_2000,dwaramkavithareddy}@yahoo.com
[2] Principal, Sri Venkateshware Institute of Technology, Anantapur, Andhra Pradesh, India
principalsvitatp@gmail.com
[3] Professor, Dept. of MCA, Sri Krishna Devaraya University
Anantapur, Andhra Pradesh, India
bachalasatya@yahoo.com

Abstract. Addressing in MANET's is of significance importance, as a mobile device cannot participate in unicast communications until it is assigned a conflict-free IP address. Allocating addresses to mobile nodes is a fundamental and difficult problem. Unlike infrastructure based networks, MANET's support autonomous and spontaneous networking and therefore, should be capable of self-organization and self-configuration. We present a new IP based address allocation protocol for MANET's based on MAC Addresses, Nonce and Duplicate Address Detection Scheme. Each node in the network is capable of assigning a unique IP address. Addresses are reclaimed automatically, as the mobile nodes leaves the network. Our approach also has support for network merging and partitioning. The proposed scheme has low communication overhead, even address distribution when applied to large scale MANET's.

Keywords: Address Allocation, Mobile Ad hoc Networks, Network initializing IP Addressing, Address Conflicts.

1 Introduction

Routing is a major part of ad hoc communication. The communication between the nodes is done through a single path established and the establishment of the path is called Routing. Most of the researchers concentrated on routing protocols and of course is a major issue to be solved earlier. But the researchers in the routing area take into assumption that the nodes in the network can be identified using their IP Addresses. A Mobile Ad hoc Network (MANET) [1] is formed by the wireless transmitting devices that communicate with each other through wireless channel and without the aid of any fixed or standard infrastructure. In general the ad hoc networks are used in some important and typical applications such as Military Field Activities, Disaster Situations, Local and Educational Requirements etc. Consider the Military Field activity where there is no network already existing and the nodes need to be communicating by initiating a new network. For the nodes to be in the network and to communicate with each other, IP Addresses are to be assigned to the mobile nodes.

V.V. Das, J. Stephen, and Y. Chaba (Eds.): CNC 2011, CCIS 142, pp. 452–457, 2011.

If these nodes want to be in a network it definitely needs to have a unique network address also called as IP addresses. Any of the nodes with the duplicate IP addresses causes malfunctioning of the network. There is no DHCP Server present which can assign the IP Addresses nor a coordinator which can assign these addresses, because the environment in MANET has no centralized control over the nodes. Each node should act as a router and should perform on its own. Where almost all the papers talk about the IP assignment/auto-configuration when the network is already established, but very few of the papers have discussed what happens when the network has been initiated. So in this paper our main intention is IP auto-configuration when the network is initiated.

2 Related Work

Stateful and stateless are the two proposals for IP address auto-configuration in ad hoc networks. Duplicate Address Detection (DAD) [4] is the stateless proposal for distributed protocol in which every node that wants to join the network, floods the network with Address Request message (AREQ). In the network if there is another node with the same address, it sends an Address Reply message (AREP) to the joining node. The joining node sees this and understands that there is another node with the same IP and again it randomly chooses another address and repeats the flooding process. Otherwise the chosen IP is assigned to the node. The disadvantage of DAD is that it cannot handle network partitions, as a result cannot fit for ad hoc networks.

The Internet Engineering Task Force(IETF) Zeroconf working group proposes a hardware based addressing [2], which assigns an IP address to a node based on the device MAC address. But, this addressing scheme has certain problems, the MAC Addresses are not unique as some manufacturers sell network adapters with non-registered MAC addresses or the MAC address may be corrupt while manufacturing. Most of the network adapters allow users to change the MAC address to arbitrary values and some of the adapters may not be having IEEE MAC addresses at all. Above all because of the privacy interface identifiers are generated randomly [3].

[5] maintains the unconfigured joining nodes as requesters and the nodes in the network that is responsible for configuring and assigning the IP for the requester is called an allocator. The IP addresses are divided into address blocks and these blocks are distributed among the allocators. The problem with this is that if the allocator is down or moves away from the network because of its mobility nature then the block of IP addresses which the allocator is assigned becomes useless and they cannot be allocated to anyone.

[6] uses two schemes for IP Auto-configuration in Mobile Ad Hoc Networks, one is Bloom Filters and the other Sequence Filters. The Bloom filters has the following disadvantages.

1. The requirement of designing k different independent hash functions can be prohibitive for large k. For a good hash function with a wide output, there should be little if any correlation between different bit-fields of such a hash, so this type of hash can be used to generate multiple "different" hash functions by slicing its output into multiple bit fields. Alternatively, one can pass k different initial values (such 0, 1, ..., k-1) to hash functions that takes an

initial value; or add these values to the key. For larger m and/or k, independence among the hash functions can be relaxed with negligible increase in false positive rate.
2. Removing an element from this simple Bloom filter is impossible. The elements maps to k bits, and although setting any one of these k bits to zero suffices to remove it, this has the side effect of removing any other elements that map onto that bit, and we have no way of determining whether any such elements have been added. Such removal would introduce a possibility for false negatives, which are not allowed.
3. Bloom filters also have the unusual property that the time needed to either add items or to check whether an item is in the set is a fixed constant, O(k), completely independent of the number of items already in the set.

Sequence Filters are the data structure to store and compact addresses based on the sequence of the addresses. There is no false-positive or false-negative in the Sequence filter. But the problem with the sequence filter is that the sequence filter can be created only, once the IP addresses are generated randomly and checked for the uniqueness by using Duplicate Address Scheme. Thus this takes more time for detecting uniqueness among the IP addresses when the network is initiated.

3 Ad Hoc IP Address Auto-configuration

Whenever a new node enters into a network, Hello messages are broadcasted to the neighbors to know who the neighbors are. As shown in the Fig.1 node 4 and node 8 are the new nodes and they are sending the broadcasts telling their presence and also to know its surrounding neighbours. In our case as soon as the network is initiated the nodes present in the network send the hello messages to the neighbors to find out who the neighbors are and also to tell their existence. The nodes generally have the Mac addresses. Embed in the hello packets the nodes send their Mac Addresses and the nonce. Nonces are the random numbers generated by the nodes only once.

This information about the neighbors is stored in the MAC-IP table which is of the format shown in Fig. 2 and further forwarded to the next neighbor through the hello

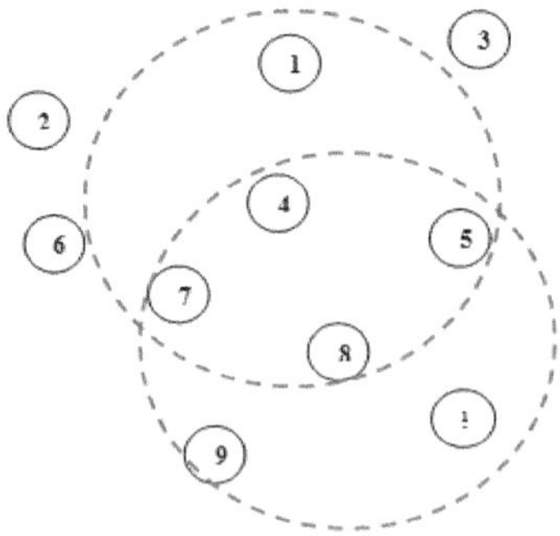

Fig. 1. Broadcasting of the hello packets by the nodes

Mac - Address	Nonce	IP-Address

Fig. 2. Format of the MAC-IP table

packets. In this way the information regarding the Mac addresses and Nonce of all the nodes spread across the network are maintained in all the nodes. Thus the MAC-IP table contains the information regarding the MAC-Address and Nonce of all the mobile nodes.

Generally the MAC Address e2-fe-2e-92-3c-39 is of the above format which has six segments separated by 5 hyphens. The MAC Addresses are sorted based on the first 8 bits and then on the next 8 bits and then so on and then finally on the Nonce. Though in some of the mobile devices MAC address does not exists then also they are sorted based on the nonce. These sorted entries are given the IP addresses in an incremental order. Thus each and every node will be maintaining the MAC-IP table which consists of the MAC Address, Nonce and IP address of all the nodes

If any of the node or set of the nodes are not within the range of the broadcast from any of the nodes which are broadcasting the hello messages, then for such node or set of nodes a separate network is formed. Later due to the mobility of the nodes if they come closer to the already existing network, then both the networks are merged.

As already mentioned before, devices now a days have higher storage capacity, more computational power, and greater wireless communication capabilities, storage of 1 row in the MAC-IP table takes 12 bytes storage space, storage of 150 rows may take a maximum of 2 KB.

Once the MAC-IP table is build, the new nodes which want to enter into the network can just know from its neighbours, the max IP present in the MAC-IP table, maintained by the periodic broadcasts. The new node assigns itself with the next IP, makes an entry in the MAC-IP table and stores the MAC-IP with itself and inform its neighbours about the change. If two nodes enter into the network at the same time then such a situation needs to be handled differently.

3.2 Reason Why Nonce Is Used

Nonce are the Random numbers they are used because the MAC Addresses are not unique as some manufacturers sell network adapters with non-registered MAC addresses or the MAC address may be corrupt while manufacturing. Most of the network adapters allow users to change the MAC address to arbitrary values and some of the adapters may not be having IEEE MAC addresses at all. Above all because of the privacy interface identifiers are generated randomly [2]. Because of the above reasons the random number is used to recognize the uniqueness.

4 Experimental Results

The entire simulation was run on ns-2 version 2.30 simulator installed in Intel Core 2 CPU of 2.00 GHz Processor and 256 MB of RAM. Table 1, Table 2, Table 3 shows the time required and the number of broadcasts required for building the MAC-IP table in 100x100, 150 X 150 and 200 X 200 terrain area respectively. If the table I, table 2 and table 3 are compared as the number of nodes remains the same and the change in the terrain range increases the number of broadcasts. Also if the terrain area is the same and the number of nodes increases the broadcasts also increases as the density of the nodes in the network increases. This is because the transmission power of the node decreases as the number of nodes in the network increases. In this case We have considered the situation where the number of nodes are limited and not more than 50.

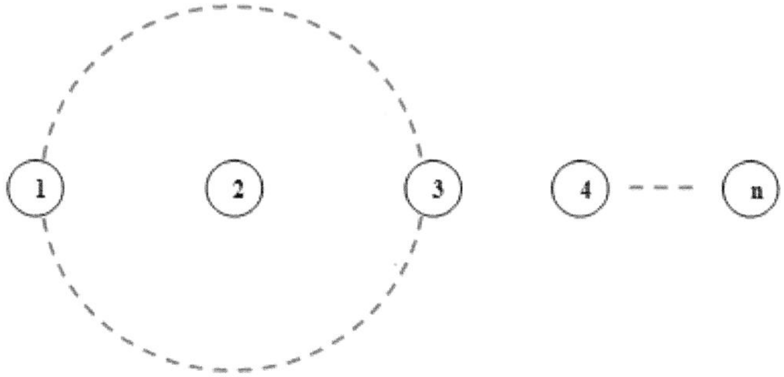

Fig. 4. Linear arrangement of Nodes

Consider the worst case shown in Fig. 4 where the nodes are linear and distant apart and by the broadcast of the hello packets only the nodes of 1 hop away can be reached in such a case also the nodes try to build a MAC-IP table using n-1 broadcasts otherwise they form a separate network due to mobility when they come closer they merge as a single network.

Table 1. Time & Number of broadcasts required when the terrain range is 100 X 100

No. of Nodes	Time Required in seconds	No. of Broadcasts required
10	0.055	2
20	0.085	3
30	0.115	4
40	0.115	4
50	0.175	6

Table 2. Time & Number of broadcasts required when the terrain range is 150 X 150

No. of Nodes	Time Required in seconds	No. of Broadcasts required
10	0.055	2
20	0.115	4
30	0.115	4
40	0.145	5
50	0.175	6

Table 3. Time & Number of broadcasts required when the terrain range is 200 X 200

No. of Nodes	Time Required in seconds	No. of Broadcasts required
10	0.055	2
20	0.115	4
30	0.145	5
40	0.175	6
50	0.175	6

5 Conclusion

The reason for using the hello packets for transfer of information regarding the MAC Addresses and the Nonce is that the hello packets are definite to exchange as the nodes in the network wants to know who the neighbors are and they does not load the network. Thus we have effectively build a MAC-IP table which is used to auto-configure the IP addresses of the nodes in the network, which takes less than 1 second of the time.

References

[1] Mobile Ad-hoc Networks (MANET), `http://www.ietf.org/html.charters/manet-charter.html`
[2] Cheshire, S., Aboba, B.: Dynamic Configureation of IPv4 Link-Local Addresses. Internet draft, Internet Engineering Task Force, Zeroconf Working Group (March 2001)
[3] Weinger, K.A.: Passive Autoconfiguration of Mobile Ad hoc Networks. IEEE Journal of Selected Areas in Communications 23(3), 507–519 (2005)
[4] Perkins, C.E., Royers, E.M., Das, S.R.: IP address autoconfiguration for ad hoc networks. Internet Draft (July 2000)
[5] Taghiloo, J., Berangi, R., Taghiloo, M., Gholami, M.: An Anti-Strom Approach for IP Address Auto-Configuration in Mobile Ad Hoc Networks. In: Mobile Ad Hoc and Sensor Systems, IEEE 2008, pp. 583–588 (2008)
[6] Fernandes, N.C., Moreira, M.D.D., Duate, O.C.M.B.: An Efficient Filter-based Addressing Protocol for Autoconfiguration of Mobile Ad Hoc Networks. In: Procedings of IEEE INFOCOM 2009 (2009)

Design of a Low Power High Speed ALU in 45nm Using GDI Technique and Its Performance Comparison

Manish Kumar[1], Md. Anwar Hussain[1], and L.L.K. Singh[2]

[1] Department of ECE, NERIST (Deemed University), Arunachal Pradesh, India
[2] Department of ECE, Mizoram University (Central University), Mizoram, India
mkr@nerist.ac.in, er.manish.k@gmail.com

Abstract. The paper presents a low power high speed Arithmetic Logic Unit (ALU) in 45 nm technology using Gate Diffusion Input (GDI) technique and its performance comparison with CMOS and nMOS Pass Transistor Logic (PTL) techniques. The simulated results revealed better performance characteristics of various logic and arithematic functions of a 1-bit ALU using GDI technique as compared to conventional CMOS and nMOS PTL techniques. GDI technique allows reducing power dissipation and delay while maintaining low complexity of logic design. MICROWIND and DSCH 3.1 EDA tools were used for the schematic layout and simulation of ALU using BSIM4 model.

Keywords: power dissipation, delay, power - delay product, GDI, CMOS, PTL.

1 Introduction

The demand for increasing speed and low power dissipation triggers numerous research efforts with the increasing demand of portable electronic appliances driven by batteries. The wish to improve the performance characteristics of VLSI circuits resulted in the development of many logic design techniques during the last two decades [1]. Pass transistor logic technique is better than the conventional CMOS technique in high speed and low power logic design because of reduced number of transistors [2], [3]. However, the major drawback of nMOS PTL technique is the reduced drive current and hence slower speed of operation at the reduced supply voltage [4].

GDI technique is superior over other design techniques in low power and high speed VLSI design as this technique uses a simple GDI cell consisting of only two transistors, to implement various complex logic functions. This technique improves the logic level swing, characteristic performances and also allows a simple design of any logic circuit using a small GDI cell. GDI technique enables simpler gates, lower transistor count, and lower power dissipation in many implementations, as compared with standard CMOS and PTL design techniques [5].

2 Low Power High Speed Design Methology using a GDI Cell

A basic GDI cell shown in Fig. 1 contains four terminals. P, N and G are the input terminals while D is used as the output terminal. The bulk of nMOS transistor is

V.V. Das, J. Stephen, and Y. Chaba (Eds.): CNC 2011, CCIS 142, pp. 458–463, 2011.

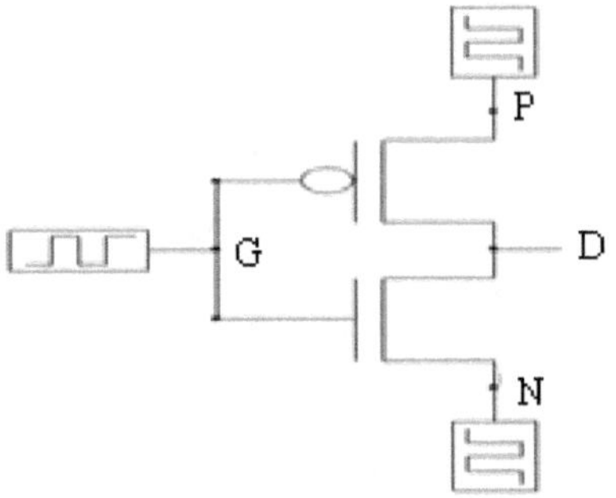

Fig. 1. A GDI cell

Table 1. Logic functions using a GDI cell

N	P	G	D	Function
0	B	A	A'B	F1
B	1	A	A'+B	F2
1	B	A	A+B	OR
B	0	A	AB	AND
C	B	A	A'B+AC	MUX
0	1	A	A'	NOT

connected to the bulk of pMOS transistor. Table 1 shows implementation of various logic functions using a single GDI cell.

A logic function Z with inputs $\{x_1, x_2, \ldots\ldots x_n\}$ can be expressed using Shannon expansion as:

$$Z(x_1, x_2,..x_n) = x_1R(x_2, x_3,..x_n) + x_1'S(x_2, x_3,...x_n) \quad (1)$$

where $R = Z|x_{1=1}$ and $S = Z|x_{1=0}$ (2)

The output function of a GDI cell shown in Fig. 1 can be written as:

$$D = A'B + AC \quad (3)$$

Where A, B and C are the inputs applied to G, P and N terminals respectively of a GDI cell.

If $A = x_1$, $B = R(x_2, x_3,..x_n)$ and $C = S(x_2, x_3,...x_n)$, then

$$D = x_1R(x_2, x_3,..x_n) + x_1'S(x_2, x_3,...x_n) \quad (4)$$

In equation 4, the input x_1 is used as a select line that selects either R or S for a given value of x_1, so that the power dissipation of the circuit reduces significantly. Shannon expansion is a very useful technique for low-power design due to its multiplexing property [6]. GDI technique is suitable for designing low power high speed VLSI circuits using very less number of transistors as compared to other design techniques.

3 Performance Characteristics Comparison of a 1-Bit ALU

The schematic diagram of a 1-bit ALU is shown in figure 2[7]. DSCH was used to draw its schematic diagram. The layout of ALU was simulated using GDI, CMOS and nMOS PTL techniques and the performance characteristics were measured in 45 nm technology using BSIM4 model. Layouts using different techniques were simulated at temperature of 27 °C, Supply voltage, V_{dd} of 1.00 V and Clock frequency, f_{clk} of 100 MHz.

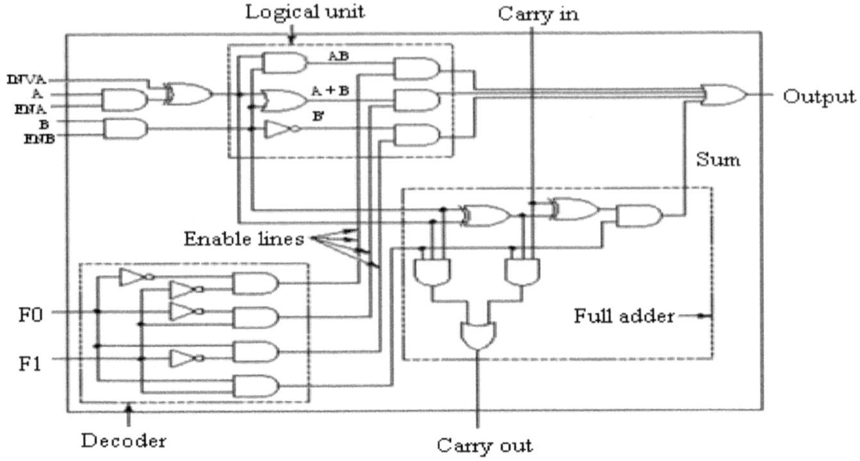

Fig. 2. 1- bit Arithmetic Logic Unit

3.1 Number of Transistors Comparison

The GDI technique used only 58 numbers of transistors while 148 and 104 numbers of transistors were required in CMOS and nMOS PTL techniques respectively to design a 1- bit ALU. Among all the design techniques, GDI technique used minimum number of transistors because each GDI cell was implemented using only two transistors.

3.2 Delay Comparison

Total delay of various logic functions of a 1- bit ALU using CMOS, nMOS PTL and GDI techniques are shown in Fig. 3. Minimum delay was achieved using GDI technique because of the use of minimum number of transistors while maximum delay was seen using nMOS PTL technique because of the reduction in the drive current at lower supply voltage.

3.3 Power Dissipation Comparison

Power dissipation of a 1- bit ALU using CMOS, nMOS PTL and GDI techniques of various logic operations are shown in Fig. 4. Among all the logic functions, increment after addition operation showed the largest power dissipation while logic zero operation consumed the least power. GDI technique proved to be the most power efficient technique.

3.4 Power - Delay Product Comparison

The performance characteristic of any logic circuit in terms of power dissipation and delay is measured by the power - delay product. The power-delay product of various logic functions of a 1-bit ALU are shown in Fig. 5. GDI technique proved to be the best among other techniques in terms of power- delay product.

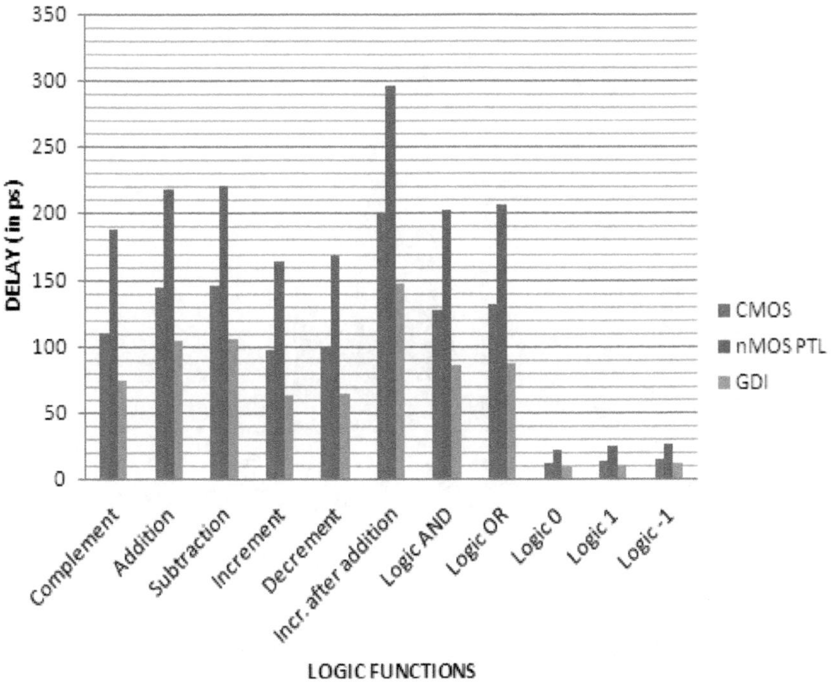

Fig. 3. Delay comparison of a 1-bit ALU

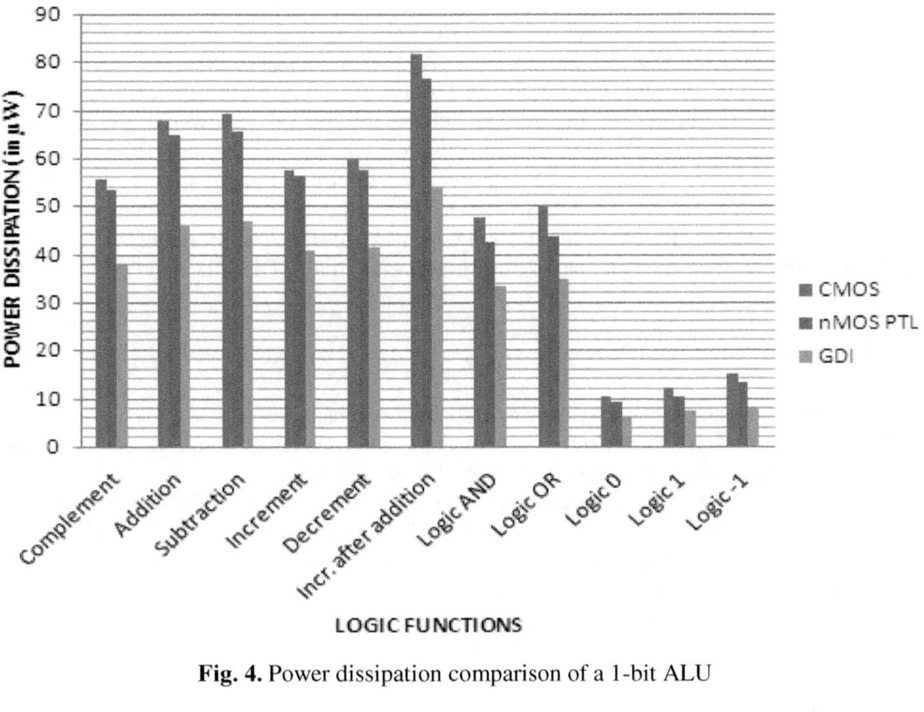

Fig. 4. Power dissipation comparison of a 1-bit ALU

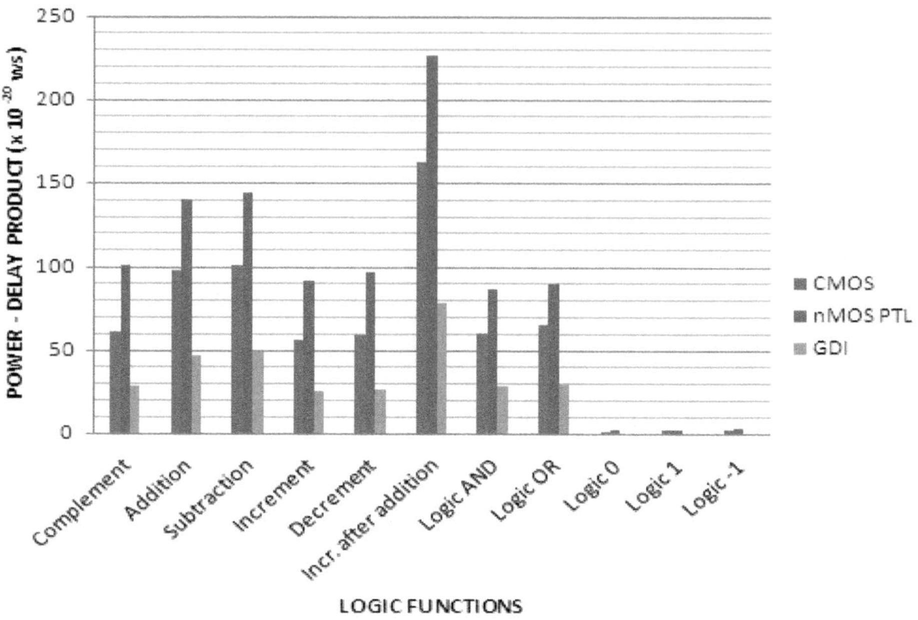

Fig. 5. Power - delay product comparison of a 1-bit ALU

4 Conclusion

Power dissipation, delay, power delay product and number of transistors of 1-bit ALU were compared using CMOS, nMOS PTL and GDI techniques using BSIM4 model in 45 nm technology. Among all the design techniques, GDI technique proved to be the best technique in terms of performance characteristics. It allowed reducing power dissipation and delay while maintaining low complexity of logic design. This method is suitable for designing fast and low power VLSI circuits using very less number of transistors as compared to CMOS, nMOS PTL and other design techniques. This technique improves the logic level swing, power dissipation characteristics and also allows a simple design of any logic circuit using a small GDI cell.

References

1. SIA Roadmap, http://www.sematech.org/public/roadmap/doc
2. Sakurai, T.: Closed-form expressions for interconnection delay, coupling, and crosstalk in VLSI's. IEEE Trans. Electron Devices 40, 118–124 (1993)
3. Adler, V., Friedman, E.G.: Delay and power expressions for a CMOS inverter driving a resistive-capacitive load. Analog Integrat. Circuits Signal Process 14, 29–39 (1997)
4. Al-Assadi, W., Jayasumana, A.P., Malaiya, Y.K.: Pass-transistor logic design. International Electron Journal 70, 739–749 (1991)
5. Morgenshtein, A., Fish, A., Wagner, I.A.: Gate-Diffusion Input (GDI) – A Power Efficient Method for Digital Combinatorial Circuits. IEEE Trans. on VLSI 10, 566–581 (2002)

6. Alidina, M., Monteiro, J., Devadas, S., Ghosh, A., Papaefthymiou, M.: Precomputation-Based Sequential Logic Optimization for Low Power. IEEE Transactions on Very Large Scale Integration (VLSI) Systems 2(4), 426–435 (1994)
7. Tenenbaum, A.: Structured Computer Organization, 4th edn. Prentice-Hall of India (1999)
8. Chandrakasan, A.P., Sheng, S., Brodersen, R.W.: Low- power CMOS digital design. IEEE J. Solid-State Circuits 27, 473–484 (1992)
9. Chandrakasan, A.P., Brodersen, R.W.: Minimizing power consumption in digital CMOS circuits. Proc. IEEE 83, 498–523 (1995)
10. Morgenshtein, A., Fish, A., Wagner, I.A.: Gate-Diffusion Input (GDI) – A Technique for Low Power Design of Digital Circuits: Analysis and Characterization. In: ISCAS 2002, USA (May 2002)

Normalization of Semantic Based Web Search Engines Using Page Rank Algorithm and Hypergraph Based Clustering

G. Archana, B. Muruganantham, and J. Jayapradha

SRM University

Abstract. In recent years, with the massive growth of web, there is an explosion of information accessible to internet users. The search logic usually tries to recover this information by exploiting many text-matching techniques. Also, traditional search engines do not have the necessary infrastructure for exploiting relation-based information that belongs to the semantic annotations for a web page. In the semantic web, each page possesses semantic metadata that record additional details concerning the web page itself. Annotations are based on classes of concepts and relations among them. We propose that relations among concepts embedded into semantic annotations can be effectively exploited to define a ranking strategy for semantic based web search engines. This approach relies on the knowledge of the user query, the underlying ontology, the hypergraph based clustering, and the web pages to be ranked. Thus, it allows us to effectively manage the search space and to reduce the complexity associated with the ranking task.

Keywords: Hypergraph, Ontology, Semantic Annotations, Text Matching.

1 Introduction

With the tremendous growth of information available to end users through the Web, search engines come to play more critical role. Most renowned search engines return result sets including many pages that are definitely useless for the user. However, when the query was send to the search engine logic, the hidden details were lost, because of the text matching technique. It also lacks in relation based information retrieval. In semantic web based on the relations, semantic annotations are used for the effective retrieval of information. The annotation is usually expressed by means of an ontology that provides a common understanding of terms within a given domain [1]. In this paper, we will prove that the relation based page ranking strategy along with hypergraph based clustering [2] which clusters high-dimensional data that retrieve the results for semantic based web search engines with better accuracy.

2 Related Works in Semantic Web Search

The aim of this paper is to show, how to make use of relations in semantic web page in ordering result set, where pages that best fit the user query are displayed first. It

V.V. Das, J. Stephen, and Y. Chaba (Eds.): CNC 2011, CCIS 142, pp. 464–467, 2011.

relies on graph-based[1] description to reduce the region of interest, but it deals with number of graphs. Our approach uses the hypergraph based clustering, for clustering the high dimensional data [2] for better accuracy and easy retrieval of data for heterogeneous environment.

3 Overview of Ranking Strategy (Proposed Approach)

The crawler application collects annotated web pages.OWL parser interprets the metadata and stores it in knowledge database. The GUI allows for the definition of a query and ordered result generated is presented to the end user. (Fig 1).

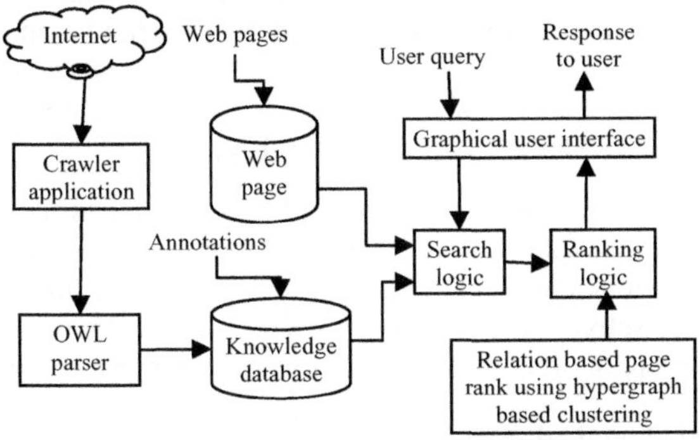

Fig. 1. Semantic Web Infrastructure

3.1 Starting Point -- Query Definition Process

In search engine like Google, a query is specified by giving a set of key words linked with logic operators and document type, language etc. But semantic search engines are capable of exploiting concepts hidden behind each keyword.

3.2 Introduction to Relation Based Ranking

The traditional search engines, returns pages without considering the information provided by the semantic mark. But semantic search engine would keyword concepts and would return a page only if they are related to associated concepts.

4 Basic Idea for Proposed Aproach

4.1 User Interaction

User interaction is the process of getting the request from the user. The searching of pages is based upon, the keyword for the retrieval of information. The fig.2 is shows the process flow.

4.2 Clustering of High Dimensional Data

The hypergraph based clustering [2] will have number of clusters such that, the items in each cluster are highly related with each other. The weight of each link is assigned on the basis of support value which is generated from each frequent item. Indexing and labeling will be done for the generated clusters for the fast access of the data.

4.3 Query Processing

In the query processing the keywords are extracted by eliminating, “is, was, the” etc. It starts with mapping between keywords and concepts. The searching is done on the basis of ontology, from the related information about the query which has been given by the user. After the searching of keyword, the ontology description and design of ontology for that particular domain is extracted from semantic annotations and web-based information.

4.4 Calculation of Relevance Score for Each Page

The Spanning Forest Generation technique [1] helps in calculating the relevance score for a single page by considering all the pages. It uses hypergraph based clustering technique for identifying more related pages and Spanning Forest Generation technique is used for effective relation based ranking, with reference to the user query. Finally the web pages are ordered with the relevance score as a result set.

4.5 Example with Stastical Data

Let the query given by the user be “List the Hotels near Guindy”. The query processing starts by searching the keywords from the query. Here the keywords are ‘Hotels’,

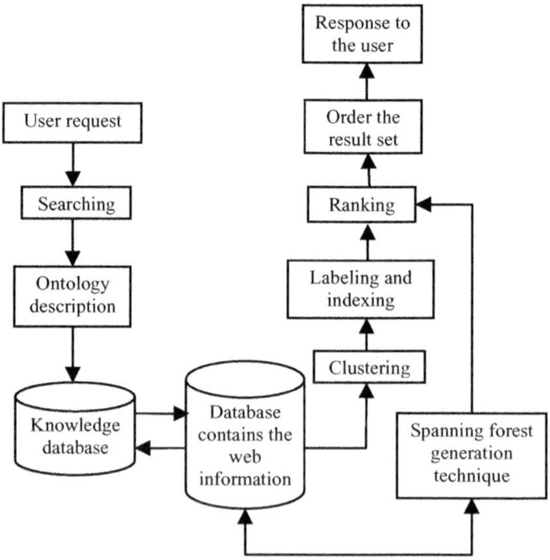

Fig. 2. Process flow

'Guindy' are taken and the web pages related to the keyword are extracted. Then the ontology description for particular domain is sent to knowledge base. The hypergraph based clustering will dynamically cluster the similar the keywords such as, "hotels with rooms, economic hotels, 3-star hotels, 5-star hotels, hotels with party halls, etc" as one cluster and "Guindy railway station, bus stand, commercial area, residential area" as another cluster. Now the relations between the keywords are identified from clusters. The ranking has to be done with the help of spanning forest generation technique. The page with high rank on the basis of relations will be displayed to the user as response to user.

5 Conclusions and Future Work

In this work, we propose a relation based ranking strategy that is capable of providing a relevance score for a web page into an annotated result set. Page relevance is measured through a hypergraph based clustering which also includes user preference questionnaires to get more accurate results. By neglecting the contribution of the remaining annotated resources, a reduction in the cost of the query answering phase could be expected. Further efforts will be requested to foster scalability into future semantic web repositories based on multiple ontologies, characterized by billions of pages, and possibly altered through next generation "semantic" spam techniques.

References

1. Lamberti, F., Sanna, A., Demartini, C.: A Relation-Based Page Rank Algorithm for Semantic Web Search Engine. IEEE Trans. Knowledge and Data Eng. 21(1) (January 2009)
2. Han, E.-H.S., Karypis, G., Kumar, V., Mobasher, B.: Clustering In A High-Dimensional Space Using Hypergraph Models. IEEE, Los Alamitos (1997)

A New Buyer-Seller Watermarking Protocol with Discrete Cosine Transform

Ashwani Kumar, Mohd Dilshad Ansari, Jabir Ali, and Kapil Kumar

Department of Computer Science, Jaypee University of Information Technology (JUIT)
Waknaghat, Distt. Solan, H.P., India
{ashwani.kumarcse,m.dilshadcse,javedkhancse,
kapil.cs89}@gmail.com

Abstract. Digital water marking is a promising technology to embed information as unperceivable signals in digital contents. Buyer-seller watermarking protocols with Discrete Cosine Transform (DCT) integrate multimedia watermarking and fingerprinting with cryptography, for copyright protection, piracy tracing, and privacy protection. It uses the public key encryption schemes. In this paper we propose a New Buyer Seller water marking protocol with DCT which is secure and flexible and gives more security from previous watermarking protocols to both buyer and seller. In this we use Public Key Infrastructure (PKI), arbitrator and watermarking certificate authority (WCA) for better security.

Keywords: Copyright Protection; Digital Watermarking; Public key infrastructure; Discrete Cosign Transform (DCT).

1 Introduction

The rapid growth of computer networks increased use of multimedia data via the Internet have resulted in fast and convenient exchange of digital information. The first known buyer-seller watermark protocol was introduced by Memon et al. [2], and it was improved by Ju et al. [3]. Since the first introduction of the concept, several alternative design solutions have been proposed in [1, 6]. In general, a buyer-seller watermarking scheme for traitor tracing involves three steps. First, a seller embeds a watermark that identifies the buyer into a digital product, such as an image. Second, when a pirated copy is found, the seller will detect the watermark of the pirated copy. At last, once the watermark of a specific buyer is identified, the seller will take the case to a court. Watermarking techniques can be broadly classified into two categories: such as spatial domain methods and transform domain methods. Spatial domain methods are less complex as no transform is used, but are not robust against attacks. A complete and sound buyer-seller watermarking protocol is expected to solve the following problem: Certification authority obligation problem, the conspiracy problem, the customer's rights problem, the unbinding problem, the anonymity problem.

V.V. Das, J. Stephen, and Y. Chaba (Eds.): CNC 2011, CCIS 142, pp. 468–471, 2011.

2 Proposed Protocol

In our proposed protocol we use robust watermark technique proposed by L Qiao [5] with the RSA cryptosystem. In our proposed protocol we use DCT to provide more security for the buyer and the seller during the transmission of digital content. In the rest of paper we defined assumption and two sub protocols which are registration protocol and watermarking generation and extraction protocol defined in [4].We cover our entire proposed protocol with figure 2 & 3.which explain the transaction in proposed protocol. The seller provides the watermark embedding operation and sells the watermarked product to the buyer. The WCA device is integrated into the seller's computer system and it will generate the watermark with the help of DCT for the buyer. In the proposed protocol we use watermarking embedding with DCT, and group signature (GS), arbiter (ARB) and watermarking certificate authority (WCA). We considered that all messages are transferred in a secure manner and digital content is still image.

2.1 Watermark Generation and Extraction Protocol with DCT

The watermark generation and extraction protocol using DCT is defined in fig. 1 (a) & (b).

Watermark Insertion using DCT-

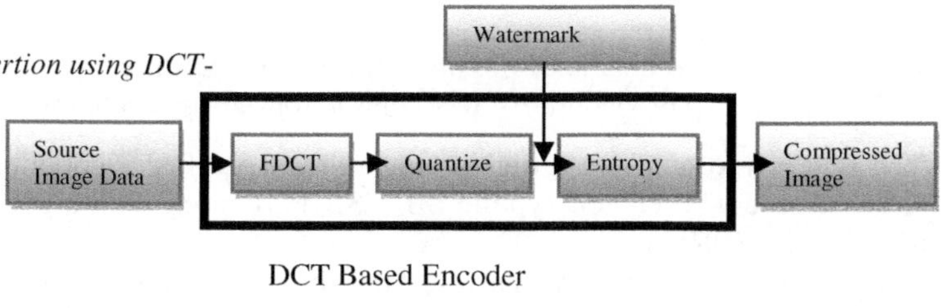

Fig. 1(a)

Watermark Extraction using DCT-

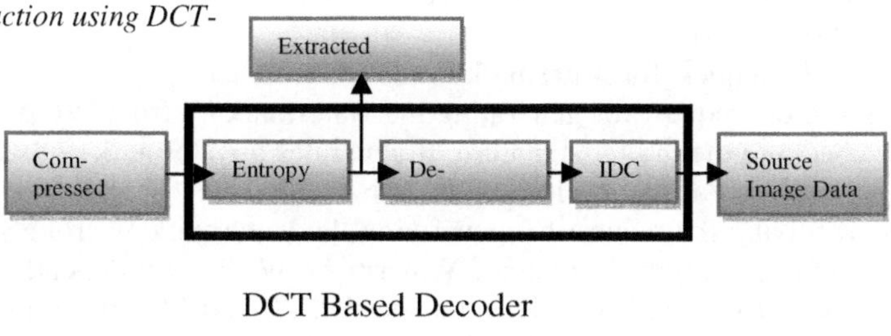

Fig. 1(b)

2.2 Registration Protocol

The registration protocol performed between the buyer B and the CA. The watermark generation and embedding protocol performed between the Seller S and the buyer B. Full description is given in reference [4].

2.3 Watermarking Protocol

Figure 2 shows the proposed new buyer-seller watermarking protocol with DCT, which is secure than its predecessors. It contains following terms.
S→ seller *B*→ buyer *WCA*→ watermark certificate authority *ARB*→ arbiter *DCT*→ discrete cosign transform.

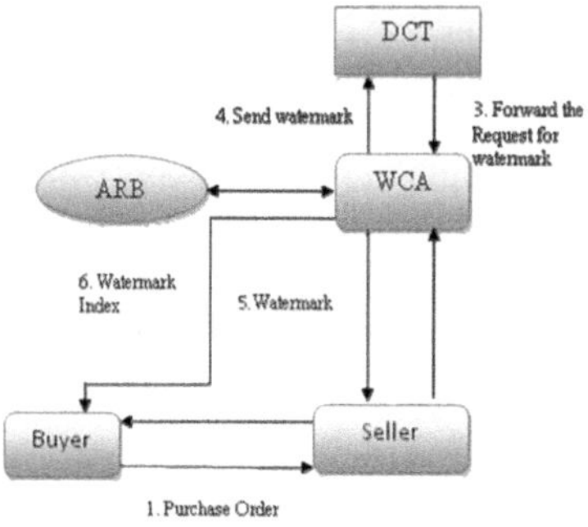

Fig. 2. New Buyer Seller Watermarking protocol with DCT

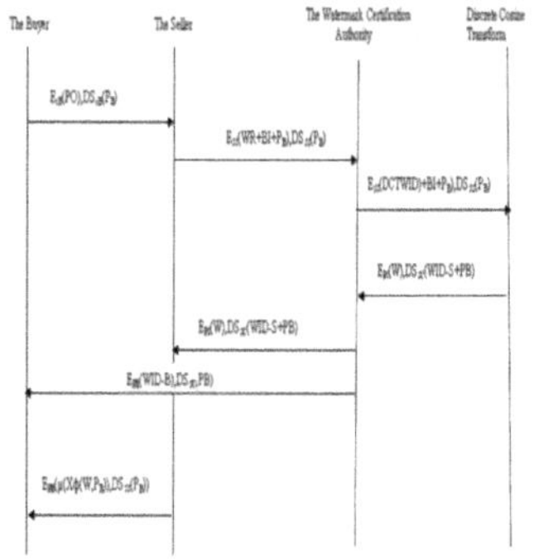

Fig. 3. Transactions in the proposed watermarking protocol

The figure 3 shows the details of possible transactions in the proposed watermarking protocol. Step-by-step procedures of the Transactions are given below:

1. When B wants to purchase a digital content from S, after negotiation B will place the purchase order (PO) by encrypting SB, ESB(PO), along with digitally signing the PB using SB, DSSB(PB).
2. S checks the authenticity by analyzing the digital signature using B's public key and decrypts the PO using PB. S places request for valid digital watermark (WR), buyer contact information (BI) such as IP address and other details to CA, along with encrypted PB using SS key, ESS(WR+BI+PB), and with digitally signing the PB using SS, DSSS(PB).
3. Forward the request for watermark to the DCT device.
4. Receiving the request for generating the watermark W from WCA, to DCT will do so after checking the digital signature. It generates a watermark with the help of DCT which is DCTWID and sends it to a WCA device.
5. Upon receiving the request for generating the watermark W from S, CA will do so after checking the digital signature. Watermark W directly depends on B's and S's public keys and Product-Copy ID number (PB, PS and PCID). Following conditions are satisfied when the watermark W is generated:

Γ (PB, PS PCID) →W (1) f (W, SB) →PS (2) μ(X'Φ (W, PB)) →X' (3)
μ' (X'© (PB)) →W (4) μ''(X'Ø (SB)) →W' (5)

Where Γ, f, μ, μ', and μ'' are homomorphism functions to generate or check the digital content or watermark. This homomorphism binary operator © and function μ' is only known to the WCA. δ (W, SS) _WID-S (6) δ' (WIS-S, SS) _W' (7)

σ (W', SB)_WID-B (8) σ(WID-B, SB)_W (9)

6. Generated watermark, W, is encrypted, EPS(W), by CA is send to S along with digitally signing the WID-S and PS, DSSC(WID S+PB).

7. The encrypted Watermark Identification number for B, EPS (WID-B), and DSSC (PB) are sending to B. This will help B to check the originality of the purchased digital content in a later stage of the transaction.

8. Upon receiving (EPS (W), DSSC (WID-S+PB)) from CA, S checks the authenticity by decrypting the digital signature using PC to get WID-S and PB. The equation (6) is applied to generate WID-S to crosscheck, with the one that is received from CA, originality of the watermark. S decrypts, DSB (EPS (W)), to get the watermark W, which will be embedded using the secrets homomorphism function μ.

9. Digitally signed PB, DSSS (PB), and encrypted watermarked digital content, EPB (μ (XΦ (W, PB)) or X', is now forwarded to B.

10. Upon receiving (EPB (μ (XΦ (W, PB)), DSSS (PB)) from S, B checks the authenticity by analyzing the digital signature. W' is generated from X¢ and SB from the equation (5) and WID-B is from the equation (8) to crosscheck, with the one that is received from CA, the originality of the digital content purchased. The briefly description of these 10 step is defined in reference [6].

3 Conclusion

This paper proposed a Watermarking Protocol solve many problems which are existing in previous watermarking protocol. This protocol provides a property that the seller does not get to know the exact watermarked copy that the buyer receives. This is designed for public key crypto system and the operations of watermark insertion are performed by the seller rather than the certification authority. The protocol has higher efficiency, and it is very secure. The security of this protocol lies on the security and robustness of the encryption standard and different group signature, and technique used in this protocol, which can further be modified and developed to improve the efficiency of the protocol.

References

1. Memon, N., Wong, P.W.: Protecting Digital Media Contents. Commun. ACM 41(7), 35–43 (1998)
2. Memon, N.D., Wong, P.W.: A buyer-seller watermarking protocol. IEEE Transactions on Image Processing 10(4), 643–649 (2001)
3. Ju, H.-S., Kim, H.-J., Lee, D.-H., Lim, J.-I.: An anonymous buyer-seller watermarking protocol with anonymity control. In: Lee, P.J., Lim, C.H. (eds.) ICISC 2002. LNCS, vol. 2587, pp. 421–432. Springer, Heidelberg (2003)
4. Zhang, J., Kou, W., Fan, K.: Secure buyer-seller watermarking protocol. IEE Proceedings Information Security 153, 15–18 (2006)
5. Qiao, L., Nahrstedlt, K.: "Watermarking Schemes and Protocols for Protecting Rightful Ownership and Customer's Right. Journal of Visual Communication and Image Representation 9, 194–210 (1998)
6. Das, V.V.: An Enhanced Secure Buyer-Seller Watermarking Protocol. In: International Conference on Advances in Computing (ICAC3 2009), pp. 498–502 (2009)

Efficient Non-chronological Dynamic Video Abstraction Using the Rack-Through Method

E. Jayabalan[1] and A. Krishnan[2]

[1] Assistant Professor of Computer Science, Government Arts College (Autonomous), Salem-7, Tamilnadu, India
[2] Dean (Academic), KSR College of Engineering, Tiruchengode, Tamilnadu, India
ejksrcas@rediffmail.com, ejksrcas@gmail.com, ejksrcas@yahoo.co.in

Abstract. Today surveillance cameras are used in many places like hospitals, museums, ATM centres etc for security purpose. Since videos captured by such cameras are time consuming, most of them are never watched. Inorder to conserve time, an abstract of the video is generated using the Rack-through Method. In the abstract video, most of the activity of the long video is condensed by simultaneously showing several actions, even when they originally occurred at different times. The abstract video also gives an index into the original video by pointing to the original time of each activity.

Keywords: Image Processing, Background Subtraction, Object Detection, Rack-through Method.

1 Introduction

Surveillance has always been playing an important role in the world. Endless videos are captured by the surveillance cameras and browsing and retrieval of such videos is time consuming. Existing approaches for video abstraction addressed mostly the temporal redundancy by selecting representative key-frames or time intervals. Motion segmentation was a major problem in such approaches. In [1], there was a focus on the two main visual cues - color and motion. Human visual system has different sensitivity to color and is also very sensitive to the boundary information. The color weight was introduced in the Bayesian clustering process. In the final phase, moving region was detected by thresholding the magnitude of motion vector. It is difficult to set the optimal threshold value. Also, region merging was performed by using only motion information. So, some parts of the object were sometimes missed out. In another approach [2], the motion segmentation problem was considered as an image-labeling process in which every pixel in the image was assigned a label describing its motion between consecutive frames.

1.1 Abstract Video

Consider an example of a video (Fig.1) in which the first man walks from right to left and after a period of inactivity, the second man walks in other direction i.e. left to right.

V.V. Das, J. Stephen, and Y. Chaba (Eds.): CNC 2011, CCIS 142, pp. 472–474, 2011.

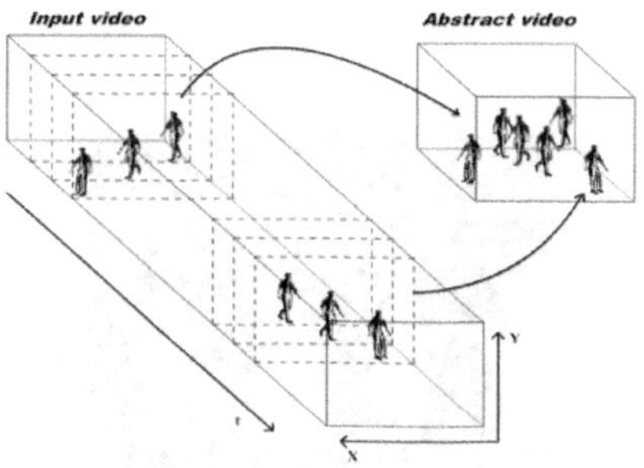

Fig. 1. Generation of a non-chronological Video

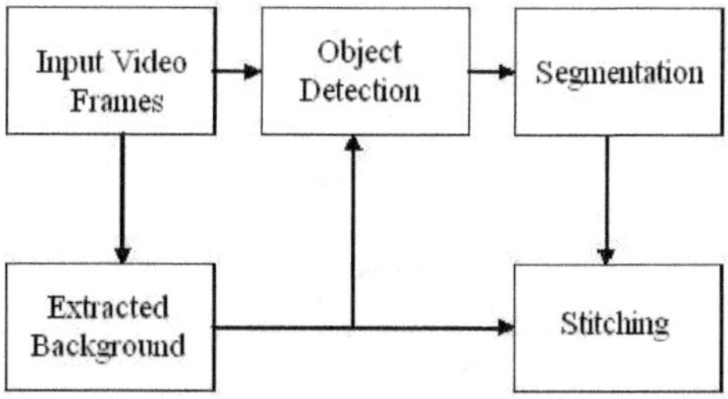

Fig. 2. Various Steps for Abstract Video

The abstract video generated in Fig.1 discards the inactive frames of the video and presents the objects in a short video by relaxing the chronological consistency of the events. Thus the abstract video preserves the dynamics of the objects. It is a substantially more compact video, playing two men simultaneously. This makes an optimal use of image regions by shifting events from their original time intervals to other time intervals when no other activities take place at these spatial locations. The block diagram shown in Fig. 2 shows the various steps involved to generate an abstract video.

2 Proposed System

The proposed Rack-through Method is used to generate a non-chronological dynamic video abstract which aims to present a compact representation of the long videos. The abstract video presents the activities of all the objects, ignoring the time at which they originally occurred. Motion segmentation problem is greatly reduced here.

3 Experimental Results

The Generated Binary mask frames are used to generate the *synopsis mask,* M_S (1). In this comb-like structure, there is a regular interval of time between two successive points. A comb-like structure is used to segment the objects from the binary mask.

Fig.3 represents the Binary Mask video and the comb-like structure used to select the objects merged to generate the synopsis mask. The Binary Mask consists of all the moving objects of the input video. The successive points of the comb-like structure point to different positions or time intervals in the binary Mask video. The objects present at these different positions are taken into consideration and are combined using logical OR operation to get synopsis mask. Fig.4 shows the synopsis mask generated by Rack-through method.

$$M_s(x,y,t_s) = \left\{ \begin{matrix} |N/(C_b)| \\ j=0 \end{matrix} \right. [M_s(x,y,t_s) | M(j*C_b)] \quad for\ all\ 0 \le t_s \cdot \tag{1}$$

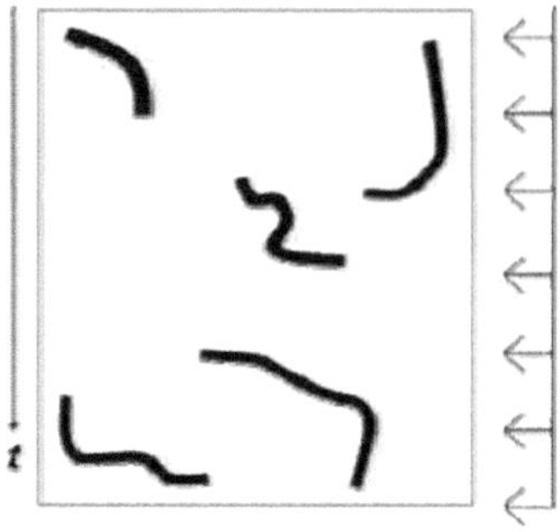

Fig. 3. Comb-like Structure used in Rack-throuh method

Fig. 4. Synopsis Mask

4 Conclusion

In this paper, the proposed Rack-through Method is used to segment the objects of the input video. This method is an efficient method. If the time interval between two successive points of the comb-like structure is very small, then we get a dense activity in the output video. If the time interval between two successive points of the comb-like structure is very large, then we have chances to miss out some of the activities of the input video. Hence a moderate time interval between the two successive points of the comb is to be considered for an effective abstract video.

References

1. Yoon, K., Kweon, I.: Moving Object Segmentation Algorithm for Human-like Vision System. In: 1st International Workshop on Human-friendly Welfare Robotic Systems, Taejon, Korea, pp. 885–890 (2000)
2. Wang, J., Lu, H., Eude, G., Liu, O.: Moving Object Segmentation Using Graph Cuts. In: 7th International Conference on Signal Processing, vol. I, pp. 777–780 (2004)
3. Pritch, P., Rav-Acha, A., Peleg, S.J.: Non-Chronological Video Synopsis and Indexing. IEEE Transactions on Pattern Analysis and Machine Intelligence (PAMI) 30(11), 1971–1984 (2008)
4. Pritch, P., Rav-Acha, A., Peleg, S.: Making a Long Video Short: Dynamic Video Synopsis. In: IEEE Computer Society Conference on Computer Vision and Pattern Recognition, New York, pp. 435–444 (2006)

Target Tracking in Aerial Videos Using Image Registration Techniques

E. Jayabalan[1] and A. Krishnan[2]

[1] Assistant Professor of Computer Science, Government Arts College (Autonomous), Salem-7, Tamilnadu, India

[2] Dean (Academic), KSR College of Engineering, Tiruchengode, Tamilnadu, India

ejksrcas@rediffmail.com, ejksrcas@gmail.com, ejksrcas@yahoo.co.in

Abstract. Image Registration is the basic step used for the target tracking in aerial videos. Image Registration is a process, which finds the location where optimal matching is obtained by matching a template image called the reference image over the searching region of an input image using a suitable similarity measures. The method although computationally intensive, is simple, straightforward and robust and requires no apriori information about the two images. A number of image registration algorithms are studied on images with varying size of search space and reference image. All these algorithms are based on similarity measure between the reference image and search space. The key issues in image registration are the time required for registration and the accuracy of registration viz., measure of how close is the match between sub image in the search space and the reference image.

Keywords: Image registration, Reference image, Search region, Similarity measures, Sub image.

1 Introduction

In the registration process three types of correlation techniques namely normalized area correlation (NC), Sequential Similarity Detection Algorithm (SSDA) and Frequency domain based *Fast Fourier Transform* (FFT) based approach are implemented. All these algorithms are based on similarity measure between the reference image and search space [1][2][3][4]. The key issues in image registration are the time required for registration and the accuracy of registration viz., measure of how close is the match between sub image in the search space and the reference image.

2 Proposed System

2.1 Accuracy of Registration

Accuracy of registration is usually calculated using the following formula: Let R(i,j) is the reference image and S(i,j) is the matched sub image in the search space of $l \times m$. Then the accuracy is calculated as

V.V. Das, J. Stephen, and Y. Chaba (Eds.): CNC 2011, CCIS 142, pp. 475–477, 2011.

$$\frac{\sum_{i=1}^{l}\sum_{j=1}^{m}\left(R(i,j)-S(i,j)\right)}{l\times m} \tag{1}$$

Matching is referred as misregistration, which is the sum of the residual differences at the best match point exceeds the specific threshold.

3 Experimental Results

3.1 Matching Error Analysis of Image Registration Algorithms

The accuracy of registration is tested and analyzed for the three algorithms as SSDA, Cross Correlation method and FFT based correlation.

Table 1. (a),(b),(c) and (d) shows the observed values of average Error Analysis

Video	Algorithm	Error in the matching
	SSDA	4.02
	Cross Correlation	9.6
	FFT Based Correlation	37.24

(a)

Video	Algorithm	Error in the matching
	SSDA	2.54
	Cross Correlation	14.26
	FFT based Correlation	42.49

(b)

Video	Algorithm	Error in the matching
	SSDA	26.75
	Cross Correlation	22.41
	FFT based Correlation	41.40

(c)

Video	Algorithm	Error in the matching
	SSDA	12.54
	Cross Correlation	12.06
	FFT based Correlation	58.49

(d)

The accuracy of registration is tested and analyzed for the three algorithms as SSDA, Cross Correlation method and FFT based correlation as shown in Fig.1 and Fig.2.

Its clearly shown that FFT based correlation is very efficient in time for searching the target window but it gives less accuracy of registration when compared to the SSDA and cross correlation methods. It's proven that there is a trade off between the time taken for registration and error in registration for each correlation algorithm.

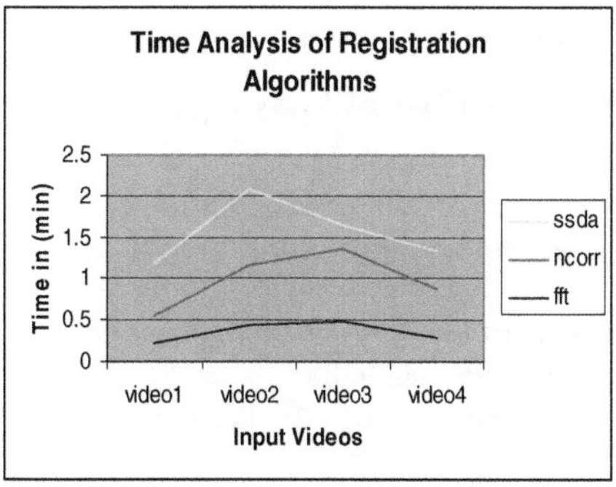

Fig. 1. Time analysis of Registration Algorithms

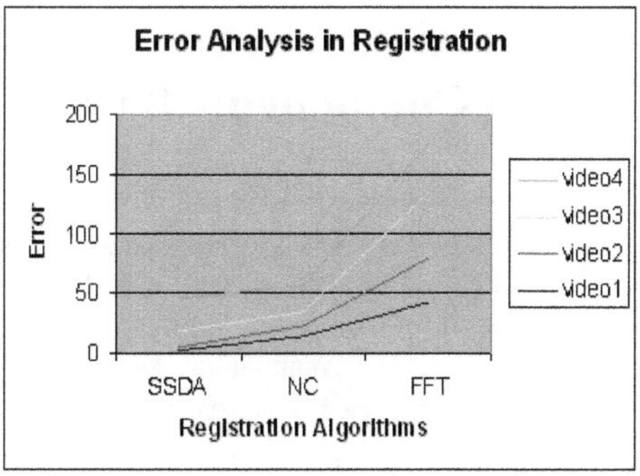

Fig. 2. Error analyses in Registration

4 Conclusion

In cases of tracking along multiple shots, Image Registration algorithms are implemented and tested for aerial videos. Image Registration is a process, which finds the location where optimal matching is obtained by matching a template image called the reference image over the searching region of an input image using a suitable similarity measures. For an efficient matching Normalized area Correlation, SSDA and FFT based correlation methods are used as similarity measures. The method although computationally intensives, is simple, straightforward, and robust requires no apriori information. In this paper, still there is a scope to improve the efficiency in time and matching of objects at the preprocessing stage of registration in real time aerial videos where camera is moving along with the target object.

References

1. Zhang, D., Lu, G.: Segmentation of Moving Objects in Image Sequence: A Review. J. Circuits, Systems and Signal Processing 20(2), 143–183 (2001)
2. Yilmaz, A., Javed, O., Shah, M.: Object Tracking:Survey. J. ACM Computing Surveys 38(4), Article 13 (2006)
3. Zitova, B., Flusser, J.: Image Registration Methods: A Survey. J. Elsevier, Image and Vision Computing 21, 977–1000 (2003)
4. Wyawahare, M.V., Patil, P.M., Abhyankar, H.K.: Image Registration Techniques: An Overview. J. Image Processing and Pattern Recognition 2, 11–28 (2009)

Cost Control and Quality Maintenance Metrics for Component- Based Software Development

Rachit Mohan Garg*, Kapil Kumar, and Yamini Sood

Deptt. of Comp. Sc. & Engg., Jaypee University of Information Technology, Waknaghat, Distt. Solan, Himachal Pradesh, India
{rachit.mohan.garg,kapil.cs89,eryaminisood}@gmail.com

Abstract. Component Based Software Development tries to develop a new system based on the integration of components designed previously. In performing this integration main problem encountered are errors & defects that affect the proper functioning of the software. Deriving a quality measure for reusable components has proven to be challenging task now a days. The main aim of this research is to derive quality metrics for the effective integration of the reusable component & thus increasing the quality of the overall developed software and thereby reducing cost.

Keywords: Component Based Software Development, Cost Metrics, Quality Metrics.

1 Introduction

Shifting to commercial off-the-shelf (COTS) [1], [2] components appears predictable, necessitating drastic changes to current software development and business practices. New approaches to quality and risk management are needed to handle the growth of CBSs. With software development proceeding at extraordinary speed, in-house development of all system components may prove too costly in terms of both time and money. Large-scale component reuse and COTS component acquirement can generate savings in development resources, which can then be applied to quality improvement, such as enhancements to reliability, availability, and ease of maintenance.

Reliability is a foremost choice as without reliability the product is of no use. Reliability is hindered when defects occur in the product. So to increase reliability a check on defects is to be kept.

Cost is an obvious choice for the focus group. Time to Market is closely related to cost and directly impacts the market viability of the software product. Some of the others metric that can be considered [6], [7] are adaptability, complexity, fault profile, customer satisfaction etc.

1.1 Benefits of Reuse on Productivity of Software

Reuse has been advocated as a means for reducing development cost. It improves productivity, because the life cycle now requires less input to obtain the same output

* Corresponding author.

V.V. Das, J. Stephen, and Y. Chaba (Eds.): CNC 2011, CCIS 142, pp. 478–481, 2011.

or productivity may increase simply because fewer work products are created from scratch. In general, reuse improves productivity by reducing the amount of time and labor needed to develop and maintain a software product.

1.2 Benefits of Reuse on Quality of Software

The accumulated defect fixes result in a higher quality work product because work products are used multiple times; Moreover, Reuse prevents and removes defects earlier in the life cycle because the cost of prevention and debugging defects can be amortized over a greater number of uses.

2 Proposed Scheme

2.1 Fault Detection Metric for CBSD

Implementation of Fault Detection in CBSD. When CBSD Technique is used for software development, we define

$$FDM_i = \frac{Err_i}{Err_i + Err_{i+1}} \tag{1}$$

where Err_i is the number of errors found in the ith component and Err_{i+1} is the number of errors found after integrating i+1th component with the ith component. In this context, to obtain the high quality product, FD for the whole product should approach 1. That means finding all errors of each component ensures the ideal value for FDM. FDM is calculated before integration of each component.

Planning a Defect Profile: For each stage of the CBSD, the following parameters are calculated, the number of defects likely to be inserted (n), removal efficiency (y), the number of defects likely to be added in next stage of component integration. So the number of defects likely to be removed is y*n & number of defects remaining will be n-(y*n). Finally calculate collective removal efficiency.

2.2 Cost Estimation Metric for CBSD

We can categorize the type of reuse in the context of cost estimation as follows:

Component Reuse without Modification. In this case, the average cost of developing using reusable components can be formulated as follows:

$$Cost_{search} + (1 - p) * Develop_{noreuse} \tag{2}$$

Where $Cost_{search}$ is the cost of performing a search operation, $Develop_{noreuse}$ is the cost of developing without reuse and p is the probability that the component is found in the component library. It is observed that the reuse option would be preferable only if:

$$Cost_{search} + (1 - p) * Develop_{noreuse} < Develop_{noreuse} \tag{3}$$

Component Reuse with Modification. The average cost of developing using reusable components can be formulated by equation 4.

$$Cost_{search} + Cost_{adapt} + (1 - p) * Develop_{noreuse} \tag{4}$$

In both the cases, the cost saving due to reuse can be formulated using a simple equation 5.

$$Cost_{saved} = Cost_{reuse} + Cost_{noreuse} \tag{5}$$

3 Analysis

3.1 Analysis of Cost Estimation Metric

The reduction in cost in different software companies is depicted by the figure 1. On the x-axis we have plotted the % of reduced cost & on y-axis we have plotted no. of components reused.

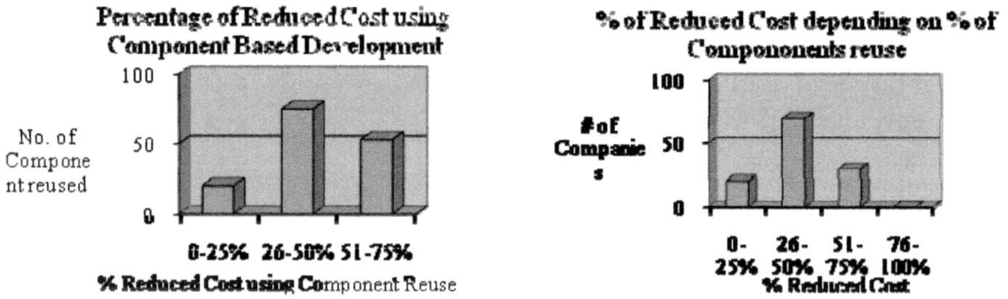

Fig. 1. (a)(b) Cost reduction on the basis of number of component reused in the organization

3.2 Analysis of Defect Removal Metric

In figure 2 the curve A from START to CODE represents the numbers of defects that were created or entered into the production processes during requirements analysis (REQ), High-level design (HLD), Low-level design (LLD), and Code in a software project. Curve A shows no pre-test defect removal activities. Consequently each test activity; e.g., (Unit Test (UT), Integration Test (IT), and System Test (ST)), as shown in curve B, finds and removes varying volumes of the generated or injected defects.

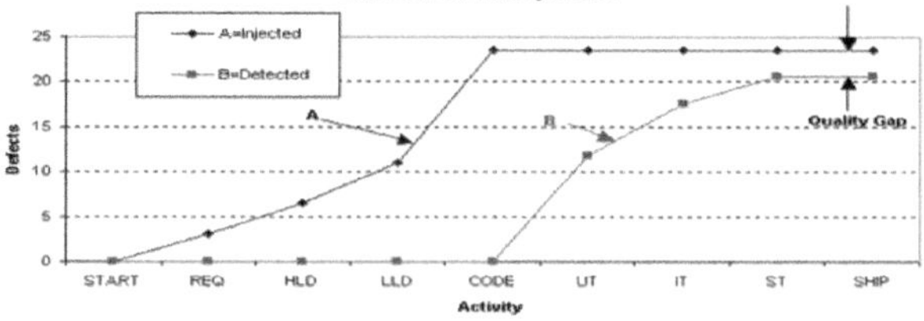

Fig. 2. Analysis of Defect removal

4 Conclusion

This paper provides metric for cost and fault estimation in using a component at various stages of a software development. The reuse orientation provides many advantages, but it also requires systematic approach in design, planning, development, support of a more complex maintenance process, and foremost, more consideration being given to quality i.e. cost estimation & error density of using components in a software.

So we conclude that using cost metrics the cost of software development is reduced considerably but not fully. Using Fault Detection Method the no. of defects that occur while integrating the components is detected using the relation provided.

References

1. Pressman, R.S.: Software Engineering-A practitioner's approach, 6th edn. McGraw-Hill, New York (2001)
2. Sommerville: Introduction to Software Engineering, 8th edn. Addison-Wesley, Reading (2007)
3. Crnkovic, I.: Component-based Software Engineering – New Challenges in Software Development
4. Futrell, R.T., Shafer, D.F., Shafer, L.I.: Quality Software Project Management, 1st edn. Pearson Education, London
5. Szyperski, C.: Component Software: Beyond Object Oriented Programming, 2nd edn. Addison-Wesley, Reading (2002)
6. Cai, X., Lyu, M.R., Wong, K.-F., Ko, R.: Component-Based Software Engineering: Technologies, Development Frameworks, and Quality Assurance Schemes. IEEE, Los Alamitos (2000)
7. Sedigh-Ali, S., Ghafoor, A., Paul, R.A.: Metrics and Models for Cost and Quality of Component-Based Software. In: Proceedings of the Sixth IEEE International Symposium on Object-Oriented Real-Time Distributed Computing (ISORC 2003). IEEE, Los Alamitos (2003)
8. Poulin, J.: Measuring Software Reusability. In: Proceedings of the 3rd International Conference on Software Reuse: Advances in Software Reusability. IEEE, Los Alamitos (1994)
9. Basili, V., Briand, L., Melo, W.: Measuring the impact of reuse on quality and productivity in object-oriented systems. Technical Report CS- TR-3395, University of Maryland, Computer Science Department (1995)

Study of Process Variations on f_t in 30 nm Gate Length FinFET Using TCAD Simulations

B. Lakshmi and R. Srinivasan

Department of Information Technology
S.S.N College of Engineering
Chennai, India
laxmi.balu@yahoo.com, srinivasanr@ssn.edu.in

Abstract. This paper investigates the effect of process variations on unity gain frequency (f_t) in 30 nm gate length FinFET by performing extensive 3D TCAD simulations. TCAD tools are used to study device sensitivities on process variations. Sensitivity of f_t on five different process parameters is studied. It is found that f_t is more sensitive to gate length, underlap and corner radius, and less sensitive to hardmask height. Sensitivity of f_t to fin width depends upon channel doping levels.

Keywords: f_t, FinFET, sensitivity, process variations, TCAD.

1 Introduction

The intensive downscaling of CMOS transistors has been the major driving force behind the growth of the semiconductor industry for the past 20–30 years [1]. Many novel device structures have evolved to overcome the scaling limitations of conventional bulk MOSFETs. The FinFET, in particular, has received tremendous attention and is one of the most likely candidates to replace planar-based MOSFET for the sub 32-nm technology node, because of their quasi-planar structure and the compatibility with CMOS. For RF applications, unity gain frequency (f_t) is one of the important metric and it is given by $g_m/2\Pi C_{gg}$, where g_m is the transconductance, C_{gg} is the combination of gate-source capacitance (C_{gs}), gate-drain capacitance (C_{gd}) and overlap capacitance (C_{ov}).

This paper investigates the effect of various process parameters of FinFET on f_t, using 3D TCAD simulations. Five different process parameters are varied over a range of values to capture their sensitivity on f_t (Refer Table 1). Next section describes the simulation environment and the various device structure parameters of interest. Section 3 discusses the simulation results. Finally section 4 provides conclusions.

2 Simulation Environment

Sentaurus TCAD simulator from Synopsys [2] is used to perform all the simulations.

- Sentaurus structure editor (SDE): To create the 2D and 3D device structure, to define doping, to generate mesh and to define contacts.

V.V. Das, J. Stephen, and Y. Chaba (Eds.): CNC 2011, CCIS 142, pp. 482–486, 2011.

- Sentaurus device simulator (SDEVICE): To perform all DC and AC simulations and to extract the device related parameters.
- Tecplot and Inspect : The former is used to view the structure of the device and the latter is used to extract the results.

The 3-D device structure generated from SDE is shown in Fig. 1. Figure 2 shows a 2D cut of the 3-D structure which depicts the fin cross section. Figure 3 is a 2D schematic diagram which shows various parameters of this study. Table 1 gives the dimensions of the nominal device and also tells the range of the various parameters studied in this paper. Throughout this study a uniform channel doping of $1\text{x}10^{16}/\text{cm}^3$ and source/drain doping of $1\text{x}10^{20}/\text{cm}^3$ are used. The gate electrode work function of 4.337 eV is used with gate oxide thickness of 1 nm. Device simulator includes the appropriate models for band to band tunneling, quantization of inversion layer charge, doping dependency of mobility, effect of high and normal electric fields on mobility, and velocity saturation. Supply voltage (Vdd) used in this study is 0.8 V. Standard AC simulations are done in SDEVICE and f_t is extracted from these results. At various gate biases f_t is calculated and the maximum of them is taken as f_t. The simulator was calibrated against the published results on FinFETs [3] and [4].

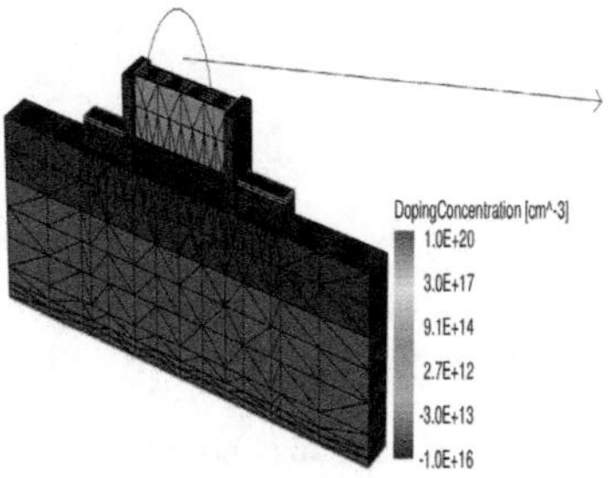

Fig. 1. 3-D structure of dual-gate FinFET

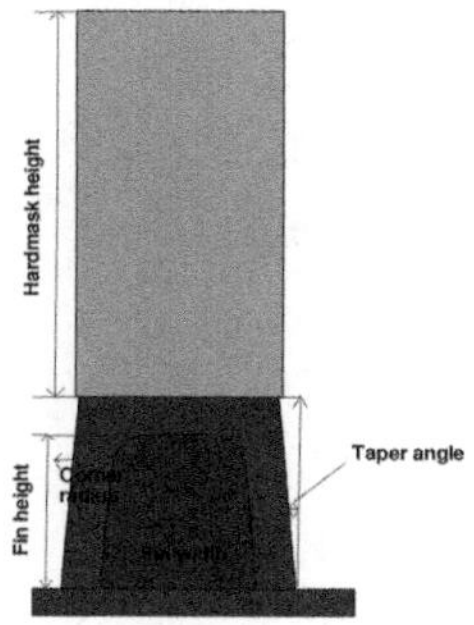

Fig. 2. Enlarged portion of rounded region

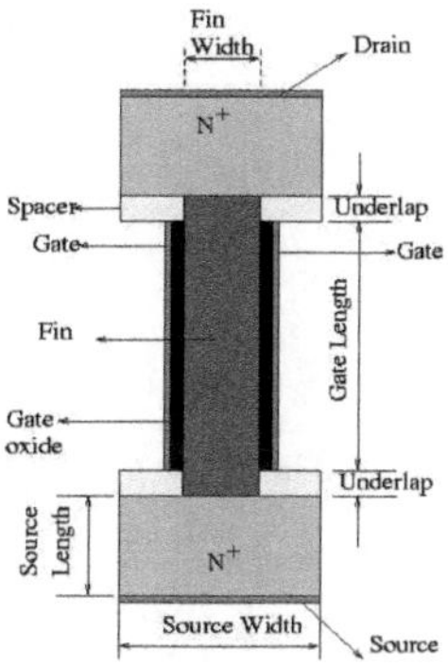

Fig. 3. Schematic of 2-D structure

Table 1. Dimensions of the device and their range of values

Process parameters	Nominal Value	Range of values
Gate length (L_g)	30 nm	20 nm - 40 nm
Fin width (W)	4 nm	3 nm - 10 nm
Underlap (L_{un})	3 nm	1 nm - 8 nm
Corner radius	1 nm	0 nm - 2 nm
Hard mask height	10 nm	0 nm- 100 nm

3 Results and Discussion

3.1 Variation in Gate Length

Figure 4 shows the variation of f_t against gate length (L_g). It can be observed from Fig. 4 that f_t initially increases and then gets decreased. The ratio C_{gd}/C_{gs} decreases with L_g [5] and this in turn affects f_t [6]. This is the cause for the initial increase in f_t with respect to L_g. As seen earlier, f_t is decided by both g_m and C_{gg}. The continuous degradation of g_m with L_g causes f_t to decrease from some point.

3.2 Variation in Underlap

Figure 5 depicts the plot between f_t and underlap (L_{un}). As L_{un} increases from 1 nm to 8 nm, g_m decreases from 48 μ-mho to 26 μ-mho due to increased series resistance whereas C_{gg} decreases from 9.1 aF to 5.5 aF because of fringing capacitance [7]. Since, g_m is more dominant f_t is expected to decrease when L_{un} gets increased. This can be observed from Fig. 5 which gives f_t versus L_{un} plot.

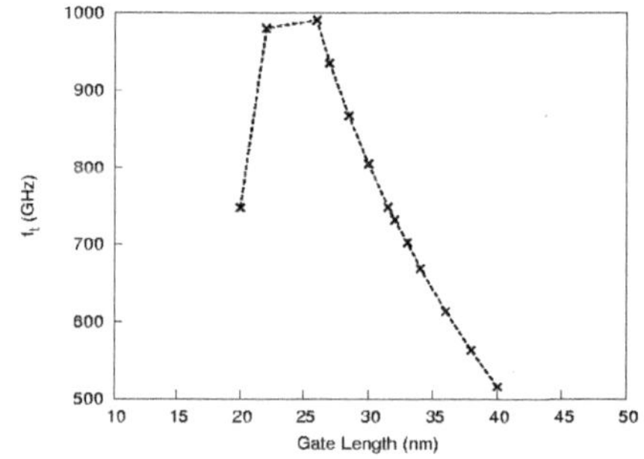

Fig. 4. f_t versus gate length

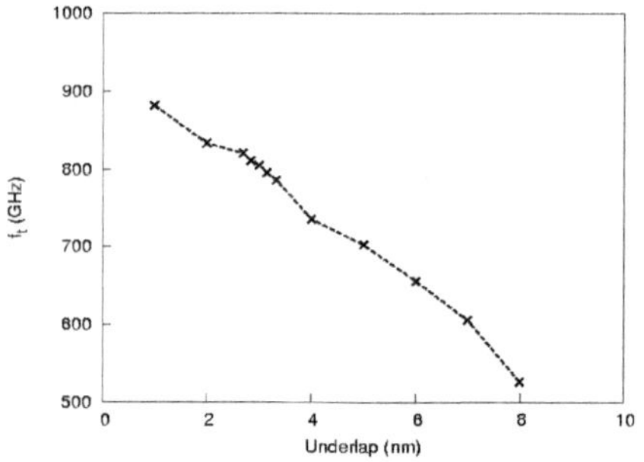

Fig. 5. f_t versus underlap

3.3 Variation in Fin Width

When fin width (W) is varied we may either face volume inversion or may not, depending upon the channel doping levels. When the channel doping is $1 \times 10^{16}/cm^3$ volume inversion is not seen. Therefore, the increase in W results in the increase in current and thereby g_m and f_t [8]. Figure 6 shows this kind of behavior between f_t and W. For the channel doping around $1.5 \times 10^{18}/cm^3$, volume inversion effect is seen which causes f_t to decrease initially and then to increase w.r.t W which is depicted in Figure 7.

3.4 Variation in Corner Radius

Figure 8 shows variation of f_t w.r.t to corner radius. The corner rounding ratio is defined as r/R, where r is the corner radius and R is the center radius (i.e., 2R = W). Higher r/R leads to a more circular cross-sectional shape, and r/R = 0 (rectangle), 1 and 2 (circular) have been used in the simulations. However, r/R should be no less

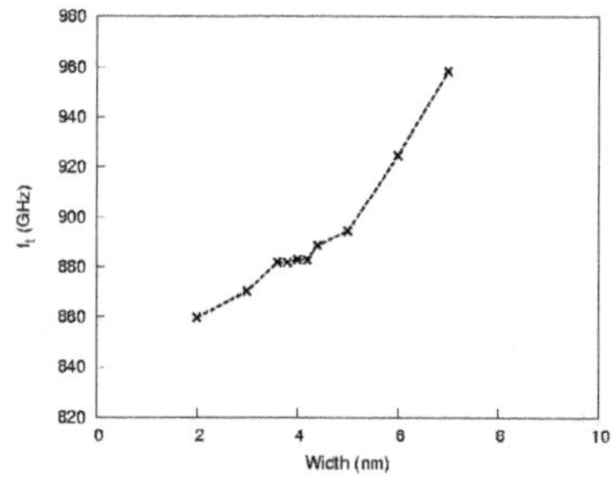

Fig. 6. Variation of f_t w.r.t W without volume inversion

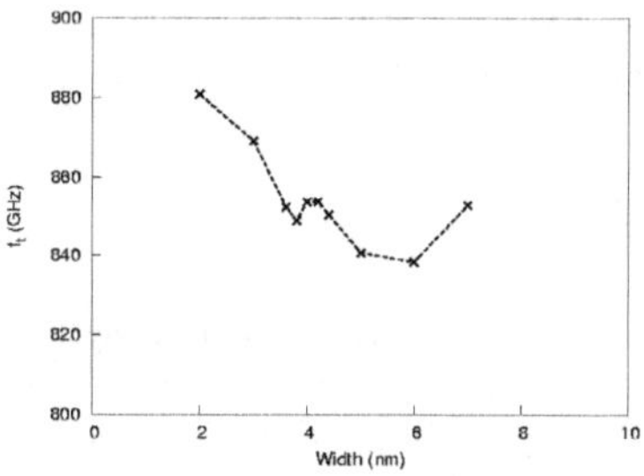

Fig. 7. Variation of f_t w.r.t W with volume inversion

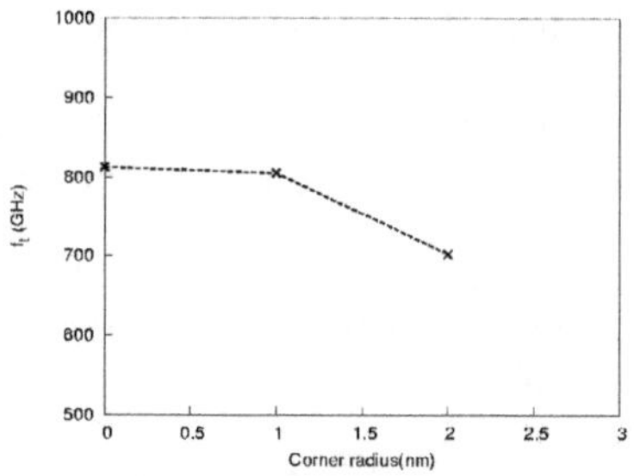

Fig. 8. f_t versus corner radius

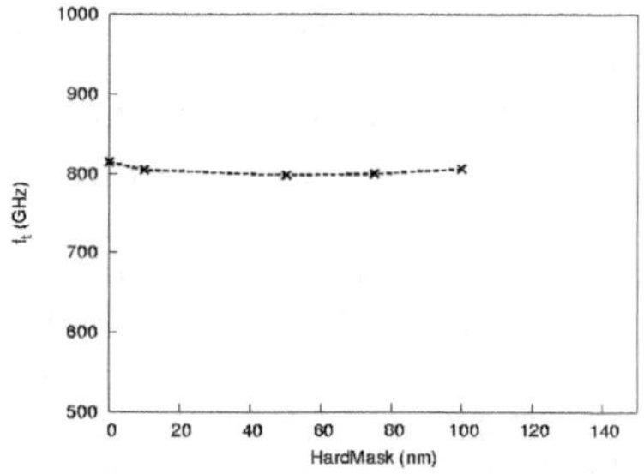

Fig. 9. Variation of f_t with respect to hardmask height

than 75% to obtain fluctuations of f_t of less than 10% [9].When the corner is rounded with bigger radius the fin has larger amount of gate oxide in those regions. Therefore the current and g_m are expected to go down and thus decreasing f_t.

3.5 Variation in Hardmask Height

Figure 9 depicts variation of f_t w.r.t hardmask height. As hardmask height increases, C_{gg} gets decreased due to reduced fringing field. This effect can be seen at the initial stage. Further increase in hardmask height makes g_m and C_{gg} to become insensitive [10] thus making f_t to become insensitive w.r.to hardmask height.

4 Conclusion

Five different process parameters gate length, underlap, fin width, corner radius, hard mask height were considered in this study to find out their effect on f_t. It was found that gate length, underlap, and corner radius are more sensitive parameters whereas hardmask height is less sensitive parameter. Depending on the channel doping levels, fin width may be a more or less sensitive parameter.

Acknowledgments. This work is supported by Department of Science & Technology, Government of India under SERC scheme.

References

1. Skotnicki, T., Hutchby, J., King, T., Wong, H., Boeuf, F.: The end of CMOS scaling. IEEE Circuits Devices Mag. 21(1), 16–26 (2005)
2. Synopsys Sentaurus Device User Guide (2008-2009)
3. Sabry, Y.M., Abdolkader, T.M., Farouk, W.F.: Quantum Transport Based Simulation and Design Optimization of a 10 nm FinFET. In: 4th International Conference on Design & Technology of Integrated Systems in Nanoscale Era, DTIS 2009, April 6-7, pp. 125–129 (2009)
4. Nuttinck, S., Parvais, B., Curatola, G., Mercha, A.: Double-Gate finFETs as a CMOS Technology Downscaling Option: An RF Perspective. IEEE Transactions on Electron Devices 54(2) (2007)
5. Kim, H.-S., Park, K., Oh, H., Jung, E.S.: Importance of Vth and Substrate Resistance Control for RF Performance Improvement in MOSFETs. IEEE Electron Device letters 30(10) (October 2009)
6. Dambrine, G., Raynaud, C., Lederer, D., Dehan, M., Rozeaux, O., Vanmackelberg, M., Danneville, F., Lepilliet, S., Raskin, J.P.: What are the Limiting Parameters of Deep-Submicron MOSFETs for High Frequency Applications. IEEE Electron Device Letters 24(3) (March 2003)
7. Morteza, F., Hamed, N., Fatemeh, K.: The Impact of Structural Parameters on the electrical characteristics of Nano scale DC-SOI MOSFETs in sub-thereshold region. In: SETIT 2007 4th International Conference: Sciences of Electronic, Technologies of Information and Telecommunications, March 25-29 (2007)
8. Curatola, G., Nuttinck, S.: The Role of Volume Inversion on the Intrinsic RF Performance of Double-Gate FinFETs. IEEE Transactions On Electron Devices 54(1) (January 2007)
9. Wang, R., Zhuge, J., Huang, R., Tian, Y., Xiao, H., Zhang, L., Li, C., Zhan, X., Wang, Y.: Analog/RF Performance of Si Nanowire MOSFETs and the Impact of Process Variation. IEEE transactions on Electron Devices 54(6) (June 2007)
10. Kim, K.Y., Park, S.R., Seo, S.H., Kang, G.C., Roh, K.S., Lee, S., Song, K.J., Choi, C.M., Jeon, K., Park, J.H., Lee, J., Kim, T.Y., Kim, D.M., Kim, D.H.: Comparative Study on Circuit Performance Depending on Double-Gate (DG) and Triple-Gate (TG) FinFET Structure. In: The 15th Korean Conference on Semiconductors (February 2008)

A Cluster Based Multi-Radio Multi-Channel Assignment Approach in Wireless Mesh Networks

Ghosh Saurav[1], Das Niva[1], and Sarkar Tanmoy[2]

[1] A.K. Choudhury School of Information Technology, University of Calcutta
[2] School of Mobile Computing and Communication, Jadavpur University
Kolkata, India
{sauravghoshcu,niva.cu,tanmoysarkar1}@gmail.com

Abstract. Wireless mesh networks (WMNs) are receiving increasing attention as an effective means to provide broadband internet. Throughput is a major QoS in WMNs keeping in view of their perceived application areas and others being connectivity and reliability. WMNs use multiple radios and orthogonal communication channels to reduce interference and increase throughput and at the same time providing path redundancy, reliability and connectivity. In our work we propose a cluster based channel assignment scheme for WMNs and assume that the wireless radio interfaces are equipped with IEEE 802.11 network interface cards (NICs). We have extensively simulated our work using ns-2 network simulator and compared it against well-known channel assignment techniques and our result exhibits a significant increase in throughput.

Keywords: Multi-Radio wireless mesh networks, algorithm, clustering, channel assignment, throughput, and simulation.

1 Introduction

Wireless Mesh Networks[1,2] are dynamically self-organized and self-configured with a high degree of fault tolerance having low deployment and maintenance cost. They are an alternative to the traditional Wi-Fi networks, which requires extensive infrastructure and suffers from the last mile connectivity problem. WMNs provide a platform to integrate various types of networks like WSN, Wi-Max, and Cellular and also to provide broadband internet connectivity to mobile nodes or to remote residential areas. All the perceived applications of WMNs require a high throughput, low latency and adequate path redundancy as its major QoS requirements. Multihop wireless communication in WMNs is beset with several problems such as high interference, increased collisions due hidden/exposed terminals and high levels of congestion with all aggregating to produce an extremely low end-to-end throughput. Due to the rapid advancement in electronics technology, it is feasible to equip each wireless node with multiple radios with each operating simultaneously over different orthogonal channels. The resulting network known as multi-radio multi-channel mesh (MR-MC WMNs) reduces wireless interference as adjacent radios operate over different orthogonal channels and the throughput increases. The mapping between a radio interface and an orthogonal channel is known as the channel assignment problem (CA)

V.V. Das, J. Stephen, and Y. Chaba (Eds.): CNC 2011, CCIS 142, pp. 487–491, 2011.

[3,7] and is critical to the proper functioning of the MR-MC WMNs. Various formulations for CA have been proposed in literature and it is proved that CA is a NP-Hard problem. As such, no deterministic polynomial bounded method exists for CA and we have to concentrate on developing heuristic methods.

2 Related Work

A number of channel assignment methods are proposed in literature and a good overview can be found in [7]. One such method BFS-CA [4], proposed by K. Ramachandran et al., models the MR-MC WMNs by Multi-Radio Conflict Graph (MCG), in which all the possible edges between radio interfaces are modeled as nodes in the MCG. Then a breadth first search is executed to visit and color each node such that two adjacent nodes are assigned different colors. It is a centralized approach in which all the computations are done by the gateway node. A distributed channel assignment approach in WMN proposed by Raniwala and Chiueh called `Hyacinth' which jointly performs channel assignment and its associated routing. Here a tree like topology is extracted on which the method runs. In both [4, 5] above both network connectivity and topology is preserved. Naveed et al. propose an all-in-one scheme called the CoMTaC [6] to jointly address the topology control, channel assignment and routing.

3 Proposed Work

A cluster-based approach is employed in two steps to divide the network into clusters and localize the channel assignment problem within each cluster maintaining network connectivity. In the first step, the network nodes are grouped into clusters and the cluster heads are selected. The second step focuses on the channel assignment. We assume that every node has knowledge of its hop distance from its nearest gateway node. The function Hopcount (v) is hop distance of node v from its nearest Gateway having id GID (v).The clustering algorithm is as follows:

Procedure Cluster:
Input: Connectivity Graph (V, E), number of nodes n, set of Gateway nodes V_G where $|V_G| = m$.
Output: Set χ containing cluster set $X_{i=1,2,3...}$ where X_i is the ith cluster head.
Step 1: Initially there are no cluster heads in the network.
Set $\chi=\varphi$, $m=|V_G|$ and $p=m$: initial number of clusters.
for i = 1 to m do // initially each gateway node considered as a cluster head.
$X_i = v_i$, where $v_i € V_G$ and $\chi = X_i$.
end for
Step 2: Each node joining a cluster with a gateway node as cluster head.
for i = 1 to n do x=GID (v_i): v_i € V
$X_x = X_x \cup \{v_i\}$, clusterdist (v_i) = hopcount (v_i)
CHID (v_i) = GID (v_i): CHID is cluster head ID
end for

Step 3: The number of Gateway nodes are limited, so the radius of the cluster is large. To restrict the size of a cluster to a radius $r \leq 2$ procedure ConstructCluster is called.
Set r=2. while clusterdist(x) > r for any $x \in \{X_1 \cup X_2 \ldots \cup X_p\}$ do
ConstructCluster (χ, p, n, V)
end while
End Procedure Cluster
Procedure ConstructCluster (χ, p, n, V)
h =x such that clusterdist(x) is maximum
p = p+1
$X_p = \{h\}$
CHID (h) = h
$X_{CHID(h)} = X_{CHID(h)} \setminus \{h\}$
for i =1 to n do if clusterdist(v_i) > dist(h, v_i) then
$X_p = X_p \cup (v_i)$
$X_{CHID(vi)} = X_{CHID(vi)} \setminus \{v_i\}$
CHID (v_i) = h
end if
end for
end **Procedure ConstructCluster**

The cluster radius, *r*,(maximum hop distance from the cluster) is a design parameter. We have set r=2. The cluster head selects other cluster head based on maximum hop distance from itself. Then the cluster is constructed around newly selected cluster head by adding nodes to 2 hop distances, as our interference range is two hops because of the protocol interference model as depicted in fig 1. The cluster head then checks among its one hop neighbor, who has the maximum neighboring nodes and selects them as control nodes (CN). The CN finds its child nodes from its one-hop neighbors and these nodes are controlled only by that CN.

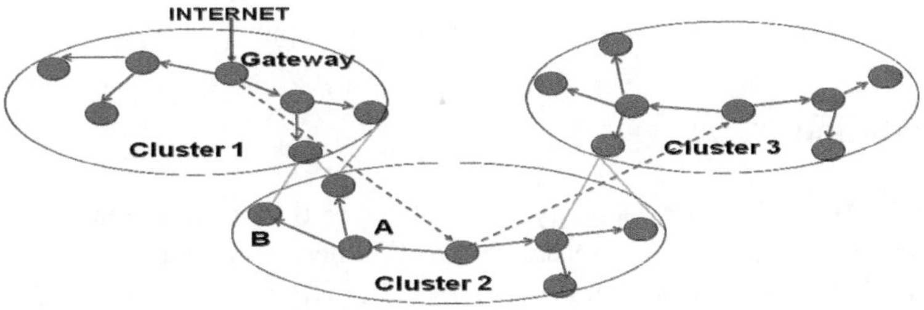

Fig. 1. Formation of Cluster

Procedure Control-Child Node Selection within a Cluster
Step 1. CH will sends “HELLO” packets to its one hop neighboring nodes (say node A in Fig 1.)
Step 2. The neighboring nodes (i.e.A’s) send an “ACK” packet to the CH.

Step 3. The neighboring nodes (i.e.A's) will send a "HELLO" packet to its 1-hop neighbors (say node B in Fig 1.).
Step 4. Two hop neighbors of the CH (i.e.B's) will send an "ACK" packet to one hop neighbors of CH (i.e.A's).
Step 5. One hop nodes (i.e.A's) Counts the number of "ACK" packets received from two hop neighbors (i.e.B's) and send it to CH.
Step 6. The CH selects the Control Nodes from among its one hop neighbors (i.e.A's) based on the basis of maximum number of children (i.e.B's) it has.
Step 7. CH sends a "JOIN" request to those selected Control Nodes.
Step 8. The selected Control Nodes sends "JOIN" request to its neighboring nodes (i.e.B's).
Step 9. On receiving the "JOIN" request, the nodes (i.e.B's) will send a "JOIN-ACK" message to the Control Node from which it gets the first "JOIN" request.
Step 10. Then the Control Node adds that node in its list of Child Node and sends an "ACCEPT" message to the child node.

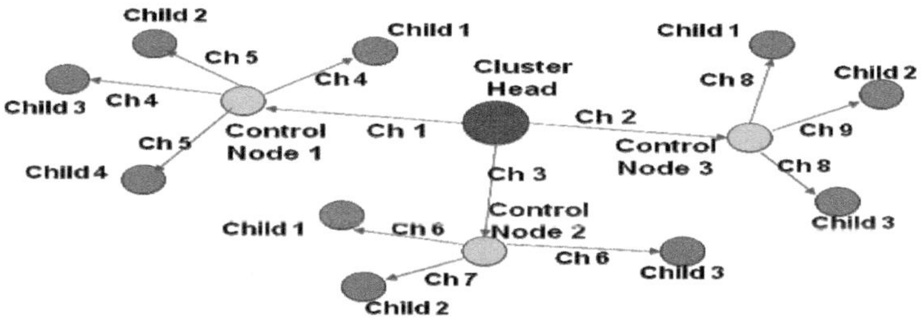

Fig. 2. Channel assignment to nodes within a cluster

A proportionate set of mutually orthogonal channels are being allocated to each CH based on the number of nodes in that cluster with adjacent clusters being allocated mutually non-interfering sets of orthogonal channels. The CH selects its communication channels to its control nodes and a set of non-interfering channels to each of its control nodes to communicate with their children as depicted in fig 2.

4 Simulation

A 25-node MR-MC WMN topology operating over twelve orthogonal channels and having four gateway nodes is created in NS-2.33 network simulator. Network interface cards (NICs) are divided into two classes: (1) UP-NICs that are used to connect to the parent and (2) DOWN-NICs that are used to connect to the children. We have assumed that each node runs the DSR routing protocol. We have compared our results i.e. scenario 3, with BFS-CA[4] i.e. scenario 2 and the naïve (single channel) approach i.e. scenario 1. The packet delivery ratio is of our method is approximately the same as BFS-CA[4] as shown in fig 3 but our method outperforms the latter with respect to throughput as shown in fig 4.

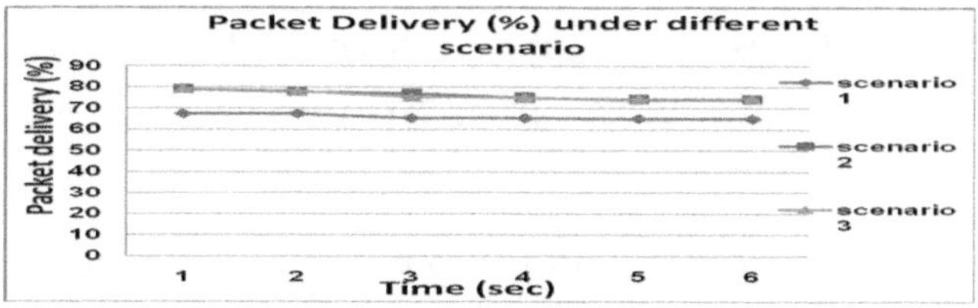

Fig. 3. Packet Delivery

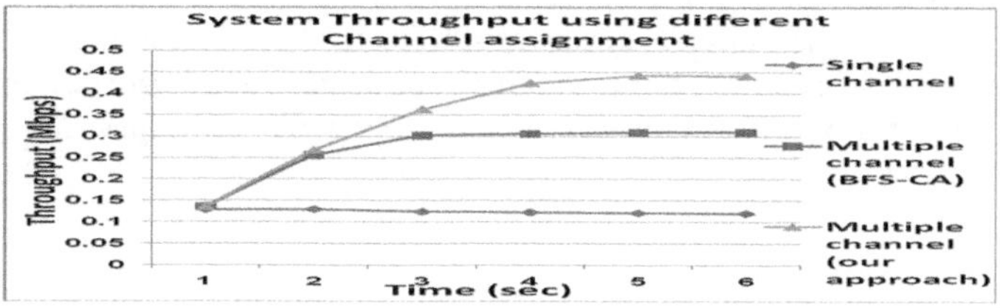

Fig. 4. System throughput

5 Conclusion

Our method does not consider network traffic or interferences from external networks working in the similar frequency ranges. A statistical analysis of the network traffic is an important pre-requisite prior to channel assignment and after that some soft computing techniques can be applied for channel allocation. Therefore, we shall try to make this possible for multiple gateway WMNs scenario and a better performance.

References

[1] Ishmael, J., Race, N.: Wireless Mesh Networks. In: Middleware for Network Eccentric and Mobile Applications (2009)

[2] Akyildiz, I.F., Wang, X., Wang, W.: A Survey on Wireless mesh networks. IEEE Radio Communications (September 2005)

[3] Conti, M., Das, S.K., Lenzini, L., Skalli, H.: Channel Assignment Strategies for Wireless Mesh Networks. In: Wireless Mesh Networks: Architectures and Protocols. Springer, Heidelberg (2007)

[4] Ramachandran, K., Belding, E., Almeroth, K., Buddhikot, M.: Interference aware channel assignment in multi-radio wireless mesh networks. In: IEEE INFOCOM (2006)

[5] Raniwala, A., Chiueh, T.-C.: Architecture and algorithms for an IEEE 802.11-based multichannel wireless mesh network. In: IEEE INFOCOM (2005)

[6] Naveed, A., Kanhere, S.S., Jha, S.: Topology control and channel assignment in multiradio multi-channel wireless mesh networks. In: IEEE MASS (2007)

[7] Si, W., Selvakennedy, S., Zomaya, A.Y.: An overview of Channel Assignment methods for multi-radio multi-channel wireless mesh networks. Journal of Parallel and Distributed Computing (2009)

VSPACE: A New Secret Sharing Scheme Using Vector Space

Prasenjit Choudhury[1], Alokparna Banerjee[2], V. Satyanarayana[2], and Gourish Ghosh[3]

[1] Department of Computer Applications, NIT Durgapur, West Bengal, India
[2] Department of Information Technology, NIT Durgapur, West Bengal, India
[3] Tata Consultancy Services, Kolkata, West Bengal, India
prasenjit0007@yahoo.co.in, speak2parna@yahoo.co.in, btech.sandeep@gmail.com, gourishghosh@gmail.com

Abstract. In this paper, a new secret sharing scheme utilizing the vector space is proposed. Initially, the basis vectors of each character of the secret are found on the basis of a random partition value, *N*. Next, the basis vectors of each partition are distributed in separate files and kept well distributed throughout the network. Only an authenticated person who knows the location of the shares can collect all the shares, distributed throughout the network, remove the redundancy and take original basis vectors to reconstruct the actual secret. The scheme is well secured against Brute Force attacks and partial discloser problem. Malicious behaviors are restricted by assigning dummy shares as long as they are not trustworthy.

Keywords: Secret sharing, vector space, basis vectors, MANET, security, cryptosystem.

1 Introduction

Secret sharing (threshold cryptography) is used to divide a cryptographic action, such as the generation and management of a secret key or the computation using secret keys, among several parties, in such a way that the action can be performed if and only if at least a certain number of parties collaborate.

In this paper, a new secret sharing scheme using vector space is proposed. It needs less computation and hence is suitable to meet the security requirements [2] of MANET [1]. It is more secure against eavesdropping, Brute force attack [8] and partial discloser problem.

The rest of the paper is organized as follows. In Section 2, background of the proposed scheme is discussed, followed by a brief introduction about vectors and vector space in Section 3, and vector space associated with a character in Section 4. In Section 5, a new secret sharing scheme is proposed with example. Security of the scheme is analyzed in Section 6, followed by concluding remarks in Section 7.

V.V. Das, J. Stephen, and Y. Chaba (Eds.): CNC 2011, CCIS 142, pp. 492–497, 2011.

2 Background

In Shamir's secret sharing scheme [3], a secret (S) is divided into n pieces in such a way that it is easily reconstructable from any k pieces, but even complete knowledge of k-1 pieces reveals absolutely no information about secret. Here, the main goal is to divide S into n pieces S_1, ..., S_n in such a way that knowledge of any k or more S_i pieces of share makes S easily computable. It is not possible to reconstruct S from any k-1 or fewer S_i pieces. Such a scheme is called a (k, n) threshold scheme [4].

Problems with Shamir's scheme. Shamir's secret sharing paradigms suffers with the following problems:

- In (k, n), the value of threshold, k is fixed. If the adversary gains the knowledge of 'any k' number of secrets out of n, he can easily reconstruct the original secret.
- Secret sharing scheme does not depend on individual character of the secret.
- The computations are not feasible.
- The scheme suffers from partial disclosure problem.

In Shamir's (k, n) threshold scheme, the value of k was fixed, but in our approach it is variable. We have distributed shares depending on the reputation [5] [6] of each node. Our scheme is similar to (n, n) secret sharing scheme, although it satisfies the properties of threshold secret sharing scheme.

3 Vectors and Vector Space

A point in a Euclidean plane can be represented by an ordered pair of numbers $X = (X_1, X_2)$ [7]. Point X can also be regarded as a vector emanating from the origin $O = (0, 0)$ to the point (X_1, X_2). In GCF(2), every number can be either 0/1.

Definition. An n-dimensional vector space over the field F consists of:

1. A field F (with its element set S and two operations * and •)
2. A set W of k-tuples (numbers are taken from F)
3. A binary operation '$\oplus$' called vector sum between the elements of the set W, such that W is an abelian group under this operation $\oplus$
4. A binary operation '$\otimes$' called the scalar multiplication between the elements of the set W, such that W is an abelian group under this operation $\otimes$

4 Vector Space Associated with a Character

An ASCII character, C can be represented by 8 tuples, $C= (c_1, c_2, c_3, c_4, c_5, c_6, c_7, c_8)$, such that c_i is either 0 or 1[7]. For instance, the character A is represented by 01000001. Altogether, 2^8 or 256 such 8 tuples are possible. For example, say $J \leftarrow C_1$ and $r \leftarrow C_2$. J is represented as 01001010 and r is represented as 01110010.

The ring sum $C_1 \oplus C_2$ is represented by 00111000, which is the ASCII value of '8'.

5 Our Proposal of Secret Sharing

Motivation of our scheme. In the proposed secret sharing scheme, graph theory and basis vectors were used for the distribution of shares. Initially, the binary ASCII value of each character of the share was found. This binary representation of the secret string was then grouped into partitions, depending on a random partition value N, such that each partition P_i has N bits. P_i was then broken down into basis vectors. Let k_i be the number of basis vectors in each P_i. As the number of basis vectors in each P_i is variable, hence k_i is also variable. The basis vectors of each P_i were used to construct the secret shares. Hence, to reconstruct a secret, each partition, P_i must be restored, for which all the k_i basis vectors are needed. Thus, k_i is the threshold of each P_i. But k_i being variable for different partitions, it was more difficult for the adversary to guess all the values of k_i. Moreover, without knowing the value of N (which is a random number known by the dealer only), the adversary could not retrieve the individual characters of the secret. Thus Brute Force attack [8] became very difficult to achieve. Secondly, to reconstruct the shares into original secret, it needed less computation because basis vectors were introduced.

The shares are distributed depending on the reputation [5] [6] of each node. The shares which have more number of basis vectors are distributed among the set of trustworthy nodes, so that they can reconstruct the secret whenever required. The new nodes of the network, whose reputation is, yet to be determined, get the shares with more number of redundant vectors. Moreover, we have used partitioning so that malicious users cannot get partial information from his share or other shares.

Our scheme has the following 3 phases:

5.1 Initial Phase

In this phase, the individual characters of the secret were partitioned based on a random partition value N and then the basis vectors of each partition were found out. It consisted of two functions, *partition()* and *findbvarray()*. The function *partition()* takes *str* as input and divides it into T number of partitions, such that T=*str.length*. The function *findbvarray()* finds the secret shares (basis vectors along with redundancy) of each partition, P and stores them in an array.

Importance of Partitioning

- To eliminate partial disclosure problem a partition value was introduced to convert each character to an un-known format. So, malicious user cannot get partial information from his share or other shares.
- Maximum number of basis vectors in an ASCII character is only 7. So a maximum of 7 shares can be obtained without using the partition technique. By introducing the partition mechanism, any number of shares can be obtained based on the size of the random partition value. Partition value being random, number of shares for the same secret but different partition value is different.

Input. Suppose the secret string: alo & the partition value, N: 10
str:= alo. Fig 1 shows each character of *str* along with its binary ASCII value.

Each character of *str*:=alo	a	l	o
Binary ASCII Value	01100001	01101100	01101111

Fig. 1. ASCII value of each character of *str* for *str*:=alo

Taking N:=10, we find the partitions P_1, P_2 and P_3. We make the size of P_3 equal to 10 (N) by padding it with extra '1's. Fig 2 shows each partition P_i along with its binary ASCII value.

Each Partition P_i	P_1	P_2	P_3
Binary ASCII Value	0110000101	1011000110	1111111111

Fig. 2. ASCII Value of each partition, P_i

Maximum number of Basis Vectors in P_1, P_2 and P_3 is 10. So, M=10= Number of shares. Now we find the basis vectors of each partition. Fig 3 shows the secret shares (basis vectors and redundancy vectors) of each partition P_i.

Each Partition P_i		P_1	P_2	P_3	
Secret Shares of P_i	1	0000000001	0000000010	0000000001	→ To *File1*
	2	0000000100	0000000100	0000000010	→ To *File2*
	3	0010000000	0001000000	0000000100	→ To *File3*
	4	0100000000	0010000000	0000001000	→ To *File4*
	5	0000000101	1000000000	0000010000	→ To *File5*
	6	0010000100	0000000110	0000100000	→ To *File6*
	7	0110000000	0001000100	0001000000	→ To *File7*
	8	0100000101	0011000000	0010000000	→ To *File8*
	9	0010000001	1010000000	0100000000	→ To *File9*
	10	0100000100	1000000110	1000000000	→ To *File10*

Basis Vectors (light) · Redundant Vectors (shaded)

Fig. 3. The Secret Shares corresponding to each partition, P_i

5.2 Distribution Phase

In this phase, the basis vectors of each partition was distributed in separate files and kept well distributed throughout the network. It consisted of the function *distribution()*, which takes *str* as the input and puts the secret shares in files.

The secret shares are then distributed in 10 various files all over the network. Fig 4 shows the contents of *File1* to *File10*.

5.3 Reconstruction Phase

In this phase, only an authenticated person, knowing the location of the shares and having the right authority can collect all the shares, distributed throughout the network, remove the redundancy and take the original basis vectors to reconstruct the

Contents of *File1*:	0000000001	0000000010	0000000001
Contents of *File2*:	0000000100	0000000100	0000000010
Contents of *File3*:	0010000000	0001000000	0000000100
Contents of *File4*:	0100000000	0010000000	0000001000
Contents of *File5*:	0000000101	1000000000	0000010000
Contents of *File6*:	0010000100	0000000110	0000100000
Contents of *File7*:	0110000000	0001000100	0001000000
Contents of *File8*:	0100000101	0011000000	0010000000
Contents of *File9*:	0010000001	1010000000	0100000000
Contents of *File10*:	0100000100	1000000110	1000000000

Fig. 4. Contents of the Secret Share Files

partitions. He can retrieve the individual characters of the secret using *N*. This phase consisted of two functions, *isBasisVector()* and *reconstruct()*. The *isBasisVector()* function takes an integer value and checks whether it is a basis vector or not. If the integer is a basis vector, it returns 1, otherwise it returns 0. The *reconstruct()* function reads the basis vectors from the secret files created by the *distribution()* function and outputs the secret message.

Output. The secret is: alo

6 Security Analysis

As the secret is shared between several users, each can get only a single secret key. Hence it is hard to attack by Brute Force [8]. Moreover, value of *k* being variable for different partitions, it becomes more difficult for the adversary to guess all the values of *k*. But there is chance of Brute Force attack by acquiring partial information from a shared secret. A malicious user can try to get another shared secret with the help of his share of secret. To overcome this problem, we have introduced a random partition value, *N*. Without knowing the value of *N* (which is a random number known by the dealer only), the adversary cannot retrieve the individual characters of the secret. Thus Brute Force attack is very difficult to achieve.

The scheme is also secured against partial discloser problem of secret sharing. Introduction of the random partition value, *N* converts each character of the secret string to an unknown format. So, malicious users cannot get partial information from his share or other shares. Moreover, we can restrict behaviour of malicious users by assigning dummy shares as long as they are not trustworthy.

7 Conclusions

In this paper, we have introduced a new secured scheme for sharing secrets in the network. The secret string was broken down into *shares* with the help of basis vectors. The sharing technique is well secured against partial discloser problem, as well as Brute Force attacks. We have also restricted the behaviour of malicious users by assigning dummy shares to them on the basis of multi-degree or weighted entropy, as long as they are not trustworthy. The scheme takes less computation, compared to the remaining approaches of secret sharing. Thus, it is very useful in MANET.

References

1. Corson, S., Macker, J.: Mobile Ad hoc Networking (MANET): Routing Protocol Performance Issues and Evaluation Considerations. RFC 2501 (January 1999)
2. Yang, H., Luo, H., Ye, F., Songwulu, Zhang, L.: Security in Mobile ADHOC NETWORKS: Challenges and Solutions. In: IEEE Wireless Communications and Networking conference (2004)
3. Shamir, A.: How to Share a Secret. Communications of ASM 22 (11) (November 1979)
4. Beimel, A., Tassa, T., Weinreb, E.: Characterizing Ideal Weighted Threshold Secret Sharing. SIAM Journal on Discrete Mathematics 22(1) (February 2008)
5. Michiardi, P., Molva, R.: CORE: A Collaborative Reputation Mechanism to Enforce Node Cooperation in Mobile Ad hoc Networks. In: Proceedings of the IFIP TC6/TC11 Sixth Joint Working Conference on Communications and Multimedia Security: Advanced Communications and Multimedia Security, pp. 107–121 (September 2002)
6. Bonnefoi, P.-F., Sauveron, D.: MANETS: An Exclusive choice between use and security? Computing and Informatics 22, 1001–1013 (2003) (October 22, 2007)
7. Graph Theory by Narshing Deo [PHI]
8. Brute Force Attack, http://en.wikipedia.org/wiki/Brute_force_attack

Secure Group Deffie-Hellman Key Exchange with ID Based Cryptography

Om Pal, Anupam Saxena, Uttam Kumawat, Ravi Batra, and Zia Saquib

Centre for Development of Advanced Computing, Mumbai, India
{ompal,anupam,uttamkumawat,ravibatra,saquib}@cdacmumbai.in

Abstract. Multicast gives large-scale content distribution by providing an efficient transport mechanism for one-to-many and many-to-many communications. There are various security issues in multicast communication, related to the nature of multicast. Group Diffie-Hellman key exchange protocol is a well known protocol for establishing common key on the open channel; however, due to middle-man-attack, the confidentiality can be compromised. To remove the man-in-middle attack in Group Diffie-Hellman, we propose a secure identity-based key management scheme, which combines the principles of Group Diffie-Hellman and of trusted third party; scheme uses the identity of the nodes to establish the common key for a group.

Keywords: Key Distribution, Key Management, Broadcast, Multicast, Identity based cryptography.

1 Introduction

Due to exponential growth in the use of Internet technology, life has become easier, and we have benefited by the use of this technology; but maintaining the complete secrecy of data is an open issue. One critical security requirement is that the communication channels need to be secured. For providing confidentiality on open channel, encryption and decryption are needed. Key management schemes are used for achieving confidentiality, authentication, non-repudiation etc. Confidentiality ensures that no one can read the message except the intended receiver. Integrity ensures that the data has not been changed. Authenticity ensures that the origin of an object is what it claims to be. Non-repudiation ensures that the sender cannot deny about sending a message later. Considering these security issues, we can not ignore the role of Key management and key distribution. Key management implies different services such as generation, registration, distribution, installation, storage, retrieval, renewal, revocation, archiving or destruction of the key. Strength of encryption and decryption is always dependent on encryption algorithm and key which is used for encryption and decryption. Secure keys are needed to be established before applying any cryptographic technique to secure the communication.

Multicast enables efficient large-scale content distribution, by providing an efficient transport mechanism for one-to-many and many-to-many communications. Today, the applications that are of multicast nature have increased (as video conference, distance learning, pay per view TV, financial stock quote distribution, etc).

V.V. Das, J. Stephen, and Y. Chaba (Eds.): CNC 2011, CCIS 142, pp. 498–503, 2011.

One of the important tasks for the use of multicast communication is ensuring the authenticity. Receiver must be able to verify the identity of sender. First level of functionality for the receiver is to verify that the data is coming from a group member. The next level of functionality for the receiver is to verify that it is from an authorized sender. Functionality may be précised to determining the exact identity of the sender. In case of multicast communication, the authentication is main problem, as it requires that a large number of recipients must verify the sender.

In this paper, we present a scheme for establishing a common key for the group and the scheme also eliminates the possibility of man-in-middle attack among the communicating parties. Rest of paper is organized as follows: in next section related work is presented, in section 3, proposed scheme is given. section 4 concludes the paper and references are given in the last section.

2 Related Work

Key management is a well known research topic in the cryptographic research community. A lot of work has already been done in field of key management but still this field is not fully saturated.

Emmanuel Bresson [11] presented multicasting approach for dynamic group. The good feature of Emmanuel's scheme is that it builds group key dynamically, and time complexity of the group key generation is linear. If there are 'n' number of nodes is in the group, then only 'n' numbers of transmissions are required. In this scheme first node sends some parameters (g, g^a) to next node to initiate the process; where 'g' is the generator of the group and 'a' is the secrete parameter of first node. After getting parameters from first node, second node raises the power of its secret number on all received parameters. Second node passes all modified parameters and the last parameter of first node to third node. This chain remains continuous up to the second last node of the group. After raising the secret number, last node of the group broadcast all parameters. From the broadcast values, each node takes its respective part and then builds the group key. This scheme provides the linear time complexity but it suffers from man-in-middle attack. Any attacker can also participate during the construction of group key and attacker can get all confidential information of other members.

In the paper "one round protocol for tripartite Diffie-Hellman", Antoine Joux A proposed a three participants variation of the Diffie-Hellman protocol [3], based on the Weil and Tate pairings on elliptic curves. It facilitated reducing discrete logarithm problem on some elliptic curves to the discrete logarithm problem in a finite field.

Authors [1] presented group key distribution protocol. In this protocol, numbers of messages are high, so that the computational time is also high. If any member leaves the group then it requires to setup the group key again, otherwise forward secrecy will become a problem. Scheme is also vulnerable to man-in-middle attack. The Group Key Distribution protocol using Diffie-Hellman (DH)[1] constructs the group key by using two stages flow which is called as up flow and down flow.

In 1984, Adi Shamir, described "Identity-Based Cryptosystems and Signature Schemes"[2], where any publicly-known string (e.g. someone's email address) could be used as public key. The Identity-based cryptography uses public-key cryptography in a different manner, where the public key of a user gives unique information about the identity of the user. This can use the text name, or domain name, or the physical IP address as a key.

In the paper "Identity-Based Encryption from the Weil Pairing" [5] Dan Boneh and Matthew Franklin proposed a fully functional identity-based encryption scheme and compared the performance with ElGamal encryption. Scheme has chosen ciphertext security in the random oracle model assuming a variant of the computational Diffie-Hellman problem, and is based on bilinear maps (like Weil pairing on elliptic curves) between groups. They proposed the PKG as distributed so that the master-key is never available at a single location.

Our scheme uses concept of Group Deffie-Hellman key exchange and Id based cryptography to establish the common key for the group and eliminates the possibility of man-in-middle attack. In next section proposed scheme is presented.

3 Proposed Scheme

Scheme assumes that there is one centralize server (Private Key Generator) which is used to generate the private key for each node. Firstly, PKG generates private key for each node by using identity of nodes and the master private key (Private Key of PKG). After the generation of private key for each node, the PKG distributes the private keys to the corresponding nodes manually. For the future use, PKG can store private key of any node in its database or it can generate at run time. To reduce the run time overhead, PKG can store private keys of all nodes in its database but in this case extra storage is needed. Any node can generate the public key of any other node by using identity of that node and master public key (public key of PKG).

Let Id_a is the identity of node A, Id_b is the identity of node B. M_{pvt} is the master private key, M_{pub} is the master public key. F is a function which is used for generating private and public key of any node, p is a large prime number, G is a group of order p, g is the generator of the group.

Generation of Private Key: Let A_{pvt} is the private key of node A. PKG generates Apvt by using function F as

$$A_{pvt} = F(Id_a, M_{pvt}) \tag{1}$$

In same way PKG generates Bpvt for node B using

$$B_{pvt} = F(Id_b, M_{pvt}) \tag{2}$$

Generation of Public Key*:* Any node can generate public key of any other node by using master public key M_{pub}, function F and Id of corresponding node like-

$$B_{pub} = F\,(Id_b, M_{pub}) \tag{3}$$

Generation of Group Key*:* Let node A want to establish group key with nodes B,C,D then A sends signed request (message 1) with list of group nodes (B,C,D) to PKG after getting message1 from A, PKG sends a message which includes generator g, time stamp t, member list A,B,C,D to A on open channel. A sends message 3 to B, B sends message 4 to C, C sends message 5 to D as shown in figure-1.

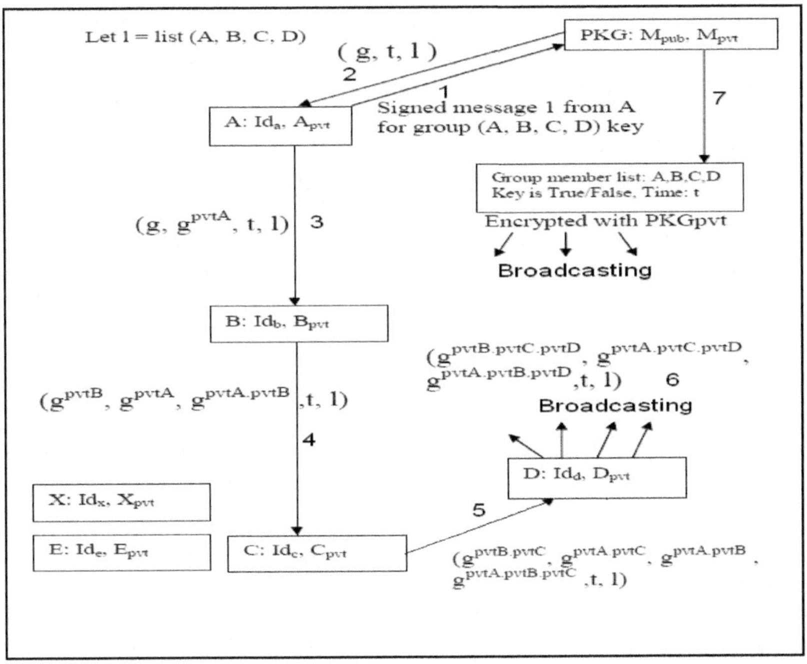

Fig. 1. Key Setup Diagram

After getting message 5 from C, D computes common key $g^{pvtA.pvtB.pvtC.pvtD}$ and D does not broadcast this common key. D broadcasts $g^{pbtB.pvtC.pvtD}$, $g^{pvtA.pvtC.pvtD}$, $g^{pvtA.pvtB.pvtD}$, t, l. In broadcast message 6, first value belongs to A, second to B, third to C and so on. Each node takes it respective part from broadcast message 6 and applies its private key to obtain group key. A applies its private key on first value; B applies its private key on second value and so on. A computes group key $(g^{pbtB.pvtC.pvtD})^{pvtA}$ and B computes group key $(g^{pbtA.pvtC.pvtD})^{pvtB}$ and so on.

In scheme like [11] PKG is not there so any attacker can also participate in the group and he/she can read common messages easily.

In our proposed solution, PKG takes the first value (which belongs to A) from broadcast message and applies the private key of A to obtain group key k_1 and then computes the group key k_2 by using generator g and private keys of all group members. After getting k_1 and k_2, PKG compares these values, if k_2 matches with k_1 then group key is attack free. PKG broadcasts the trueness/falseness of group key through message 7. Contents of message 7 are group member list, time stamp, key verification information and before broadcasting PKG encrypts message 7 by its private key. Through message 7 every node gets the information about trueness of key, if key is true then they can use this group key for communication.

Key Revocation: Any group can revoke its key by sending fresh request to PKG. Nodes can use this fresh group key as a long term key or they can also use it only for one session.

Security Analysis against Man-in-Middle Attack: In our scheme key verification is done by PKG, so if an attacker participates in the group, then PKG will catch the attacker through verification mechanism. If key verification is false, then every node will get this false status and it will not use the newly generated group key, and a fresh request for group key will be sent to the PKG. Time stamp is also associated with each message, so that there is no possibility of replay attacks.

Time Complexity: Time complexity of our scheme is linear. If there are 'n' number of nodes in the group, then 'n+2' number of transmissions are expected. So complexity is O(n).

4 Conclusion

Key management is the back bone of the cryptography. For speeding up the encryption/decryption, symmetric key is a favorable approach. In this paper, first we discussed the key management approaches, followed by issues in the available key management approaches, like man in middle attack in Group Deffie-Hellman Key exchange. The proposed protocol distributes processing among different communication nodes and provides authentication and security by using the features of ID based cryptography. Proposed scheme eliminates the man-in-middle attack in Group Deffie-Hellman Key Exchange and establishes a secure group key among nodes.

References

1. Steiner, M., Tsudik, G., Waidner, M.: Diffie-Hellman Key Distribution Extended to Group Communication
2. Shamir, A.: Identity-Based Cryptosystems and Signature Schemes. In: Blakely, G.R., Chaum, D. (eds.) CRYPTO 1984. LNCS, vol. 196, pp. 47–53. Springer, Heidelberg (1985)
3. Joux, A.: A one round protocol for tripartite diffie-hellman. In: Bosma, W. (ed.) ANTS 2000. LNCS, vol. 1838, pp. 385–394. Springer, Heidelberg (2000)
4. Menezes, A.: An Introduction to Pairing-Based Cryptography: Mathematics Subject Classification. Primary 94A60 (1991)
5. Boneh, D., Franklin, M.: Identity-Based Encryption from the Weil Pairing. SIAM J. of Computing 32(3), 586–615 (2003)
6. Oliveira, L.B., Scott, M., López, J., Dahab, R.: TinyPBC: Pairings for Authenticated Identity-Based Non-Interactive Key Distribution in Sensor Networks
7. Naureen, A., Akram, A., Riaz, R., Kim, K.-H., Farooq Ahmed, H.: Performance and Security Assessment of a PKC Based Key Management Scheme for Hierarchical Sensor Networks. IEEE, Los Alamitos (2008)
8. Kapil, A., Rana, S.: Identity-Based Key Management in MANETs using Public Key Cryptography. International Journal of Security (IJS) 3(1)
9. Mokammel Haque, M., Pathan, A.-S.K., Hong, C.S., Huh, E.-N.: An Asymmetric Key-Based Security Architecture for Wireless Sensor Networks. KSII Transactions on Internet and Information Systems 2(5) (October 2008)

10. Mahalanobis, A.: Deffie-Hellman Key Exchange Protocol, Its Generalization and Nilpotent Groups. PhD thesis, Florida Atlantic University, Boca Raton, Florida (August 2005)
11. Bresson, E., Chevassut, O., Pointcheval, D., Quisquater, J.-J.: Provably Authenticated Group Deffie-Hellman Key Exchange. In: Computer and Communications Security-Proc of ACM CCS 2001, November 5-8 (2001)
12. Mollin, R.A.: An Introduction To Cryptography, 2nd edn. Hand Book. Chapman & Hall / CRC, Taylor & Francis Group

Computing with Enhanced Test Suite Reduction Heuristic

S. Selvakumar[1], N. Ramaraj[2], R.M. Naaghumeenal[1],
R. Nandini[1], and S. Sheeba karthika[1]

[1] Department of Information Technology
Thiagarajar College of Engineering
Madurai, India
[2] Department of Computer Science & Engineering
G.K.M College of Engineering
Chennai, India
ssit@tce.edu, prof.ramaraj@yahoo.in,
{naaghumeenal,nandini.nan,skarthika635}@gmail.com

Abstract. Software computing is a goal-oriented activity requiring, benefiting from, or creating applications for commercial, network communications etc. The process of development includes designing, building and testing software systems for a wide range of purposes, processing, structuring, and managing various kinds of information. To validate the developed one, time and resource constraints are taken into account; therefore test suite reduction technique is the important stage to be followed to test the developed software. Several test suite reduction techniques are present. Our approach is a new type of test suite reduction heuristic in which we propose an algorithm for test suite reduction based on changed condition/coverage criteria (CC/CC) to test the software on satisfying complete user's requirement with less time, cost and more coverage of requirements this increases computing efficiency.

Keywords: CC/CC pair, Software testing, testing constraints, regression testing, test suite reduction, computing efficiency.

1 Introduction

Software testing is essential in the development of any software system for efficient computing. These test objectives vary in granularity. Therefore, it is necessary to reduce the test suite to improve computing efficiency and save test cost. A test suite is a set of test cases. A test case consists of execution conditions, input data, and expected results. It requires seven weeks to run all its test cases and costs several hundred thousand dollars to execute them. In order to overcome this problem, it's better to choose test suite minimization ,which, mainly is used to identify reduced test suite that provide same coverage of the test requirements as the original suite. In this paper, we present a new approach for test case reduction heuristic based on CC/CC pair which deals with both requirement satisfaction and coverage criteria.

V.V. Das, J. Stephen, and Y. Chaba (Eds.): CNC 2011, CCIS 142, pp. 504–507, 2011.

2 Background

In the literature, almost all the approaches to test case generation will consider how to avoid generating redundant test cases. Typical test suite reduction techniques include heuristic approaches and MC/DC criterion is a more rigorous criterion which was advanced by Chilenski et. al [2]. It can be used to reduce the size of test suite to economize on testing time. If a test suite is said to satisfy MC/DC criterion, all executing test cases in the test suite will guarantee that [3]. Although multiple-condition coverage (MCC) and MC/DC are both sensitive, the growth rate of the amount of test cases is different. In general, n conditions within an expression require n+1 test cases but MCC requires 2n test cases for the same Boolean expression [4].The [5] reported that the selected test cases are valid for MC/DC against each gate.

2.1 CC/CC Pair and Test Suite Reduction

In our proposed method it makes use of the concept of changed condition/coverage criteria in which higher weight test case is checked for both MC/DC pair concept and for the requirement coverage conditions.

Definition
The CC/CC a pair of test cases that must contain at least one input same and the other inputs different (replace one by another), but it must always be producing the different outputs and also find the requirement coverage of selected test case.

3 Proposed Test Suite Reduction Technique

3.1 Objectives

i) To reduce the number of all test cases: Generally, the larger the input domain, the more exhaustive the testing would be. To avoid this problem, a minimum set of test cases needs to be created using an algorithm to select a subset that represents the entire input domain.
ii) To find the technique for automatic generation of test cases: To reduce the high cost of manual software testing while increasing reliability of the testing processes.
iii) To keep a minimum number of test runs: The best technique must be able to reduce test cases from minimum test oracles in test run.

4 Experimental Study

4.1 Subject Programs, Faulty Versions and Test Case Pools

We experimented with objects that are retrieved from software infrastructure repository [1] and the programs developed as an academic project .We examined the types of errors introduced in the faulty versions of each subject program and identified six distinct categories of seeded errors. For each program there is a test pool which contains test cases developed for different black-box and white-box testing objectives.

```
Algorithm:
Input:
        Set of requirements:   SR={req[i] / i∈ N}
        Set of test cases:     Stc = {tc [i] / i∈ M}
        Set of test suite:     STs = {Ts[i] / i∈ N}

Output:
      Rs={ } : reduced set of test cases which satisfies complete coverage
 Begin
      Step(1):
            Set Rs={}   as a empty set initially
      Step(2):
            Find all possible test cases (a1(n), b1(n), c1(n), .......n1(n))
            /* where 'n' is the extension of each inputs
            and a1,b1,c1,.....n1 represents the total number of inputs a, b, c*/
     Step(3):
          Assign appropriate test case to req(Req1....TS[])
     Step(4):
                For each i=1 to ntc
                Wtci=calculate Weight(tc[i]);
     Step(5):
                Mwtc=select MAX (Wtc);
                mtc[]=select all tc with Mwtc;
      Step(6):
                For each i=1 to mtc
                If( mtc >1)
               Wtci = Randomly Select One (Mwtc)
               Req.cov(Wtci)=calculate CC/CC pair value [i] for selected Wtci;
      Step(7):
            For each i=1 to n
            If( ( tci∈ Req.cov(Wtci)) and ( tci∈ Tsi) )
            {
               DELETE (Tsi) ;
            }
       Step(8):
    Repeat the process from step (6) to step (7) until all Tsi get deleted indicates that the requirement get satisfied
```

Fig. 1. SuiteRed Algorithm for test suite reduction

Name	Lines of code	No. of Classes
Triangle	123	2
Sample	66	1
average	131	1
area testnumber	186	1
Gcd	142	2
Binary-Search-Tree	130	3
Array-Partition	13	1
Doubly-Linked-List	277	1
Sorting	130	1
Vector	254	1
Binary-Heap	72	2
Disjoint-Set	35	1
Stack	114	5
Elevator	934	12
OrdSet	229	2
deadlock	24	4
accountsubtype	89	6
account	66	3
producer_consumer	99	8
alarm_clock	125	6
linkedlist	121	5

Table 1. Software Infrastructure Repository [1] and Academic Subject Programs

4.2 Experiment Instrumentation

The studied test suite reduction techniques were implemented using JDK 1.4. All the implemented techniques were executed on a PC with an Intel Pentium Dual CPU T3400 @ 2.16 GHz 2.17 GHz CPU and 2 GB memory running the Windows 2000 Professional operating system.

5 Experimental Analysis

We evaluate four characteristics of the reduction algorithm: size reduction of test-suites, time to perform the reduction, cost reduction of test-suites, and requirement coverage.

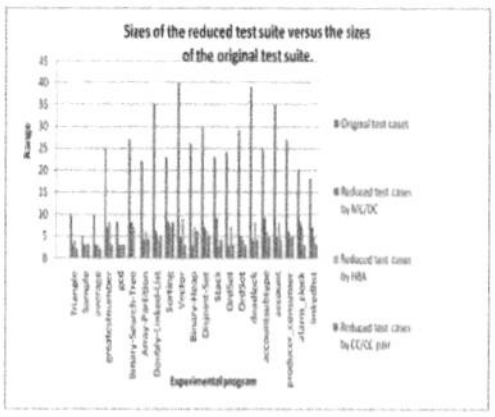

Fig. 2. Size of test suites

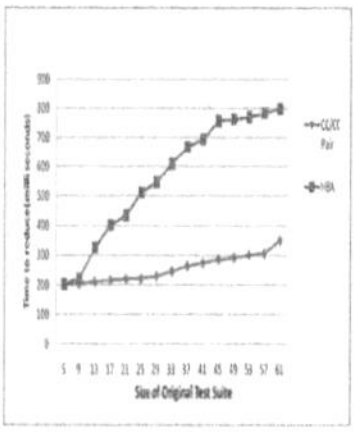

Fig. 3. Time of test suite

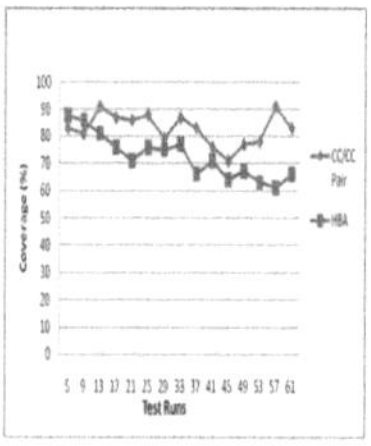

Fig. 4. Coverage information

The results show that HBA is slightly more competent for size reduction than MC/DC, and when test-execution cost is taken into account, the ability of CC/CC to reduce size is considerably improved.

6 Conclusion

In this paper we proposed, a new type of test case reduction for effective computing based on CC/CC PAIR with the simple implementation part which indicates that the test case reduction factors namely execution time, cost, requirement coverage are taken into account. Finally, we empirically evaluated the CC/CC, HBA, and MC/DC on suites of different sizes, comparing our reduced suites to those generated HBA reduction and MC/DC reduction, respectively.

Acknowledgments. We thank Dr. Gregg Rothermel, Department of Computer Science, University of Nebraska for providing the Siemens Suite of programs.

References

1. `http://www.cse.unl.edu/~galileo/sir`
2. Chilenski, J.J., Miller, S.P.: Applicabiliy of Modified Condition/Decision Coverage to Software Testing. Software Engineering Journal 9(5), 193–200 (1994)
3. Chang, J.-R., Huang, C.-Y.: A Study of Enhanced MC/DC Coverage Criterion for Software Testing. In: 31st Annual International Computer Software and Applications Conference(COMPSAC 2007), pp. 457–464 (2007)
4. Aidemark, J., Folkesson, P., Karlsson, J.: Path-Based Error Coverage Prediction. Journal of Electronic Testing: Theory and Applications 18(3), 343–349 (2002)
5. Hayhurst, K.J., Veerhusen, D.S., Chilenski, J.J., Rierson, L.K.: A Practical Tutorial on Modified Condition/Decision Coverage. Report NASA/TM-2001-210876, NASA (May 2001)

Robotic Arm with Finger Movement

Dinesh Mittal*, Ashish Prashar, Vaibhav Gupta, Rahul Yadav, and Rachana Nagal

Amity School of Engineering and Technology, New Delhi, India
{dinesh.2185,ashish.prshr,vaibhav.gupta17oct,ry1558,
rachana.nagal}@Springer.com

Abstract. Robotic arm is an electro-mechanical system with functions similar to that of human arm. The proposed robotic arm is intelligent in the sense that it senses the movement of human arm with the help of the sensors attached to the hand and then move in accordance with it. Many exoskeletal robotic arms have been developed for teleoperation having force reflection. They can measure operator's arm motion and apply reactive force to the operator as well [1]. The movement of robotic arm is basically due to the two main components which are servomotor along with microcontroller. Since, this design includes servomotor for the movement of robotic arm, this will give very precise movements. Movement of servomotor is based on width of the rectangular pulse generated by the combined working of the monostable and astable multivibrator. As the pulse width changes servomotor detects the changes in it and then moves accordingly.

1 Introduction

Robotics is the engineering science and technology of robots, and their design, manufacture, application, and structural disposition. Robotics is related to electronics, mechanics, and software. Robot is a virtual or mechanical artificial agent. In practice, it is usually an electro-mechanical machine which is guided by computer or electronic programming, and is thus able to do tasks on its own. Robots are generally classified for two main applications, which are industrial robots and service robots. In the past, the industrial robot has been extensively applied in industrial manufacturing and construction, such as the steel, automobile, electronic, and semiconductor industries, and has been the main focus of research development. Robotic arms were among the first robots developed and are by far the most common. Various components of robotic arm are structure, actuation, sensing, manipulation, power source etc. The proposed wearable robotic arm is designed to have not only high wear-ability, but also simple design and operation.

1.1 Related Work

Although much work has been done in robotic technology (see [1] [5] [6]), the proposed arm is different from others as the operator has to wear only the sensors and not the whole arm. This arm also includes the fingers and their individual movement

* Corresponding author.

V.V. Das, J. Stephen, and Y. Chaba (Eds.): CNC 2011, CCIS 142, pp. 508–513, 2011.

according to the human fingers. This arm can be used in workplaces where more human casualties occur. The operator can operate the arm from a distance of 2-3 metres, so fewer chances of causalities will be there.

2 Description of Design

Intelligent Robotic arm employed one up and down shoulder movement, one up and down elbow movement and two wrist movements namely rotate and up-down. It can also move its fingers at there base. Servomotors are used for the movement of portion of robotic arm from shoulder to wrist and micro DC motors are used for the movement of fingers. Circuit used for driving servomotor is servodriver. This arm is different from all other robotic arms that have been so far developed as follows – 1. It senses the motion of human hand with the help of sensors in the form of potentiometers attached to it and then moves in accordance with that. 2. Since this design is using servomotor, so it can move at angle ranging 0-180degrees and its movement is very accurate and precise. 3 It senses the motion of fingers with the help of reed switches and then moves accordingly.

2.1 Servodriver

Servomotor works on the principle of Pulse Width Modulation (PWM).Circuit used for driving servomotor is servodriver which is shown in fig. 1. Servodriver was made using few basic components like 555 timer IC, resistors and capacitors. The circuit consisted of two parts namely monostable and astable multivibrator. The first part was 555 timer IC operating in astable multivibrator mode which provides continuous triggering pulses for the second part which was 555 timer IC operating in monostable multivibrator mode. The astable multivibrator circuit consists of external resistors R1, R2, potentiometer RV1 and external capacitor C3. These resistors, potentiometers and capacitors were used to control the pulse width as follows [2]

$$W = 0.693*(R1+R2+RV5)*(C1+C2). \tag{1}$$

When 5V was applied to the pin number 8 and 4 of the 555 timer IC operating in astable multivibrator mode, it produced continuous triggering pulses from output pin 3. This pulse goes to the input of second 555 timer IC operating in monostable mode through capacitor C4. Due to C4, only the ac component passed to second 555 timer input. R3 acted as a pull up resistor for trigger.The second 555 IC helped in formation of Pulse Width Modulation which was controlled by the resistor R3, pot. RV2 and capacitor C4 attached to the IC. Hence, a signal which was PWM in nature was generated and can be used to drive the servomotor. These resistors, potentiometers and capacitors were used to control the pulse width as follows [2]

$$W = 1.1*(R4+ RV2)*(C6). \tag{2}$$

RV1, R5 and C5 are used for voltage control. The output of monostable multivibrator passes through RV3 and R6 which were used to control the amplitude of the output signal. The nature of operation of monostable mode was same as that of astable mode except from the fact that it required external trigger for producing PWM. As stated

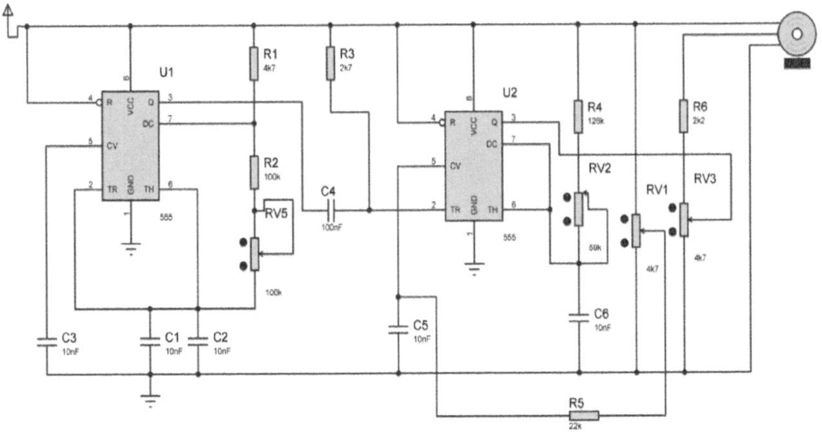

Fig. 1. Circuit diagram used for driving servomotor[2]

earlier the movement of servomotor depends on the change in the pulse width of triggering pulse. Next, when the value of resistance varied through potentiometer, then the pulse width of triggering pulse changed. As a result of this servomotor rotated through different angles according to the width of the triggering pulse at that time. Rotation of the servomotor depends upon the pulse width as follows – 1.When the pulse width is anywhere between 0 and 1.25ms, then there is no rotation in the servomotor. Even if the pulse width reached 1.25ms mark, then also there is 0 degrees of rotation [4]. 2. When the pulse width is further increased and it reaches the 1.50 ms mark, then the servomotor rotates through 90 degrees [4].3. Increasing the pulse width of triggering input again beyond the 1.50 ms up to 1.75 ms, then the servomotor rotates through 180 degrees of angle [4].

2.2 Finger Driving Circuit

Servodriver was used to control the movement from shoulder to wrist part of the arm. Movement of fingers was done using micro DC motors using circuit as shown in fig.2.Movement of fingers was based on the microcontroller and dual H-Bridge IC. Since H-bridge is a dual IC, so it can be used to drive 2 micro DC motors at once. Output from microcontroller goes to the input of H-Bridge IC which was used to move the micro DC motor. The microcontroller uses the program which was stored in its internal RAM, so the pin number 29 and 31 were made high through attaching them to V_{cc}.

The basic working of the circuit depends on the working of the push-to-off reed switches which were attached at the base of human fingers. Magnets were attached to the upper part of the human finger. When the finger was moved (forward), the reed switch was opened, which set the bits of Port 3, which acted as input port for controller. Through port 3, the controller received the input from reed switch and generated a pulse from port 2 of microcontroller which acted as a signal for motor driving IC L293D.

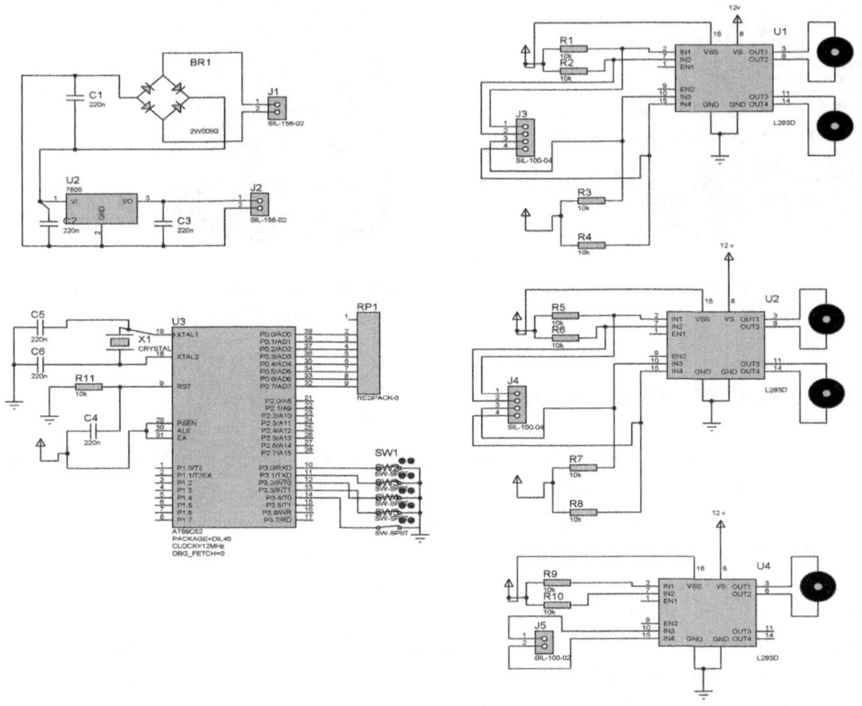

Fig. 2. Circuit diagram for running DC Motor

2.3 Logic for the Code

For delay generation, timer 0 has been used in mode 1. Due to push-to-off reed switches, initially the bits 3.0-3.4 are reset. In MAIN sub module, bit 3.0 is checked. If it is not set then control jumps to bit 3.1 to check it and so on. If any one of the pins is set, then the control will be transferred to a subroutine KEY for that particular input pin. Subroutines are used to set pin 2 and clear pin 7 of L293D. As a result, the motor will move in forward direction. After some delay, the motor will stop.

It will again go to the MAIN routine and check whether the corresponding bit is reset or not. If it is reset, the control will be transferred to a subroutine which makes pin 2 clear and pin 7 set of L293D. Now the motor will move in backward direction for same time and same speed as it did for the forward movement. Now the control is transferred to MAIN subroutine and it will check the next bit.

2.4 Experimental Results

Figure 3 shows an experimental result when operator's movement was not allowed and robotic arm was showing no movement as the pulse width is 1.25ms. Figure 4 shows another experimental result when operator's movement was allowed. In this case, due to the movement of operator's arm sensor generated a pulse width which was more than 1.25ms, hence arm showed a movement. Here, the arm showed maximum movement as pulse width was 1.75ms.

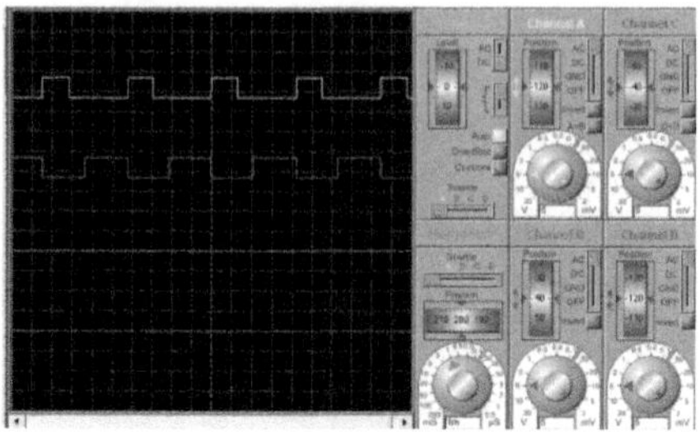

Fig. 3. Sensor driven performance test (without input)

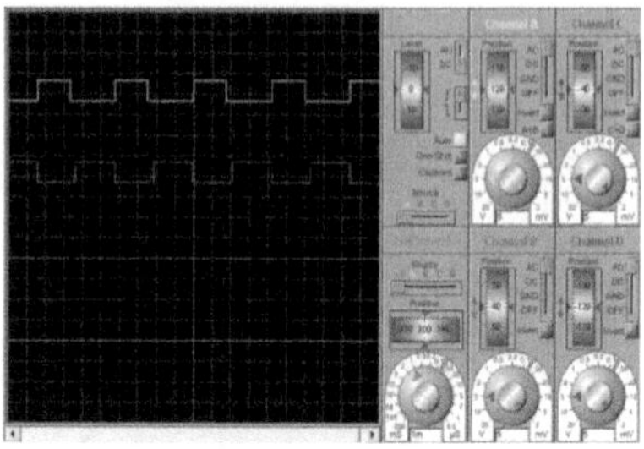

Fig. 4. Sensor driven performance test (with input)

3 Implementation and Result

The Formation of Intelligent Robotic Arm was done in a number of stages which includes –1. The Formation of individual frames for holding the servomotor which were responsible for arm movement.2. Formation of individual frames for holding the micro DC motors which were responsible for the movement of fingers. 3. Assembling the individual modules thus formed to make a complete robotic arm from shoulder to fingers.4. Realizing the finally tested circuit either on general purpose board or printed circuit board which will be used finally.5. Placing sensors on hand as well as on the base of the fingers so that they can sense the motion of human hand and fingers and then moved accordingly.6. Finally assembled robotic arm is as shown in the fig 5.

In this way the 'Intelligent Robotic Arm' was implemented. The design of robotic arm can be extended to make it to show all the movements i.e. the movement of each and every joint of human hand. Intelligent Robotic Arm can also be extended to work in wireless mode instead wired mode using infrared technology which will prove to be of great advantage for future use.

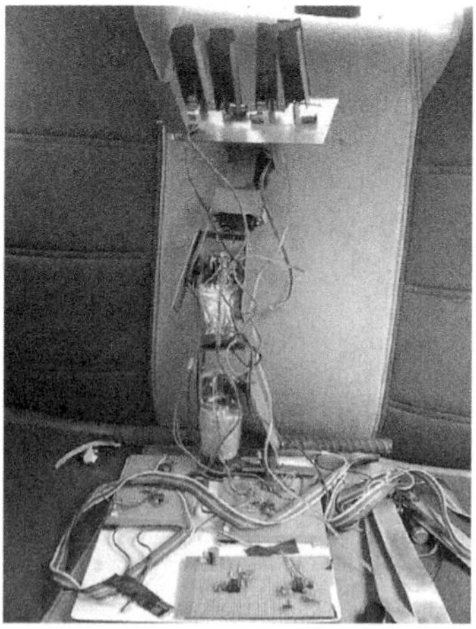

Fig. 5. Complete robotic arm from shoulder to palm

In this arm, servomotor is used which is light in weight as compared to the pneumatic cylinder. Also, servomotor expects to see a pulse every 20 milliseconds so they are very sensitive for any change in input as compared to other motors and pneumatic cylinders.

References

1. Jeong, Y., Lee, D., Kim, K., Park, J.O.: A wearable robotic arm with high force-reflection capability. In: Proceedings of 9th IEEE International Workshop on Robot and Human Interactive Communication, RO-MAN 2000, September 27-29, pp. 411–416 (2000)
2. Malvino: Electronic Principles, 6th edn. Tata Mcgraw-Hill Publishing Company Limited
3. Niku, S.B.: Introduction to Robotics Analysis, Systems, Applications. Pearson Education Asia (First Indian Reprint 2002)
4. http://www.seattlerobotics.org/guide/servos.html
5. Kang, C.-G., Park, E.-J., Son, I.-X., Kim, Y.-W., Ryu, K.-S.: Conceptual design for robotic arm wrestling. In: Conference on Robotics, Automation and Mechatronics, December 1-3, pp. 1072–1076 (2004)
6. Nakai, A., Ohashi, T., Hashimoto, H.: 7 DOF Arm Type Haptic Interface for Teleoperat ion and Virtual Reality Systems. In: Intl. Conference on Intelligent Robots and Systems, pp. 1266–1271 (October 1998)

Impulse Noise Removal in Digital Images Using Image Fusion Technique

J. Harikiran, B. Saichandana, P.V. Lakshmi, V. SivaPrasad, and A. Aishwarya

Department of IT, GIT, GITAM University, Visakhapatnam
jharikiran@yahoo.co.in, bschandana@yahoo.co.in

Abstract. This paper introduces the concept of image fusion of filtered noisy image for impulse noise reduction. Image fusion is the process of combining two or more images into a single image while retaining the important features of each image. Multiple image fusion is an important technique used in military, remote sensing and medical applications. Five different filtering algorithms are used individually for filtering the image captured from the sensor. The filtered images are fused to obtain a high quality image compared to individually denoised images. In-order to better appraise the noise cancellation behavior of our fusion technique from the point of view of human perception, an edge detection is performed using canny filter for the fused image. Experimental results show that this method is capable of producing better results compared to individually denoised images.

Keywords: Impulse Noise, Image Enhancement, Image Restoration, Image Processing, Image Fusion.

1 Introduction

Order statistics filters exhibit better performance as compared to linear filters when restoring images corrupted by impulse noise. Impulse noises are short duration noises which degrade an image. They may occur during image acquisition, due to switching, sensor temperature. They may also occur due to interference in the channel and due to atmospheric disturbances during image transmission. The goal of the filtering action is to cancel noise while preserving the integrity of edge and detail information. In this paper, a novel technique for impulse noise reduction is proposed. In our technique, first an image in captured by a sensor. The captured image is converted from RGB to YIQ domain. The image is filtered in this domain with five different smoothing filters. The filtering is applied only to the chrominance (I and Q) component of YIQ domain and the filtered image is converted from YIQ into RGB domain. The denoised images obtained from five different filters are fused them to obtain a high quality image free from impulse noise.

The rest of the paper is organized as follows. Section 2 presents the conversion of color image from RGB to YIQ domain and YIQ to RGB domain, Section 3 presents the image fusion technique with five different smoothing filters, Section 4 presents the experimental results and finally Section 5 report conclusion.

V.V. Das, J. Stephen, and Y. Chaba (Eds.): CNC 2011, CCIS 142, pp. 514–518, 2011.

2 Conversion from RGB to YIQ and YIQ to RGB

The early approaches to color image processing are performed by processing each RGB components separately. A disadvantage of these methods is the loss of correlation between the color channels resulting in color shifts [4][5]. That is a substitution of a noisy pixel color through a new false color, which does not fit into the local neighbourhood. This means that a noisy pixel is replaced by another noisy pixel. In our work, the YIQ system is used to process color image. The principle advantage of this space in image processing is that the color information components (I and Q) are processed leaving the luminance component (Y) It will be a need to convert RGB to YIQ system for this purpose.

3 Image Fusion Technique

The process of combining two or more images into a single image while retaining the important features of each image is called image fusion. In this paper, the filtered images from five different smoothing algorithms are fused to obtain a high quality denoised image. The block diagram of the process is shown in figure 3.

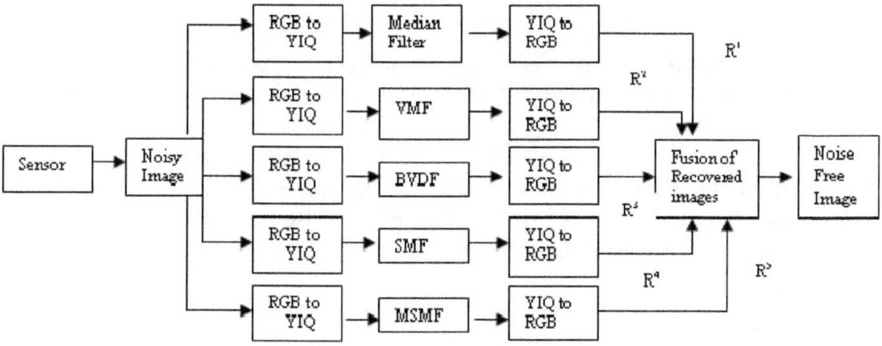

Fig. 3. Block diagram of image fusion technique

The five different smoothing algorithms used in our technique are described as follows: Order-static filters are nonlinear filters whose response is based on the ordering (ranking) the pixels contained in the image area encompassed by the filter, and then replacing the value of the center pixel with the value determined by the ranking result. The best known filter in this category is the median filter, which as the name implies, replaces the value of the pixel by the median of the intensity values in the neighborhood of that pixel defined in (3). The pixel with the median magnitude is used to replace the pixel in the signal studied.

$$\text{MEDIANFILTER}(x_1,x_2,\ldots\ldots x_N) = \text{MEDIAN}(x_1,x_2,\ldots\ldots x_N) \tag{3}$$

The median filter is more robust with respect to the presence of noise.

In the **Vector median filter** (VMF) [3] for the ordering of the vectors in a particular kernel or mask a suitable distance measure is chosen. The vector pixels in the window are ordered on the basis of the sum of the distances between each vector pixel and the other vector pixels in the window. The sum of the distances is arranged in the ascending order and then the same ordering is associated with the vector pixels. The vector pixel with the smallest sum of distances is the vector median pixel. The vector median filter is represented as

$$X_{VMF} = \textit{vectormedian}\ (\text{window}) \tag{4}$$

If δ_i is the sum of the distances of the ith vector pixel with all the other vectors in the kernel, then

$$\delta_i = \sum_{j=1}^{N} \Delta(X_i, X_j) \tag{5}$$

where $(1 \le i \le N)$ and X_i and X_j are the vectors, N=9.

$\Delta(X_i, X_j)$ is the distance measure given by the L_1 norm or the city block distance which is more suited to non correlated noise. The ordering may be illustrated as

$$\delta_1 \le \delta_2 \le \delta_3 \le, \ldots, \le \delta_9 \tag{6}$$

and this implies the same ordering to the corresponding vector pixels i.e.

$$X_{(1)} \le X_{(2)} \le, \ldots, \le X_{(9)} \tag{7}$$

where the subscripts are the ranks. Since the vector pixel with the smallest sum of distances is the vector median pixel, it will correspond to rank 1 of the ordered pixels,

$$\text{i.e.,}\quad X_{VMF} = X_{(1)} \tag{8}$$

For **Basic Vector Directional Filter (BVDF) [6]**, Let W be the processing window of size n, and let x_i, i=1,2,…n be the pixels in W. Let also the vector valued image function at pixel x_i is denoted as f_i. Let α_i correspond to f_i.

$$\alpha_i = \sum_{j=1}^{n} A(f_i, f_j) \tag{9}$$

where $A(f_i, f_j)$ denotes the angle between f_i and f_j.

An ordering of the α_i's

$$\alpha_{(1)} \le \alpha_{(2)} \le \ldots\ldots \le \alpha_{(m)} \le \ldots\ldots \alpha_{(n)} \tag{10}$$

implies the same ordering to the corresponding f_i's

$$f^{(1)} \le f^{(2)} \le \ldots\ldots \le f^{(m)} \le \ldots\ldots f^{(n)} \tag{11}$$

The first term in (11) constitutes the output of the BVDF

$$\text{BVDF}\ [f_1, f_2, \ldots\ldots f_n] = f^{(1)} \tag{12}$$

The **Spatial median filter** (SMF) [4] is a uniform smoothing algorithm with the purpose of removing noise and fine points of image data while maintaining edges around larger shapes. The SMF is based on the spatial median quantile function which is a L_1 norm metric that measures the difference between two vectors. The spatial depth between a point and a set of points is defined by

$$S_{depth}(X,x_1,x_2,\ldots\ldots x_N)=1-\frac{1}{N-1}\left\|\sum_{i=1}^{N}\frac{X-xi}{\|X-xi\|}\right\| \tag{13}$$

Let $r_1,r_2,\ldots r_N$ represent $x_1,x_2,\ldots x_N$ in rank order such that

$$S_{depth}(r_1,x_1,x_2,\ldots\ldots x_N) \geq S_{depth}(r_2,x_1,x_2,\ldots\ldots x_N) \geq \ldots \geq S_{depth}(r_N,x_1,x_2,\ldots\ldots x_N) \tag{14}$$

and let r_c represent the center pixel under the mask . Then

$$SMF(x_1,x_2,\ldots\ldots x_N)= r_1 \tag{15}$$

In the **Modified Spatial Median Filter** (MSMF) [4], we first calculate the spatial depth of every point within the mask and then sort these spatial depths in descending order. After the spatial depth of each point within the mask is computed, an attempt is made to use this information to first decide if the mask's center point is an uncorrupted point. If the determination is made that a point is not corrupted, then the point will not be changed. If the point is corrupted, then the point is replaced with the point with the largest spatial depth. We can prevent some of the smoothing by looking for the position of the center point in the spatial order statistic. Let us consider a parameter P (where $1\leq P \leq N$, where N represents numbers of points in the mask), which represents the estimated number of original points under a mask of points. If the position of the center mask point appears within the first P ranks of the spatial order statistic, then we can argue that while the center point is not the best representative point of the mask, it is likely to be original data and should not be replaced. The MSMF is defined by

$$MSMF(T,x_1,x_2,\ldots\ldots x_N) = \begin{cases} r_c & c \leq P \\ r_1 & c > P \end{cases} \tag{16}$$

We generated five denoised images using the above smoothing algorithms. By fusing the best features of these five denoised images, a high quality image is obtained. The fusion of the filtered images is given by

$$F = \sum_i \zeta R^i \quad \text{where i= 1,2,3,4,5} \tag{17}$$

R^1 is the median filtered image, R^2 vector median filtered image, R^3 vector directional filtered image, R^4 spatial median filtered image and R^5 modified spatial median filtered image. ζ is the fidelity factor. The fusion criterion depends on the fidelity factor of the image. For a heavily noised image the fidelity factor ζ will be smaller compared to a lightly noised image and hence we fuse the images, each image contributes to the recovered image depending on its noise density. A lightly noised image contributes more compared to a heavily noised image. This helps to obtain a high quality fused image. We have taken fidelity factor ζ=0.2.

4 Experimental Results

This section presents the simulation results illustrating the performance of the proposed fusion technique. The test image employed here is the true color image "parrot" with 290×290 pixels. The salt and pepper noise is added into the image with noise density 0.4. All filters considered operate using 3×3 processing window shown in figure 4.

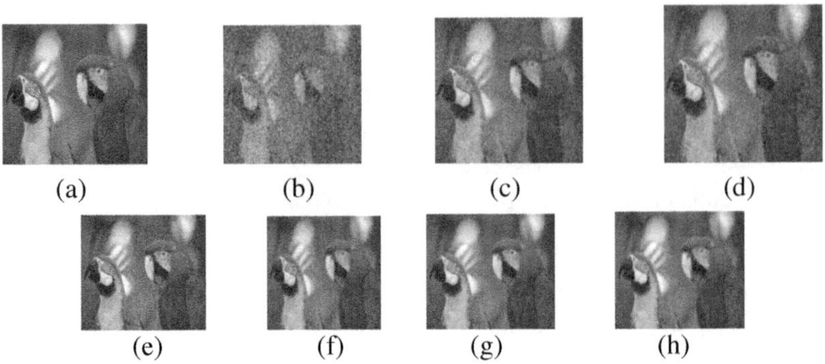

Fig. 4. a) original image, b) impulse noise image corrupted by noise density 0.4, c) median filter, d) VMF, e) BVDF, f) SMF g) MSMF with P=4 and h) Fused image.

Edges define the boundaries between regions in an image, which helps with segmentation and object recognition. We used canny filter [5] for detection of edges in our "parrot" image shown in figure 5.

(a) original image (b) Noise image (c)Resultant image by our method

Fig. 5. Edge detection using canny filter

5 Conclusion

We proposed fusion technique for removing impulse noise in digital images. The image captured by the sensor undergoes filtering by different smoothing filters and the resultant images are fused to attain high quality image. This restoration of image data is very likely to find potential applications in a number of different areas such as electromagnetic imaging of objects, medical diagnostics, remote sensing, robotics, etc.

References

1. Kao, O.: A Parallel Triangle Operator for Noise Removal in True Colour Images, Department of Computer Science, Technical University of Clausthal, 38678 Clausthal –Zellerfeld Germany
2. Baker, M.N., Al-Zuky, A.A.: Color Image Noise Reduction Using Fuzzy Filtering. Journal of Engineering and Development 12(2) (June 2008) ISSN 1813-7822
3. Laskar, R.H., Bhowmick, B., Biswas, R., Kar, S.: Removal of impulse noise from color image. IEEE, Los Alamitos (2009)
4. Nadernejad, E.: Edge Detection Techniques: Evaluations and Comparisions. Applied Mathematical Sciences 2(31), 1507–1520 (2008)
5. Indu, S., Ramesh, C.: A noise fading technique for images highly corrupted with impulse noise. In: Proceedings of the ICCTA 2007. IEEE, Los Alamitos (2007)
6. Karakos, D.G., Trahanias, P.E.: Generalized Multichannel Image-Filtering Structures. IEEE Transactions on Image Processing 6(7) (July 1997)

Faster Arithmetic and Logical Unit CMOS Design with Reduced Number of Transistors

Rachit Patel, Harpreet Parashar, and Mohd. Wajid

Deptt. of Electronics and Communication Engineering,
Jaypee University of Information Technology, Solan, India
{rachit05081gece,harpreet123100,shamsi.shamsi}@gmail.com

Abstract. The Arithmetic and Logic Unit (ALU) is a combination circuit that performs a number of arithmetic and logical operations. Over the past two decades, Complementary Metal Oxide Semiconductor (CMOS) technology has played important role in designing high performance systems because of the advantages that CMOS provides: an exceptionally low power-delay product, the ability to accommodate millions of devices on a single chip. To take the benefits of CMOS technology a novel ALU Circuit is proposed in this paper. An improved and efficient adder circuit called mirror adder [3] is used which helps in decreasing the RC delay. Also a programmable logic circuit is included to configure mirror adder circuit to subtractor circuit depending upon programmable input; this implementation helps in reducing the transistor count and power dissipation, decreasing parasitic capacitance hence increasing speed.

Keywords: ALU, CMOS, VLSI, Circuit.

1 Introduction

High performance systems and a variety of real time Digital Signal Processing systems derive their performance from VLSI solutions. Since fast arithmetic units are critical to all such high-performance applications [1]. Delay and power dissipation for a circuit also have emerged as the major concerns of designers and depend on the number of transistors are used in the circuit. When the number of transistors is more the capacitance is more due to which the delay is more so here our aim is to reduce the delay and power dissipation [3] [4]. In this paper authors reduced the number of transistor using a programmable adder and subtractor [2]. The factors consider in this paper: number of gates used, performance in terms of speed, delay and parasitic capacitance. The section 2 explains the ALU, the section 3 explains how the mirror adder can be used as subtractor, section 4 gives proposed ALU operational and selection circuits, and section 5 discusses on proposed ALU circuit and design techniques used in it. Finally section 6 concludes the paper.

2 Arithmetic and Logical Operations

The arithmetic logic unit (ALU) is the core of a CPU in a computer. An eight function instruction set CMOS ALU Fig.-1 performs Addition, Subtraction, AND, NAND,

V.V. Das, J. Stephen, and Y. Chaba (Eds.): CNC 2011, CCIS 142, pp. 519–522, 2011.

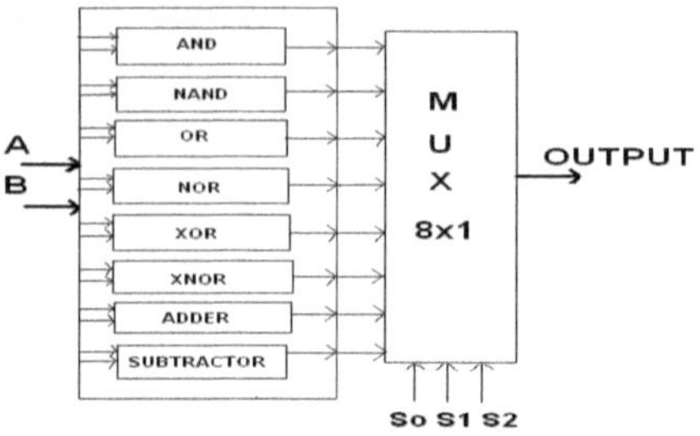

Fig. 1. Block diagram of an ALU

OR, NOR, XOR and XNOR. Each of these functions is performed on two 1-bit inputs and each of the single bit building blocks can be cascaded together to form a four bit ALU [2]. ADDITION operation is performed using mirror adder which depends upon a programmable input i.e., P when it is equal to 0. SUBTRACTION operation is performed using the same circuitry i.e., mirror adder when the programmable input P = 1. Each of the CMOS ALU functions is performed on a single bit input.

3 Mirror Adder as Subtractor

Number of transistors used to design a circuit is an important parameter as it influences overall delay and power dissipation. Also when the number of transistors is more the capacitance is more due to which the delay increases further [4]. Therefore, in CMOS ALU design addition and subtraction is performed using same circuitry in order to reduce the transistor count [3]. The adder block used here is Mirror Adder whose operation i.e. addition or subtraction depends upon a programmable input pin i.e. *P*. In Fig.-2, SUM = (B`.P + P`.B) + A + P, where $P = C_{in}$, *P* is programmable input. When *P*=0 then output of adder will be A+B. But when P=1 then output A+B`+1 which is equivalent to A-B [3]. In this way we can use the adder in both ways thereby saving the extra number of gates.

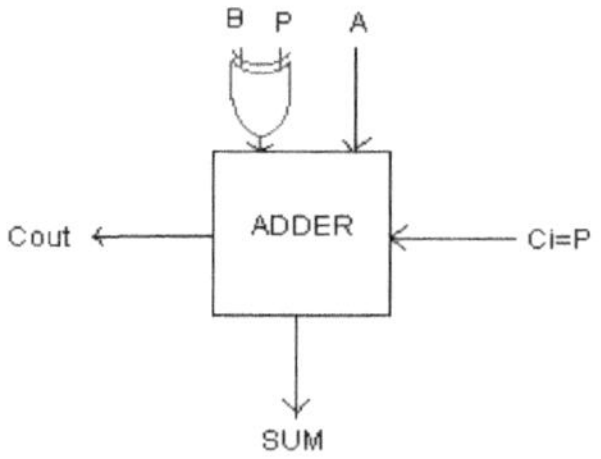

Fig. 2. Block diagram of a programmable adder

4 Proposed ALU Circuitry

Figure 3 shows ALU realization using CMOS gates. Left upper part of the figure 3 shows the implementation of various logic gate operations like *AND, NAND, OR,*

NOR, XOR and *XNOR* whereas left down part shows adder and subtractor implementation using mirror adder with a XOR gate as programmable logic. Mirror adder performs both addition and subtraction on the basis of the value of programmable logic input *P* [4] [3]. Figure 3 right part shows the selection circuitry implemented using an 8x1 MUX with 3 select lines viz. S_0, S_1 and S_2 which help in selecting the operation that will be performed by the ALU like AND, NAND, addition etc. The 3 control signals S_0 (=*P*), S_1 and S_2 show the desired output in the order as shown below:

Operations	$S_2S_1S_0$		Operations	$S_2S_1S_0$
AND	0 0 0		XOR	1 0 0
NAND	0 0 1		XNOR	1 0 1
OR	0 1 0		ADDITION	1 1 0
NOR	0 1 1		SUBTRACTION	1 1 1

5 Discussion on Proposed Circuit

5.1 Progressive Transistor Sizing

The propagation delay calculated using Elmore Delay Model shows that in the MUX circuitry resistance of M_{96} (R_{96}) in fig. -3 appear 4 times, the resistance of M_{95} appears 3 times, etc. Consequently progressive scaling of the transistors is beneficial: M_{96}>M_{95}>M_{94}>M_{93} [3]. This approach reduces the dominant resistance like M_{96} (R_{96}) and thus the delay. Similarly, same approach can be use for other resistances like M_{100}, M_{104} and so on.

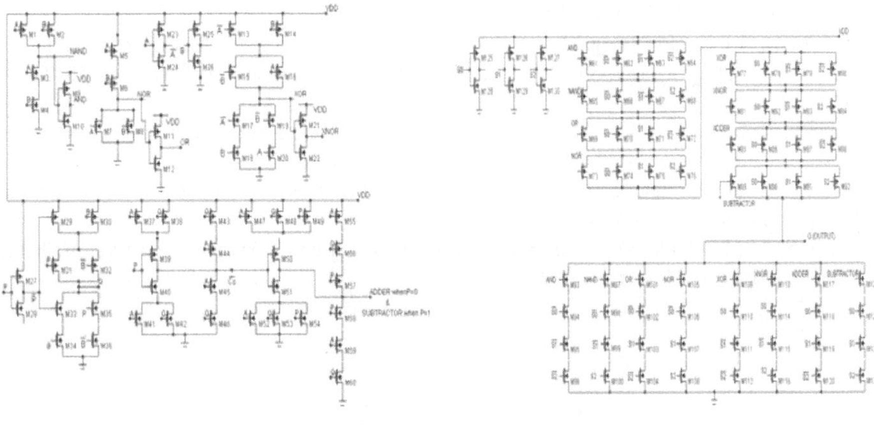

Fig. 3. ALU Operational circuitry (left) and Selection Circuitry (right)

5.2 Input Reordering

In the MUX circuitry, SUBTRACTOR signal is a critical signal as compared to select lines input S_0, S_1 and S_2 as it is coming after passing through a number of logic gates so there will be delay before it gets stable [4]. Putting the critical path transistors closer to the output of the gate can result in a speed up. Therefore, SUBTRACTOR signal is placed closer to the output signal. Similarly other critical signals like AND, NAND, OR, NOR etc. are also placed closer to the output to reduce delay [3].

5.3 Performance

With the reduction of number of gates the overall capacitance is reduced. Progressive transistor sizing & input reordering helped in reducing delay. Thus, hence improve the performance [4] [3].

6 Conclusion

ALU being the core of any processor has been designed using CMOS because it uses various flexible design techniques to reduce delay, power, cost and to increase speed. Also, CMOS approach helps in accommodating large number of transistors on a single chip. All this improves the performance of the ALU. Mirror adder with programmable logic circuitry is an efficient method used to reduce the transistor count by performing dual function of addition and subtraction using single adder circuitry. Design techniques like Progressive transistor sizing and input reordering is used to reduce the overall delay which in turn increase the performance.

References

1. Chau, K.T., Chan, C.C.: A Spice compatible model of permanent magnet dc motor drives. In: Proc. 1995 IEEE Int. conf. Power Electronics and Drive Systems, Kowloon, Singapore, vol. 1, pp. 477–482 (February 1995)
2. Vahidi, B., Beiza, J.: Using PSpice in Teaching Impulse Voltage Testing of Power Transformers to Senior Undergraduate Students. IEEE Transactions on Education 48(2) (May 2005)
3. Rabaey, J.M., Chandrakasan, A., Nikolic, B.: Digital Integrated Circuits – A Design Perspective, 2nd edn. Pearson Educational Publishers, London (2008)
4. Weste, N., Eshraghian, K.: Principles of CMOS VLSI Design. Addison-Wesley Publishing Company, Menlo Park (2004)
5. Kang, S.-M., Leblebici, Y.: CMOS digital integrated circuits: Analysis and Design, 3rd edn. Tata Mcgraw-hill, New York (2003)

Dimensionality Reduction of Data Warehouse Using Wavelet Transformation: An Enhanced Approach for Business Process

Ardhendu Tripathy[1], Kaberi Das[2], and Tripti Swarnakar[3]

[1] I.T.E.R (S.O.A University),
Jagamohan Nagar, Bhubaneswar
ardhendutripathy2007@gmail.com
[2] Department of Computer Applications, S.O.A University,
Jagamohan Nagar, Bhubaneswar
kaberidas@iter.ac.in
[3] Department of Computer Applications, S.O.A University,
Jagamohan Nagar, Bhubaneswar
tswarnakar@iter.ac.in

Abstract. Dimensionality reduction is an essential data preprocessing technique used for reducing the number of random variables under consideration. Traditional dimensionality reduction approaches fall under two categories one is Feature Extraction and Feature Selection. Many data warehouse contain massive amount of data, accumulated over long period of time. This is done in order to maintain the privacy concerns in which the data are aggregated to a certain levels. Another reason is to maintain a balance between the use of data that changes as data age and the size of data, thus avoiding over large data warehouse. In this paper an enhanced approach for business process has been carried out using dimensionality reduction by implementing wavelet transformation, which will help us in automated selection of most relevant independent attributes in a data warehouse.

Keywords: Dimensionality Reduction, Data Warehouse, Wavelet Transformation.

1 Introduction

Dimensionality reduction is termed as Data encoding or transformation that can be applied so as to obtain a reduced or compressed representation of original data. The dimension of data is the number of variables that are measured on each observation. High dimensional databases present many mathematical challenges as well as some opportunities and are bound to give rise to new theoretical developments. We distinguish two major types of dimensionality reduction methods [1, 2, 3]: linear and non-linear.

In this paper we present methodology for automated selection of the most relevant independent attributes in a data warehouse. Taking assumptions that in most data warehouse projects the process of reducing data warehouse dimensionality can be

V.V. Das, J. Stephen, and Y. Chaba (Eds.): CNC 2011, CCIS 142, pp. 523–525, 2011.

supported by feature selection procedures. The main advantage of reducing data warehouse dimensionality is that most data mining algorithm can be applied to the detail data in data warehouse.

1.1 Popular Techniques for Dimensionality Reduction

SVD (Singular Value Decomposition) or PCA (Principal Component Analysis) is a classic technique for dimension reduction which keeps the Euclidean distance between the time series with singular values [4, 5]. SVD is a global transformation technique (i.e.) the entire data set is examined to compute the axes. Popular feature extraction techniques are DFT (Discrete Fourier Transform) and DWT (Discrete Wavelet Transform) where the sequence is projected into the frequency domain (DFT) or tiling of the time frequency plane (DWT).

1.2 Wavelet Transformation

A wavelet is a little piece of a wave. Where a sinusoidal wave is used by Fourier Transforms carries on repeating itself to infinity unless and until a wavelet exists within a finite domain and also with a zero-valued. A wavelet transform involves convolving the signal against the particular position.

We have proposed a dimensionality reduction technique called Wavelet transform. A wavelet was borne out of a need for further developments from Fourier transforms.

2 Procedure

Our main aim is to remove noise and abnormal data using dimensionality reduction. So, here we will be considering an approach called Wavelet Transformation [5, 9].

When we actually perform the discrete wavelet transform, the common approach involves using the dyadic scheme. This is where we increase the scale and step spacing between wavelets by a factor of two at each step. e.g., in the first pass, we would use wavelets of scale 1 and space them 1 sample apart. Then in the second pass, we'd scale them to two samples, and space them apart by two sample points. Everywhere you see an x is a point where the convolution is performed at that scale. The smallest scale we need to consider is where the wavelet takes up two sample points. In short, we treat the wavelet transform as if it is a filter bank.

3 Implementation

Wavelets are a mathematical tool for the hierarchical decomposition of functions. Wavelets represent a function in terms of a coarse overall shape. To start the wavelet decomposition procedure, first we need to choose the wavelet basis functions. Haar wavelets are conceptually the simplest wavelet basis functions, and for purposes of exposition in this paper, we focus our discussion on Haar wavelets. They are fastest to compute and easiest to implement.

Suppose we have a one-dimensional signal": [2, 2, 7, 11]: We perform a wavelet transform on it. We first average the signal values, pairwise, to get the new lower-resolution signal with values [2, 9]. Haar wavelets store the pairwise differences of the original values as detail coefficients. In the example, the two detail coefficients are 2-2 = 0 and 11-7 = 4.

Resolution	Averages	Detail Coefficients
4	[2, 2, 7, 11]	
2	[2, 9]	[0, 4]
1	[5.5]	[7]

Fig. 1. Implementation result

4 Conclusion

In this paper, we present an I/O-efficient technique based upon a multiresolution wavelet decomposition that yields an approximate and space-efficient representation of the data cube. We build our compact data cube on the partial sums of the raw data values. One drawback of our current approach is that the construction of the wavelet decomposition is performed on the dense data cube, which may be very large. Developing automated approaches to the design of data warehouses requires further investigation of several issues.

References

1. Gupta, V.R.: A Introduction to Data Warehousing, System Services Corporation, Chicago, Illinois (1997)
2. Inmon, W.H., Hackathorn, R.D.: Using the Data Warehouse. John Wiley & Sons, New York (1994)
3. Korth, H.F., Silberschatz, A.: Database System Concepts. McGraw-Hill, Inc., New York (1991)
4. Schouten, H.: Analysis and Design of Data Warehouses. In: Proc. International Workshop on
5. Carreira-Perpinan, M.A.: A review of dimension reduction techniques. Technical report CS-96-09, Department of Computer Science, University of Sheffield (1997)
6. Karypis, G., Han, E.H.: Concept indexing: A fast dimensionality reduction algorithm with applications to document retrieval and categorization. In: Proceedings of the Conference on Information and Knowledge Management (2000)

Analysis of Adaptive Floating Search Feature Selection Algorithm

D. Devakumari[1] and K. Thangavel[2]

[1] Assistant Professor, Department of Computer Science,
Government Arts College, Dharmapuri
[2] Professor & Head, Department of Computer Science,
Periyar University, Salem
ramdevshri@gmail.com, drktvelu@yahoo.com

Abstract. In feature selection, a search problem of finding a subset of features from a given set of measurements has been of interest for a long time. This paper proposes an Adaptive Floating Search feature selection method (AFS) based on contribution entropy with flexible backtracking capabilities. Experimental results show that the proposed method performs well in selecting an optimal size of the relevant feature set. Using common indices such as Rand Index and Jaccard Coefficient it is found that the clusters formed with selected features are as good as clusters formed with all features.

Keywords: Data Mining, Unsupervised Feature Selection, Contribution Entropy (CE), Adaptive Floating Search (AFS), Clustering.

1 Introduction

Unsupervised feature selection is a process of selecting features without referring back to the data classification or any other target function [1, 2, 3].

An intuitive, efficient and deterministic principle, depending on authentic properties of the data, which serves as a reliable criterion for feature ranking is based on SVD-entropy. Here, a feature is selected according to its Contribution to the Entropy (CE) calculated on a leave-one-out basis [4] . It had been demonstrated that this principle can be turned into efficient and successful feature selection methods like Simple Ranking (SR), Sequential Forward Selection (SFS) and Sequential Backward Elimination (SBE).

In this paper, another sequential search method is presented, the Adaptive Floating Search Feature Selection (AFS), which attempts to compensate for the weaknesses of SFS and SBE with some backtracking capabilities. This method is an extension of the LRS algorithm which repeatedly adds L features and removes R features. Rather than fixing the values of L and R, the floating search method allows those values to be determined from the data [11]. Hence, this method is more favorable for feature selection than the above mentioned ones.

The organization of the paper is as follows : Section 2 presents the background. The proposed work is presented in section 3. The experimental results and analysis are provided in section 4. This paper concludes in section 5.

V.V. Das, J. Stephen, and Y. Chaba (Eds.): CNC 2011, CCIS 142, pp. 526–530, 2011.

2 Background

A dataset of n instances and m features $A_{[nxm]} = \{A_1, A_2, \ldots, A_i, \ldots, A_n\}$ is considered, where each instance, or observation, A_i is a vector of m measurements or features. The objective is to obtain a subset of features of size $m_c < m$, that best represents the data.

According to Alter et al., [8], the contribution of the ith feature to the entropy (CE_i) is defined by a leave-one-out comparison and is given by

$$CE_i = E\,(A_{[nXm]}) - E\,(A_{[nX(m-1)]}) \tag{1}$$

where the ith feature was removed in $A_{[nX(m-1)]}$. The average of all CE is represented by c.

Entropy maximization had been implemented in three different ways [4].

i) Simple Ranking (SR): m_c features are selected according to the highest ranking order of their CE values.
ii) Sequential Forward Selection (SFS): Two implementations are considered here.
   a) SFS1: The first feature is chosen according to the highest CE. Then among all other features the one which, together with the first feature, produces a 2-feature set with highest entropy is selected. The iteration is continued over all m - 2 features to choose the third according to maximal entropy and so on, until m_c features are selected.
   b) SFS2: The first feature is chosen as before. The CE values of the remaining set of size m - 1 is recalculated and the second feature is selected according to the highest CE value. This is continued until m_c features are selected.
iii) Sequential Backward Elimination (SBE): The feature with the lowest CE value. Is eliminated. The CE values are recalculated and the lowest one is iteratively eliminated until m_c features remain.

3 Proposed Work

The LRS algorithm is a generalization of SFS and SBE.

If L > R, LRS starts from the empty set and repeatedly adds L features and removes R features.

If L < R, LRS starts from the full set and repeatedly removes R features followed by L feature additions.

The main limitation of LRS is the lack of a theory to help predict the optimal values of L and R. The proposed AFS method is an extension of LRS since there is no need to specify any parameters such as L and R. The number of forward(adding) or backward(removing) steps is determined dynamically during the method's run so as to maximize the criterion function.

In this paper two inequalities are proposed based on which the number of features to be added and removed can be determined respectively.

The first inequality is as follows:

$$\text{Max CE} - \text{Average CE} \leq \text{CE(m)} \leq \text{Max CE} \tag{2}$$

Each feature m which satisfies inequality (2) will be selected in one step.

The second inequality is as follows:

$$\text{Min CE} \leq \text{CE(m)} \leq \text{Min CE} + \text{Average CE} \tag{3}$$

If any feature m which satisfies inequality (3) has been included in the selected list, it is removed. Both set of selected and removed features are eliminated from the full set of features. CE values are calculated for the remaining features and the process is repeated till m_c features are selected.

AFS method can be implemented in any one of the following ways:

i) AFS1: It starts from the empty set and performs forward selection. After each forward step, it performs backward steps as long as the objective function increases.
ii) AFS2: It starts from the full set and performs backward selection. After each backward step, it performs forward steps as long as the objective function decreases.

In this paper, the AFS1 method is experimented and the pseudo code is given below:

Pseudo code for Adaptive Floating Search Method

Step 1 : Start with $M_{AFS} = \phi$ and
$M'_{PLUS} = M'_{MINUS} = M$.
Step 2 : Calculate the CE score of each element in M using equation (1)
Step 3 : Select each element m from M'_{PLUS} which satisfies the inequality (2)
Step 4 : Remove it from M'_{PLUS} and insert into M_{AFS}.
Step 5 : Select each element m from M'_{MINUS} which satisfies the inequality (3)
Step 6 : Remove it from M'_{MINUS} and also from M_{AFS} (if it is already included in M_{AFS}).
Step 7 : While size of $M_{AFS} < m_c$
a) For each element m in M'_{PLUS} and M'_{MINUS} (that is not included in M_{AFS}) recalculate its CE score.
b) Go to step (3).
Step 8 : End.

---

4 Experimental Results and Analysis

The AFS1 method of feature selection is experimented with five different benchmark data sets obtained from the UCI machine learning repository [www.archive.ics.uci.edu], namely ionosphere, lung cancer, SPEC TF heart, sonar and web log data. Experiments show that the AFS1 method eliminates some of the features selected by SFS2, thus reducing the number of features. The results are tabulated in Table 1:

Table 1. Features selected by the SR, FS2 and AFS1 methods using contribution entropy for different datasets

Data set	Total Number of Features	Number of Features Selected		
		SR	SFS2	AFS1
Ionosphere	34	16	16	16
Lung Cancer	56	24	24	23
SPEC TF Heart	44	17	17	9
Sonar	60	32	32	18
Web Log Data	145	113	113	96

The quality of the clusters constructed with all features and with selected features are assessed using some common indices like Rand Index and Jaccard Coefficient.

Since AFS1 is an unsupervised feature selection method, the class labels that were originally present in the datasets are omitted. Now, for the purpose of cluster validation, the classification details are considered. Clustering results are compared with actual classification to check the reliability of the clusters.

The following figure presents the graphical representation of cluster validations using Jaccard Coefficient for chosen datasets.

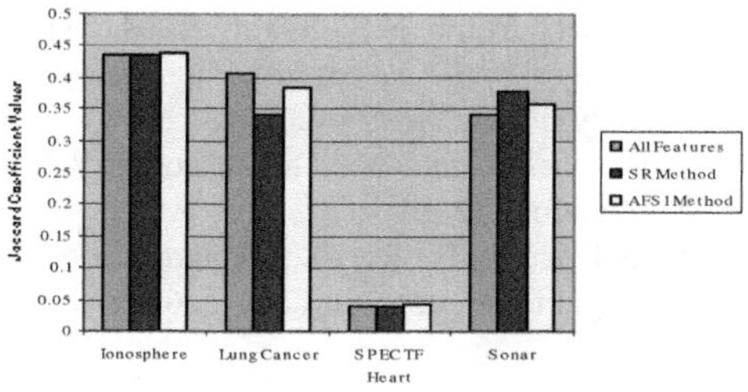

Fig. 1. Graphical representation of Jaccard Coefficient values for clusters formed with all features and features selected with SR and AFS1 methods

It is illustrated that, an optimal set of features selected using AFS1 method is sufficient to maintain the quality of clusters equivalent to those clusters formed with all the features. Hence, it is concluded that the proposed method effectively reduces the number of features while at the same time preserving the classification accuracy.

5 Conclusions

Based on the principle of Contribution Entropy, four feature selection methods had already been implemented. This paper had proposed the adaptive floating search feature selection method which is an extension of the LRS algorithms. Rather than fixing the values of L and R, the floating search method allows those values to be determined from the data. The proposed method was experimented and proved to select fewer and better features as compared with the other existing methods.

References

1. Handl, J., Knowles, J.: Feature Subset Selection in Unsupervised Learning via Multi objective Optimization. International Journal of Computational Intelligence Research 2(3), 217–238 (2006)
2. Guyon, I., Elisseeff, A.: An Introduction to Variable and Feature Selection. Journal of Machine Learning Research 3, 1157–1182 (2003)
3. Liu, H., Li, J., Wong, L.: A Comparative Study on Feature Selection and Classification Methods Using Gene Expression Profiles and Proteomic Patterns. In: Lathrop, R., Miyano, K.N.S., Takagi, T., Kanehisa, M. (eds.) 13th International Conference on Genome Informatics, pp. 51–60. Universal Academy Press, Tokyo (2002)
4. Varshavsky, R., Gottlieb, A., Linial, M., Horn, D.: Novel Unsupervised Feature Filtering of Biological Data. Bioinformatics/bti 283, 1–5 (2005)
5. Liu, H., Yu, L.: Towards Integrating Feature Selection Algorithms for Classification and Clustering. IEEE Transactions on Knowledge and Data Engineering 17, 491–502 (2005)
6. Dy, J.G., Brodley, C.E.: Feature selection for unsupervised learning. Journal of Machine Learning Research, 845–889 (2004)
7. Mitra, P., Murthy, C.A., Pal, S.K.: Unsupervised feature selection using feature similarity. IEEE Transactions on Pattern Analysis and Machine Intelligence, 301–312 (2002)
8. Alter, O., Brown, P.O., Botstein, D.: Singular value decomposition for genome-wide expression data processing and modeling. PNAS 97, 10101–10106 (2000)
9. Kanungo, T., Mount, D.M., Netanyahu, N.S., Silverman, R., Wu, A., Piatko, C.D.: An efficient K-Means clustering algorithm: Analysis and Implementation. IEEE Transactions on Pattern Analysis and Machine Intelligence 24 (2002)
10. Halkidi, M., Batistakis, Y., Vazirgiannis, M.: On Clustering Validation Techniques. Intelligent Information Systems Journal (2001)
11. Somol, P., Pudil, P., Novovicova, J., Paclik, P.: Adaptive Floating Search Methods in Feature Selection. Pattern Recognition Letters 20, 1157–1163 (1999)

Reconfigurable Vlsi Architecture for Graph Color Allocation with Its Chromatic Number

Khushboo Mirza[1], Mohd Wajid[2], Syed Atiqur Rahman[3], Abhishek Srivastava[4], and Masoma Khatoon[5]

[1] Dept. of Electronics, AMU Aligarh
khushboo.mirza@gmail.com
[2] Dept. of ECE, JUIT, Solan
shamsi.shamsi@gmail.com
[3] Dept. of Electronics, AMU Aligarh
syed.atiq.amu@gmail.com
[4] Dept. of ECE, JIIT, Noida
abhisheks@rediffmail.com
[5] Dept. of Electronics, AMU Aligarh
masomak@gmail.com

Abstract. the graph coloring problem is an optimization problem which finds application in diverse fields, like register allocation, frequency assignment in mobile communication etc. Various algorithms [3] have been implemented to solve this problem. Greedy algorithm is one of them. The existing architectures for greedy algorithm, uses parallel processors. In this paper a novel VLSI architecture for graph color allocation with chromatic number is proposed based on reconfigurable hardware - FPGA.

Keywords: FPGA; Graph coloring; VLSI; Algorithm.

1 Introduction

Graph coloring problem involves assigning colors to vertices of a graph in such a way that adjacent vertices have distinct colors. Graph coloring problem arises when limited numbers of resources are available and multiple users are in need of these resources at the same time. For example, register allocation problem, which involves assigning variables to a limited number of hardware registers during program execution. A simple solution is to create a graph where the nodes represent variables and an edge represents conflict between its nodes. A coloring is then a conflict-free assignment.

Greedy method is the most widely used algorithm for the graph coloring problem [2]. In this paper a novel architecture is discussed to implement the Greedy algorithm.

2 Operation of the Architecture

The approach presented here to implement greedy algorithm consists of performing the simple bit wise AND operation between two matrices- one containing information regarding the connectivity of a node to other nodes ($\mathbf{D}_v$) and the other matrix

V.V. Das, J. Stephen, and Y. Chaba (Eds.): CNC 2011, CCIS 142, pp. 531–534, 2011.

containing the color assignment information($\mathbf{C}_V$). The steps involved in this algorithm are as follows:

i. Store the information of graph G (V,E) in a square matrix '$\mathbf{D}_V$'.
ii. Initialize a color square matrix '$\mathbf{C}_V$' to zero.
iii. Take the 1st row of $\mathbf{D}_V$ matrix i.e. d_i where i=1.
iv. $L_j = d_i$ & c_j, where $j \in I^+$ [1, V].
v. The lowest j which makes L_j= [0 0 0 0 ..... 0] is taken and the ith bit of c_j is set to 1.
vi. i=i+1; and repeat step iv, v and vi till i=V.
vii. $\chi(G)$ is equal to the number of non zero rows of $\mathbf{C}_V$ matrix

Rows of D_V matrix corresponds to the nodes 1 to V of the graph and the columns corresponds to the nodes V to 1 and the binary elements give connection information of the nodes (1 for connection and 0 for no connection). The above method is explained for coloring of graph shown in fig 1(a). Initially no color is assigned to any node.

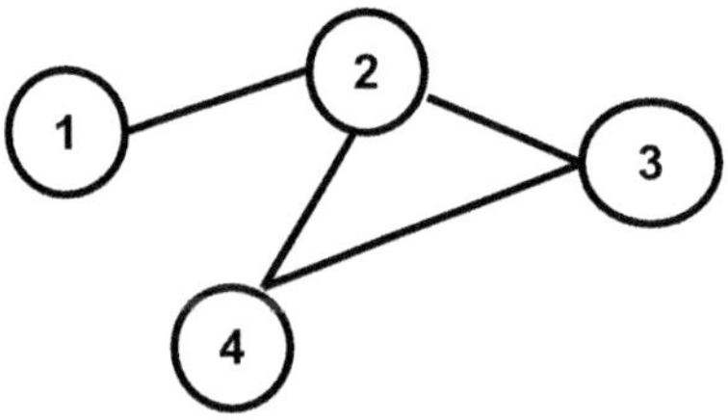

$$D_V = \begin{bmatrix} 0 & 0 & 1 & 1 \\ 1 & 1 & 1 & 1 \\ 1 & 1 & 1 & 0 \\ 1 & 1 & 1 & 0 \end{bmatrix} \qquad C_V = \begin{bmatrix} 0 & 0 & 0 & 0 \\ 0 & 0 & 0 & 0 \\ 0 & 0 & 0 & 0 \\ 0 & 0 & 0 & 0 \end{bmatrix}$$

(a) (b) (c)

Fig. 1. (a) Graph G(4,4) with 4 nodes and 4 edges, (b) The connection matrix DV , (c) Initial color matrix C_V

The graph shown in fig 1(a) can be colored using above method in four iterations. Initially no color is assigned to any node. The color matrix gets modified at each iteration and different colors are assigned accordingly. The contents of each matrix and finally the colored matrix obtained from above method are shown in figure 2.

$$\begin{bmatrix} 0 & 0 & 0 & 1 \\ 0 & 0 & 0 & 0 \\ 0 & 0 & 0 & 0 \\ 0 & 0 & 0 & 0 \end{bmatrix} \qquad \begin{bmatrix} 0 & 0 & 0 & 1 \\ 0 & 0 & 1 & 0 \\ 0 & 0 & 0 & 0 \\ 0 & 0 & 0 & 0 \end{bmatrix}$$

(a) (b)

$$\begin{bmatrix} 0 & 1 & 0 & 1 \\ 0 & 0 & 1 & 0 \\ 0 & 0 & 0 & 0 \\ 0 & 0 & 0 & 0 \end{bmatrix} \qquad \begin{bmatrix} 0 & 1 & 0 & 1 \\ 0 & 0 & 1 & 0 \\ 1 & 0 & 0 & 0 \\ 0 & 0 & 0 & 0 \end{bmatrix}$$

(c) (d)

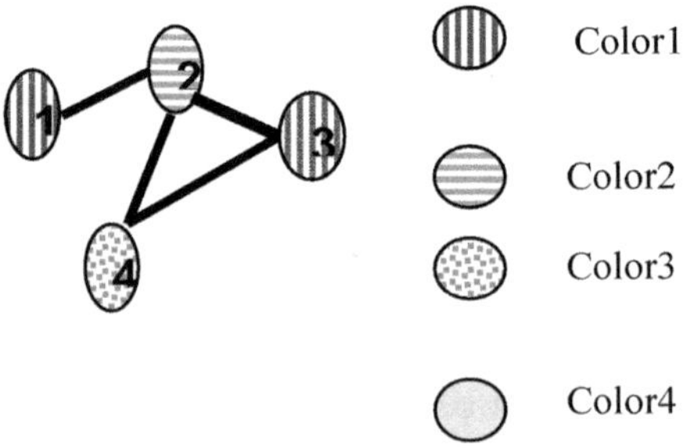

(e)

Fig. 2. (a)C_V after 1st iteration, (b)C_V after 2nd iteration, (c)C_V after 3rd iteration, (d)C_V after 4th iteration, (e)Graph G(4,4) after coloring and chromatic number is 3

3 Proposed Hardware Architecture

The hardware architecture for the algorithm explained in the section 2 is shown in Figure 3.

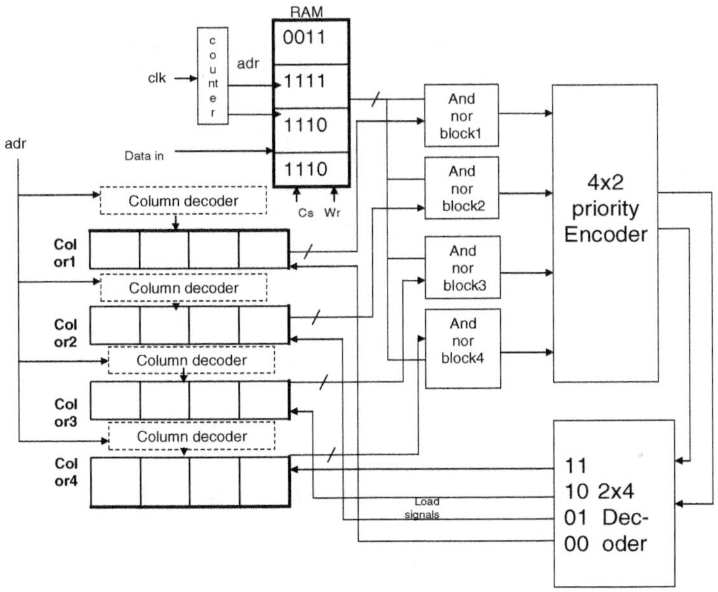

Fig. 3. Hardware Architecture of Graph G(4,4) Coloring Problem

The entire operation for this problem of four nodes takes place in eight clock cycles. In the first 4 clocks the RAM is loaded with the connection matrix D_V and the color assignment takes place in next 4 clock cycles.

The operations that are performed in these last four clock cycles is as follows - In each cycle counter points towards the RAM location from where data is AND-NORed with all the color registers. Then the four outputs of the AND-NOR blocks are applied to the input of the priority encoder. The priority encoder generates the address corresponding to the lowest activated input. This in turn activates the load input of the appropriate color register and the appropriate bit of the color register from the left is set based on the value of the address generated by the counter in that clock cycle.

Clock event 1: The reset signal sets all the color registers to zero. The counter is set to 00. The first bit from the left of color1 register is set to 1.

Clock event 2: Counter increments to 01. The 2nd bit of the color2 register from the left is set to 1.

Clock event 3: Counter increments to 10. The 3rd bit of the color1 register from the left is set to 1 based on the value of the address generated by the counter in this clock cycle.

Clock event 4: Counter increments to 11. The 4th bit of the color3 register from the left is set to 1 based on the value of the address generated by the counter in that clock cycle.

4 Simulation and Synthesis Results

The RTL simulation (using Verilog HDL) and synthesis (using Xilinx ISE 8.2i with Xilinx Spartan technology) results are shown in Figure 4 and Table 1 respectively.

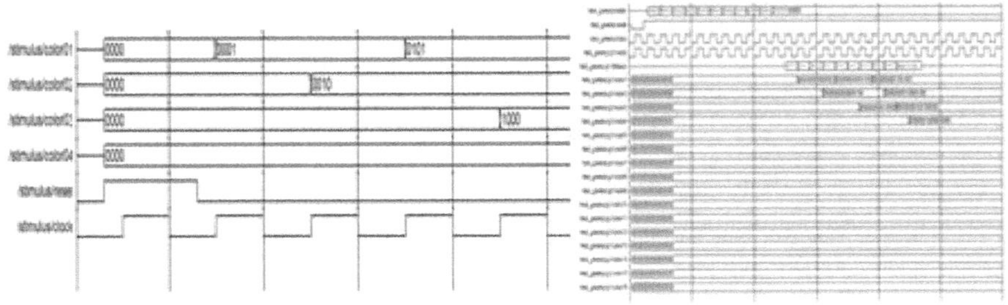

(a) (b)

Fig. 4. Simulation results (a) for G(4,4) after loading of RAM; (b) for G(11,20) graph

Table 1. The device utilization summary obtained after synthesis

Logic Utilization	Used	Available
Number of Slices	170	14752
Number of Slice Flip Flops	201	29504
Number of 4 input LUTs	290	29504
Number of GCLKs	1	24

References

[1] Takefuji, Y., Lee, K.C.: Artificial Neural Networks for Four-Coloring Map problems and K-Colorability Problems. IEEE Trans.on Circuits and Systems 38(3), 326–333 (1991)

[2] Rosen Kenneth, H., AT&T Laboratory: Discrete mathematics and its Applications. TataM-cGraw-Hill (2003)

[3] Sklyarov, V., Skliarova, I., Pimentel, B.: FPGA-Based Implementation of Graph coloring Algorithms. Studies in computational Intellegence (SCI) 76, 255–231 (2007)

[4] Ciesielski, M.J., Yang, S., Perkowski, M.A.: Multiple-valued Boolean minimization based on graph coloring. In: Proceedings of IEEE International Conference on Computer Design: VLSI in Computers and Processors, ICCD 1989 (1989)

Retracted Chapter: Hop-Count Filtering: A Successful Security Alongside Spoofed Network Traffic

N.B. Pokale[1] and Samuel S. Chopde[2]

[1] Asst. Professor - Department Information Technology
Sinhgad Academy of Engineering, Pune, India
nbpokale@gmail.com

[2] Research Scholar - Department Information Technology
Sinhgad Academy of Engineering, Pune, India
samuel.chopde@gmail.com

Abstract. IP spoofing has been exploited by Distributed Denial of Service (DDoS) attacks[2] to conceal flooding sources and localities in flooding traffic, and coax legitimate hosts into becoming reflectors, redirecting and amplifying flooding traffic. Although any field in the IP header can be forged, one cannot falsify the number of hops an IP packet takes to reach its destination. This hop-count[1] information can be inferred from the Time-to-Live (TTL)[3] value in the IP header. Base on this observation, we present a novel filtering technique that is immediately deployable to weed out spoofed IP packets. Through analysis using network measurement data, we show that *Hop-Count Filtering* (HCF) can identify close to 90% of spoofed IP packets, and then discard them with little collateral damage.

Keywords: DDoS, TTL, HOP-COUNT.

1 Introduction

1.1 Hop Count

In computer networking, a hop represents one portion of the path between source and destination. When communicating over the Internet, for example, data passes through a number of intermediate devices (like routers) rather than flowing directly over a single wire. Each such device causes data to "hop" between network connection. Network utilities like **ping**[4] can be used to determine the hop count to a specific destination. Ping generates packets that include a field for the hop count. Each time a device receives these packets, that device modifies the packet, incrementing the hop count by one. Both routers and bridges can managing hop counts, but intermediate devices (like hubs) cannot.

1.2 TTL (Time to Live)

TTL stands for "Time To Live". This is a piece of data in a network packet that specifies how many routers the packet can pass through before the packet expires and is

---

This paper has been retracted because a considerable amount of its content was plagiarized from the following publication:
"Hop-Count Filtering: An Effective Defense Against Spoofed DDoS Traffic" byCheng Jin, Haining Wang, and Kang G. Shin, Copyright 2003 ACM 1-58113-738-9/03/0010.
An erratum to this chapter can be found at http://dx.doi.org/10.1007/978-3-642-19542-6_132

V.V. Das, J. Stephen, and Y. Chaba (Eds.): CNC 2011, CCIS 142, pp. 535–538, 2011.

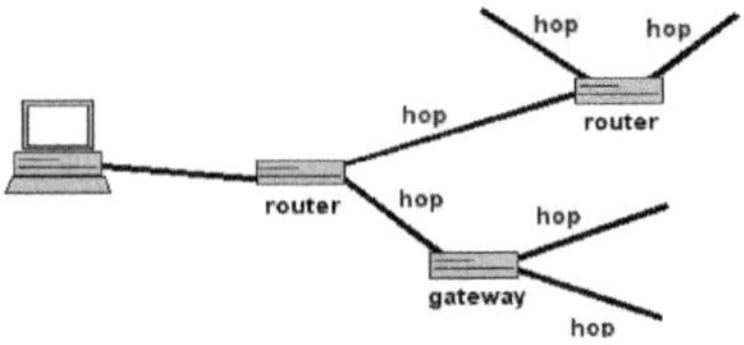

Fig. 1. Hop in Networks

thrown away. Every router that the packet travels through subtracts one from the TTL counter. When it reaches zero, the packet expires. The router will drop the packet that has expired. We can configured one or more routers such that a packet may circulate between three routers. Without the TTL counter, this packet would circulate endlessly, taking up bandwidth until someone fixes the routing.

1.3 TTL Based HopCount Computation

Since hop-count information is not directly stored in the IP header, one has to compute it based on the TTL field. TTL is an 8-bit field in the IP header. Each intermediate router decrements the TTL value of an in-transit IP packet by one before forwarding it to the next-hop. The final TTL value when a packet reaches its destination is therefore the initial TTL subtracted by the number of intermediate hops.

Since our filter starts to discard packets only upon detection of a DDoS attack, such end-systems would suffer only during an actual DDoS attack. The study shows that the OSs that use "odd" initial TTLs are older OSs. Thus, the benefit of deploying HCF should out-weight the risk of denying service to those end-hosts during attacks.

2 Related Work

Researchers have used the distribution of TTL values seen at servers to detect abnormal load spikes due to DDoS traffic . The Razor team at Bindview built Despoof, which is a command-line antispoofing utility. Despoof compares the TTL of a received packet that is considered "suspicious," with the actual TTL of a test packet sent to the source IP address, for verification. However, Despoof requires the administrator to choose packet, and to manually verify. Thus, the per-packet processing overhead is prohibitively high for weeding out spoofed traffic in real time.

3 Feasibility of Hop Count Filtering

The feasibility of HCF hinges on three factors: (1) stability of hop-counts, (2) diversity of hop-count distribution. In this section, we first examine the stability of hop-counts. Then, we assess if valid hop-counts to a server are diverse enough, so that matching the hop-count with the source IP address of each packet suffices to recognize spoofed packets with high probability.

3.1 Hop Count Stability

The stability in hop-counts between an Internet server and its clients is crucial for HCF to work correctly and effectively. Frequent changes in the hop-count between the server and each of its clients not only lead to excessive mapping updates, but also greatly reduce filtering accuracy when an out-of-date mapping is in use during attacks. According to the study of end-to-end routing stability in, the Internet paths were found to be dominated by a few prevalent routes, and about two thirds of the Internet paths studied were observed to have routes persisting for either days or week.

3.2 Diversity of Hop Count Distribution

Because HCF cannot recognize forged packets whose source IP addresses have the same hop-count value as that of a zombie, a diverse hop-count distribution is critical to effective filtering. It is necessary to examine hop-count distributions at various locations in the Internet to ensure that hop-counts are not concentrated around a single value. If 90% of client IP addresses are ten hops away from a server, one would not be able to distinguish many spoofed packets from legitimate ones using HCF alone.

Table 1. Diversity of traceroute gateway locations

Type	Sample Number
Commercial site	11
Educational site	4
Non profit site	2
Foreign site	18
.net site	12

4 Staying “Alert” to Drdos Attacks

In DRDoS attacks, an attacker forges IP packets that contain legitimate requests, such as DNS queries, by setting the source IP addresses of these spoofed packets to the actual victim's IP address. The attacker sends these spoofed packets to number of reflectors. The usual intrusion detection methods based on the ongoing traffic volume or access patterns may not be sensitive enough to detect such spoofed traffic. In contrast, HCF specifically looks for IP spoofing.

4.1 Blocking Bandwidth Attacks

To protect server resources such as CPU and memory, HCF can be installed at a server itself or at any network device near the servers, i.e., inside the ‘last-mile’ region, such as the firewall of an organization.This is not effective against DDoS attacks that target the bandwidth of a network to/from the server.

HCF at the web server will notice this sudden rise of spoofed traffic and inform the network administrator via an authenticated channel. The administrator can have the ISP install a packet filter in the ISP’s edge router, based on the HCF tableOnce the

HCF table is enabled at the ISP's edge router, most spoofed packets will be intercepted, and only a very small percentage that slip through HCF, will consume bandwidth. In this case, having two separable states is crucial since routers usually cannot observe established TCP connections and use the safe update procedure.

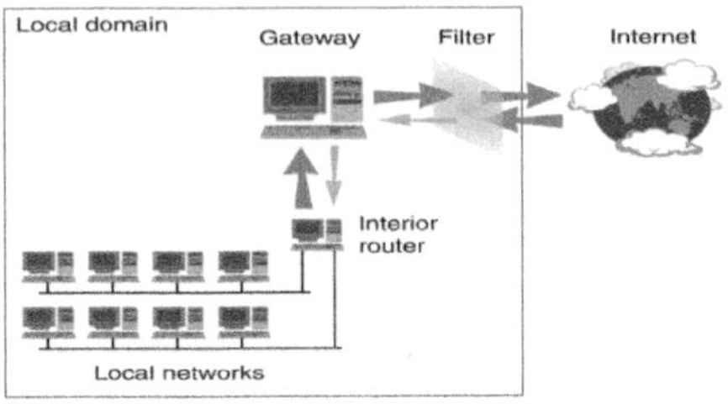

Fig. 2. Packet Filtering at a router to protect bandwidth

5 Conclusion

Thus Hop Count Filtering is a very effective method to monitor traffic, analyze by analyzing the packets and has proved to be very effective as even by effective means no one can alter the Hop counts for a packet. thus it is a very effective security alongside the network traffic.

References

[1] http://www.pcmag.com/encyclopedia_term/0,2542,t=hop+count&i=44344,00.asp
[2] http://en.wikipedia.org/wiki/Denial-of-service_attack
[3] http://www.techterms.com/definition/ttl
[4] http://searchnetworking.techtarget.com/definition/ping

Impulse Noise Reduction Using Mathematical Morphology

J. Harikiran, R. Usha Rani, P.V. Lakshmi, Y. Anil Bharath, and A. Sindhura

Department of IT, GIT, GITAM University, Visakhapatnam
jharikiran@yahoo.co.in, ushamurthy06@rediffmail.com

Abstract. This paper introduces a new class of filter for the removal of impulse noise in images using mathematical morphology. Basic mathematical morphological theory and operations are introduced at first, and then the self adaptive median morphological filter is proposed for the removal of impulse noise. The noise density (ND) is calculated for the image, which is used as the estimation function to select the size of the structuring element unit [SEU]. The subsequent algorithm for edge detection based on morphology is designed to better appraise the noise cancellation behavior of our filter from the point of view of human perception. Simulation results show that the proposed filter can effectively resolve the gap between detail-preserving and noise removing.

Keywords: Mathematical Morphology, Image Noise Reduction, Noise Density, Structuring Element Unit, Edge Detection.

1 Introduction

Image noise is the random variation of brightness or color information in images produced by the sensor and circuitry of a scanner or digital camera. The development of techniques for noise removal is of paramount importance for image-based measurement systems [1]. In-order to smooth out from noise many filtering architectures have been proposed in the literature [2]. Indeed, noise can significantly decrease the accuracy of very critical operations such as feature extraction and object recognition. The goal of the filtering action is to cancel noise while preserving the integrity of edge and detail information, non-linear approaches generally provide more satisfactory results than linear techniques. The mathematical morphological filter is an important nonlinear filter for signal processing. Recently people introduced number of morphological algorithms. In this paper, the self adaptive median morphological filter is proposed which overcomes the weakness of conventional morphology and maintains details of the original image when removing image noise. The paper is organized as follows: The basic mathematical morphological operations are presented in section 2, Noise density (ND) calculation is presented in section 3, self adaptive median morphological filter is presented in section 4, morphological edge detection method in section 5, Experimental results are presented in section 6 and finally section 7 reports conclusion.

V.V. Das, J. Stephen, and Y. Chaba (Eds.): CNC 2011, CCIS 142, pp. 539–543, 2011.

2 Basic Morphological Operations

In morphology, basic operations include erosion, dilation, opening and closing. The erosion of an image I by a flat structuring element (SE) b at any location (x,y) is defined as the minimum value of the image in the region coincident with b when the origin of b is at (x,y). In equation form, the erosion at (x,y) of an image I by a structuring element b is given by

$$[\, I \ominus b] (x,y) = \min_{(s,t)\in b} \{ I(x+s, y+t) \} \tag{1}$$

The dilation of an image I by a flat structuring element (SE) b at any location (x,y) is defined as the maximum value of the image in the window outlined by $\hat{b}$ when the origin of $\hat{b}$ is at (x,y). In equation form, the erosion at (x,y) of an image I by a structuring element b is given by

$$[\, I \oplus b] (x,y) = \max_{(s,t)\in b} \{ I(x-s, y-t) \} \tag{2}$$

The opening of image I by structuring element b, denoted I o b is

$$I \circ b = (I \ominus b) \oplus b \tag{3}$$

Similarly, the closing of I by b, denoted $I \bullet b$, is

$$I \bullet b = (I \oplus b) \ominus b \tag{4}$$

Usually the opening and closing operations are used to the morphological filter. They smooth the signal with different behaviors.

3 Noise Density (ND) Calculation

Algorithm I

Step 1: Let I be the noisy image of size NXN of an object or scene captured by sensor
Step 2: The noise boundaries of noisy image I are computed by spike detection technique [7][9]. Let L_1 and L_2 be the lower and upper noise boundaries for the noisy image
Step 3: The binary map(BM) of the noisy image is developed using the noise boundaries L_1 and L_2. If the image pixel 'y' lies within the noise boundaries, then it is uncorrupted and represented by a '0' in the binary map. The corrupted pixel is represented by a '1'in binary map.

$$BM = \begin{cases} \text{'0' if } L_1 < y < L_2 \\ \text{'1' if } y < L_1 \text{ or } y > L_2 \end{cases} \tag{7}$$

step 4: Compute the noise density ND of the noisy image.

$$ND = \frac{\text{sum of '1's in BM}}{N * N} \tag{8}$$

The value of ND ranges from 0 to 1.

4 Self Adaptive Median Morphological Filter

In morphology, the erosion operation is a basic operation that removes the image characteristic value which enlarged by noise, but retains the image characteristic value which reduced by noise. Analogously, the dilation, an inverse of erosion. In the conventional morphological filtering, the noisy image is processed sequentially by an opening operation and two erosion operations. However, the residual noise of the image remains obviously. Therefore, extreme value operation in conventional morphology restricts the application of morphology in image noise reduction. In this paper, the extreme value operation is changed to the median operation in the erosion and dilation according to the median principle, and builds the SEU based on the zero square matrix.

The median erosion and median dilation of image signals with structuring element b can be expressed as

Median Erosion:

$$[I \ominus b] (x,y) = \underset{(s,t)\in b}{\text{Median}} \{ I(x + s, y + t) \} \tag{9}$$

Median Dilation:

$$[I \oplus b] (x,y) = \underset{(s,t)\in b}{\text{Median}} \{ I(x - s, y - t) \} \tag{10}$$

where Median denotes median operation of image signals. Median morphological operations are built on a combination of median erosion and median dilation operations. The structuring element unit (SEU) is built based on the zero square matrix using Algorithm II. If the SEU size is too big, image noises can be removed, but image details will lose seriously. Inversely if the size is too small, image noise cannot be removed efficiently. The ND of image is looked as the estimation function to select the size of SEU.

Algorithm II

Step 1: Initialize the variable, and make n=0;
Step 2: Using the SEU based on the zero square matrix to implement median erosion and median dilation for the image, get the output image I_n, calculate the ND_n of I_n ; and make n=n+1;
Step 3: Implement the median erosion and median dilation for the image again, and get the new output image I_{n+1}. Calculate ND_{n+1} of I_{n+1}; and make n=n+1 again;
Step 4: If $ND_n > ND_{n+1}$, then make the SEU size equal to n ; if not, jump to step 3 and continue to calculation.

Step 5: Using the SEU based on n orders zero square matrix to implement median erosion operation and median dilation operation for the image, get the output image I_{out} .

5 Edge Detection Algorithm

Edge detector is one of the most important tools in computer vision. The edge detection process serves to simplify the analysis of images by drastically reducing the amount of data to be processed, while at the same time preserving useful structural information about object boundaries [8]. Morphological edge detection algorithm selects appropriate structuring element of the processed image and makes use of the basic theory of morphology including erosion, dilation, opening and closing operation and the synthesization operations of them to get clear image edge.

In this paper, a mathematical morphology edge detection algorithm is proposed. First, smooth the image by first closing and then dilation. The perfect image edge will be got by performing the difference between the processed image by above process and the image before dilation. The following is our algorithm:

$$(M \bullet b) \oplus b - M \bullet b \tag{14}$$

where $M = (I \bullet b) \circ b$

6 Experimental Results

This section presents the simulation results illustrating the performance of the proposed filter. The test image employed here is the lungs CT scan image with 351×294 pixels. The salt and pepper noise is added into the image with noise density 0.4. The noise model was computer simulated. Figure 1 shows the results of filtering the lungs CT scan image which is corrupted by impulse noise. Edges define the boundaries between regions in an image, which helps with segmentation and object recognition. In order to better appraise the noise cancellation behavior of our filter from the point of view of human perception, we perform edge detection mechanism for the filtered image by using the algorithm described in section 4. Figure 2 shows that our method significantly reduces impulse noise and the image details have been satisfactorily preserved.

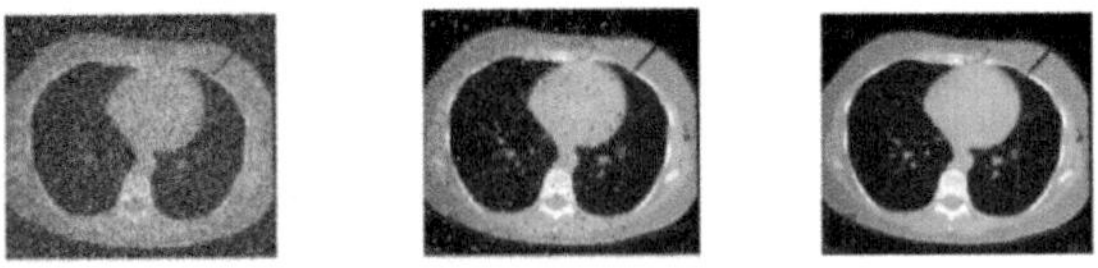

(a) (b) (c)

Fig. 1. (a) original image (b) conventional morphological filter (c) filtering by our method

(a) (b)

Fig. 2. a) edge detection on conventional morphological filtered image. b) edge detection on our proposed filtered image.

7 Conclusion

Based on median operation in erosion and dilation, a new technique for the restoration of images corrupted by impulse noise has been presented. The proposed filtering approach maintains the characters of morphological operations and the flexibility of structuring element, overcomes the weakness of information missing in the conventional morphological image noise reduction. The structuring element unit (SEU) is built based on the zero square matrix. The ND of image is used as the estimation function to select the size of SEU. A method for edge detection is also presented to better appraise the noise cancellation behavior of our filtering models. This restoration of image data is very likely to find potential applications in a number of different areas such as electromagnetic imaging of objects, medical diagnostics, remote sensing, robotics, etc. Experimental results show that the proposed method yields very satisfactory results.

References

1. Van der Heijden, F.: Image Based Measurement Systems. Wiley, NewYork (1994)
2. Jain, K.: Fundamentals of Digital Image Processing. Prentice-Hall, Englewood Cliffs (1989)
3. Zhao, C.: Research on digital morphological filter theory and its algorithms. Harbin Institute of technology, Harbin (1998)
4. Liu, J.: Research on morphological filter algorithms based on mathematical morphology. Harbin Institute of technology, Harbin (2006)
5. Ge, P., Su, P., Wang, X., Zhang, Y.: Hybrid method in noise removing based on mathematical morphology. Computer Engineering and Design 28(7), 1584–1585 (2007)
6. Peng, B., Wang, Y.: Research and Implementation of denoising algorithm for low-illimunation image. Journal of Computer Applications 27(6), 1455–1457 (2007)
7. Indu, S., Ramesh, C.: A noise fading technique for images highly corrupted with impulse noise. In: Proceedings of the ICCTA 2007. IEEE, Los Alamitos (2007)
8. Nadernejad, E.: Edge Detection Techniques: Evaluations and Comparisions. Applied Mathematical Sciences 2(31), 1507–1520 (2008)
9. Indu, S., Ramesh, C.: Image fusion algorithm for impulse noise reduction. In: Proceedings of the ARTCom. IEEE, Los Alamitos (2009)

Improving the QoS Routing for Mobile Ad Hoc Networks

Mamatha Balachandra[1], K.V. Prema[1], and M. Krishnamoorthy[2]

[1] Dept. of Computer Science and Engg., Manipal Institute of Technology
[2] Dept. of Master of Computer Applications, Manipal Institute of Technology
Manipal University, Manipal, India
mamtha.bc@manipal.edu

Abstract. Quality of Service support for Mobile Ad-hoc Networks is a challenging task due to dynamic topology and limited resource. In this paper initially systematic performance study of two routing protocol for ad hoc networks Ad hoc On Demand Distance Vector (AODV)[1], Ad hoc On Demand Multipath Distance Vector (AOMDV)[2] based on QoS parameters throughput, delay, packet delivery ratio is made. Then a novel QoS routing algorithm called Quality of Service – Ad hoc On-demand Multipath Distance Vector Routing (QoS-AOMDV) based on Adhoc On-demand Multipath Distance Vector Routing (AOMDV) is proposed to support QoS constraints.

Keywords: Mobile Ad hoc Networks (MANETs), Quality of Service (QoS), Ad-hoc On demand Distance Vector (AODV), Ad-hoc On demand Multipath Distance Vector (AOMDV).

1 Introduction

In recent years, on-demand routing protocols have attained more attention in mobile ad hoc networks as compared to other routing schemes due to their abilities and efficiency. There exist many on- demand routing protocols for mobile ad hoc networks (MANETS). Most of the protocols, however, use a single route and do not utilize multiple alternate paths. Multipath routing allows the establishment of multiple paths between a single source and single destination node and when a path breaks an alternate path is used instead of initiating a new route discovery. QoS is very important especially in the case of multimedia traffic. Our goal is to carry out a systematic performance study of two routing protocol for ad hoc networks Ad hoc On Demand Distance Vector (AODV)[1], Ad hoc On Demand Multipath Distance Vector (AOMDV)[2] Algorithms based on QoS parameters and the idea is to improve these QoS parameters for these routing protocols.

2 Unipath and Multipath Routing in MANETs

Two main classes of ad hoc routing protocols are table-based and on-demand protocols [3]. In table-based protocols [4] [5], each node maintains a routing table containing routes to all nodes in the network. In on-demand protocols [6] [7], nodes

V.V. Das, J. Stephen, and Y. Chaba (Eds.): CNC 2011, CCIS 142, pp. 544–546, 2011.

only compute routes when they are needed. Multipath routing consists of finding multiple routes between a source and destination node[8].

3 Simulation

The performances of 2 on-demand routing protocols , ie. AODV and AOMDV are compared using NS-2 simulation. The following 2 metrics are used to compare the performances of 2 routing protocols:

i. Packet Delivery Ratio (PDR) ii. Delay

(i) Delay vs. number of nodes

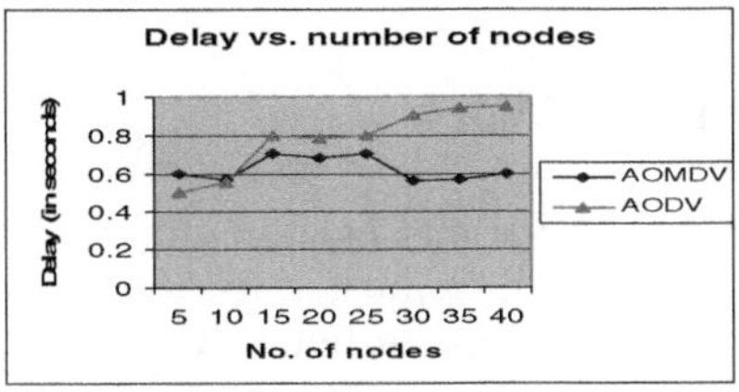

Fig. 1. Delay versus number of nodes

(ii) PDR vs. number of nodes

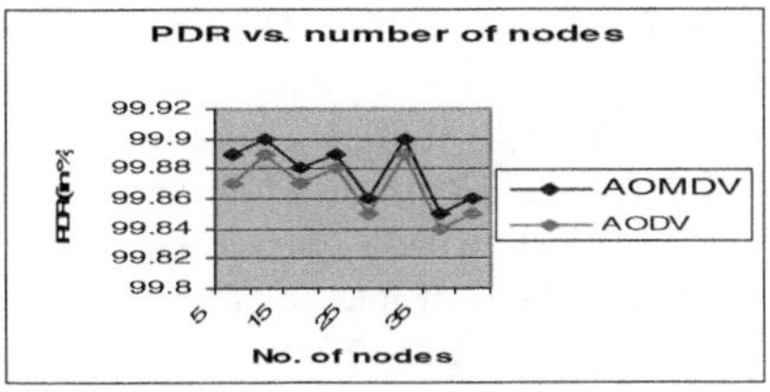

Fig. 2. PDR versus number of nodes

4 Method to Improve AOMDV

Step (i): The source node first checks its routing table to find a route to the destination. If route (or routes) found, it selects route according to fresh routes or the optimal multiple metrics' values and sends the data to the destination.

Step (ii): If no route exists, it initiates a route discovery process. The source constructs a RREQ packet with the destination node.

Step (iii): If the intermediate node has a route entry, it will send a route reply RREP packet back to the source along the current metrics values based on the following outcomes:

a) If multiple entries to the same destination are found, it will select the route based on the QoS metrics.
b) If it doesn't have a route entry then it will rebroadcast the RREQ packet into the network.

Step (iv): After receiving one Route Request packet, the destination node will construct a Route Reply packet by appending the additional fields of RREP packet i.e., Delay and PDR values.

Step (v): Each intermediate node along the forward path to the source will update the multiple metrics fields of RREP packet by comparing the multiple metrics value on the link on which the RREP was received The source node waits for a timeout period to receive multiple route replies. Storing all the routes in the buffer. Subsequently, it selects the best path based on the multiple metrics' values.

5 Conclusion and Future Work

Using NS-2 the performance of various most commonly used routing protocols AODV and AOMDV are compared. Compared to AODV the overall performance of AOMDV is better in all the aspects. In future, we implement this algorithm to compute the multipath route to reach to the destination which meets the QoS metrics.

References

1. Perkins, C., Belding-Royer, E., Das, S.: Adhoc On-Demand Distance Vector (AODV) Routing. RFC 3561 (2003)
2. Marina, M.K., Das, S.R.: On-demand Multipath Distance Vector Routing in Ad Hoc Networks. Department of Electrical & Computer Engineering, USA
3. Biradar, R.V., Patil, V.C.: Classification and Comparison of routing Techniques in Wireless Ad-hoc Networks. In: Proceedings of international Symposium on Ad-hoc Ubiquitous Computing (ISHUC 2006), pp. 7–11 (2006)
4. Murthy, S., Garcia-Luna-Aceves, J.J.: An Efficient Routing Protocol for Wireless Networks. Mobile Networks and Applications 1(2) (1996)
5. Perkins, C.E., Bhagwat, P.: Highly Dynamic Destination-Sequenced Distance-Vector Routing (DSDV) for Mobile Computers. In: ACM SIGCOMM, pp. 234–244 (1994)
6. Perkins, C.E., Royer, E.M.: Ad-hoc On-Demand Distance Vector Routing. In: Proceedings of the 2nd IEEE Workshop on Mobile Computing Systems and Applications
7. Johnson, D.B., Malt, D.A.: Dynamic Source Routing in Ad Hoc Wireless Net works. Mobile Computing, 153–181 (1996)
8. Miller, M.J., So, J.: Improving Fault Tolerance in AODV, University of Illinois

An Approach towards Promoter Database Search Using Hidden Markov Model

A. Meera[1,*] and Lalitha Rangarajan[2]

[1] BMS College of Engineering, Bengaluru
meeramesh01@gmail.com
[2] Department of Computer Science, University of Mysore, Mysore

Abstract. A common task in bioinformatics is the comparison of biological sequences to probabilistic models in order to evaluate their similarity. In the proposed work we have made an attempt to classify the given query promoter sequence by comparing with the set of promoter sequences belonging to different pathways. The promoter sequences are extracted from NCBI database and converted to motif (TFBS) sequences by using 'TF SEARCH' tool. A probabilistic model is developed for each motif sequence of the pathways considered and a database is created. Given query motif sequence is compared with all the sequences in the database and classified based on the log probability score obtained.

Keywords: Database, Expectation Maximization, Hidden markov model, maximum likelihood, promoter sequence.

1 Introduction

Bioinformatics is the science of managing, mining, and interpreting information from observations of biological processes. Various genome projects have contributed to an exponential growth in DNA and protein sequence databases. This growth has prompted scientists to develop faster and more sophisticated algorithms to keep pace with the increasing size of the databases [1]. Software improvements combined with state-of-the-art hardware have allowed computational biologists to enter a new era of comparative genomics [2], [3], [4].

Gene consists of a sequence of nucleotides from which RNAs (Ribo Nucleic Acid) can be transcribed to give either non-coding regulatory RNAs or m RNAs or rRNAs. The temporal, spatial and quantitative expression of RNA from genes is decided by the sequence existing before the sequence of a gene. This region is the promoter region. Conserved sequences in promoter regions (non coding regions) are called motifs [5]. Transcriptional factors are enzymes that bind to specific sequences (motifs or TFBS) in the promoter and regulate expression of the gene. Promoter regions of genes with similar expression patterns may not show sequence similarity, even though they may be regulated by similar configurations of TFs. As a result, similar functions are frequently encoded by diverse sequences [6]. It is assumed that genes exhibiting

* Corresponding author.

V.V. Das, J. Stephen, and Y. Chaba (Eds.): CNC 2011, CCIS 142, pp. 547–552, 2011.

similar expression patterns would also share similar configurations of TFs in their promoter. Comparison of promoter sequences may give insight into finding evolutionary distance and gene order [7].

A common task in bioinformatics is the comparison of biological sequences to probabilistic models in order to evaluate their similarity. Hidden Markov Models (HMMs) are one of the most popular models for analysis of sequential processes taking place in a random way, where "randomness" may also be an abstraction covering the fact that a detailed analytical model for the internal matters are unavailable. Such a sequential process can be observed from outside by its emission sequence (letters, sounds, measures of features, all kinds of signals) produced over time, and a HMM postulates a hypothesis about the internal machinery in terms of a finite state automaton equipped with probabilities for the different state transitions and single emissions. Decoding or prediction for a given observed sequence means to compute the most probable state transitions that the HMM can go through to produce the sequence, and thus this represents a best hypothesis for the internal structure or "content" of the sequence. HMMs are widely used in speech recognition and biological sequence analysis. Usually, for decoding Viterbi algorithm is used. The original Viterbi algorithm involves multiplications with probabilities. One way to avoid this is to take the logarithm of the model parameters, thus replacing the multiplications with additions. This is called Fast Viterbi or alternate Viterbi.

In the present work, we have developed a method for finding similarity between the query motif sequence and the database consisting of motif sequences of metabolic pathways using maximum likelihood procedure.

2 Methodology

The promoter sequences are extracted from NCBI database and converted to motif (TFBS) sequences using 'TF SEARCH' tool. As a case study we have considered promoter sequences of two metabolic pathways namely, Central Metabolic Pathway (CMP) and amino acid pathway.

Step 1: Motif sequences are converted into numerical sequences.
We have assumed N=Number of states in the model=5

Step 2: Initially model λ=(A,B,π) is assumed.
A=a_{ij} is state transition matrix
B is output/ Observation / emission matrix.
Π is initial state distribution matrix.
Initially, we assume

$$A = a_{ij} = \begin{bmatrix} 0.2 & 0.2 & 0.2 & 0.2 & 0.2 \\ 0.2 & 0.2 & 0.2 & 0.2 & 0.2 \\ 0.2 & 0.2 & 0.2 & 0.2 & 0.2 \\ 0.2 & 0.2 & 0.2 & 0.2 & 0.2 \\ 0.2 & 0.2 & 0.2 & 0.2 & 0.2 \end{bmatrix} \quad \{\text{ NXN ie. 5X5}\}$$

$$\pi = \begin{bmatrix} 1 \\ 0 \\ 0 \\ 0 \\ 0 \end{bmatrix} \quad \{NX1 \text{ ie. } 5X1\}$$

$$B=b_j(o_t) = \begin{bmatrix} 5/210 & \cdots & 5/210 \\ \vdots & \ddots & \vdots \\ 5/210 & \cdots & 5/210 \end{bmatrix}$$

{ NXM ie. 5X210}
Where M=210(for example), is the number of observables (motifs) in the given motif sequence.

Step 3: Using Equation $\alpha_1(i) = \pi_i b_i(o_1), \quad 1 \leq j \leq N.$ (1)
the initial value of forward variable α_1, is calculated $\alpha_1(1)=(1)(5/210)=5/210$
Then we obtain forward variable $\alpha_t(i)$, (probability of partial observation of sequence) using Equation

$$\alpha_{t+1}(j) = b_j(O_{t+1}) \sum_{i=1}^{N} \alpha_t(i) a_{ij}, \quad 1 \leq j \leq N \tag{2}$$

Step 4: The backward variable $\beta_{t(i)}$ (the probability of being in state S_i, given the partial observation $o_{t+1},\ldots,o_T$) is calculated using equation

$$\beta_t(i) = \sum_{j=1}^{N} \beta_{t+1}(i) a_{ij} o_j(O_{t+1}), \quad 1 \leq i \leq N \tag{3}$$

$$t = T-1 \ldots . N$$

e.g. $\beta_{1(1)} = a_{1,1} * b_1(O_2) * \beta_2(1)$ [at i=1,j=1,t=1] = (0.2)(5/210)(1)=1.0/210

Step 5: Using an initial parameter instantiation $\lambda=(A,B,\pi)$, the Baum-welch algorithm or Expectation Maximisation[8] [9] (EM method) re-estimates three parameters π_i, $a_{i,j}$ and $b_i(o_t)$ where $\pi_i =$ initial state distribution
$a_{i,j}$ =Transition probabilities
$b_i(o_t)$=Emission probabilities
Here we re-estimate1) Transition Probabilities:

$\xi_t(i,j)$ is the probability of being in state s_i at time t and going to state s_j, given the current model and parameters

$$\xi_t(i,j) = P(q_t = i, q_{t+1} = j \mid O, \lambda) . \tag{4}$$

$$= \frac{P(q_t = i, q_{t+1} = j \mid O, \lambda)}{P(O \mid \lambda)}$$

$$= \frac{\alpha_t(i) a_{ij} b_j(\boldsymbol{o}_{t+1}) \beta t + 1(j)}{\sum_{i=1}^{N} \sum_{j=1}^{N} \alpha_t(i) a_{ij} b_j(\boldsymbol{o}_{t+1}) \beta_{t+1}(j)}$$

$$a_{ij} = \frac{\textit{expected number of transitions from state } i \textit{ to state } j}{\textit{expected number of transitions from state } i}$$

$$a_{ij} = \frac{\sum_{t=1}^{T-1} \xi_t(i,j)}{\sum_{t=1}^{T-1} \gamma_t(i)}. \tag{5}$$
$$= \frac{\sum_{t=1}^{T-1} \alpha_t(i) a_{ij} b_j(\boldsymbol{o}_{t+1}) \beta_{t+1}(j)}{\sum_{t=1}^{T-1} \alpha_t(i) \beta_t(i)}$$

2) Emission probabilities $b_i(k)$ =

$$\frac{\textit{expected number of times in } s_{ii} \textit{ and observed symbol } vk}{\textit{expected number of times of times in state } si}$$

^

$$b_j(k) = \sum_{t=1}^{t} \delta(o_t, v_k)\ \gamma_t(i) / \sum_{t=1}^{T} \gamma_t(i)\ b_i(k). \tag{6}$$

Where $\delta(o_t, v_k) = 1\ if\, o_t = v_k\ otherwise\ 0$

and $\gamma_t(i)$ (state probability), the probability of being in state s_i, given the complete observation $o_1 \ldots\ldots\ldots o_T$.

3) Initial distribution probability

$$\hat{\pi} = \gamma_1(i)$$

The Updated Model is

$\lambda' = (\hat{A}, \hat{B}, \hat{\pi})$ by the following update rules: $\hat{a}_{ij} = \frac{\sum_{t=1}^{T-1} \xi_t(i,j)}{\sum_{t=1}^{T-1} \gamma_t(i)}$

$$\hat{b}_{i(k)} = \sum_{t=1}^{t} \delta(o_t, v_k)\ \gamma_t(i) / \sum_{t=1}^{T} \gamma_t(i)$$

$$\hat{\pi} = \gamma_1(i) \text{ Where } (i) = \sum_{j=1}^{N} \xi_t\ (i,j).$$

For each values of updated model λ, P(o/ λ) is calculated and above step is repeated until P(o/ λ) reaches local maxima.

Step 6: Testing the motif sequence (query sequence): Using Viterbi algorithm, we find P(o | λ) for optimal sequence.

1. Initialization

$$\delta_1(i) = \pi_i b_i(\boldsymbol{o}1), \quad 1 \leq i \leq N$$

$$\delta_t(j) = b_j(\boldsymbol{o}_t) \max_{0 \leq x \leq 1} \delta_t - 1(i) a_{ij}$$

2.Induct

$$1 \leq j \leq N$$
$$P^* = \max_{0 \leq x \leq N}[\delta t - 1(i)a_{ij}],$$
$$1 \leq j \leq N$$

The probability value is generated for all models.

3 Result and Discussion

The query promoter sequence which is in the form of motif sequence is compared with all the sequences in the database and the log probability score with each sequence is displayed. The sequence in the database which has highest similarity with the query sequence is highlighted (in blue and underlined). Here, a section of the result has been shown.

pathway- CMP(citrate synthase)and Organism - BOS-5:	-2571.002094
pathway- CMP(citrate synthase) and Organism - BOS-10:	-2604.309499
pathway-CMP(citrate synthase)and Organism - Can-10:	-3319.908145
pathway- CMP(citrate synthase)and Organism - HS-19:	-1365.157139
pathway -CMP(citrate synthase)and Organism - HS-12:	-4079.434347
pathway- CMP(citrate synthase)and Organism - HS-2:	-2743.449172
pathway- CMP(citrate synthase)and Organism - HS-3:	-2527.987613

Genes coding for proteins is a very important pattern recognition problem in bioinformatics. Promoters, the upstream sequences of gene coding regions, are one of the keys to understanding the underlying hypotheses of gene expression and regulation which is mainly controlled by transcription factors (TFs), i.e., proteins that bind to promoter regions at specific sites (TFBSs) and regulate the process of transcription initiation.

Hidden Markov models (HMMs) are a well-developed technology for classification of multivariate data that have been used extensively in speech recognition, handwriting recognition and even sign recognition. HMMs have been used widely for classification such as protein structures[10], autonomous terrain mapping and Classification [11], etc. In the present work, we have made an attempt to classify the given query motif sequence by comparison with the given set of motif sequence database using maximum likelihood procedure. Each comparison gives a score reflecting the degree of similarity between query and the motif sequence in the database. Higher the score, greater is the degree of similarity.

References

1. Benson, D.A., Karsch-Mizrachi, I., Lipman, D.J., Ostell, J., Wheeler, D.L.: GenBank. Nucleic Acids Research 34, D16–D20 (2006)
2. Altschul, S.F., Gish, W., Miller, W., Myers, E.W., Lipman, D.J.: Basic Local Alignment Search Tool. J. Molecular Biology 215, 403–410 (1990)

3. Altschul, S., Madden, T.L., Schaffer, A.A., Zheng, J., Zhang, Z., Miller, M., Lipman, D.L.: Gapped BLAST and PSI-BLAST: A New Generation of Protein Database Search Programs. Nucleic Acid Research 25, 3389–3402 (1997)
4. Pearson, W.R., Lipman, D.J.: Improved Tools for Biological Sequence Comparison. Proc. Nat'l Academy of Science 85, 2444–2448 (1988)
5. Brutlag, D.: Multiple sequence alignment and Motifs. Bioinformatics methods and Techniques. Stanford University, Stanford center for Professional development (2002)
6. Blanco, E., Messeguer, X., Smith, T.F., Guigó, R.: Transcription factor map alignment of promoter regions. PLoS Comput. Biol. 2(5), 49 (2006)
7. Meera, A., Rangarajan, L., Bhat, S.: Computational Approach Towards Finding Evolutionary Distance And Gene order Using Promoter Sequences Of Central Metabolic Pathway" Inter disciplinary sciences-computational life sciences doi: 0.1007/s12539-009-0017-3 (2009) (Spriger link)
8. Baum, L.E., et al.: A maximization technique occurring in the statistical analysis of probabilistic functions of Markov chains. Ann. Math. Statist. 41, 164–171 (1970)
9. Lawrence, E., Reilly, A.: An expectation maximization algorithm for the identification and characterization of common sites in unaligned biopolymer sequences. Proteins 7, 41–51 (1990)
10. Mirceva, G., Davcev, D.: HMM based approach for classifying protein structures. International Journal of Bio- Science and Bio- Technology 1(1) (December 2009)
11. Wolf, D.F., Sukhatme, G.S., Fox, D., Burgard, W.: Autonomous Terrain Mapping and Classification Using Hidden Markov Models. In: Proceedings of the 2005 IEEE International Conference on Robotics and Automation, vol. 2005, pp. 2026–2031. IEEE, Los Alamitos (2005)

Computational Characterization of CNT-Papain Interactions for Developing a Biosensor

C.R. Athira[1], S. Shilpa[2], C.M. Parvathy[2], S. Nithya[2], V. Vinothaa[2], S.P. Subashini[2], K. Varun Gopal[2], Bharath Parimalam[2], and P.K. Krishnan Namboori[2]

[1] Amrita School of Biotechnology, Amrita Vishwa Vidyapeetham University, Amritapuri, Kollam - 690 525, India
[2] Computational Chemistry Group, Computational Engineering and Networking, Amrita Vishwa Vidyapeetham University, Ettimadai, Coimbatore - 641 105, India
athirarajeev@gmail.com, n_krishnan@cb.amrita.edu, varungopal19@gmail.com

Abstract. THC and its metabolites are detectable in body fluids and could be used to detect marijuana use. This research work reveals the possibility of using a carbon nano tube (CNT) based enzyme sensor for the detection of 9-carboxy tetrahydrocannabinol (9 THC-COOH) in urine using a cysteine protease obtained from papaya plant called papain. A comparative study of armchair and zigzag conformation of CNTs with the protein was carried out and the suitable conformer of CNT has been identified. The electrostatic energy variations and the thermodynamic energy differences of CNTs have been computed. A characteristic gradient in electrostatic potential has been identified for the CNT-protein complexes which could be effectively sensed using electrochemical sensors. The computational analysis of CNT-papain interaction supports the possibility of developing a biosensor using papain coated carbon nano tubes for the detection of THC-COOH in urine of the suspected marijuana consumed persons. A comparative analysis of the interaction energies of arm-chair and zigzag conformations of carbon nano tubes with the cysteine protease enzyme papain has been carried out. The arm chair conformation with configuration index 18, 18 (m, n) has been identified as keeping the highest interaction energy of 45.218 kcal/mol. With this conformer designing of an electrochemical sensor has been suggested.

Keywords: Carbon Nanotube, Electrochemical Sensor, MD simulation.

1 Introduction

The main active chemical in marijuana is delta-9-tetrahydrocannabinol (THC) [1]. THC or its metabolites 9-carboxy THC stays in body fluids as markers for detection of cannabis consumption. Marijuana testing in the work-place has been adopted widely by employers in the United States to determine the use of drug by the employee and to ensure workplace safety, monitor medication compliance and to promote 'drug-free' work-places [2]. Studies show that the acute effects of smoking cannabis impair performance for a period of about 4 hours. Cannabis detection is also made mandatory for

V.V. Das, J. Stephen, and Y. Chaba (Eds.): CNC 2011, CCIS 142, pp. 553–556, 2011.

athletics, defense and also driving personals [3, 4]. Prevailing commercially available cannabis detecting kits like quick screen and rapid check utilizes the method of gas chromatography and HPLC techniques [5, 6], which are highly expensive.

Carbon nanotubes (CNTs) have many distinct properties for the development of next generation of sensors [7]. The application of CNTs in next-generation of sensors has the potential of revolutionizing the sensor industry due to its inherent properties such as small size, high strength, high electrical and thermal conductivity, and high specific surface area [8]. High sensitivity and low power consumption make CNT based sensors suitable, especially, for high accuracy and battery-powered applications. The purified CNTs have very large surface area and extremely high conductivity [9]. Thus, the sensors using CNT composite as sensing materials have higher sensitivity [10].Papain is a cysteine hydrolase that is stable and active under a wide range of conditions like at elevated temperatures. Papain has been reported to have a degradative effect in the inactive metabolite of the active component THC in marijuana, the 9-carboxy THC. In an experiment conducted checking the possibility of papain as a potential adulterant in the urine test for detection of cannabis use, it was reported that there was a direct pH, temperature, and time-dependent correlation between the increase in papain concentration and the decrease in THC-COOH concentration from the untreated control groups.

This research paper includes the possibility of using papain coated carbon nano tubes as biosensors for the detection of marijuana consumption using computational techniques. Although the use of CNT leads into a great revolution in the sensor development, certain challenges keep the progress under a slow pace. The challenges include the high cost for obtaining pure and uncontaminated nanotubes, lack of detailed understanding of the growth mechanism of carbon nano tubes and also the difficulties in growing defect-free nanotubes continuously to macroscopic length. Computational methods could be a better substitute for the prevailing scenario of carbon nano tubes research ensuring a reduced cost, time and effort in the research and providing an initial screening and designing step. Another major reason for the adoption of computational method is due to the unavailability of Marijuana for experimental research, being an illicit hallucinogen.

2 Computational Methodology

The structure of 9-caarboxy tetrahydrocannabinol (THC-COOH) was obtained from pubchem and was subjected to geometrical optimization. 36 different single walled carbon nano tubes were designed using Material studio [11] with varying m and n values and were subjected to modeling in the quantum mechanical level. Materials Studio is a flexible client-server software environment that brings some of the most advanced material simulation and modeling technology. Materials Studio makes it easy for creating and studying models of molecular structures, exploiting stunning graphics capabilities to present the results. Materials Studio draws on well-validated and widely applied simulation methods from the leader in high quality materials simulation software. Protein molecule, papain was collected from the repository Protein Data Bank (PDB) [12] and was subjected to computational modeling. All the modeling and simulations was performed using Accelrys Discovery studio with 'smart minimiser' algorithm and CHARMm force field [13]. The input for the modeling was the PDB file collected from the repository Protein Data Bank. The collected protein structures were subjected to geometry optimization and the corresponding minimum interactional potential energy, RMS gradient, van der Waals energy, electrostatic energy and kinetic energy values were computed. Most of the structures attained convergence at about 2000 steps of iteration. Using the cdocker tool [14] in the Discovery studio the CNT-papain interaction analysis was carried out and the best interaction results were selected based on the cdocker score. The interaction analysis between the papain and 9-carboxy tetrahydrocannabinol and CNT-papain complex with tetrahydrocannabinol was also found out using cdocker tool. The electrostatic potential energy of papain and papain-CNT with THC-COOH was calculated using the MGL tool [15].

3 Results and Discussions

The carbon nanotube structures have been subjected to quantum mechanical modeling and molecular dynamic simulation. The enthalpy values for all the nanoparticles are found to be negative indicating them as exothermic compounds and explains the thermodynamic stability of nanoparticles. The interactional potential energy and the van der Waals energy of the papain proteins have been found to be negative. The RMS gradient value was very close to zero indicating them to be in stable state. The docking results, obtained for various poses have been analyzed based on the interactional energies. An interaction is thermodynamically feasible if and only if the reaction is exothermic. From the results of the docking analysis done using the cdocker algorithm, the armchair conformation and the papain protein complex has been found to be showing high interactions with the THC-COOH.

Table 1. Interaction Energy of Sensor Components with THC-COOH

Components Involved	Interaction Energy
Papain	35.826
CNT-Papain Complex	45.218

Arm chair CNT with m and n configuration of 18, 18and the papain protein showed the maximum interaction with THC-COOH with an energy of 45.218 kcal/mol. Results from the MGL tool indicates the difference in the electrostatic potential energy of papain and papain with THC-COOH, which predicts the possibility of potential difference that could be measured using an electrochemical sensor.

Table 2. Result of Electrostatic Analysis

Components	Electrostatic energy
Papain	-32.82
Papain With THC-COOH	-47.35

4 Conclusion

The computational analysis results of CNT-papain interaction strengthen the possibility of a biosensor using papain coated carbon nano tubes for the detection of THC-COOH in urine which indicates the marijuana consumption. A comparative study of the interaction energy of arm-chair and zig-zag conformations of carbon nano tubes which show metallic and semiconducting properties respectively were done with the cysteine protease enzyme papain and the arm chair conformation with 18,18(m, n) was identified with the highest stable interactional potential energy. The selected CNT also showed a lesser total energy which indicates a stable combination of papain with 18, 18 CNT and also higher interaction energy with THC-COOH. The electrostatic potential energy analysis suggests the possibility of using electrochemical sensors which when applied with required voltage produces current proportional to the concentration of the substrate.

References

1. Herkenham, M., Lynn, A., Little, M.D., et al.: Cannabinoid receptor localization in the brain. Proc Natl. Acad. Sci. 87(5), 1932–1936 (1990)
2. Baselt, R.: Disposition of Toxic Drugs and Chemicals in Man, 8th edn., pp. 1513–1518. Biomedical Publications, Foster City (2008)
3. Pate, D.W.: Possible role of ultraviolet radiation in evolution of Cannabis chemotypes. Economic Botany 37, 396–405 (1983)
4. Lydon, J., Teramura, A.H., Coffman, C.B.: UV-B radiation effects on photosynthesis, growth and cannabinoid production of two Cannabis sativa chemotypes. Photochem. Photobiol. (1987)
5. Solinas, M., Scherma, M., Fattore, L., et al.: Nicotinic alpha 7 receptors as a new target for treatment of cannabis abuse. J. Neurosci. 27(21), 5615–5620 (2007)
6. Fellermeier, M., Zenk, M.H.: Prenylation of olivetolate by a hemp transferase yields cannabigerolic acid, the precursor of tetrahydrocannabinol. FEBS Lett. 427 (1998)
7. Wang, X., Li, Q., Xie, J., Jin, Z., Wang, J., et al.: Fabrication of Ultralong and Electrically Uniform Single-Walled Carbon Nanotubes on Clean Substrates. Nano Letters 9(9), 3137–3141 (2009)
8. Seetharamappa, J., Yellappa, S., D'Souza, F.: Carbon Nanotubes: Next Generation of Electronic Materials (2006)
9. Sinha, N., Ma, J., Yeow, J.T.W.: Carbon Nanotube-Based Sensors. Journal of Nanoscience and Nanotechnology 6, 573–590 (2006)
10. Saito, R., Dresselhaus, G., Dresselhaus, M.S.: Physical properties of carbon nanotubes (1998)
11. Accelrys Software inc., Materials Studio Release Notes, Release 4.3, San Diego: Accelrys Software Inc. (2008)
12. Berman, H.M., Westbrook, J., Feng, Z., Gilliland, G., Bhat, T.N., Weissig, H., Shindyalov, I.N., et al.: The Protein Data Bank. Nucleic Acids Research 28(1), 235–242 (2000)
13. Accelrys Software Inc., Discovery Studio Modeling Environment, Release 2.1, San Diego: Accelrys Software Inc. (2007)
14. Wu, G., Robertson, D.H., et al.: Detailed Analysis of Grid-Based Molecular Docking: A Case Study of CDOCKER - A CHARMm-Based MD Docking Algorithm. J. Comp. Chem. 24, 1549 (2003)
15. Sanner, M.F.: Python: A Programming Language for Software Integration and Development. J. Mol. Graphics Mod. 17, 57–61 (1999)

A Fast Sentence Searching Algorithm

Rohit Kamal Saxena, Kamlendra Pratap Singh, and U.C. Jaiswal

M. M. M. Engineering College, Gorakhpur, U.P. India - 273010
{rohit_kamal2003,ucj_jaiswal}@yahoo.com,
22kamlendra@gmail.com

Abstract. With ever increasing digital data in form of files, there is a need of efficient text searching algorithms with their applications. This paper describes an algorithm for searching any sentence in the given pool of files. The efficiency of the algorithm depends on the fact about sentences in any language that, there can be no such condition where a sentence may be a part of some similar sentence (as in case of patterns). Keeping this property of the languages in mind, we are presenting a linear order search algorithm for sentences with no preprocessing time.

Keywords: text search, sentence searching, pattern, searching in files application.

1 Introduction

In this computerized world, data is increasing at a rate of fifty percent per annum. Therefore, it is quite common that a firm may add up a terabyte of data per year. As the size of data is increasing at such a tremendous rate, we should focus on some efficient searching algorithms so to fulfill our purpose of storing data.

At present, the existing text searching algorithms search the pattern/words in some text. **Brute-force search**, otherwise known as naïve algorithms, use the simplest method of the searching through the search space. Other than the Brute-Force technique, all other pattern-searching algorithms have some preprocessing time over the pattern to reduce the searching time. The most common searching technique employed by text editors or some other searches is the Boyre-Moore string-searching algorithm.

The major challenge of today's world is the type of search i.e. the search can be for a sentence, a paragraph, a pattern etc and it was observed that the time required of all the three searches using a common search algorithm is not the best in all cases. For e.g. KMP and Boyre-Moore algorithms are efficient for searching a pattern while in case of searching for paragraph or sentence their efficiency may be increased or some other techniques can be used with efficient search times. Considering this, we use different search algorithms for different types of search.

2 Related Work

Amongst the several text-searching algorithms designed until now, the simplest one is the Naive or Brute-Force Algorithm. Rabin-Karp is another searching technique that makes use of elementary number-theoretic notations such as equivalence of two

V.V. Das, J. Stephen, and Y. Chaba (Eds.): CNC 2011, CCIS 142, pp. 557–561, 2011.

numbers modulo a third number. Other algorithm is the Knuth-Morris—Pratt algorithm that is a linear time string-matching algorithm [2]. This algorithm uses a prefix function π that encapsulates knowledge about how the pattern matches against shifts or itself. Now the most commonly used text-searching algorithm is the Boyre-Moore Algorithm that takes a *sub-linear* searching time [1]. It uses two functions i.e. a bad character and a good prefix functions that require certain preprocessing. Let m be the length of the sentence and let n be the length of the search space (file). The Table-1 gives the comparison of the asymptotic time analysis of various text searching algorithms.

Table 1. Comparative asymptotic time analysis

Algorithm	Preprocessing time	Matching time
Naïve String Search Algorithm	0(no preprocessing)	Θ((n-m+1) m)
Rabin-Karp string search algorithm	Θ(m)	average Θ(n+m), worst Θ((n-m+1) m)
Finite state automaton based	Θ(m |Σ|)	Θ(n)
KMP algorithm	Θ(m)	Θ(n)
Boyer–Moore search Algorithm	Θ(m + |Σ|)	Ω(n/m), O(n)
Our Algorithm	*0(no preprocessing)*	*Θ(n)*

3 Semantic Differences between Pattern and Sentences/Paragraph

Searching for a pattern or word is quite different from searching a sentence or a paragraph in a text or a file. This difference is due to the grammatical construct of the English language. We can explain it better with an example.

For searching pattern/word:
Pattern: abbabb

In case of patterns, there may be symmetry or there may be a case where in text a pattern may contain a pattern. Consider a case if text is ***abbabbabb,*** in this case the text contains a pattern twice where one is the part of other pattern.

While in case of sentences, there is no such condition where a pattern might have symmetry or a case where one sentence may be a part of similar sentence.

Second drawback of employing pattern search for searching sentences or paragraph is the length of the sentence/paragraph. Both the searching techniques i.e. KMP and Boyre-Moore algorithm have $\Theta(m)$ and $\Theta(m + |\Sigma|)$ preprocessing times respectively. In case the length of sentence increases, the preprocessing time also increases while all the preprocessing would go in vain due to the grammar construct.

The algorithm presented as a solution does not have any preprocessing time while the running time is *$\Theta(n)$*. Under all cases i.e. the best, average and worst case, the running time remains invariant. Whether the sentence is present in the file or not, we need to traverse the text till the end, character by character.

4 Algorithm

We now specify the algorithm. The search for a sentence in a pool of files includes reading of the given set of files. Initially we had implemented this algorithm using C-Compiler.

Not considering the way of reading files at present, we may specify our algorithm as follows:

```
1.  While(!EndOfFile)
2.   Do read a single character from file, x
3.   pos←pos+1
4.   If sentence[i]=x then
5.   i←i+1
6.   Else
7.     i←0
8.     If sentence[i]=x then
9.      i←i+1
10.  If i = LengthOfSentence then
11     c←c+1
12     i←0
13 Return c
```

The above algorithm returns *'c'*, i.e. the number of times the sentence to be searched occurs in a single file. It can scan more than one file; one by one and thus help in distinguishing between the set of files that contain a sentence or paragraph and the ones that do not contain it.

The above algorithm works by scanning the file character by character and comparing each character of the file with the ones in the sentence to we wish to search. We may see the algorithm in two phases as described in the following lines.

(a) Initially, we compare the first character of the file is with the first character of the sentence to be searched. If there is a match, we increment ***i*** else we set the pointer again to 0 and check for the first character of the sentence.

(b) We now check the value of ***'i'*** if it is equal to length of the sentence or not. Value of ***'i'*** will be equal to the length of sentence only in a condition if the sentence is found thus we increment the value of ***'c'***

5 Theoretical Analysis

Considering the illustrated algorithm, we can see that the complete complexity of searching a sentence in a file is equal to $\boldsymbol{\Theta(n)}$ without having any pre-processing time where n is the number of characters in the file.

Lines 1-12 show that this particular loop continues until the end of file i.e. iterates 'n' (no. of characters in file) times. *Line 2* reads a single character at a time thus having O(1) complexity. Similarly, *line 3* also executes once in a loop. *Lines 4-9* check whether the character read from file is present in the sentence or not and accordingly, the respective lines execute. In case we find a mismatch, we check it for the first character of the sentence we are searching. *Lines 10-12* check if the sentence is found in the file and accordingly increment the counter of the number of sentences by 1. Finally, line 13 returns the number of times the sentence is present in the file.

This clearly shows that there is a single loop iterating **'n'** times. Thus, the complexity of the algorithm is **Θ(n),** under all circumstances as the loop continues till the last character of the file whether or not the sentence is present in the file.

6 Implementation

To demonstrate our algorithm we implemented it using C-Compiler and tested the time of execution using a text file (e-book of Harry Potter). We used a text file for testing instead of .pdf or .doc as it is easier to read a text file in C with respect to other formats. We were even able to search a sentence in multiple files simultaneously and display the result accordingly (Fig 1). The same algorithm has been implemented using C#.NET to render an interactive interface for the user.

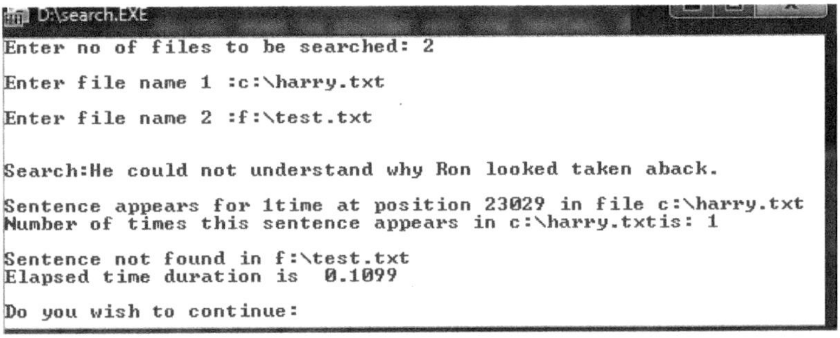

Fig. 1. Result of the search

7 Conclusion and Future Scope

At a tremendous rate of increase in digital files, there is a need of application that can help in searching those files for required data that too at an efficient rate. We are thus providing such application with an efficient sentence-searching algorithm for the above purpose. The theoretical complexity of our algorithm is **Θ(n)** with **no preprocessing**

Some of the extra features that can be included in our application are

Normal Searching:
Normal Searching allows the use of the question mark (?) and asterisk (*) to match one and one or more characters respectively. All white space is treated the same and multiple white space characters are treated as one.

Search a drive, path or multiple drives and paths:
Such as *C:\ / \\Corp-backup\C\Accounting*

References

1. Boyer, R.S., Moore, J.S.: A Fast String Searching Algorithm. Carom. ACM 20(10), 262–272 (1977)
2. Knuth, D.E., Morris, J.H., Pratt, V.R.: Fast pattern matching strings. TR CS-74-440, Stanford University, Stanford California (1974)
3. Karp, R.M., Rabin, M.O.: Efficient randomized pattern-matching algorithms (March 1987)
4. Sunday, D.M.: A very fast substring search algorithm. Commun. ACM 33(8), 132–142 (1990), http://doi.acm.org/10.1145/79173.79184, doi:10.1145/79173.79184
5. Cormen, T.H., Leiserson, C.E., Rivest, R.L., Stein, C.: String Matching. In: Introduction to Algorithms, 2nd edn., ch. 32, pp. 906–932. MIT Press/McGraw-Hill (2001), ISBN 0-262-03293-7
6. Sedgewick, R.: String Searching. In: Algorithms, ch. 19, pp. 241–256. Addison-Wesley Publication Company, Reading, ISBN O-201 -06672-6
7. Cole, R.: Tight Bounds on the complexity of the Boyer-Moore Algorithm. In: Proceedings of the 2nd Annual ACM-SIAM Symposium on Discrete Algorithms, http://portal.acm.org/citation.cfm?id=127830

Early Congestion Detection and Self Cure Routing in Manet

T. Senthil kumaran and V. Sankaranarayanan

BSA Crescent Engineering College, Vandalur, Chennai, Tamilnadu, India
senthilkumaran@bsauniv.ac.in,
sankarammu@bsauniv.ac.in

Abstract. Ad hoc mobile networks are composed of mobile nodes communicating through wireless medium, without any fixed backbone infrastructure. In these networks, congestion occurs in any intermediate node when data packets travel from source to destination incurring high packet loss and long delay, which cause the performance degradation of a network. This paper presents an early detection of congestion and self cure routing protocol for wireless Ad-hoc networks called as EDCSCAODV. All the primary path nodes periodically calculate its queue_status at node level. While using early congestion detection technique, node detects congestion that is likely to happen and send warning message to its neighbors. All neighbors listen to the node's warning message and apply automatic cure algorithm to get alternate path to it, if and when a non-congested local alternate path is available for transmitting data. Thus, EDCSCAODV improves performance in terms of reducing delay, routing overhead and increases packet delivery ratio without incurring any significant additional cost. The performance of EDCSCAODV was compared with AODV using the Ns-2 simulator. The result reveals that EDCSCAODV has significant improvement over conventional AODV routing schemes.

Keywords: Ad hoc Networks, Congestion, AODV, EDCSCAODV.

1 Introduction

Ad-hoc wireless networks are usually defined as an autonomous system of nodes connected by wireless links communicating in a multi-hop fashion [11], [12]. One of the fundamental tasks that an ad hoc network should perform is congestion control and its main objective is to limit the delay and buffer overflow caused by network congestion and provide better performance of the network. In wireline networks, congestion control is implemented at the transport layer and is often designed separately from functions of other layers [1], [2]. However, these results do not apply directly to wireless networks because, ad hoc networks result in large amount of packet loss, high delay, unfair scenarios and low throughputs [1],[3]. A new perspective on this problem might be to realize congestion control in the MAC or network layer [1], [3]. The Ad Hoc On-Demand Distance Vector (AODV) routing protocol is an adaptation of the DSDV protocol for dynamic link conditions [8], [9], [10]. Since AODV has no congestion control mechanisms, it may also lead to the following

V.V. Das, J. Stephen, and Y. Chaba (Eds.): CNC 2011, CCIS 142, pp. 562–567, 2011.

problems (i) Long delay (ii) Many packet losses (iii) Low throughput [1]. Our motivation is to clear that congestion which is a dominant cause for packet loss in MANETs. Typically, reducing packet loss involves congestion control running on top of a mobility and failure adaptive routing protocol at the network layer [1]. We use early detection techniques to detect the congestion well in advance [Section 2.1]. We present a method to find an alternate path to control the congestion [Section 2.2]. Ns2 simulation is used to analyze the results and prove the performance [Section 3.0].

2 Early Detection of Congestion and Self Cure Routing Protocol

2.1 Early Congestion Detection Techniques

Early detection of congestion and self cure routing protocol (EDCSCAODV) is a unicast routing protocol for MANET. The source discovers the route to the destination, it broadcasts an RREQ packet toward the destination, the destination responds to the first arrived RREQ and sends back an RREP packet. The RREP will travel back in the path the RREQ previously travelled and adds a new entry in its routing table. Congestion in a network may occur at any interval, if the load on the network is greater than the capacity of the network, the node becomes congested and starts losing packets. Congestion metric can be used at a node to detect the congestion well in advance otherwise it may be harmful when the buffer is almost full. We used base design of Random Early Detection queue model for constructing buffer [6], [7]. The expression (1) and (2) are used to set the Minimum threshold and Maximum threshold values

$$\text{Minth} = 25\%\text{buffer\_size} \quad (1)$$

$$\text{Maxth} = 3 * \text{Minth} \quad (2)$$

To detect the congestion well in advance, compute the average queue size as

$$\text{Avgque} = (1 - w_q) * \text{Avgque} + \text{Inst\_Que} * w_q \quad (3)$$

Where wq, the queue weight is a constant parameter (wq=0.002 from RED queue experimental result [11]) and Inst_Que is an instantaneous queue size.

In our early detection model, we introduce Queue_status of actual queue size over average queue size given by Equation (4), which reflects the heaviness of the incoming traffic. Based on the Queue_status, the mobile node can get useful information about the incoming traffic. If the Queue_status value is large, the incoming traffic becomes bursty. The continuous growth of the Queue_status indicates that the incoming heavy traffic is beyond the mobile node's buffer capacity and buffer overflow is imminent.

$$\text{Queue\_status} = \text{Inst\_que} - \text{Avgque} \quad (4)$$

If Queue_status < minimum threshold, the incoming traffic is low and queue is in safe zone. If Queue_status > minimum threshold and Inst_Que < maximum threshold, the incoming is above average and queue is in likely to be in congested zone. If Inst_Que > maximum threshold, the incoming traffic is heavy and queue is in congested zone.

2.2 Self Cure Routing

A primary path of a node predicts its congestion status and periodically broadcasts a congestion status packet (CSP) with TTL = 1. The CSP packet P contains the node's congestion status and a set of parameters (Source S, Destination D, Hop Count hop, Sequence Number Seq, Congestion Status Cong and Neighbors information N_list), each for a destination appearing in the routing table. S-D (P) is used to represent the source-destination pair of the flow that the packet belongs to a routing table maintained at each node. When the neighbors receive a CSP packet from its primary path node X regarding destination Dst, Neighbors will be aware of the congestion status of X. When neighbor overhears (CSP) a packet P, it checks its source and destination information in their routing table. If P belongs to a flow that has not been recorded by this node, a new entry is added in its table. Else let E be the existing routing table entry representing the flow, we use Src(E),Dst(E), Seq_no(E), and Neighbors_list(E) are used to represent four fields such as the source, destination, sequence number and Neighbors list. The Neighbors_list(E) only collects packets of the flow carrying the same sequence number as Seq_no(E), if Seq(P) is less than Seq_no(E) then the packet P is ignored. If Seq(P) is greater than Seq_no(E) then Seq_no(E) is updated to Seq(P) and reset Neighbors_list(E). Now the neighboring nodes are used to identify the a 2-Step alternate sub-path [4], [5].the Neighbors_list(E) now had three entries (A, B, C) meeting conditions The current node's congestion status <= the minimum level of the congestion status of A, B and C. If condition is met, then activates the sub-path and deletes entry E from the Neighbors table, then it has found a better sub-path. The neighboring node is used to identify Upstream Node of a 3-Step alternate sub-path [4],[5].If Neighbors_list (E) now had two entries A and B, meeting condition The current node's congestion status <= the minimum level of the congestion status of A and at least two times less than the congestion status of B, If condition is met, then it adds node B to the Waiting Indicator list, and then it may be able to seek a better sub-path with a downstream node. Now the neighbor node is used to Identify Downstream Node of a 3-Step alternate sub-path [4], [5]. If Neighbors_list (e) now had two entries B and C, meeting special condition the current node's congestion status is at least two times less than the congestion status of B and less than the congestion status of C, then it may be able to seek a better sub-path with an upstream node. If condition is met, then the current node broadcasts a special hello message with the information (node B is a candidate for a 3-Step Alternate Sub-path). When cooperating node receives the special hello message, it checks special hello message fields with an entry in the current node's Waiting Indicator list, both do not math then ignore it. Otherwise, special hello message does match a Waiting Indicator entry, so a better sub-path has been found. Activate the new sub-path.

3 Performance Study

A comparison of EDCSCAODV's performance to AODV routing protocols in MANET is made using the Network Simulator (Ns2.33) [13].

3.1 Simulation Configuration and Performance Metrics

The network consisted of 100 nodes in a 1,400m *1400 m rectangular field. The MAC layer was based on IEEE 802.11 DCF. The channel propagation model we used combines 2-ray ground reflection models. A routing buffer at the network layer could store up to 50 data packets. Each simulation ran 600 seconds.

3.2 Varying Number of Connections

In this simulation, the number of connection (source and destination) was varied from 10 to 50, CBR sending rate 4 packets per second, maximum node speed 10 *m*/*s* and pause time 30*s*. Fig.1 (a) and Fig.1 (b) had shown the packet delivery ratio, the End-to-End delay for EDCSCAODV and AODV respectively. When the numbers of flows were less than 20, EDCSCAODV did not show much improvement over AODV. When the number of flows increases from 20 to 50, the packet delivery ratio of AODV had a sudden fall from 87% to 14 %. When compared with the above packet delivery ratio, the figures of the EDCSCAODV fall only a gradual fall from 91% to 59% as shown in Fig.1 (a). One observed that the End-to-End delay between AODV and EDCSCAODV as shown in Fig.1 (b), when the number of flows increased from 20 to 50. The End-to-End delay had increased from 0.7 seconds to 7.9 seconds. The corresponding variation for EDCSCAODV was from 0.5 to 4.7.

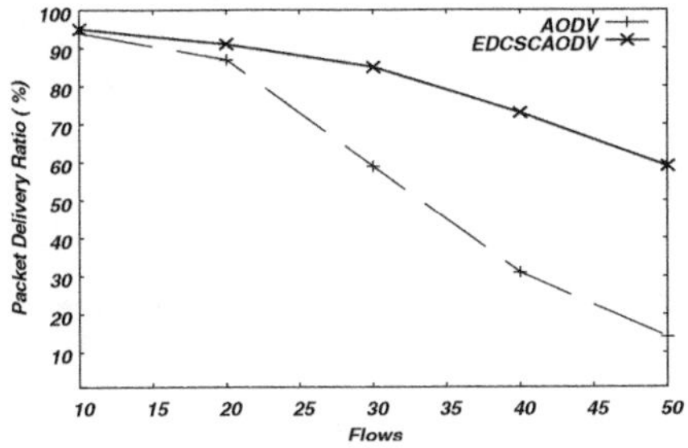

(a)

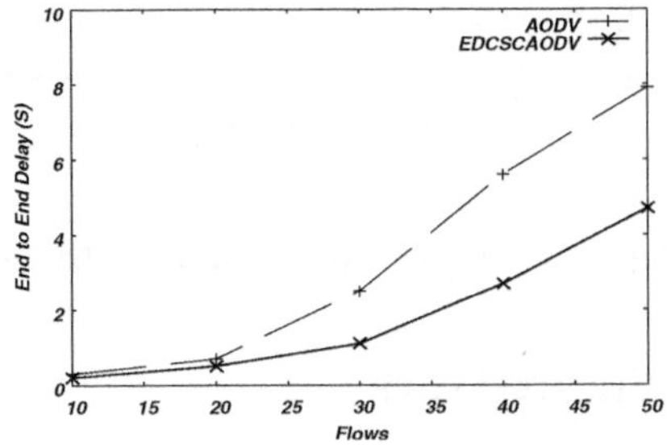

(b)

Fig. 1. Performance when no of Connection change (source and destination)

3.3 Varying CBR Load

In this simulation, the number of connection (different source and different destination) was kept at 20. The CBR sources sent data packets to the destinations at different rates, varying from 4 packets/s to 16 packets/s. Fig.2 (a) and Fig.2 (b) had shown the packet delivery ratio, the End-to-End delay between EDCSCAODV and AODV respectively. When the data packet-sending rate was greater than 4 packets per seconds, the packet delivery ratio of AODV had a sudden fall from 55% to 10 %. When compared with the above packet delivery ratio, the figures of the EDCSCAODV was fallen only a gradual fall from 85% to 34% as shown in Fig.2 (a). The End-to-End delay between AODV and EDCSCAODV was shown in Fig.2 (b). When the data packet-sending rate was greater than 4 packets per seconds, the End to End delay was increased from 1.58 seconds to 5.2 seconds the corresponding variation for EDCSCAODV was from 0.9 to 3.5.

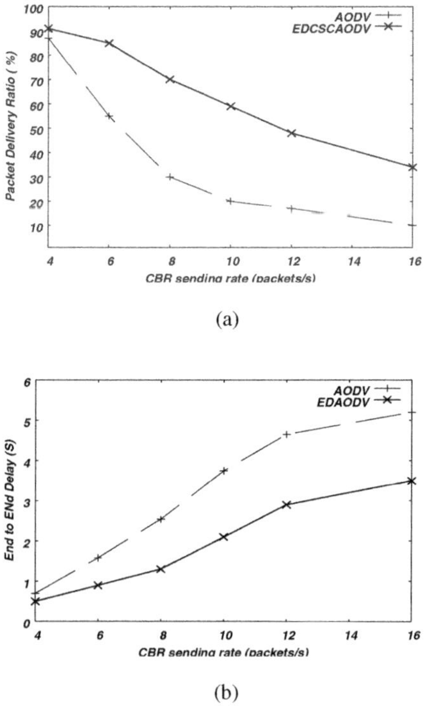

Fig. 2. Performance when CBR load changes

4 Conclusion and Future Work

Thus the proposed EDCSCAODV is early congestion detection and self cure routing protocol in MANETs. EDCSCAODV has lost fewer packets than general AODV. This is because EDCSCAODV tries to detect congestion in advance from occurring in the first place, rather than dealing with it reactively. A key in EDCSCAODV design is to cure congestion using local alternate path concept. This concept tries to find out non-congested alternate path automatically and tries to avoid congestion in the network. The ns-2-based simulation was confirmed that the advantages of EDCSCAODV and demonstrated a significant improvement End-to-End delay and packet delivery ratio over AODV. As future work we will be extended to other networks like WMN and WSN and study its performance.

References

1. Tran, D.A., Raghavendra, H.: Congestion Adaptive Routing in Mobile Ad Hoc Networks. IEEE Transactions on Parallel and Distributed Systems 17(11), 16–28 (2006)
2. Yu, Y., Giannakis, G.B.: Cross-Layer Congestion and Contention Control for Wireless Ad Hoc Networks. IEEE Transactions on Wireless Communications 7(1), 37–42 (2008)
3. Lochert, C., Scheuermann, B., Mauve, M.: A Survey on Congestion Control for Mobile Ad-Hoc Networks. Wiley Wireless Communications and Mobile Computing 7(5), 655–676 (2007)
4. Gui, C., Mohapatra, P.: A framework for self-healing and optimizing routing techniques for mobile ad hoc networks. ACM Transaction on Wireless Networks 14(1), 29–46 (2008)
5. Yen, Y.-S., Chang, H.-C., Chang, R.-S., Chao, H.-C.: Routing with adaptive path and limited flooding for mobile ad hoc networks. Elsevier Transaction on Computers and Electrical Engineering 36(2), 280–290 (2010)
6. Floyd, S., Jacobson, V.: Random early detection gateways for congestion avoidance. IEEE/ACM Transaction on Networking 1(4), 397–413 (1993)
7. Feng, G., Agarwal, A.K., Jayaraman, A., Siew, C.K.: Modified RED gateways under bursty traffic. IEEE Communications Letters 8(5), 323–325 (2004)
8. Perkins, C.E., Belding-Royer, E.M., Chakeres, I.: Ad Hoc on Demand Distance Vector (AODV) routing. IETF Internet draft (2003)
9. Broch, J., Johnson, D., Maltz, D.: The Dynamic Source Routing Protocol for Mobile Ad Hoc Networks. IETF Internet draft (1999)
10. Perkins, C.E.: Highly Dynamic Destination-Sequenced Distance-Vector Routing (DSDV) for Mobile Computers. In: Proc. ACM SIGCOMM, pp. 234–344 (1994)
11. Siva Ram Murthy, C., Manoj, B.S.: Ad hoc Wireless Networks–Architectures and Protocols. Pearson Education, London (2007)
12. Johnson, Maltz, D.: Ad Hoc Networking. Addison-Wesley, Reading (2001)
13. NS2 Network Simulator, `http://www.isi.edu/nsnam/ns/` (last accessed on February 3, 2009)

WCDMA-Third Generation Radio Interface

Mridula S. Korde[1] and Anagha P. Rathkanthiwar[2]

[1] Department of Electronics Engineering,
Shri Ramdeobaba Kamla Nehru Engineering College,
Nagpur, India
mridula_korde@yahoo.com
[2] Department of Electronics Engineering,
Priyadarshini College of Engineering and Architecture,
Nagpur, India
anagharathkanthiwar@yahoo.com

Abstract. Cellular networks and mobile phone use continues to grow at a rapid pace around the world. To satisfy ever-increasing demands for higher data rates, as well as to allow more users to simultaneously access the network, an interface known as wideband code division multiple access (WCDMA) is used. In this paper, we discuss the basic characteristics of WCDMA that make them attractive for high data rate transmission over wireless and mobile channels. This paper also presents the wide-band code-division multiple-access (WCDMA) radio interface chosen by ETSI as the basic radio-access technology for the universal mobile telecommunications system (UMTS). A detailed description of the physical layer of ETSI WCDMA is given together with an overview of the higher layers of the WCDMA radio interface.

Keywords: IMT-2000, radio access, UMTS, WCDMA, cdma2000, 3G PP.

1 Introduction

The systems that go under the ITU name of IMT-2000 and within ETSI as the universal mobile telecommunications system (UMTS) extend the services provided by current second-generation systems (GSM, PDC, IS-136, and IS-95) with high-rate data capabilities. The main application for these high-rate data services is wireless packet transfer, e.g., for wireless access to the Internet. However, UMTS will also support high-rate circuit-switched services such as video. Wide-band code-division multiple-access (WCDMA) has been chosen as the basic radio access technology for UMTS/IMT-2000. Compared to second-generation narrow-band CDMA, the WCDMA radio interface offers significant improvements, in addition to the support of higher rate services. These include: improved coverage and capacity due to a higher bandwidth and coherent uplink detection; support of interfrequency handover necessary for high-capacity hierarchical cell structures (HCS's); support for capacity-improving technologies such as adaptive antennas and multiuser detection; and a fast and efficient packet-access protocol.

V.V. Das, J. Stephen, and Y. Chaba (Eds.): CNC 2011, CCIS 142, pp. 568–573, 2011.

2 Parameters of WCDMA

WCDMA has two modes characterized by the duplex method: FDD (frequency division duplex) and TDD (time division duplex), for operating with paired and unpaired bands, respectively [1]. The chip rate of the system is 3.84 Mcps. The frame length is 10 ms and each frame is divided into 15 slots (2560 chip/slot at the chip rate 3.84 Mcps). Spreading factors range from 256 to 4 in the uplink and from 512 to 4 in the downlink. Thus, the respective modulation symbol rates vary from 960 k symbols/s to 15 k symbols/s (7.5 k symbols/s) for FDD uplink. For separating channels from the same source, orthogonal variable spreading factor (OVSF) channelization codes are used. In the downlink, Gold codes with a 10-ms period (38400 chips at 3.84 Mcps) are used to separate different cells, with the actual code itself length 218-1 chips. In the uplink, Gold codes with a 10- ms period, or alternatively short codes with a 256-chip period, are used to separate the different users. For the channel coding three options are supported: convolutional coding, turbo coding, or no channel coding. Channel coding selection is indicated by upper layers. The modulation scheme is QPSK. The carrier spacing has a raster of 200 kHz and can vary from 4.2 to 5.4 MHz. [2] [3].

3 Protocol Architecture

The protocol architecture is similar to the current ITU-R protocol architecture, ITU-R M.1035. The air interface is layered into three protocol layers:

- The physical layer (layer 1, L1);
- The data link layer (layer 2, L2);
- Network layer (layer 3, L3).

The physical layer interfaces the medium access control (MAC) sublayer of layer 2 and the radio resource control (RRC) layer of layer 3. The physical layer offers different transport channels to MAC. A transport channel is characterized by how the information is transferred over the radio interface. Transport channels are channel coded and then mapped to the physical channels specified in the physical layer. MAC offers different logical channels to the radio link control (RLC) sublayer of layer 2.

3.1 Logical Channels

The MAC layer provides data transfer services on logical channels. A set of logical channel types is defined for different kinds of data transfer services as offered by MAC. Each logical channel type is defined by the type of information that is transferred. Logical channels are classified into two groups: (Table 1)

Control channels for the transfer of control plane information (Table 2) Traffic channels for the transfer of user plane information (Table 3).

Table 1. Logical Channel Structure

Channel Name	Subchannels
Control channel (CCH)	1.Broadcast control channel (BCCH) 2.Paging control channel (PCCH) 3.Dedicated control channel (DCCH) 4.Common control channel (CCCH) 5.Shared channel control channel (SHCCH) 6.ODMA dedicated control channel (ODCCH) 7.ODMA common control channel (OCCCH)
Traffic channel (TCH)	1.Dedicated traffic channel (DTCH) 2.ODMA dedicated traffic channel (ODTCH) 3.Common traffic channel (CTCH)

Table 2. Logical Control Channels

Channel Name	Function
Broadcast channel (BCH)	Downlink transport channel that is used to broadcast system- and cell-specific information. The BCH is always transmitted over the entire cell with a low fixed bit rate.
Paging control channel (PCCH)	Downlink channel that transfers paging information and is used when: Network does not know the location cell of the mobile station; The mobile station is in the cell connected state (utilizing sleep mode procedures).
Common control channel (CCCH)	Bidirectional channel that transfers control information between network and mobile stations. This channel is used: By the mobile stations having no RRC connection with the network; By the mobile stations using common transport channels when accessing a new cell after cell reselection.
Dedicated control channel (DCCH)	Point-to-point bidirectional channel that transmits dedicated control information between a mobile station and the network. This channel is established through RRC connection setup procedure.
ODMA common control channel(OCCCH)	Bidirectional channel for transmitting control information between mobile stations.
ODMA dedicated control channel (ODCCH)	Point-to-point bidirectional channel that transmits dedicated control information between mobile stations. This channel is established through RRC connection setup procedure.

Table 3. Traffic Channels

Channel Name	Function
Dedicated traffic channel (DTCH)	Point-to-point channel, dedicated to one mobile station, for the transfer of user information. A DTCH can exist in both uplink and downlink.
ODMA dedicated traffic channel (ODTCH)	Point-to-point channel, dedicated to one mobile station, for the transfer of user information between mobile stations. An ODTCH exists in relay link. A point-to-multipoint unidirectional channel for transfer of dedicated user information for all or a group of specified mobile stations

3.2 Transport Channels

A transport channel is defined by how and with what characteristics data is transferred over the air interface. There exist two types of transport channels:

- Dedicated channels;

There is one dedicated transport channel, the dedicated channel (DCH), which is a downlink or uplink transport channel.

3.3 Physical Channels

The transport channels are channel coded and matched to the data rate offered by physical channels. Thereafter, the transport channels are mapped on the physical channels. Physical channels consist of radio frames and time slots. The length of a radio frame is 10 ms and one frame consists of 15 time slots. A time slot is a unit, which consists of fields containing bits. The number of bits per time slot depends on the physical channel. Depending on the symbol rate of the physical channel, the configuration of radio frames or time slots varies. The basic physical resource is the code/frequency plane. In addition, on the uplink, different information streams may be transmitted on the I and Q branch. Consequently, a physical channel corresponds to a specific carrier frequency, code, and, on the uplink, relative phase (0 or p/2).

- Uplink Physical Channels

There are two uplink dedicated physical and two common physical channels: The uplink dedicated physical data channel (uplink DPDCH) and the uplink dedicated physical control channel (uplink DPCCH). The physical random access channel (PRACH) and physical common packet channel (PCPCH). The uplink DPDCH is used to carry dedicated data generated at layer 2 and above (i.e., the dedicated transport channel (DCH)). The uplink DPCCH is used to carry control information generated at layer 1. Control information consists of known pilot bits to support channel estimation for coherent detection, transmit power-control (TPC) commands, feedback information (FBI), and an optional transport-format combination indicator (TFCI). [4]

Figure 1 show the principle frame structure of the uplink dedicated physical channels. Each frame of length 10 ms is split into 15 slots, each of length *T*slot = 2560 chips, corresponding to one power-control period. The parameter *k* in Figure 1 determines the number of bits per uplink DPDCH/DPCCH slot. It is related to the spreading factor (SF) of the physical channel as SF = 256/2*k*. The DPDCH spreading factor may thus range from 256 down to 4.

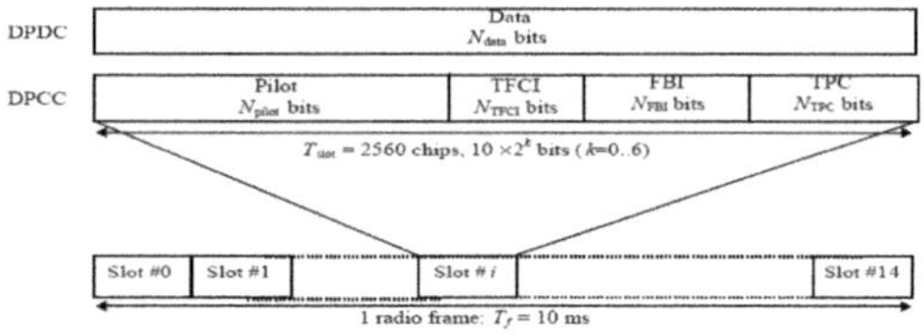

Fig. 1. Frame structure for uplink DPDCH/DPCCH

- Downlink Physical Channels

There is one downlink dedicated physical channel, one shared and five common control channels:

- Downlink dedicated physical channel (DPCH);
- Physical downlink shared channel (DSCH);
- Primary and secondary common pilot channels (CPICH);
- Primary and secondary common control physical channels (CCPCH);
- Synchronization channel (SCH).

Figure 2 shows the frame structure of the DPCH. On the DPCH, the dedicated transport channel is transmitted time multiplexed with control information generated at layer 1 (known pilot bits, power-control commands, and an optional transport-format combination indicator). DPCH can contain several simultaneous services when TFCI is transmitted or a fixed rate service when TFCI is not transmitted. The network determines if a TFCI should be transmitted. The PCPCH is used to carry the CPCH transport channel. The CPCH transmission is based on DSMA-CD approach with fast acquisition indication. The mobile station can start transmission at a number of well-defined time-offsets, relative to the frame boundary of the received BCH of the current cell. The CPCH random-access transmission consists of one or several access preambles of length 4096 chips, one collision detection preamble (CD-P) of length 4096 chips, a DPCCH power control preamble (PC-P) (which is either 0 slots or 8 slots in length), and a message of variable length *N*x10 ms.

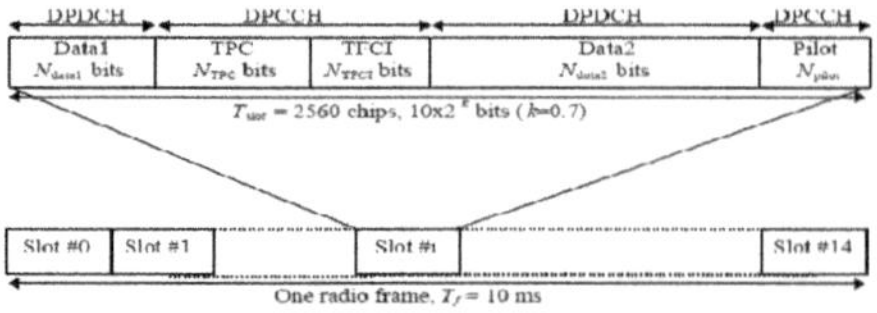

Fig. 2. Frame structure for downlink DPCH

Common pilot channel (CPICH) is a fixed-rate (30 Kbps, SF=256) downlink physical channel that carries a predefined bit/symbol sequence. There are two types of common pilot channels, the primary and secondary CPICH. The primary CCPCH is a fixed-rate (30 Kbps, SF=256) downlink physical channels used to carry the BCH. Common control physical channels are not inner-loop power controlled.

The page indicator channel (PICH) is a fixed-rate (SF=256) physical channel used to carry the page indicators.

4 Summary and Proposed Simulation

In this paper, a discussion of some of the essential features of WCDMA has been presented. The WCDMA radio interface chosen by ETSI as the basic radio-access technology for UMTS extends the services of second-generation systems with wide-area coverage of high-rate data transmission and efficient packet access. Especially, compared to second-generation narrowband CDMA, the WCDMA technology provides improved capacity and coverage due to wider bandwidth and coherent uplink. In this paper, a detailed description of the physical layer of ETSI WCDMA has been given together with an overview of the structure of the higher layers of the WCDMA radio interface (layers 2 and 3). Proposed simulation for the WCDMA system will be based on the performance analysis of various channels and synchronization aspects between base station and mobile station.

References

1. Milstein, L.B.: Wideband Code Division Multiple Access. Invited Paper, IEEE Journal on Selected Areas in Communications 18(8), 1344–1354 (2000)
2. Holma, H., Toskala, A.: WCDMA for UMTS: Radio Access for Third Generation Mobile Communications. John Wiley, Chichester (2000)
3. Gudmundson, M.: WCDMA - The Third Generation Radio Access. In: 29th European Microwave Conference, Munich, pp. 7–11 (1999)
4. Dahlman, E., Beming, P., Knutsson, J., Ovesjö, F., Pearson, M., Robol, C.: WCDMA—The Radio Interface for Future Mobile Multimedia Communications. IEEE Transactions on Vehicular Technology 47(4), 1105–1118 (1998)
5. Adachi, F., Sawahashi, M., Suda, H.: Wideband DS-CDMA for Next-Generation Mobile le Communications Systems. IEEE Communications Magazine, 56–69 (September 1998)

Image Registration Using Mexican-Hat Wavelets and Invariant Moments

Jignesh Sarvaiya[1], Suprava Patnaik[1], and Hemant Goklani[2]

[1] ECED, SVNIT-Surat-7, India
jns@eced.svnit.ac.in, ssp@eced.svnit.ac.in
[2] ECED, SVMIT-Bharuch, India
hsgoklani@gmail.com

Abstract. Image registration is the process of determining the transformation which best matches, according to some similarity measure, two images or more of the same scene taken at different times or from different viewpoints so that corresponding pixels in the two images correspond to the same physical region of the scene being imaged. It is an essential step for a great variety of applications such as computer vision, stereo navigation, medical image analysis, pattern recognition etc. In this paper, an algorithm is proposed which consists of Feature point extraction, based on scale-interaction of Mexican-hat wavelets, Feature correspondence between the extracted features points using invariant Zernike moments of neighborhoods centered on feature points, and transformation parameter estimation and mapping.

Keywords: Image Registration, Mexican-hat wavelet, Zernike Moment.

1 Introduction

Given two or more different images to be registered, image registration estimates the parameters of the geometrical transformation model that maps the sensed images back to its reference image [1]. In image registration, robustness of the algorithm is the main and required goal. But due to diversification of images acquired their contents and purpose of their alignment, it is almost impossible to design universal method for image registration that fulfill all requirements and suits all types of applications [2].

Many of the image registration techniques have been reviewed [1] [2] [3]. Image registration techniques broadly classified as Intensity and feature based [14].

The proposed algorithm utilizes the new approach, which exploits a Mexican Hat Wavelet (MHW), to extract significant image features from reference and sensed image. Then correspondence is established using Zernike moment-based similarity measure [5] [9]. Experimental results show that the proposed algorithm leads to acceptable accuracy and robustness against several image deformations and image processing operations. Wavelet analysis allows the use of long time intervals where more precise low-frequency information, and shorter regions where high-frequency information. So, the resolution is the main issue for using Wavelet Transform [12] [13].

V.V. Das, J. Stephen, and Y. Chaba (Eds.): CNC 2011, CCIS 142, pp. 574–577, 2011.

2 Feature Point Extraction Using Scale Interaction of MHW

Extracted feature points must have to fulfill some basic objectives like invariant to various geometrical deformations, robust enough to some level of variations in scale, etc. The proposed method satisfies all the above objectives, which is based on finding the local maxima of the response of a feature detection operation, which involves convolving the image with the Mexican-hat wavelets. Feature point extraction using scale interaction of Gabor wavelets was proposed in [6]. The Mexican hat is a Laplacian of a Gaussian and its isotropic property makes it insensitive to orientation and a good choice for feature extraction [7]. A Mexican-hat wavelet has the shape of a signal with a positive peak in a negative dish. Mexican hat based feature extraction methods such as the ones in [8] and its further development in [9]. For feature point extraction convolve the image with the Mexican-hat wavelet. It is given by below equation [10]

$$MHW(S) = \frac{1}{\sigma}\left(2 - \frac{x_1^2 + x_2^2}{\sigma^2}\right) e^{-\frac{\left(\frac{x_1^2+x_2^2}{\sigma^2}\right)}{2}} \quad (1)$$

First of all reference image is taken as input, we call it im_ref(x,y). After taking its Fourier transform, its convolution is done with the Mexican hat wavelet. This represented in the equation form as

$$R(S) = \text{im\_ref}\,(x, y)_f \otimes MHW(S) \quad (2)$$

Next step is to get the scale interaction output or absolute difference between the obtained responses. It is given as

$$R(S_1, S_2) = |\,R(S_1) - R(S_2)\,| \quad (3)$$

The response R (S_1, S_2) can be obtained in the frequency domain using:

$$= |\text{IFFT}\,\{\,\text{im\_ref}(x, y)_f \times MHW(S_1)\,\} - \{\,\text{im\_ref}(x, y)_f \times MHW(S_2)\,\}| \quad (4)$$

Where im_ref(x, y)$_f$ denotes the Fourier transform of im_ref (x, y). FFT and IFFT represent the Fourier transform and its inverse respectively [4] [11]. The second stage of the feature extraction process localizes the feature points of the image by finding the local maxima of the response R (S_1, S_2). A local maximum is a point with maximum value that is greater than a specified threshold T_n in a disk shaped neighborhood of radius r_n.

3 A Proposed Registration Algorithm

The main objective of the proposed algorithm is accurately registering images which are geometrically distorted and may have undergone various degradations. The proposed algorithm consists of three main steps: (i) Feature point extraction: In this proposed algorithm scale interaction Mexican hat Wavelet is used for the feature point extraction from reference and sensed image. (ii) Obtaining the correspondence: To find the correspondence between the feature points of both images, invariant Zernike moments of neighborhoods centered on the feature points is used. (iii)Transformation parameter estimation: To estimate the transformation parameters between the two images using an iterative weighted least squares algorithm. (iv)

Resampling and Registration: After successful estimation of the transformation parameters, resample the sensed image and register it with the reference image.

4 Experimental Results

Here an 8-bit gray level 'living room' image is taken as a reference image. Some parameters were manually adjusted. Scales S_1 and S_2 taken 2 and 4, respectively, T_n= 55, l_n=32, r_n= 16. Feature points which give best correlation marked as control points (red circle) as shown in Fig. 1 (a) and (b). Then after apply transformation to the sensed image to register it on the reference image. The registered image is shown in Fig.1 (c) & 1 (d). Fig.2 shows the registered image with rotated by 330^0 and scaled up to 3. Fig.3 shows Gaussian noise as a distortion and register that sensed image with reference.

(a) (b) (c) (d)

Fig. 1. (a) Correspondence between feature points in the reference (b) Sensed image (Rotated by 140^0) (c) Registered image (d) Registered image overlaid on Reference image

(a) (b) (c) (d)

Fig. 2. (a) Reference image (b) Sensed image (rotated by 330^0 Scaled by 3) (c) Registered image (d) Registered image overlaid on Reference image

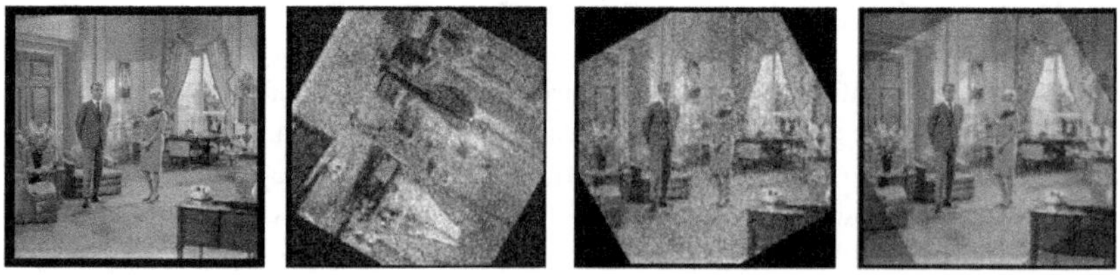

(a) (b) (c) (d)

Fig. 3. (a) Reference image (b) Sensed image (rotated by 240^0 with Gaussian noise, mean 0 and variance 0.02) (c) Registered image (d) Registered image overlaid on Reference image

5 Conclusion

Performance of the proposed algorithm was evaluated and found to be robust against rotation, image compression and expansion, noise contamination as well as other geometric transformations. The experimental results clearly indicate that the registration accuracy and robustness is very acceptable. This confirms the success of the proposed feature based algorithm for image registration.

References

1. Brown, L.G.: A survey of Image Registration Techniques. ACM Computing Surveys 24(4), 325–376 (1992)
2. Hong, G., Zhang, Y.: Combination of feature-based and area-based image registration technique for high resolution remote sensing image. In: IEEE International Conference (2007)
3. Zitova, B., Flusser, J.: Image Registration methods: A Survey. Image Vision Computing 21(11), 977–1000 (2003)
4. Collins, D.L., Holmes, C.J., Peters, T.M., Evans, A.C.: Automatic 3D model-based neuroanatomical segmentation. In: Human Brain Mapping, pp. 190–208 (1995)
5. Sarvaiya, J., Patnaik, S., Goklani, H.: Image Registration Using NSCT and Invariant Moments. International Journal of Image Processing, ISSN (1985-2304) 4(2), 119–130 (2010)
6. Manjunath, B., Shekhar, C., Chellappa, R.: A new approach to image feature detection with applications. Pattern Recognition 29(4), 627–640 (1996)
7. Bhattacharjee, S., Kutter, M.: Compression tolerant image authentication. In: Proceedings of the IEEE International Conference on Image Processing (ICIP), Chicago, USA, pp. 435–439 (1998)
8. Kutter, M., Bhattacharjee, S.K., Ebrahimi, T.: Toward second generation watermarking schemes. In: Proceedings of the IEEE International Conference on Image Processing (ICIP), Kobe, Japan, pp. 320–323 (1999)
9. Yasein, M.S., Agathoklis, P.: A robust, feature-based algorithm for aerial image registration. In: Proceedings of the IEEE International Symposium on Industrial Electronics (ISIE 2007), Vigo, Spain, pp. 1731–1736 (2007)
10. Yasein, M.S., Agathoklis, P.: Automatic and robust image registration using feature points extraction and zernike moments invariants. In: Proceedings of the Fifth IEEE International Symposium on Signal Processing and Information Technology, Athens, Greece, pp. 566–571 (2005)
11. Tang, C.-W., Hang, H.-M.: A Feature-Based Robust Digital Image Watermarking Scheme. IEEE Transactions on Signal Processing 51(4), 950–959 (2003)
12. Gonzalez, Woods: Digial Image Processing, 3rd edn. Prentice Hall Publicaion, Englewood Cliffs (2008)
13. Mallat, S.: A theory of multiresolution signal decomposition, The wavelet representation. IEEE Trans. Pattern Analysis Machine Intell. PAMI-11(7) (1989)
14. Reddy, B.S., Chatterji, B.N.: An FFT-Based technique for translation, rotation, and scale-invariant image registration. IEEE Trans. on Image Processing 5 (1996)

Centralized Management Approach for WLAN

Ashwini Dalvi[1,*], PamuKumar Swamy[2], and B.B. Meshram[3]

[1] Computer Department, VJTI, Matunga Mumbai-19
ashwinidalvi@gmail.com
[2] IT Consultant, NewTech Computer Services, Mumbai
Pamu_kumar@rediffmail.com
[3] Computer Department, VJTI, Matunga Mumbai-19

Abstract. Wireless Local Area Networks (WLAN) have proliferated within the enterprise. Implementation of WLAN started with standalone AP or independent AP. The independent AP is configured individually by administrator. But with growing scenario of enterprise management of individual AP become difficult. The central management of AP solves management problem of individual AP. There are various approach proposed for central management of WLAN. This paper discusses the central management of WLAN of leading enterprises such as Intel, Cisco, Meraki enterprises. The paper concludes with comparative analysis of these approaches of central management use in enterprises and proposed hybrid approach for WLAN management.

Keywords: 802.11, FAT AP, Thin AP, Centralized Management.

1 Introduction

The WLAN technologies offer benefits of flexibility, reduced installation time, and reduced cost. WLAN implementation step consists of prepare, plan, design, implement, operate and optimize [1]. WLANs typically consist of Access Points (APs) distributed across the enterprise. Initially, the implementation of WLAN is achieved with individual AP, also called as FAT AP. Each Access Point operates as an independent device. These APs often provide high speed wireless connectivity but access the Internet via independent, relatively low-speed DSL or cable modem links [2]. The independent AP architecture suffers from scalability and management problem. With thousands of access points, it simply isn't possible to configure and alone manage such a large wireless network. On the other hand, the scarcity of APs in conventional enterprise WLANs limits their performance in a variety of ways [3]. Managing scalability and managerial issues of independent AP is first step to centralized management of WLAN. This centralized approach is also called as Thin AP or dependent AP architecture. A thin or controller-based access point is usually part of a centrally managed enterprise WLAN network.

The paper is organized as follows section 2 discuses centralized management approaches of three leading enterprises such as Intel, Meraki Enterprise and Cisco. In section 3 we have done comparative analysis of discussed centralized approach.

[*] Corresponding author.

V.V. Das, J. Stephen, and Y. Chaba (Eds.): CNC 2011, CCIS 142, pp. 578–580, 2011.

2 Centralized Approaches for WLAN

Enterprise WLANs have made a dramatic shift towards centralized architectures in the recent past [4]. Easy management and better control are reasons behind this shift. The Thin AP architecture studied in literature to meet the challenges of its implementation [5, 6].

2.1 WLAN Network Management Server by Intel

Intel IT developed a new management and support structure for wireless LANs (WLANs) [7]. Intel IT team found traditional WLAN management approach is time consuming and labor intensive. Intel IT has implemented a distributed wireless infrastructure that uses multiple designs and suppliers. The idea behind this infrastructure is to focus management of WLAN as service rather than collection of disparate component. Intel's central WLAN management system uses a main console and management server clusters to monitor and control enterprise WLAN.

Central Manager handles administrative issues of regional network engineer, global service desk engineer, end users.

2.2 Meraki Enterprise Cloud Controller

In the Meraki architecture, there is only one hardware component: the access points. All control, configuration, optimization, and mobility control is centralized and delivered as service by the Meraki Cloud Controller (MCC) from Meraki's data centers [8]. By eliminating separate controllers, and moving intelligence into the cloud, hosted wireless LANs reduce deployment time and complexity while enabling multi-site, scalable wireless LANs. In this configuration administrator logs into the controller system through web browser and provide access to wireless network of his account. All management is done remotely through a Web browser. The bottleneck due to centralized approach is avoided in case of Meraki Cloud Controller. Meraki's cloud-based architecture provides significant advantages over legacy hardware-based solutions. This configuration provides opportunity for cost reduction for WLAN management.

2.3 Cisco WLAN Controller

The Cisco Wireless LAN Controller Modules allow small and medium-sized businesses (SMBs) and enterprise branch offices to cost-effectively deploy and manage secure wireless LANs. Cisco WLAN controller is responsible for system wide security, scalability, control and reliability that is required to build secure enterprise level wireless network. The Cisco Wireless LAN Solution is designed to provide 802.11 wireless networking solutions for enterprises and service providers [9]. Cisco controller has advantages of simple deployment and management.

3 Analysis of Centralized Approaches for WLAN

Recently the centralized model, commonly known as the thin AP or centralized AP architecture, has gained popularity. As discussed earlier most of the enterprises have their own approach for centralization of WLAN. The following table evaluates Intel, Meraki Enterprise and Cisco for their centralized WLAN management approach. Intel model uses network management server. This server manages administrator console to control and maintain different WLAN sites. With this cost factor for Intel centralized management of WLAN is manageable factor. Whereas Meraki enterprise consider centralized approach as software service rather than hardware implementation. This addresses scalability issue of access point effectively. The bottleneck due to centralized approach is avoided in case of Meraki Cloud Controller. Cisco uses innovative clustering technology between wireless LAN controllers to help ensure mobility across an entire wireless network. Cisco approach minimize need for operational support and increase security of WLAN.

This analysis concludes that above discussed approaches addressess individual organization need as per its own requirement, but there is need to have globally accepted solution for centralized management for WLAN environment.

4 Conclusion

Dense deployment of individual AP in the enterprise environment has necessitated the introduction of thin Access Points controlled by a central wireless controller. This central controller can be implanted with different approaches in different enterprises. In this paper we discuss such three approach adapted by leading enterprise and give comparative evaluation of these approach.

References

[1] Intel and Cisco WLAN Deployment Guide

[2] Kandula, S., Ching-Ju Lin, K., Badirkhanli, U., Katabi, D.: FatVAP: Aggregating AP Backhaul Capacity toMaximize Throughput. In: 5th USENIX Symposium on Networked Systems Design USENIX Association and Implementation, NSDI 2008 (2008)

[3] Murty, R., Padhye, J., Chandra, R., Wolman, A., Zill, B.: Designing high- performance enterprise Wi-Fi networks (August 2008)

[4] Shrivastava, V., Ahmed, N., Rayanchu, S., Banerje, S., Keshav, S., Papagiannaki, K., Mishra, A.: CENTAUR: Realizing the Full Potential of Centralized WLANs through a Hybrid Data Path. In: MobiCom 2009, Beijing, China, September 20–25 (2009)

[5] Ahmed, N., Keshav, S.: SMARTA: A SelfManaging Architecture for Thin AccessPoints. In: CoNext 2006 (2006)

[6] Murty, R., Wolman, A., Padhye, J., Welsh, M.: An Architecture for ExtensibleWireless LANs

[7] Mauch, M., Veum, G., Vallejo, M.: Managing a Global Wireless LAN, Intel White Paper (April 2010)

[8] Meraki Hosted Architecture, Maraki White Paper (March 2009)

[9] Cisco Wireless LAN Controller Configuration

A New Metric Based Cluster Head Selection Technique for Increasing the Lifetime of Energy Aware Wireless Sensor Network

Sanghita Bhattacharya[1], Arpan Sarbadhikari[2], and Subhansu Bandyapadhyay[3]

[1] Department of Computer Science & Engineering, National Institute of Technology, Durgapur-713 209, India
sanghita.b@gmail.com
[2] A.K. Choudhury School of Information Technology, University of Calcutta, Kolkata-700 009, India
arpan1983@yahoo.co.in
[3] Senior member IEEE, Department of Computer Science & Engineering, University of Calcutta, Kolkata-700 009, India
subhansu@computer.org

Abstract. The sensor nodes in a Wireless Sensor Network (WSN) have to work with limited energy and a major issue in designing such kind of network is energy efficiency. In existing cluster based routing protocols such as LEACH, cluster heads are selected randomly which may cause unbalanced distribution of energy among nodes and would shorten the life time of the network. In the scheme proposed here, cluster heads are selected on the basis of a *weighted energy-distance* metric instead of selecting them randomly. Computer simulation shows that the proposed scheme improves life time of a network to a large extend.

Keywords: WSN, Cluster-head, Energy efficiency, Network lifetime.

1 Introduction

Wireless sensor networks (WSN) consists of large number of low cost, limited energy, application specific sensor nodes distributed in an ad-hoc manner [2], with a specific *base station* (BS) for involved in the node to node transmission with wireless communication. Sensor nodes are small in size and hold small amount of power-hence energy consumption is an important issue in designing such networks. Since wireless communications consume significant amounts of battery power, it is apparent that sensor nodes should expend as little energy as possible for receiving and transmitting data. Any conventional technique like direct communication of data from one node to another has to be avoided because the energy loss incurred can be large and this factor is also dependent on the location of the sensor node relative to the *base station* (BS). It is necessary for communication protocols to maximize nodes' lifetimes [1], reduce bandwidth consumption by using local collaboration among the nodes, and tolerate node failures.

V.V. Das, J. Stephen, and Y. Chaba (Eds.): CNC 2011, CCIS 142, pp. 581–586, 2011.

A large number of papers address the issue of energy-aware routing from different angles. For example, use of conventional multi-hop routing protocols such as Minimum Transmission Energy protocol (MTE) [3] results in an undesirable effect - nodes closest to the base station will rapidly drain their energy resource and may ultimately become dead, as, they are involved in relaying continuously a large number of messages to the base station. To overcome this problem, clustering techniques have been proposed to reduce the energy consumption and consequently the network lifetime is improved. According to this methodology, a set of cluster heads are selected at regular intervals and rest of the regular nodes act as member nodes. Member nodes are clustered around the head and send their data to head - the head nodes aggregate the data with its own data and send the aggregated packet directly to base station.

There exists several cluster based routing protocols [4, 5] for minimizing energy consumption. Clustering of nodes is an elegant solution - clusters are formed to fuse data before transmitting to the base station. Using data fusion the amount of data transmitted between sensor nodes and the BS can be reduced. A designated node in each cluster collects and fuses data from nodes in its cluster and transmits the result to the BS. LEACH [4] is a classic hierarchical cluster based routing protocol where cluster heads are selected using randomization in each round. However, it may lead to uneven energy consumption and can shorten network life time. HEED [6] elects cluster heads based on residual energy and node connectivity. PEGASIS [7] use chain topology with data fusion to reduce energy consumption. However, for such kind of networks, nodes far away from the base station exhaust their energy rapidly and would die soon.

In this paper a new method for cluster head selection is proposed - they are not selected randomly as done earlier. Instead, a cost function based on *'weighted energy–distance'* metric is used for such selection. Computer simulation reveals that the proposed scheme extends the life time of the network by a large amount.

2 System Model

A sensor network consisting of N homogeneous nodes randomly deployed over a terrain to monitor continuously the targeted environment. The sensor node set is denoted as $S = \{s_1, s_2,\ldots, s_N\}$ where $|S|=N$ and each node has an unique ID. The base station (BS) is assumed to be located far from the sensing field - BS and all sensors nodes remain static after deployment and the links are considered to be symmetric. A sensor is also capable of adjusting its transmission power using the distance from the receiver.

A model shown in [4] is used in this work for the communication energy dissipation. Both the free space (d^2 power loss) and the multi-path fading (d^4 power loss) channel models are used, depending on the distance between the transmitter and receiver. The energy spent for transmission of an l-bit packet over distance d is

$$E_{Tx}(l, d) = lE_{elec} + l \ d^{\alpha} = lE_{elec} + l \quad d^2, d < do \quad = lE_{elec} + l\varepsilon_{amp} d^4, d \geq do \tag{1}$$

The electronics energy, E_{elec}, depends on factors such as the digital coding, and modulation, whereas the amplifier energy $\varepsilon_{amp} d^4$ or d^2, depends on the transmission distance and the acceptable bit-error rate. We have $do = \sqrt{\varepsilon fs / \varepsilon amp}$ To receive this message, the radio expends energy amounting to:

$$E_{Rx}(l,d)=lE_{elec}. \tag{2}$$

3 The Proposed Scheme

In this section, the proposed scheme for two level cluster based sensor network is presented. Sensor nodes collect topological information of the network through localized interactions, which is then transferred to the base station during the network initialization phase. The subsequent cycle of data collection is defined as a round and each round consists of three phases: (1) cluster head selection (2) cluster formation (3) data transmission and cluster head rotation. Here, each cluster head compresses the data received from all the nodes within its cluster and relays the compressed data to the base station for further processing. In designing the routing protocol, the essential primary goal is to improve the life time of the network. In order to achieve this, energy consumption of the nodes need to be well balanced so that no node drains its battery power rapidly. In the scheme proposed, the cluster heads are selected based on three parameters viz.,

(i) *Node Residual Energy:* Selected nodes must have a high residual energy because cluster heads are involved in high energy consumption operations.

(ii) *Centrality:* The node centrality defines the node's importance based on how central the node is to the cluster. It is the sum of squad distance $d_{(si,sj)}$ from node s_i to other nodes s_j in cluster of s_i and distance from s_i to base station $d_{(si,BS)}$

$$\text{Centrality}=d_{(si,BS)}^{\ 2} + \sum_{j=1}^{n(si)} d(si,sj)^{\ 2}, \tag{3}$$

Where n(si) is number of all live nodes in cluster of s_i. A low centrality value is desired.

(iii) *Cluster head (CH) frequency:* A record of the number of times each node served as cluster head is kept to avoid electing the same nodes repeatedly.

The algorithm ensures that the selected cluster heads have high residual energy, low centrality and low cluster Head Frequency.

3.1 Initialization Phase

During the time of initialization, BS sends initialization message to all the nodes. After receiving this, each node sends its distance from neighbor nodes, amount of residual energy available and the ID to BS.

3.2 Cluster Head Selection

In every round a node decides whether it will become the cluster head for the current round or not. The cluster head selection probability ($CHPs_i$)

$$CHPs_i = p(\gamma RE.s_i/E_o.s_i + (1\text{-}\gamma)(1/(d_1+d_2)+1/FQ.s_i)) \tag{4}$$

is calculated using the three factors such as node residual energy, centrality and cluster head frequency. A node then generates a random number T where T= [0,1] as mentioned in [4]. If CHP_{si} is less than T then that node is selected as cluster head.

Here, $RE.s_i$ is residual energy of s_i, $E_o.s_i$ is initial energy at the time of deployment $FQ.s_i$ is cluster head frequency, $d_1= \sum_{j=1}^{n(si)} d(si,sj)^{2}$ where $d(s_i,s_j)$ is distance from s_i to other nodes s_j in cluster of s_i and n(si) is number of all live nodes in cluster of s_i ; $d_2=d_{(si,BS)}^{2}$. γ ($0\leq \gamma \leq 1$) is the weighing parameter and p is initial clustering probability and its value is 5%[4].

In this proposed scheme, cluster head selection probability of nodes per round decreases or increases with regard to (i) residual energy of node, (ii) distance of node with neighbors as well as (iii) cluster head selection frequency.

3.3 Cluster Formation

In most of the algorithms, nodes choose the cluster heads based on the *Received Signal Strength (RSS)* to reduce their energy consumption. However, as discussed in [8], this approach may result in cluster heads exhausting their energy rapidly during data transmission (since nodes further away from the base station consume higher level of energy).

1. Cost function based on Distance-Energy Metric (DEM)

To overcome this problem stated above, a new cost function based on a distance-energy metric (DEM) is introduced.

$$Cost(s_i,CH_j)=f(s_i,CHj)-RE_{CHj} \tag{5}$$

Here, $f(s_i,CH_j)= d^2_{si,CHj} / d_{f\_max} + d^2_{CHj,BS}/g_{total\_dis}$; $d_{f\_max}=\max(d_{si,CHj})$;

$$g_{total\_dis=} = \sum_{j=1}^{N} d(si,BS)$$

RE_{CHj} is residual energy of jth cluster head, $d_{(CHj,BS)}$ and $d_{(si,BS)}$ denote distances from cluster head CH_j to base station BS and node s_i to BS respectively and, $d_{(si,CHj)}$ is distance between node s_i and cluster head CH_j. Each node then selects the cluster head with minimum cost and sends the '*JOIN*' message containing its ID and residual energy to that cluster head.

4 Performance Evaluation

The scheme described was implemented in MATLAB simulator. Simulation work was undertaken with 100 nodes distributed over 100m * 100 m area and the base station is located at (50,175). The data packet size is 4000 bits and control packet size is 200 bits. Initial energy of each node is 0.5 joules while E_{ele}=50nJ/bit,$\varepsilon_{fs=}$10pJ/bit/m^2, ε_{amp}=0.001pJ/bit/m^{4}and *do=87m*.

A Performance Metrics

Three important performance metrics are selected to analyze the performance of the proposed algorithm against LEACH [4] and Direct Communication:

1. *Network Life Time i.e., First Node Die (FND):* The number of rounds that have elapsed before energy content of the first node is exhausted.
2. *Standard deviation of Total Energy Consumption:* This metric reflects how uniform energy consumption between the nodes. Lower value of this metric means better energy distribution among the nodes. It can be calculated using the following equation:

$$E_{cost}=\sqrt{\sum_{i=1}^{N}\frac{1}{N}(Esi-E\alpha)^{2}},\ E_a=E[E_{si}]$$

   where E_{si}= Total energy consumption of the node s_i , ; N = total number of operational nodes
3. *Packet delivery ratio* (PDR): It is the fraction of the successfully delivered data packets to the base station divided by those generated and transmitted by sensor nodes before first node dead.

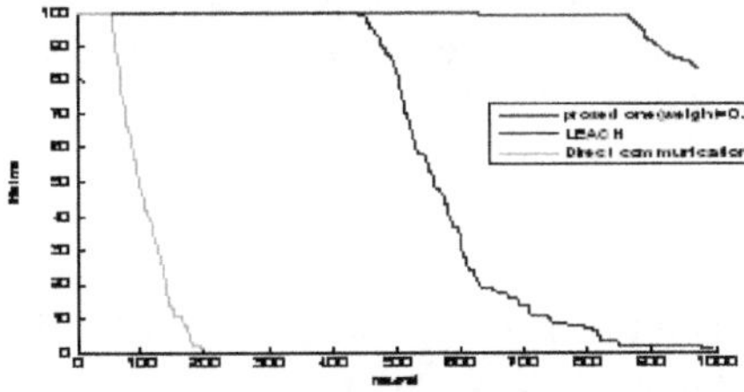

Fig. 1. Live nodes comparison between routing protocols

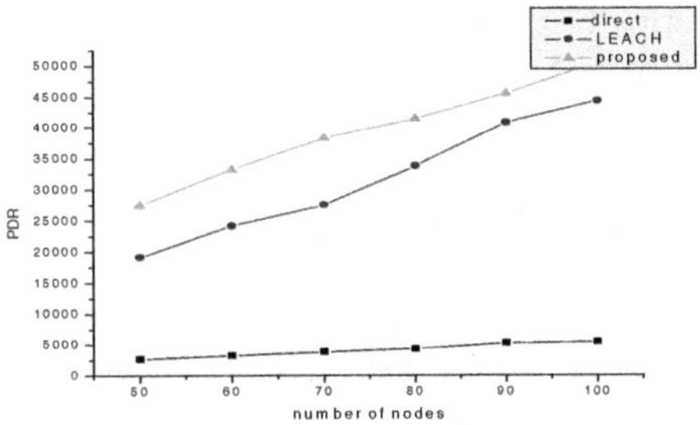

Fig. 2. PDR vs. Number of nodes

Figure 1 shows number of live nodes as rounds proceed for three schemes. In [4], same node may have the probability of being selected as the cluster head again even through its residual energy has become low. Consequently, it would drain its energy very fast and become dead. In the proposed scheme, as the weighted energy-distance metric is considered for head selection, lifetime of the nodes is improved considerably and energy becomes distributed evenly.

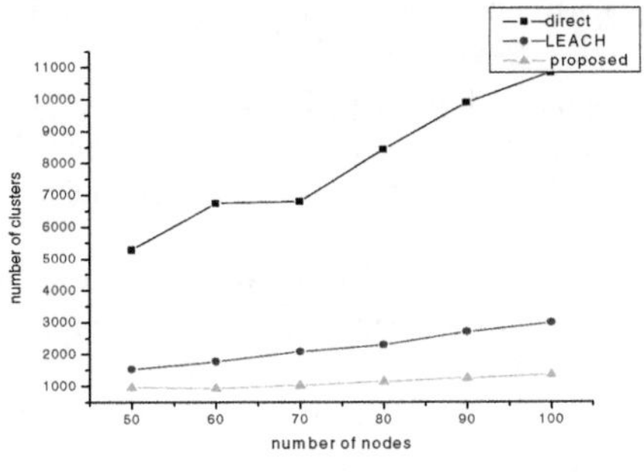

Fig. 3. Number of nodes vs. number of clusters

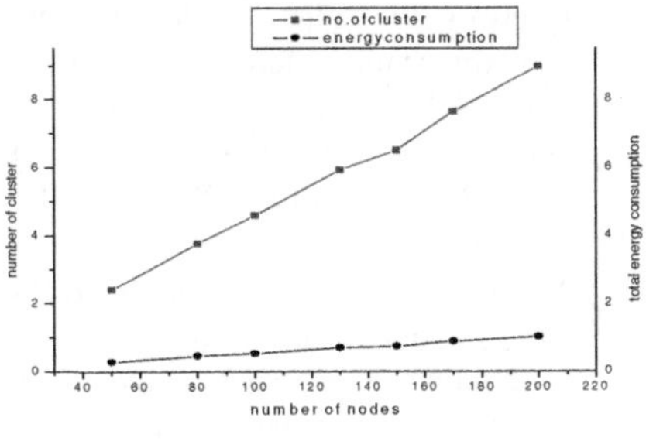

Fig. 4. Number of nodes vs. transmission cost

From fig. 2 it can be seen that as number of nodes increases linearly PDR increases at a much higher rate in the proposed method than [4] and direct method. In fig 3, In case of number of clusters parameter, it is much lower in proposed one than the others. It implies that clustering overhead is much lower in the proposed scheme and thus life time improves significantly. In fig.4 as the number of nodes increases, total energy consumption of the nodes increases extensively. As number of nodes increase, small numbers of clusters are formed which subsequently increases total energy consumption as well as number of clusters.

5 Conclusion and Future Work

In this paper, for solving the problem of unbalance in energy consumption of the sensor nodes, a hierarchical cluster based weighted energy- distance routing protocol has been introduced. It has been shown through simulation that its performance, in terms of network life time, is much better than other well known approaches. Here, maximum hop count is restricted to 2. Work is being carried out for extending our model for multi hop communication to improve life of the sensor network.

Acknowledgements

The first author gratefully acknowledges the facilities and support provided by the Director, National Institute of Technology, Durgapur 713 209, India.

References

1. Singh, S., Woo, M., Raghavendra, C.S.: Power-Aware Routing in Mobile Ad Hoc Networks. In: Proc. of ACM/IEEE Mobicom 1998 (1998)
2. Sohrabi, K., Gao, J., Ailawadhi, V., Pottie, G.J.: Protocols for Self-Organization of a Wireless Sensor Network. IEEE Personal Communication 7(5), 16–27 (2000)
3. Ettus, M.: System Capacity, Latency and Power Consumption in Multi hop-Routed SS-CDMA Wireless Networks. In: Proce of IEEE Radio and Wireless Configuration (August 1998)
4. Heinzelman, W.R., Chandrakasan, A., Balakrishnan, H.: Energy Efficient Communication Protocol for Wireless Micro Sensor Networks. In: Proc. of Hawaii International Conference on System Sciences, pp. 1–10 (January 2000)
5. Kim, K.T., Youn, H.Y.: Energy-Driven Adaptive Clustering Hierarchy (EDACH) for Wireless Sensor Networks. In: Enokido, T., Yan, L., Xiao, B., Kim, D.Y., Dai, Y.-S., Yang, L.T. (eds.) EUC-WS 2005. LNCS, vol. 3823, pp. 1098–1107. Springer, Heidelberg (2005)
6. Younsis, O., Fahmy, S.: Distributed Clustering in Ad hoc Sensor Networks: A hybrid, energy-efficient approach. In: Proc. of 13th Joint Conference on IEEE Computer and Communications Societies, INFOCOMM (March 2004)
7. Lindsey, S., Raghavendra, C.S.: PEGASIS: Power Efficient Gathering in Sensor Information Systems. In: Proc. of IEEE Aerospace conference, vol. 3, pp. 1125–1130 (March 2002)
8. Ye, M., Li, C.F., Chen, G.H., Wu, J.: EECS: An Energy Efficient Clustering Scheme in Wireless Sensor Networks. In: Proc. of the IEEE Int. Performance Computing and Communications Conf. (IPCCC), pp. 535–540 (2005)

A Feature Extraction Algorithm with Minimum Entropy for Images

M. Sindhu

Department of Information Technology,
Indira Institute of Engineering & Technology, Thiruvallur, TN, India
sindhumoha@gmail.com

Abstract. The increasing volume of data generated by imaging modalities which justifies the use of different compression techniques. To decrease the storage space and efficiency of transfer the images over network for access to electronic record. In this paper, method for feature level image for images in second-generation wavelet domain is proposed. The proposed algorithm yield feature extraction method to produce the optimum features with high confidence in producing best results of compression.

Keywords: Entropy function, Feature Extraction, Wavelet-Domain.

1 Introduction

Image compression plays a fundamental role for the efficient and cost-effective use of digital medical imaging technology and applications. Image compression reduces the storage space required by an Image and the bandwidth needed when streaming that image across network [1]. Image compression algorithms can be divided into two groups: algorithms, which compress images without data loss and an exact reproduction of the original image from the compressed file. However, this lossless compression scheme achieves relatively low compression rates of about the ratio of 3:1. The other algorithm, which compress images with data loss by eliminating the redundant and high frequency data from an image, which is usually outside the range of human visual perception. This lossy compression results in much higher compression ratio, typically 20:1 or greater, but with irretrievable data loss.

The compressed image has two implications: fidelity and intelligibility. Fidelity describes how the reconstructed image differs from the original one. Intelligibility shows the ability through which the image can offer information to people with classification accuracy [2].

The basic idea of the Discrete Wavelet-Domain (DWT) is that of successive approximation, together with that of "added detail". At each stage, the input signal is decomposed into a coarse approximation signal (which can be considered a low pass version of the input) and an "added detail" signal (which can be considered a high pass version). In the wavelet domain, by comparing the energies of different wavelet transform levels of the different pixels with a varying in frequency. This transform technique currently provides the most promising approach to high-quality image

V.V. Das, J. Stephen, and Y. Chaba (Eds.): CNC 2011, CCIS 142, pp. 587–590, 2011.

compression, which is essential for tele-radiology. Wavelet coders for images have been implemented both with scalar quantization [3] and vector quantization [4].

In this paper, we investigate the image compression in the wavelet-domain for encoding along with the optimum features of an image with minimum entropy to achieve better compression to save memory and bandwidth. The proposed method performs two level wavelet decomposition then the root pixels are chosen to apply the extraction of high confidence features, with that optimum set of features to increase in compression ratio and storage space significantly.

2 Proposed Method

In the proposed algorithm, a standard test image is decomposed with 2-level DWT and applying the entropy functions to the root subband that is corresponds to the approximation image. By applying DWT, the image is actually divided i.e., decomposed into four subbands and critically sub sampled as shown in Fig.1. These four subbands arise from separable applications of vertical and horizontal filter. The subbands labeled (H,L), (H,H) and (L,H) represent the finest scale wavelet coefficients, i.e., detail images while the subband (L,L) corresponds to coarse level coefficients, i.e., approximation image. To obtain the next coarse level wavelet coefficients, the subband (L,L) alone is further decomposed and critically sampled. This results in two-level wavelet decomposition as shown in Fig. 2.

After analyzing the transformed pixels, most of the root level pixels (LL subband) found to be significant and most of the pixels lie at the bottom of the pyramid found to be insignificant. The region or sets, henceforth referred to as (LL,LL) can be varying dimensions. The dimension of a set (LL,LL) depends on the dimension of the original image and the subband level of the pyramidal structure at which the set lie. The root level pixels are only considered for features extraction to further applying the compression. The feature extraction with Minimum Entropy method involves in choosing minimum of entropy functions representing the population entropy of classification classes. The confidence of the compression result is an important issue in image compression that lies in feature extracted from raw input data. The simple way to quantify the confidence of the features is to calculate the entropy of the predictions using equation (4). H is called as entropy of the predictions using uncertainty in the system. A large entropy measure (very uncertain) is produced by a system where the probabilities are equal.

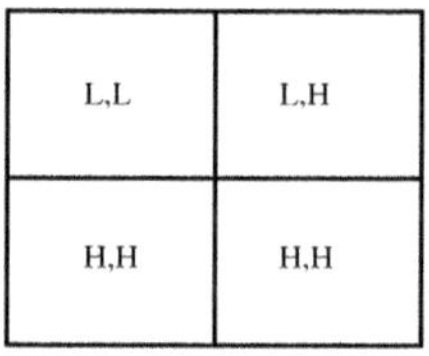

Fig. 1. One Level Decomposition

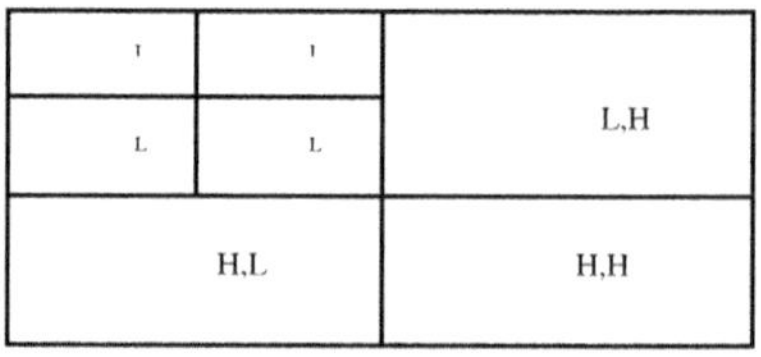

Fig. 2. Two Level Decomposition

The eigen vectors (column vectors) having repeated data will give maximum entropy value which is not useful in representing the unwanted pixels. The larger the entropy (H), the smaller the confidence. Hence, eigen vectors having minimum entropy value is chosen as feature vectors. The steps involved in the proposed algorithm is as follows:

Procedure Process (LL, LL)

1. Initialization

   Partition the image into different subbands by two level decomposing.

   Set I = root

2. If I is (LL, LL) subband

   Load the Input pixel elements x.

   Calculate the mean $(x - \bar{x})$ with

$$\bar{x} = \frac{1}{n} \sum_{i=1}^{n} \sum_{j=1}^{n} x_{ji} \tag{1}$$

3. Compute the covariance matrix

$$C = \frac{1}{n} \sum_{i,j=1}^{n} (x_{ij} - \bar{x})(x_{ij} - \bar{x}) \tag{2}$$

4. For each covariance matrix compute eigen values and eigen vectors.

5. Form feature matrix. Arrange Eigen values by their Eigen values from highest to lowest.

   Feature Vector = [$evect_1$, $evect_2$, ..., $evect_n$] (3)

6. Compute entropy. For each column of feature vector matrix and rearrange the matrix according to their entropy value from minimum to maximum.

$$H = -\sum_{i} p(y_i) * \log(p(y_i)) \tag{4}$$

7 Choose only first d eigen vectors as principle components in such a way that d<n.

8 If (d<n) Step 2. Else search for the subband (LL, LL).

After the selection of features (symbols) are coded in a bit stream using any compression algorithm the compressed data can be obtained.

3 Simulation Results

For evaluation standard test image Lenna is used. By using the proposed algorithm, the main feature of an image has been obtained. The wavelet-domain transformation is applied to the image with the features. The proposed algorithm yield high confidence measure for features by computing entropy-using equation (4) of each column of Feature Vector. The Feature Vector matrix has been rearranged according to their entropy value from minimum to maximum [5]. This algorithm has reduced data by choosing the first d minimum entropy's eigenvectors as principle components. The final data set is derived by using equation (5).

$$[\text{Final Data}] = [\text{Reduced Feature Vector}]^T * [\text{mean adjusted Input Pixel}] \quad (5)$$

The average entropy $\overline{H}$ for the entries feature data set can be calculated as

$$\overline{H} = \frac{1}{d} \sum_{j=1}^{d} H_j \quad (6)$$

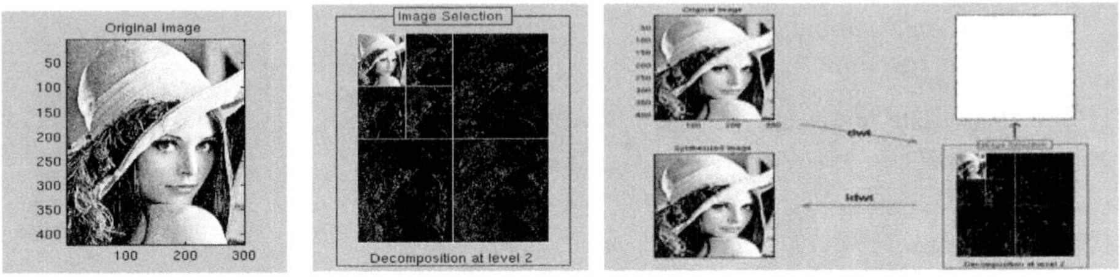

Fig. 3. Compression and Decompression of Standard Test Image

4 Conclusion

It is observed that the proposed algorithm produce the high confidence features which can achieve better compression ratio that can save memory and bandwidth in wavelet domain decomposition. This may also helps in medical field to handle, store and transmit the medical images such as Magnetic resonance imaging (MRI), Computer Tomography (CT), Positron Emission Tomography (PET) and Signal Photon Emission Computed Tomography (SPECT) effectively.

References

1. Sindhu, M., Rajkamal, R.: Image and Its Compression Techniques – A Review. International Journal of Recent Trends in Engineering 2(4), 71–75 (2009)
2. Ghrare, S.E., Ali, M.A.M., Ismail, M., Jumari, K.: The Effect of Image Data Compression on the Clinical Information Quality of Compressed Computed Tomography Images for Teleradiology Applications. European Journal of Scientific Research 23(1), 6–12 (2008)
3. Mallat, S.G.: A theory for multiresolution Signal decomposition: The wavelet representation. IEEE Trans. Pattern Anal. And Mach. Intef. 1, 674–693 (1989)
4. Antonini, M., Barlaud, M., Mathieu, P., Daybechies, I.: Image Coding using wavelet transform. IEEE Trans. Image Proc. IP 1, 205–220 (1992)
5. Rajkamal, R., Vanaja Ranjan, P.: Soft Computing Feature Extraction Algorithm with Minimum Entropy. International Journal of Distributed Energy Resources 6(3), 253–261 (2010)

Shot Boundary Detection Algorithm Based on Color Texture Moments

B.H. Shekar[1], M. Sharmila Kumari[2], and Raghuram Holla[1]

[1] Department of Computer Science, Mangalore University, Karnataka, India
[2] Department of Computer Science and Engineering, P A College of Engineering, Mangalore, Karnataka, India
bhshekar@gmail.com, sharmilabp@gmail.com, raghu247@gmail.com

Abstract. In this paper, a shot boundary detection method based on color texture moments is proposed. Here we apply the Local Feature Transform (LFT) on each frame of a video data for feature extraction purpose. Then the first and second moments are computed for each channel of the color space and the feature vector is formed. The dissimilarity measure higher than a predefined threshold indicates the presence of a shot cut. Experiments have been conducted on the standard TRECVID video database to evaluate the effectiveness of the proposed method.

1 Introduction

A sequence of frames captured by one camera with continuous action in time and space is referred to as a video shot. A video shot boundary detection, the first step in the process of indexing, aims to detect the boundaries by computing and comparing the similarity or difference between adjacent frames. The various approaches differ concerning the used features. However, it is quite difficult to give a precise definition of a video shot transition since many factors such as camera motions may change the video content significantly.

We have seen several models which were developed based upon statistical properties of pixels [3], edge differences [9] or Histogram differences [10] etc. Zhang et al. [10] used Histogram differences method with two thresholds in order to detect gradual transitions. Audio and motion features were used to improve shot boundary detection. Saraceno et al. [6] classify audio according to silence, speech, music, or noise and use this information to verify shot boundaries hypothesized by image-based features.

The proposed model uses color texture moments as features introduced by Hui et al. [2] in the frequency domain. In the proposed work, we compute the color texture moments in the (SVcos(H), SVsin(H), V) color space to overcome some short comings of the HSV color space. The rest of the paper is organized as follows. Section 2 presents the proposed model. Experimental results and comparative analysis are shown in section 3. Conclusion is given in section 4.

V.V. Das, J. Stephen, and Y. Chaba (Eds.): CNC 2011, CCIS 142, pp. 591–594, 2011.

2 Proposed Methodology

This section presents the proposed model developed using color texture moments. Given a video containing many frames, we compute the color texture moments for each frame and the Euclidean distance measure is used to measure the similarity between the frames. A shot boundary is said to be detected when the dissimilarity measure between the frames exceeds the threshold.

The proposed model is based on Local Fourier Transform (LFT) developed for image segmentation by [1]. As suggested in [1], to overcome soft color changes between red and magenta in HSV space, modified color space defined below is used to extract features.

$$x = (x_1, x_2, x_3) \tag{1}$$

Where,

$$\begin{aligned} x_1 &= S*V*cos(H) \\ x_2 &= S*V*sin(H) \\ x_3 &= V \end{aligned} \tag{2}$$

Let $\{I(x,y) \mid x=0,..,L-1,\ y=0,\ldots,M-1\}$ denote the image of dimension LxM obtained from the video frame. The 8-neighbourhood of pixel (x, y) is $\{P_0\ P_1\ P_2\ P_3\ P_4\ P_5\ P_6\ P_7\}$ in anti-clockwise order. Assuming it is a periodic sequence with a period of 8, we denote it as $I(x, y, n) = P_n$, $0 \le n \le 7$. Intuitively, similar local parts of texture have similar series of $I\ (x, y, n)$ and their Fourier transform coefficients in frequency field are similar correspondingly. Hence, the LFT can be represented by 8 unique templates operating on the image respectively. We utilize the eight templates defined in [1] to extract the local Fourier coefficients directly. Then the eight characteristic maps can be obtained as follows:

$$FI(K) = \{ F(x,y,k) \mid x = 0,\ldots,L-1,\ y=0,\ldots,M-1,\ 0 \le k \le 7\} \tag{3}$$

where $F(x,y,k) = \frac{1}{8}\sum_{n=0}^{7} I(x,y,n)e^{-j\pi kn}$

So, FI(k) represents the co-occurrence of gray levels and their spatial distribution.

From the probability theory, we know that a probability distribution is uniquely characterized by its moments. If we interpret the color distribution of a characteristic map as a probability distribution, then the color distribution can be characterized by its moments [4]. We compute the first and second moments of each characteristic map FI(k) for feature extraction. The moments are calculated independently in each of the eight maps for every color channel. Then, we obtain a 16-dimensional feature vector for each channel to form 48-dimensional feature vector for every frame. The feature vector for frame j is given by:

$$F_j = \begin{bmatrix} FI_{x1}^1(0),\ldots,FI_{x1}^1(k),\ FI_{x2}^1(0),\ldots,FI_{x2}^1(k),\ FI_{x3}^1(0),\ldots,FI_{x3}^1(k), \\ FI_{x1}^2(0),\ldots,FI_{x1}^2(k),\ FI_{x2}^2(0),\ldots,FI_{x2}^2(k),\ FI_{x3}^2(0),\ldots,FI_{x3}^2(k), \end{bmatrix} \tag{4}$$

where $FI_x^{\,i}(k)$ represents i^{th} moment computed for the kth characteristic map for the color space x.

Then the difference between the frames f_i and f_{j-1} is computed as follows:

$$D = d(F_j - F_{j-1}) - d(F_{j-1} - F_{j-2}) \tag{5}$$

where,

$$d\left(F_j, F_{j-1}\right) = \sum_{h=1}^{48} \left\| F_j(h) - F_{j-1}(h) \right\|^2 \tag{6}$$

When q is set to 2, $d(i, j)$ is the Euclidean distance between the frame i and frame j. The value of D indicates the change tendency of consecutive frames. If $|D| > T_1$, then a cut is detected, where T_1 is the cut threshold.

3 Experimental Results

This section presents the results of the experiments conducted to corroborate the success of the proposed model. We have conducted experimentation on TRECVID video database.

The performance of the proposed model is evaluated using precision and recall as evaluation metrics. The precision and recall data obtained for the proposed model on three different video segments along with recently proposed Local histogram based algorithm [8] is reported in Table 1.

Table 1. Precision and Recall metrics for TRECVID video segments

Video Segment	Metrics	Proposed Model ($T_1 = 0.24$)	Local histogram based model[8] ($T_1 = 4 \times 10^4$)
BOR03	Precision	1.00	0.917
	Recall	0.9286	0.846
	Combined measure	0.963	0.880
BOR08	Precision	0.69	0.330
	Recall	0.9	0.700
	Combined measure	0.7826	0.452
SENSES111	Precision	1.00	1.000
	Recall	1.00	1.000
	Combined measure	1.00	1.000

In order to exhibit the computational efficiency, computing time taken by the proposed model for feature extraction for any single frame is given in Table 2 along with the other algorithms proposed for shot detection. One can notice here that the proposed model consume less time when compared to any other model for feature extraction purpose.

Table 2. Comparison of feature extraction time of the proposed model with other methods

Shot detection method	Per frame feature extraction time (sec.)
Temporal slices analysis [5]	0.878
Color-SIFT [7]	0.836
Local Histogram [8]	0.288
Colour Texture Moments based model (Proposed model)	0.277

4 Conclusion

We have presented an efficient texture features based model for identifying shot boundaries of a video segment. We have developed colour texture moments based model for accurate detection of cuts in a shot. Experimental results on standard TRECVID video database reveal that the proposed model best suits for shot detection purpose.

References

1. Zhou, F., Feng, J., Shi, Q.: Image Segmentation Based on Local Fourier Coefficients Histogram. In: Proc. SPIE - 2nd Int. Conf. on Multispectral Image Processing and Pattern Recognition, Wuhan, China (November 2001)
2. Yu, H., Li, M., Zhang, H.-J., Feng, J.: Color texture moments for content-based image retrieval. In: Proceedings of 2002 International Conference on Image Processing, June 24-28, vol. 3, pp. 929–932 (2002)
3. Kasturi, R., Jain, R.: Dynamic Vision. In: Kasturi, R., Jain, R. (eds.) Computer Vision: Principles. IEEE Computer Society Press, Washington (1991)
4. Stricker, M., Orengo, M.: Similarity of color images. In: Proc. SPIE on Storage and Retrieval for Image and Video Databases, San Jose, USA, vol. 2420, pp. 381–392 (February 1995)
5. Ngo, C.-W., Pong, T.-C., Zhang, H.-J.: On Clustering and Retrieval of Video Shots through Temporal Slices Analysis. IEEE Trans. On Multimedia 4(4), 446–458 (2002)
6. Saraceno, C., Leonardi, R.: Audio as a Support to Scene Change Detection and Characterization of Video Sequences. In: Proc. Int. Conf. Acoustics, Speech, and Signal Processing, Munich, Germany, pp. 2597–2600 (April 1997)
7. Sharmila Kumari, M., Shekar, B.H.: Color-SIFT model: a robust and an accurate shot boundary detection algorithm. In: Proc. SPIE, vol. 7546, p. 75463S (2010)
8. Fu, X., Zeng, J.-x.: Local Features Based Image Sequence Retrieval. Journal of computers 5(7) (July 2010)
9. Zabih, R., Miller, J., Mai, K.: A Feature-based Algorithm for Detecting and Classifying Scene Breaks. In: Proc. ACM Multimedia 1995, San Fransisco, CA, pp. 189–200 (November 1995)
10. Zhang, H., Kankanhalli, A., Smoliar, S.: Automatic Partitioning of Video. Multimedia Systems 1(1), 10–28 (1993)

Secured and Authentic Communication by Combined Approach of Digital Watermarking and Steganography

Satya Prakash Sahu[1] and Satya Verma[2]

[1] Asst. Prof., Deptt. of I.T., NIT Raipur, (C.G.), India
spsahu.it@nitrr.ac.in
[2] PG Student, Deptt. of Elect. Engg., NIT Raipur (C.G.), India
satyaritu@yahoo.co.in

Abstract. The rapid evolution of the Internet makes the transmission of digital multimedia content such as text, audio, images, and video easier. Digital media can be accessed or distributed through the network. As the communication over internet is easier and convenient, the security is the main concern. And one of the issues is authenticity of contents. There are two different approaches to achieve security and authenticity. One of the approaches for secured transmission is steganography in which secrete message is hidden behind the image. For authenticity, digital watermarking is a technique which embeds a digital signature or digital watermark that asserts the ownership or intellectual property rights of the media creator or owner, in the digital media such as text, audio, image, and video. In this paper a hybrid approach is being proposed, in which watermarking and steganography techniques have been combined, which exploits the advantages of both techniques resulting more robustness against different types of attacks.

Keywords: Information Hiding, Payload, Opacity, Watermarked-steganographic Image.

1 Introduction

For redundancy, copying is simple with no loss of fidelity, that is, the copy of a digital medium is identical to the original one. An unlimited number of identical copies of digital media can be illegally produced; this is a serious threat to the copyright of the media owner. Therefore, to protect and enforce intellectual property rights of the media owner is an important issue in the digital world [1]. The digital watermarking technique embeds a digital signature or digital watermark, which asserts the ownership or intellectual property rights of the media creator or owner, in the digital media such as text, audio, image, and video. The watermark can then be extracted from the watermarked media to identify the author or distributor of the media. The main objective of Steganography is to hide the information under a cover media so that the outsiders may not discover the information contained in the said media. This is the major distinction between Steganography and other methods of hidden exchange of information [2]. Information hiding, steganography, and watermarking are three closely related fields that have a great deal of overlap and share many

V.V. Das, J. Stephen, and Y. Chaba (Eds.): CNC 2011, CCIS 142, pp. 595–599, 2011.

technical approaches as shown in figure 1. The techniques involved in such applications are collectively referred to as information hiding [3]. The key difference between steganography and watermarking is the absence (in steganography) of an active adversary mainly because usually no value is associated with the act of removing the information hidden in the host content. Nevertheless, steganography may need to be robust against accidental or common distortion like compressions or color adjustment (in this case we will talk about active steganography) [4].

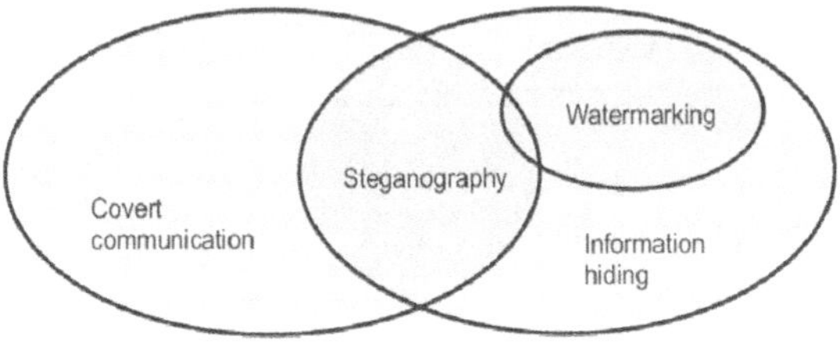

Fig. 1. Relationship between Steganography, watermarking and related fields

All steganographic and watermarking algorithms have to comply with a few basic requirements. The most important requirement is that an algorithm has to be imperceptible. These requirements are analyzed as [5], [6], [7]-Security, Imperceptibility, Capacity and Robustness.

2 Proposed Work

In this section, the combined approach has been proposed by the use of Watermark & Steganographic technique as illustrated in figure 2(a) at sender end. The final output is *watermarked-steganographic image*. Now the final image (*watermarked-steganographic image*) can be used for the secured communication over the non-secure network. At the receiving end, the extraction of secrete message is carried as illustrated in figure 2(b). The implementation of work can be carried out by following tasks:

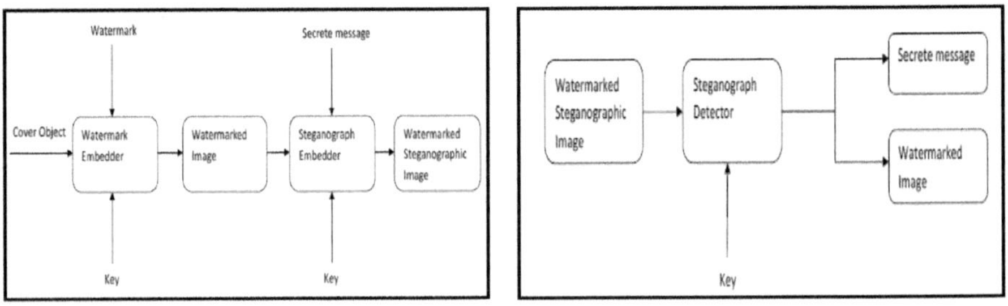

Fig. 2. (a) Embedding of watermark and message, (b) Extraction of secrete message

2.1 Watermark Embedding

1. Because watermark is visible, so first it is needed to set the transparency of the watermark.
2. The embedding of the watermark in cover image is done by changing the RGB components of the cover image. Here we make use of opacity factor.
   Let, C(x, y) - pixel of cover image, W (x', y') - is the pixel of watermark, CW (x, y) - is the pixel of watermarked image and OF- is opacity factor.
   The pixel of watermarked image:

$$CW = C \times (100 - OF) / 100 + W \times OF/100$$

2.2 Message Embedding with Encryption

1. Before employing steganography, message is encrypted with a key. For the encryption message is XORed with the key.
2. The embedding is done in following manner:
   i. Read the RGB (color) value of the pixel.
   ii. If the color value of the pixel is odd then
      a) If bit of the message is 0,
         Then increment the color value of the pixel by 1.
      b) If bit of the message is 1,
         Then leave the color value of the pixel as it is.
   iii. If the color value of the pixel is even, then
      a) If the bit of the message is 0,
         Then leave the color value of the pixel as it is.
      b) If bit of the message is 1,
         Then increment the color value of the pixel by 1.

2.3 Message Extraction with Decryption

1. For the extraction of message from watermarked image is done in following manner:
   i. Read the RGB color value of the pixel.
   ii. If the color value of the pixel is odd then interpret 1 for the character bit.
   iii. If the color value of the pixel is even then interpret 0 for the character bit.
2. For the decryption of extracted message, the private key is XORed with the extracted message.

3 Experimental Results

The images used in the experiment were color images. A typical result for *Lena* image of size 256x256 is given in Figure 3. The original, watermarked and watermarked-steganographic image is shown in Figure 3(a), (b) & (c) respectively. This final image figure 3(c) is having message of size 53 words. And similarly the work is also tested for other images with varying the size of image & message and opacity.

(a) (b) (c)

Fig. 3. (a) Original (b) Watermarked (c) Watermarked-Steganographic

3.1 Perceptibility

Perceptibility expresses amount of distortion caused by watermark embedding. In other words, it indicates how visible the watermark is. It is measured by peak signal-to-noise ratio (PSNR). The less the value of PSNR is the more perceptible the watermark is.

$$PSNR = 10 * \log_{10} \frac{\sum_{i=1}^{M} \sum_{j=1}^{N} [C'(i,j)]^2}{\sum_{i=1}^{M} \sum_{j=1}^{N} [C(i,j) - C'(i,j)]^2}$$

Next table gives the PSNR between original image and watermarked-steganographic image.

S.No	Image	Image Size	Opacity	Length of message (no of words)	PSNR
1	Leena	256x256	50%	53	44.8982
2	Baby	650x643	50%	589	62.6175
3	Tree	1176x1256	75%	921	52.7571
4	Zebra	600x401	25%	645	46.5895
5	Tiger	690x461	100%	3264	53.6662
6	Teddy	412x467	10%	276	50.2488

3.2 Robustness

Experiment shows that the approach is robust against some geometrical transformations. If rotation or scaling is done in the watermarked-steganographic image and then the contents of secrete message remains unaltered (in some cases it is extracted in jumbled form), in order to get the exact message, it is required to set the original position of the image at receiver end. If an image is rotated by an angle Θ then the pixels of rotated image will be

$$X' = X*COS(\Theta) - Y*SIN(\Theta) \ \& \ Y' = X*SIN(\Theta) + Y*COS(\Theta)$$

Where (X, Y) is the pixel of watermarked-steganographic image & (X', Y') is the pixel of rotated image.

And if the image is vertically flipped then the pixel of the flipped image is given by

$$X' = -X \ , \ Y' = Y;$$

3.3 Editing

If some editing is done in the watermarked-steganographic image, then also it is possible to extract the original secrete message. In experiments, images are modified & edited and the message had been successfully extracted.

4 Conclusions

In this paper, the different aspects & issues of these techniques such as security, imperceptibility, payload capacity and robustness are studied. Subsequently a combined approach of watermarking and steganography had been proposed. The work had been implemented for a textual secrete message with the help of watermark embedder & steganography embedder using encryption with secret key. The watermarked-steganographic image is then communicated over the non secure network. And similarly at receiver end, the extraction of secrete message has been achieved by steganograph detector using decryption technique. It has been observed that, by this approach, not only the secured & authentic mode of transmission for textual message is achieved but also the robustness against general geometric transformation attacks is shown. The method has been exposed to several simulation tests checking up their resistance to various types of attacks. It can further be extended for the invisible watermark containing the secrete message and also for the different file & image formats.

References

[1] Wei, H., Yuan, M., Zhao, J., Kou, Z.: Research and Realization of Digital Watermark for Picture Protecting. In: First International Workshop on Education Technology and Computer Science (2009)

[2] Judge, J.C.: Steganography: Past, Present, Future. SANS white paper (November 30, 2001), `http://www.sans.org/rr/papers/index.php?id=552`

[3] Kharrazi, M., Sencar, H., Memon, N.: Image Steganography: Concepts and Practice. Lecture Note Series, Institute for Mathematical Sciences. National University of Singapore (2004)

[4] Anderson, R.J., Petitcolas, F.A.P.: On the limits of steganography. IEEE Journal of selected Areas in Communications (May 1998)

[5] Moerland, T.: Steganography and Steganalysis. Leiden Institute of Advanced Computing Science, `http://www.liacs.nl/home/tmoerl/privtech.pdf`

[6] Cummins, J., Diskin, P., Lau, S., Parlett, R.: Steganography and Digital watermarking. Student Seminar Report, School of Computer Science, University of Birmingham (2004), `http://www.cs.bham.ac.uk/~mdr/teaching/modules03/security/students/SS5/Steganography.htm`

[7] Sharma, R.K., Decker, S.: Practical challenges for digital watermarking applications. In: Proceedings of IEEE Fourth Workshop on Multimedia Signal Processing, pp. 237–242 (2001)

Building and Controlling a Ball and Plate System

Erikson Ferry Sinaga, Edward Boris Manurung, Vahdat A. Chee, and Arko Djajadi

Department of Mechatronics,
Swiss German University,
BSD City, Indonesia
erikson.sinaga@sgu.ac.id

Abstract. The Ball and Plate system is a ball balancing system which uses a closed loop control system. The ball balancing system is used for laboratorial purposes and also in automation systems. The closed loop control which is used is a digital Proportional Integral Derivative (PID) control. The ball and plate system will be used as an illustration to understand the basic concepts and principles of control system and algorithm for beginner students.

Keywords: Ball and plate, Control System, PID Controller, ball balancing.

1 Introduction

Balancing is a challenging topic and can be used effectively to ilustrate the role of control system, especially in mechatronics application. The ball and plate system consist of a number of parts which interact and support one another. This paper is concerned with the overall and detail functions of these parts and products. The touch screen panel senses where the ball is and where it is moving to, while the servo motor acts as the actuator which tilts the table to move the ball. A PID controller performs its controlling function to archieve the desired balancing objective.

2 Ball and Plate System

2.1 Mechanical Part Design

The mechanical design of the ball and plate system is to physically place all the different components of the system as efficiently as possible so as to be able to get a working system of the Ball and Plate system. The mechanical parts drawn in this project are the base plate, table plate, servo holders & shafts, ball joints as given in Figure 1.

The various parts; the touchscreen sensor, the servo motors, were measured to fit in the mechanical design.

V.V. Das, J. Stephen, and Y. Chaba (Eds.): CNC 2011, CCIS 142, pp. 600–608, 2011.

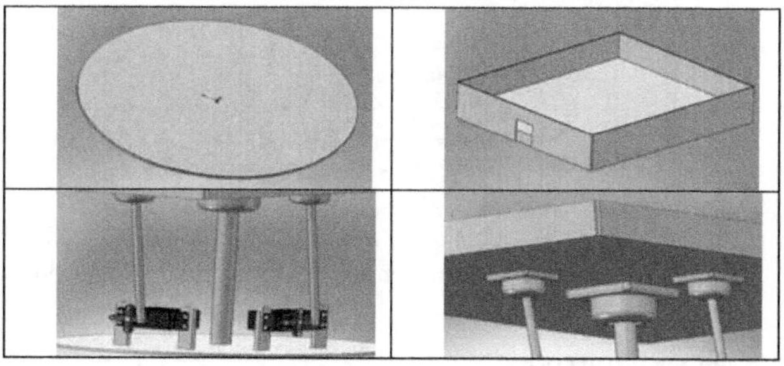

Fig. 1. Mechanical Part Design

2.2 Electrical and Control Parts

2.2.1 Touch Screen Panel and Controller

The touchscreen technology is used in a wide variety of applications, for example in room automation or touch buttons on control panels. There are several types of touchscreen technology available to be used, such as infrared touchscreen, resistive touchscreen, and capacitive touchscreen. In this project the author uses resistive touchscreen panel. In this project, a resistive touchscreen is used, since it works in a very simple way by using two very thin layers of electrically conductive material, sometimes transparent.

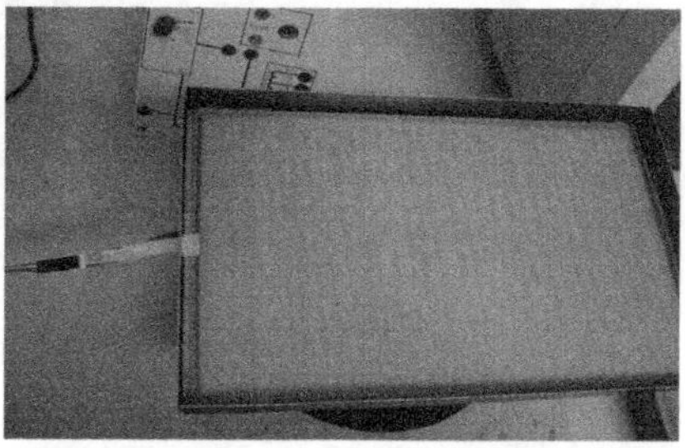

Fig. 2. Touchscreen panel on the table

As can be seen on Figure 2, the touchscreen is placed to fit on the table. The touchscreen detects the position of the ball which is on the table. Once the ball is placed on the table, the touchscreen directly sends the position of the ball to the computer through a touchscreen controller. And wherever the ball slides to, the change of position will be sent to the computer.

The touch screen technology is very efficient but it is only able to calculate one point at a time. If two points on the screen are touched, the sensor will not able to know which input to calculate. It will create an error on sensing positions.

The touchscreen has four wires, two of which are for the power. It is supplied with a 5V DC voltage. This power is obtained from the touchscreen controller as shown in Figure 3, which is connected to a personal computer through its USB serial port. The oter two wires from the touchscreen are for each layer for position signals.

Fig. 3. Touchscreen controller and its four wire connection

2.2.2 DT Basic Servo Controller

The DT Basic BS2sx is used in this project as a servo controller and not as the main program processor. The DT Basic is supplied with a 5V or 9V DC voltage which can be from a battery. The basic stamp is then connected to the computer through a serial port RS 232 for controlling and program downloading. The PBASIC library already includes one to control parallax servo motors easily, hence making it suitable for this project.

2.2.3 Servo Motor

The parallax servo motor consists of three wires, one is connected to the power, 5V-dc (red cable), one is connected to ground (black), and the final one is the control wire (white). The controller gives the desired angle of the motor through this white wire, shown in Figure 4.

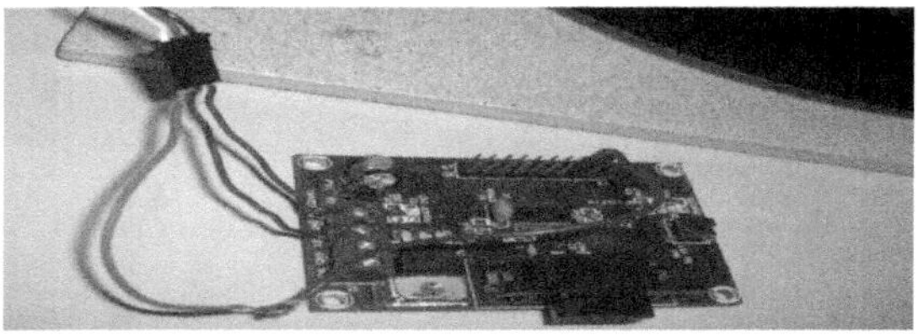

Fig. 4. Pins of servo motor and connection to the Basic Stamp

The angle of the motor rotation is determined by the duration of a pulse with an amplitude of 5V-dc that is applied to the control wire. This is called pulse width modulation. The servo expects a pulse every 20ms, and it is the length of that pulse that the servo motor gets which will determine to what angle the motor rotates. So, for every 20ms the PWM will be set to 1 or high for as long as the input value, and to 0 or low until the rest of the 20ms.

2.2.4 Servo Motor Calibration

The PWM of the servo motor needs an ON value of 0.4ms to 2.4ms to get from 0° to 210°. The basic stamp BS2sx uses its own units of duration in its program. Each unit equals to 0.8μs. So to get the value of 0.4ms for the servo motor to get to the extreme left position it needs 500 pulses and to get the servo to be at the position of the extreme right is needs 3000 pulses.

2.3 Programming

2.3.1 Serial Communication Programming

The main processor for the system will be in the computer using application Visual Basic. After the main program gets its input from the sensors, it processes the input and gives the output to the servo controller. The output is given to the servo controller through serial communication.

On the servo controller side (basic stamp BS2sx) it is asked to input data from the serial port with 9600 baud rate, 8 data bits and no parity bit. After setting the baud rate, the command is to get two decimal values; each value is for each position of the servo motors.

2.3.2 Programming and Algorithm

The main program gets its input from the touchscreen sensor as a position of the mouse. To detect the X and Y coordinate of where the ball is the mouse move event is used. Once the X and Y coordinate of the ball is known, wherever the ball rolls to, the PID controler in the main program will calculate how much the servo needs to tilt. See Figure 5 and 6 for detail.

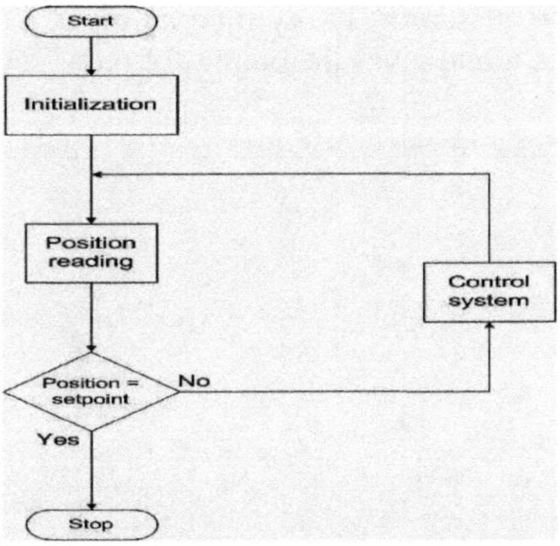

Fig. 5. Flowchart of Visual Basic program

Figure 6a is the initialization process includes defining several variables which will be later used in the program. And also opening and choosing the serial portsare also part of it.

Figure 6b explains that after the control system does its calculation, a positioning calibration like stated earlier is needed to send the right value for the servo motors. And after this process, the value will be sent to the servo controller, thus changing the angle of the servo motor arm and the position of the ball.

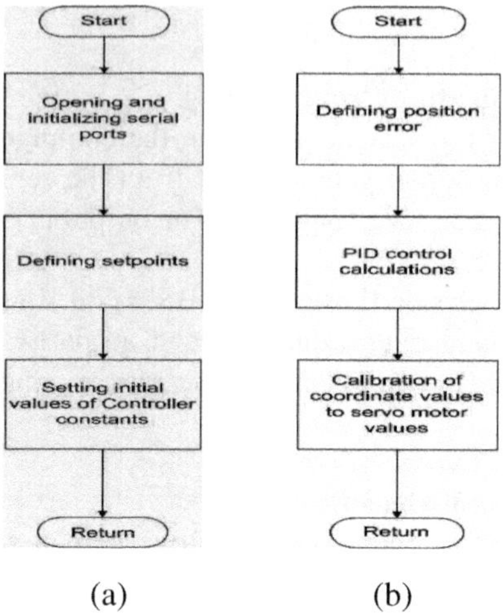

Fig. 6. (a) Flowchart of initialization; (b) Flowchart of control system

2.3.3 PID Control System Programming

The PID control system is a closed loop system which means it continuously needs a feedback from some sort of sensor. Every time an input is given to the system, the control system calculates it and gives the output for the system dynamics to act based on it.

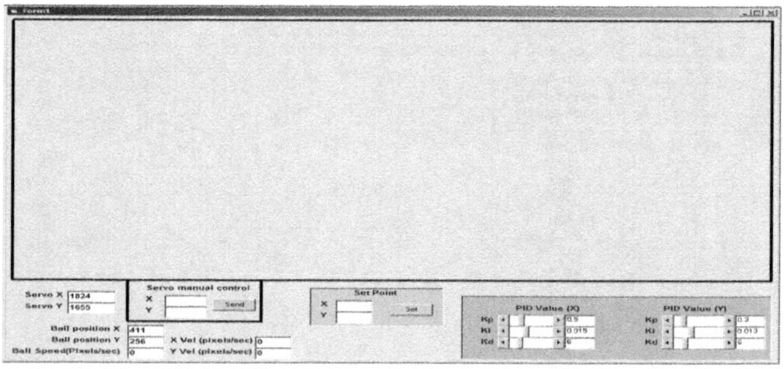

Fig. 7. Main program display form

Figure 7 shows text boxes, command buttons, and labels, which are often called events, that is used in the program.

3 Testing, Result and Discussion

3.1 Servo Motor and DT Basic Servo Controller Testing

The angle of the servo motor was measured with a protractor from SGU's workshop with a tolerance of ±1°. The extreme left of the servo position is regarded as 0° and adding the angle would mean rotating it to the right. The measurements were the same for both servo motors.

Table 1. Servo motor testing

Value given to the servo(n)	Value of PWM (μs)	Actual angle of servo (°)	Theoretical Angle of servo(°)
500	400	0.00	0
1000	800	41.00	42.01
2000	1600	125.00	126.05
3000	2400	209.00	210.08

From table 1, we can see that the servo motors are rotating with the right calibration as in the theory with a small amount of error.

The last testing was to confirm whether the servo motors have enough torque to tilt the table with the ball above them. The ball was placed on every corner of the table and then the servo motor tilted the table to the opposite direction or holds it steady on the position. The testing showed that the ball could roll freely to every direction. This means the servo motors do have enough torque for the system. Figure 8 shows the actual ball on the plate of touchscreen.

Fig. 8. Billiard ball on the table plate

3.2 Control System Testing and Discussion

The PID control system was tested both by individual terms and by combining the terms. The control system was tested whilst the system was fully operational. The purpose of this testing was to see if the PID control is sufficient to control a ball to be immobile on the desired positions on a plate. The testing will also show that a close loop control is required for the Ball and Plate system. The testing was divided into four sections; first three terms would be done by individual terms and their effect of the system. And the last testing would test the PID altogether.

3.2.1 Proportional Control Testing and Discussion

The first testing was done with a Kp value of 1 both for X and Y axis. Figure 9 shows the response of the ball with that value of Kp. The upper graph is the response in the X axis, and the lower graph is for the Y axis. The red line is the set point given to the system. And the blue line shows the position of the ball. The text box underneath the graphs shows the PID constant value. A 59g ±1g ball was placed on one of the corners of the table every time the program runs.

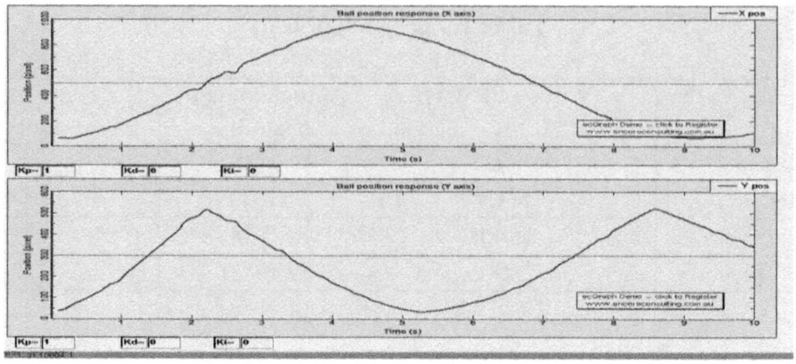

Fig. 9. Ball position response with Kp = 1

When the Kp value is then increased to 2 for both axes, the response is faster than when the Kp was 1.

3.2.2 Integral Control Testing and Discussion

The integral term is used to eliminate steady state error. The problem with this control term is that it increases overshoot. For the reason of the integral control having not much effect when applied by itself, the next testing were done by combining the integral control with the proportional control.

3.2.3 Derivative Control Testing and Discussion

The derivative term testing started with a small value of the derivative constant (Kd) until the desired value was obtained. Individual testing of the derivative term was not done because it did not have an effect to the output when the ball is immobile.

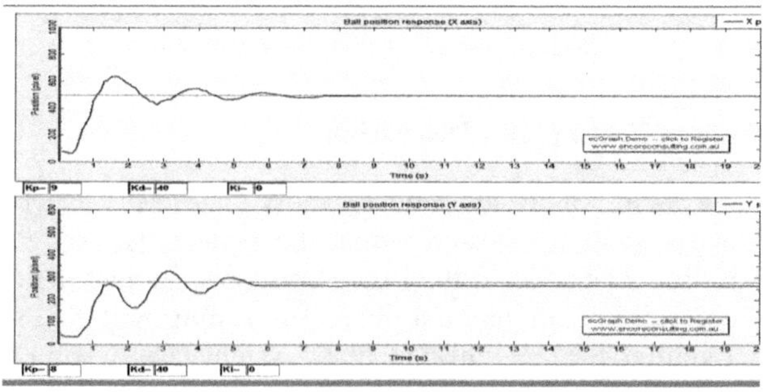

Fig. 10. Ball position response with Kd = 40

The Kd value was set to 40 (Figure 10).The graph shows that finally the ball was able to rest immobile in the set point although it was not exactly in the set point; there is a steady state error.

It is seen that the overshooting that was caused by the proportional term is finally reduced. The overshooting, which was the system's primary problem can now be reduced with the right Kd value. The time of the testing was doubled to 20 seconds in this graph and the coming graphs.

3.2.4 PID Control Testing and Discussion

The next testing was to combine all three control terms as a whole system. The Kp and Kd which is used for this testing were obtained from earlier testing in the derivative control. These constant are KpX = 9, KpY = 8, KdX = 40, and KdY = 35. The controller to be added to this system is the integral term. Because the integral term did not show much response on its individual testing, a random value of Ki = 0.3 was used on the first testing.

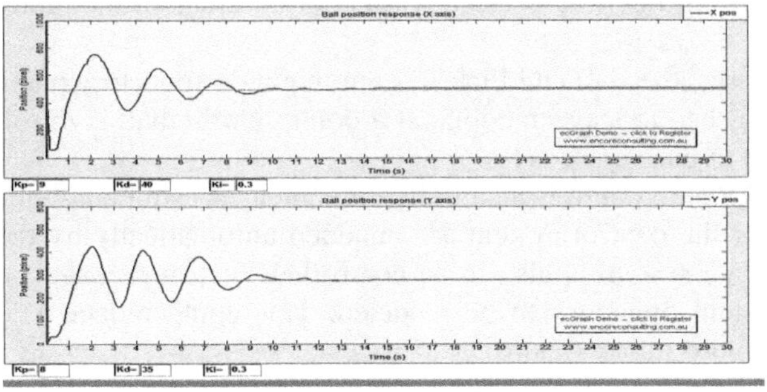

Fig. 11. Ball position response with KpX, Y = 9, 8, KdX, Y = 40, 35, Ki = 0.3

The result of this first testing is shown in Figure 11. It is seen that because of the Ki, the ball does not stop when it nearly reached the set point at the 11th second. It moved around a little and finally reached the set point. This proofs that the Ki has an effect to the system, it tries to eliminate the steady state error.

The overshooting of the ball response was much betters when KiX value of 0.15 and KiY = 0.13 show that the overshooting was much better than of Ki = 0.3. See Figure 12.

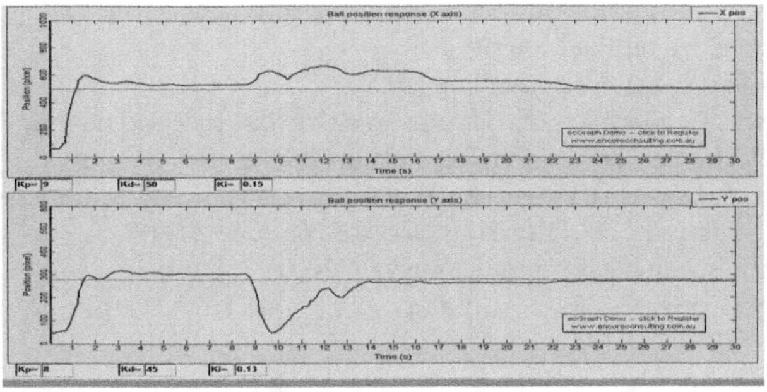

Fig. 12. Chosen PID constant with steady state disturbance

When testing the derivative controller, the Ball and Plate system was again tested to deal with a steady state disturbance. The new control where the Integral term was added was given the same disturbance. The base plate was tilted to the same degree as before. Figure 12 shows that after the ball has reached the set point, the base plate was tilted on the 9th second. And it is seen, the ball tries to make its way back to the set point. And on the 24th second, the ball reached the set point. This proofs that the steady state error was indeed eliminated by the integral term.

3.3 Conclusion and Recommendation

In conclusion, the overall project objective was achieved. The control system was able to stabilize the ball in any given point on the table plate. The PID control system enables the ball to reach the given point with a fast speed yet still being smooth. Some problem might occur due to the mechanical shortcomings of the table rotating in the yaw axis.

For future works, the Ball and Plate system might do trajectory planning, where the ball is not only asked to stay immobile at a point, but the ball is to follow a path, for example, to turn in the same circles.

The Ball and Plate system is also to be implemented using auto tuning PID, where the constants for the control system are obtained automatically by the system itself. The Ball and Plate system is also to be controlled by a more exact controller which requires the system dynamics to be modeled. This could reduce a lot of error and creates a much more stable system.

References

1. Ogata, K.: Modern Control Engineering, 4th edn. Prentice hall, New Jersey (2002)
2. Bolton, W.: Mechatronics, 3rd edn. Pearson Education Limited, England (2003)
3. Edwards, S.: Programming and Customizing the Basic Stamp computer, 2nd edn. (2006)
4. Rizzoni, G.: Principles and Applications of Electrical Engineering, Revised, 4th edn. McGraw Hill, New York (2004)
5. Ritonga, C.: Designing and Controlling a Ball and Beam Control Using Microcontroller. Swiss German University (December 007)
6. Electrotest , Resistive touchscreen (2009), `http://www.allproducts.com/manufacture98/electrot/product3.html` (last accessed on June 2009)
7. Ermicro, Basic servo motor controlling (2009), `http://www.ermicro.com/blog/?p=771` (last accessed on June 2009)
8. Innovative electronics, DT-Basic micro system (2009), `http://www.innovativeelectronics.com/innovative_electronics/pro_dtbasicmicrosys.htm` (last accessed on June 2009)
9. Parallax, Basic stamp microcontroller module (2009), `http://www.parallax.com/tabid/295/Default.aspx` (last accessed on June 2009)

A VLSI Based Scheme for Implementation of BIBD

Ravi R. Ranjan, Vadana Gupta, Monjul Saikia, and Jayanta Bora

North Eastern Regional Institute of Science and Technology, Arunachal Pradesh, India
rvbunty@gmail.com, van_nerist@rediffmail.com,
monjuls@gmail.com, jba@nerist.ac.in

Abstract. Combinatorial design theory is an area of combinatorics that is devoted to studying the problem of selecting subsets of object from a larger set of objects such that certain relationships between these subsets are satisfied. BIBD pattern is useful in vast areas like agriculture, education, and consumer product evaluation experiment, error control coding, digital finger-printing, water-marking and also they are used in organ metallic and biological field too. Very-large-scale integration (VLSI) is the process of creating integrated circuits by combining thousands of transistor-based circuits into a single chip, which involves packing more and more logic devices into smaller and smaller areas. This chip gives relax to the software by minimizing a part of execution of software. Here VLSI design methodology is used to fit a circuit for balanced incomplete block design (BIBD). Our proposed scheme generates and simulates a BIBD patterns.

Keywords: Combinatorics, Balanced incomplete block design, VLSI, Verilog.

1 Introduction

Combinatorial design theory is an area of combinatorics that is devoted to studying the problem of selecting subsets of object from a larger set of objects such that certain relationships between these subsets are satisfied. The theory of combinatorial designs is a field of mathematics that has found application to a variety of applied fields, ranging from the construction digital finger-printing [1], water-marking [2] and also they are used in organ metallic and biological field, error-correcting codes [12] to the design of statistical experiments. Design of bigger valued balance incomplete block design takes a large complexity of computation.

VLSI has been around for a long time, there is nothing new about it but as a side effect of advances in the world of computers, there has been a dramatic proliferation of tools that can be used to design VLSI circuits [3]. Alongside, obeying Moore's law, the capability of an IC has increased exponentially over the years, in terms of computation power, utilization of available area. The combined effect of these two advances is that people can now put diverse functionality into the IC's, opening up new frontiers. Examples are embedded systems, where intelligent devices are put inside everyday objects, and ubiquitous computing where small computing devices proliferate to such an extent that even the shoes you wear may actually do something useful like monitoring your heartbeats. These two fields are kind a related and getting

V.V. Das, J. Stephen, and Y. Chaba (Eds.): CNC 2011, CCIS 142, pp. 609–614, 2011.

into their description can easily lead to another article. Digital VLSI circuits are predominantly CMOS based. The way normal blocks like latches and gates are implemented is different from what students have seen so far, but the behavior remains the same. All the miniaturization involves new things to consider. A lot of thought has to go into actual implementations as well as design. Let us look at some of the factors involved:

Circuit Delays: Large complicated circuits running at very high frequencies have one big problem to tackle - the problem of delays in propagation of signals through gates and wires. The operation speed is so large that as the delays add up, they can actually become comparable to the clock speeds.

Power: Another effect of high operation frequencies is increased consumption of power. This has two-fold effect - devices consume batteries faster, and heat dissipation increases.

Layout: Laying out the circuit components is task common to all branches of electronics. What's so special in our case is that there are many possible ways to do this; there can be multiple layers of different materials on the same silicon, there can be different arrangements of the smaller parts for the same component and so on.

2 Combinatorial Design

A (v, k, λ) balanced incomplete block design (BIBD) [4] is a pair (X, A), where Λ is a collection of k-element subsets (blocks) of a v-element set X, such that each pair of elements of X occurs together in exactly λ blocks. The main two properties of BIBD are discussed as follows

Property 1: Each object will belong to the same amount of blocks as any other object. Therefore, if r is the amount of blocks that an arbitrary object belongs to then at the same time, n would be the amount of blocks that would form for a (v, k, λ)-BIBD. The relationship between v, k, r, and n is given by equation 1.

$$vr = nk \tag{1}$$

That is if the repetitions are counted, as each of the v objects belongs to r blocks, a total of vr objects are there. Similarly, by observing that each of the n blocks have k objects in it, i.e. a total of nk objects.

Property 2: It deals with another relationship between r, k, v, and λ. For a particular object y contained by block B along with another object. When defined in pairs the (x, B) corresponds to blocks B containing an object $x \neq y$. There are $(v - 1)$ ways, we can select an $x \in X$ such that $x \neq y$. Thus each of these x appears in λ blocks, giving a total of $\lambda\ (v - 1)$ pairs (x, B). Starting with the block B, there are r ways to choose a block B with y contained in B. For this block, there are $(k - 1)$ ways to choose x different from y, giving a total of $r(k - 1)$ pairs (x, B). Thus these can be set equal to get the relationship $\lambda(v - 1) = r(k - 1)$. With these two relationships, r can be solved as in equations 2 and 3.

$$r = \lambda(v - 1)/(k - 1) \tag{2}$$

$$n = \lambda(v2 - v)/(k2 - k) \tag{3}$$

These relationships indicate the possibility for a particular (v, k, λ)-BIBD to exist. As r and n should necessarily be integers, the above equations can be used to check whether for a certain existing (v, k, λ)-BIBD, r and n result to an integer on calculation. But the most tough problem to be solved is devising a systematic method in combinatorial design to determine the existence of a (v, k, λ)-BIBD precisely. Moreover there are a number of constructions for generating infinite families of BIBDs with different (v, k, λ) values.

A simple example: We consider a set X containing seven objects:

$$X = \{1, 2, 3, 4, 5, 6, 7\}$$

Now, if we choose a subset of three of them at a time, there are a total of $\left\{ {7 \atop 3} \right\} = 21$ different ways to select these subsets. So far, we have not required any incidence relationships. Now, suppose that we impose a constraint that any pair of objects must appear only once. For instance, if the combination of (1, 2, 3) has been selected, then we do not allow (1, 2, 4) to be selected as the pair (1, 2) has already been used. The triplet (3, 4, 5), however, is still possible since it does not contain any pair of objects already appearing in (1, 2, 3).With this constraint, the number of different valid combinations is reduced to seven. The selection is not unique, though. One possible set of combinations is

$$\{123, 145, 246, 167, 347, 257, 356\};$$

And another possible set of combinations is

$$\{124, 136, 157, 235, 267, 347, 456\}.$$

$$\begin{bmatrix} 1 & 1 & 1 & 0 & 0 & 0 & 0 \\ 1 & 0 & 0 & 1 & 1 & 0 & 0 \\ 0 & 1 & 0 & 1 & 0 & 1 & 0 \\ 1 & 0 & 0 & 0 & 0 & 1 & 1 \\ 0 & 0 & 1 & 1 & 0 & 0 & 1 \\ 0 & 1 & 0 & 0 & 1 & 0 & 1 \\ 0 & 0 & 1 & 0 & 1 & 1 & 0 \end{bmatrix}$$

Either of them satisfies the rule laid out above, so we consider the second one and denote it by A. What we have obtained is known as a balanced incomplete block design (BIBD), and is typically referred to as a (7, 3, 1)-BIBD. Here, the 7 refers to the total number of objects, while the 3 refers to the number of objects that we are allowed to choose at a time, and 1 indicates that each pair of objects is allowed to appear once. For a BIBD (v,k,λ) total number of blocks $n = \lambda(v^2-v)/(k^2-k)$

3 VLSI Design Process

A typical digital design flow [5, 6] is as follows: Specification, Architecture, RTL Coding, RTL Verification, Synthesis and Backend. All modern digital designs start with a designer writing a hardware description of the IC (using HDL or Hardware Description Language) in Verilog/VHDL. A Verilog or VHDL program essentially describes the hardware (logic gates, Flip-Flops, counters etc) and the interconnection of the circuit blocks and the functionality. Various CAD tools are available to synthesize a circuit based on HDL. Without going into details, we can say that the VHDL, can be called as the "C" of the VLSI industry. VHDL stands for "VHSIC Hardware Definition Language", where VHSIC stands for "Very High Speed Integrated Circuit". This language is used to design the circuits at a high-level, in two ways. It can either be a behavioral description, which describes what the circuit is supposed to do, or a structural description, which describes what the circuit is made of. There are other languages for describing circuits, such as Verilog, which work in a similar fashion.

Both forms of description are then used to generate a very low-level description that actually spells out how all this is to be fabricated on the silicon chips. This will result in the manufacture of the intended IC. There are a number of directions a person can take in VLSI, and they are all closely related to each other. Together, these developments are going to make possible the visions of embedded systems and ubiquitous computing.

Reconfigurable computing

Reconfigurable computing is a very interesting and pretty recent development in microelectronics. Reconfigurable computing involves specially fabricated devices called FPGA's, that when programmed act just like normal electronic circuits. They are so designed that by changing or "reprogramming" the connections between numerous sub-modules, the FPGA's can be made to behave like any circuit we wish.

This fantastic ability to create modifiable circuits again opens up new possibilities in microelectronics. Consider for example, microprocessors which are partly reconfigurable. We know that running complex programs can benefit greatly if support was built into the hardware itself. We could have a microprocessor that could optimize itself for every task that it tackled! Or then consider a system that is too big to implement on hardware that may be limited by cost, or other constraints. If we use a reconfigurable platform, we could design the system so that parts of it are mapped onto the same hardware, at different times. One could think of many such applications, not the least of which is prototyping - using an FPGA to try out a new design before it is actually fabricated. This can drastically reduce development cycles, and also save some money that would have been spent in fabricating prototype IC's.

Software Engineers taking over hardware design

ASIC's provide the path to creating miniature devices that can do a lot of diverse functions. But with the impending boom in this kind of technology, what we need is a large number of people who can design these IC's. This is where we realize that we cross the threshold between a chip designer and a systems designer at a higher level.

The need for the hardware compilers

Before we go further let us look at why we need this kind of technology that can convert high-level languages into hardware definitions. Silicon compiler [7] is a software, that takes user's requirement and specification and automatically generates an integrated circuit(IC). It converts hardware-description language such as Verilog or VHDL into logic called "netlist" and places equivalent logic gates on the IC.

4 Algorithm and Flow of the Design

The algorithm for BIBD(7,3,1) design is as follows.

Step 1: Input a words in registers of size 7-bit each having 1's count=3
Step 2: From k = 0, Calculate numbers of 1's in each word up to k = 127
Step 3: Put the words in look-up table having number of 1's count = 3
Step 4: Take the 1st entry of look-up table as first input
Step 5: Assign the input word as 1st output register
Step 6: xor it with each entry of look-up table and check for the xor sequence having 1's count= 4
Step 7: If found, assign the entry as 2nd output register
Step 8: Repeat step 6 for remaining entries and then check it for 1's count=4 for each of previous outputs
Step 9: If success then assign as next output registers else moves to next entry
Flow Chart is shown in figure1

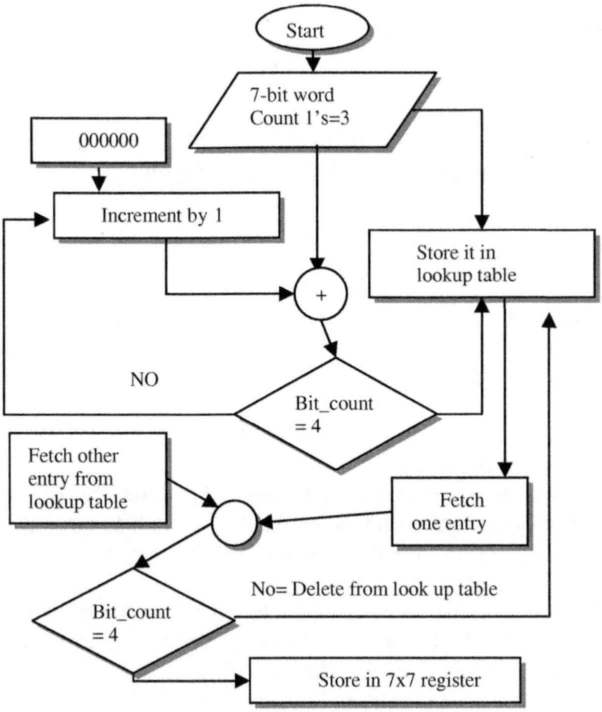

Fig. 1. Flow Chart

5 Experimental Results

The implementation of VLSI based scheme for BIBD(7,3,1) was successfully implemented using XILINX ISE-10.1 using the language VERILOG keeping in mind the requirement of the required patterns needed to be generated. Experiment is done for various 35 possible input sequences. Figure 2 shows the result of register content when input sequence 1110000 is set as input. Output patterns are the patterns of other BIBD block patterns.

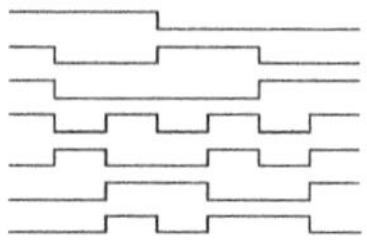

Fig. 2. Output BIBD patterns

6 Conclusion

The project has been a great learning experience about VLSI systems and Verilog. This has no direct application but can be used very efficiently in different fields as mentioned earlier. There exist other methods for implementing a BIBD patterns. Our discussed method is pettarn generation for BIBD(7,3,1) which can be generalized for a BIBD(n,k,λ). In our future work, we plan to use these generated patterns to hardware watermarking.

Acknowledgement. We'd like to thank Prof. Md. A. Hussain, Mr. Swanirbhar Majumder and the anonymous referees for valuable inputs.

References

[1] Boneh, D., Show, J.: Collusion-Secure fingerprinting for digital data. IEEE Trans. Inf Theory 44(5), 1897–1905 (1998)

[2] Cox, I., Bloom, J., Miller, M.: Digital Watermarking: Principles and Practice. Morgan Kaumann, San Mateo (2001)

[3] Malik, J., Ojha, A.: Design of a VLSI FPGA Integrated Circuit, pp. 12–15. IEEE, Los Alamitos, ISBN:0-7803-8898-4

[4] Stinson, D.R., Wei, R.: Combinatorial prperties and constructions of traceability schemes and frameproff codes. SIAM J. on Discrete Math. 11(1), 41–53 (1998)

[5] Rubin, S.M.: Computer Aids for VLSI Design (1994), `http://www.rulabinsky.com/cavd`

[6] Lee, J.M.: Verilog quickstart, A practical guide to simulation and synthesis in Verilog

[7] Johannsen, D.L.: Bristle Block: A Silicon Compiler. In: Proceedings 16th Design Automation Conference, pp. 310–313 (June 1979)

[8] Xilinx ISE Design Suite 10.1 Software Manuals

[9] Palnitkar, S.: Verilog® HDL: A Guide to Digital Design and Synthesis, 2nd edn.

[10] Brown, S., Vranesic, Z.: Fundamentals of Digital Logic with Verilog Design

[11] Gallager, R.G.: Low-density parity-check codes. IRE Transactions on Information Theory IT-8, 21–28 (1962)

[12] Saikia, M., et al.: Coded Fingerprinting Based Watermarking to Resist Collusion Attacks and Trace Colluders. In: ACE 2010 (2010), ISBN - 978-1-4244-7154-6

On 3D Streaming over Peer-to-Peer Networks

Vishaldeep Garg and Sonal Bora

Jaypee Institute of Information Technology,
Deemed University, Noida, India-201010
{garg.vishaldeep,sonal.bora}@gmail.com

Abstract. 3D streaming over peer to peer networks is gaining great importance with increase in internet advancements. After a huge success in linear streaming over peer to peer networks, many nonlinear media streaming solutions has been proposed recently. In this work, we first give brief survey on some proposed techniques in the domain of Non Linear Multimedia content streaming. Secondly, analysis on key challenges like scalability, content authenticity etc. and their proposed solutions for 3D streaming over P2P channel is made. The major objective is to give a broad picture about the current scenario of P2P 3D streaming models in order to guide and motivate the ongoing research in a better direction.

Keywords: 3D streaming, Multimedia, Peer-to-Peer, Non Linear Streaming.

1 Introduction

There has been enormous research work going on in the field of streaming media for the past two decades. The streaming models have witnessed huge success. Earlier media streaming was limited to linear streaming (2D streaming) but quite recently non-linear streaming (3D streaming) has become popular such as virtual environments and video games. In addition to the 2D uses, the 3D users can have different transmission sequence and calculations depending on their users and their interests or visibility.

This paper is organized broadly in two sections as - In section 2, an overview about 3D streaming types and the conceptual model. In section 3, major emphasis is done on key challenges and their proposed solutions, to finally conclude.

2 Non-Linear (3D) Streaming Techniques

3D streaming refers to the continuous and real-time delivery of 3D content (e.g., meshes, textures, animations, etc.) over networks to allow user interactions without a prior download [20]-[21]. 3D streaming techniques can mainly be categorized into four main types: single 3D object streaming, scene streaming, visualization streaming, and image-based streaming [3].

FLoD [4] formulated a conceptual model for P2P-based 3D scene streaming by identifying its main tasks and presented the first related framework where a P2P-VE overlay is used to discover neighbors for content exchange.

V.V. Das, J. Stephen, and Y. Chaba (Eds.): CNC 2011, CCIS 142, pp. 615–617, 2011.

Extending the work done in [15] on navigation over 3D cities in real time J. Royan et. al. proposed another design for P2P based 3D streaming [16], where primary focus was on client server architecture models and incremental level-of-detail (LOD) models Botev et al. [23] proposed HyperVerse design, where Clients are notified directly by the servers for both object and source discoveries.

3 Key Challenges and Proposed Solutions

There were several challenges prior to formulation of FLoD framework; like Distributed visibility determination, Dynamic group management, Peer and piece selection piece and peer selection. In order to deal with the challenges, FLoD [4] separated the main client side into two layers i.e. the graphics layer and the networking layer. But then a problem was identified in case of FLoD [4] that neighbor relationship was purely based on logical relationship but without the knowledge of physical networks. DPN [10] (Discovery of Physical Neighbor) technique was proposed in order to deal with the issue, according to which groups peers with physical proximity together, reduces the latencies among peer communications.

The server bottleneck problem in case of client/server framework was another important issue. This issue was solved by building a peer to peer connected architecture to become decentralized, making every client a service receiver and meanwhile a provider [17]. Another proposed solution was to use central server architecture to deliver the important base layer content to each peer [18].

Building streaming frameworks based on P2P creates security concerns for commercial vendors [7]. In applications such as online shopping, Integrity verification and authentication has become imperative [8]. One Solution to this issue can be generating a message authentication code (MAC), taking the contents to be verified as and a secret key known only to the content owner and receiver as inputs. The Content and the MAC are sent together. The receiver can recalculate the MAC with the content received and the secret key and verify it with the MAC received. Only the intended recipients who possess the key can verify the MAC. Another method of content authentication is use of Digital Signature (DS) which is a piece of information based on both the content and the owner's private key. The receiver can use the owner's public key for verification of the content. Since the public key is publicly accessible so it is useful for public authentication [6].

4 Conclusion

Recently multimedia streaming has gained greater importance with the ever-increasing needs for video conferences, video programs, long-distance learning, on-line games and entertainment etc. Despite of great achievement in the domain of 2D streaming over P2P networks, 3D streaming over P2P networks is more complex.

After a brief introduction about 3D streaming techniques, the key challenges and their proposed solutions in the domain of 3D streaming over P2P networks were summarized in this paper with an objective to guide the ongoing research in a better direction. Our future work will focus on the broadcasting technologies with an extension towards how IPTV will be able to cope up with 3D content streaming.

References

1. Garg, V., et al.: Issues concerned with Multimedia Content Streaming over a Peer-to-Peer Network. In: Proc. ITC 2010, March 12-13, pp. 122–126 (2010)
2. Hu, S.-Y., et al.: A Case for Peer-to-Peer 3D Streaming. ASCEND (November 2006)
3. Huang, T.: 3D Scene Streaming over P2P Network. Master's thesis, Taiwan (2007)
4. Hu, S., et al.: FLoD: A Framework for Peer-to-Peer 3D Streaming. In: Proc. INFOCOM (2008)
5. Sung, W., et al.: Selection Strategies for Peer-to-Peer 3D Streaming. In: Proc. NOSSDAV (2008)
6. Wu, H.-T., et al.: Public authentication of 3d mesh models. In: International Conference on Web Intelligence, pp. 940–948 (2006)
7. Chan, M.-C., et al.: Secure Peer-to-Peer 3D Streaming. Multimedia Tools & Applications (2008)
8. Chan, M.-C., et al.: Group-Based Peer-to-Peer 3D Streaming Authentication. In: Proceeding of the 15th Int. Conference on Parallel and Distributed Systems (2009)
9. Cheng, W.: Streaming of 3D progressive meshes. In: Proceeding of the 16th ACM International Conference on Multimedia (2008)
10. Wu, C., et al.: Discovery of Physical Neighbors for P2P 3D Streaming. In: Proc. ICUMT (2009)
11. Chien, C.: Bandwidth Aware Peer-to-Peer 3D Streaming. Master's thesis, Taiwan (2009)
12. Hu, S.-Y.: Peer-to-Peer 3D Streaming. Ph.D Dissertation, Taiwan (November 2009)
13. Mayer-Patel, K., et al.: Adaptive streaming for nonlinear media. IEEE Multimedia (2007)
14. Royan, J., et al.: Hierarchical representation of virtual cities for progressive transmission over networks. In: Proc. 3DPTV (2006)
15. Royan, J., et al.: A new progressive and hierarchical representation for network based navigation in urban environments. In: Proc. VMV, pp. 299–307 (2003)
16. Royan, J., et al.: Network-Based Visualization of 3D Landscapes and City Models. IEEE Computer Graphics and Applications 27, 70–79 (2007)
17. Gao, W., et al.: Recent Advances in Peer-to-Peer Media Streaming Systems. China Communications 3(5), 52–57 (2006)
18. Chen, Z., et al.: 3D-wavelet based Secure and Scalable Media Streaming in a Central controlled P2P Framework. In: Int. Conf. on Advanced Networking and Application (2007)
19. Cheng, L., et al.: Real-time 3d graphics streaming using mpeg-4. In: Proc. IEEE/ACM Wksp. On Broadband Wireless Services and Appl. (2004)
20. Teler, E., et al.: Streaming of complex 3d scenes for remote walkthroughs. In: EUROGRAPHICS (2001)
21. Tran, D., et al.: A peer-to-peer architecture for media streaming. IEEE JSAC (2004)
22. Cohen, B.: Incentives build robustness in bittorrent. In: Proc. Economics of P2P Systems (2003)
23. Botev, J., et al.: The HyperVerse - Concepts for a Federated and Torrent-Based 3D Web. Int. J. Advanced Media and Communication (2008)

Secure Identity-Based Key Establishment Protocol

Om Pal[1], Anupam Saxena[1], Zia Saquib[1], and Bernard L. Menezes[2]

[1] Centre for Development of Advanced Computing, Mumbai, India
{ompal,anupam,saquib}@cdacmumbai.in
[2] Indian Institute of Technology Bombay, Mumbai, India
bernard@it.iitb.ac.in

Abstract. Diffie-Hellman key exchange protocol is a well known protocol for establishing the key on the open channel but due to man-in-middle attack confidentiality of the message might be compromised. Public Key Infrastructure (PKI) requires a central authority or trusted third party to provide fundamental security services. To address above issues like trustiness in PKI and man-in-middle attack in Diffie-Hellman, we propose a secure identity-based key management scheme which combines the principles of Diffie-Hellman, third party and uses the identity of the receiver to establish the common key between two nodes.

Keywords: Key Distribution, Key Management, Session Key, Deffie-Hellman, Man-in-Middle, Key Exchange.

1 Introduction

Due to exponential growth in the use of Internet technology, life has become easier. It is true that we have benefited by the use of this technology but we have failed to maintain complete secrecy of our data. One critical security requirement is that communication channels need to be secured. For providing confidentiality on open channel, encryption and decryption is needed. Key management schemes are used for Confidentiality, Authentication, Non-repudiation etc. Confidentiality ensures that no one can read the message except the intended receiver. Integrity will ensure that data has not been changed. Authenticity ensures that the origin of an object is what it claims to be. Non-repudiation proves that the sender cannot deny about sending a message later. Availability is the ability to use the information or resources when it is desired. After considering security issues of confidentiality, authentication and integrity of data we can not ignore role of Key management and key distribution. Key management implies different services such as generation, registration, distribution, installation, storage, retrieval, renewal, revocation, archiving or destruction of the key. Strength of encryption and decryption is dependent on encryption algorithm and key which is used for encryption and decryption. Secure keys need to be established before cryptographic techniques can be used to secure the communication.

There are two areas of key management: Asymmetric key management and symmetric key management. In asymmetric key management, public and private key pair is used. In symmetric key management secret key between two or more than two

V.V. Das, J. Stephen, and Y. Chaba (Eds.): CNC 2011, CCIS 142, pp. 618–623, 2011.

parties is used. Main hurdle in symmetric key management is the distribution of secret key between the communicating parties. Symmetric key can be distributed manually at each party but in this case trustiness is the issue. Symmetric key can be also established at both communicating partiers by exchanging some parameters on open channel. Deffie-Hellman is the one of the approach which establishes the symmetric key at both ends by sending some parameter on the open channel. But in Deffie-Hellman key exchange an attacker can play role of middle man between two communicating parties.

In this paper we establish the symmetric key between two parties and also our scheme eliminates the possibility of man-in-middle attack between the communicating parties.

Rest of paper is organized as follows: in next section related work is presented. In section 3, proposed scheme is given. Section 4 concludes the paper and in last reference section is given.

2 Related Work

Key management is a well known research topic in the cryptographic research community. A lot of work has already been done in field of key management but still this field is not fully saturated.

Michael Steiner Gene Tsudik Michael Waidne [12] presented three new group Deffie-Hellman key distribution protocols. In each protocol, number of messages is more, so the computational time is high. If any member leaves the group then there is the need to setup the group key again otherwise forward secrecy will be problem. Schemes are also vulnerable to man-in-middle attack.

Emmanuel Bresson [7] presented multicasting approach for dynamic group. The good feature of Emmanuel's scheme is that it builds group key dynamically, and time complexity of the group key generation is linear. If there are 'n' number of nodes is in the group, then only 'n' numbers of transmissions are required. In this scheme first node sends some parameters (g, g^a) to next node to initiate the process; where 'g' is the generator of the group and 'a' is the secrete parameter of first node. After getting parameters from first node, second node raises the power of its secret number on all received parameters. Second node passes all modified parameters and the last parameter of first node to third node. This chain remains continuous up to the second last node of the group. After raising the secret number, last node of the group broadcast all parameters. From the broadcast values, each node takes its respective part and then builds the group key. This scheme provides the linear time complexity but it suffers from man-in-middle attack. Any attacker can also participate during the construction of group key and attacker can get all confidential information of other members.

In the paper "Performance and Security Assessment of a PKC Based Key Management Scheme for Hierarchical Sensor Networks" [8], Ayesha Naureen, Attiya Akram, Rabia Riaz, Ki-Hyung Kim, and H. Farooq Ahmed, presented a public key cryptography based key management scheme where cluster based architecture is used, in which the base station serves as the key distribution centre and maintains public keys for all the nodes in the network. In all the nodes, the private keys are pre-installed and the public key exchanges are done via broadcasts.

In 1984, Adi Shamir, described "Identity-Based Cryptosystems and Signature Schemes"[11], where any publicly-known string (e.g. someone's email address) could be used as public key. The Identity-based cryptography uses public-key cryptography in a different manner, where the public key of a user gives unique information about the identity of the user. This can use the text name, or domain name, or the physical IP address as a key.

Our scheme uses concept of Deffie-Hellman key exchange and Id based cryptography to establish the common key between two parties and eliminates the possibility of man-in-middle attack. In next section proposed scheme is presented.

2.1 Man-in-Middle Attack in Deffie-Hellman Key Exchange

Let Alice and Bob are two parties who want to communicate with each other. Let 'a' is the secret of Alice, 'b' is the secret of Bob, 'g' is a generator of the group G, 'p' is the order of the group. Both 'g' and 'p' are open parameters. There is a man-in-middle, say Eave.

In presence of Eave, two keys (K and K') are established. By using key K, Eave can communicate with Alice and by using K', he can communicate with Bob; and both Alice and Bob are not aware about the presence of Eave. Alice thinks that she is talking to Bob but in actual she is talking to Eave.

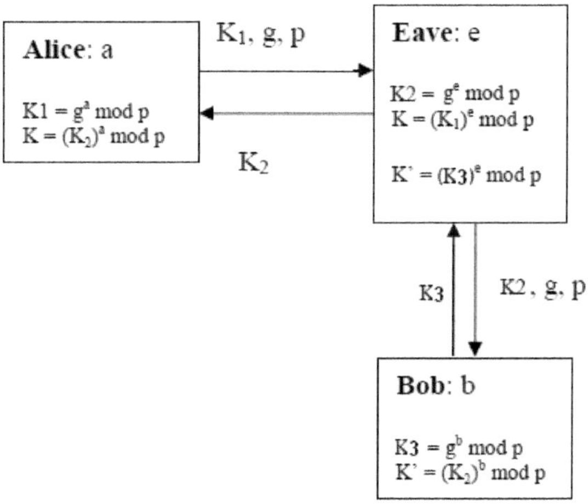

Fig. 1. Key setup in the presence of Eave

3 Proposed Scheme

Scheme assumes that there is one centralize server (Partial Key Generator) which is used to generate the partial key for communicating pairs.

First PKG generates secret key of any node by using the identity of that node and the master key (secret key of PKG). Generated secret key is the common key between PKG and node for which this secret key is generated. After generation of secret key for each node, it distributes the secret keys to corresponding nodes manually.

Each node stores its secret key for communication with PKG. PKG never stores any secret key of any node. In case of need of secret key of any node, PKG calculates the secret key at run time by using identity of node and its master key. Partial key of any node is the random number (r) which is sent by PKG for common key generation. PKG encrypts the partial key by secret key of receiving node before transmitting.

Let Id_a is the identity of node A, Id_b is the identity of node B. M is the master key (secret key) of PKG. F is a one way function which is used for generating secret key M_i of any node i, p is a large prime number, G is a group of order p, g is the generator of the group, a is a secret parameter of node A, b is a secret parameter of node B.

Generation of Secret Key: Let M_a is the secret key of node A. PKG generates Ma by using function F as

$$M_a = F(Id_a, M) \tag{1}$$

Similarly PKG generates Mb for node B by using function F as

$$M_b = F(Id_b, M) \tag{2}$$

After establishment of secret key with PKG, any node can get partial key from PKG for generating common key with any other node.

Generation of Common Key: Let node A and node B want to generate a common key then it will be generated in this way-

Request to PKG for partial key: Let A wants to communicate with B then A will request to PKG for partial key (random number). After getting request from A, PKG will send common partial key (r) to A and B by using secrets key Ma of A and secret key Mb of B. Now both nodes have common partial key of PKG and they can use it for common key generation.

Creation of partial key of A
Let K_1 is partial key of node A then $K_1 = g^a \bmod p$

Now A transmits K_1 to node B

Creation of partial key of B
Let K_2 is partial key of node B then $K_2 = g^b \bmod p$
Now B transmits K_2 to node A

Generation of common key for A and B: Now both nodes A and B have values K_1, K_2 and 'r', where 'r' is a partial key of PKG for node A and B.

A will generate the common key by using

$$K = (K_2)^{ar} \bmod p \tag{3}$$

B generates common key by using

$$K = (K_1)^{br} \bmod p \tag{4}$$

Now both parties have common key, hence they can communicate with each other.

Key Revocation: Any pair of nodes can revoke their key by sending fresh request to PKG for obtaining partial key of PKG. By using partial key of PKG they can construct a fresh common key. Nodes can use this new key as a long term key or as a session key.

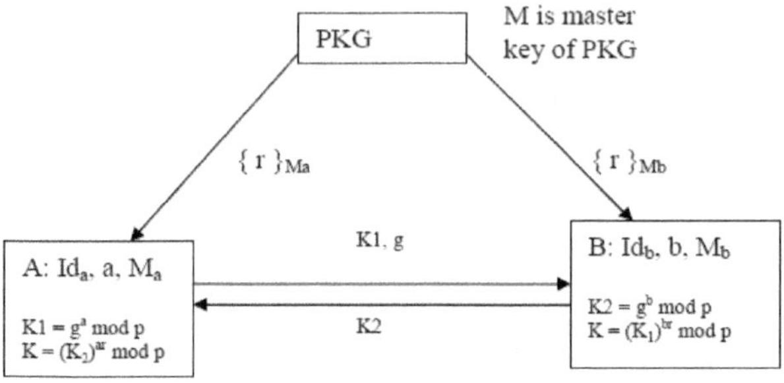

Fig. 2. Proposed scheme

Security Analysis against Man –in- Middle Attack: As scheme assumes secret 'r' is transferred from PKG to any node by using the secret key of respective node. Only parameters 'g', K_1 and K_2 passes on open channel so attacker can only get these parameters. Partial key (r) of PKG is secret from attacker so attacker will not be able to drive the common key of nodes A and B. Hence man-in-middle attack is not possible in the proposed scheme.

4 Conclusion

Key management is the back bone of the cryptography. For speeding up the encryption/decryption, symmetric key is the favorable approach. In this paper first we discussed the some key management approaches, followed by issues in key management approaches like man in middle attack in Deffie-Hellman Key exchange. Finally we proposed our solution which eliminates the man-in-middle attack in Deffie-Hellman Key Exchange and establishes a secure common key between any two nodes. In the proposed scheme node stores secret key with PKG only and on demand basis, node establishes common key with any other node of the network. If there is very large number of nodes in the network, it is not feasible for any node to store a common key for each node. Hence the presented solution is an appropriate solution for establishing a common key on demand basis.

References

1. Menezes, A.: An Introduction to Pairing-Based Cryptography: Mathematics Subject Classification. Primary 94A60 (1991)
2. Boneh, D., Franklin, M.: Identity-Based Encryption from the Weil Pairing. SIAM J. of Computing 32(3), 586–615 (2003)
3. Bojkovic, Z.S., Bakmaz, B.M., Bakmaz, M.R.: Security Issues in Wireless Sensor Networks. International Journal of Communications 2(1) (2008)
4. Mokammel Haque, M., Pathan, A.-S.K.: An Efficient PKC-Based Security Architecture for Wireless Sensor Networks. IEEE, Los Alamitos (2007), ISBN:1-4244-1513-06/07/$25.00
5. Ma, J., Zhang, S., Zhong, Y., Wu, Y.: PEAN: A Public Key Authentication Scheme in Wireless Sensor and Actor Network. In: Proceedings of The Sixth IEEE International

Conference on Computer and Information Technology, CIT 2006 (2006)
6. Oliveira, L.B., Scott, M., López, J., Dahab, R.: TinyPBC: Pairings for Authenticated Identity-Based Non-Interactive Key Distribution in Sensor Networks
7. Bresson, E., Chevassut, O., Pointcheval, D., Quisquater, J.-J.: Provably Authenticated Group Deffie-Hellman Key Exchange. In: Computer and Communications Security-Proc.of ACM CCS 2001, November 5-8 (2001)
8. Naureen, A., Akram, A., Riaz, R., Kim, K.-H., Farooq Ahmed, H.: Performance and Security Assessment of a PKC Based Key Management Scheme for Hierarchical Sensor Networks. IEEE, Los Alamitos (2008) ISBN: 978-1-4244-1645-5/08/$25.00
9. Zhang, Y., Liu, W., Lou, W., Fang, Y.: Securing Sensor Networks with Location-Based Keys. IEEE, Los Alamitos (2005), ISBN: 0-7803-8966-2/05/$20.00
10. Steiner, M., Tsudik, G., Waidner, M.: Diffie-Hellman Key Distribution Extended to Group Communication
11. Shamir, A.: Identity-Based Cryptosystems and Signature Schemes. In: Blakely, G.R., Chaum, D. (eds.) CRYPTO 1984. LNCS, vol. 196, pp. 47–53. Springer, Heidelberg (1985)
12. Diffie-Hellman Key Distribution Extended to Group Communication by Michael Steiner Gene Tsudik Michael Waidne, IBM Ziirich Research Laboratory CH-8803 Riischlikon, Switzerland

RSA Based Threshold Cryptography for Secure Routing and Key Exchange in Group Communication

Rajkumari Retoliya[1,*], Swapnil Soner[2], Ati Jain[3], and Anshu Tripathi[1]

[1] Mahakal Institute of Technology, RGPV University, Ujjain, India
retoliya.rajkumari@gmail.com, anshu.tripathi@yahoo.com
[2] Truba College of Engineering, RGPV University, Indore, India
swapnil.soner@gmail.com
[3] Swami Vivekanand College of Engineering, RGPV University, Indore, India
jain.ati@gmail.com

Abstract. Purpose of this paper is to provide reliable data transmission in group communication. We here combine RSA and Threshold Cryptosystem to securely deliver messages in group. As long as the all receivers receive secure data, it can recover the original message. We worked on RSA for secure data forwarding; we consider both splitting plaintext before encryption, and splitting ciphertext after encryption. Also we suggest exchanging keys between a Diffie-Hellman by which group communication is more secure.

Keywords: Threshold Cryptography, Group Communication, Attacks, Cryptanalyst, Diffie Hellman.

1 Introduction

A number of processes which cooperate to provide a services and an abstract identity to name a collection of processes for coordination among processes of a group. Scenario may be out of n users in a network, some t ($t<<n$) of them would like to discuss on a common concern. These t parties termed *privileged users* must communicate themselves and others must not be able to listen to the conversation between these t parties. Hence, there is a need to find new technology for such confidential communication termed as *Secure Group Communication* [3]. A solution is to have a shared key between every pair of users, which leads to storing $(n\ 1)$ keys with each user. Also, to send a message, sender must encrypt the message to each user in the group separately. Hence, the general goal of Secure Group Communication is to establish a *common secret key*, also called *Secure Group key* among privileged users for confidential communication. In any secure group communication scenario, since all the group members can send and receive messages, the sender of a message must be able to indicate his identity and the receivers must be able to verify the authenticity of the message [4]. A group of users, termed malicious parties may collude with each other and try to derive the common group key. The security of the group key management scheme is based on number of colluding parties. Scheme is termed as k

* Corresponding author.

V.V. Das, J. Stephen, and Y. Chaba (Eds.): CNC 2011, CCIS 142, pp. 624–627, 2011.

secure, if it is not possible to derive the common key even after *k* non-group (non-privileged) members collude with each other, where *k* is termed as *threshold.*

2 Threshold Cryptography

Threshold cryptography allows a party of say n people to share the ability of performing a cryptographic operation. Any t parties ($t < n$) can perform this operation jointly. This t is calling the threshold. If is infeasible for nay t-1 parties to do so, even by collusion. Also the secret cannot be recovered by any subset of parties [1].

However, key management in group communication has to be distributed service as there is no fixed infrastructure to provide centralized service. Another major challenge is to deliver reliable data transmission when some nodes may be compromised. Attackers can disrupt data transmission and incur significant data loss by tampering with, fraudulently redirecting, or even dropping data traffic.

3 RSA Algorithm

Rivest-Shamir-Adleman (RSA) scheme has become the most widely accepted and implemented general-purpose approach to public-key encryption. The RSA scheme is a block cipher. Each plaintext block is an integer between 0 and $n - 1$ for some n, which leads to a block size $\leq \log2(n)$. The typical block size for RSA is 1024 bits. Steps of this algorithm are as:

Key Generation: Pick two large prime numbers p and q, $p \neq q$;
Calculate $n = p \times q$ Calculate $\Phi(n) = (p - 1)(q - 1)$
Pick e, so that $\gcd(e, \Phi(n)) = 1$, $1 < e < \Phi(n)$
Calculate, so that $d. e \bmod \Phi(n) = 1$, d is the multiplicative inverse of e in mod $\Phi(n)$
Get public key as $PUK = \{e, n\}$ Get private key as $PRK = \{d, n\}$
Encryption: For plaintext block $M < n$, its ciphertext $C = M^e \bmod n$.
Decryption: For ciphertext block C, its plaintext is $M = C^d \bmod n$.

3.1 Why RSA Works

As we have seen from the RSA design, RSA algorithm uses modular exponentiation operation. For $n = p \cdot q$, e which is relatively prime to $\Phi(n)$, has exponential inverse in mod n. Its exponential inverse d can be calculated as the multiplicative inverse of e in mod $\Phi(n)$. The reason is illustrated as follows. Based on Euler's theorem, for y which satisfies

$y \bmod \Phi(n) = 1$, the following equation holds.
$x^y \bmod \Phi(n) = x \bmod n$

As $d .e \bmod \Phi(n) = 1$, we have that $M^{ed} \equiv M \bmod n$. So the correctness of RSA cryptosystem is shown as follows.

Encryption: $C = M^e \bmod n$;
Decryption: $M = C^d \bmod n = (M^e)^d \bmod n = M^{ed} \bmod n = M \bmod n = M$.

The premise behind RSA's security is the assumption that factoring a big number (n into p, and q) is hard. And thus it is difficult to determine Φ (n). Without the knowledge of Φ (n), it would be hard to derive d based on the knowledge of e. However factoring n is not the only way to break RSA.

4 Diffie Hellman Key Exchange Algorithms

Diffie–Hellman establishes a shared secret that can be used for secret communications by exchanging data over a public network [6]. Here is explanation the encryption's mathematics:

All users agree on global parameters:
Large prime integer or polynomial q, α a primitive root mod q
Each user (**A**) generates their key. Here let us take:
Public Key **d as** = $\mathbf{x_A}$, Private Key **e** *as* = $\mathbf{y_B}$ and secret key (number): x_A
Compute their **public key**: $y_A = \alpha^{xA} \bmod q$
Each user (**B**) generates their key : secret key (number): y_B
Compute their **public key**: $y_B = \alpha^{xB} \bmod q$
Shared session key for users A & B is K_{AB}:

$$K_{AB} = \alpha^{xA.xB} \bmod q \quad = y_A^{xB} \bmod q \text{ (which } \mathbf{B} \text{ can compute)}$$
$$= y_B^{xA} \bmod q \text{ (which } \mathbf{A} \text{ can compute)}$$

RSA is known algorithm so it is easily decoded. In RSA if d is compromised then p, q, n, e and d must be regenerated and another deficiency is to select e (public key) for the individual. Diffie-Hellman key Exchange algorithm does not protect against replay attacks and another deficiency is to man-in-the-middle-attack. This concept work in group communication with threshold cryptography gives more secure key exchange and date security. In the group communication all member belong to group exchange key using the threshold concept all are these combination initially secure with RSA and after that Diffie Hellman work [2].

5 Examples for New Concept

Example 1: P =7, q = 11, n = 77, Φ (n) = 60
The factors of 60 is 2, 2, 3, and 5.Therefore, our public key e must not have a factor of 2, 3, and 5.Let us choose the public key value of e as 13. Then select the private key d such that d.e mod Φ (n) =1.Let us choose d as 37, because we can see that (13*37) mod 60= 481 mod 60=1, which satisfies our condition.

Now, RSA for encryption, $C = 5^{13} \bmod 77 = 26$
RSA For decryption, $M = 26^{37} \bmod 77 = 5$
But we are taking Public Key **d as** = $\mathbf{x_A}$ Private Key **e** *as* = $\mathbf{x_B}$

We use a prime number n=77 and base α =5. We assumes X_A is 37 then calculate $y_A = \alpha^{xA} \bmod q$ i.e $y_A = 5^{37} \bmod 77 = 47$

Then we assumes X_B is 13 then calculate $y_B = \alpha^{xB}$ mod q and $y_B = 5^{13}$ mod 77 = 26

Shared session key for users A & B is K_{AB}: $K_{AB} = \alpha^{xA.xB}$ mod q that is 5^{13*37}mod 77= 5 and y_A^{xB} mod q is 47^{13} mod 77= **5**and y_B^{xA} mod q is 26^{37}mod 77=**5**

Example 2: P =7, q = 17, n = 119, Φ (n) = 96

The factors of 96 is 2,22,2,2, and 3.Therefore , our public key e must not have a factor of 2,and 3.Let us choose the public key value of e as 5.

Then select the private key d such that d.e mod Φ (n) =1.Let us choose d as 77, because we can see that (5*77) mod 96= 385 mod 96=1,which satisfies our condition.

Now, RSA for encryption, $C = 10^5$ mod 119 = 40

For RSA decryption, $M = 40^{77}$ mod 119 = 10

But we are taking Public Key **d as = x_A** Private Key **e *as* = x_B**

We use a prime number n=119 and base α =10. We assumes X_A is 77 then calculate $y_A = \alpha^{xA}$ mod q , $y_A = 10^{77}$ mod 119 = 96

Then we assumes X_B is 5 then calculate $y_B = \alpha^{xB}$ mod q, $y_B = 10^5$ mod 119 = 40

Shared session key for users A & B is K_{AB}:

$K_{AB} = \alpha^{xA.xB}$ mod q that is, 10^{77*5} mod 119 = 10 and y_A^{xB} mod q is 96^5 mod 119= **10** and y_B^{xA} mod q is 40^{77}mod 119 =**10**

Above all examples, all the answers are same as follow diffi-hellman concept. But here we provide this new algorithm is more security than RSA and Diffi-Hellmna because public key of Diffi Hellman from RSA's public and private key.

Acknowledgements

We would like to thank the all faculty members, Prof. J. K. Khatwani and Prof. Lalji Prasad of the institute, who helped us lot in calculating the facts and figures and the anonymous reviewers who provided helpful feedback on our manuscript.

References

1. Ertaul, L.: Elliptic Curve Cryptography Threshold Cryptography. IJCSNS International Journal of Computer Science and Network Security 7(4), 48 (2007)
2. Sun, Y., Trappe, W., Liu, K.J.R.: Network-Aware Security for Group Communications. Springer, Heidelberg (2008) ISBN: 978-0-387-68846-6
3. Xu, S.: On the Security of Group Communication Schemes. Department of Computer Science, University of Texas at San Antonio
4. Aparna, R., Amberker, B.B.: Dynamic Authenticated Secure Group Communication. World Academy of Science, Engineering and Technology 34 (2007)
5. Desmedt, Y., Frankel, Y.: Threshold cryptosystems. In: Brassard, G. (ed.) CRYPTO 1989. LNCS, vol. 435, pp. 307–315. Springer, Heidelberg (1990)
6. Bresson, E., Chevassut, O., Pointcheval, D.: Dynamic group Diffi-hellman key exchange under standard assumption. In: Knudsen, L.R. (ed.) EUROCRYPT 2002. LNCS, vol. 2332, p. 321. Springer, Heidelberg (2002)
7. Rivest, R.L., Shamir, A., Adleman, L.: A Method for Obtaining Digital Signatures and Public-Key Cryptos

Cooperative Caching Strategy in Mobile Ad Hoc Networks

Naveen Chauhan, Lalit K. Awasthi, and Narottam Chand

Department of Computer Science and Engineering
National Institute of Technology, Hamirpur (H.P.)-177005
{naveen,lalit,nar}@nitham.ac.in

Abstract. We propose a cooperative caching scheme, called global cluster cooperation (GCC) in mobile ad hoc networks. In this scheme information regarding the contents of all the mobile nodes (MNs) within a cluster, is maintained at each node. Simulation experiments show that the GCC caching mechanism achieves significant improvements in average query latency and reduces the cache management overheads in comparison with other caching strategies.

Keywords: Ad hoc networks, cooperative caching, cluster, consistency.

1 Introduction

Caching is one of the most attractive techniques that improve data retrieval performance in wireless mobile environment [2, 3]. With caching, the data access delay is reduced since data access requests can be served from the local cache, thereby obviating the need for data transmission over the scarce wireless links. However, caching techniques used in one-hop mobile environment may not be applicable to multi-hop ad hoc environment since the data or request may need to go through multiple hops. Variable data size, frequent data updates, limited client resources, insufficient wireless bandwidth and clients' mobility make cache management a challenging task in mobile ad hoc networks. As mobile nodes in ad hoc networks may have similar tasks and share common interest, cooperative caching, which allows the sharing and coordination of cached data among multiple nodes, can be used to reduce the bandwidth and power consumption. In this paper, we investigate the data retrieval challenge of mobile ad hoc networks and propose a novel scheme, called global cluster cooperation (GCC) for caching.

2 System Model

In GCC, when a client suffers from a cache miss (called local cache miss), the client will look up the required data item from the cluster members. Only when the client

V.V. Das, J. Stephen, and Y. Chaba (Eds.): CNC 2011, CCIS 142, pp. 628–630, 2011.

cannot find the data item in the cluster members' caches (called cluster cache miss), it will request the CSN which keeps the global cache state (GCS) and maintains the information about the node in the network which has copy of desired data item. If a cluster other than requesting nodes' cluster has the requested data (called remote cache hit), then it can serve the request without forwarding it further towards the server. Otherwise, the request will be satisfied by the server.

Based on the above idea, we propose a cache discovery algorithm to determine the data access path to a node having the requested cached data or to the data source. Assume that MH_i denotes mobile node/client i. In Figure 1, let us assume MH_i sends a request for a data item d_x and MH_k is located along the path through which the request travels to the data source MH_s, where $k \in \{a, c, d\}$.

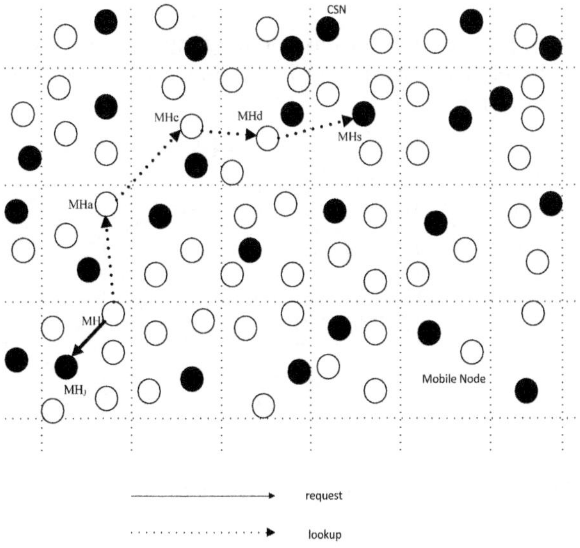

Fig. 1. Request packet from client MH_i is forwarded to the data source MH_s

3 Simulation Results

3.1 Effects of Cache Size

Figure 2 shows the effect of cache size on message overhead by varying the cache size from 200 KB to 1400 KB. It shows that GCC performs much better than NC (No Cooperation) and CacheData [1] in terms of message overhead.

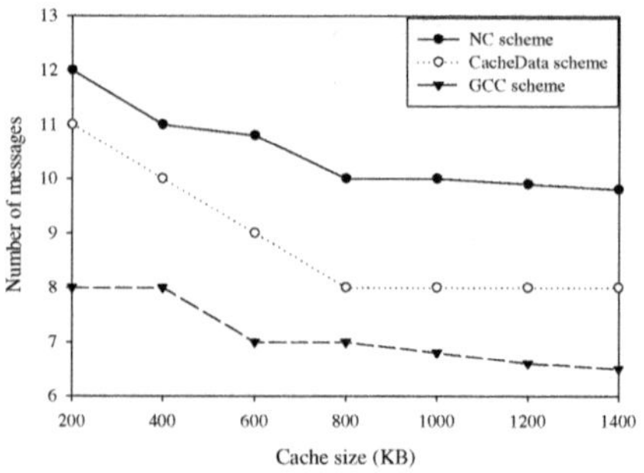

Fig. 2. Effects of cache size on message overhead

3.2 Effects of Transmission Range

Figure 3 shows that increase in transmission range increases the expected progress of the packet towards its final destination but at the expense of a higher energy consumption per transmission.

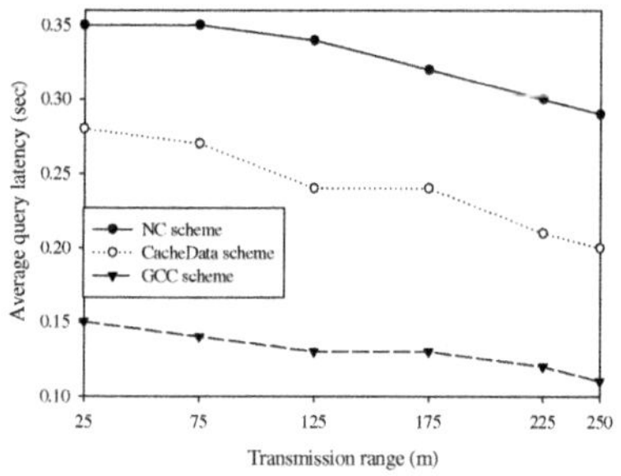

Fig. 3. Effects of transmission range on average query latency

References

1. Yin, L., Cao, G.: Supporting Cooperative Caching in Ad Hoc Networks. In: IEEE INFOCOM, pp. 2537–2547 (2004)
2. Chand, N., Joshi, R.C., Misra, M.: A Zone Cooperation Approach for Efficient Caching in Mobile Ad Hoc Networks. International Journal of Communication System 19(9), 1009–1028 (2006)
3. Tang, B., Gupta, H., Das, R.: Benefit-Based Data Caching in Ad Hoc Networks. IEEE Transactions of Mobile Computing 7(3), 289–304 (2008)

Enhancement of Color Image Using Forth Order Partial Differential Equation Based on S-Type Enhancement

Awanish Kumar Kaushik[1], Anubhav Kumar[1], R.L. Yadava[1], and Divya Saxena[2]

[1] Department of Electronics and Communication, Galgotia's college of Engineering and Technology, G. Noida (U.P.), India
[2] Department of Mathematics, Vishveshwarya Institute of Engineering and Technology, G.B. Nagar (U.P.), India
kaushik2feb@gmail.com, rajput.anubhav@gmail.com, divya_saxena070@yahoo.co.in

Abstract. The color images generally contain data which may be corrupted by noise, and the subject under study is normally blurred because of improper focusing during image acquisition. Thus, this paper, proposes the use of fourth order Partial Differential Equation and some enhancement parameter to remove this problem. Our purposed method makes the use of information of different intensity from normal image and noisy images, so that, the performance of our proposed method is better than that of existing color image enhancement algorithms. Various experiments are used to evaluate and analysis the performance of our proposed algorithm. Experimental results tested from a large data set of image have demonstrated that the proposed method is effective and practical. Some parameters like as PSNR are analyzed for both existing and proposed method to determine the success and limitation of our method.

Keywords: Fourth Order Partial Differential Equations, Image Enhancement, S type transform, PSNR.

1 Introduction

Color image enhancement develops very fast and plays an important role in the presentation of an image in recent years. Because of wide application of color image enhancement peoples are more attracted to this process, for example, in multimedia, biomedical science and internet etc. field. The generalization of these techniques to color images is not straight forward. Unlike grey scale images, there are some factors in color images like hue, saturation, intensity, which need to be properly taken care of for image enhancement. Hue is the attribute of the color which defines what kind of color it is i.e. blue or an brown.

Various algorithms are available for contrast enhancement in grey scale images, which change the grey values of pixels depending on the criteria for enhancement. George et al. [1] Color image enhancement and denoising operation is performed by using an optimized set of filters. Scaling and shifting are two hue preserving techniques which are proposed by Yang et al. [7] for the processing of luminance and

V.V. Das, J. Stephen, and Y. Chaba (Eds.): CNC 2011, CCIS 142, pp. 631–636, 2011.

saturation components. Strickland et al. [6] proposed an enhancement technique which is based on the fact that objects can exhibit variation in color saturation with little or no corresponding luminance variation. Various distinct methods for color contrast enhancement are proposed in [8, 9] using 3-D histogram of colored images.

In particular, mathematical frameworks define a powerful tool of partial differential equations and functional analysis has been extensively studied to provide the fundamental answer in image processing. The PDE models can be widely used in a broad range of image restoration tasks like as denoising, image enhancement [10,11] and color image processing [12,13].

In this paper we focused on Enhancement of Color Image using Forth order Partial Differential Equation. The paper is characterized in following sections: Section 2 gives the theoretical foundation, Section 3 gives the proposed model, section 4 gives result and analysis and section 5 gives conclusions respectively.

2 Paper Preparation

Our proposed model for color medical image enhancement using fourth order PDE (Partial Differential Equation) consists of four stages and which is shown in block diagram in figure 1. It is a task in which the value of a set of pixel of one image is transformed to a new value of a set of pixel so that the produce new image is visually pleasing and is also more suitable for analysis. The block diagram of proposed method describe below.

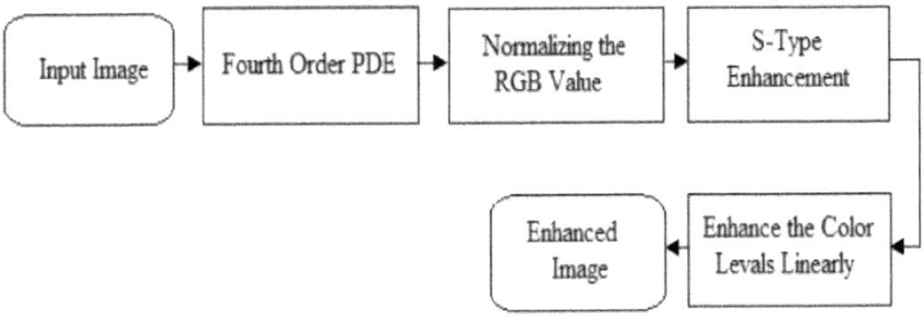

Fig. 1. Block Diagram of Proposed Method

2.1 Fourth Order PDE

One of the most commonly used PDE based denoising technique is introduced by the Perona-Malik method. The Perona - Malik equation for an image u is given by

$$du/dt= div[c (\nabla u) \nabla u],\ u(x, y) \mid t=0 = u0 (x, y) \tag{1}$$

Where ∇u *is* the gradient of the image u, *div* is the divergence operator and c is the diffusion coefficient. The diffusion coefficient c is a non decreasing function and diffuses more on plateaus and less on edges and thus edges are preserved. Two such diffusion coefficients suggested by Perona and Malik [3] are

$$C(s) = 1/(1+(s/k)\ 2\ ) \tag{2}$$

And

$$C(s) = \exp\ [-(s/k)^2 \tag{3}$$

In this work we implemented and tested the fourth order PDEs proposed by *You* and *Kaveh* . *You* and *Kaveh* [14] proposed an *L2 – curvature* gradient flow method which is given as

$$du/dt = -\nabla^2[c\,(\nabla^2 u)\,\nabla^2 u] \tag{4}$$

Where $\nabla^2 u$ is the Laplacian of the image u.The discrete form of non-linear fourth order PDE described in equation (4)

$$u^{n+1}_{i,j} = u^{n}_{i,j} - \Delta t\, \nabla^2 g^{n}_{i,j} \tag{5}$$

2.2 Normalized the RGB Value

In general, a pixel in a color image has a color vector with three components R, G, and B and the color images are saved and viewed using RGB color space. The process of an image enhancement in any of the above mentioned spaces (i.e., CMY, etc.), the image require to be transformed to that space. The transformations involved in transforming the color image from RGB space to other mentioned spaces and again the inverse coordinate transformation has to be implemented for displaying the images.

2.3 S-Type Enhancement

S-type enhancement technique is generally and widely used contrast enhancement technique for grey scale images. The methodologies of S-type enhancement techniques can be found in the enhancement theory. Generally the grey scale contrast enhancement techniques is used to enhance the intensity of the color image and they are hue preserving. Hue is that attribute of a color which defines what kind of color it is, i.e., a yellow or a green. The main objective is to keep the transformed values within the range of the RGB space, i.e., the transformations should be free from gamut problem.

2.4 Enhance the Color Level Linearly

At last the image is recovered by enhance the color level linearity and this is possible by contrast enhancement of gray scale images by performing the same operation on the RGB plane changes the hue of the pixel values of the image. This enhances the image without affecting the hue of the pixels. But, over enhancement of certain pixels leads to gamut problem when the image is transformed back to RGB plane. So by using this above method, we can avoid this problem.

3 Proposed Algorithm

Step-1 Numerate the Laplacian of the image intensity function as:

$$\nabla^2 u^{n}_{i,j} = u^{n}_{i+1,j} + u^{n}_{i-1,j} + u^{n}_{i,j+1} + u^{n}_{i,j-1} - 4\,u^{n}_{i,j} \tag{6}$$

Step-2 Numerate the value of the following function:

$$g(\nabla^2 u) = f'(\,|\nabla^2 u|\,)\,[\nabla^2 u/(\,|\nabla^2 u|\,)] = c(\,|\nabla^2 u|\,)\,\nabla^2 u \tag{7}$$

As

$$g^{n}_{i,j} = g\,(\nabla^2 u^{n}_{i,j}) \tag{8}$$

Step-3 Numerate the Lapcaian of, $g^{n}_{i,j}$:

$$\nabla^2 g^{n}_{i,j} = g^{n}_{i+1,j} + g^{n}_{i-1,j} + g^{n}_{i,j+1} + g^{n}_{i,j-1} - 4\, g^{n}_{i,j} \tag{9}$$

Step-4 Numerate the: $u^{n+1}_{i,j}$:

$$u^{n+1}_{i,j} = u^{n}_{i,j} - \Delta t\, \nabla^2 g^{n}_{i,j} \tag{10}$$

Step-5 Transform the RGB color vector of image to CMY :

$$\tilde{x}\,' = (\alpha\, x_1+\beta,\ \alpha\, x_2+\beta,\ \alpha\, x_3+\beta) \tag{11}$$

Where $\tilde{x}\,'$ represent a transformation which is combination of scaling and shifting, α is a scaling factor and β is a shifting factor. In this equation x_1, x_2, x_3 are the normalized red, green and blue pixel values respectively.

Step-6 Perform the transform in the CMY space :

A generalized transformation changing $\tilde{x}$ to $\tilde{x}\,'$ is given as

$$x'_k = \alpha\,(x_k)x_k + \beta(x_k),\ \text{for}\ \tilde{x} \in u,\ k=1,2,3 \tag{12}$$

where x_k is define the value of pixels , k = 1,2,3 for red, green and blue respectively. If taking the shifting function $\beta\,(x_k)$ to be zero, the transform would be-

$$x'_k = \alpha\,(x_k)x_k\,,\ \text{for}\ \tilde{x} \in I,\ k=1,2,3 \tag{13}$$

Step-7 Revert back to the RGB space :

The above equation modifies the three components of the color vector by three different scales. This condition leads to change in hue of the color vector, which is against our aim. So the objective solution to this is to transform the color vector to CMY space and process it there. This will be dealt with in two separate cases.

Case I) $\alpha\,(l\tilde{x}) \le 1$, where $l\tilde{x} = (x_1+ x_2+ x_3)$. And $x'_k = \alpha\,(l_{x'})x_k$, for k=1,2,3
Case II) $\alpha\,(l\tilde{x}) > 1$

i) Transform the RGB color vector $\tilde{x}$ to CMY color vector $\tilde{y}$ by $\tilde{y} = (y_1+ y_2+ y_3)$ where $y_k = 1-x_k$, k=1, , 2, 3., ii) Find $l\tilde{y} = (y_1+ y_2+ y_3) = 3- l\tilde{x}$. , iii) Find $g(l\tilde{y}) = 3-f(l\tilde{x})$, $\alpha(l\tilde{y}) = g(l\tilde{y}) / (l\tilde{y})$., iv) $y'_k = \alpha(l\tilde{y})\, y_k$ for k=1,2,3. , v) Back to RGB space by the transformation- $\tilde{x}' = (1-y'_1, 1-y'_2, 1-y'_3)$

4 Result and Analysis

Figure (6) show the comparing between the enhanced image and original image. After comparison, it is show that the detail information of enhance image is more accurate and resolving power is increased in the modified image. For further analyzing and diagnosing purpose, the enhancement processing is very effective and important preparation. The test set for this evaluation experiment randomly selected from the internet with 256*256 and 512*512 sizes. Matlab 7.0 software platform is use to perform the experiment. The PC for experiment is equipped with an Intel P4 2.4GHz Personal laptop and 2GB memory. Figure 2, 3,4,5,6, show that our proposed method for image enhancement has excellent performance with wide variety of set of images. So that we can say our proposed method has a strong and impressive approach to images enhancement in normal and noisy environment.

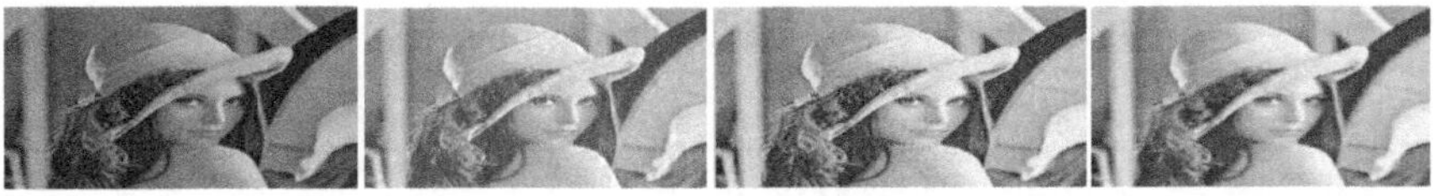

(a) (b) (c) (d)

Fig. 2. Lena noisy Image (a) Original image (b)Enhanced from Mukherjee method (c) Enhanced from Murtaza method (d) Enhanced from Proposed method

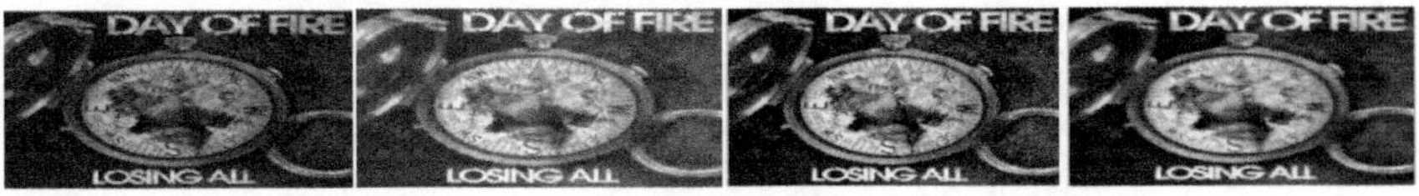

(a) (b) (c) (d)

Fig. 3. Watch Image (a) Original image (b)Enhanced from Mukherjee method (c) Enhanced from Murtaza method (d) Enhanced from Proposed method

Table 1.

	PSNR		
Images	Proposed Method	Mukherjee et al.[2]	Murtaza et al. [3]
Lena noisy	13.41	13.16	3.27
Watch	17.90	14.20	7.43

Table 1 show the comparison of performance of our proposed method for enhancement with various existing method, where our proposed method has a better performance in contrast enhancement and peak signal to noise ratio (PSNR).

5 Conclusion

Color image denoising and image enhancement is performed by using our proposed method. This research represents an image enhancement processes which was perform by image filtering process and enhancement process. From the results, the forth order Partial differential equation is use for filtering with and enhancement parameter performed the most accurate result of all the processes. This was proved by the value of signal to noise ratio which was considered a great amount of value. Experiments show that our proposed method obtains better enhancement result, and it is more effective for color image enhancement in normal and noisy environment.

References

1. George, J., Indu S P.: Color Image Enhancement and Denoising Using an Optimized Filternet Based Local Structure Tensor Analysis. In: 9th International Conference on Signal Processing, ICSP 2008, Beijing, pp. 236–239 (October 2008)
2. Mukherjee, J., Mitra, S.K.: Enhancement of Color Images by Scaling the DCT Coefficients. IEEE Transactions on Image Processing 17(10), 1783–1794 (2008)

3. Murtaza, S.M., Ahmad, J., Ali, U.: Efficient Generalized Colored image Enhancement. In: IEEE Conference on Cybernetics and Intelligent Systems, Bangkok, pp. 1–5 (June 2006)
4. Naik, S.K., Murthy, C.A.: Hue-Preserving Color Image Enhancement Without Gamut Problem. IEEE Transactions on Image Processing 12(12), 1591–1598 (2003)
5. Wang, X., Yang, S., Xu, X.: An Effective Method to Colour Medical Image Enhancement. In: 2007 IEEE/ICME International Conference on Complex Medical Engineering (2007)
6. Strickland, R.N., Kim, C.S., McDonnel, W.F.: Digital color image enhancement based on the saturation component. Opt. Eng. 26(7), 609–616 (1987)
7. Yang, C.C., Rodriguez, J.J.: Efficient luminance and saturation processing techniques for bypassing color coordinate transformations. In: Proc. IEEE Int. Conf. on Systems, Man, and Cybernetics, Vancouver, BC, vol. 1, pp. 667–672 (October 1995)
8. Buzuloiu, V., Ciuc, M., Rangayyan, R.M., Vertan, C.: Adaptive neighborhood histogram equalization of color images. J. Electron. Imag. 10(2), 445–459 (2001)
9. Zhang, Q., Mlsna, P.A., Rodriguez, J.J.: A recursive technique for 3- D histogram enhancement of color images. In: Proceedings of the IEEE Southwest Symposium on Image Analysis and Interpretation, San Antonio, TX, pp. 218–223 (April 1996)
10. Marquina, A., Osher, S.: Explicit algorithms for a new time dependent model based on level set motion for nonlinear deblurring and noise removal. SIAM Journal on Scientific Computing 22, 387–405 (2000)
11. Strong, D., Chan, T.: Edge-preserving and scale-dependent properties of total variation regularization. Inverse Problems 19(6), 165–187 (2003)
12. Blomgren, P.V., Chan, T.F.: Color TV,; Total variation methods for restoration of vector valued images. IEEE Transactions on Image Processing 7(3), 304–309 (1998)
13. Chan, T.F., Shen, J.: Variational restoration of nonflat image features: Models and algorithmms. SIAM J. Appl. Math. 61(4), 1338–1361 (2000)

Log Based Recovery with Low Overhead for Mobile Computing Systems

Awadhesh Kumar Singh[1,*] and Parmeet Kaur[2]

[1] Department of Computer Engineering, National Institute of Technology, Kurukshetra, Haryana, India
aksinreck@rediffmail.com

[2] Department of Computer Science, Jaypee Institute of Information Technology, Noida, UP, India
parmeet.kaur@jiit.ac.in

Abstract. The article proposes a recovery protocol for applications in mobile computing environment by combining movement based checkpointing with message logging. The focus of the scheme is to have a low overhead to the normal application execution due to the recovery scheme. The cost to locate the mobile host, the number of messages exchanged over the wireless media, and the size of recovery related data stored with a mobile host, have been reduced. We also include the correctness proof and performance analysis of our protocol in the presentation.

Keywords: checkpointing; message logging; mobile host.

1 Introduction and System Model

We consider a typical mobile computing environment consisting of mobile hosts (MH) and a wired network of some fixed computers, which may be regular hosts or Base Transceiver Stations (BTS). The MHs and the wired network communicate through the BTSs. A BTS manages the MHs within its cell. Mobile hosts may move between cells, thus, necessitating a transfer of control from one BTS to another. A single switch Mobile Switching Centre (MSC) controls a number of BTSs in a region.

The recovery algorithms for mobile hosts need to consider additional constraints than those for fixed hosts. The recovery procedure needs to be aware of the location of the mobile device to reduce the *search cost* [2]. Also, the recovery of the application should not be affected by the unavailability of some nodes. The data storage on the mobile hosts is limited, so usually the storage at Base Transceiver Station (BTS) is used for storing the message logs. But the bandwidth of the wireless link between the Mobile Host (MH) and the BTS is constrained than that of fixed links in the static network. A recovery protocol that reduces the size of messages stored on the MHs or decreases the number of messages passed over the wireless media will be efficient [2].

[*] Corresponding author.

V.V. Das, J. Stephen, and Y. Chaba (Eds.): CNC 2011, CCIS 142, pp. 637–642, 2011.

Considerable work has been done in the past to achieve an efficient scheme for recovery of applications in the mobile computing systems. Some of the work involves only checkpoints [1][4], while others employ message logging[3][5] [7]. In our work, we employ movement based checkpointing and message logging for recovery. A fixed host keeps track of the mobile hosts in a region to reduce the cost to locate any mobile device. We utilize the storage at a BTS to store the message logs and tentative checkpoints of the mobile hosts. However, the number of messages exchanged over the wireless media is low. The concept of a region is defined and used.

2 Proposed Scheme

A fixed host is introduced as the regional coordinator system (RCS) for a region. This RCS may be a MSC or any other fixed host specifically designated for this purpose. A region will have a number of BTSs. A MH entering the cell of a BTS is registered with that BTS. However, the BTS with which the mobile host first registers when it enters a region is designated as the InitialBTS (IBTS). The IBTS will manage the message log of the MH till the MH is in the same region.

The RCS maintains a list of the mobile hosts in the region at any given time by using a data structure, CURRENT_REGISTRATIONS; of the form < MH, current BTS of the MH>.This data structure has to be updated by a BTS when a MH registers with it. A MH identifies itself to any BTS with <own MH id, id of its initial BTS of the region, id of its previous BTS>. This information is stored by the BTS. The size of this data stored at the MH is less than storage required at the MH in protocols like [3, 7] where the entire log set, containing a list of all the previous BTSs after the last checkpoint, is stored at the MH.

2.1 Message Logging

- *When a MH enters a network for the first time*

MH registers with a BTS and updates its own data structure with the id of the current BTS as the IBTS. The IBTS updates the CURRENT_REGISTRATIONS data structure at the RCS to inform the location of MH. Further messages being sent by the MH are logged at the IBTS.

- *When there is a handoff from the IBTS to another BTS within the same region*

The MH updates its own data structure to show the id of the previous BTS, which is the IBTS in this case. A tentative checkpoint is taken at the IBTS. The current BTS updates the entry at the RCS to show that the MH is now registered with it. Further messages sent by the MH are logged at its current BTS using sender based logging.

- *When there is a handoff from any BTS to another BTS within the same region*

The MH moves from BTS_j to BTS_k in the same region

$IBTS \rightarrow BTS_i \rightarrow BTS_j \rightarrow BTS_k$

The current BTS, BTS_k, updates the CURRENT_REGISTRATIONS at the RCS to inform the presence of the MH with it. Once BTS_j knows that the MH is with BTS_k, it sends a *delete_log* message to the BTS_i which will then append its log for the MH to IBTS and delete its own copy of the log.

In case, the next BTS with which a mobile host is registering is the same as its previous BTS, the present log is appended to previous' log and then own copy is deleted. Thus, whatever be the movement pattern of a MH within a region, only the message log at the current and previous BTS may not be available at the IBTS at a given instant. This handles the storage management issue at the BTS.

- *When there is a handoff from a BTS in one region to a BTS outside the region*

The first BTS with which it registers becomes the initial BTS. The MH updates its data structure to show that the id of the IBTS for the region. Previous BTS is set to null. A tentative checkpoint is taken at this BTS and this stores further messages. The current IBTS stores the record of this MH at the RCS of this region. The previous IBTS deletes its tentative checkpoint for the MH and the RCS of the previous region deletes the record of this MH in its CURRENT_REGISTRATIONS.

Hence, at any time at most one tentative checkpoint is maintained for a MH.

- *On disconnection of a MH from the current BTS*

When a MH disconnects or switches off voluntarily, it sends a *disconnect* message to its BTS which sets the status of the MH as disconnected. If the MH reconnects with the same BTS later, the BTS changes its status to connected. Further messages are appended to the previous message log. If later the MH connects with a different BTS in the same region, the RCS has the required information about the MH. But if the MH reconnects in a different region, the RCS of this region sends a *locate_ MH* message to the other RCSs. The RCS with the latest record of the MH informs the current RCS. In both cases, the respective handoff procedure from the previous to current BTS takes place. The previous BTS deletes the record of the MH after the log unification at the new BTS. Further log is maintained by the new BTS. While the MH is disconnected, if a checkpoint request reaches the BTS, it participates in the message log unification for the MH irrespective of the fact that the MH is currently disconnected.

2.2 Global Checkpoint

A global checkpoint is taken in a coordinated fashion and can be initiated by any RCS. A RCS sends a checkpoint request to all the other RCSs and fixed hosts $F_1 \ldots F_n$ Fixed hosts take tentative local checkpoints and return a message to inform the initiator RCS. Every RCS sends a message to its BTSs. The current BTS of the MH sends a message to previous BTS of the MH to append the log for this MH at the initial BTS of the MH. The present BTS also unifies its own message log at the initial BTS. A reply is then sent back to the initiator RCS. The RCS asks the hosts to finalize the checkpoints which then make their checkpoints permanent.

3 Recovery

If the MH recovers in the same region in which it crashed, the RCS of the region has a record of the MH and its BTS before the crash in its region. In case, a MH recovers in a region different from the one in which it crashed, then the RCS of the current region sends a *locate_MH* message to other RCSs. The RCS with the latest record of the MH informs the current RCS using a message over the fixed media for the recovery process. The log at that BTS and its previous BTS are unified at the IBTS of the MH in the region. The MH can recover by using the tentative checkpoint at its IBTS along with the unified log at the IBTS.

4 Evaluation

To evaluate the overhead caused to the application execution because of the recovery protocol, we define two parameters. Checkpoint Expense (Chk_exp) is the increase in execution expense of a process due to a single checkpoint taken at a BTS. Control Expense is the increase in execution expense of a process due to the control messages exchanged during the checkpointing process.

Calculation of Chk_exp
A tentative checkpoint is taken at the IBTS when there is a handoff from the IBTS to another BTS in the same region or if the MH moves to a new region

The expense for taking a single tentative checkpoint

= $1*E_w$;where E_w is the expense of sending a message on the wireless media from a MH to its BTS

The expense for unification of log at the IBTS by the any other BTS $=1*E_f$; where E_f is the cost of sending a message across a fixed network

Hence, for a single MH in one region, say R_x

$$Chk\_exp = E_w + (N_x-1)*E_f \tag{1}$$

where N_x is the number of registrations for a MH in the region R_x

Calculation of Con_exp
The following messages contribute to the *Con_exp*

- The updating of current BTS of a MH at the RCS = N in a region
- *Delete_log* messages passed = N - 3 in a region
- control messages from the RCS to the BTSs in the region to take a global checkpoint = N in a region
- The message passed from the current to previous BTS in a region for the unification of log at the IBTS at the time of global checkpoint = 1
- The search by a RCS of the last RCS of a MH at the time of recovery = (number of RCSs - 1)* E_f

If k is the number of regions in which the MH moves, then

$$Con\_Exp=\sum_{x=1}^{k} ((3N_x\text{-}2)^*E_f)+(\text{no.of RCSs-}1)^*E_f \quad (2)$$

We observe that, due to the recovery procedure, only one additional wireless message is passed that too only at the time of taking a tentative checkpoint at a BTS i.e. is $O(k)$ where k is the number of regions in which the MH moves. This number is much less than the number of checkpoints if the checkpoints are taken periodically [7] or if checkpoints are taken based on mobility rates [3] or if a checkpoint is taken each time a MH moves out of a cell [1].

5 Proof of Correctness

Theorem 1: No orphan messages can be generated by the system.

Proof: For the nondeterministic event *e*, of the receipt of a message at the process *p*, define the following sets:

Depend(*e*) consists of *p*, and any process whose state depends on the event *e* according to Lamport's happened before relation [9].

Log(*e*) is the set of processes that have logged a copy of *e*'s determinant in their volatile memory.

Stable(*e*), is true if *e*'s determinant is logged on stable storage.

A process *p* becomes an orphan when *p* itself does not fail and *p*'s state depends on the execution of a nondeterministic event *e* whose determinant cannot be recovered from stable storage or from the volatile memory of a surviving process. Formally,

$$\forall\, e : \neg\, Stable\,(e) \Rightarrow Depend(e) \subseteq Log(e).$$

This property is called the *always-no-orphans* consistency condition [6]. In our scheme, if a BTS_j has logged the receipt of a message for a MH registered with it, then some BTS_i must have logged the message content and send event in its stable storage. This is because sender based pessimistic logging is utilized is utilized along with the communication protocol itself. Hence, the determinant of a nondeterministic event is always available upon recovery. Thus, our protocol satisfies the *always-no-orphans* condition.

Theorem 2: The recovery of a mobile host is consistent assuming reliable BTSs.

Proof: A MH can recover in the same region as it crashed or in a different region. In both cases, the record of the latest BTS when the MH crashed can be found with the RCS of the corresponding region. This BTS will have the record of the IBTS and the previous BTS in the region for the MH. Assuming reliable BTSs, every message, that has been delivered to the MH, after the last checkpoint, is logged at the corresponding BTS. The message logging scheme ensures that the message log gets unified at the IBTS of the MH in the correct sequence as per the handoff pattern of the MH The latest checkpoint for the MH at the IBTS can be retrieved on the recovery of MH. All messages can be replayed in the correct sequence according to the unified log after restoring the latest checkpoint. Hence, the MHcan reconstruct the same sequence of state intervals as before the failure and the lost state interval becomes empty. Therefore, there can be no orphan state interval and the recovery of MH is consistent among reliable BTSs.

6 Conclusion

In this paper, we have proposed a recovery protocol for the mobile computing environment which incurs very low overhead in the normal application execution. A MH can be located by passing messages only to the fixed RCSs, which are fewer in number than the MHs, over the fixed media. A checkpoint can be taken only by initially passing message to the current BTS of the MH. A small number of messages are passed over the wireless media and the size of recovery related data stored at the MH is lower than in many other contemporary protocols. At any given time, only one tentative checkpoint is kept for a MH. Therefore, the presented protocol implies a low overhead due to recovery process.

References

1. Acharya, A., Badrinath, B.R.: Checkpointing distributed applications on mobile computers. In: 3rd Int'l Con. on Parallel and Distributed Information Systems, pp. 73–80 (October 1994)
2. Badrinath, B.R., Acharya, A., Imielinski, T.: Designing distributed algorithms for mobile computing networks. Computer Communications 19, 309–320 (1996)
3. George, S.E., Chen, I., Jin, Y.: Movement based checkpointing and logging for recovery in mobile computing systems. In: Proc. of MobiDE 2006, pp. 51–58 (June 2006)
4. Neves, N., Fuchs, W.K.: Adaptive recovery for mobile environments, vol. 40(1), pp. 68–74. ACM Press, New York (1997)
5. Higaki, H., Takizawa, M.: Checkpoint recovery protocol for reliable mobile systems. In: 17th IEEE Symposium on Reliable Distributed Systems, pp. 93–99 (October 1998)
6. Alvisi, L., Marzullo, K.: Message logging: pessimistic, optimistic, causal, and optimal. IEEE Transactions on Software Engineering, 149–159 (February 1998)
7. Park, T., Woo, N., Yeom, H.Y.: An efficient optimistic message logging scheme for the recoverable mobile computing systems. ISAS-SCI (1) (2001)
8. Elnozahy, E.N., Alvisi, L., Wang, Y.M., Johnson, D.B.: A survey of rollback-recovery protocols in message-passing systems. ACM Computing Surveys (CSUR) 34(3), 375–408 (2002)
9. Lamport, L.: Time, clocks, and the ordering of events in a distributed system. Commun. ACM 21(7), 558–565 (1978)

Custom Processor Implementation of a Self Clocked Coding Scheme for Low Power RF Communication Systems

Femy John[1] and P. Jagadeesh Kumar[2]

1 Assistant Professor, Electronics and Communication Engineering Department, Viswajyothi College of Engineering and Technology, Ernakulam, Kerala, India
femy.john@rediffmail.com
[2] Senior Lecturer, Department of Electronics Engineering, Model Engineering College, Ernakulam, Kerala, India
jagadeeshkumarp@mec.ac.in

Abstract. The paper describes the implementation of Manchester Encoder and Decoder modules that is suitable for RF communication systems. The architecture has been designed for low power which is an important parameter for implantable devices. The PLL based clock recovery circuitry presented here can be used to decode data of frequency 2 MHz to 500 MHz. It can decode the data correctly even in the presence of jitter upto ¼ T. The encoder – decoder modules use no analog parts which make it less sensitive to process variations. Since the design is digital, it is insensitive to fluctuations in the process parameters, power supply etc.

Keywords: Implants, Manchester encoding, clock recovery, digital PLL.

1 Introduction

Technological advances are making it possible for communication links to carry more and faster signals. Biomedical implants utilize short distance wireless communication to avoid the infection associated with long wires connected between the implant and the external controller[1]. An example of the nature of data communication between external controller and the implant may include information about patterns for micro-stimulation and feedback for deep brain stimulation etc.

Implantable devices can be programmed by means of wireless digital communications. For synchronization, data and clock are combined at the transmitter and are sent to the implant. A technique that has received considerable acceptance is called Manchester coding[2]. This coding provides strong timing information by providing a transition for every bit, whether it be a one or a zero. It also eliminates the residual DC problem, regardless of the message sequence, by providing both a positive and negative polarity for every bit[3].

Phase Locked Loops can be used to regenerate the desired clock at the receiver and to synchronize it to the rate of the received bit stream. One of the key advantages of digital PLLs over their analog counterparts is that they remove the need for large

V.V. Das, J. Stephen, and Y. Chaba (Eds.): CNC 2011, CCIS 142, pp. 643–648, 2011.

capacitors within the loop filter by utilizing digital circuits to achieve the desired filtering function[4].

2 Manchester Encoder

Manchester data encoding is the process of a logical combining of the serial data to be encoded and the clock used to establish the bit rate. Fig 1 shows the block diagram of a Manchester encoder.

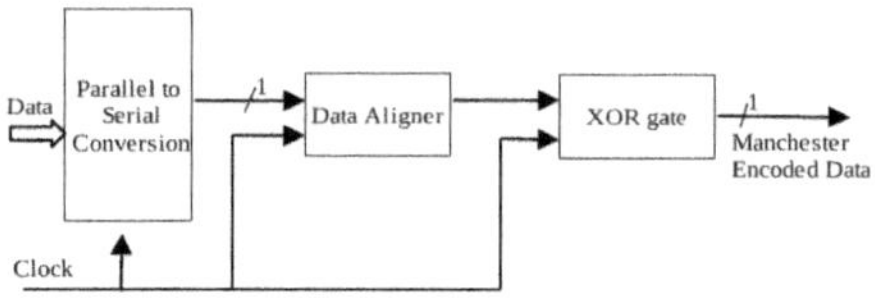

Fig. 1. Manchester Encoder

3 PLL Based Manchester Decoder

A serial data stream contains a series of 1's and 0's. In the case of Manchester Encoding, the clock has to be recovered from data at the receiver. In the hardware clock recovery method, Phase-Locked Loops are used to regenerate the desired clock, and synchronize it to the rate of the received bit stream. Block diagram of the proposed DPLL based Manchester Decoder is shown in Fig 2. It consists mainly of a frequency generator (the digital controlled oscillator), a phase-frequency comparator, and a filter interconnected.

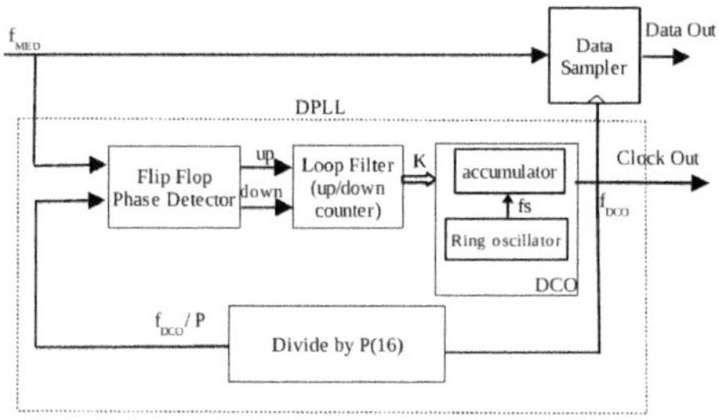

Fig. 2. Digital PLL based clock recovery circuit

The DCO frequency divided by an integer P is compared with the incoming Manchester encoded data. Assuming a proper design, the feedback mechanism converges and the DCO output frequency reaches steady state, where

$$f_{DCO} = P * f_{MED} \quad (1)$$

The actual search for the value of K that directs the system to an "in frequency-phase window" state is performed by a low-pass filter which consists of a successive approximation register (SAR). The SAR provides a value of K at the input of the accumulator type DCO after each frequency-phase comparison. At the first iteration, the SAR proposes the integer 2N-1 to the accumulator, and waits for the Loop Filter in order to propose to the accumulator either a larger value of K (if f_{DCO} is too slow) or a smaller value of K (if f_{DCO} is too fast). This iterative search for a K solution proceeds until the locked conditions are met by the system.

3.1 Digitally Controlled Oscillator (DCO)

An accumulator-type Digital Controlled Oscillator (DCO) proposes a DPLL architecture where faster locking can be attained [5]. Compared to the conventional digital PLL, overall hardware complexity can be decreased for a given high jitter performance and timing resolution. With the accumulator-type DCO, the DCO output is generated by successively adding the value of an integer K to itself at the high frequency rate fs, of a system clock as indicated in Fig. 3. The DCO output is given by the most significant bit of the output of the latch, so that the DCO output frequency is given by the formula

$$f_{DCO} = K\ (\ fs / 2^N\ )\ . \tag{2}$$

The oscillator for the digital PLL can be implemented with a ring oscillator with one inverter replaced by a NAND gate for shutting down the ring oscillator during idle mode.

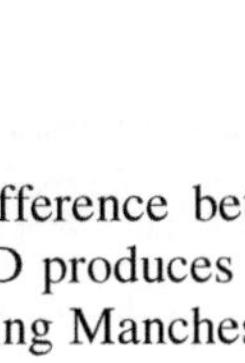

Fig. 3. Accumulator Type DCO

3.2 Phase Detector

This module monitors the phase difference between the clocks. A sequential-logic PFD is shown in Figure 4. This PFD produces UP and DOWN signals depending on the phase difference between incoming Manchester encoded data, f_{MED} and feedback clock, f_{DCO}

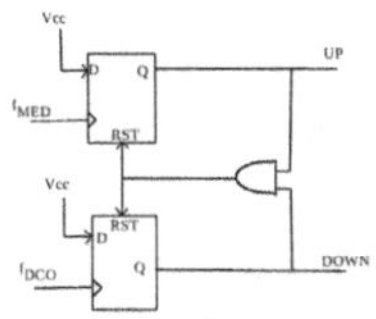

Fig. 4. Flip Flop Phase Detector

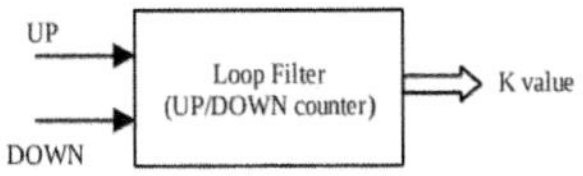

Fig. 5. Loop Filter

3.3 Loop Filter

The Successive Approximation Register (SAR) and its controller (Fig.5) serve as a low pass filter supplying the DCO with frequency and phase correction data [5]. The SAR provides a value of K at the input of the accumulator-type DCO after each frequency-phase comparison. For an N-bit accumulator-type DCO, it is sufficient to choose an N-bit SAR since the DCO output frequencies that can be generated by such an arrangement require that the corresponding K values, which define f_{DCO} , to be less than 2^N .

4 Implementation Results

The modeling of Manchester encoder-decoder modules are done using VHDL. The simulation results for the encoder module is given in Fig 6. 'md' is the data that has to be encoded. 'med' is the Manchester encoded data obtained.

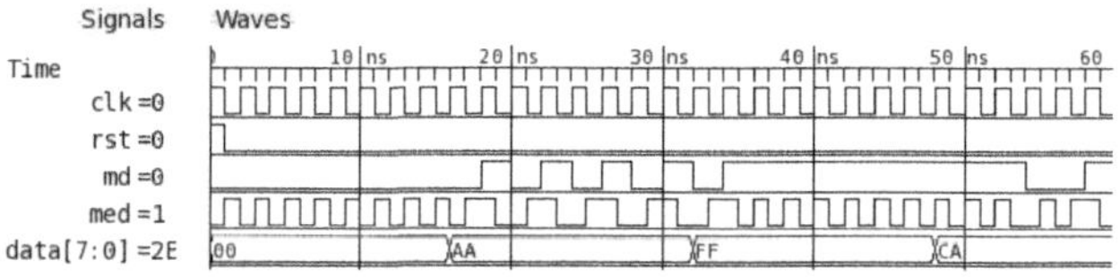

Fig. 6. Simulation Results for Manchester Encoder

Fig 7 shows the simulation result for PLL based Manchester decoder. First a stream of preamble bits is send for the PLL to get locked. In the locked condition, the up and down signals of the phase detector will be zero indicating zero phase error. After recovering the clock using the PLL, the data can be decoded using this clock. 'rxc' is the recovered clock and rxd is the recovered data. The recovered clock, 'rxc' can be used as the clock for the receiver part.

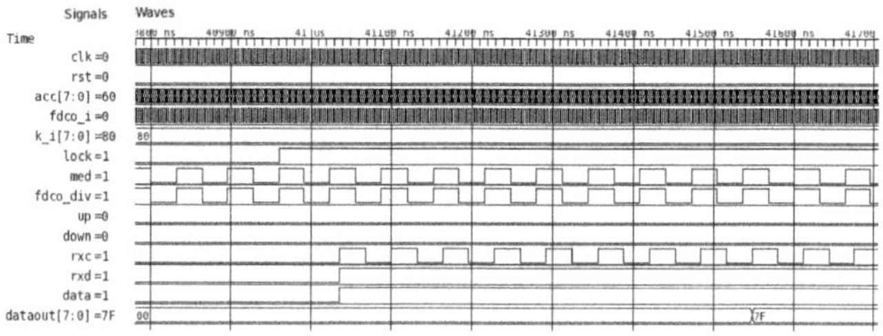

Fig. 7. Simulation Results for PLL based Manchester Decoder

Verification is done by implementing the encoder and decoder modules in Spartan Xc3s400 FPGA. Netlist and custom layout of each blocks were obtained using Alliance digital design tool[8]. The standard cell library used was sxlib which is based on 0.13nm technology. The custom processor implementation details are given in Table 1.

Table 1. Custom Processor implementation details of Manchester encoder-decoder modules

Module	No of Standard Cells	Static Power Consumption (nW)	Area (μm^2)	No of metal layers used	Extracted Capacitance (pF)
Manchester Encoder	342	448.36	11150.15	4	68.9
Oversampling Manchester Decoder	140	141.84	6488.63	4	20.9
Ring Oscillator (15 stages)	16	9.86	181.5	3	1.1
DCO	41	57.28	1503.43	4	8
Phase Detector	21	18.72	635.25	3	3.3
Loop Filter	261	295.78	5717.25	4	42.7
Frequency Divider	51	50.28	1355.2	4	7.6
Digital PLL	355	415.66	8279.43	4	59.6
Digital PLL based Manchester Decoder	1543	2012.82	42083.8	4	295.1

5 Conclusion

Encoder – decoder modules for a low power, self clocked encoding scheme (Manchester), that is suitable for RF communication systems has been successfully designed and implemented. The traditional analog loop filter and VCO in a PLL are replaced by their digital counterparts. The PLL based clock recovery circuitry presented here can be used to decode data of frequency 2 MHz to 500 MHz. It can decode the data correctly even in the presence of jitter upto ¼ T where T is the bit period. Layout for encoder and decoder are done using 130 nm process technology. The encoder module dissipates 448.36 nW power. Decoder module implemented using PLL based clock recovery module dissipates 2.01 μW power. PLL based method is superior to oversampling method [7] since it can decode data in a wide range of frequencies. The encoder – decoder modules use no analog part which makes it less sensitive to process variations.

The Manchester encoding presented here can be extended for multiple channels and also for 2 way communication. As a future enhancement in the PLL, further improvements can be done to reduce the capture time and to improve the resolution of the oscillator.

References

1. Arabi, K., Sawan, M.A.: Electronic Design of a Multichannel Programmable Implant for Neuromuscular Electrical Stimulation. IEEE Transactions on Rehabilitation Engineering 7, 204–214 (1999)
2. Karim, A., Mohamad, S.: A Novel CMOS Digital Clock and Data Decoder for implanted systems. IEEE-Engineering in Medicine and Biology Society 2, 1663–1664 (1995)
3. Ayoub Khan, M., Sharma, M., Brahmanandha Prabhu, R.: FSM based Manchester Encoder for UHF RFID Tag Emulator. In: Proceedings of the 2008 International Conference on Computing, Communication and Networking, pp. 1–6 (2008)
4. Chen, D.-L.: A Power and Area Efficient CMOS Clock/Data Recovery Circuit for High Speed Serial Interfaces. IEEE Journal of Solid State Circuits 31 (1996)
5. Walters, S.M., Troudet, T.: Digital Phase-Locked Loop with Jitter Bounded. IEEE Transactions on Circuits and Systems 36, 980–987 (1989)

6. Razavi, B.: Monolithic Phase-Locked Loops and Clock-Recovery Circuits. In: Collection of IEEE PLL papers. IEEE Press, Los Alamitos (1996)
7. Lin, Y.-H., Tu, S.H.-L.: Implementation of an Oversampling Data Recovery Receiver for Serial Link Communications. In: Proceedings of Seventh International Symposium on Signal Processing, vol. 1, pp. 613–616 (2003)
8. Alliance, `http://www-asim.lip6.fr/recherche/alliance/doc/designflow/tools.html`

A Novel Message Routing in Unstructured P2P Using CIS and Ant Search Algorithm

M. Sadish Sendil[1], N. Nagarajan[5], U. Kaleelurrahaman[2], M. Kavitha[3], and S. Karthik[4]

[1] Research Scholar and Associate Professor, [2] II M.Tech (IT), [3] II M.E (CSE), [4] Prof. and Head
SNS College of Technology, Coimbatore - 641035, Tamilnadu, India
[5] Principal and Research Supervisor,
Coimbatore Institute of Engineering and Technology,
Coimbatore - 641109, Tamilnadu, India

Abstract. Peer-to-Peer (P2P) overlay systems offer a substrate for the construction of large scale and distributed applications. P2P applications are differentiated in to two major forms, structured P2P and unstructured P2P. Unlike structured P2P, unstructured P2P does not maintain any technique to keep track of the other peers in the network. Communications between the peers are carried out by either flooding or random walk mechanism. But these mechanisms consume lot of bandwidth during communication. In this paper, we propose a novel message routing mechanism which uses the CIS and Ant algorithm to provide efficient communication between peers.

Keywords: Peer-to-Peer, Structured P2P, Unstructured P2P, Ant Search algorithm, CIS algorithm, Message Routing.

1 Introduction

A peer-to-peer, commonly abbreviated to P2P, is any distributed network architecture composed of participants that make a portion of their resources (such as processing power, disk storage or network bandwidth) directly available to other network participants, without the need for central coordination instances (such as servers or stable hosts).Since Peers in Unstructured P2P is unaware of the other peer's information in the network, the messages have to be flooded in the network. I.e. the sender peer has to flood the message in the network to communicate with the designated peer. Since there is no correlation between a peer and the content managed by it, there is no guarantee that flooding will find the desired peer in the network. Flooding also causes a high amount of signaling traffic in the network and hence such networks typically have very poor efficiency [4][5][6].

To overcome this, an efficient routing mechanism has to be introduced. In this paper we propose a novel routing mechanism which combines the advantages of Cluster based Intelligent Search algorithm (CIS) and Ant Search algorithm [3] which communicates between peers efficiently. i.e. Peers in the Unstructured P2P are clustered and each cluster is assigned with a cluster head. The cluster head consist of information about the

V.V. Das, J. Stephen, and Y. Chaba (Eds.): CNC 2011, CCIS 142, pp. 649–652, 2011.

peers in its cluster. The messages are routed through the cluster heads. The route information is stored in the intermediate peers as in ant search algorithm.

2 Related Works

In [7] it is proposed that the peers maintains routing table to keep track of the information about the searched files and the information about peer that has the file. The information about the peer refers to the peer id associated with the peer.

In [8] we find that to cluster a certain group of peers there used a peer's interest identical table which gives information about the peers of specific cluster.

In [2] we find an efficient message routing technique implemented in structured Peer-to-Peer which can be used in our proposal.

3 Proposed Methodology

The proposed message routing methodology can be illustrated in the Fig.1. There are three major steps to be introduced in the proposed method. They are, 1. Assign Ids for Each Peers in the network. It can be done by using a CA as in [9]. 2. Divides the network into disjoint clusters. 3. Assign Cluster heads and gateways as in [1] and route the messages through the cluster head.

In unstructured peer to peer information about one peer is unknown to the other. i.e. to enable communication between the peers, the peers in the network should know some information about the other peer in the network. To enable this we assign peer ids to each peer that enters the network. i.e. as in [9] we use a CA which assigns the ids for each peer.

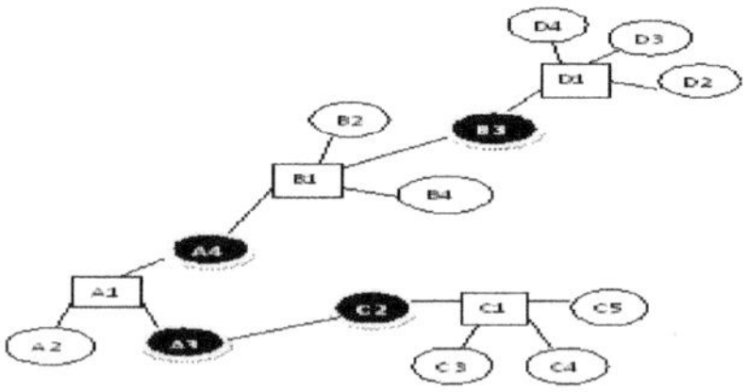

Fig. 1. An Overview of Cluster based intelligent search

The peer ids are unique and it helps the peers to communicate with each other. Usage of CA in the routing mechanism has two advantages:

1. Used in Assigning Peer ID (unique) to each peer in the network.
2. Used to Provide Secure communication between the peers [9]

After assigning the peer ids network is divided into number of disjoint clusters. Each cluster has an elected cluster head (CH). CHs communicate with their neighbor CHs via Gateways. Two CHs are neighbors of each other if their distance is at most 3 hops. The cluster head consist of information about the other peers in its own cluster. i.e. the peer ids of each peer in the cluster are placed in its cluster head.

The communication is carried out through the cluster head. When a non cluster head node initiates message communication, the cluster head process the transaction of message between the sender and receiver. Sender uses the peer id of the receiver to send message. The cluster head look up for the peer id in the information stored in it. If it is available in it then it identifies that the peer is in its own cluster and forwards it to the peer. Else it forward it to the neighbor cluster heads. When the receiver receives the message it sends acknowledgement to the sender.

During the acknowledgement process the intermediate cluster heads are updated with information of the receiver's id. i.e. the id of the receiver is updated in the intermediate cluster heads along with a parameter called hit ratio as used in ant search. The peer id of the receiver and its hit ratio are updated in the route path table that we update in the intermediate cluster head. The hit ratio is incremented when the acknowledgement received to the intermediate cluster. The hit ratio is used to make the cluster head intelligently forward message to the cluster heads that takes it to the designated destination.

For example: From Figure 1 we try to send message to peer D1 from peer A1. Peer A1 send message to its cluster head A. The cluster head check the information of peer ids stored in it. The peer id D1 is not present in A. Hence it forwards the Packet to B and C. Still the peer is not in cluster B and C which are given in the cluster heads B and C. hence they forward the message to their neighbor cluster. Now the cluster head D receives the message. The cluster head D checks its information and finds that the peer is in its cluster and forwards the message to the peer.

After receiving the message the peer sends acknowledgement through its cluster head. During acknowledgement the intermediate cluster head B update the route path table which saves the information about the peer id (D1) and increments the hit ratio value of it. This is updated to make cluster intelligently forward message to designated neighbor cluster heads. I.e. during the next communication the cluster head forwards the message to the cluster head which has the positive hit ratio. For example if the message is again transferred between A1 and D1, the route path selected will be B. The cluster head C will not have any positive value for D1. Hence message is not forwarded through it

Also to avoid any failure in the intermediate peer communication we use the tracer routing mechanism as shown in [2]. The message is forwarded and acknowledgement is send back immediate to the former cluster head. So that the former cluster heads assumes that the next cluster head have received the message.To improve security in our routing mechanism we follow the same mechanism that is used in [9].

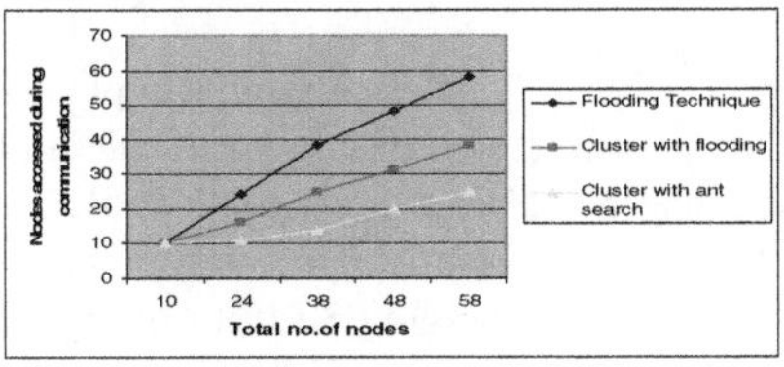

Fig. 2. Comparison of proposed method with other techniques

4 Conclusion

Peers in Unstructured Peer-to-Peer are unaware of the information about the other peers in the network. Hence communication between them is carried out by flooding the messages in the network. But this mechanism is inefficient. In this paper we propose a novel routing mechanism which uses the CIS and Ant search algorithm to intelligently transfer message between the peers. Our approach partitions the entire network into disjoint clusters with one cluster head (CH) per cluster. CHs form a MIS and are connected through gateway nodes. Each node takes one role, a CH, a gateway, or an ordinary node. The CH consists of information (peer id) about the other peers in its cluster. Using the information the CH intelligently forward message to the next CH until the desired peer reached. Our method significantly reduces the paths used for communication. i.e. it uses only a fewer routing path for communication when compared with other existing techniques. The result is compared and shown in Fig.2.

References

1. Li, X., Wu, J.: Cluster-based Intelligent Searching in Unstructured Peer-to-Peer Networks. In: 25th IEEE International Conference on Distributed Computing Systems, pp. 642–645. IEEE Press, Los Alamitos (2005)
2. Xiang, X., Jin, T.: Efficient secure message routing in structured P2P. In: International Conference on Networks Security, Wireless Communications and Trusted Computing, pp. 354–357. IEEE Press, Los Alamitos (2009)
3. Michlmayr, E.: Ant Algorithms for Search in Unstructured Peer-to-Peer Networks. In: 22nd International Conference on Data Engineering Workshops, pp. x142–x142. IEEE Press, Los Alamitos (2006)
4. Lin, T., Lin, P., Wang, H., Chen, C.: Dynamic search algorithm in unstructured peer to peer Networks. IEEE Trans. Parallel Distrib. Syst. 19, 654–666 (2009)
5. Dorrigiv, R., Lopez-Ortiz, A., Prałat, P.: Search Algorithms for Unstructured Peer-to- Peer Networks. In: 32nd IEEE Conference on Local Computer Networks, pp. 343–352. IEEE Press, Los Alamitos (2007)
6. Reynolds, P., Vahdat, A.: Efficient Peer-to-Peer Keyword Searching. In: Endler, M., Schmidt, D.C. (eds.) Middleware 2003. LNCS, vol. 2672, pp. 21–50. Springer, Heidelberg (2003)
7. Chen, G., Low, C.P., Yang, Z.: Enhancing Search Performance in Unstructured P2P Networks Based on Users' Common Interest. IEEE Trans. Parallel Distrib. Syst. 19, 821–836 (2008)
8. Tan, Y., Lin, Y., Chen, Z., Dong, T.: Research and Implementation on Routing Scheme Based on Interest Mining in unstructured P2P Systems. In: 7th International Conference on Web-Age Information Management Workshops, pp. 2–2. IEEE Press, Los Alamitos (2006)
9. Kraxberger, S., Payer, U.: Secure Routing Approach for Unstructured P2P Systems. In: 3rd IEEE International Conference on Emerging Security Information, Systems and Technologies, pp. 210–216. IEEE Press, Los Alamitos (2009)

Image Fusion Framework

Ujwala Patil[1], Uma Mudengudi[1], K. Ganesh[1], and Ravikiran Patil[2]

[1] BVBCET Hubli, Karnataka, India
ujwalapatil@bvb.edu
[2] Mind Tree, Bangalore, Karnataka, India

Abstract. In this paper we propose image fusion framework, which includes image registration, demonstration of different image fusion techniques, and a new quality metrics to evaluate the performance of different fusion algorithms. Image fusion is a process of combining two or more images of same scene to get the more informative image. Registration, the coordinate transformation of one image with respect to other is the preprocessing step for the fusion process, we propose an algorithm for automatic registration of two images which are related by affine motion model in a computationally efficient manner using normalized mutual information (NMI). Registered images are fused using fusion algorithms like pyramid, wavelet and principle component analysis (PCA). We propose new quality metrics and analyze the performance of fusion algorithms for variety of dataset.

Keywords: Image Registration, image fusion, quality analysis.

1 Introduction

Image fusion is a process of generating a single fused image using a set of input images. The input images must be registered, we propose an algorithm for automatic registration using iterative hierarchical frame work with NMI as a similarity metric. Registered images from proposed robust registration method are fused using different algorithms. We propose new common quality metric for multi modal and multisensor images and separate metric for multi focal image.

2 Image Registration Using NMI

There have been several different approaches to image registration. The authors in [1] propose a hierarchical framework for motion estimation. The main criterion for choosing the hierarchical framework is to improve the computational efficiency, however they use sum of square difference (SSD) as similarity metric. Authors in [2], [3] give the transformation that aligns reference image and target image by maximizing mutual information, however mutual information is much sensitive to the variation in correlation and entropy. Most of the above methods do not use hierarchical model in the registration process. We use iterative hierarchical model with NMI as a similarity metric as it is almost invariant to the changing of entropy.

V.V. Das, J. Stephen, and Y. Chaba (Eds.): CNC 2011, CCIS 142, pp. 653–657, 2011.

The steps involved in the proposed robust registration algorithm are

1) Construct the image pyramid for each input, initialize the transformation matrix *P* (we use identity matrix as the initial estimates of *P*).
2) For each resolution level of pyramid:

    a) warp the target image towards the reference image using the current transformation parameters, $S_1'=warp(S_2{:}P)$, where S_1 is reference image S_2 is target image, S_1' is estimated image.
    b) refine *P* according to the residual misalignment between reference image and warped image.
    c) measure the similarity between the reference and the estimated image using NMI as the similarity metric.
    d) steps (a) to (c) if difference in the NMI in the previous and current iterations is above threshold θ (we use $\theta = 0.001$)

3) Propagate *P* to the next pyramid level and repeat steps (a) to (d) till all levels of pyramids are completed.
4) Warp the target image towards the reference image using estimates of *P* Results of proposed registration are shown in Fig 1.

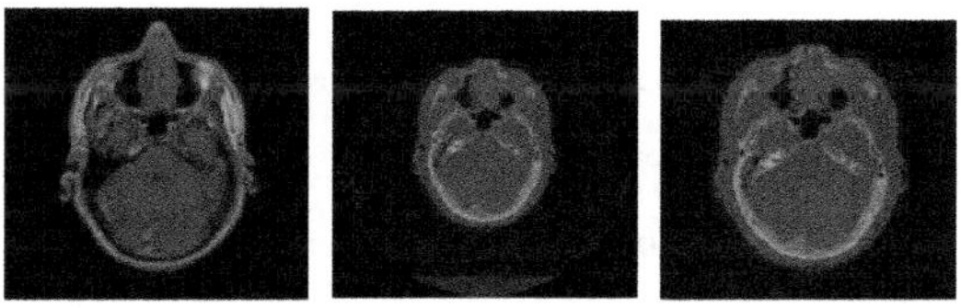

(a) (b) (c)

Fig. 1. Image registration. Images are taken from Visible Human Project data set, National library of medicine, USA (a) MRI image. (b) CT image. (c) Registered CT image, using proposed method. (Algorithm stopped after 4 iterations at L3 with 0.6789 NMI, at L2 it stopped after 5 iterations with 0.5479 NMI, at L1 it stopped after 2 iterations with 0.1569 NMI.

We demonstrate proposed registration algorithm for different synthetic and real dataset and the results are shown in Table 1.

Table 1. Number of Iteration and NMI at different levels of proposed hierarchical model for registration

Images	Levels			NMI			Error
	L1	L2	L3	L3	L2	L1	
CT and MRI	4	5	2	0.6789	0.5479	0.1569	-
Real	7	3	3	0.5977	0.5764	0.3104	6.730
Synthetic-1	17	3	2	0.7381	0.6962	0.3780	2.733
Synthetic-2	7	3	2	0.7345	0.6897	0.3496	3.181
Synthetic-3	11	3	3	0.7247	0.6090	0.3808	2.524

2.1 Image Fusion

Several fusion algorithms starting from simple pixel based to sophisticated wavelets and PCA based are available [4], [5]. Table 2 shows the different fusion algorithms for different dataset.

Table 2. Image fusion using different fusion algorithms

Image-1	Image-2	Fused using PCA	Fused using pyramid	Fused using Wavelet
CT 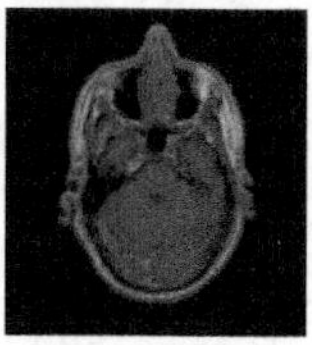	MRI 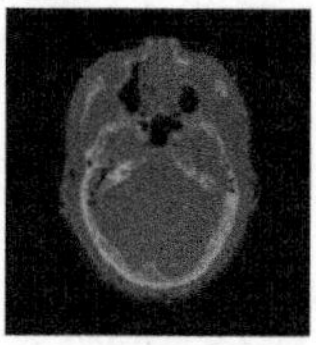	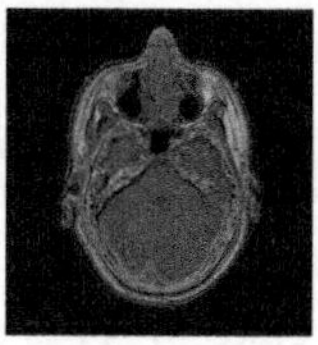	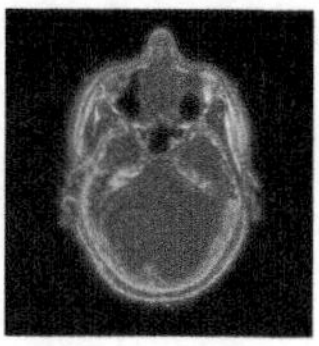	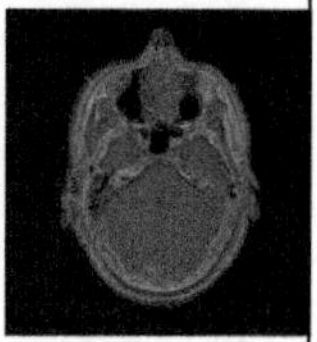
Visiblerange image 	IR range image 			
Focused clock 	Focused man 			

2.2 Quality Analysis of Image Fusion

There is an increasing need for performance or quality assessment tools in order to compare the results obtained with different image fusion algorithms. We used spatial frequency (SF), entropy (ENTR), standard deviation (SD), cross entropy (CE), fusion mutual information (FMI) and structural similarity index (SSIM) for the analysis of fusion algorithms [6].

Proposed quality metrics: Existing quality parameters are not sufficient to analyze the performance of fusion algorithms, we propose new common quality metric Q_{M1} for multi sensor and multi modal images and Q_{M2} separate metric for multi focal images

$$Q_{M1} = e^{\alpha\,(FMI+CE+SSIM)} + \beta * \ln\,(SF) + \gamma * \ln\,(VI + SD) \tag{1}$$

$$Q_{M2} = e^{\alpha\,(FMI+CE+SSIM)} + \beta * \ln\,(SF) + \gamma * \ln\,(VI) + \theta * \ln\,(SD) \tag{2}$$

Where α, β, γ, θ are tuning parameters. The first term of Q_{M1} depends on FMI, CE and SSIM. We assign exponential function for this term to increase the value since mutual information and cross entropy have small values. SF has relatively large value compared to other metric so it is minimized by applying logarithmic function. Sum of SD and VI has relatively large value than others, so we apply logarithmic compression to the sum of SD and VI. For multi focal images we reduce VI and SD by applying separate logarithmic in Q_{M2}. Table 3 shows the analysis using existing quality parameters and proposed quality parameters

(Q_{M1}, α=0.6, β = γ = 1. Q_{M2}, α=0.6, β = θ = 1, γ=1.5)

Table 3. Quality analysis of different fusion algorithms

Input images	Fusion algorithms	Quality metrics							
		SD	SSIM	VI	SF	MI	CE	Q_{M1}	Q_{M2}
Multi modal	PCA	38.5	0.88	11710.6	10.22	4.87	0.33	51.46	58.47
	Pyramid	34.27	0.91	12801.1	5.00	4.06	0.9	46.84	53.18
	Wavelet	36.28	0.89	11134.7	7.05	4.24	0.59	44.03	50.64
Multi sensor	PCA	52.23	0.46	611.28	3.46	4.72	0.40	38.94	43.38
	Pyramid	51.12	0.92	571.40	4.02	5.71	0.7	91.52	96.10
	Wavelet	56.56	0.61	539.83	5.75	5.61	0.65	71.87	76.76
Multi focal	PCA	46.85	0.88	682	10.34	0.07	0.01	12.18	17.74
	Pyramid	41.67	0.59	585.87	9.70	0.37	0.16	12.08	17.52
	Wavelet	46.98	0.83	11134.7	9.88	0.21	0.04	15.09	22.03

3 Conclusion

We proposed image fusion framework including registration, demonstration of fusion algorithms and new quality metrics to analyze the performance of fusion algorithms. Registration is computationally efficient with hierarchical model and NMI as similarity metric. RMS error is used to compare the target image and the warped image. Registered images are fused and quality analysis is done for the different fusion algorithms. Analysis shows that PCA works better for multi modal dataset, we can use pyramid for multi sensor and wavelet fusion is better for multi focal images.

References

1. Bergen, J., Anandan, P., Hanna, K., Rajesh, H., Zhiyong, G., Gu, B., Lin, J.: Hierarchical model based motion estimation. In: Sandini, G. (ed.) ECCV 1992. LNCS, vol. 588, pp. 436–440. Springer, Heidelberg (1992)
2. Zhiyong, G., Gu, B., Lin, J.: Monomodal image registration using mutual information based methods. Image and Vision Computing archive, 164–173 (2008)
3. Viola, P., Wells, W.M.: Alignment by maximization of mutual information, pp. 16–23 (1995)

4. Piella, G., Heijmans, H.: A new quality metric for image fusion. In: Proceedings International Conference on Image Processing, pp. 173–176 (2003)
5. Kong, J., Zheng, K., Zhang, J., Feng, X.: Multi-focus image fusion using spatial frequency and genetic algorithm. International Journal of Computer Science and Network Security, 2220–2224 (May 2008)
6. Naidu, V.P.S., Raol, J.R.: Pixel-level image fusion using wavelets and principle component analysis. Defence Science Journal, 338–352 (2008)

An Enhancement to BioPPISVMExtractor Using Multidimensional Support Vector Machine

Mahapatra Saswati and Swarnkar Tripti

Department of Computer Applicatons, ITER
Siksha O Anusandhan University, Bhubaneswar
S_aswati@yahoo.co.in, tswarnakar@iter.ac.in

Abstract. Protein–protein interactions play a key role in various aspects of the structural and functional organization of the cell. Automatically extracting protein–protein interaction information from increasing biomedical literature can help to build protein relation network, predict protein function and design new drugs. This paper adds an improvement to the protein-protein extraction system BioPPISVMExtractor. This system applies a new SVM based method called multi dimensional support vector machines (MD-SVMs) that generates multiple orthogonal axes, which correspond to decision functions, from high dimensional space based on margin between two classes. The multiple orthogonal axes approach of MD-SVM is efficient and useful for both the visualization and class prediction. The system improves the precision with an acceptable recall.

Keywords: protein-protein interactions, conditional random fields, multi dimensional support vector machines (MD-SVM).

1 Introduction

Protein–protein interactions (PPI) play a key role in various aspects of the structural and functional organization of the cell. However most of this information are hidden in different articles. To date, more than 16 million citations of such articles are available in the MEDLINE database [1]. Many hand-curated databases, such as DIP [2], BIND [3], IntAct [4] and STRING [5], have been built to store PPI information. Manually retrieving such information is very complex due to the lack of formal structure in the natural language narrative in these documents. Thus, automatically extraction of protein interaction information from biomedical literature has become an important research area.

Existing PPI extraction works can be roughly divided into three categories: manual pattern engineering approaches, grammar engineering approaches and machine learning approaches. Machine learning approaches such as Maximum entropy[7], HVS[8], SVM[6] have gained interest for extracting protein interaction information. BioPPISVMExtractor[9] is a SVM based fully automated PPI extraction system. This system achieves a recall of 71.83% and precision of 49.28%. However SVMs predict a class

V.V. Das, J. Stephen, and Y. Chaba (Eds.): CNC 2011, CCIS 142, pp. 658–662, 2011.

of test samples by a decision function. As a result, information loss of original data is enormous.

The main purpose of this paper is to boost the classification accuracy of the classifier by introducing multidimensional SVM(MD-SVM)[10].This approach overcomes the shortcomings of SVM by introducing multiple orthogonal axes for better visualization and class prediction. Our experimental results show that by introducing MD-SVM the precision of extraction system can substantially be increased with an acceptable recall.

2 Methods

The system architecture of MD-SVM based extraction system is shown in Fig. 1. The trained MD-SVM classifier is used to identify which protein pairs in a sentence have a biologically relevant relationship between them. MD-SVM based extraction system consists of the following processing stages to extract interaction information: pronoun resolution, protein name recognition, features extraction and MD-SVM classification. The details are described in the following sections.

2.1 Pronoun Resolution

In our pronoun resolution module, noun and noun phrase in text are identified using GENIA Tagger which is specifically tuned for biomedical text such as MEDLINE abstracts and achieves an F-score of 98.20% on GENIA corpus [12,13].

2.2 Protein Name Recognition

Conditional Random Fields(CRF) based method is applied to find gene and protein mentions with promising early results [14].

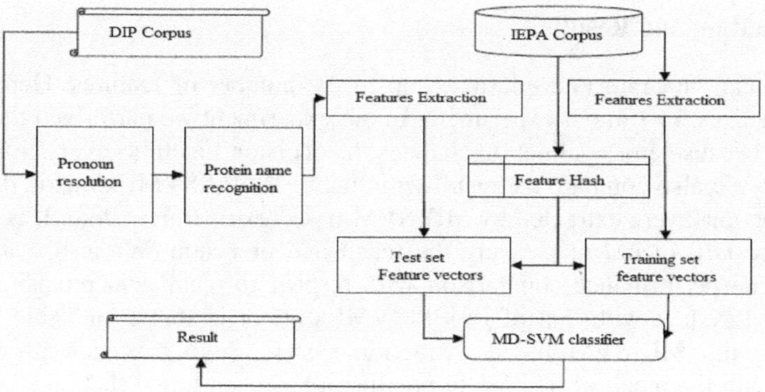

Fig. 1. System architecture of MD-SVM based extraction system

2.3 MD-SVM Model

MD-SVM model is trained to recognize protein protein interactions by generating more than one decision functions. It gives multiple optimal solution sets (ω^k, b^k, ξ^k), k=1,2,......l with l<=n so that all the directions ω^k are orthogonal to each other.

For a set of training data (x_i, y_i) i=1,2....n $x_i \in R^n$ yi$\in$$\{-1,+1\}$.$x_i$ is a feature vector of ith training example and yi is class label associated with ith training example. The decision function of kth step is of the form:

$$f^k(x)=\text{sign}\ (\sum \alpha_i^k y_i K(x_i,x)) + b^k \text{ for } i=1...m.$$

Here we have chosen k=1,2,3 to simplify our experiment. We have taken linear kernel $K(x_i,x_j) = x_i.x_j$ for classification. In our experiment, the SVM-light package [15] was used. The penalty parameter C in setting the SVM is a very important parameter since it controls the trade off between the training error and the margin.

2.4 Feature Selection

In our experiments the words (words from protein names, words between and surrounding protein names), protein name distance, key word, link path features are used for MD- SVM classification.

3 Experiment and Discussion

3.1 Corpus

The training and test set in our experiments is the same dataset that was used for the BioPPISVMExtractor evaluation so that the results are comparable. i.e. The Interaction Extraction Performance Assessment (IEPA) corpus[11] is used as training set and for evaluation, 394 interactions were identified from the DIP database. All protein names have been tagged correctly in the IEPA corpus.

3.2 Evaluation and Results

MD-SVM can generate multiple axes, up to the number of features. Here we have taken three axes, k=3 in our experiment. In our experiment we performed the hold-out validation because cross validation changes the decision functions every time the data set is split. We also compare the result with that of BioPPISVMExtractor.1982 candidate interactions were extracted by MD-SVM based extraction system. It is compared with source DIP record to measure the recall and precision. We also evaluated the Fscores. The performance comparison with respect to recall and precision of MD-SVM based system with that of BioPPISVMExtractor is shown in Table 1. By implementing the MD-SVM based extraction system, high precision of 70.79% is achieved with the recall of 67.26%.In practice it has been found that recall and precision are inversely related to each other.Fig:2 shows the relationship between precision and recall for MD-SVM based Extraction system.

Table 1. Recall and Precision comparison of MD-SVM based extraction system with BIioP-PISVMExtractor

Recall results	MD-SVM based extraction system		BioPPISVMExtractor		Precision results	MD-SVM based extraction system		BioPPISVMExtractor	
	cases	percentage	cases	percentage		cases	percentage	cases	percentage
Match	265	67.26	283	71.83	Correct	1403	70.79	894	49.28
no match	129	32.74	111	28.17	Incorrect	579	29.21	920	50.72
Total	394	100	394	100	Total	1982	100	1814	100

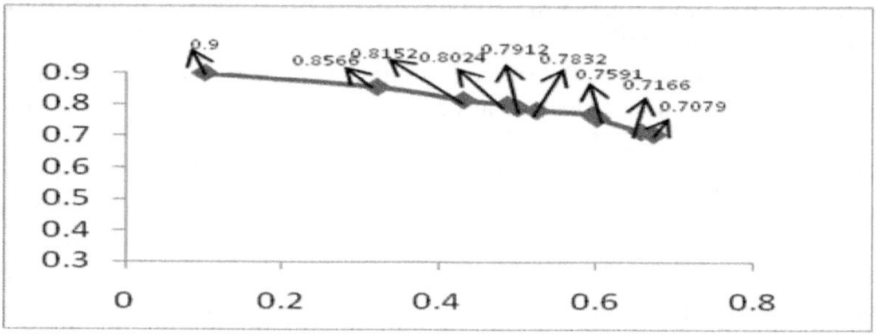

Fig. 2. Relationship between precision and recall

4 Conclusion

The purpose of our work is to further enhance the precision performance of the SVM based PPI extraction system. In this paper we have presented a MD-SVM based PPI extraction system. Experimental evaluation of MD-SVM based PPI extraction system with SVM based system-BioPPISVMExtractor shows that our system achieves a higher precision with a permissible recall. Also it as found that MD-SVMs are useful for outlier detection.

References

[1] Pubmed-overview, `http://www.ncbi.nlm.nih.gov/entrez/query/static/overview.html`

[2] Xenarios, I., Salwinski, L., Duan, X.J., Higney, P., Kim, S.-M., David, E.: DIP, the database of interacting proteins: a research tool for studying cellular networks of protein interactions. Nucleic Acids Res. 30(1), 303–305 (2002)

[3] Bader Gary, D., Doron, B., Houge Christopher, W.V.: BIND: the biomolecular interaction network database. Nucleic Acids Res. 31(1), 248–250 (2003)

[4] Henning, H., Luisa, M.-P., Chris, L., Sugath, M.: IntAct: an open source molecular interaction database. Nucleic Acids Res. 1(32 Database issue), 452–455 (2004)

[5] von Mering, C., Jensen Lars, J., Snel, B., Hooper Sean, D.: STRING: known and predicted protein–protein associations, integrated and transferred across organisms. Nucleic Acids Res. 33(Database issue), 433–437 (2005)

[6] Donaldson, I., Martin, J., de Bruijn, B., Wolting, C., Lay, V., et al.: Pre-BIND and textomy-mining the biomedical literature for protein–protein interactions using a support vector machine. BMC Bioinform. 4(1), 11 (2003)
[7] Xiao, J., Su, J., Zhou, G.D., Tan, C.L.: Protein–protein interaction extraction: a supervised learning approach. In: Proc. the First International Symposium on Semantic Mining in Biomedicine, Hinxton, UK, April 10-13 (2005)
[8] Zhou, D., He, Y., Kwoh, C.K.: Extracting protein–protein interactions from the literature using the Hidden Vector State Model. In: Proc. International Workshop on Bioinformatics Research and Applications, Reading, UK (2006)
[9] Zhihao, Lin, H., Li, Y.: BioPPISVMExtractor: A protein–protein interaction extractor for biomedical literature using SVM and rich feature sets. Journal of Biomedical Informatics 43(2010), 88–96 (2008)
[10] Komura, D., Nakamura, H., Tsutsumi, S., Aburatani, H., Ihara, S.: Multidimensional support vector machines forvisualization of gene expression data. Bioinformatics-Original Paper 21(4), 439–444 (2005), doi: 10.1093/bioinformatics/bti188
[11] Ding, J., Berleant, D., Nettleton, D., Wurtele, E.: Mining MEDLINE: abstracts, sentences, or phrases? In: Proc. Pacific Symposium on Biocomputing, Hawaii, USA (2002)
[12] Tsuruoka, Y., Tateishi, Y., Kim, J.D., Ohta, T., McNaught, J., Ananiadou, S., et al.: Developing a robust part-of-speech tagger for biomedical text. In: Proc. The 10th Panhellenic Conference on Informatics, Volos, Greece (November 2005)
[13] Kim, J.D., Ohta, T., Tateisi, Y., Tsujii, J.: GENIA corpus – a semantically annotated corpus for bio-text mining. Bioinformatics 19(suppl. 1), i180–i182 (2003)
[14] McDonald, R., Pereira, F.: Identifying gene and protein mentions in text using conditional random fields. BMC Bioinform. 6(suppl. 1), S6 (2005)
[15] Joachims, T.: Making large-scale SVM learning practical. In: Advances in Kernel Methods – Support Vector Learning, ch.11, pp. 169–184. MIT Press, Cambridge (1999)

Industrial Monitoring Using Zigbee Network

M. Shanmugaraj[1], C. Muthu Ramya[1], and R. Prabakaran[2]

[1] M.E. Pervasive Computing Technologies,
[2] Assistant Professor (EEE)/CCT
Anna University of Technology, Tiruchirappalli, India
{innoraj,ramyacpct,hiprabakaran}@gmail.com

Abstract. We have developed a Zigbee network in real time using Zigbee nodes. Discussed Wireless Sensor Network (WSN), Zigbee Technology, and its Applications. Derived results from the network created. Characteristics and network parameters such as data rate, packet delay, Received Signal Strength Indication and packet loss are derived. It is found that this Zigbee network is more suitable for industrial monitoring and control Applications.

Keywords: Industrial Monitoring, Wireless Sensor Network, Zigbee Technology, Received Signal Strength Indicator (RSSI), Packet Delay.

1 Introduction

Wireless Sensor Networks have emerged as a growing research area where most research scholars and industrialists focus on. Many people believe Wireless Sensor Network would become more important than internet in future. WSN consists of small nodes with sensing, computation, and wireless communications capabilities. WSN can also be defined as a network of devices, which can sense the environment and communicate the information gathered from the monitored field through wireless communications.

A typical sensor node consists of the following components: Sensing Element, Battery, Processor, Memory, and Communication Equipment. Sensor is a device which converts a physical phenomenon into an electrical signal. Sensors act as the interface between the physical world and the world of electrical devices, such as computers. Actuators represent the other part of the interface, which convert electrical signals into physical phenomena. [1] Sensing Material may be physical, chemical and biological. Physical sensing material consists of Magnetic, Light, Sound and so on. Battery represents the heart of the sensor unit as it decides the lifetime of the system. Processor used in the sensor node performs several important functions such as memory management, maintaining the resources of the system, handling the interrupts, turning off communication equipment when not in use to save power.

V.V. Das, J. Stephen, and Y. Chaba (Eds.): CNC 2011, CCIS 142, pp. 663–665, 2011.

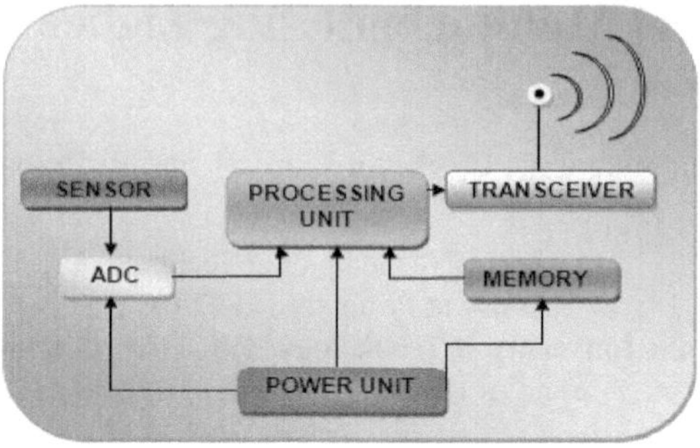

Fig. 1. Components of a Sensor Node

2 Zigbee Technology

Zigbee is the global wireless technology connecting dramatically various devices to work together and enhance everyday life. Zigbee was introduced by IEEE and the Zigbee Alliance provides the standards for various applications. Zigbee Alliance includes companies such as Texas Instruments, Free scale, Philips, Cisco, Digi, Honeywell, Intel, Siemens, Sony, Ember, Emerson, Reliant Energy, Itron, Kroger [2].

Zigbee is a low power, low cost, low data rate wireless standard for WPAN featured with security, reliability, large network capacity, easily deployed, short delay, long transmission range. Zigbee protocol stack defines four layer namely PHY layer, MAC layer, Network layer, Application layer. Zigbee defines Application & Network layer whereas PHY & MAC layer are defined by IEEE 802.15.4. Zigbee found applications in variety of fields such as Smart energy, Home automation, Health care, Remote control and monitoring, Telecommunication services, Retail services, Building automation, Environmental monitoring, Industrial sensing and diagnostics, Agriculture, Military applications and so on.

3 Experimental Details

Our design adopts the mesh topological network structure. Before deploying the node on to the network, the hardware component is tested using X-CTU software provided by Digi. Range test has been performed for various distances in our campus. The experiments were carried out in our lab without any noise interference. To analyse the network characteristics Zigbee Dongle (sniffer) is a simple, plug and play device connected to computer via USB port or USB extension cable. The IEEE 802.15.4 USB Dongle from Integration provides a complete 802.15.4 interface that can be quickly and easily connected to a host computer. The dongle provides a simple method of integrating 802.15.4 or Zigbee into computers, gateways or bridge devices.

Wireless Protocol Analyser (WPA) is s graphical protocol analyser which scans a channel over the network and displays the various statistics, packet information and etc. WPA provides real time 802.15.4 packet sniffer with sub microsecond timestamps, essential tool for application debug [4]. X-CTU is windows-based application provided by Digi. This program is designed to interact with the firmware files on Digi's RF products and to provide a simple-to-use graphical user interface to them [5].

The Zigbee node can be configured based on the function set required for the network. We have proposed a simple model, where three Zigbee nodes have been used with different function performing namely Zigbee End device, Zigbee Router and Zigbee Coordinator. Various physical phenomena such as temperature, pressure, noise, light, humidity, gas, oil, water level, Chemical substances can be monitoring in an industrial environment using this model. There have been many wired system proposed in literatures both theoretically and practically. We have developed these using Zigbee Technology. The sensing unit consists of sensor and power supply which may be either external supply or battery operated. The sensing unit consists of sensor and power supply which may be either external supply or battery operated. The sensed data from the sensor is fed to the processing unit where the controller processes the data which is then passed to the node for transmission over wireless network. The Zigbee node receives the data and scan for available node to transmit the data to the coordinator node. The coordinator receives the data, process it and transmit over wired media to monitor and control the system. In our experiment we calculated the propagation delay, data rate and throughput by capturing the packets using Wireless Protocol Analyser.

4 Conclusion

We have given overview about Zigbee technology, developed a Zigbee network in real time with Zigbee nodes. Network parameters such as packet delay and received signal strength indication are derived. It is found that increase in distance range would affect the overall performance of the network. It is observed that Zigbee technology would be a better one to be implemented in monitoring and control application.

References

[1] Sensor, http://www.stanford.edu/class/me220/data/lectures/lect02
[2] Zigbee Alliance, http://www.zigbee.org
[3] Yong, S., Yi, Z., Jian, W., Tinggao, Q.: Research and Implementation of Zigbee Networking. In: Proceedings of the 2009 IEEE International Conference on Mechatronics and Automation, Changchun, China, August 9-12, pp. 3992–3996 (2009)
[4] Wireless Protocol Analyser, http://www.integration.com
[5] X-CTU, http://www.digi.com/support

Erratum: A New Metric Based Cluster Head Selection Technique for Increasing the Lifetime of Energy Aware Wireless Sensor Network

Sanghita Bhattacharya[1], Arpan Sarbadhikari[2], and Subhansu Bandyapadhyay[3]

[1] Department of Computer Science & Engineering, National Institute of Technology, Durgapur-713 209, India
sanghita.b@gmail.com
[2] A.K. Choudhury School of Information Technology, University of Calcutta, Kolkata-700 009, India
arpan1983@yahoo.co.in
[3] Senior member IEEE, Department of Computer Science & Engineering, University of Calcutta, Kolkata-700 009, India
subhansu@computer.org

V.V. Das, J. Stephen, and Y. Chaba (Eds.): CNC 2011, CCIS 142, pp. 581–586, 2011.

DOI 10.1007/978-3-642-19542-6_131

In the original version, the name of the first author was spelled incorrectly by mistake. It should be "Sanghita Bhattacharjee".

The original online version for this chapter can be found at
http://dx.doi.org/10.1007/978-3-642-19542-6_114

Retraction Note to: Hop-Count Filtering: A Successful Security Alongside Spoofed Network Traffic

N.B. Pokale[1] and Samuel S. Chopde[2]

[1] Asst. Professor - Department Information Technology
Sinhgad Academy of Engineering, Pune, India
nbpokale@gmail.com
[2] Research Scholar - Department Information Technology
Sinhgad Academy of Engineering, Pune, India
samuel.chopde@gmail.com

DOI 10.1007/978-3-642-19542-6_132

The paper starting on page 535 of this volume has been retracted because a considerable amount of its content was plagiarized from the following publication:

"Hop-Count Filtering: An Effective Defense Against Spoofed DDoS Traffic" by Cheng Jin, Haining Wang, and Kang G. Shin, Copyright 2003 ACM 1-58113-738-9/03/0010.

The original online version for this chapter can be found at
http://dx.doi.org/10.1007/978-3-642-19542-6_104

V.V. Das, J. Stephen, and Y. Chaba (Eds.): CNC 2011, CCIS 142, p. E2, 2016.

Author Index

GPSR Compliance

The European Union's (EU) General Product Safety Regulation (GPSR) is a set of rules that requires consumer products to be safe and our obligations to ensure this.

If you have any concerns about our products, you can contact us on ProductSafety@springernature.com

In case Publisher is established outside the EU, the EU authorized representative is:

Springer Nature Customer Service Center GmbH
Europaplatz 3
69115 Heidelberg, Germany

Batch number: 10844611

Printed by Printforce, the Netherlands